THE OXFORD HANDBOOK OF

THE ECONOMICS OF NETWORKS

CONSULTING EDITORS

THE OXFORD HANDBOOK OF
THE ECONOMICS OF NETWORKS

Edited by

YANN BRAMOULLÉ, ANDREA GALEOTTI,

and

BRIAN W. ROGERS

OXFORD
UNIVERSITY PRESS

OXFORD
UNIVERSITY PRESS

Oxford University Press is a department of the University of Oxford. It furthers the University's objective of excellence in research, scholarship, and education by publishing worldwide. Oxford is a registered trade mark of Oxford University Press in the UK and certain other countries.

Published in the United States of America by Oxford University Press
198 Madison Avenue, New York, NY 10016, United States of America.

Library of Congress Cataloging-in-Publication Data
Names: Bramoullé, Yann, editor. | Galeotti, Andrea, editor. | Rogers, Brian W., editor.
Title: The Oxford handbook of the economics of networks / edited by Yann Bramoullé, Andrea Galeotti, Brian W. Rogers.
Description: Oxford ; New York : Oxford University Press, [2016] | Includes bibliographical references and index.
Identifiers: LCCN 2015034929 | ISBN 9780199948277 (alk. paper)
Subjects: LCSH: Social networks–Economic aspects. | Information networks–Economic aspects. | Social sciences–Network analysis.
Classification: LCC HM741 .O94 2016 | DDC 302.3–dc23 LC record available at http://lccn.loc.gov/2015034929

9 8 7 6 5 4 3
Printed on acid-free paper
Printed in the United States of America

Contents

PART IV EMPIRICS AND EXPERIMENTS

PART V DIFFUSION, LEARNING, AND CONTAGION

PART VI COMMUNITIES

PART VII ORGANIZATIONS AND MARKETS

List of Contributors

Daron Acemoglu Elizabeth and James Killian Professor of Economics, MIT

Sinan Aral David Austin Professor of Management, Sloan School of Management, MIT

Lori Beaman Assistant Professor, Department of Economics, Northwestern University

Francis Bloch Professor of Economics, Université Paris 1 and Paris School of Economics

Vincent Boucher Assistant Professor, Department of Economics, Laval University

Yann Bramoullé CNRS Research Fellow, Aix-Marseille School of Economics, Aix-Marseille University

Emily Breza Assistant Professor, Finance and Economics, Columbia Business School

Antonio Cabrales Professor, Department of Economics, University College London

Arun G. Chandrasekhar Assistant Professor, Department of Economics, Stanford University

Thomas Chaney Professor of Economics, Toulouse School of Economics

Syngjoo Choi Associate Professor, Department of Economics, Seoul National University and University College London

Daniele Condorelli Lecturer, Department of Economics, University of Essex

Wouter Dessein Eli Ginzberg Professor of Finance and Economics, Columbia Business School

Marcin Dziubiński Assistant Professor, Institute of Informatics, University of Warsaw

Nicholas Economides Professor of Economics, Stern School of Business, New York University

Bernard Fortin Professor Department of Economics, Laval University

Douglas Gale Professor of Economics and Director of Research in the Brevan Howard Centre for Financial Research, Imperial College London

Andrea Galeotti Professor, Department of Economics, University of Essex

Edoardo Gallo University Lecturer, Faculty of Economics, University of Cambridge, and Official Fellow, Queens' College, Cambridge.

Benjamin Golub Assistant Professor, Department of Economics, Harvard University

Piero Gottardi Professor, Department of Economics, European University Institute

Sanjeev Goyal Professor of Economics and Fellow of Christ's College, University of Cambridge

Matthew O. Jackson William D. Eberle Professor of Economics, Stanford University

Shachar Kariv Benjamin N. Ward Professor of Economics, University of California, Berkeley

Alan Kirman Emeritus Professor of Economics, Aix-Marseille University

Rachel Kranton James B. Duke Professor of Economics, Duke University

P. J. Lamberson Assistant Professor, Department of Communication Studies, University of California, Los Angeles

Mihai Manea Associate Professor, Department of Economics, MIT

Ana Mauleon Professor of Economics, Saint-Louis University, Brussels

Dina Mayzlin Associate Professor of Marketing, Marshall School of Business, University of Southern California

Markus Mobius Principal Researcher, Microsoft Research

Kaivan Munshi Frank Ramsey Professor of Economics, University of Cambridge

Francesco Nava Lecturer in Economics, London School of Economics

Asuman Ozdaglar Professor, Department of Electrical Engineering and Computer Science, MIT

Paolo Pin Associate Professor, Department of Decision Sciences, Bocconi University

Andrea Prat Richard Paul Richman Professor of Business and Professor of Economics, Columbia University

Brian W. Rogers Associate Professor of Economics, Washington University in St. Louis

Tanya Rosenblat Associate Professor of Information, School of Information and Associate Professor of Economics, University of Michigan

Evan Sadler Prize Fellow in Economics, History, and Politics, Harvard University

Alireza Tahbaz-Salehi Daniel W. Stanton Associate Professor of Business, Columbia Business School

Vincent Vannetelbosch Professor of Economics, University of Louvain

Fernando Vega-Redondo Professor, Department of Decision Sciences, Bocconi University

Adrien Vigier Assistant Professor, Department of Economics, University of Oslo

Duncan J. Watts Principal Researcher, Microsoft Research

Yves Zenou Professor of Economics, Stockholm University

THE OXFORD HANDBOOK OF

THE ECONOMICS OF NETWORKS

PART I

INTRODUCTION

CHAPTER 1

INTRODUCTION TO THE HANDBOOK

YANN BRAMOULLÉ, ANDREA GALEOTTI, AND BRIAN W. ROGERS

1.1 Foreword

During the past 25 years, economics has undergone a kind of quiet revolution. In the 1990s, when we were first being introduced to economics at the graduate level, most economists viewed social networks as largely outside the realm of economics. This situation has changed radically, and for a number of separate reasons.

First, game theory was expanding its reach, and thinking of social ties from a strategic perspective was a natural step in this development. Second, new applications have arisen for which networks are of clear importance, including, to name a few, the Internet, Facebook and other social media, cell phones, and communication networks. A related third factor is increased access to relevant data from both new and more traditional applications, along with the computing power to analyze it. Finally, the growth of the study of social networks in economics is part of an interdisciplinary growth, with counterparts in computer science, physics, finance, and sociology.

Economists' views have evolved in roughly three phases. In the first, social networks entered the lexicon of economics and the study of networks became accepted as a legitimate, albeit perhaps esoteric, subject of research. This is the period of the pioneering, high-impact papers of, for example, Jackson and Wolinsky (1996), Bala and Goyal (2000), and Kranton and Minehart (2001).

In the second stage, the literature has grown exponentially and on both the extensive and intensive margins. Major economic journals have increasingly published papers related to networks, and this research has expanded along all key dimensions: in the types of questions asked, the methods used, and the applications studied. At some point in perhaps the last 10 years, the economics of networks became a field of research in its own right with dedicated JEL codes, massive online courses, workshops and conferences, and best-selling textbooks.

We believe we have now reached a third phase, where researchers appreciate that there are many important questions and issues that, in order to be properly studied, require a network dimension. It might be said that the field of network economics has thereby matured into an established discipline both within the community of economists, and at the boundaries of several interdisciplinary efforts.

We hope that the publication of this handbook at this critical and exciting juncture will help to catalyze this last evolution. Researchers are placing networks at the heart of many of their current reflections, including the study of exchange markets, the recent financial crisis, international trade, migration, development, and the labor market, all of which are represented in this volume.

We have focused the content on economic applications of networks. Most of the chapters are written by economists, but there are contributions that incorporate complementary views from related disciplines. Additionally, the contributions come both from the most senior established voices in the field, as well as from prominent young researchers, whose current work is shaping the frontier of the science. We have organized the handbook more topically than methodologically. In what follows, we sketch the contents of the handbook and summarize its areas of concentration.

1.2 Overview

Section II offers perspectives on the state of the field from three of its founders: Alan Kirman, Sanjeev Goyal, and Matthew Jackson. Alan Kirman and Sanjeev Goyal both discuss the thought-provoking possibility that networks may provide a paradigm shift for economics.

In Chapter 2, Alan Kirman argues cogently that a focus on the structure of interactions between agents should become the new benchmark in economic modelling. He shows that this focus is needed, for instance, to understand patterns of transactions in the Marseille wholesale fish market or the origins and unfolding of the recent financial crisis. He also observes that this shift in paradigm has not yet taken place in macroeconomics.

In Chapter 3, Sanjeev Goyal argues that the economic treatment of networks has experienced the three classical phases of a true paradigm shift. From a major shift in modeling, illustrated by the introduction of networks in the literature on social learning in the 1990s, to the study of new questions, like network formation, to the current integration into normal economic science and the study of markets. Alan Kirman's and Sanjeev Goyal's points of view could be reconciled by observing that while networks seem to have reached the core of microeconomics, they still lie at the periphery of macroeconomics, although this too may be changing rapidly.

In Chapter 4, Matthew Jackson provides a concise summary of the achievements in the field and a wide-ranging discussion of open problems and areas that are ripe for further investigation. These include new applications in development, labor, international trade, and international relations; the development of statistical and econometric methods for studying network formation; and theoretical studies of strategic influence, dynamic interactions, and the multiple dimensions of social relationships.

Section III gathers contributions on games played on networks and on models of network formation. These two kinds of models represent, in a sense, the historical core of the economic theory of networks.

In Chapter 5, Yann Bramoullé and Rachel Kranton review the literature on games played on fixed networks. In these models, while agents interact with their direct neighbors only, strategic interactions imply that Nash equilibria depend on the whole network. They develop a common analytical framework to study a wide class of games, establish new connections between models, and introduce a notion of interdependence to analyze how a shock on one agent eventually affects the action of another agent.

In Chapter 6, Francesco Nava surveys a recent and growing literature on repeated games with local monitoring and local interactions. He shows that classical Folk Theorems generally extend independently of the network structure. By contrast, the network has a strong impact on cooperation and equilibrium payoffs obtained for fixed discount rates. These models help capture important features of diverse phenomena, including favor exchange, risk-sharing, and lending.

In Chapter 7, Paolo Pin and Brian Rogers argue that incorporating random components into network formation models is crucial for reconciling theoretical predictions with observed patterns in real networks. Consequently, essentially all empirical work in the field is based largely on random network formation. They focus also on the role of homophily in network formation; that is, the consequences for network structure of the tendency for agents to connect with others who are similar.

In Chapter 8, Ana Mauleon and Vincent Vannetelbosch discuss the literature on strategic network formation under mutual consent. They present and discuss various notions of stability and illustrate the distinctive impact of these notions in applications. In particular, they show how to operationalize the idea that agents are farsighted, and they study the implications of farsightedness in models of R&D, trade, and criminal networks.

In Chapter 9, Fernando Vega-Redondo reviews a recent and stimulating literature that studies the simultaneous determination of actions and links. He describes how such joint determination operates in contexts of coordination, cooperation, intermediation, bargaining, public goods, learning, and conflict. A broad lesson is that endogenizing both kinds of decisions often allows one to obtain much sharper predictions than when either actions or links are treated as fixed.

In Chapter 10, Marcin Dziubiński, Sanjeev Goyal, and Adrien Vigier survey a recent but potentially wide-ranging literature on conflict and networks. They discuss the robustness of infrastructure networks, cybersecurity, and criminal networks and how networks can be optimally attacked, defended, and designed. They also review conflicts and strategic alliances and the nascent economic literature aiming at understanding the network aspects of these issues.

In Chapter 11, Yves Zenou provides an overview of the literature of key players on networks. Key players solve specific planners' problems in network contexts. For example, which node should be removed to obtain the highest reduction in aggregate activity? Which set of nodes should be targeted to quickly diffuse information or attitudes? Empirical implementations of key-player policies are discussed and are shown to outperform standard targeting policies across diverse circumstances.

Section IV adopts an empirical focus, and explores the following kinds of questions: How should one estimate the effects of networks? What are the main statistical features of networks and how should their formation be analyzed empirically? How can experiments help answer these challenging questions?

In Chapter 12, Vincent Boucher and Bernard Fortin discuss recent developments and challenges in the empirics of the effects of networks. They show that social multipliers may not be identifiable without *a priori* assumptions on the underlying microeconomic model. They discuss the possibility of testing for the endogeneity of the network and the difficulties raised by the common use of proxies. A broad conclusion is that greater care should be taken to ground econometric models in sound microeconomic foundations.

In Chapter 13, Arun Chandrasekhar surveys a recent and fast-growing body of work on the econometrics of network formation. He identifies two key difficulties: how to achieve consistency in estimation when a single network is observed and how to account for sparseness and clustering. He discusses the microeconomic underpinnings and statistical properties of the two main classes of models: those with conditional edge-independence and those allowing for higher-order dependence across links.

In Chapter 14, Duncan Watts focuses on a main empirical feature of networks: the small-world property, expressing the idea that any two individuals can be connected via short chains of intermediaries. He distinguishes between a topological and an algorithmic version of this property and surveys the empirical and theoretical literature on the topic. He shows that this robust feature of social networks has important economic, epidemiological, and social implications.

In Chapter 15, Sinan Aral discusses the design and analysis of networked experiments. He shows how large-scale experiments have grown following the development of digitization and online interactions, and that this has led to a corresponding increase in the complexity of research methods involved. He reviews innovations in design, such as randomization at different levels, and advances in analysis, including the modeling of treatment response and interference, estimation, and inference in network contexts.

In Chapter 16, Emily Breza surveys the fast-growing area of field experiments in developing countries. In contexts where formal institutions are not well-functioning, social networks generally play a central role. Field experiments have thus allowed researchers to better understand the impact of networks on learning and diffusion, other-regarding preferences, peer monitoring and enforcement, risk-sharing, and the formation and evolution of networks.

In Chapter 17, Syngjoo Choi, Edoardo Gallo, and Shachar Kariv provide an overview of the literature utilizing laboratory experiments to study network contexts. Researchers can impose an exogenous network structure in the lab, which facilitates causal estimations of network effects. They first discuss how the network structure affects behavior in games played on networks and in games where information flows through the network. They then look at experiments on networks and markets in which the network determines trading opportunities, and also when it can be used to circulate information about traders' reputations.

Section V reflects on the role of networks in determining outcomes in the key applications of diffusion, learning, and contagion. The four chapters in this section are closely related in their underlying motivation, but they depart from each other in the way in which diffusion and linkages are modeled.

In Chapter 18, P. J. Lamberson focuses on diffusion processes developed in the epidemiology literature, for example, the SIR model and its variants. These models are based on simple rules that mirror disease spread, and they were originally studied under the assumption of uniform and random interactions. In the last two decades, the literature in computer science, economics, and statistical physics has extended the analysis of these models to more complex patterns of social interactions, using in particular random graphs. This chapter, in addition to providing an account of a literature that is less well known to economists, also provides an introduction to a set of mathematical tools to study diffusion processes in networks that, we believe, can be fruitfully embedded in economics models.

In Chapter 19, Ben Golub and Evan Sadler consider diffusion of information and opinions. The first part of the chapter reviews models where agents are Bayesian and learn from the observation of the actions of previous agents. The set of agents that an individual observes before taking her action is specified by the underlying social network. The second part of the chapter focuses on models where, in every period, agents update linearly their opinions based on the opinions of their neighbors. The linearity in the updating process implies that agents neglect possible correlations in the information that they receive from their social contacts.

In Chapter 20, Antonio Cabrales, Douglas Gale, and Piero Gottardi review models of financial contagion. In this literature links across institutions capture financial obligations. Microeconomic volatility and agents' limited liability trigger contagion. The chapter provides a unifying framework to study how default by some agents can spread in a connected system. It discusses the important question of how networks can be designed in order to limit contagion processes and, at the same time, help to smooth consumption and share risk across institutions.

In Chapter 21, Daron Acemoglu, Asu Ozdaglar, and Alireza Tahbaz-Salehi study the role that network topology plays in determining how microeconomic shocks propagate, and possibly get amplified, through an economic system. The chapter provides a general framework that embeds recent models studying shock propagation in different contexts, including games in networks, financial interactions, supply chains, and macroeconomic models.

Section VI provides an account of the economic consequences of social ties within a community.

In Chapter 22, Markus Mobius and Tanya Rosenblat review some of the most recent theoretical and empirical work on how social ties facilitate informal lending and risk-sharing. This is particularly important in economies where formal financial institutions are not well developed. They describe, in particular, an approach whereby social ties can be leveraged as social collateral so as to mitigate moral hazard.

In Chapter 23, Kaivan Munshi describes the interplay between social ties in a community and migration. It first provides a fascinating historical account of migrant networks, highlighting the benefits that community networks can have in facilitating migration. The chapter then discusses the difficulties in providing credible statistical evidence that networks support migration, and suggests possible solutions.

In Chapter 24, Lori Beaman discusses an active literature on labor market and networks. She reviews the theoretical literature, exploring the three main roles played by networks in the labor market: social ties can be used to transmit information on jobs, on qualities of employers and employees, and to exert peer pressure. While the importance of networks is well-established empirically, the applied literature is still struggling to understand and identify the precise mechanisms at work.

Section VII offers a survey on recent developments of the theory of networks in the areas of organizational economics, trade and market functioning, and industrial organization.

In Chapter 25, Wouter Dessein and Andrea Prat discuss the role of information in organizations. Organizational economics has tended to model communication as exogenous flows. The key role that communication plays in coordinating activities and creating new opportunities motivates a cost-benefit analysis of the design of communication flows. The intersection of network economics and organizational economics is largely unexplored. This chapter argues that incorporating a network approach to organizational economics generates new questions, new insights, and new empirical predictions on organizational trends.

The next two chapters survey strategic models of exchange in networks. In these models the network describes "who can trade with whom." The absence of a connection models a trading friction that prevents direct exchange between the two parties. How does network location affect the terms of trade and profit? What are the types of inefficiencies that emerge in trading networks?

In Chapter 26, Mihai Manea surveys the literature on bilateral trade. Linked agents are, for example, buyers and sellers. Their position in the network determines their bargaining power as it defines the opportunities that they have to trade with other agents.

In Chapter 27, Daniele Condorelli and Andrea Galeotti survey strategic models of intermediation. In this case, buyers and sellers are not necessarily connected, but there are intermediaries who can buy and resell. The resale value of an intermediary is endogenous to the trading protocol and to the intermediary's network location.

Understanding the way in which resale values form and are shaped by the network provides insights on how resale markets affect market outcomes.

In Chapter 28, Thomas Chaney surveys recent advances in international trade and the economics of networks. Connections may represent trade flows between countries, export-import relationships across firms, and information diffusion via, for example, migration. This chapter shows how a network approach to international trade can help uncover micro and macro properties of the structure of international trade. It also shows how an explicit formalization of production networks can help in rationalizing data on supply chains, and, possibly, quantifying their fragility and the consequences for macroeconomics outcomes.

The next two chapters summarize a vibrant and multidisciplinary literature on marketing strategies in the presence of social effects. Social effects include both word-of-mouth communication processes and consumption externalities across consumers. The general question is how firms can design marketing strategies to tip and to leverage such social effects.

In Chapter 30, Dina Mayzlin studies possible actions that firms can take to stimulate social interactions and to increase the efficacy of social effects. In the process she reviews a large body of literature, mostly in marketing, that studies the management of social interactions and their effects.

In Chapter 29, Francis Bloch considers the complementary problem where firms take the network of social interaction as given and consider how to optimally leverage social effects to stimulate demand and increase profits.

In Chapter 31, Nicholas Economides discusses the economic features of the Internet with a focus on the current network neutrality debate After describing the general structure of the Internet, the chapter introduces the concept of network neutrality. It then uses the theory of pricing in two-sided markets in order to highlight possible consequences of abolishing network neutrality on the functioning of the Internet.

References

Bala, Venkatesh and Sanjeev Goyal (2000). "A noncooperative model of network formation." *Econometrica* 68(5), 1181–1229.

Jackson, Matthew O. and Asher Wolinsky (1996). "A strategic model of social and economic networks." *Journal of Economic Theory* 71(1), 44–74.

Kranton, Rachel E. and Deborah F. Minehart (2001). "A theory of buyer-seller networks." *The American Economic Review* 91(3), 485–508.

PART II

PERSPECTIVES

CHAPTER 2

NETWORKS: A PARADIGM SHIFT FOR ECONOMICS?

ALAN KIRMAN

> Yet it is likely that one day we will know much more about how economies work—or fail to work—by understanding better the physical structures that underlie brain functioning. Those structures—networks of neurons that communicate with each other via axons and dendrites—underlie the familiar analogy of the brain to a computer—networks of transistors that communicate with each other via electric wires. The economy is the next analogy: a network of people who communicate with each other via electronic and other connections. The brain, the computer, and the economy: all three are devices whose purpose is to solve fundamental information problems in coordinating the activities of individual units—the neurons, the transistors, or individual people. As we improve our understanding of the problems that any one of these devices solves—and how it overcomes obstacles in doing so—we learn something valuable about all three.
>
> —Robert Shiller 2011

It was Thomas Kuhn in his 1962 book *The Structure of Scientific Revolutions* who developed the idea of a "paradigm shift" in a discipline. He argued that such a shift occurs when scientists encounter anomalies that cannot be explained by the universally accepted paradigm which has provided the framework for the existing theory. He suggested that when such a change occurs what he referred to as the "world view" of the scientists in the discipline changes. A number of economists and policy-makers have suggested that, with the experience of the current economic crisis we have reached the point at which such a shift will occur in the discipline of economics.

Yet, it might seem pretentious to suggest that the development of network theory and its applications would be at the heart of such a change in economics. Underlying the development of economics in the last two centuries has been a commitment to the principles of what might be called "classical science" (for example, Morin 2006), the most relevant of which, for this discussion, can be summarized as follows: *The principle of reduction*, that consists in knowing any composite from only the knowledge of its basic constituting elements.

While accepting that the behavior of the aggregate is the result of the behavior of the components, the important idea, and one which has pervaded economic theory, is that it is enough to know how those components behave without worrying about the direct interaction and feedbacks between them. This is the core of methodological individualism, and indeed Robert Lucas went as far as to say that it is, for him, an article of faith that one should only make assumptions about individuals. However, this leads us directly to the basic problem with recent events. What has been the major source of dissatisfaction with current economic theory, and macroeconomic theory in particular? It has been the absence of major endogenous upheavals in the models. Major changes are attributed to exogenous shocks but do not emerge from the internal dynamics of the system. This is not surprising given the structure of modern macroeconomic models, which do not allow for direct interactions among the individuals in the economy or for the consequences of the network that governs those interactions, and thus exclude one of the major sources of intrinsic volatility. One gets a glimpse of how understanding the network of linkages between asymmetric firms may generate large systemic shocks in the work of Acemoglu et al. (2011) and Gabaix (2011), and this is discussed in Acemoglu's chapter in this book. But this has not penetrated modern macroeconomics other than as justification for a more fat-tailed distribution of aggregate shocks. The importance of the direct interaction between firms, which is an essential ingredient of the evolution of the economy, is not recognized. Macroeconomic models remain in the tradition of a set of actors taking essentially independent decisions with the mutual influence of these actions relegated to the role of inconvenient externalities.

But to understand why we have come to this point, it is worth just casting a brief glance at the path that economics has followed from Adam Smith to the present. Our current theory, particularly in macroeconomics, is the result of a long evolution in which economics has tried to produce a coherent model to underpin the laissez faire liberal, social, and philosophical approach, which has become dominant. This suggests that, left to their own devices, participants in an economy will self-organize into a satisfactory state. Yet, within the standard General Equilibrium framework, this cannot be proved. Thus, theoreticians and policy-makers are confined to studying equilibria and changes in the latter without any convincing argument as to why the economy should be in such a state. Suppose instead that we base our paradigm on a view of the economy as a complex interactive system. This would mean taking as a starting point the direct interaction between market participants who may vary widely in their characteristics and their information. Such a system may show no tendency to self-equilibrate, and endogenous crises and aggregate changes will be a standard feature of our models.

Where then do networks come into play? As Lucas (1988) said, somewhat paradoxically given his insistence on restricting assumption to those on individual characteristics, "Applications of economic theory to market or group behavior require assumptions about the mode of interaction among agents as well as about individual behavior."

If we are to understand aggregate economic behavior, we have to not just make assumptions about the individual characteristics and motives of the actors in isolation and the economy, but also the structure that links them together. Sociologists have long adopted the idea that the individual is embedded in a social network and that his or her choices and actions are influenced, directly and indirectly, by those to whom he is linked. This has been pushed in many directions, such as the development of the Actor Network Theory (ANT) in sociology. Whittle and Spicer (2008) believe that ANT prefers to seek out complex patterns of causality rooted in connections between actors. In other words, what generates causal relations cannot be understood by looking at individuals in isolation. Furthermore, as Vega-Redondo emphasizes in his chapter in this book, the links between actors are also chosen by the actors themselves, and this further complicates the analysis.

James Coleman noted, with regret, the efforts by some empirical sociologists to move to an approach more akin to that of economists, taking samples of independent purposeful individuals and leaving to one side the conflict between the micro and macro level:

> There was no comparable development of tools for analysis of the behavior of interacting systems of individuals or for capturing the interdependencies of individual actions as they combine to produce a system-level outcome. The far greater complexity required of tools for these purposes constituted a serious impediment to their development and continues to do so (though some methods such as those generally labeled "network analysis" move in that direction). The end result is extraordinarily elaborated methods for analysis of the behavior of a set of independent entities (most often individuals), with little development of methods for characterizing systemic action resulting from the interdependent actions of members of the system.
>
> —Coleman 1986, p. 1316

These remarks were made some 30 years ago but should still resonate with economists. The tradition on which modern economic analysis is based goes back a long way, at least as far as Hobbes. Hobbes' view (1651) is now referred to as atomism and is a more extreme form of methodolological individualism. He envisioned society as a group of unconnected individuals whose current characteristics were unaffected, now, or in the past, by those to whom they are linked. He proposed to "consider men as if but even now sprung out of the earth, and suddainly (like Mushromes) come to full maturity without all kind of engagement to each other."

Later economists did not reject interaction out of hand, though it was often limited to market interaction over which the actors involved had no strategic control, and where agents were essentially negligible price takers. This is, of course, the basic vision of the Walrasian General Equilibrium model.

To caricature, if we take the Walrasian model seriously, it consists of agents linked to some central but unspecified agent who calls out prices for all goods. Thus the basic network is a star, and the agent in the center is often referred to as the "Walrasian auctioneer."[1] The Walrasian road does not seem to offer a vision of the economy that would place network theory in the forefront.

The evolution of game theory can be considered as the natural sequel to a development in economic theory that is very different from that followed by general equilibrium theory and which opened the door for network theory to play a central role. To caricature again, one could think of a full blown game theoretic model of the economy as made up of agents, all of whom interact consciously and strategically with all the others. In this case the network of links between the agents would correspond to the complete graph. One might argue that, from a macroeconomic point of view, a fully game-theoretic analysis would seem to attribute knowledge and calculating capacity to agents, which is so far beyond what can reasonably be postulated that it is best to assume that agents simply act in isolation. Yet this would seem to negate the interest of a great deal of economic analysis. Indeed, the natural reaction to this would be to suggest that agents interact strategically with only a limited number of other agents, and once we admit this then we can use network theory to specify who is "playing with whom," as do Bramoullé and Kranton in their chapter. But, at least for the moment, the use of game-theoretic analysis puts quite severe restrictions on the sort of models we can analyze; perhaps the most important of these that the network effects are small. If we are to put network theory at center stage, then this must be because the interaction effects can sometimes be large.

The other extreme, the pure Walrasian approach, would suggest that individuals essentially only interact through large, impersonal markets, and that who trades with whom, can safely be ignored. Indeed, some network models have been criticized precisely because they assume that agents only trade or interact with a limited number of others, and as Durlauf (2012) has rightly indicated, there are large, even anonymous markets on which many individuals trade without being specifically linked to the other participants. However, the purpose of this chapter is to suggest that such markets are, in fact, the exception, and that closer empirical inspection of most markets shows that the networks of interactions between individuals do play an important role. Indeed, to assert that what individuals choose to do, on the apparently large, anonymous markets, is not influenced by those with whom they are linked through family, social, or other connections is unrealistic. In fact, individuals, firms, and collective entities are all embedded in many different sets of relations. To obtain a reasonably complete picture of how aggregate activity emerges, we cannot afford to ignore the influence of individuals on each other, and the important fact that not only do these influences change over time, but the structures of the networks themselves are constantly being endogenously modified.

We have to acknowledge that the direct interaction between agents and the way in which that interaction is organized has fundamental consequences for aggregate

[1] Walras never mentioned the auctioneer himself, as Walker (1996) has indicated, but as De Vroey (1999) has pointed out, the auctioneer seems to be the only device consistent with Walras' arguments.

outcomes. To reiterate, when agents are directly linked to and influence each other, the relationship between the behavior of individuals and the behavior of aggregate variables will be different than in the anonymous market situation in which all agents are linked to each other only through the price system. What we observe at the aggregate level will not mimic what we observe at the individual level, nor will it correspond to the behavior of some "representative individual." Moreover, that rationality we attribute to economic individuals, in order to justify and analyze the behavior of aggregates, may have to be modified. Just as neurologists would not think of explaining behavior by studying the changes in a representative neuron, neither should economists try to explain aggregate phenomena in this way. This does not mean that one should not be interested in what happens at the micro-level, but rather that the passage to the aggregate level is mediated by the network structure in which individuals find themselves. Neurologists will continue to examine what happens at the molecular level but would not argue that there is some simple passage from that level to the aggregate activity of the brain that does not involve the network of interactions between neurons. Of course, as economists, unlike neurologists, we do not usually descend as far as the level of the neurons of economic agents, but as interest in so-called neuro-economics has developed, it has been argued that economic behavior is very much determined by the network of neurons that is activated in a certain situation, and that as the situation changes another network may become active. Thus even at this level it is the network structure of the neurons that is important (see Oullier et al. 2008).

To use another analogy, we would not expect to be able to explain how much food is stored by a colony of ants by looking at the behavior of individual ants in isolation. The organization of the ants plays an essential role. This example raises an important point. Far from complicating things, taking direct account of interaction and the networks which organize it, actually makes life simpler for the economic theorist. This is because the reasoning and calculating capacities we need to attribute to economic agents may be of a lesser order than in standard models.

Two examples of direct interaction are, perhaps, worth evoking at this stage. The first concerns the vexed question of the stability of general equilibrium. Here, most of the attention has focused, as I have remarked, on the idea of some central market clearing mechanism that might lead an economy to equilibrium. However, many economists from Edgeworth (1881)[2] to Hayek (1945)[3] have insisted on the fact that direct interaction between agents plays a crucial role in determining aggregate

[2] Indeed Walras (*Letter no. 927 to Von Bortkiewicz*, in Jaffe 1965) complained that Edgeworth considered all his work on price adjustment as worthless and that the only realistic assumption was that prices emerged as the result of haggling between individuals.

[3] Hayek (1945), for example, explicitly argued that information is never available to all the participants in the economy but that it is transmitted by means of transactions between individuals, each of whom possesses some small part of that information. He argued that prices would emerge that would encapsulate all this information. However, he was unable to specify the precise mechanism through which this would happen.

outcomes. Edgeworth (1881) was dismissive of the Walrasian idea of a market as a set of isolated individuals:

> You might suppose each dealer to write down his demand, how much of an article he would take at each price, without attempting to conceal his requirements; and these data having been furnished to a sort of market-machine, the price to be passionlessly evaluated.
>
> —Edgeworth 1881, p. 30

The natural setting within which to study Edgeworth's own bargaining approach would seem to have come with the development of game theory. Nevertheless, frustratingly, the various discussions of price formation mechanisms in which individuals trade directly with each other have almost invariably left the structure of trade to one side. The various non-tatonnement processes, or trading mechanisms, a la Feldman (1973, 1974) assume that individuals are drawn at random to trade with each other. If one thinks of bargaining processes such as that of Rubinstein and Wolinsky (1985), for example, nothing is said about who bargains with whom, just that there are individuals in a bargaining situation. There has, of course, been a later literature on bargaining in networks (see, e.g., Abreu and Manea 2012 and Manea 2011). However, the general question as to how individuals through their direct interaction might generate equilibrium prices did not involve the explicit use of networks. Manea, in his chapter in this handbook, looks at a number of the questions associated with bargaining in networks and highlights the importance of the discount factor that players use when evaluating outcomes as bargaining proceeds. This is an interesting problem, since it is not clear to what extent, in empirical situations where the market functions for a limited and specified period of time that discounting plays a real role; for example, it might be that the discount factor is a proxy for the uncertainty as to how many units are left to be sold.

It might also be worth observing that a great deal of the game-theoretic literature on this subject has concentrated on the convergence of the process to an equilibrium, and often a Nash equilibrium. But, as Von Neumann and Morgenstern suggested, the nature of the game theory that they developed was fundamentally unsatisfactory because of its static nature, and Morgenstern said that Von Neumann's aversion to the Nash equilibrium was precisely that it avoided the non-equilibrium dynamics involved. Manea also emphasizes the fact that we need to take into consideration the extent to which individuals know the structure of the network, and what the matching mechanism is.

A second example of how local interaction has long been envisaged as playing a part in the evolution of the economy is that involving the simple idea that individuals rarely obtain information independently and then act on it, but rather tend to be influenced by what others are doing. It was Poincaré (1908) who put his finger firmly on this. He was critical of Bachelier's (1900) thesis for which he was the external arbiter. Bachelier laid the foundations of what has come to be called the efficient markets hypothesis, by assuming that individuals observe information independently of each other and by acting upon it, that information is incorporated into prices. Yet, Poincaré objected to this and argued that in fact markets and financial markets are characterized by "herd behavior." He explained that he there is a natural trait in human nature which is to herd, like Panurge's sheep, and this undermined the mathematical structure that Bachelier developed.

Thus, there was imbedded in Poincaré's vision of how markets functioned a notion of contagion. Yet, nothing was specified about how this contagion spread, other than to say that individuals have a direct influence on each other. The idea of isolated individuals receiving idiosyncratic information and, somehow, through the way in which they interact, transmitting this information to others, was very different from a network of mutually observing participants.

However, if one wants to be more specific about the consequences of direct interaction, one has, as Lucas observed, to say more about how this interaction is organized and, in particular, who interacts with whom. To analyze this, we need to know about the network of links between the individuals, whether these are consumers, firms, or other collective entities. As I have said, networks and network analysis have played a central role in many disciplines, both in the social sciences and in the "hard sciences," but for a long time their role in economics was ambiguous. Direct interaction and the results of that interaction were classified as "externalities" and were considered as "imperfections" from the point of view of the benchmark model in which the participants only interact through the price system.

For many economists the study of networks was limited to analysis of the functioning of physical networks such as the railway, the telephone system or, now, the Internet. Yet, more recently it has been recognized that networks, in fact, are much more fundamental and pervasive than this, and this is well illustrated by the work of Goyal (2007), Jackson (2008), Vega-Redondo (2007), and Ioannides (2013) on economic and social networks. Indeed, almost any serious, detailed analyis of economic activity leads to the conclusion that network structures both within and between organizations are important. Nevertheless, a consensus seems to have developed among economists that networks and graph theory are useful tools to complement standard economic analysis but do not constitute the basis for rethinking the model to which we have become accustomed. These authors are regarded with great respect but are not generally thought of as the basis for a new paradigm. The whole purpose of this chapter is to suggest, to the contrary, that founding economic theory on a network-based benchmark model would provide insights and answers to enigmas which the standard model does not let us handle. One effort to develop such an approach is that of Easley and Kleinberg (2013), but there are few signs that their textbook will be commonly used at the undergraduate level.

To make the basic point, I will present some illustrations as to how networks can produce phenonema that are often difficult to generate with standard models. The suggestion then is that direct interaction is the benchmark and pure competition but an exceptional and very special case that does not necessarily provide the best basis for the analysis of empirical phenomena. As a Chinese philosopher said, "Water that is too pure has no fish in it."

2.1 Markets

In the first example that I mentioned, that of price adjustment in the general equilibrium model, the mechanism can be characterized as akin to an auctioneer who calls out the prices for all goods. But what happens if, as Edgeworth (1881) envisaged, trade takes place between individuals—who sets the prices and why should the same price prevail for units of the same good? Those who have studied the *theory of value* seriously would argue that if we observe different prices, these are for units at different times and places even if they have identical physical characteristics. This means that technically they are different goods. But, leaving this objection to one side for a moment, would one then expect to find different units of the same good at the same place being traded over a limited period of time at the same price? Consider Cournot's definition of a market:

> . . . not any particular market place in which things are bought and sold but the whole of any region in which buyers and sellers are in such free intercourse with each other that the prices of the same goods tend to equality easily and quickly.
>
> — A. Cournot, *Recherches sur les Principes Mathématiques de la Théorie des Richesses*, Chapter IV

He does not explain how this might happen but does suggest that, unlike in a network view, all participants interact with all the others and somehow this eliminates price discrepancies. However, if this were to be essentially the case, the vast literature on searching for the lowest price or wage when there is a distribution of prices for the same good would be of no empirical interest. To see why this is not the case, consider the following example of a real market, that of the wholesale fish market in Marseille, where the conditions for such a process to work would seem to be ideally fulfilled.

This market is situated at Saumaty on the coast at the northern edge of Marseille, which for the period for which the data was collected was open every day of the year from 2 a.m. to 6 a.m. At this market, over 500 buyers and 45 sellers come together, although they are not present every day and transact more than 130 types of fish. Prices are not posted. All transactions are pairwise. There is little negotiation and prices can reasonably be regarded as take it or leave it prices given by the seller.

The data set consists of the details of every individual transaction made over a period of three years. The data was systematically collected and recorded by the Chamber of Commerce, which managed the market. The following information was provided for each transaction:

(i) The name of the buyer
(ii) The name of the seller
(iii) The type of fish
(iv) The weight of the lot
(v) The price per kilo at which it was sold
(vi) The order of the transaction in the daily sales of the seller.

The data runs from January 1, 1988 to June 30, 1991. The total number of transactions for which we have data is 237,162.

Two specific questions can be asked about the functioning of this market. Do all traders trade with each other, or do networks of buyers and sellers form? Does this result in different units of the same fish being transacted at the same price as Cournot seemed to suggest? In other words, does the market self-organize into a state characterized by what is traditionally assumed, or does it exhibit rather different features?

2.1.1 A Simple Model

It is often useful to build on a very simple theoretical model for which we have analytical results, and then to add more realistic features that make exact results more difficult to obtain. Then the model can be simulated, to see whether the results obtained in the simple case still hold.

In the market, there are n buyers indexed by i and m sellers indexed by j. The buyers update their probability of visiting sellers on the basis of the profit that they obtained in the past from them. If we denote by $J_{ij}(t)$ the cumulated profit, up to period t, that buyer i has obtained from trading with seller j where

$$J_{ij}(t) = \Pi_{ij} + (1 - \gamma)J_{ij}(t - 1) \tag{2.1}$$

and where Π_{ij} is the profit that the buyer i makes if he visits seller j and the latter still has fish available (assume for the time being that the profit does not vary over time). Then the probability $p_{ij}(t)$ that i will visit j in that period is given by

$$p_{ij}(t) = \frac{e^{\beta J_{ij}(t)}}{\sum\limits_{k} e^{\beta J_{ik}(t)}} \tag{2.2}$$

where β is a reinforcement parameter which describes how sensitive the individual is to past profits. This nonlinear updating rule will be familiar from many different disciplines and is also widely used in statistical physics. It is known as the "logit" rule or, in game theory, as the "quantal response" rule. The rule is based on two simple principles. Agents make probabilistic choices between actions. Actions that have generated better outcomes in the past are more likely to be used in the future.[4]

To simplify matters at the outset we will start with a continuous approximation of our model which is actually in discrete time. Furthermore, we will replace the random variables by their expected values. This is referred to as the "mean field" approach. In

[4] Such a process has long been adopted and modeled by psychologists (see, e.g., Bush and Mosteller 1955). It is a special form of reinforcement learning. It has also been widely used in evolutionary and experimental game theory (see Roth and Erev 1995), and a more elaborate model has been constructed by Camerer and Ho (1999).

this way it is easy to see that the change in cumulated profit for the buyer is given by

$$\frac{dJ_{ij}}{dt} = -\gamma J_{ij} + E(\Pi_{ij}). \tag{2.3}$$

Using the learning rule that I have given, we know the probability for agent i to visit seller j and can therefore calculate the expected gain from that visit. Recall that there are two things involved here—the probability that the seller j still has fish available when buyer i arrives, and the probability that the latter chooses seller j. So the expectation is given by

$$E(\Pi_{ij}) = \Pr(q_j > 0)\Pi_{ij}\frac{\exp(\beta J_{ij})}{\sum_k \exp(\beta J_{ij})}. \tag{2.4}$$

Now consider an even simpler case where the seller is sure to have fish, in which case we have

$$\Pr(q_j > 0) = 1. \tag{2.5}$$

Now simplify even further and look at the case where there are just two sellers (and furthermore each time a buyer visits one of the sellers he receives a fixed profit of Π) and find the equilibrium level for the cumulated profit for a buyer from seller 1. This will, of course, be when

$$\frac{dJ_1}{dt} = 0. \tag{2.6}$$

Substituting this gives

$$\gamma J_1 = \Pi\frac{\exp(\beta J_1)}{\exp(\beta J_1) + \exp(\beta J_2)}. \tag{2.7}$$

Now take the difference between the profits from the two sellers and we have

$$\Delta = J_1 - J_2. \tag{2.8}$$

If we now substitute we have the following expression

$$\Delta = \frac{\exp(\beta\Delta - 1)\Pi}{\exp(\beta\Delta + 1)\gamma}. \tag{2.9}$$

We now have simply to solve this equation for Δ and this gives two cases. First, consider

$$\beta < \beta_c = \frac{2\gamma}{\Pi}. \tag{2.10}$$

In this case, when the importance attached to previous experience is below the critical value β_c we have

$$\Delta = 0 \quad J_1 = J_2\frac{\Pi}{2\gamma}. \tag{2.11}$$

There is a single solution and the cumulated profits from both sellers, and hence the probabilities of visiting them, are the same.

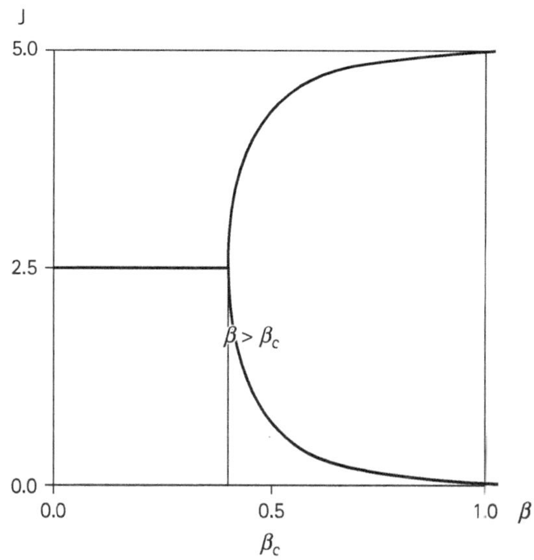

FIGURE 2.1 Source: Weisbuch et al. (2000).

However, when $\beta > \beta_c$ then there are three solutions and $\Delta = 0$ is unstable and there is a rapid transition at $\beta > \beta_c$. By this we mean that as soon as β passes above the critical value, the probabilities of visiting each seller rapidly become very different. All of this is illustrated in Figure 2.1.

Note that even in this most basic case, which would seem to have simplified away all the interesting features of the market, a small change in one parameter, β, the sensitivity to past profits, leads to a radical change in the behavior of the buyer who will, when β is high, not treat the two identical sellers identically but will attach himself with high probability to one of them. Furthermore, it is easy to show that there is considerable inertia, or hysteresis, in the system. Once attached, the other seller would have to have a considerably lower price to be able to attract the buyer back (see Weisbuch et al. 2000). However, suppose that we start to introduce some more realistic features into the model. For example, sellers no longer have a fixed predetermined amount of fish, but decide each day in the light of their experience how much to buy, while buyers reinforce on their previous experience. Initially, suppose that sellers keep their price fixed, and adjust the amount of fish that they buy and then put on offer each day. Each seller forecasts his sales on the basis of past sales. Suppose that we then simulate the evolution of the purchases of the individuals in the market. As shown in Weisbuch et al. (2000), we observe that when buyers have a β below their critical value β_c, they keep shopping around. Since sellers are faced with varying numbers of clients, they find it difficult to anticipate demand. As a result, considerable amounts of fish remain unsold at the end of the day, and some sellers find themselves unable to satisfy the demand. If, on the other hand, buyers have coefficients β above their critical value, β_c then buyers become

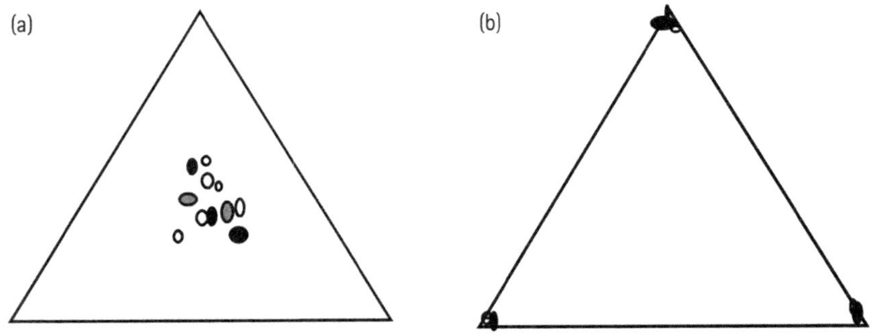

FIGURE 2.2 Source: Kirman (2010).

loyal to their sellers and the latter can then forecast the demand correctly and much less fish is wasted.

This would seem to be in contradiction with the overall efficiency of the market since buyers who become locked on to a particular seller can no longer benefit from the fact that another seller may provide more profit. Nevertheless, it became clear from the simulations that the overall efficiency of the market, in terms of the percentage of fish purchased which was then sold, was higher in the presence of loyal buyers. Thus, an informational loss was compensated for by an efficiency gain.

The buyers and sellers in the market structured themselves into a network which improved the throughput of fish, even though this might seem to have deprived some of the actors of information in the process. The seeds of the explanation can be found in the probabilistic rule, the "logit rule" (equation 7), used to determine the probability of a buyer visiting a seller on the basis of past profits. This rule can be derived (see Weisbuch et al. 1998) from the maximization of a convex combination of the information gain, "exploration" from visiting new sellers, measured by entropy and the benefit from using past experience, "exploitation." Individuals who, in the market, have a low critical value β of the key parameter β_c i.e., who attach considerable importance to previous experience, will rapidly converge to visiting a single seller, and thus give up the benefits of exploring new opportunities.

Proceeding in this way, by examining the evolution of the structure of relationships in the market, provides a very different picture from the standard one of anonymous individuals trading at equilibrium prices.

One further point is worth mentioning. When we simulated the model with, for example, 3 sellers and 30 buyers, where the latter all had β values below their critical value, β_c individuals continued to "shop around," as in Figure 2.2a, where each buyer is a dot in the 3 simplex, his vector of probabilities of visiting the various sellers. However when the buyers had values of β higher than the critical value β_c they rapidly became loyal to particular sellers, as in Figure 2.2b. Now a natural question arises. What if the population of buyers is mixed with differing critical values β_c. The answer might be expected to be that the structure of the graph would be at least partially demolished,

FIGURE 2.3 Source: Kirman (2010).

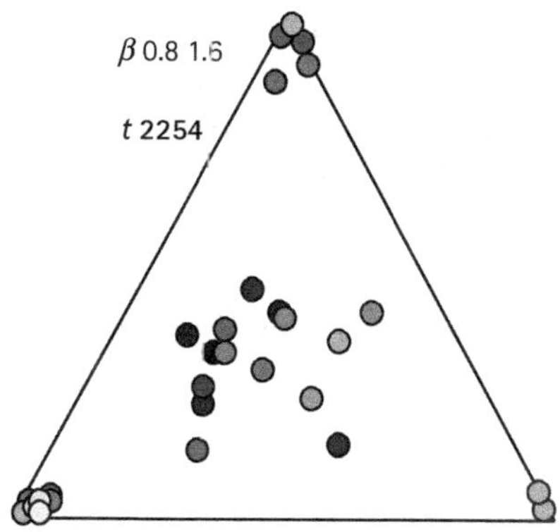

since sellers would no longer be able to predict accurately the number of buyers they would receive and this might undermine the loyalty links. However, simulating the model in this case showed that the presence of low, below the critical β_c buyers did not prevent the higher value buyers from becoming loyal, as can be seen in Figure 2.3 (Kirman 2010).

In that figure there are two groups of buyers—those with a low threshold β_c and those with a high threshold β_c. Even after over 2000 iterations, the separation between the loyalists and the shoppers persists. The empirical evidence from the Marseille fish market is clear, as shown in Weisbuch et al. (2000). There is a clear bimodal distribution of loyalty, by which I mean the number of sellers visited by a buyer in a month. There is a concentration of buyers on one, that is, who only visit one seller, and then a distribution of other buyers with a mean of 5. This reflects precisely the prediction of the theoretical model.

However, when trying to build a more realistic model of a market, at each step there is some important feature of the market that has been omitted but which could safely be ignored in a more conventional model. In the case of the Marseille fish market, as soon as one understands that networks will evolve at least partially into clusters of buyers around different sellers, and that there is inertia in the system, the question naturally arises as to why the sellers do not exploit this structure by charging higher prices to their loyal customers.

To examine this question, we built an "agent-based model" (ABM), in Kirman and Vriend (2001), to see what would happen if buyers, using simple rules, not only decided which sellers to visit but also which prices to accept, and sellers decided which prices to charge to which buyers at each point in time. Such a model is not analytically tractable but can be simulated and robustness checks made on the relevant parameters. In the model in question, each individual agent uses a Classifier System (an approach to learning developed by John Holland 1976), for each decision and this means that each agent has four such systems "in his head." A classifier system consists of a set of rules. Each rule has a condition "if . . . " and an action "then . . . " and in addition each rule is assigned a certain strength. The classifier system decides which of the rules will be active at a given point in time. It checks the conditional part of the rule and decides among all of those rules for which the condition is satisfied which to choose. This is done by a simple auction procedure. Each rule makes a "bid" to be the current rule and this bid = current strength $+\varepsilon$ where ε is white noise, a normal random variable with mean 0 and fixed variance. The rule with the highest "bid" in this auction becomes the active rule. The white noise means that there was always some experimenting going on, and there was always some probability that a rule, however bad, will be chosen.

The classifier system updates the strength s of a rule that has been active and has generated a reward s_{t-1} at time $t - 1$ as follows. $s_t = s_{t-1} - cs_{t-1} + creward_{t-1}$ where $0 < c < 1$. Hence, as long as the reward generated by the rule on day $t - 1$ is greater than its strength at $t - 1$ the strength will increase. The strength of each rule converges to the weighted average of the rewards generated by that rule. What the reward is will depend on the rule in question. Supposing that in our market example the rule for the buyer is of the form, *"if the price proposed by the seller for one unit of fish in the morning is 10 euros then accept."* The reward for using this rule would then be the profit that is generated by using it. In this case that would be the price at which the unit of fish is sold on the retail market minus the price paid. When the model is started, the strengths of all rules are equal.

The question now is how does the market structure itself? Do the networks of buyers linked to sellers persist? Will price discrimination occur or will buyers move on to cheaper suppliers? The answers are interesting. The networks do indeed persist, and loyal buyers pay higher prices than their less loyal counterparts, but paradoxically make higher profits. This is because sellers learn to serve their loyal customers first but to charge them higher prices. The advantage of better service more than offsets the higher prices paid. The market self-organizes into a stable situation in which the networks become well established and change very little. The empirical evidence confirms these findings. Buyers who visit several sellers pay lower prices than those who concentrate all their purchases on one seller but are, in general, less profitable.

The picture that emerges from all this is far from that of the perfectly competitive market but the importance of the network of buyer-seller relations is clear. This is still a simplified model, and there are a number of other features that one might want to take account of, such as family ties or homophily considerations. Graddy (1995) in her investigation of the Fulton Fish Market, for example, found that ethnic considerations played a role in determining how much buyers paid. Geertz (1978), in his famous studies of North African bazaars, studied the impact of the relationship between buyers and sellers, and Mclean and Padgett (1997) emphasized the role of family relations in the textile markets in Rennaissance Florence. The latter is particularly striking with regard to what we observed on the Marseille fish market. As they say:

> . . . Both from their perspective as buyers and from their perspective as sellers, banking and wool firms engaged in significantly more repeat-business and "relational trading" than perfect competition theory permits. This was true not only internal to the banking and wool sectors themselves, but, more importantly, between the banking and wool firms

and a majority of all other sectors. Thus, for a two hundred year period between 1300 and 1500 the structure of the Florentine markets bore many of the hallmarks we still observe in modern markets. Networks of relations are omnipresent and the anonymous competitive market is a rare exception.

2.2 Networks in the Financial Sector

Let me now turn to another example, that of the financial sector, which has too often been described as one in which most of the action takes place between numerous relatively anonymous actors and where the structure of the links between these actors is of little importance. It is in fact clear from recent events that this is far from the case.

Consider for a moment the crisis of confidence in the world economy after 2008. Here we see the role of networks. Bad risks from the American mortgage market had been bundled with good risks into derivatives, and these had diffused through the international banking system. Up to that point the financial system, thought of as a network of banks, had become larger and more connected and it was argued that the resultant diversification of risk was a stabilizing influence. Yet there was little analysis of the effect of the increasing connectivity of the network on the probability of an epidemic of negative impacts. The problem is that as risk is diversified into different instruments, those who buy these instruments lose track of the underlying assets. Thus, while the risk is diversified, the information is not. When this happens, an epidemic of mistrust can develop as each bank in the network is wary of lending to another which may have inherited risks that turned out to be bad. Worse, not only may banks find themselves with assets which may turn out to be toxic but the market may also revise its valuation of these assets so that the prices of even those derivatives that do not contain toxic elements may fall. Thus, the fact that banks are highly interdependent through, in part, the exchange of these assets, is crucial to our understanding of how a market or an economy functions or may fail to function.

There is by now a considerable literature on networks and contagion in financial networks and an excellent survey is given by Cabralles et al. (2015), in their chapter of this book. They give an account of both the theoretical developments and of some of the more empirical work which is the focus here. However, the theoretical models that they discuss are still very highly simplified and stylized. For the banking sector, for example, almost all theoretical models take, as the primary function of banks, that of direct financial intermediation. That is, banks simply match those with funds available with those who need those funds to invest in economic projects (see, e.g., Gertler and Kyotaki 2011). Yet it is known that such activities are a small proportion of the total activites of banks. Funds transit through a multitude of operations and transformations as they are packaged and repackaged and this trade in derivatives is now the principal activity of the sector. Figure 2.4 illustrates this.

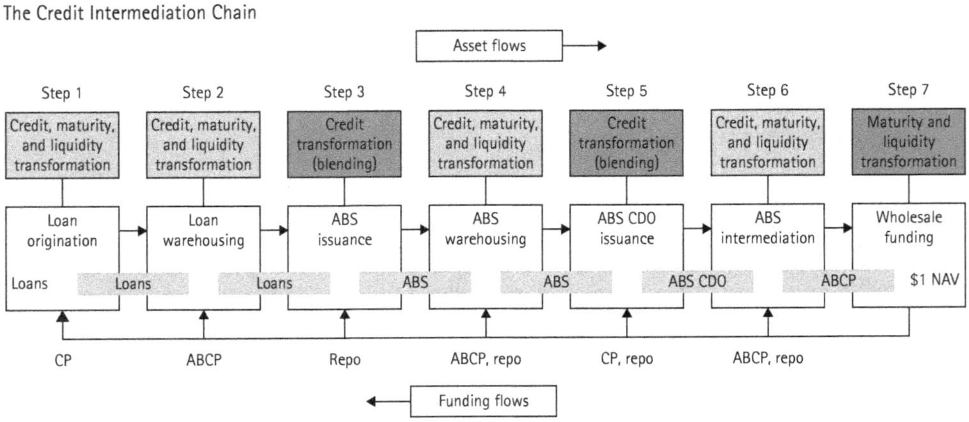

FIGURE 2.4 Source: Pozsar (2010).

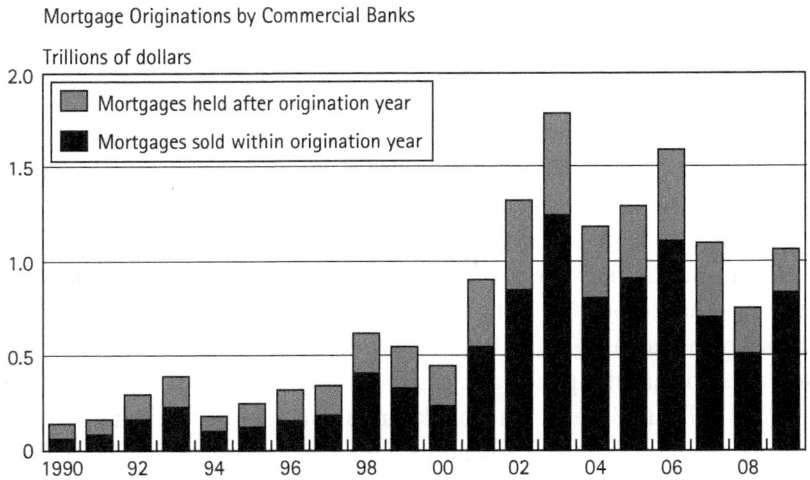

FIGURE 2.5 Source: Cetorelli et al. (2012).

To see how far we have moved from a system where banks issued loans to their clients and then received the associated interest and capital payments, it suffices to examine how many mortgages remain on the books of the institution that issued them. This is illustrated in Figure 2.5. As is clear the majority of mortgages were sold to other institutions.

The transformation of the banking sector and of financial intermediation was far from specific to the United States. An indication of this is given by the expansion of the derivatives market in the United Kingdom, as illustrated in Figure 2.6.

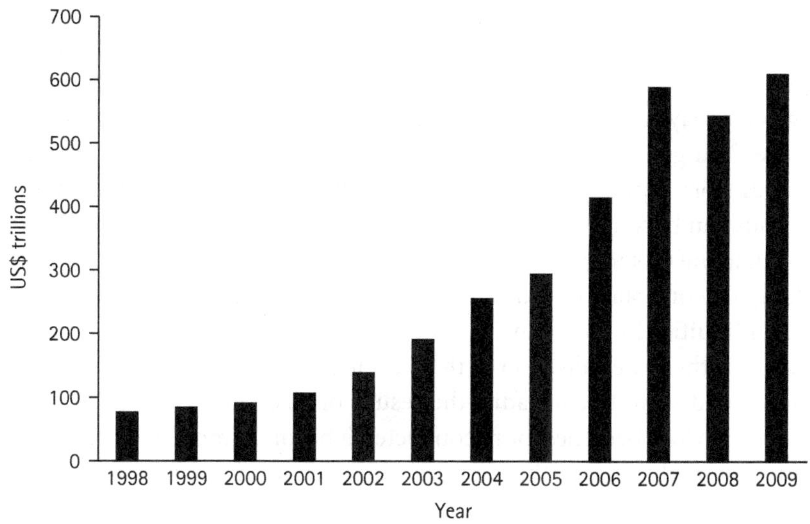

FIGURE 2.6 Notional principal value of outstanding derivative contracts, recorded at year end. These include foreign exchange, interest rates, equities, commodities, and credit derivatives. Data from UK Department for Business, Innovation and Skills, International Monetary Fund and Bank of England calculations (Source: Haldane and May 2011).

This reflects the fact that the linkages between banks were rapidly expanding. What happened subsequently was that various banks were obliged to reassess their losses as a result of the subprime episode and its consequences. These losses were due not only to their discovering the true nature of their assets, but also to the revaluation of these downward by the market. The resultant losses of the banks enhanced the epidemic of mistrust. It is possible that it is simply the increased connectivity of the network that favored the development of such an epidemic. But, in fact, the problem is more subtle than this. Propagation of information or of shocks may be more likely, and the effect may be bigger in networks that are much less than fully connected.

This problem has already been discussed by Allen and Gale (2000). Using a network structure involving four banks, they showed that the spread of contagion depends crucially on the pattern of interconnectedness between banks; that is, on their exposure to each other. When the network is completely connected, with all banks having exposures to each other, such that the amount of interbank deposits held by any bank is evenly spread over all other banks, the impact of a shock is easily absorbed. Every bank suffers a small loss and there is no contagion.

By contrast, when the connectivity of the network is lower, with banks only having exposures to a few counterparties, the system is more fragile. The initial impact of a shock is concentrated among neighboring banks. Once these succumb, the premature liquidation of long-term assets and the associated loss of value bring previously unaffected banks into the front line of contagion. Thus, the structure of the network heightens rather than dampens the effect of a shock. Indeed, there is evidence that even in large, apparently anonymous markets, participants trade or interact with a rather small group of other traders. Hence, the fact that the participants are clustered into limited groups may cause the propagation of a shock which was not particularly large at the outset.[5] In general, what we want to know is whether it is true that a network that emerges from a particular evolution of trading relationships which are mutually advantageous can become fragile without those who participate in it realizing what is going on. Haldane (2009) of the Bank of England, when talking about the evolution of the banking network, stated, "This evolution in the topology of the network meant that sharp discontinuities in the financial system were an accident waiting to happen. The present crisis is the materialisation of that accident" (p. 4).

Yet, this would seem to contradict the results of Allen and Gale (2000), for in their model as the network becomes more connected it becomes more robust, while what we have witnessed might suggest the contrary. However, it is important to realize that it is not only the connectivity of the network, but other characteristics also that come into play when considering systemic risk. It is worth examining the relation between these two phenomena when the scope of the model is enlarged. For example, Iori et al. (2006) find that when the network of banks is heterogenous, an inter-bank market can have ambiguous consequences for systemic risk. As usual, the benefits of mutual insurance need to be contrasted with the possibility of contagion. They also show, however, that as the connectivity increases the system becomes more stable, echoing the finding of Allen and Gale (1998). Similar conclusions in relation to the degree of network connectivity are reached by Gai and Kapadia (2010), Montagna and Lux (2013), and Nier et al. (2007). However the fact that it is not enough to consider just connectivity is also shown by Gai and Kapadia (2010). They introduce idiosyncratic and aggregate shocks and compare the impacts of the two. They show that when banks have relatively high capital to total asset ratios, the system can survive the former type of shock but, with aggregate shocks, contagion risk significantly increases. Efforts such as these to model the financial system using complex networks analysis have been promoted by ecologist Robert May (see Haldane and May 2011), who sees clear analogies between the collapse of ecosystems and that of the banking network.

Over the last 10 years, many researchers have followed this promising line of research by using these analytical tools to improve our understanding of the financial sector in general and banking systems in particular. Yet the literature has often had to focus on simple models to obtain analytic results. Allen and Gale (2000) use a very simple four-bank structure, for example, and Cabralles et al. in this handbook compare a ring structure for the network of institutions with a completely connected graph. However, the banking network tends to have a more "core-periphery" structure, and it would be interesting to extend their analysis to such a case.

[5] See Allen and Babus (2009) for a review of various applications of network theory to the study of financial issues.

In parallel to the literature in economics is econophysics, which has used complex systems to analyze various economic systems, including the behavior of asset prices (for a review, see Varela et al. 2013). This field has not focused on analytical results and often uses simulations.

Battiston et al. (2012) characterize the evolution over time of a network of credit relations among financial agents as a system of coupled stochastic processes. Each process describes the dynamics of individual financial robustness, while the coupling results from a network of liabilities among agents. Here again the role of the network of interdependencies between entitities is crucial. The average level of risk diversification of the agents coincides with the density of links in the network. In addition to a process of diffusion of financial distress, they also consider a discrete process of default cascade due to the re-evaluation of agents' assets. They then investigate the probability of individual defaults as well as the probability of systemic default as a function of the network density. They underline the reasons for doubting the idea that diversification of risk always leads to a more stable financial system, since in their model, as in other network models, a tension emerges between individual risk and systemic risk. As the number of counterparties in the credit network increases beyond a certain value, the default probability, both individual and systemic, starts to increase. They explain that this tension originates from the fact that agents are subject to a financial accelerator mechanism. In other words, individual financial fragility feeding back on itself may amplify the effect of an initial shock and lead to a full-fledged systemic crisis. Once one accepts the crucial role that the network of linkages between agents or entities plays, their results offer a simple possible explanation for the endogenous emergence of systemic risk in a credit network.

Again, the network structure as it transmits information and shocks can generate sudden large and endogenous changes in the overall performance of the system. Understanding the dynamics of this complex system depends on the recognition of a certain number of factors. Among the most important is the heterogeneity of the actors involved. This in turn involves several considerations, the size of the entitities, their different scope, and the geographical extent of their operations. These aspects are well illustrated in the graph of Australian banking interdependencies (large exposures) in Figure 2.7. Casual visual inspection reveals that this particular network is far from the simple forms such as rings or complete networks often used for analytical tractability.

What we observe is that this network has a number of rather special features. The big Australian banks are connected both to foreign banks and the smaller credit institutions, but the latter have few connections with banks outside Australia while the foreign banks are linked to each other and to the major Australian banks. Analyzing the consequences of a difficulty in some part of this system is not possible without understanding the structure of the network. In other words, both the nature of the nodes and the structure of the graph in which they are embedded play an important role.

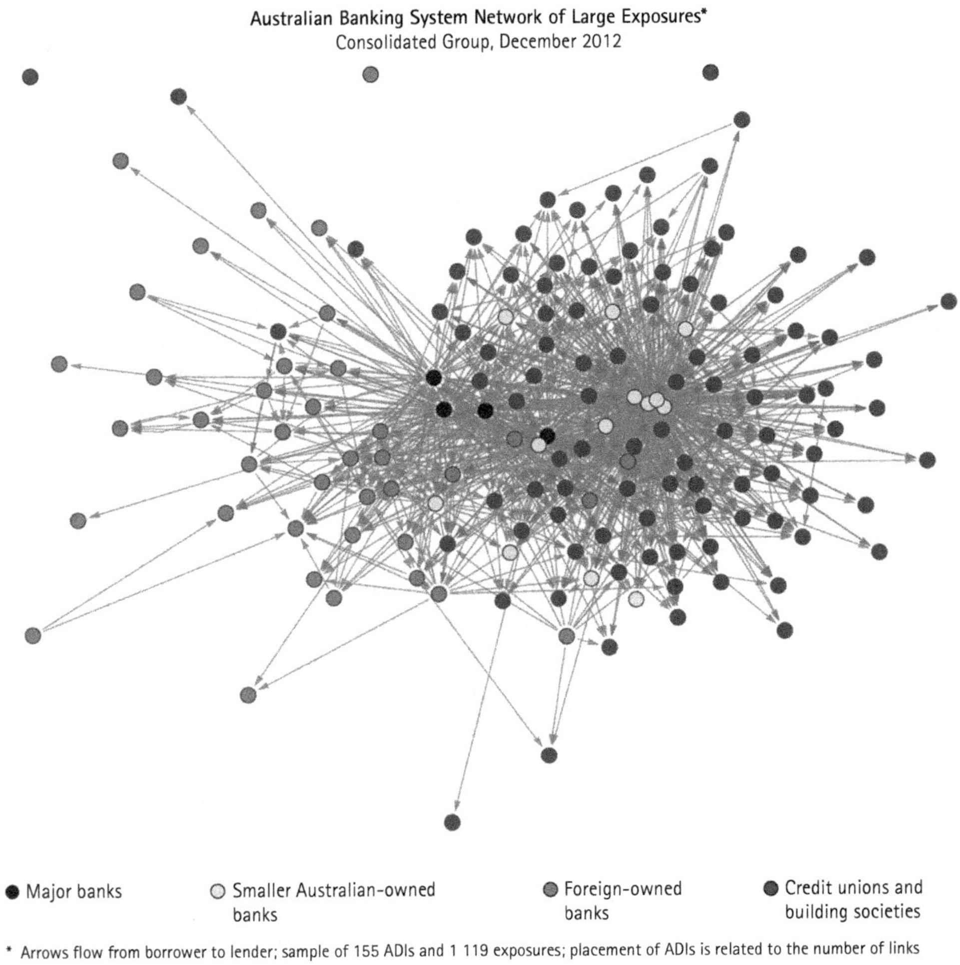

FIGURE 2.7

2.2.1 Derivatives

As mentioned earlier, an important part of the financial puzzle is the role of derivatives. These instruments package tranches of assets held by one financial institution into new assets, which are then sold to other institutions. The resultant instruments can be repackaged and resold. This, it was thought, would diminish risk, since the risk in the instruments in question was diversified. The opposite turned out to be true. As Warren Buffett (2002) stated even before the crisis:

> The derivatives genie is now well out of the bottle, and these instruments will almost certainly multiply in variety and number until some event makes their toxicity clear.

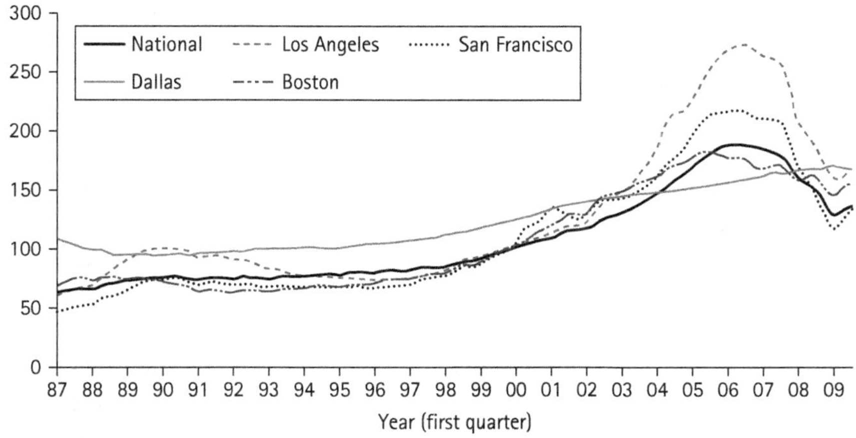

FIGURE 2.8 The evolution of house price indices in different cities in the United States, 1987–2009 (Source: Case Shiller).

> Central banks and governments have so far found no effective way to control, or even monitor, the risks posed by these contracts. In my view, derivatives are financial weapons of mass destruction, carrying dangers that, while now latent, are potentially lethal.

What is the relation between these assets and the network of banks that issue and sell them? A first point is that the actors in the financial sector network have a tendency to adopt the rules or norms that currently prevail in that network. To illustrate this approach, I will now give an example of a model that portrays a market crash and in which individuals have local and limited knowledge, but whose actions are governed by the choices made by those with whom they are in contact.

2.2.2 An Example

To see how the interactions of entities or individuals with whom they are linked can produce an endogenously generated catastrophic change in a market, it is worth considering an empirical example. The dramatic collapse of the prices of MBS (mortgage-backed securities) at the beginning of the current crisis is such an example and had profound ramifications for the economy as a whole (Anand et al. 2013). This collapse occurred rapidly despite the fact that the weakness of the assets underlying the derivatives had increased progressively over time. Each of the instruments in question had a single overall rating but, in fact, consisted of tranches of mortgages with different ratings. Figure 2.8 shows the evolution of housing prices over 20 years. The developments from 2006 onward should have been a first indication that defaults on the mortgages involved in the MBS were likely to rise.

The principal reason for this was that the percentage of loans that represented "positive equity"—that is, where the value of the house was greater than the outstanding amount of the house loan—diminished as the increase in house prices slowed down and was reversed.[6]

In Figure 2.9 the default rate on mortgages in the United States for various years is shown. The increase that occurred in that rate for mortgages issued in 2006 and 2007 can be explained by two factors. First, house prices increased more slowly than previously, and the gain from defaulting offset any capital gain. Second, more mortgages were issued on easier terms. To see this in retrospect, of mortgages issued in 2004, 10% were delinquent after 30 months, whereas for those issued in 2007, 10% were delinquent after only 8 months. The evidence was public and available. Yet the evolution of the prices of MBS did not reflect this steady increase, as can be seen in Figure 2.10.

Prices of similarly rated assets remained stable and then suddenly collapsed. There was a different date of collapse for each asset class, and one could argue that at each of these dates some relevant news was revealed and that this caused the downturn. This is hardly convincing as there was a whole chain of actors from the mortgager to the investor in MBS that would have simultaneously received the news and reacted similarly. However, many have argued that the constancy of the prices before the collapse reflected the rational expectations of all these actors and that their interaction then had no influence over the prices of these assets. Yet at the onset of the crisis, Ashcraft and Schuermann (2008) had already pointed out a number of important informational frictions arising in links in these chains of actors, each of which could potentially lead to a breakdown in the system. These frictions ranged from problems of moral hazard involving the payment of the credit rating agencies by asset managers, to principal agent difficulties between the investor and the asset manager, to moral hazard problems between the issuers of mortgages and the arranger who set up a special vehicle to hold the mortgages and securitize them for eventual purchase by investors. What the authors observed is that the presence of all these frictions led to the initial breakdown of this market. The market was a network of interacting agents and was very far from transparent.

What is particularly important to observe is that despite the fact that the delinquency of differently rated assets was evolving similarly, as seen in Figure 2.9, it is clear that the collapse of the better-rated assets came later. This suggests that the estimates of the probability of default on these instruments were influenced by the ratings, whether or not this was justified by the underlying fundamentals. Thus there were other actors, the rating agencies, in the network who influenced the evolution of prices. Pozsar et al. (2010) show that prices were strongly correlated with ratings but that ratings were very poorly correlated with default rates, indicating clearly that the information provided was far from perfect. The essential question is, why did the behavior of the actors in the

[6] In a number of states in the United States (where loans are "non-recourse") the owners of a property on which they have a loan can simply turn the house over to the bank that issued the loan without having any further financial contribution to make.

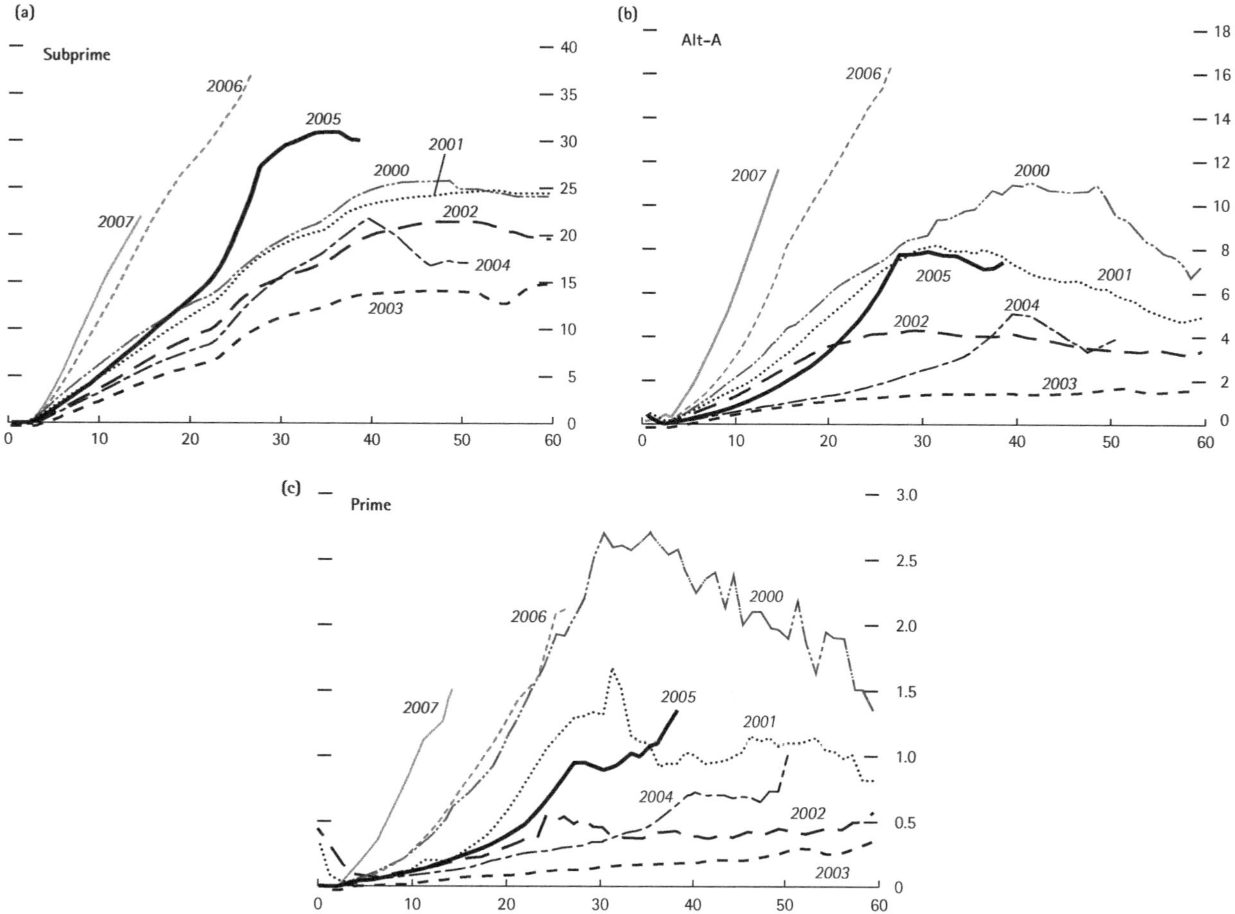

FIGURE 2.9 Delinquency rates on mortgages (months from origin) in the United States originating from 2004 to 2007 (Source: Merrill Lynch and Loan Performance).

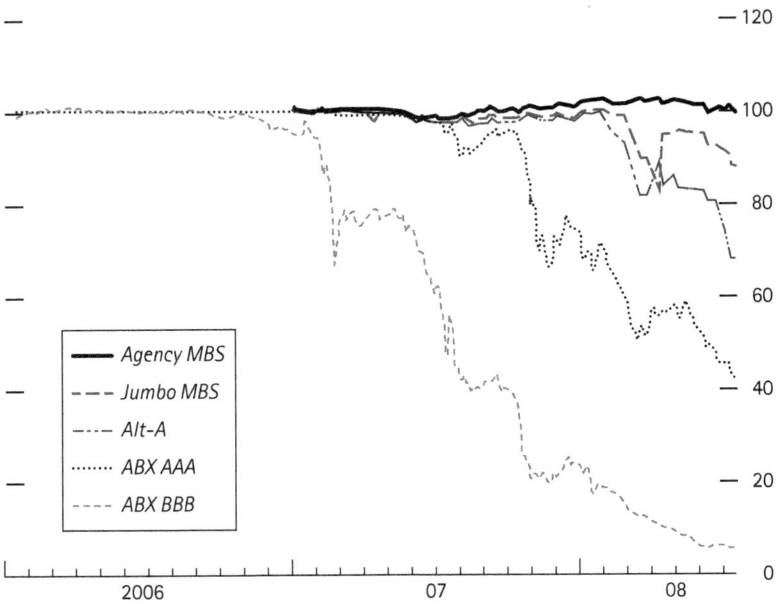

FIGURE 2.10 The prices of MBS with different ratings.

market and their relation to each other prevent the available evidence from translating into a steady decline in the prices of MBS rather than into the sudden collapses that actually occurred?

To understand this, in Anand et al. (2010) we constructed a model of the banking system in which the participants securitize their loans and then sell them via Special Purpose Entities (SPE) to each other. The derivatives thus constructed were made up of bundles of parts of different underlying assets with varying degrees of riskiness. These were frequently rebundled and sold again. It is not obvious why the issuers would have an interest in doing this. The reason lies in the so-called "recourse rule" which puts a risk weight of 50% on individual mortgages but only 20% on highly rated mortgage-backed securities. The amount of capital banks have to hold is conditioned by the risk weighting of their assets. Thus banks bundled their assets to free up capital. Here one sees immediately the role of the network. There is obviously considerable pressure on the rating agencies from those who employ them to overrate derivatives. Worse, even if a rational investor might doubt the rating assigned to a derivative, to reliably estimate the risk entailed checking on the current status of each of the assets in a bundle and involved considerable costs. In other words, the banks were holding derivatives—in particular, MBS—the content of which was extremely costly to evaluate. Not only this, but some of the banks knowingly misrepresented the value of the underlying assets they were selling to their investor clients, as J. P. Morgan admitted in a $13 billion settlement.[7]

[7] November 22, 2013.

These investors had neither the means nor the sophistication to evaluate whether the mortgages in the MBS did, in fact, meet the underwriting standards as was claimed. Finally, the banks when trading among themselves were not doing due diligence. Thus, although these assets were being actively traded, it is difficult to argue that the prices of the associated transactions reflected the decisions of fully informed rational agents. What I have just described is an oversimplified version of a market in which some agents had an interest in hiding or misrepresenting information and others had no incentive to go to the expense of obtaining the full information. Many of the incentives were, contrary to the conventional view of such markets, in the wrong direction. The network structure of the actors and the incentives with which they were faced led to a loss of information as the interlinkages created by securitization developed.

It was worth considering this example in some detail, since it is only by looking at the organization, incentives, and interactions between the actors that one can understand that trying to model such a framework as an anonymous market inhabited by fully informed agents with rational expectations does not capture the essence of what is happening. This market is, to repeat, a complex interactive network, and using a standard model would not have reflected this. The model we proposed focuses on the interaction between the actors in the banking network and provides an explanation as to how a sudden price collapse can occur despite the fact that the underlying fundamentals were gradually changing over a considerable period of time.

Our simple model, which reflects the concerns expressed by Haldane and May (2011), describes the behavior of the participants on the market. This behavior was not irrational in the normal sense of the word but was not fully rational in the sense that that term is understood in economics. The agents in the model have short horizons and condition their evaluation of an ABS not by always examining the fundamentals underlying the instrument, or by analyzing the general evolution of the housing market. Rather, they depended on the ratings of that MBS by the credit rating agencies, and their willingness to buy depended on how much checking was being done by those with whom they traded.

2.2.3 A Simple Model

In this model developed by Anand et al. (2013), the system consists of $i = 1,\ldots,N$ agents, which, in the case of the sub-prime crisis, for simplicity, we can think of as the banks who were both the issuers—via SPEs—and the investors in these ABS. These banks or agents are linked in a network corresponding to the over-the-counter market (OTC), and at each period an agent draws at random another agent among her neighbors. Each agent i is characterized by a variable $z \in \{0, 1\}$which specifies whether she follows ($z = 1$) or not ($z = 0$), the following behavioral rule: purchase an ABS relying on signals from the rating agencies, without independently evaluating the fundamental value of the underlying assets.

Succinctly, we write

$$z_i = \left[\begin{array}{ll} 1 & \textit{if agent } \mathrm{i} \textit{ follows the rule} \\ 0 & \textit{if the agent } \mathrm{i} \textit{ does risk analysis} \end{array} \right. \tag{2.12}$$

The rationale for adopting the rule, as we will see, is not based on the fundamental quality of the asset but rather on the fact that others also follow the rule. If, in fact, enough other participants do so, the agent becomes convinced, not irrationally that the ABS is highly liquid and hence easy to trade.

Assume that the ABS is toxic with probability p. By toxic we mean, for example, that the underlying asset was too favorably rated by a rating agency and either that the original borrower of the loan has already defaulted, or has a higher probability of defaulting, as he is delinquent in his payments. Assume that the cost of purchasing a security is p_0 whereas the payoff from successfully re-selling the security is p_1 where $p_1 - p_0 > 0$. However, if the buyer checks and finds the ABM to be toxic, the price now becomes a "fire sale" price p_2 where $p_2 - p_0 < 0$. The buyer can be sure to avoid this outcome by checking at a cost of X drawn from a p.d.f. $\Phi(X)$ Now one can rescale and reduce the number of parameters by normalizing such that: $p_1 - p_2 = 1$ and $p_0 - p_2 = c$. The agent is then faced with the following problem:

Table 2.1

	Check and toxic	Don't check
$Z(i) = 0$	$-X_i$	$1 - c - X_i$
$Z(i) = 1$	$-c$	$1 - c$

The columns represent the strategy of the buyer and the rows those of the seller. Now consider the expected payoff to the seller of each strategy.

$$u_i(z_i = 1) = E[-p(1 - z_j)c] + [1 - p(1 - z_j)](1 - c) = 1 - p(1 - \bar{z}_i) - c \tag{2.13}$$

where $\bar{z}_i = E(z_j)$ for $j \in N_i$. That is, agent i can correctly estimate the average choice of rule by his neighbors but not the choice of each individual. Thus we have

$$z_i = \frac{1}{k_i} \sum_{j \in N_i} z_j. \tag{2.14}$$

Now the expected payoff from not following the rule and choosing $z_i = 0$—that is, from checking the value of the underlying assets, is:

$$u_i(z_i = 0) = (1 - p)(1 - c) - X_i. \tag{2.15}$$

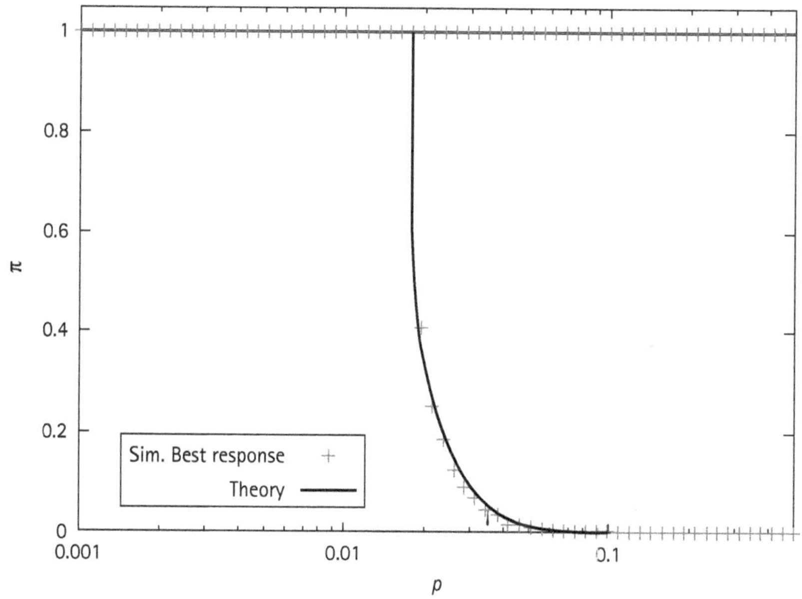

FIGURE 2.11 The coexistence of two equilibria, either all $z_i = 1$ or $z_i = 0$ (Source: Anand et al. 2013).

Thus if the agent checks and finds the assets to be toxic, he simply incurs the cost of checking, while if the asset is not toxic he obtains the difference between the selling and buying price less the checking cost. The strategy, which constitutes the best reply to the strategies of the neighbors, is then given by:

$$z_i = \Theta[u_i(1) - u_i(0)] = \Theta\left[p\left(\frac{1}{k_i}\right)\sum_{j \in N_i} z_j - c\right] \tag{2.16}$$

where the function Θ is defined as $\Theta(x) = 1$ if $x > 0$ and 0 otherwise. Note that the agents are assumed to know the probability of default of the underlying assets. However, in reality, the common perception of p reflected the over-optimistic evaluation of the rating agencies. For low values of p there is one equilibrium in which all agents choose not to check, but once a critical value of the commonly perceived p is passed, another equilibrium emerges in which all agents check. This is illustrated from numerical simulations in Figure 2.11.

When there are two equilibria there is no reason to believe that one or the other will be necessarily realized. However, we can introduce, as in our first example, some noise and assume that the agents only make the best response with a certain probability. What we impose is that, as the superiority of one strategy over the other increases, the probability of choosing that strategy increases. Given this, we can examine whether they coordinate on one equilibrium as p increases. We now use the logit rule, already

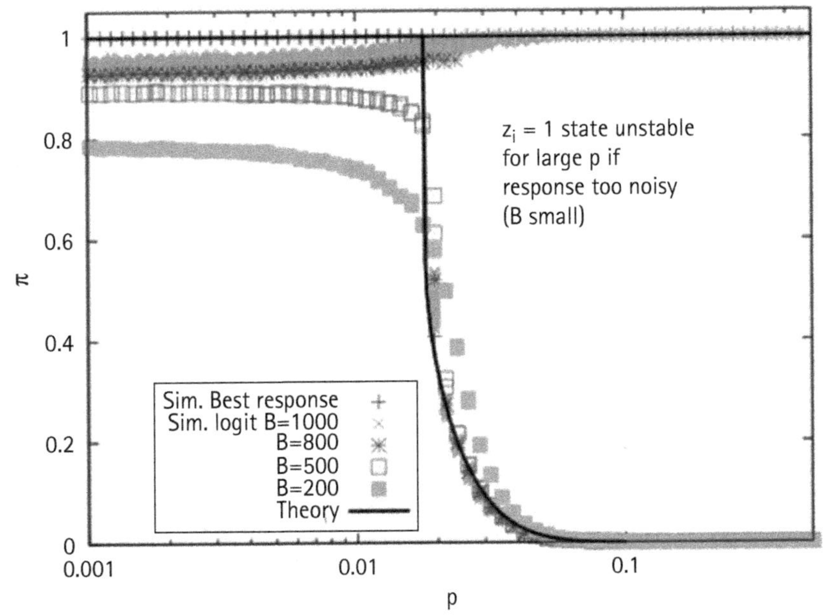

FIGURE 2.12 The evolution of the equilibrium state as p increases (Source: Anand et al. 2010).

discussed earlier, which has the required property. Thus, the probability of choosing $z_i = 1$ is given by:

$$P(z_i - 1) = \frac{e^{\beta u_i(1)}}{e^{\beta u_i(1)} + e^{\beta u_i(0)}} \tag{2.17}$$

where β is a parameter indicating the sensitivity of the agent to the difference between the payoffs from the two strategies. If $\beta = 0$ the agent chooses one of the two strategies at random, whereas if $\beta \to \infty$ then the probability of choosing the best response goes to one.

With the noise in the decisions of the agents, the system switches suddenly from one equilibrium to another. There are two things to observe here. First, a continuous evolution of p, the perceived probability of default or toxicity, leads to a sudden and large change in the equilibrium state. This, in turn, provokes a sharp decline in the prices of the asset-backed security, which is just what was observed and shown in Figure 2.12. The fact that the collapse occurred later for better rated MBS reflects the effect of the ratings on perceived probabilities rather than any real differences in those probabilities across assets. In fact, Pozsar et al. (2010) found that MBS ratings did not fully reflect publicly available data.

The second important observation is that the existence of a certain amount of noise in the decisions of agents leads to the selection of a particular equilibrium. In the context, it is a very partial equilibrium since the evolution of p has been taken as exogenous, and to fully model the process this would also need to be modeled.

However, in a situation where agents are influenced by each others' decisions and where their decision making is not fully "rational," we capture some important empirical facts. The individuals involved are far from the infinitely farsighted optimizers of standard models and are making relatively simple binary decisions, based on the actions of their partners. This can lead to major changes in the aggregate state of the market. Again this is clearly not a comprehensive model of what is, in reality, a very complicated system, but it does capture some of the characteristics that lead to major aggregate shifts without any specific major exogenous shock.

A more ambitious goal would be to build a model in which there are no equilibria and in which the market, its organization, and the behavior of the agents are constantly and simultaneously evolving.

As should be clear by now, analyzing the financial sector makes little sense without taking into account its network structure(s). The very definition of the nodes and the links between them already poses problems. Yet, most theoretical models of the financial network impose a very simple structure on the nodes and their relations. As soon as one tries to define carefully what the nature of a link is, and furthermore one tries to model how these links might be built or broken, and in addition one models the different nature of the financial entities, the situation looks very different from that of the simple but tractable theoretical models which have been built so far. This is why many of the more empirically focused studies of financial networks are often based on simulations.

2.2.4 The Evolution of the System and its Regulation

In looking at the evolution of the banking system, we see why the simple market paradigm, with banks acting as intermediaries between lenders and borrowers is wholly inadequate as a model of the actual transactions system.

> The demonstrated ability of regulated banking institutions to adapt to the changing environment suggests that there may be much to learn about the future evolution of intermediation directly from the observation of banks. Risks are still likely to be concentrated in other parts of the system—that is, outside of banks' balance sheets—but there is a good chance a bank will be involved in new mutations of the intermediation system, either directly or indirectly. This observation thus suggests a new role for bank supervisors: In addition to carrying out their main mandate of monitoring the health of banking firms, supervisors could contribute to dynamic and forward-looking oversight of the whole system of financial intermediation as it continues to evolve.
>
> —Cetorelli et al. 2012, pp. 10–11

What is being argued here is that the system still serves an overall purpose of financial intermediation, but that it is far from being the simple process described in most economic models. As the authors suggest, we ignore the evolution of the structure of the system and all the feedbacks involved and the consequent potential fragility at our peril. When integrating the financial sector into macroeconomic models, account has to be taken of the network structure of that sector, and the way in which contagion can occur between entitities which might seem, at first sight, to be robust to shocks. In fact, the banking sector reflects many of the network features to be found elsewhere in the economy. It is very difficult to find a sector that operates in the way that our basic models describe. As soon as one starts to examine in detail any particular market, firm, or sector, one is forced to admit that the characteristics of the relations within and between different entitities plays a major role in determining the aggregate outcomes.

2.3 Conclusion

The basic message of this chapter is that we should start from networks rather than the standard model of the isolated optimizing individual as our benchmark for economic models. But this does not mean that the sort of network models, particularly the theoretical ones, currently being built will provide easier and simpler answers to economic problems. Copernicus' ideas did not immediately conquer the world of astrophysics, partly because the predictions of his "model" were not demonstrably superior to those of the much-modified Ptolomean construct. Indeed, we see echoes of this in today's discussions in economics, where it is claimed by some that the old model can be recovered if certain limitations are imposed on the underlying interactions. Acemoglu et al., in this handbook, show how important the nature of the interaction and aggregation functions are in this regard. Until the analysis of the cases in which the limitations do not hold progresses further, economists will prefer to work within the more Ptolomean setting. Recall that the Copernican shift had to wait for the underlying theory to be more fully developed. It needed Galileo and Kepler to consolidate the paradigm shift. It would be foolish to put the controversies in economics on that level. Still, it is interesting to see that Goyal, a leading contributor to the development of network theory in economics, in his perspectives chapter in this book, feels that networks in economics have become, in Kuhnian terms, "normal science." For Kuhn, this would mean that all the scientists in the discipline have shifted their perspective and have taken networks as the central focus of attention, or put more bluntly consider networks to be the benchmark model in economics. It is not clear to me that this is the case. As Goyal rightly points out, economics has basically developed along two lines. The first, and still the standard benchmark, is that of a large number of agents interacting through some central mechanism but without much consideration for the structure of individual interactions. The second is the game theory approach, which typically deals with rather small numbers of sophisticated and strategic agents and studies the equilibria in such situations. Goyal believes there is a clear and unresolved tension between the two and perhaps the most empirically relevant situation is one in which there are many individuals who are locally linked but largely unaware of the way in which the system as a whole develops. A full paradigm shift would not endeavor to make progress in either the competitive model or the pure game-theoretic framework but would rather take the more realistic intermediate sort of situation as a starting point. It would not relegate networks and network theory to the status of a tool which has been used to give useful insights into those situations that economics has typically struggled with. Yet, it takes time for a paradigm shift to consolidate and it may be the case that network theory will simply consolidate our previous science. I suspect that the change will be more radical than that as the network approach continues to influence economics. But it is important to understand that it is unlikely to lead to models with the simple causal relationships so sought after by macroeconomists.

In this regard, it is worth noting the caveats expressed by Johnson, a leading expert in complex networks.

> In models of complex systems and networks, tiny changes in the model's assumptions—or changes in what it means to be a node, a link or 'infectious'—can inadvertently invert the emergent dynamics, for example by turning a stable output into an unstable one. Such changes can therefore amplify the inherent risk in any resulting policy suggestions. There is already substantial consensus that policy-makers need to embrace financial market risk within the framework of complex dynamic systems. However, markets contain many heterogeneous objects, the interactions of which may change in any number of ways in the blink of an eye (or the click of a mouse). This new dynamic regime, in which the character of both the links and nodes can change on the same timescale, lies well beyond standard models of ecological food webs, disease spreading and networks. The resulting dynamic interplay can generate unexpectedly large market fluctuations—and it is these that invalidate the financial industry's existing approach to the pricing of financial derivatives and the management of risk.
>
> —Neil Johnson 2011, p. 302

Although he rightly suggests that we should avoid the hubris of thinking that the sort of network models currently being introduced can be immediately and risklessly applied to analyze economic problems and suggest policy measures, two things are worth noting. In the first place his general criticism of the applicability of models applies just as much to standard economic models as to network-based models. Secondly, the fragility he describes is due to the intrinsically dynamic nature of networks and as Jackson observes in his chapter in this book, this is one of the major challenges for the use of networks in economics. Taking this dynamic aspect into account makes prediction more difficult, but it cannot be worse than working with mechanical models in which no such dynamics are present and arriving at totally erroneous conclusions.

References

Abreu, Dilip, and Mihai Manea (2012). "Bargaining and efficiency in networks." *Journal of Economic Theory* 147(1), 43–71.

Acemoglu, Daron, Vasco M. Carvalho, Asuman Ozdaglar, and Alireza Tahbaz-Salehi (2012). "The network origins of aggregate fluctuations." *Econometrica* 80(5), 1977–2016.

Allen, Franklin, and Ana Babus (2009). "Networks in finance." In *The Network Challenge: Strategy, Profit, and Risk in an Interlinked World*, Paul R. Kleindorfer and Yoram Wind, eds., 367–382. Philadelphia: Wharton School.

Allen, Franklin, and Douglas Gale (2000). "Financial contagion." *Journal of Political Economy* 108, 1–33.

Anand, Kartik, Alan Kirman, and Matteo Marsili (2013). "Epidemics of rules, rational negligence and market crashes." *The European Journal of Finance* 19, 438–447.

Ashcraft, A. B., and T. Schuermann (2008). "Understanding the securitization of subprime mortgage credit." *Federal Reserve Bank of New York Staff Report* 318, March.

Battiston, S., D. Delli Gatti, M. Gallegati, B. Greenwald, and M. Gallegati (2012). "Liaisons dangereuses: Increasing connectivity, risk sharing, and systemic risk." *Journal of Economic Dynamics & Control* 36, 1121–1141.

Bougheas, Spiros, and Alan Kirman (2014). "Complex financial networks and systemic risk: A review." *CESifo Working Paper Series* No. 4756.

Buffett, Warren (2002). "Berkshire Hathaway annual report."

Bush, R. R. and F. Mosteller (1955). *Stochastic Models for Learning*. New York: Wiley.

Cetorelli, Nicola, Benjamin H. Mandel, and Lindsay Mollineaux (2012). "The evolution of banks and financial intermediation: Framing the analysis." *Federal Reserve Bank of New York Economic Policy Review* 18(2), 1–12.

Coleman, James S. (1986). "Social theory, social research, and a theory of action." *The American Journal of Sociology* 91(6), 1309–1335.

Cournot, A.-A. (1838). *Recherches sur les principes mathématiques de la théorie des richesses*. Paris: L Hachette.

De Vroey, M. (1999). "Transforming Walras into a Marshallian economist: A critical review of Donald Walker's *Walras's Market Models*." *Journal of the History of Economic Thought* 21, 413–435.

Durlauf, Steven N. (2012). "Complexity, economics, and public policy." *Politics, Philosophy & Economics* 11(1), 45–75.

Easley, David, and Jon Kleinberg (2010). *Networks, Crowds, and Markets. Reasoning about a Highly Connected World*. Cambridge: Cambridge University Press.

Edgeworth, F. Y. (1881). *Mathematical Psychics: An Essay on the Application of Mathematics to the Moral Sciences*. London: Kegan.

Elliott, Matthew, Benjamin Golub, and Matthew O. Jackson (2014). "Financial networks and contagion." *American Economic Review* 104(10), 3115–3153.

Feldman, Allan (1973). "Bilateral trading processes, pairwise optimality, and Pareto optimality." *The Review of Economic Studies* 40(4), 463–473.

Feldman, Allan (1974). "Recontracting stability." *Econometrica* 42(1), 35–44.

Gabaix, Xavier (2011). "The granular origins of aggregate fluctuations." *Econometrica* 79, 733–772.

Gai, Prasanna, and Sujit Kapadia (2010). "Contagion in financial networks." *Proceedings of the Royal Society A* 466(2120), 2401–2423.

Geertz, C. (1978). "The bazaar economy: Information and search in peasant marketing." *American Economic Review* 68, 28–32.

Gertler, M. and N. Kyotaki (2011). "Financial intermediation and credit policy in business cycle analysis." In *Handbook of Monetary Economics*, Volume 3, B. Friedman and M. Woodford, eds. Amsterdam: North-Holland.

Goyal, S. (2007). *Connections: An Introduction to the Economics of Networks.* Princeton, NJ: Princeton University Press.

Graddy, K. (1995). "Testing for imperfect competition at the Fulton fish market." *Rand Journal of Economics* 26, 75–92.

Graddy, K., and G. Hall (2010). "A dynamic model of price discrimination and inventory management at the Fulton fish market." *Journal of Economic Behavior and Organization* 80(1), 6–19.

Haldane, A. (2009). "Rethinking the financial network." Speech delivered at the Financial Student Association, Amsterdam.

Haldane, A. G., and R. M. May (2011). "Systemic risk in banking ecosystems." *Nature* 469, 351–355.

Hayek, Friedrich A. (1945). "The use of knowledge in society." *American Economic Review* 35(4), 519–530.

Hobbes, Thomas (1651). *De Cive, or the Citizen.* New York: Appleton-Century-Crofts, 1949.

Holland, J. H. (1976). "Adaptation." In *Progress in Theoretical Biology IV*, R. Rosen and F. M. Snell, eds., 263–293. New York: Academic Press.

Ioannides, Y. (2013). *From Neighborhoods to Nations: The Economics of Social Interactions.* Princeton, NJ: Princeton University Press.

Iori, Giulia, Saqib Jafarey, and Francisco G. Padilla (2006). "Systemic risk on the interbank market." *Journal of Economic Behavior & Organization* 61(4), 525–542.

Jackson, Matthew O. (2008). *Social and Economic Networks.* Princeton, NJ: Princeton University Press.

Johnson, N. (2011). "Proposing policy by analogy is risky." *Nature* 469, 301.

Kuhn, Thomas (1962). *The Structure of Scientific Revolutions.* Chicago, IL: University of Chicago Press.

Latour, B. (2005). *Reassembling the Social: An Introduction to Actor Network Theory.* Oxford: Oxford University Press.

Lux, T. (2011). "Network theory is sorely required." *Nature* 469, 302.

Manea, Mihai (2011). "Bargaining in stationary networks." *American Economic Review* 101(5), 2042–2080.

McClean, P., and J. F. Padgett (1997). "Was Florence a perfectly competitive market? Transactional evidence from the Renaissance." *Theory and Society* 26(2), 209–244.

Montagna, Mattia, and Thomas Lux (2013). "Contagion risk in the Interbank Market: A probabilistic approach to cope with incomplete structural information," FinMaP-Working Papers no. 8.

Morin, Edgar (2007). "Restricted complexity, general complexity." In *Worldviews, Science, and Us: Philosophy and Complexity*, Carlos Gershenson, Diederik Aerts, and Bruce Edmonds, eds. Singapore: World Scientific.

Nier, E., J. Yang, T. Yorulmazer, and A. Alentorn (2007). "Network models and financial stability." *Journal of Economic Dynamics and Control* 31(6), 2033–2060.

Oullier, O., A. P. Kirman, and J. A. S. Kelso (2008). "The coordination dynamics of economic decision-making: A multi-level approach to social neuroeconomics." *IEEE Transactions on Neural and Rehabilitation Systems Engineering* 16(6), 557–571.

Poincaré, H. (1908). *Science et Méthode*. Paris: Flammarion.

Pozsar, Z., T. Adrian, A. Ashcraft, and H. Boesky (2010). "Shadow banking." *Federal Reserve Bank of New York Staff Reports*, no. 458, July.

Roth, A. E., and I. Erev (1995). "Learning from extensive-form games: Experimental data and simple dynamic models in the intermediate term." *Games and Economic Behavior* 8, 164–212.

Rubinstein, A., and A. Wolinsky (1985). "Equilibrium in a market with sequential bargaining." *Econometrica* 5, 1133–1150.

Shiller, Robert (2011). "The neuroeconomics revolution." Project *Syndicate*, November 21.

Varela, L. M., G. Rotondo, M. Ausloos, and J. Carrete (2015). "Complex network analysis in socioeconomic models." In *Complexity and Geographical Economics: Topics and Tools*, Pasquale Commendatore, Saime Suna Kayam, and Ingrid Kubin, eds. Berlin, New York: Springer, Verlag.

Vega-Redondo, F. (2007). *Complex Social Networks*. Econometric Society Monograph Series. Cambridge: Cambridge University Press.

Walker, D. A. (1996). *Walras's Market Models*. Cambridge: Cambridge University Press.

Weisbuch, G., O. Chenevez, A. Kirman, and J.-P. Nadal (1998). "A formal approach to market organization: Choice functions, mean field approximations and maximum entropy principle." In *Advances in Self-Organization and Evolutionary Economics*, J. Lesourne and A. Orlean, eds., 149–159. Paris: Economica.

Weisbuch, G., A. Kirman, and D. Herreiner (2000). "Market organisation and trading relationships." *Economic Journal* 110, 411–436.

Whittle, Andrea, and André Spicer (2008). "Is actor network theory critique?" *Organization Studies* 29, 611.

CHAPTER 3

NETWORKS IN ECONOMICS A PERSPECTIVE ON THE LITERATURE

SANJEEV GOYAL

3.1 Introduction

The study of networks is one of the most dynamic fields of research in economics today. Research on networks is regularly published in the top journals of the profession. The best university presses bring out research monographs on networks written by economists. Researchers in networks win prizes and are awarded prestigious grants from research funding bodies. Some of the best graduate students in economics choose to work on networks. Governments, nongovernmental organizations, and private companies all express an interest in the findings of the networks research community.

When the editors of this handbook invited me to write a perspective piece, I felt that it was a good moment to pause, stand back, and take stock.

In this chapter, I will argue that at the start of the 1990s, the introduction of networks in economics marked a major break with existing practice. Halfway through the third decade of its life, the study of networks has come of age. It is now "normal science" in the sense of Kuhn (1962).

I start by drawing out the ways in which networks constitute a radical departure from standard models in economics in the 1980s. This will be illustrated with a case study: the research on social learning, before and after the introduction of networks. The case study will draw attention to a major methodological innovation: locating individuals within a general framework of social connections. The analysis yields a number of insights that relate social structure to individual behavior, learning dynamics, diffusion, and welfare.

I would like to thank Nizar Allouch, Matthew Elliott, Julien Gagnon, Edoardo Gallo, Manuel Mueller-Frank, Anja Prummer, and the editors for their comments on an earlier version of the paper.

The second step in my argument shows how initial breakthroughs led to the investigation of major new questions. The crucial role of social structure motivated a study of the origins of networks: where do they come from and what do they look like? This is a question that was not viewed as especially important or interesting prior to the research on social learning. I present a case study on the theory of network formation to sketch its key ingredients and its main insights. A distinctive feature of the research on network formation is a very wide range of applications. I illustrate the scope of the theory of network formation by showing that it provides an elegant resolution to two long-standing puzzles in the social sciences: one, racial homophily patterns in friendship; and two, the law of the few in social communication.

I then argue that as the research on behavior in fixed networks and on network formation has matured, it has taken on progressively more ambitious themes. This sets the stage for a discussion of the nature of research on networks today. I will argue that this work now encompasses the classical notions of preferences, individual choice, strategic behavior, competition and prices. As a result, networks are now becoming central to improving our understanding of macroeconomic volatility and cycles, patterns of international trade, contagion and risk in financial and social systems, resilience of infrastructure and supply chains, economic development, unemployment and inequality, and a host of other important phenomena. The present handbook offers a panoramic view of the tremendous vitality of this research.

By way of conclusion, I discuss some tensions between traditional models in economics and models in networks and how these tensions shape current research practice.

3.2 Networks in Economics: A Significant Departure

This section develops the first step in the argument. I show that networks mark a radical departure and go beyond the traditional model in economics in the late 1980s. The theoretical framework of economics was then organized around concepts for the study of interaction in small groups (game theory) and interaction among large groups (competitive markets and general equilibrium). A number of phenomena appear to arise in between these two extremes. By way of illustration and for concreteness, I take up and discuss the case of learning and diffusion. The received models in the 1980s appeared to be inadequate both from the viewpoint of introspection and failing to account for empirical patterns. This led to the first attempts at modeling decision making by individuals embedded in a social network. I sketch the main ingredients of these first set of models and then show how they depart from received models. I discuss the main insights and relate them to subsequent developments in the field.

The diffusion of new ideas, opinions, products, and technologies is key to understanding social change, growth, and economic development. Traditionally, economists have focused on a combination of individual heterogeneity and unknown profitability to explain patterns in diffusion; a prominent early contribution is Griliches (1957).[1] Through the 1970s and 1980s the study of information imperfections and asymmetries occupied center stage in economics. As information has economic value, it was natural to ask if starting from incomplete or imperfect information, individuals would "acquire" information and "learn" the "optimal" action. In an early paper, Rothschild (1974) showed that a patient and dynamic optimizing agent will stop learning and hence will get locked into a suboptimal action, with positive probability. This observation was followed by a large and technically sophisticated literature on single agent learning that continued through the 1980s and well into the 1990s. Alongside this work, there emerged a parallel line of work on multi-agent learning: do individuals learn rational expectations and do individuals learn to play Nash equilibrium (Fudenberg and Levine 1998; Evans and Honkapohja 2001)?

The research focused on a single individual or on groups where interaction was uniform and homogenous. However, the early work of Lazarsfeld, Berelson, and Gaudet (1948), Katz and Lazersfeld (1955), Coleman (1966), Granovetter (1973, 1974), and Ryan and Gross (1943) in sociology, Hagerstrand (1967) in economic geography, and Rogers (1983) in communications points to a world that lies somewhere in between these two extremes: individuals, be they farmers or consumers or doctors, typically interact with only a small subset of the group. These small subsets—the neighborhoods—are stable and overlap in rich and complicated ways with the neighborhoods of others. The implicit argument in Coleman (1966) and Hagerstand (1968) is that these connections are the conduit through which information and influence flow and this shapes the diffusion of new ideas and practices.

The analysis of many actors located within complex networks strains the plausibility of delicate chains of strategic reasoning, which are typical in game theory. Similarly, the "local" interaction in social networks makes anonymous competitive equilibrium analysis implausible. This tension between the small and the large is fundamental to understanding behavior and it posed a challenge to the theory of the 1980s and motivated a major advance: a framework with multiple agents, making repeated choices and embodied in a stable social network. Early papers in this tradition include Bala and Goyal (1998, 2001) and Ellison and Fudenberg (1993).

I focus on Bala and Goyal (1998, 2001) as they present a framework that combines a general model of networks with individual choice and learning dynamics. The key innovation is embedding rational individuals in a (directed) network. Individuals do not know the true value of different actions. So they experiment individually and use their own past experience, but also use the experience of their friends and neighbors to guide their decision making. Moreover, their neighbors in turn use the experience of their friends, and so forth. In this way, information flows across the connections of a network. The interest is in understanding how the social structure shapes individual choice, belief dynamics, and the diffusion of actions.[2]

[1] For a brief summary of the state of the literature in the 1980s, see Feder, Just, and Zilberman (1985).

[2] For a survey of other early attempts to embed individual choice within social structure, see Kirman and Zimmerman (2001).

Bala and Goyal (1998, 2001) develop a number of general results. Their first result draws out an important implication of network connections: if two individuals are directly or indirectly linked then, in the long run, they will learn from each other, make similar decisions, and earn the same amount. Thus local learning ensures that all agents in a *connected* society obtain the same utility, in the long run. Second, they show that highly connected *"hub"* players can hinder learning and the diffusion of desirable practices. Thus societies with influential hub nodes may get locked into inferior practices. Third, they develop sufficient conditions on social structure—a combination of connectedness and local autonomy (that allows space and time for individual experimentation)—that guarantee convergence to the optimal action. Fourth, the simulations of the model generate patterns of diffusion through local spread that are consistent with observed empirical patterns.

Furthermore, the methodological innovations in the framework are worth noting. These papers introduced a number of concepts from graph theory—directed graphs, connectedness, inequalities in connectivity (hubs and spokes), and the combination of node heterogeneity and network structure. Second, they located incomplete information, individual beliefs revision, and dynamics of choice within a directed graph. This combination of individual choice and graphs is central to subsequent work on networks in economics.

These papers mark a radical departure with the traditional model in economics, as practiced in the 1980s. At that time, as pointed out earlier, economists focused their efforts at understanding single-agent learning or public learning. The work in the early 1990s introduced graphs—with their attendant concepts. This permits the study of large social groups with overlapping neighborhoods. It goes beyond the established models for small groups as well as those that dealt with large groups. These early papers on social learning also foreshadow some of the tensions in the literature that shape current research practice. I briefly turn to this tension now.

The goal of the papers was to understand how Bayesian/fully rational individuals learn in a social network. The complications involved in making inferences about what neighbors observe and learn obliged the authors to make the assumption that an individual ignores the information about actions and experience of neighbors possibly contained in the choice of action of the neighbor. This simplifying assumption reflects a tension that is a recurring theme in the research in networks. The tension arises from problems of tractability: models with fully rational agents and general network structures are difficult to analyze, especially in terms of deriving a clear relation between the network structure and individual behavior. It is also difficult to incorporate heterogeneity in a tractable way within a network model with fully rational agents.

As interest in social networks has grown, interest has moved to the study of a tighter characterization of how social networks shape learning. These questions posed difficult technical challenges and held up progress for a number of years; it is only in recent years that there has been a revival of interest in learning and diffusion in social networks. This new research has proceeded broadly along two lines: one strand of work has simplified the individual decision-making ingredient of the model and focused attention on rich social networks and individual heterogeneities. The other strand maintains (and generalizes) the assumptions on individual beliefs and decision making but works within a special class of networks.

The potential of the first line of attack is well reflected in the papers that build on the DeGroot model (1974) of opinions and consensus; see DeMarzo, Vyanos, and Zweibel (2003), Golub and Jackson (2010, 2012). In their recent work, Golub and Jackson provide a number of important results on this subject. In their 2010 paper, building on the work of DeMarzo, Vyanos, and Zweibel (2003), they show that vanishing social influence is both necessary as well as sufficient in a context of naive learning to guarantee that complete learning obtains in a large society. This result nicely complements the analysis of incomplete learning in the Bala and Goyal (1998) framework, and it illustrates the value of moving to a simpler, reduced-form model of beliefs and choice. In their 2012 paper, Golub and Jackson study how the speed of learning in a society depends on homophily: the tendency of agents to associate disproportionately with those having similar traits. When agents' beliefs or behaviors are developed by averaging what they see among their neighbors, convergence to a consensus is slowed by the presence of homophily, but is not influenced by network density. The key contribution is a new measure of homophily based on the relative frequencies of interactions among different groups. These findings are very important in view of the presence of homophily in social relations (reported below in the network formation section).

The second strand of the literature maintains the assumption of Bayesian/rational decision makers: as the belief dynamics are generally nonlinear, there are technical hurdles to a full understanding of the role of social networks in shaping learning and rates of convergence. This challenge has motivated an interesting program of research; recent contributions include Acemoglu et al. (2011), Jhadbabaie, Molavi, and Tahbaz-Salehi (2013), Mossel, Sly, and Tamuz (2015), and Mueller-Frank (2013), and others.

The work of Acemoglu et al. (2011) reflects the potential in this line of work. This paper builds on the pioneering work of Banerjee (1992) and Bikhchandani, Hirshleifer, and Welch (1992) and combines it with the theory of random graphs. There is a single sequence of privately informed individuals who take one action each. Before making his choice an individual gets to observe the actions of all the people who have made a choice earlier. The actions of his predecessors potentially reveal their private information. An individual can therefore use the information revealed via the actions of others along with his own private information to make decisions. The principal question is: do individuals eventually learn and choose the optimal action?

In their work, Banerjee (1992) and Bikhchandani, Hirshleifer, and Welch (1992) showed that there was a possibility of herding on the inefficient action. In their model all individuals observe everyone who has gone before them before making a choice. Acemoglu et al. (2011) relax this assumption. They propose that an individual draws a sample from the past individuals. The focus is on properties of this sample. They show that learning is complete and the optimal action is chosen with probability 1 if the sample is "expanding": expanding observations implies that the probability of an agent being observed goes to 0 as we move across time. This in turn means that there is a bound on the influence an individual can have in the long run. This is sufficient to ensure that privately generated information is not blocked out and that agents eventually choose the optimal action in the long run.

The details of the arguments and the technical methods differ across Bala and Goyal (1998), Golub and Jackson (2010) and Acemoglu et al. (2011), but they point to a similar idea: in an information-rich world, social learning obtains when no single individual exercises social influence.

The initial motivation underlying the study of social learning in networks came from the empirical patterns on spatial and temporal diffusion of ideas, products, and technologies. As the theoretical work has progressed, economists have examined the empirical implications of networks for learning and diffusion of networks. Important contributions to this line of work include Conley and Udry (2010) and Banerjee et al. (2013). For a survey on the role of networks in economic development, see Munshi (2014).

Conley and and Udry (2010) investigate the role of social learning in the diffusion of a new agricultural technology in Ghana. The novelty of their work is a detailed description of each individual farmer's information neighborhood. They find evidence that farmers adjust their inputs to align with those of their information neighbors who were particularly successful in previous periods. As a check on the interpretation of these social learning effects they also study input choices for another crop, of known technology: there are no social learning effects in the latter case.

Banerjee et al. (2013) study the diffusion of micro-finance in Indian villages. They exogenously vary the "injection points" of information across villages and ask how differences in network characteristics of the initial seeding nodes affects eventual adoption of micro-finance. Their main finding is that micro-finance is significantly higher when the injection points have higher eigenvector centrality. The authors also estimate a model of information diffusion and adoption using detailed data on demographic and network variables. These papers mark a departure from the earlier tradition of empirical work in its explicit treatment of network architecture. It also implicitly hints at the substantial advances in data availability and computing power. Big data is a broad-scale technological change, and this empirical work reflects an important point of contact with economics.

By way of summary, this case study brings out three general points. The first is methodological: around the early 1990s we see the first set of models that incorporate individual choice and network structure within a common framework. This marks a major advance in the conceptual framework and it brings the concepts and the insights of graph theory and networks into the mainstream of economics. This expansion also points to problems of tractability that shape future research. The second is the collection of analytical results: the role of social structure, in particular the role of hubs in shaping learning and creating blockages and lock-ins. These results motivate a large and flourishing current research program on how structure shapes behavior; a prominent instance of this is the large body of work on games on networks.[3] The third point is the move toward the study of large data sets on networks and of dynamics of behavior on these networks. This hints to the impact of big data on research in economics; we will return to this point again in subsequent sections. Finally, the case study highlights how an economic approach combines with networks to deliver insights in areas that were previously the reserve of sociologists and geographers.[4]

3.3 The Origins of Networks

The finding that the social structure has large effects on individual behavior and well-being motivates an examination of the structure of social networks and very quickly leads to a study of their origins. This section develops the second part of the argument—the emergence of entirely new research questions—presents a case study of the theory of network formation.

At the very outset, it is worth emphasizing the novelty of the approach: the traditional approach in sociology and related social sciences focuses on the effects of social structure on behavior (Granovetter 1985; Smelser and Swedberg 2005). The economic approach to network formation locates the origins of networks in individual choice. It therefore puts the traditional approach firmly on its head!

The second and equally interesting point to note is that a "framework," once in place, begins to formulate entirely new questions, questions which would have appeared meaningless or without any great interest prior to its emergence. The new questions on network formation deepen the original investigations—on the effects of networks on individual behavior and learning dynamics—and they expand and reshape the research program in a profound way.

The key innovation is the idea that the social (and economic) structure itself is created through purposeful individual activity. This is a powerful idea and it has had far-reaching effects: it has deepened our understanding of classical questions in economics and the other social sciences and motivated altogether new lines of enquiry.

The beginnings of the theory of network formation can be traced to the work of Boorman (1975), Aumann and Myerson (1988), and Myerson (1991). The general framework and a systematic theory of network formation was first presented in Bala and Goyal (2000) and in Jackson and Wolinsky (1996). The two papers present complementary approaches to the process of network formation. Over the past two

[3] For early work in this field, see Goyal and Moraga (2001), Bramoulle and Kranton (2007), and Ballester, Calvo-Armenogol, and Zenou (2006); for surveys of this work see Goyal (2007) and Jackson and Zenou (2014).

[4] One of the distinctive aspects of research on networks in economics is that it is concurrent with a very broad interest in networks across the social and the information sciences. For a discussion of the distinctiveness of the economic approach and the many points of contact and overlap between the different disciplines, see Goyal (2007).

decades, this has been a very active field of research; one of its great successes has been a wide and expanding range of applications. As the applications have progressed, empirical analysis and experimental investigations have also gathered momentum.

In this section I briefly sketch the building blocks and some of the main insights of the theory of network formation. I then illustrate the scope of the theory through a discussion of two classical questions in the social sciences.

I first take up the approach of **unilateral link formation**. This approach was introduced in Goyal (1993) and systematically studied in Bala and Goyal (2000). Consider a collection of individuals, each of whom can form a link with any subset of the remaining players. Link formation is unilateral: an individual can decide to form a link with another individual by paying for the link. It takes time and effort to sustain a link. A link with another individual allows access, in part and in due course, to the benefits available to the latter via his own links. Thus individual links generate externalities whose value depends on the level of decay/delay associated with indirect links. As links are created on an individual basis, the network formation process can be analyzed as a noncooperative game. The paper allows for general payoffs—increasing in number of people accessed and declining in number of links formed.

There are interesting practical examples of this type of link formation—hyperlinks across webpages, citations, "following" relations on Twitter, and gifts. But the principal appeal of this model is its simplicity. This simplicity allows for a systematic study of a number of central questions concerning social and economic networks.

Bala and Goyal (2000) provide a characterization of the architecture of equilibrium networks. The equilibrium networks have simple architectures: star (hub-spoke) networks and the cycle are salient. This prediction of great heterogeneity in connections and the presence of highly connected hub nodes is an important theoretical contribution. In the star network, the central hub node will generally earn much larger payoffs as compared to the peripheral nodes. Thus, the theory provides a foundation for the idea that social structures may sustain great inequality. This part of the theory ties in closely and gains significance in the context of the earlier discussion on patterns of social learning in networks with highly connected hubs.

Bala and Goyal (2000) also introduce the study of dynamics of linking: at every point, individuals may revise their current links. Their main insight is that individual efforts to access benefits offered by others lead, rapidly, to the emergence of an equilibrium social network, under a variety of circumstances. This provides a foundation to the idea that certain social structures—including those that are very unequal—may be dynamically stable.

One virtue of the unilateral link formation model is that it allows us to combine graphs and the tools of noncooperative game theory. This approach has subsequently been used in the study of a variety of economic problems. For a survey of this work, see Bloch and Dutta (2012) and Goyal (2007).

I turn next to **two-sided** or **bilateral link formation**. This approach was introduced and developed in Jackson and Wolinsky (1996). A link between two players requires the approval of both players involved. This is the natural way to think about link formation in a number of social and economic contexts, such as the formation of friendship ties, co-authorship, collaborations between firms, trading links between buyers and sellers, and free-trade agreements between nations.

The simplest way to think of two-sided link formation is to imagine an announcement game along the lines of the game sketched by Myerson (1991). Each player announces a set of *intended* links. A link between two individuals A and B is formed if both A and B announce an intention to create a link. In a context where links are two sided there are elements of "cooperation" involved in the formation of a link, and so solving such games calls for new concepts.

It is useful to begin the discussion with the familiar notion of Nash equilibrium as this will illustrate some of the conceptual issues that arise in the study of network formation with two-sided links. If every player announces that she wants to form no links then a best response is to announce no links. In other words, the "empty" network is a Nash equilibrium for any network formation game. To overcome this type of coordination failure, Jackson and Wolinsky (1996) propose the concept of *pairwise stable networks.*

A network is said to be pairwise stable if no individual wishes to delete a link and if no two unlinked individuals wish to form a link: pairwise stability looks at the attractiveness of links in a network *g*, *one at a time*. Formally, every link present in a stable network must be profitable for the players involved in the link. For every link not present in the network it must be the case that if one player strictly gains from the link then the other player must be strictly worse off.

The second important contribution of the Jackson and Wolinsky (1996) paper was the result of a conflict between stability and efficiency. This highlights the pervasive externalities in linking activity and is a recurring theme in the subsequent research in this area.

The great attraction of pairwise stability is its simplicity. For any network it is relatively easy to check whether the two conditions are satisfied. The theoretical properties of this solution concept have been developed systematically and the solution concept has been widely applied; for a survey of this work, see Bloch and Dutta (2012) and Jackson (2008).

The network formation framework has motivated a vibrant theoretical, empirical, and experimental literature. For book-length overviews, see Goyal (2007) and Jackson (2008). For a recent overview on network formation models and applications, see the handbook on social economics edited by Bisin, Benhabib, and Jackson (2012).

One of the great successes of this research has been an extensive set of economic applications. Examples of this work include formation of research collaboration networks (Goyal and Joshi 2003), core and periphery in networks (Hojman and Szeidl 2008), structural holes in trading networks (Goyal and Vega-Redondo 2007), networks and coordination (Goyal and Vega-Redondo 2005; Jackson and Watts 2002), co-author networks (Jackson and Wolinsky 1996), collusion networks (Belleflamme and Bloch 2004), information networks (Galeotti and Goyal 2010), peer networks (Cabrales, Calvo-Armengol, and Zenou 2011), labor market networks (Calvo-Armengol 2004; Calvo-Armengol and Jackson 2004), buyer-seller networks (Kranton and Minehart

2001), risk-sharing networks (Ambrus, Mobius and Szeidl 2014; Ambrus, Chandrashekhar, and Elliott 2014; Bloch, Genicot, and Ray 2008; Bramoulle and Kranton 2007), financial networks (Babus, 2008; Cabrales, Gottardi, and Vega-Redondo 2013; Farboodi 2014), cyberattack and network design (Goyal and Vigier 2014; Acemoglu, Malekian, and Ozdaglar 2014), and free-trade agreement networks (Goyal and Joshi 2006; Furasawa and Konishi 2007).

I now discuss two applications of this approach that highlight its ability to address important open questions in the social sciences.

Homophily and Social Structure: An important strand of the research examines the effect of heterogeneity on social networks. This work asks how differences in costs and benefits of linking affect linking behavior and shape the overall social network. Early work in this field extended the basic network formation model and looked for equilibrium or pairwise stable networks (for an overview, see Goyal 2007). In more recent years, interest has shifted to dynamic models of linking, and the key element explored is *homophily*.

It is generally agreed that human relations exhibit homophily: individuals prefer to be friends with others like themselves. There is a long and distinguished literature on patterns of homophily in friendships. For an influential early discussion, see Lazarsfeld and Merton (1954); for a recent survey, see McPherson, Smith-Lovin, and Cook (2001).

Three empirical regularities have been highlighted in this literature. One, larger groups tend to form more same-type ties and fewer other-type ties than small groups. Two, that larger groups form more ties per capita. Three, all groups exhibit an inbreeding bias: there are more friendships within a group relative to fraction in population (with the greatest bias being in middle-sized groups).

Currarini, Jackson, and Pin (2009) develop a model of friendship formation that helps to explain these homophily patterns. Individuals have types and access type-dependent benefits from friendships. The innovation here is the model of friendships: individuals form friendships taking into account their preferences and the relative proportions of different types in the population. The authors examine the properties of a steady-state equilibrium of a matching process of friendship formation. They show that the three empirical regularities arise as properties of the steady–state social relations if preferences exhibit biases.

The study of homophily and its consequences for inequality and segregation is a very active field of research (see, e.g., Bramoulle et al. 2014).

The Law of the Few: The classical early work of Lazarsfeld, Berelson, and Gaudet (1948) and Katz and Lazersfeld (1955) investigated the impact of personal contacts and mass media on voting and consumer choice with regard to product brands, films, and fashion changes. They found that personal contacts play a dominant role in disseminating information, which in turn shapes individuals decisions. In particular, they identified 20% of their sample of 4,000 individuals as the primary source of information for the rest. Moreover, there exist only minor differences between the observable characteristics of the influencers and the others. Recent empirical work on virtual social communities reveals a similar pattern of communication. How do we account for this pattern of specialization?

Galeotti and Goyal (2010) propose a model to study this question. This model combines the Bala and Goyal (2000) model of network formation with a model of local public goods developed by Bramoulle and Kranton (2007). They study a setting in which individuals choose to personally acquire information and to form connections with others to access the information these contacts acquire. Their main finding is that every (strict) equilibrium of the game exhibits the law of the few. The network has a core-periphery architecture; the players in the core acquire information personally, while the peripheral players acquire no information personally but form links and get all their information from the core players. The core group is small relative to number of individuals. They also show that a small heterogeneity in costs of acquiring information has strong effects: the individuals who have slightly lower costs of acquiring information constitute the core and acquire all the information. Thus strategic forces amplify small initial differences to create large differences in behavior and location in social structure.

A number of subsequent papers have explored models that combine behavior and formation of structure (see, e.g., Baetz 2014 and Hiller 2013).

The theoretical research on network formation has been accompanied by a growing sophistication in empirical investigations on the structure and dynamics of networks. It is useful to view this work as dealing with small and large networks. Small networks may involve oligopolistic firms or countries (and range from a few nodes to a hundred nodes), while large networks involve scientists, individual consumers, or webpages (and may contain from hundreds of thousands to millions of nodes). The theoretical models in economics provide a good account for the stylized features of small and medium-sized networks, but these models do not account well for the complexities of large and massive networks. As computational power has grown the interest and ability to map very large networks has also grown. This gap between the successful theoretical game theoretic models and the properties of large empirical networks now presents a major challenge to the received theory.

Empirical contexts appear to be rich in heterogeneity and in dynamics: this suggests that individuals will have limited information on the identity of nodes and on the structure of the network. An early response to this complexity is a paper by Jackson and Rogers (2007). They present a dynamic model of linking where newly born nodes use a combination of random linking and network-based linking to form new networks. The linking follows plausible rules of thumb, and individuals do not take into account the effects of their actions or the evolution of the network in making their choices. Nevertheless, the dynamics of linking generate a concrete set of predictions on network structure as a function of key parameters, such as the ratio of random versus network-based linking. The model provides us with a mechanism that shapes network evolution. In this sense it is close to the models in physics and mathematics. However, as the microstructure is kept minimal—preferences, information, and choice are not modeled explicitly—it is unclear how one can make use of the model for normative or policy experimentation. There is thus a tension between the standard approach in economics and the need to account for large-scale empirical networks.[5]

Empirical work also suggests that a relation between two individuals have different facets: a link may typically perform a variety of functions and they are interrelated. Existing theoretical work, on the other hand, assumes that a network performs a single function and studies this role in isolation. Moving from single-role networks to multiplex networks is an important challenge for future work.

To summarize, the theory of network formation provides an account for how individual action shapes social and economic structure. In founding a theory of social structure upon individual choice, it presents a major departure from earlier traditions in sociology and the other social sciences. The theory combines elements of the theory of games with probability theory and graph theory. It offers insights into a variety of real world phenomena, especially in the context of small and medium-sized networks. There, however, remains a tension between the stark predictions of the successful models of network formation and the complex properties of large and evolving networks.

3.4 The Route to Normal Science: Prices, Competition, and Networks

Throughout the 1990s, research on networks focused on behavior in given networks and on the theory of network formation. I would say that this research proceeded at a "pure" and "general" level with relatively little connection to applications. But, by the end of the decade, economists working in networks began to address specific economic questions and, as they did so, they began to develop models that included competition, prices, and markets. Indeed, the last decade has seen the emergence of a very rich and thriving research program in which these standard economic concepts play a central role. In this respect, there is a close analogy with the spread of game theory during the 1980s and 1990s in one applied field after another in economics. We are beginning to see a similar trend with networks.

The aim of this section is to illustrate this development through a discussion of a few notable examples. I find it useful to distinguish between "social" and "economic" networks; while this distinction is a little artificial in some applications, it is helpful in organizing the discussion.

[5] Galeotti et al. (2010) present one resolution to this tension. They assume that individuals have limited and local information of the complex network in which they are embedded. This leads them to develop a model of incomplete private information and examine how variations in the strategic structure of the game, the depth of network knowledge, and the underlying network structure jointly shape behavior.

3.4.1 Economic Networks and Markets

In the classical Walrasian model it is assumed that individuals are anonymous, that they can all trade with each other and that this trade takes place at a common price. In real-world markets buyers and sellers develop durable relations of exchange; these relations are personal and there are definite limitations on who can trade with whom. Moreover, terms of trade differ across traders and they depend on the network structure of relations; for empirical evidence, see, for example, Uzzi (1993) and Kirman and Vignes (1991), among others. These findings lead us to view markets as networks. We are then led to ask what are the incentives of buyers and sellers to form durable relations?

I start with a discussion of the **market as a network**. Kranton and Minehart (2001) consider a model with two stages. In stage 1, buyers unilaterally choose to form links with sellers. These links enable buyers to procure goods or inputs. Buyers trade off expected gains from trade against costs of link formation.[6] In stage 2, the valuations of buyers are realized; they then engage in trade with sellers restricted by the network structure defined in the first stage. The trading in stage 2 takes place through a centralized auction where at each price efficient matches are determined.

The paper establishes two major results: networks arise as a mechanism to pool uncertainty in demand and trade it off with the costs of establishing bilateral ties. The analysis also reveals, somewhat surprisingly, that an efficient allocation mechanism (ex-post competitive environment) is sufficient to align the buyers' incentives to form ties with the social incentives. For a systematic recent study of inefficiencies in networked markets, see Elliott (2014).

The Kranton-Minehart (2010) model pertains to direct buyer-seller ties: the modern economy is characterized by an extensive and complicated array of supply chains, that span industries and countries. A more recent strand of the literature studies pricing, contracting and competition in these supply chains (Choi, Galeotti and Goyal 2014; Gale and Kariv 2009; Goyal and Vega-Redondo 2007; Manea 2013; Kotowoski and Leister 2012; and Nava 2014). This work builds on the earlier tradition of games on networks. The study of supply chains is now also a major focus of study in macroeconomics: the interest here is in understanding how the network structure of production connections amplifies shocks and thereby generates more or less aggregate volatility (Acemoglu et al. 2012; Carvalho 2014).

I now turn to the other standard model of markets: **oligopolistic competition**. The standard model of oligopoly assumes that firms noncooperatively set prices (or quantities). Empirical work suggests that collaboration between firms is common. This collaboration takes a variety of forms, which include creation and sharing of knowledge about markets and technologies, the setting of market standards, and the sharing of facilities (such as distribution channels or plane capacity). Typically, collaboration ties

[6] For a related strand of the literature on buyer-seller networks with a different modeling approach—based on heuristic learning rules and random linking decisions—see Weisbuch, Kirman, and Herreiner (2000).

are bilateral and are embedded within a broader network of similar ties with other firms. Considerable asymmetries exist between the level of collaborative activity across firms, with some firms forming several ties whereas others are poorly linked (Powell 1990; Hagedoorn 2002; Gulati 2007). This empirical evidence motivates the study of the relation between oligopolistic market competition and collaboration networks.

Goyal and Joshi (2003) present a framework for the study of these issues. The two ingredients are a model of bilateral link formation and the standard textbook model of oligopoly. The innovation here is a general model of bilateral linking. Prior to competing in the market, firms can form pairwise collaborative links with other firms. These pairwise links involve a commitment of resources and lead to lower costs of production of the collaborating firms.

Goyal and Joshi (2003) show that incentives to form links are closely related to market competition: in a standard homogenous good model, if competition is in prices, firms choose to form no links, while if competition is in quantities, then firms typically form dense but asymmetric networks of collaboration. Thus collaborations are used by firms to generate competitive advantage. Moreover, market competition shapes networks, and these networks in turn define the competitiveness of firms in a market. This two-way flow of influence between markets and networks is central to understanding economic activity.

Collaboration networks in markets remains an active field of research; recent papers explore socially optimal networks and the scope of public policy (Koenig et al. 2014; Westbrock 2010).

In the applications above, I have considered direct ties between buyers and sellers or between firms. I now turn to **intermediation**. The financial sector embodies intermediation in a pure form—that between the sources and the eventual users of savings. Traditional models of the banking sector generally pay little attention to the rich patterns of intermediation within the sector. Following the financial crises of 2008, there was renewed interest in the role of interconnections among financial institutions as a source for the transmission and possible amplification of shocks. Consequently, a number of papers have documented the structure of the inter-bank lending network (see, e.g., Bech and Atalay 2010; Afonso and Lagos 2012; Van Lelyveld I and t' Veld 2012). The broad consensus is that this network has a core-periphery structure: there is a core of large banks that are densely interconnected, and a large number of smaller banks at the periphery. There is a net inflow of funds from the peripheral banks to the core banks. These empirical findings motivate the study of economic mechanisms underlying the formation of core-periphery financial networks.

In a recent paper, Farboodi (2014) develops a model of the financial sector in which banks choose to form endogenous intermediation links with each other. There are banks that have links with depositors and banks that have links with potential investors. A link between two banks is a durable relationship. Links are unilateral: a link from A to B constitutes a commitment from A to honor any loan demand from B. A bank has an incentive to form multiple links and be the intermediary between a source bank and a destination bank as it can then earn "rents." Farboodi (2014) shows that a core-periphery network endogenously emerges as an equilibrium outcome. An important result is that the network is inefficient as banks who lend to investors "over-connect," exposing themselves to excessive counter-party risk, while (depositor linked) banks who mainly provide funding end up with too few connections. This creates excessive risk in the system at large.

This paper builds on the early work of Allen and Gale (2000) and Babus (2006) in financial networks and Goyal and Vega-Redondo (2007) in the theory of network formation. It is also part of the recent flurry of activity in the area of financial networks: other recent papers include Allen, Babus, and Carletti (2012); Acemoglu, Ozdaglar, and Tahbaz-Salehi (2015); Cabrales, Gottardi, and Vega-Redondo (2012); and Elliott, Golub, and Jackson (2014).

3.4.2 Social Networks and Markets

In this section, I discuss the interaction between social ties between individuals—examples include friendships, family ties, street neighborhoods, and caste, race, and religious affiliations—and markets. As a first approximation, these ties may be taken as given and external to the specific economic problem being studied.[7]

I start with a discussion of **social networks in product markets**. In the standard product market model a firm chooses prices, advertising strategy and quality taking as given heterogenous consumer preferences (Tirole 1994). The roles of friends, neighbors, and colleagues in shaping consumer choice have been brought out in a number of studies over the years. In the past, the practical use of such social influences for advertising or pricing was hampered by a lack of good data. The growth of the Internet and the large amounts of data on online social networking along with the other advances in information technology have made it considerably easier to gather data on social networks and have led to an explosion in interest on how firms and governments can harness the power of social networks to promote social and private goals. Practical interest has centered on questions such as: for which product categories are networks important and when are they unimportant; what are the relevant aspects of networks; how should a firm use social networks to promote its product; how much should a firm be willing to pay to acquire information about social networks; and how should a firm compete with other firms on social networks.

Galeotti and Goyal (2009) studied the optimal strategy of a firm that wishes to reach maximum number of consumers located in a social network. The firm chooses whom to target with advertisements, under the assumption that advertisements travel through connections. They showed that the optimal strategy involves targeting more

[7] The effect of social networks on economic life is the subject matter of economic sociology; for an overview of this work, see Smelser and Swedberg (2005). There are overlaps between the work in sociology and research on networks in economics and a deeper engagement would be mutually beneficial. But a detailed discussion of the relations will not be attempted here, as it will take me too far afield.

connected nodes in some cases and poorly connected nodes in other cases. The use of social networks always leads to higher sales and greater profits. However, an increase in the level and dispersion of social interaction can have non-monotonic effects on level of optimal strategy. Finally, they also showed that the returns to investing in market research on social networks are greater in more unequal networks.

These results are obtained in a setting with one firm, with one step spread of advertisement, and with advertising strategy only. Current research extends and deepens the scope of the analysis significantly to include multiple firms, dynamics of spreading information, and to allow for pricing in addition to advertising choices (Fainmesser and Galeotti 2014; Goyal and Kearns 2012; and Campbell 2013). The use of social networks for optimal diffusion of information remains a very active field of research in economics and in other social and information sciences.

In a related line of work, researchers have explored the use of targeted pricing strategy in social networks. In the standard network externalities, literature pricing was conditional on the size of aggregate consumer base (Farrell and Saloner 1986; Katz and Shapiro 1985). Network externality often arises through the use of common products or services in personal interaction. So the value of adopting a product to a consumer would depend on how many of her neighbors adopt the same product. In other words, the value depends on adoption patterns in the local neighborhood. This motivates the study of optimal pricing that is sensitive to the network properties of individuals. In an important paper, Bloch and Querou (2013) analyzed the problem of optimal monopoly pricing in social networks where agents care about consumption of their neighbors. They showed that optimal prices are related to consumer centrality in the social network. This relation depends on the market structure (monopoly vs. oligopoly). They identified two situations where the monopolist does not discriminate between nodes in the network: linear monopoly with consumption externalities and local monopolies with price externalities. Aoyagi (2014) extended this work to allow for more general externalities and competition among several firms.

I turn next to the role of **social networks in financial markets**. This is one setting where the standard market model of anonymous traders and common prices that reveal information of traders has been very dominant. In recent years this standard view has been the subject of much attention, both empirically and theoretically. By way of motivation, I start with some empirical findings. Cohen, Frazzini, and Malloy (2008) showed that portfolio managers place larger bets on firms when they went to school with senior managers or a board member and perform significantly better on these holdings. Similarly, Hong, Kubik, and Stein (2005) documented that U.S. fund managers located in the same city commit to correlated investment decisions. The authors argued that such correlated choices may be due to peer-to-peer communication or because fund managers in a given area base their decisions upon common sources of information. The empirical evidence motivates a more systematic study of the relation between information social networks and trader behavior and aggregate outcomes on volume and prices. Ozsoylev and Walden (2011) and Colla and Mele (2010) studied asset pricing in markets where traders are located in information networks. They studied trader behavior on a fixed network and derived a relation between equilibrium outcomes (on prices and trading volume) and the network topology.

I turn finally to the role of **social networks in labor markets**. Workers like jobs that suit their skills and location preferences, while firms are keen to hire workers who have the right ability for the job. However, workers do not know which firms have vacancies, and finding the right job takes time and effort. Similarly, firms do not know which workers are looking for a job. Faced with this lack of information, workers look for job advertisements in newspapers and magazines. They also spread the word among their friends and acquaintances that they are looking for a job, and indeed there is substantial evidence that they often get information on job vacancies via their personal connections. A second type of information problem concerns the ability of workers: a person generally knows more about his own ability as compared to a potential employer. Indeed, this asymmetry in information leads workers to invest in signals of their quality (such as educational degrees, certificates, and licenses), and it leads potential employers to ask for references and recommendation letters. Referrals—references and recommendation letters—are widely used in the process of matching workers and firms. A letter of reference is only valuable in so far as the employer can trust the writer of the letter; this suggests that the structure of personal connections is likely to play an important role in matching workers and firms.

These observations raise the question: how does the pattern of social contacts affect the flow of information about jobs? The flow of information across persons will influence how quickly workers are matched with jobs, which will in turn shape the level of employment. The patterns of social connections will also determine who gets information and when; this in turn may determine who gets a job and who is left unemployed, which will in turn have a bearing on the distribution of earnings and overall inequality in a society. The study of social networks in labor markets has a long and distinguished history; for a survey of this work, see Granovetter (1974) and Goyal (2007). The older work has an empirical orientation. In recent years, economists have made significant progress in the theoretical analysis of this issue.

Calvo-Armengol and Jackson (2004) studied a model of information transmission on job vacancies. Their analysis of this model yielded three main insights. The first insight was that the employment status of individuals in a social network is positively correlated. Empirical work suggests that there is significant correlation in employment status within social communities or geographically contiguous city districts. The second insight was that the probability of finding a job declined with the duration of unemployment. Duration dependence of unemployment has been widely documented. The reason that duration dependence arises in a context of social information sharing is that a longer spell of unemployment reveals that a person's social contacts are less likely to be employed, which in turn makes it less likely that they will pass on information concerning vacancies. The third insight was that small initial differences in employment across otherwise identical groups of individuals can have significant effects on the incentives to drop out of the labor market. However, if individuals drop out of the social network this then reduces the value to others of remaining in the network. Thus small initial differences can create a sequence of drop outs, which in turns can have long run effects on employment prospects of the group.

Montgomery (1991) studied the adverse selection problem in labor markets. The analysis of this model yielded two key insights. The first insight of the analysis was that workers with more connections will earn a higher wage and that firms who hire through contacts (of their existing high quality workers) will earn higher profits. The reason for this relation between connections and wages is simple: more connections imply a higher number of referral wage offers from firms (on average) and this translates into a higher accepted wage (on average). The second insight was that an increase in the density of social connections raises the inequality in wages. This effect is a reflection of the lemons effect: an increase in social ties means that more high-ability workers are hired via referrals, and this lowers the quality of workers who go into the open market, thereby pushing down their wage, relatively.

In Calvo-Armengol and Jackson (2003), prices and competition do not play a role, while in Montgomery (1991) there is no modeling of network topology. The integration of network topology with the usual ingredients of pricing and competition remains a challenging open problem.

3.5 Closing the Circle

In this chapter, my aim has been to locate the innovation that networks bring to economics, to assess the current state of research, and point to current challenges.

I have argued that research on networks in economics started with theoretical innovation in the early 1990s. At the start, both the research on social learning and the research on network formation emerged in relative autonomy from applications and empirical work. When the results came in contact with substantive issues in economics, this research gathered momentum. It generated new questions that built on the research and lent the program greater significance. In the last decade the innovations have matured; now they combine with preexisting concepts—such as competition, asymmetric information, and prices—to build an encompassing framework. In the words of Kuhn (1962), research on networks in economics has acquired the traits of "normal" science.

I have also identified a persistent and growing tension between (much of) the theoretical research and the empirical findings on behavior and structure of large and complex networks. This tension should not come as a surprise: traditionally, economists have been successful at developing concepts for the study of interaction in small closed groups (game theory) and interaction among large groups (competitive markets and general equilibrium). A social network typically consists of a large number of individuals, and any individual interacts only with a small subset of them. The analysis of many actors located within complex networks strains the plausibility of delicate chains of strategic reasoning that are typical in game theory. Similarly, the "local" interaction in social networks makes anonymous competitive equilibrium analysis implausible. Thus networks fall between the "small" and the "large." So, while much has been accomplished, I believe that this tension will continue to offer a fertile ground for new and exciting research in the years to come.

References

Acemoglu, D., V. Carvalho, A. Ozdaglar, and A. Tahbaz-Salehi (2012). "The network origins of aggregate fluctuations." *Econometrica* 80(5), 1977–2016.

Acemoglu, D., M. A. Dahleh, I. Lobel, A. Ozdaglar (2011). "Bayesian learning in social networks." *Review of Economic Studies* 78, 1201–1246.

Acemoglu, D., A. Malekian, and A. Ozdaglar (2014). "Network security and contagion." mimeo, MIT.

Acemoglu, D., A. Ozdaglar, and A. Tahbaz-Salehi (2015). "Systemic risk and stability in financial networks." *American Economic Review* 105(2), 564–608.

Afonso, G. and R. Lagos (2012). "Trade dynamics in the market for federal funds." *Federal Reserve of New York Staff Report.*

Allen, F., A. Babus, E. Carletti (2012). "Asset commonality, debt maturity and systemic risk." *Journal of Financial Economics* 104(3), 519–534.

Ambrus, A., M. Mobius, and A. Szeidl (2014). "Consumption risk-sharing in social networks." *American Economic Review* 104(1), 149–182.

Ambrus, A., A. Chandrashekhar, and M. Elliott (2014). "Social investments, informal risk sharing and inequality." mimeo, Duke and Caltech.

Aoyagi, M. (2014). "Bertrand competition under network externalities." mimeo, Osaka University.

Aumann, R. and R. Myerson (1988). "Endogenous formation of links between players and coalitions: An application to the Shapley Value." In *The Shapley Value*, A. Roth (ed.), 175–191. Cambridge: Cambridge University Press.

Babus, A. (2006). "The formation of financial networks." mimeo.

Baetz, O., (2015), "Social activity and network formation." *Theoretical Economics* 10, 315–340.

Bala, V. and S. Goyal (1995). "A theory of learning with heterogeneous agents." *International Economic Review* 36(2), 303–323.

Bala, V. and S. Goyal (1998). "Learning from neighbours." *Review of Economic Studies* 65, 595–621.

Bala, V. and S. Goyal (2000). "A non-cooperative model of network formation." *Econometrica* 68(5), 1181–1231.

Bala, V. and S. Goyal (2001). "Conformism and diversity under social learning." *Economic Theory* 17, 101–120.

Ballester, C., Calvo-Armengol, A., and Y. Zenou (2006). "Who's who in networks. Wanted: The key player." *Econometrica* 74, 1403–1417.

Banerjee, A. (1992). "A simple model of herd behavior." *Quarterly Journal of Economics* 107(3), 797–817.

Banerjee, A., A. Chandrasekhar, E. Duflo, and M. O. Jackson (2013). "The diffusion of microfinance." *Science* 341, 61–74.

Barabasi, A. L. (2002). *Linked.* New York: Perseus Books.

Bech, M. L. and E. Atalay (2010). "The topology of the federal funds market." *Physica A: Statistical Mechanics and its Applications* 389(22), 5223–5246.

Belleflamme, P. and F. Bloch (2004). "Market sharing agreements and collusive networks." *International Economic Review* 45(2), 387–411.

Bikhchandani, S., D. Hirshliefer, and I. Welch (1992). "A theory of fads, fashion, custom, and cultural change as informational cascades." *Journal Political Economy* 100, 992–1023.

Bisin, A., J. Benhabib, and M. Jackson (2012). *The Handbook of Social Economics: Volumes I–II*. Amsterdam: North Holland.

Bloch, F. and B. Dutta (2012). "Formation of networks and coalitions." In *Social Economics*, A. Bisin, J. Benhabib, and M. Jackson, eds. Amsterdam: North Holland.

Bloch, F., G. Genicot, and D. Ray (2008). "Informal insurance in social networks." *Journal of Economic Theory* 143(1), 2008, 36–58.

Bloch, F., and N. Querou (2013). "Pricing in social networks." *Games and Economic Behavior* 80, 263–281.

Bollobás, B. (1998). *Modern Graph Theory*. Berlin: Springer Verlag.

Boorman, S. (1975). "A combinatorial optimization model for transmission of job information through contact networks." *Bell Journal of Economics* 6(1), 216–249.

Bourdieu, P. (1977). *Outline of a Theory of Practice*. Cambridge: Cambridge University Press.

Bramoullé, Y. and R. Kranton (2007). "Public goods in networks." *Journal of Economic Theory* 135, 478–494.

Bramoullé, Y. and R. Kranton (2007). "Risk-sharing networks." *Journal of Economic Behavior and Organization* 64(34), 275–294.

Bramoulleé, Y., S. Currarini, M. O. Jackson, P. Pin, and B. Rogers (2012). "Homophily and long-run integration in social networks." *Journal of Economic Theory* 147(5), 1754–1786.

Cabrales, Gottardi and Vega-Redondo (2012). "Risk-sharing and contagion in networks." Working paper. EUI, Florence.

Cabrales, A., A. Calvo-Armengol, and Y. Zenou (2011). "Social interactions and spillovers." *Games and Economic Behavior* 72, 339–360.

Calvo-Armengol, A. (2004). "Job contact networks." *Journal of Economic Theory* 115(1), 191–206.

Calvo-Armengol, A. and M. O. Jackson (2004). "The effects of social networks on employment and inequality." *American Economic Review* 94(3), 426–454.

Campbell, A (2013). "Word-of-mouth communication and percolation in social networks." *American Economic Review* 103(6), 2466–2498.

Carvalho, V. (2014). "From micro to macro via production networks." *Journal of Economic Perspectives* 28(4), 23–48.

Coase, R. (1937). "The nature of the firm." *Economica* 4(16), 386–405.

Cohen, L., A. Frazzini, and C. Malloy (2008). "The small world of investing: Board connections and mutual fund returns." *Journal of Political Economy* 116, 951–979.

Coleman, J. (1966). *Medical Innovation: A Diffusion Study*. New York: Bobbs-Merrill.

Coleman, J. (1988). "Social capital in the creation of human capital." *American Journal of Sociology* 94, S95–S120.

Conley, T. G. and C. R. Udry (2010). "Learning about a new technology: Pineapple in Ghana." *American Economic Review* 100(1), 35–69.

Currarini, S., M. O. Jackson, and P. Pin (2009). "An economic model of friendship: Homophily, minorities and segregation." *Econometrica* 77(4), 1003–1045.

Dasgupta, P. and I. Serageldin (1999). *Social Capital: An Multifaceted Perspective.* Washington, DC: World Bank Publications.

DeGroot, M. H. (1974). "Reaching a consensus." *Journal of the American Statistical Association* 69(345), 118–121.

Doreian, P. and F. Stokman (2001). "Evolution of social networks." Special Volume: *Journal of Mathematical Sociology* 25, 1–10.

Durlauf, S. and P. Young (2001). *Social Dynamics.* Washington, DC: Brookings Institute.

Easley, D. and J. Kleinberg (2010). *Networks, Crowds, and Markets: Reasoning About a Highly Connected World.* Cambridge: Cambridge University Press.

Elliott, M. (2015). "Inefficiencies in networked markets." *American Economic Journal: Microeconomics* 7(4), 43–82.

Elliott, M., B. Golub, and M. O. Jackson (2012). "Financial networks and contagion." *American Economic Review* 104(10), 3115–3153.

Ellison, G. and D. Fudenberg (1993). "Rules of thumb for social learning." *Journal of Political Economy* 101, 612–643.

Ellison, G. and D. Fudenberg (1995). "Word of mouth communication and social learning." *Quarterly Journal of Economics* 110, 93–126.

Evans, G. and S. Honkapohja (2001). *Learning and Expectations in Microeconomics.* Princeton, NJ: Princeton University Press.

Fainmesser, I. and A. Galeotti (2014). "The value of network information." *mimeo*, Essex University.

Farboodi, M. (2014). "Intermediation and voluntary exposure to counter-party risk." Working paper, University of Chicago.

Farrell, F. and G. Saloner (1986). "Installed base and compatibility: Innovation, product preannouncements, and predation." *American Economic Review* 76, 940–955.

Feder, G., R. E. Just, and D. Zilberman (1985). "Adoption of agricultural innovations in developing countries: A survey." *Economic Development and Cultural Change* 33, 255–298.

Cohen, L., A. Frazini, and C. Malloy (2008). "The small world of investing: Board connections and mutual fund returns." *Journal of Political Economy* 116(5), 951–979.

Fudenberg, D. and D. K. Levine (1998). *The Theory of Learning in Games.* Cambridge, MA: MIT Press.

Fursawa, T. and H. Konishi (2007). "Free trade networks." *Journal of International Economics* 72(2), 310–335.

Galeotti, A., and S. Goyal (2009). "Influencing the influencers: A theory of strategic diffusion." *Rand Journal of Economics* 40(3), 509–532.

Galeotti, A. and S. Goyal (2010). "The law of the few." *American Economic Review* 100, 1468–1492.

Galeotti, A., D. Goyal, M. Jackson, F. Vega-Redondo, and L. Yariv (2010). "Network games." *Review of Economic Studies* 77(1), 218–244.

Golub, B. and M. O. Jackson (2012). "How homophily affects the speed of learning and best response dynamics." *Quarterly Journal of Economics* 127(3), 1287–1338.

Golub, B. and M. O. Jackson (2010). "Naive learning and the wisdom of crowds." *American Economic Journal: Microeconomics* 2(1), 112–149.

Goyal, S. (1993). "Sustainable communication networks." *Tinbergen Institute Discussion Paper TI 93-250.*

Goyal, S. (2007). *Connections: An Introduction to the Economics of Networks.* Princeton, NJ: Princeton University Press.

Goyal, S. (2012). "Learning in networks." In *Handbook of Social Economics*, J. BenHabib, A. Bisin, and M. O. Jackson, eds. North Holland.

Goyal, S. and S. Joshi (2003). "Networks of collaboration in oligopoly." *Games and Economic Behavior* 43(1), 57–85.

Goyal, S. and S. Joshi (2006). "Bilateralism and free trade." *International Economic Review* 47(3), 749–778.

Goyal, S. and M. Kearns (2012). "Competitive contagion in networks." *Symposium in Theory of Computing* (STOC).

Goyal, S., M. van der Leij, and J. L. Moraga-González (2004). "Economics: Emerging small world." *Journal of Political Economy* 114, 403–412.

Goyal, S. and J. L. Moraga-Gonzalez (2001). "R&D networks." *Rand Journal of Economics* 32(4), 686–707.

Goyal, S. and F. Vega-Redondo (2005). "Network formation and social coordination." *Games and Economic Behavior* 50, 178–207.

Goyal, S. and F. Vega-Redondo (2007). "Structural holes in social networks." *Journal of Economic Theory* 137, 460–492.

Goyal, S. and A. Vigier (2014). "Attack, defence and contagion in networks." *Review of Economics Studies* 81(4), 1518–1542.

Granovetter, M. (1973). "The strength of weak ties." *American Journal of Sociology* 78, 1360–1380.

Granovetter, M. (1974). *Getting a Job: A Study of Contacts and Careers.* Cambridge, MA: Harvard University Press.

Granovetter, M. (1985). "Economic action and social structure: The problem of embeddedness." *American Journal of Sociology* 3, 481–510.

Griliches, Z. (1957). "Hybrid corn: An exploration in the economics of technological change." *Econometrica* 2(4), 501–522.

Gulati, R. (2007). *Managing Network Resources.* Oxford: Oxford University Press.

Gulati, R., N. Nohria, and A. Zaheer (2000). "Strategic networks." *Strategic Management Journal* 21, 203–215.

Hagedoorn, J. (2002). "Inter-firm R&D partnerships: An overview of major trends and patterns since 1960." *Research Policy* 31, 477–492.

Hagerstrand, J. (1967). *Innovation Diffusion as a Spatial Process.* Translated by A. Pred. Chicago, IL: University of Chicago Press.

Hiller, T. (2012). "Peer effects in endogenous networks." Working paper, Bristol University.

Hojman, D. and A. Szeidl (2008). "Core and periphery in endogenous networks." *Journal of Economic Theory* 139(1), 295–309.

Hong, H., J. D. Kubik, and J. C. Stein (2005). "Thy neighbor's portfolio: Word-of-mouth effects in the holdings and trades of money managers." *Journal of Finance* 60, 2801–2824.

Jackson, M. O. (2008). *Social and Economic Networks.* Princeton, NJ: Princeton University Press.

Jackson, M. O. and B. Rogers (2007). "Meeting strangers and friends of friends: How random are social networks?" *American Economic Review* 97(3), 890–915.

Jackson, M. O. and A. Wolinsky (1996). "A strategic model of economic and social networks." *Journal of Economic Theory* 71(1), 44–74.

Jackson, M. O. and Y. Zenou (2014). "Games on networks." *Handbook of Game Theory, Vol. 4*, Peyton Young and Shmel Zamir, eds.

Jhadbabaie, A., P. Molavi, and A-R. Tahbaz-Salehi (2013). "Heterogeneity and the speed of learning in social networks." Working paper, Columbia University.

Karlan, D., M. Mobius, T. Rosenblat, and A. Szeidl (2009). "Trust and social collateral." *Quarterly Journal of Economics* 124(3), 1307–1361.

Katz, M. and C. Shapiro (1985). "Network externalities, competition and compatibility." *American Economic Review* 75(3), 424–440.

Katz, E. and P. F. Lazarsfeld (1955). *Personal Influence: The Part Played by People in the Flow of Mass Communications.* New York: Free Press.

Kirman, A. (1997). "The economy as an evolving network." *Journal of Evolutionary Economics* 7, 339–353.

Kirman, A. and J-P. Zimmerman (2001). *Economics with Heterogenous Interacting Agents*, Berlin: Springer-Verlag.

Kirman, A. and A. Vignes (1991). "Price dispersion: Theoretical considerations and empirical evidence from the Marseilles Fish Market." in Kenneth J. Arrow, ed., *Issues in Contemporary Economics: Proceedings of the Ninth World Congress of the International Economic Association.* Athens, Greece. Volume 1 : Markets and Welfare. New York: New York University Press.

Kotowoski, M. and M. Leister (2014). "Trading networks and equilibrium intermediation." mimeo, University of California, Berkeley.

Koenig, M., C. J. Tessone, and Y. Zenou (2014). "Nestedness in networks: A theoretical model and some applications." *Theoretical Economics* 9, 695–752.

Kranton, R. and D. Minehart (2001). "A theory of buyer-seller networks." *American Economic Review* 91(3), 485–508.

Kuhn, T. (1962). *The Structure of Scientific Revolutions.* Chicago, IL: University of Chicago.

Lazarsfeld, P. F., B. Berelson, and H. Gaudet (1948). *The People's Choice: How the Voter Makes Up His Mind in a Presidential Campaign.* New York: Columbia University Press.

Lazarsfeld, P. F. and R. Merton (1954). "Friendship as a social process: A substantive and methodological analysis." In *Freedom and Control in Modern Society*, M. Berger, T. Abel, and C. H. Page, eds. New York: Van Nostrand.

Loury, G. (1977). "A dynamic theory of racial income differences." In *Women, Minorities, and Employment Discrimination*, P. A. Wallace and A. Le Mund, eds. Lexington, MA: Lexington Books.

Van Lelyveld, I. and D. in t' Veld (2012). "Finding the core: Network structure in interbank markets." DNB Working Paper No. 348.

Manea, M. (2013). "Intermediation in networks." mimeo, MIT.

McPherson, M., L. Smith-Lovin, and J. M Cook (2001). "Birds of a feather: Homophily in social networks." *Annual Review of Sociology*, 27, 415–444.

Mossel, E., A. Sly, and O. Tamuz (2015). "Strategic learning and the topology of social networks." mimeo, MIT.

Mueller-Frank, M. (2013). "A general framework for rational learning in social networks." *Theoretical Economics* 8(1), 1–40.

Montgomery, J. (1991). "Social networks and labor-market outcomes: Toward an economic analysis." *The American Economic Review* 81(5), 1408–1418.

Munshi, K. (2014). "Community networks and the process of development." *Journal of Economic Perspectives* 28(4), 49–76.

Myerson, R (1991). *Game Theory: Analysis of Conflict.* Cambridge MA: Harvard University Press.

Nava, F. (2014). "Efficiency in decentralized oligopolistic markets." mimeo, LSE.
Ozsoylev, H., J. Walden, D. Yavuz, and R. Bildick (2014). "Investor networks in the stock market." *The Review of Financial Studies* 27(5), 1323–1366.
Ozsoylev, H. and J. Walden (2011). "Asset pricing in large information networks." *Journal of Economic Theory* 146(6), 2252–2280.
Polanyi, K. (1944). *The Great Transformation: The Political and Economic Origins of our Time.* New York: Rinehart.
Powell, W. (1990). "Neither market not hierarchy: Network forms of organization." *Research in Organizational Behavior* 12, 295–336.
Rogers, E. M. (1983). *Diffusion of Innovations*, 3rd edition. New York: Free Press.
Ryan, B. and N. Gross (1943). "The diffusion of hybrid seed corn in two Iowa communities." *Rural Sociology* 8, 15–24.
Rogers, E. (1983). *Diffusion of Innovations*, 3rd edition. New York: Free Press.
Rothschild, M. (1974). "A two-armed bandit theory of market pricing." *Journal of Economic Theory* 9(2), 185–202.
Schelling, T. (1975). *Micromotives and Macrobehavior*. New York: Norton.
Smelser, N. J. and R. Swedberg (2005). *The Handbook of Economic Sociology*, 2nd edition. Russell Sage Foundation.
Tirole, J. (1994). *The Theory of Industrial Organization*. Cambridge, MA: MIT Press.
Uzzi, B. (1996). "The sources and consequences of embeddedness: The network effect." *American Sociological Review* 61(4), 674–698.
Westbrock, B. (2010). *Inter-Firm Networks: Economic and Sociological Perspectives*. Ph.D Dissertation, Utrecht University.
Watts, D. (1999). *Small Worlds: The Dynamics of Networks between Order and Randomness.* Princeton, NJ: Princeton University Press.
Weisbuch, G., A. Kirman, and D. Herreiner (2000). "Market organisation and trading relationships." *Economic Journal* 110, 411–436.

CHAPTER 4

THE PAST AND FUTURE OF NETWORK ANALYSIS IN ECONOMICS

MATTHEW O. JACKSON

4.1 An Explosion of Research on Networks in Economics

Over the past two decades, research on networks in economics has grown exponentially—from a handful of papers up through the late 1990s to thousands today. This explosive growth is due to a number of factors. First and foremost, to understand many economic behaviors—from the dynamics of product adoption to financial contagions—it is necessary to account for the patterns of interactions. Failing to include network structure can lead to a deficient understanding of an observed behavior and to poor policy design. Second, there are increasingly well-understood features of networks (e.g., how densely connected a population is, how segregated it is, among others) that have specific and important implications for economic behaviors. This improved understanding has widened the collection of settings in which networks have been analyzed in conjunction with economic consequences, as it is increasingly clear how to account for network structure and relate it to behavior. Third, increasingly available data and improvements in computational capabilities are enabling the testing and application of models that could not be analyzed even a few decades ago. Moreover, data on networks of interactions is often available in conjunction with behaviors—which is essential for understanding the economic implications of network structure, and allows us to test theories and evaluate policies, and measure social learning, diffusion, and peer effects.

Thus, this handbook comes at an opportune time. We have learned an immense amount about social and economic networks and developed important new tools, and the set of applications is rapidly expanding. The literature has grown so much that taking stock of it will help researchers new to the area gain access to the large toolbox for analyzing networked interactions. In addition, there is much yet to study, and taking stock of the current knowledge base also highlights the research frontier and illuminates important open problems.

Here I highlight a few of the areas in which the literature has made substantial progress and discuss what enabled this progress, as well as where current tools are poised to make further contributions, and point out some of the most important open problems.

4.2 Successes: What Economics has Brought to the Table

Economists' forays into the study of networks over the past couple of decades have helped advance the literature in several ways.

The neoclassical economics paradigm of ("rational") individual choice has driven new modeling of network formation and of behavior on networks based on strategic interactions. While arguably narrow, that paradigm is nevertheless powerful, and it drove us to understand how individual decisions to form relationships impact the networks that emerge (see the chapters by Mauleon and Vannetelbosch; and Pin and Rogers). In particular, the early theme in the economics literature on networks was the tension between individual incentives to form relationships and the broader societal welfare from the resulting network, as individuals do not internalize the indirect impact that their relationships have on others through a multitude of factors: increased information diffusion, contagion, changes in bargaining strengths, exposure to risks, opportunities, and so on. The understanding of network externalities was greatly enhanced by those studies, and an economist's inherent welfarist perspective was essential in asking new questions and generating answers that were absent from the literature, despite the fact that there was a rich and healthy social networks literature in sociology.

Similarly, this game-theoretic approach founded another wave of the literature systematically exploring how an individual's behavior is driven by that of the individual's neighbors and hence indirectly by others in the network (see the chapters by Bramoullé and Kranton; Nava; and Zenou). This game-theoretic approach has added substantially to peer effects studies, and the understanding of delinquency, educational achievement, vaccination decisions, product adoption, participation in various programs, and other interdependent behaviors. Viewing such interactions not just as contagion processes, but ones with complementarities and substitution pressures in behaviors (e.g., choosing compatible technologies, benefiting from spillovers in knowledge, etc.) leads to a much richer mosaic of how behavior interacts with network structure.

The interest in understanding how people's decisions interact also requires a deep understanding of social learning and diffusion. Thus, there has been renewed interest in social learning and diffusion, and understanding how network structure impacts outcomes, and research in the past decade has substantially advanced that frontier. The advances include new understandings of when processes converge or reach consensus, speeds of convergence, how dynamics relate to homophily and other network features, as well as how dynamics differ according to the complexity of the learning or diffusion process and the network structure (see the chapters by Golub and Sadler; and Lamberson). Most importantly, this has resulted in a fairly systematic understanding of how dynamic processes are dictated by network structure, and how basic network characteristics, such as the degree distribution, homophily, and local patterns such as clustering, determine the outcome of social learning and diffusion processes. Similar understandings have been found in the study of networked markets (see the chapters by Condorelli and Galeotti; and Manea). This knowledge has enabled new empirical analyses and tests of the theory, and new areas of application (see the chapters by Breza; Chaney; Mobius and Rosenblat; and Munshi).

Ultimately, having such a systematic understanding that relates network structure to behavior and welfare is essential to network science having a lasting impact in economics. This is a point that I expand upon elsewhere (*Journal of Economic Perspectives*, Fall 2014), and also with Brian Rogers and Yves Zenou (in "The Economic Consequences of Social Networks"). We provide an up-to-date taxonomy of network features and their economic consequences, as well as discussion of how accounting for networks can be essential for understanding behavior and shaping policies. The longevity of network analysis in economics will derive from seeing such significant network effects in a variety of applications.

In addition to bringing the neoclassical economic paradigm to network analyses, working to develop a systematic understanding of how networks impact economic behaviors, and exploring new applications (more on that below), another important push from the economics literature has been on cleanly identifying network effects and testing many new theories. Economists' preoccupation with the inference of causation and getting clean identification of hypothesized effects (e.g., see the chapters by Boucher and Fortin; and Chandrasekhar) has increased the use of lab and field experiments in analyzing social and economic networks (e.g., see the chapters by Aral; Breza; Choi, Gallo, and Kariv; and Watts). Such techniques are certainly not new to network analysis, but the emphasis has changed. This is in part due to the progress in modeling, which has resulted in numerous new hypotheses to be tested regarding network formation, peer effects, social learning, and diffusion.[1] This is resulting in some important cross-fertilization with anthropology and sociology, which have strong traditions and expertise in field work. This is also pushing statistics and econometrics researchers to produce new techniques for analyzing data and testing network theories (more on this below).

[1] As a caution, the pressure to have clean identification should not preclude studies that uncover important correlations without establishing causation. Although one obviously has to be careful regarding what to conclude from correlations, the recent trend in natural and field experiments should not completely crowd out observational studies that unearth important relationships in the data that pave the way for further study.

4.3 A Bucket List

Despite the richness of the literature, there is so much yet to be studied. The frontiers can be thought of as relating to the recent progress in the literature in several ways. First, existing theory and techniques are enabling new applications. Next, recent theory and empirical work have made evident some particular areas where new models and techniques are needed. Finally, there is a need for some systematic meta-analyses of the tools and findings to data. Let me treat each of these in turn.

I begin with application areas that are ripe for further investigation. As mentioned above, the areas in which studies are incorporating networks are growing rapidly. There is a very natural and unusually strong interplay between theory and application in network studies. Applying the theory is helping to identify the impact of network structure on economic behaviors, and as a result is reshaping some policies, as well as testing and refining theory. One important facet of this is that incorporating network structure into the analysis gives us more precision in distinguishing various social forces, since things like diffusion of information, norms, and pressure from peers all depend in different ways on network structure, and so including networks can help us to distinguish various forms of peer effects—which can have substantial policy implications. For example, if people forego education because they are learning about its benefits through their social network, that suggests a very different policy than if they are aware but forego it due to pressures to match the behavior of their friends. For all of these reasons, a key feature of emerging, and future, studies is that they not only involve observations of network structures, but also include behavioral outcomes of the individuals or organizations involved in the networks.

Perhaps the most notable area of such exploration and growth is development economics. This is a very natural area in which to study social networks as most transactions and relations in developing countries are informal—not relying on formal contracts or exogenous institutions, but relying heavily on social interactions (see the chapters by Breza; Mobius and Rosenblat; and Munshi). This arena is important beyond its immediate welfare impact, as not only can network theory help us to understand behaviors and welfare, but small communities in developing countries tend to be relatively closed, thus giving a researcher a holistic view of the patterns of interaction and an unusual degree of control in field experiments. This helps in testing existing theories and generating new models of social learning and diffusion. Developing countries also present fairly stark instances in which to study how culture and social norms are shaped by and transmitted through networks, and can help us better understand things like corruption, collective action problems, revolutions, inequality, and growth, all of which have strong network components that are yet to be understood but are well within our current grasp.

Although labor economics is a longer-standing area of network study, it is also an arena in which many fascinating questions remain open (see the chapters by Beaman; Burt; Contractor, and Dessein and Prat). It has long been clear that networks play important roles in who has access to which opportunities, but richer data sets and new models are opening new questions and avenues for research. For example, how do individuals make decisions on how hard to study, whether to attend university, whether to engage in crime and so on? Such decisions are shaped by the social setting in which the person is embedded, and heavily influenced by family and friends. We are just beginning to unpack the many different types of interaction that are lumped into the broad category of "peer effects". Looking carefully at network patterns of interaction should help us to disentangle information spillovers from norms, and complementarities from opportunity. This continues to be a promising area for research.

Beyond these two prominent areas of application, we are also seeing increased interest in financial networks and economic fluctuations (see the chapters by Cabrales, Gale, Gottardi; and Acemoglu, Ozdaglar, and Tahbaz-Salehi). It has become increasingly obvious that a proper understanding of risk and interdependencies in an economy has to account for indirect effects and transmissions of shocks, and these are inherently network phenomena. Here tools are (rapidly) emerging, but the gap between theory and application still needs to be closed. How can one measure whether an institution is not just "too big to fail" but also "too connected to fail"? How can we properly measure counter-party risk, accounting for further connections and indirect effects? What are the implications of an increasingly global economy for economic fluctuations within industries and regions? This is an area in which network theory can have an immediate as well as lasting impact, and so it is encouraging to see new work emerging and interest from regulators and practitioners in network tools. It is also a particularly interesting area for future development, since the nodes in the network are generally financial institutions whose behavior is sophisticated and often highly strategic, which affects not only investments and production, but the evolution of the network and the resulting transmission of shocks and crises.

There are two other areas in which networks of relationships and externalities are prevalent, and yet are still quite understudied: international trade (see the chapter by Chaney) and international relations (see the chapter by Dziubiński, Goyal, and Vigier), and in fact the two areas are intertwined (a point I explore in a recent paper with Stephen Nei). Given the inherent complexity of networks, there is no reason to undertake network analyses unless there are important externalities and interdependencies across relationships. It is thus natural that many economic settings were first studied from a perspective that was either market-based with many relatively anonymous participants or bilateral with two participants. However, in international trade and international relations, there are fundamental unanswered questions that require modeling beyond a two-at-a-time or a market approach, as the actors involved are inherently networked and face large externalities. In international relations, decisions by countries of which alliances to undertake and which conflicts to enter depend on which allies other countries have. Studying international conflict without a network approach eliminates one's ability to study more than a third of conflicts, and to understand very basic interstate history. The same can be said of international trade, where at both a country level and a firm level, terms of trade depend heavily on the opportunities of partners.

Each of these applications require developing new models that capture the specific incentives at play, and also obtaining richer data sets that allow us to track networks of trade and alliances over time and see the resulting consequences.

These areas of application are clearly not the only ones in which network analyses can and should be successful over the next years, as we are also continuing to see growing numbers of network studies in political economy, marketing, and patenting, among others. Areas like development economics, international trade, and international relations are where the study of networks is relatively new and the questions are quite obvious and important, and so these are areas in which network analyses are likely to produce dramatic advances in the short run. In addition, there are overarching questions about the ultimate impact of technology that continues to make it easier to communicate quickly and cheaply throughout the world (see the chapters by Economides; and Watts).

The growing use of networked data in testing theory is also putting new pressure on the development of statistical and econometric models for studying network formation. This stems from the fact that most network studies of networked behaviors are challenged by the endogeneity of the network, which could end up correlating with behavior and unobserved factors that influence behavior (see the chapters by Aral; Boucher and Fortin; and Chandrasekhar). A major challenge in developing tractable statistical methods for analyzing network formation is that the formation of relationships are generally correlated. That is, the choice of whether to form a relationship between two parties—whether it be friendship, favor exchange, a financial transaction, sharing of risk or information—is usually substantially influenced by to whom else the two parties are connected. This leads all of the relationships in a network to be interdependent. Coupling this with the fact that many data sets consist of observations of a single network means that the data do not consist of many independent observations, but instead one large observation consisting of many dependent objects (e.g., links). Although there are some existing models that admit interdependencies (e.g., exponential random graph models), they suffer from proven computational problems, and models that are both robustly computable and admit link interdependencies are just emerging. This is an area that should experience breakthroughs and enormous advancements over the next decades.

Beyond expanding areas of application and the development of new statistical methods, there are also exciting frontiers that should be explored in network theory over the coming decades. Here is a partial list of important areas in which significant advances are likely in the near future.

First, although we now have models that are helping us to understand diffusion, games on networks, and social learning, these processes are often influenced and manipulated from outside of the networks in ways that are rapidly changing with technologies and of which we have little understanding. For instance, new products have prices and features that are designed to influence their diffusion, and they come together with marketing campaigns that are increasingly taking advantage of social media (see the chapters by Bloch; and Mayzlin). This interaction of media and word of mouth should have important implications for social learning. More generally, there are many contexts in which outside actors attempt to influence networks and the processes operating on them, such as regulators imposing restrictions on which investments financial institutions can make, and there should be some general insights that will prove useful in understanding such interactions. In addition, such processes are also influenced from within, as people may withhold or distort the information that they transmit which can have a profound impact on the diffusion of information and social learning, in ways that are only beginning to be explored (see the chapter by Golub and Sadler).

Second, although there has been substantial progress in modeling network formation and games on networks in static settings, most interactions are dynamic by nature. Although there are some dynamic models of network formation, and a few dynamic strategic analyses of favor exchange, risk-sharing, and cooperation on networks (e.g., see the chapter by Nava), this is another area where more breakthroughs loom. "Cooperative" behavior can be induced and enforced through potential changes in network structures over time: people abide by certain social norms because their social standing and relationships could deteriorate if they break with the norms.[2] While the basic concepts are fairly straightforward and we see some insights in analyses of games with various random matching technologies, we still know little about how the evolution of behavior depends on, and influences, social structure, and why some norms are robust and difficult to change and others are fragile. Developing rich dynamic models for these questions can help shed important light on growth and development, as well as persistent inequality.

Related to this point, networks are often nonstationary and highly dynamic entities, and yet standard approaches still involve either static or stationary processes, not only in modeling but also in representations. This has been driven by tractability as well as a lack of a clear picture of the extent to which the nonstationary nature of networks is a major issue. For example, social media data show highly nonstationary patterns of interaction, with some information relayed for long periods of time, and other information only for short periods, and flu contagions depend on nonstationary school and travel patterns, just to mention two obvious examples. Although this subject has received some attention, we still know little about the ultimate impact of such nonstationarities. This is a fruitful area not only for model development, but also for empirical investigations.

Third, as the set of network tools and models have expanded, we have little understanding of which tools are appropriate for which circumstances. Which are the right measures of homophily and segregation, and how does that depend on the particulars of the application? How can we decide which of the many measures of power or centrality is appropriate for a given analysis? Which algorithms for detecting underlying

[2] This becomes important in settings in which repeated interactions between any two given people are insufficient to enforce behavior via simple folk-theorem arguments, and so the broader structure of interactions becomes important.

community structures are appropriate as a function of the setting? This complicates empirical work, as trying many different measures to see which ones yield results can lead to spurious findings, and correcting for this requires complicated statistical corrections for running multiple models on the same data—which most researchers ignore.[3] Distinguishing models and techniques requires meta-theory that provides us with an understanding of which properties characterize which tools and help us pre-select among theories and form sharper hypotheses for testing. An axiomatic approach may be quite useful to dealing with these issues, as will meta-analyses of empirical projects where similar questions are explored across various case studies.

Fourth, it is clear that networks and behaviors "co-evolve": friendships influence behaviors and behaviors influence who becomes friends with whom. Although there are some models and studies of this phenomenon (see the chapter by Vega-Redondo), it is another area that is not so extensively studied, either empirically or theoretically. This phenomenon also applies beyond social interactions, having important potential implications for things such as investments by banks and other inter-linked financial entities, whose incentives regarding how to invest and whether to monitor those investments are network dependent, and influence the resulting relationships. Such co-dependencies could have far-reaching policy implications.

Fifth, a related but quite distinct observation is that people interact in many different ways at once. One might exchange favors with co-workers, share information with trading partners, and so forth. The layering of different types of relationships among individuals has not gone unnoticed, as there are many studies of multi-relational, multi-layered, or multiplexed networks. Nonetheless, we know little about when and why multiple relationships interact with each other and what the broader consequences of interactions between different types of networks might be. The literature to date has been largely observational, and this is an area where there are enormous potential gains from bringing some simple economic modeling to bear.

4.4 Closing Thoughts

It has become clear that network modeling and analysis in economics is more than a fad. The ubiquity of networked interactions in economic settings means that, if anything, network analyses should continue to grow in economics. The bucket list above provides some areas where the potential gains from new research are self-evident, and undoubtedly new areas will emerge over time as the theory and data continue to advance. The multitude of important open questions makes this a most fertile area for research, and makes this volume indispensable.

[3] This is also an issue if one splits data and runs models on one part and then checks that they still make good predictions out of sample (on the second part of the data)—as even there, considering *multiple* models can end up selecting ones that happen to spuriously provide good fits on the second data set.

Finally, as I stated at the outset, network science is inherently interdisciplinary, drawing on tools from a variety of disciplines ranging from mathematics to sociology, and including economics, statistical physics, computer science, and statistics (see the discussion by Kirman). Its breadth derives from the presence of networks throughout the social and physical world; and from the wide variety of perspectives and techniques that are useful in representing and analyzing networks as well as collecting and analyzing data. Although researchers are much more aware of each other across disciplines than ever before, and tools and techniques are crossing borders, substantial homophily remains in the research communities. This is partly driven by the historical silos and departments in which subcommunities reside, which leads to incentives and cultures that are heavily influenced by home-disciplines. This leads to prejudices within economics, just as in other disciplines, that are suspicious of research employing paradigms and methods that are unusual within the discipline. It is thus heartening to see that this handbook involves researchers from outside of economics, and that many of the chapters incorporate substantial amounts of material from outside of economics. The hope is that the trend continues, and research in network science eventually becomes seamless across historical silos and disciplines.

PART III

NETWORK GAMES AND NETWORK FORMATION

CHAPTER 5

GAMES PLAYED ON NETWORKS

YANN BRAMOULLÉ AND RACHEL KRANTON

5.1 Introduction

This chapter studies games played on networks. The games capture a wide variety of economic settings, including local public goods, peer effects, and technology adoption. Pairwise links can represent respectively, geographic proximity, peer relations, and industry ties. The collection of links form a network, and given the network, agents choose actions. Individual payoffs derive from own actions and the actions of linked parties. Because sets of links overlap, ultimately the entire network structure determines equilibrium outcomes.

The chapter develops a guide to study all these settings by nesting the games in a common framework. Start with local public goods, for example. Individuals choose some positive level of goods, which benefit themselves and their neighbors. Provision is individually costly. Players' actions are then strategic substitutes; when a player's neighbor provides more, he provides less. A peer effect game requires just two modifications. The first is a change in the sign of a parameter, so that players' actions are strategic complements rather than substitutes. The second is an upper bound on players' actions, since, for example, students can study no more than twenty-four hours in a day. Next consider a coordination game, as in technology adoption. Individuals' actions are complements; they want to choose the same action as their neighbors. Here there is one additional modification—restricting individuals to binary actions: "adopt technology A" or "adopt technology B." We show that in these games individuals have the same underlying incentives expressed in linear best replies, which are more or less constrained. The chapter systematically introduces the modifications—changes in parameter sign and/or constraints on agents' choices—and shows how they alter the analysis and affect outcomes.

The chapter has three overarching objectives. The first is to establish a common analytical framework to study this wide class of games. So doing, the chapter establishes new connections between games in the literature—particularly the connection between binary choice games, such as coordination and best-shot games, and games with continuous actions, such as public goods, peer effects, and oligopoly games. The second objective is to review and advance existing results by showing how they tie together within the common framework. The final objective is to outline directions for future research.

All the games considered in this chapter are simultaneous-move, complete information games.[1] The analysis thus employs classic solution concepts: Nash equilibrium and stable equilibrium, which is a Nash equilibrium robust to small changes in agents' actions.

The chapter studies how equilibrium outcomes relate to features of the network. In any strategic setting, researchers study the existence, uniqueness, and stability and possibly comparative statics of equilibria. In a network game, researchers strive to answer these questions in terms of the network: What features of the network determine the Nash and stable equilibrium set? How do individual network positions determine individual play? How do outcomes change when links are added to or subtracted from a network?

The chapter reviews known results and highlights open questions. Characterizations of equilibrium sets often involve conditions on the eigenvalues of the network matrix. It has been long known, for example, that in a continuous-action game with pure complementarities, such as peer effects, a contraction property based on the highest eigenvalue guarantees the existence and uniqueness of Nash equilibrium. Contraction ensures that the complementarities and network effects are sufficiently small so there is a convergence in the best replies.

In a continuous-action game with any substitutabilities, such as an oligopoly game, it is the lowest eigenvalue that appears in the equilibrium conditions. For example, a unique Nash equilibrium exists if the magnitude of the lowest eigenvalue is sufficiently small. While the highest eigenvalue is a positive number, the lowest eigenvalue is a negative number, and its magnitude captures the extent of substitutabilities in the overall network. When this condition does not hold, the computation of the Nash and stable equilibrium set is necessarily complex.

As for individual play, when network effects are sufficiently small so that there is a unique interior equilibrium, individual actions are proportional to players' Bonacich centralities. Little is known, however, about how individual play relates to network position when the network effects are larger and equilibria involve agents who are constrained in their best replies.

The chapter further engages a novel question concerning networks and equilibria: how is one agent's action affected by an exogenous shock to another agent? The

[1] We discuss network games of incomplete information in the Conclusion.

chapter develops a new notion: equilibrium *interdependence* of agents. When one firm's production cost is reduced, for example, it possibly affects equilibrium production of all firms—not just the firm's direct competitors. We study whether two players who are not directly connected are nonetheless *interdependent*, relating interdependence to changes in individual parameters. We show that path-connectedness is necessary but not sufficient for interdependence. A third player along the path could absorb the impact of one player on the other. We study comparative statics and how shocks to individuals propagate or do not propagate through a network.[2]

This consideration of individual heterogeneity opens new avenues for investigation. Individuals are characterized not only by their position in the network but also by individual costs and benefits of actions. This heterogeneity allows the study of interdependence and the connection to the empirical literature on social interaction, equilibria, and the reflection problem. This connection is a rich area for current and future research and is discussed below.

The chapter is organized as follows. Section 5.2 presents the basic simultaneous-move game on a network and constructs the common set of (linear) best replies. Section 5.3 gives examples of games in the literature that fit in the framework. Section 5.4 analyzes the simplest case—unconstrained actions—which serves as a foil for the constrained cases that follow. Sections 5.5 and 5.6 study the games where agents actions are continuous but constrained, first to be positive, then to be positive and below an upper bound. In Section 5.7, we study binary action games. Section 5.8 relates the theory of network games to empirical work on social interactions. Section 5.9 discusses related games and topics that fall outside our framework.

5.2 Games on a Network and (Modified) Linear Best Replies

This section introduces the basic game played on a network and identifies the key economic parameters. It then poses the mathematical system underlying the Nash and stable equilibria for the class of games covered in this chapter.

5.2.1 Players, Links, and Payoffs

There are n agents, and N denotes the set of all agents. Agents simultaneously choose actions; each agent i chooses an x_i in $X_i \subseteq \mathbb{R}$. Agents are embedded in a fixed network represented by an $n \times n$ matrix, or graph, $\mathbf{G}$, where $g_{ij} \in \mathbb{R}$ represents a link between

[2] The chapter by Daron Acemoglu, Asuman Ozdaglar, and Alireza Tahbaz-Salehi, in this volume, also discusses how macroeconomic outcomes are induced by microeconomic shocks under network interactions.

agents i and j. Note that g_{ij} can be weighted and positive or negative. For most of the chapter, links are assumed to be undirected; i.e., $g_{ij} = g_{ji}$. In the games below, only the actions of i's *neighbors*—the agents to whom i is linked—enter an agent i's payoff π_i.

Each agent's payoff is a function of own action, x_i, others' actions, $\mathbf{x}_{-i}$, the network, and a global parameter $\delta \in [-1,1]$, called the "payoff impact parameter," which gives the sign and magnitude of the effect of players' actions on their neighbors: $\pi_i(x_i, \mathbf{x}_{-i}; \delta, \mathbf{G})$. For a given $\mathbf{G}$ and δ, we will say a property holds *generically* (i.e., for *almost any* δ), if it holds for every $\delta \in [-1,1]$ except for possibly a finite number of values. For any square matrix $\mathbf{M}$, let $\lambda_{\min}(\mathbf{M})$ denote the lowest eigenvalue and let $\lambda_{\max}(\mathbf{M})$ denote the highest eigenvalue. Note these eigenvalues can always be written in terms of each other; that is, $\lambda_{\max}(-\mathbf{M}) = -\lambda_{\min}(\mathbf{M})$.

The signs of δ and g_{ij} determine the type of strategic interactions between i and j. As we will see, i's and j's actions are strategic complements when $\delta g_{ij} < 0$, and they are strategic substitutes when $\delta g_{ij} > 0$.[3] We say that a game has *pure complements* if $\delta < 0$ and $\forall i,j, g_{ij} \geq 0$ and *pure substitutes* if $\delta > 0$ and $\forall i,j, g_{ij} \geq 0$. The literature has paid much attention to these polar cases, at the risk of neglecting the analysis of the general case.[4] We pay careful attention to this issue in what follows.

5.2.2 Best Replies, Nash Equilibria, and Stable Equilibria

We consider pure-strategy Nash equilibria of these games.[5] Let

$$f_i(\mathbf{x}_{-i}; \delta, \mathbf{G}) = \arg\max_{x_i} \{\pi_i(x_i, \mathbf{x}_{-i}; \delta, \mathbf{G})\} \tag{5.1}$$

denote agent i's best reply to other agents' actions. The following is then the system of best replies:

$$\begin{aligned} x_1 &= f_1(\mathbf{x}_{-1}; \delta, \mathbf{G}) \\ &\vdots \\ x_n &= f_n(\mathbf{x}_{-n}; \delta, \mathbf{G}). \end{aligned} \tag{5.2}$$

A Nash equilibrium is a vector $\mathbf{x} = (x_1, \ldots, x_n)$ that satisfies this system.

The chapter considers both the full set of Nash equilibria and the subset of Nash equilibria that are *stable*. Stable, here, refers to the criterion that an equilibrium is robust

[3] An agent i's action is a strategic complement (substitute) to j's action when i's best reply is increasing (decreasing) in j's action. See Bulow, Geanakoplos, and Klemperer (1985).

[4] For example, with pure complements the Perron-Frobenius Theorem and derivative results (e.g., the addition of a link increases the largest eigenvalue) apply. But these results do not apply if there are any substitutabilities.

[5] For many continuous action games in this class, the payoff functions are concave and no mixed strategy Nash equilibria exist.

to small changes in agents' actions. Appropriate stability notions necessarily differ for continuous and binary action games. For binary actions, we use a notion of stochastic stability based on asynchronous best-reply dynamics and payoffs (Blume 1993; Young 1998).

For continuous actions, we consider a classic definition of stability which is a continuous version of textbook Nash tâtonnement.[6] Starting with a Nash equilibrium **x**, and changing agents' actions by a little bit, we ask whether the best replies lead back to the original vector. Consider the following system of differential equations:

$$\begin{aligned} \dot{x}_1 &= f_1(\mathbf{x}_{-1};\delta,\mathbf{G}) - x_1 \\ &\vdots \\ \dot{x}_n &= f_n(\mathbf{x}_{-n};\delta,\mathbf{G}) - x_n. \end{aligned} \tag{5.3}$$

By construction, a vector **x** is a stationary state of this system if and only if it is a Nash equilibrium. We say a Nash equilibrium **x** is *asymptotically stable* when (5.3) converges to **x** following any small enough perturbation.[7]

5.2.3 Class of Games and Restrictions on the Strategy Space

We consider games whose payoffs are special cases of the following generalized payoff function:

$$\pi_i(x_i,\mathbf{x}_{-i};\delta,\mathbf{G}) = v_i\left(x_i - x_i^0 + \delta\sum_j g_{ij}x_j\right) + w_i(\mathbf{x}_{-i}) \tag{5.4}$$

where v_i is increasing on $(-\infty,0]$, decreasing on $[0,+\infty)$ and symmetric around 0, so that 0 is the unique maximum of v_i, and w_i can take any shape. The individual parameter x_i^0 represents agent i's optimal action absent social interactions ($\delta = 0$ and/or $g_{ij} = 0$). A higher x_i^0 would correspond, for example, to i's greater personal benefit from actions or lower private cost. As $|\delta|$ increases, the payoff externalities of agents' actions become globally stronger.

In the base case, actions can take any real value: for each agent i, $x_i \in X_i = \mathbb{R}$. With payoffs (5.4), best replies are linear in other agents' actions:

$$f_i(\mathbf{x}_{-i}) = x_i^0 - \delta\sum_j g_{ij}x_j. \tag{5.5}$$

[6] See, e.g., Fisher (1961).

[7] Formally, following Weibull (1995, Definition 6.5, p. 243), introduce $B(\mathbf{x},\varepsilon) = \{\mathbf{y} \in \mathbb{R}_+^n : ||\mathbf{y}-\mathbf{x}|| < \varepsilon\}$ and $\xi(t,\mathbf{y})$, the value at time t of the unique solution to the system of differential equations that starts at **y**. An equilibrium **x** is *asymptotically stable* if $\forall\varepsilon > 0, \exists\eta > 0 : \forall\mathbf{y} \in B(\mathbf{x},\eta), \forall t \geq 0$, $\xi(t,\mathbf{y}) \in B(\mathbf{x},\varepsilon)$ and if $\exists\varepsilon > 0 : \forall\mathbf{y} \in B(\mathbf{x},\varepsilon), \lim_{t\to\infty}\xi(t,\mathbf{y}) = \mathbf{x}$.

The agent essentially compares his autarkic optimum, x_i^0, to the weighted sum of his neighbors' actions, $\delta \sum_j g_{ij} x_j$; his best reply is the difference between the two.

While in principle, a player's action could be any real number, all games in the literature place restrictions on players' actions which represent different real-world situations. For example, for peer effects in a classroom, there are natural lower and upper bounds—a student can study no less than zero hours and no more than twenty-four hours in a day. For technology adoption, an individual is often restricted to two actions: "adopt" or "not adopt."

For each restriction on the action space, we determine the corresponding best reply. First, agents' actions are constrained to be non-negative, which corresponds to, for example, the production of goods or services. For each agent i, $x_i \in [0, \infty)$ and the corresponding best reply is

$$f_i(\mathbf{x}_{-i}) = \max\left(0, \left(x_i^0 - \delta \sum_j g_{ij} x_j\right)\right). \tag{5.6}$$

Second, agents' actions must not be below zero nor be above some finite upper bound L: for each agent i, $x_i \in X_i = [0, L]$ with $0 < L < \infty$. The corresponding best reply is

$$f_i(\mathbf{x}_{-i}) = \min\left(\max\left(x_i^0 - \delta \sum_j g_{ij} x_j, 0\right), L\right). \tag{5.7}$$

In both cases a player's best reply is to choose, as much as possible, the difference between x_i^0, and the weighted sum $\delta \sum_j g_{ij} x_j$.

Finally, agents must choose between two discrete values: $x_i \in X_i = \{a, b\}$ with $a \leq b$. Agent i's best reply can be written in terms of a threshold value $t_i = x_i^0 - \frac{1}{2}(a+b)$. If the weighted sum of neighbors' actions is above the threshold, i's best response is a; if the weighted sum is below the threshold, agent i's best response is b; if the sum is equal to the threshold, i is indifferent between a and b. We have:

$$\begin{aligned} f_i(\mathbf{x}_{-i}) &= a \text{ if } \delta \sum_j g_{ij} x_j > t_i; \\ f_i(\mathbf{x}_{-i}) &= b \text{ if } \delta \sum_j g_{ij} x_j < t_i; \\ f_i(\mathbf{x}_{-i}) &= \{a, b\} \text{ if } \delta \sum_j g_{ij} x_j = t_i. \end{aligned} \tag{5.8}$$

The best replies for the constrained actions, (5.6), (5.7), and (5.8), can all be obtained from (5.5). Let $\widehat{x}_i(\mathbf{x}_{-i}) \equiv x_i^0 - \delta \sum_j g_{ij} x_j$ denote the *unconstrained optimum*. When agents' choices are constrained, agent i's best reply is simply the value which is closest to $\widehat{x}_i(\mathbf{x}_{-i})$ within the restricted space.

5.2.4 Game Class

The chapter studies all games whose best replies have the above form. Since the best replies are equivalent, to analyze the equilibria for all games, we can consider one payoff function that satisfies the conditions of (5.4). We make extensive use of payoffs with the following quadratic form:

$$\Pi_i(x_i, \mathbf{x}_{-i}; \delta, \mathbf{G}) = -\frac{1}{2}\left(x_i - x_i^0 - \delta \sum_j g_{ij} x_j\right)^2 + w_i(\mathbf{x}_{-i}). \tag{5.9}$$

5.3 Examples of Games in the Literature

Many games in the economics and the network literature fall in this class, with different specifications of the action spaces, X_i, the link values g_{ij}, and the payoff impact parameter δ. All games in the literature involve some restriction on the strategy space.

5.3.1 Constrained Continuous Actions: Quadratic Payoffs and Benefit/Cost Payoffs

Quadratic payoffs are common and have been used to represent a variety of settings including peer effects, consumption externalities, and oligopoly. Players choose some action $x_i \in X_i = [0, \infty)$ and payoffs have a form such as

$$\pi_i(x_i, \mathbf{x}_{-i}; \delta, \mathbf{G}) = x_i^0 x_i - \frac{1}{2} x_i^2 - \delta \sum_j g_{ij} x_i x_j, \tag{5.10}$$

which is a special case of (5.9).[8] With positive links, $g_{ij} = g_{ji} \geq 0$, and negative payoff parameter, $\delta \leq 0$, these payoffs give a pure complements game, as in peer effects. For pure substitutes, $g_{ij} = g_{ji} \geq 0$, and $\delta \geq 0$, as in a Cournot game with n firms producing substitute products. The links $g_{ij} = g_{ji} = 1$ indicate firm i and firm j compete directly, and a function of δ is the overall extent of substitutability among goods. Quadratic payoffs have also been used to model settings with both substitutes and complements, as in crime games.[9]

[8] For a prominent example see Ballester, Calvó-Armengol, and Zenou (2006).

[9] The benefit from x_i is higher when the overall crime level is lower, capturing the possibility that criminals may compete for victims or territory. The cost of x_i is lower when i's friends engage in more crime, capturing the possibility of positive peer effects. See Bramoullé, Kranton and D'Amours (2014), which refines Calvó-Armengol and Zenou (2004) and Ballester, Calvó-Armengol, and Zenou (2010).

Another type of payoffs specifies the trade-off between the benefits from own and others' actions and individual costs, as in the private provision of local public goods.[10] Each agent chooses a level $x_i \in X_i = [0, \infty)$ and earns

$$\pi_i(x_i, \mathbf{x}_{-i}; \delta, \mathbf{G}) = b_i(x_i + \delta \sum_j g_{ij} x_j) - \kappa_i x_i,$$

where $b_i(\cdot)$ is strictly increasing and strictly concave, $\kappa_i > 0$ is i's marginal cost, and $b_i'(0) > \kappa_i > b_i'(+\infty)$ for all i. The links are positive, $g_{ij} = g_{ji} \geq 0$, and $\delta \geq 0$ so that agents' payoffs increase when their neighbors provide more public goods. Best replies correspond to (5.6) for x_i^0 such that $b_i'(x_i^0) = \kappa_i$. The substitutability of own and others' goods is scaled by δ.

5.3.2 Binary Actions: Coordination, Anti-Coordination, and Best-Shot Games

In many games in the literature, players have the choice between two actions such as "buy" vs. "not buy," "vote yes" vs. "vote no," and so on. For judicious choices of a, b, δ, and x_i^0, binary games played on a network have the form of best reply (5.8). Consider the payoffs of a symmetric 2×2 game with actions A and B:

	A	B
A	π_{AA}, π_{AA}	π_{AB}, π_{BA}
B	π_{BA}, π_{AB}	π_{BB}, π_{BB}

This matrix gives payoffs for a classic coordination game between two players when $\pi_{AA} > \pi_{BA}$ and $\pi_{BB} > \pi_{AB}$. On a network with all $g_{ij} \geq 0$, a player earns the sum of bilateral payoffs: $\pi_i(x_i, \mathbf{x}_{-i}) = \sum_j g_{ij} \pi(x_i, x_j)$.[11]

Agent i's best reply has a simple form. Define $p_B = (\pi_{AA} - \pi_{BA})/(\pi_{AA} + \pi_{BB} - \pi_{AB} - \pi_{BA})$, where $0 < p_B < 1$. Let $k_i = \sum_j g_{ij}$ be the weighted links of agent i's neighbors, and let $k_{iB} = \sum_j g_{ij}(B)$ be the weighted links of i's neighbors who play B. Then i strictly prefers to play B if and only if the weighted majority of his neighbors play B: $k_{iB} > p_B k_i$. To establish the correspondence with (5.8), assign numbers a and b to actions A and B, let $k_{iA} = \sum_j g_{ij}(A)$ be the weighted links of i's neighbors who play A, and note that $\sum_j g_{ij} x_j = a k_{iA} + b k_{iB} = (b - a) k_{iB} + a k_i$. The threshold $t_i = x_i^0 - \frac{1}{2}(a + b)$ is then constructed by setting $x_i^0 = \frac{1}{2}(a + b) + |\delta| k_i (a + (b - a) p_B)$ for any $\delta < 0$.

More generally, these games include any threshold game of complements as defined in Jackson (2008, p. 270). The "majority game" is thus easily recast as a coordination

[10] Bramoullé and Kranton (2007) study the private provision of local public goods. Further studies include Galeotti and Goyal (2010) and Allouch (2014).

[11] See Blume (1993, 1995), Ellison (1993), Morris (2000), Young (1998), Jackson and Watts (2002), Goyal and Vega-Redondo (2005).

game on a network. Agents earn payoffs when they choose the same action as the majority of their neighbors. Setting $\pi_{AA} = \pi_{BB} = 1$ and $\pi_{AB} = \pi_{BA} = 0$, gives $p_B = 1/2$, which corresponds to (5.8). For any $\delta < 0$ and for $a = -1$, $b = 1$, we have $x_i^0 = 0$.

For binary choice games with substitutes, such as anti-coordination games, agents want to differentiate from, rather than conform to, their neighbors. The Hawk-Dove game is one example, where in the payoff matrix above $\pi_{AA} < \pi_{BA}$ and $\pi_{BB} < \pi_{AB}$.[12] Agent i strictly prefers to play B if and only if a payoff-weighted *minority* of his neighbors plays B: $k_{iB} < p_B k_i$. A network anti-coordination game then has best responses of the form (5.8) by setting $x_i^0 = \frac{1}{2}(a+b) + \delta k_i(a+(b-a)p_b)$ for any $\delta > 0$.

In the "best-shot" game,[13] actions are also strategic substitutes and agents' actions can represent a discrete local public good. Each agent i chooses either 0 or 1, with $c \in (0,1)$ as the individual cost of taking action 1. Agents earn a benefit of 1 if any neighbor has played 1. The best reply is then $f_i = 1$ if $\sum_j g_{ij}x_j = 0$ and $f_i = 0$ if $\sum_j g_{ij}x_j > 0$. This game gives another particular case of (5.8) with $a = 0$, $b = 1$, $\delta = 1$ and $x_i^0 \in (\frac{1}{2}, \frac{3}{2})$.

5.4 Unconstrained Actions

When agents' actions are unconstrained, the games can be analyzed with relatively straightforward linear algebra. Even so, equilibrium behavior on networks gives rise to rich and complex patterns. This complexity is amplified by the introduction of constraints, studied in the next section.

5.4.1 Nash and Stable Equilibria

For unconstrained actions, a Nash equilibrium is simply a solution to the system of linear equations defined by the best replies (5.5). For $X_i = \mathbb{R}$, the system of best replies is, in matrix notation,

$$\mathbf{x}(\mathrm{I}+\delta\mathrm{G}) = \mathbf{x}^0.$$

Generically, there exists a unique Nash equilibrium. A unique equilibrium exists if $\det(\mathrm{I}+\delta\mathrm{G}) \neq 0$,[14] and then the equilibrium actions are determined by

$$\mathbf{x} = (\mathrm{I}+\delta\mathrm{G})^{-1}\mathbf{x}^0. \tag{5.11}$$

For convenience, we will label this unique unconstrained equilibrium vector $\mathbf{x}^*$. This argument also clearly holds for any directed network.

[12] For anti-coordination games played on networks see Bramoullé (2007), Bramoullé et al. (2004).

[13] See Hirshleifer (1983) and Pin and Boncinelli (2012).

[14] Note that $\det(\mathrm{I}+\delta\mathrm{G}) \neq 0$ for almost every δ. The invertibility of $(\mathrm{I}+\delta\mathrm{G})$ is a sufficient but not necessary condition for existence of a Nash equilibrium. Continua of equilibrium can exist when $\mathrm{I}+\delta\mathrm{G}$ is not invertible.

This equilibrium is asymptotically stable according to the standard conditions for the stability of a system of linear differential equations, which is here $|\lambda_{\min}(\delta\mathbf{G})| < 1$, which can also be written $|\lambda_{\max}(-\delta\mathbf{G})| < 1$. The stability condition imposes a joint restriction on the payoff impact, δ, and the network structure, which jointly give what we call the "network effects" of players' actions. The equilibrium is stable only when these network effects are small enough. When network effects are strong, the equilibrium is unstable. Bounds on actions, which also often represent real-world situations, are necessary for the existence of stable Nash equilibria.

5.4.2 Network Position

How do individual network positions affect individual actions? Ballester, Calvó-Armengol, and Zenou (2006) first establish the connection between equilibrium action and Bonacich centrality (Bonacich 1987).[15] In their model, individuals are homogenous but for their network position ($x_i^0 = x^0$ for all i). For a network $\mathbf{M}$ and a scalar q such that $(\mathbf{I} - q\mathbf{M})$ is invertible, $\mathbf{z}(q,\mathbf{M}) = (\mathbf{I} - q\mathbf{M})^{-1}\mathbf{M}\mathbf{1}$ is the vector of Bonacich centralities. It is easy to see that equilibrium actions $\mathbf{x}^*$ can be directly written in terms of centralities: $\mathbf{x}^* = x^0(\mathbf{I} + \delta\mathbf{G})^{-1}\mathbf{1} = x^0(1 - \delta\mathbf{z}(-\delta,\mathbf{G}))$.

These centralities have an interpretation in terms of paths in the network if $|q|\,\lambda_{\max}(\mathbf{M}) < 1$, so that $\det(\mathbf{I} - q\mathbf{M}) \neq 0$ and $\mathbf{z}(q,\mathbf{M}) = \sum_{k=0}^{+\infty} q^k\mathbf{M}^{k+1}\mathbf{1}$. For agent i, $z_i(q,\mathbf{M})$ is then equal to a weighted sum of the number of paths starting from i, where paths of lengths k are weighted by q^{k-1}. Agents' actions are increasing in centrality under pure complements and decreasing under pure substitutes.

Figure 5.1 below illustrates the unique equilibria—contrasting pure complements and pure substitutes—on a line with five agents when $\mathbf{x}^0 = \mathbf{1}$. For $\delta = -0.3$, agents' actions are strategic complements. The agent with the highest Bonacich centrality is in the middle of the line, as the scalar q is positive. This agent, then, has the highest level of play and agents' centralities and actions decrease moving away from the middle of the line. The outcome is quite different for $\delta = 0.3$, when agents' actions are strategic substitutes. The scalar q is now negative, giving positive weight to an agent's neighbors, but negative weight to the neighbors of neighbors. The agent in the middle of the line is not the most central agent. The intermediate agents are more central, as Bonacich centrality weights go up and down along network paths. Since the agents on the ends of the line have no other neighbors—and hence no further substitutes for their actions—their actions are highest, which leads to a lower level for the agents in the intermediate positions, which leads to a higher level for the agent in the middle.

The simple relationship between equilibrium actions and Bonacich centrality fails to hold when individuals are heterogeneous. Yet, Bonacich centrality still affords an intuitive comparative static. Let $\mathbf{x}^*(\mathbf{x}^0)$ be the unique equilibrium for a given

[15] The chapter by Yves Zenou in this volume provides further discussion of this connection, and how it relates to key-player policies.

FIGURE 5.1 Bonacich centrality and equilibrium actions: complements vs. substitutes.

$\mathbf{x}^0 = (x_1^0, \ldots, x_1^0)$. Suppose each agent i's autarkic action, x_i^0, changes by the same amount s. An increase could represent, for instance, a policy intervention that lowers all agents' individual costs. Agents' Bonacich centralities give precisely the change in their equilibrium actions. It is readily evident that

$$\mathbf{x}^*(\mathbf{x}^0 + s\mathbf{1}) - \mathbf{x}^*(\mathbf{x}^0) = s(\mathbf{1} - \delta\mathbf{z}(-\delta, \mathbf{G})).$$

5.4.3 Interdependence

Equilibrium actions depend on how the network connects different agents. We say two agents i and j are *interdependent* if a (small, exogenous) change in agent j's autarkic action would lead to an adjustment in i's action. In $\mathbf{x}^*$, consider $\partial x_i^* / \partial x_j^0$. Intuitively, x_j^0 first affects x_j^* which then affects the action of j's neighbors and then the actions of their neighbors, and so on. Through the network, this change potentially impacts all agents. We have

$$\frac{\partial x_i^*}{\partial x_j^0} = \left[(\mathbf{I} + \delta\mathbf{G})^{-1}\right]_{ij} = \sum_{k=0}^{+\infty} (-\delta)^k \left[\mathbf{G}^k\right]_{ij},$$

where the second equality holds if $|\delta|\lambda_{\max}(\mathbf{G}) < 1$. The marginal impact of x_j^0 on x_i^* is equal to a weighted sum of the number of paths from i to j. When $\mathbf{G}$ is connected (i.e., there is a path from any agent i to any other agent j), $\partial x_i^* / \partial x_j^0 \neq 0$ for all i and j and almost every δ.[16]

The direction and magnitude of the interdependence depend on whether actions are complements or substitutes. Under pure complements ($\delta < 0$), all the terms in the infinite sum are non-negative. An increase in x_j^0 leads to an increase in x_j^*, which leads to an increase in the actions of j's neighbors, which leads to an increase in the actions of their neighbors, and so on.

Figure 5.2 illustrates the interdependence of agents in a network that represents connected communities. For $\delta = -\frac{1}{4}$, and $\mathbf{x}^0 = \mathbf{1}$, in the equilibrium $\mathbf{x}^*$ the agents with no connection to the other community play 5.33 and the two agents connecting the communities play 6.67. The figure shows the value of the partial derivatives for the play of each agent following a small positive impact to x^0 for the agent to the extreme left of the graph. This impact ultimately affects the play of all agents in the network.

[16] In the presence of substitutes, the positive and negative effects could cancel each other completely for specific values of δ.

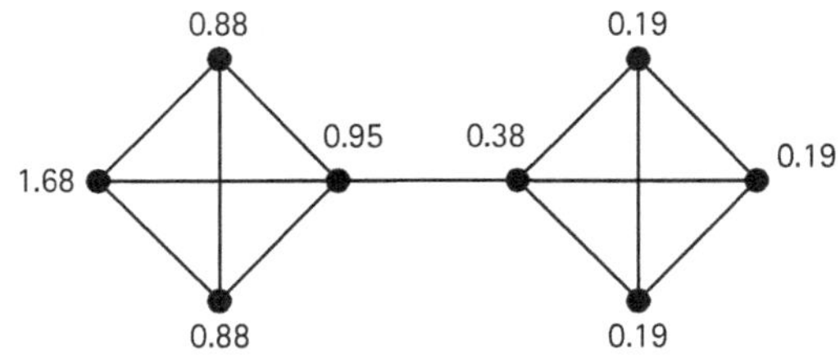

FIGURE 5.2 Measures of interdependence under complementarities.

The magnitude of the interdependence follows directly from the formula for $\partial x_i^*/\partial x_j^0$ above. With pure complements, agents are more (less) interdependent when they are connected by more and shorter (less and longer) paths.

The situation is more complex under pure substitutes ($\delta > 0$). In that case, the even terms in the infinite sum are non-negative, while the odd terms are non-positive. An increase in x_j^0 leads to an increase in x_j, which leads to a decrease in the actions of j's neighbors, which leads to an increase in the actions of their neighbors, and so on with alternating signs. Generically, the aggregate could be positive or negative.

While there are few results that relate the aggregate impacts under substitutes to the structure of the network, bipartite graphs serve as a benchmark. Bipartite graphs are the only networks where direct and indirect effects are all aligned. A graph is bipartite, by definition, when the agents can be partitioned in two sets U and V such that $g_{ij} = 0$ if $i,j \in U$ or $i,j \in V$. We can show that a graph is bipartite if and only if the length of all paths connecting two agents is exclusively even or odd. The length is even when the two agents belong to the same set (U or V) and odd when they belong to different sets. In a connected bipartite graph, $\partial x_i/\partial x_j^0 > 0$ if i and j belong to the same set, and $\partial x_i/\partial x_j^0 < 0$ if i and j belong to different sets.[17] It would be interesting to try to extend these results to more elaborate structures and interactions.

5.5 Tools for Constrained Actions: Potential Function

When players' actions are constrained, as in all economic games, the analysis of the equilibrium set is more complex. When $x_i^0 > 0$ for all players and $|\delta|$ is sufficiently small, the constrained and unconstrained equilibrium coincide.[18] All players choose actions in the interior of the action space, and the analysis of Section 5.4 applies. However, as

[17] See Appendix of Bramoullé, Kranton, and D'amours (2011).

[18] This is the case studied by Ballester, Calvó-Armengol, and Zenou (2006). In our notation, their sufficient condition for the existence of a unique interior equilibrium is $\delta < 1/(\overline{g} + \lambda_{\max}(\overline{g}\mathbf{C} - \mathbf{G}))$, where $\overline{g}$ is the value of the strongest substitute link in $\mathbf{G}$ (i.e. $\overline{g} = \max_{ij}(0, g_{ij})$) and $\mathbf{C}$ is the complete graph.

$|\delta|$ becomes larger, network effects become important, and some players will be driven to actions at the boundaries. The constraints then affect the analysis quite deeply.

To construct and analyze equilibria, we develop the following terminology. Agents who choose actions which are strictly positive and strictly lower than the upper bound are called *unconstrained.* Agents who play 0 or play L are called *constrained.* A constrained agent i is *strictly constrained* if i would remain constrained even with a small change in neighbors' actions. That is, i is strictly constrained if $x_i^0 - \delta \sum_j g_{ij} x_j > L$ when $x_i = L$ and $x_i^0 - \delta \sum_j g_{ij} x_j < 0$ if $x_i = 0$. For a network $\mathbf{G}$ and a subset of agents S, let $\mathbf{G}_S$ denote the subgraph that contains only links between the agents in S and $\mathbf{x}_S$ denote the actions of agents in S.

5.5.1 Potential Function

To analyze these games and solve for the Nash equilibria, we use the theory of potential games developed by Monderer and Shapley (1996).[19] A function $\varphi(x_i, \mathbf{x}_{-i})$ is a *potential function* for a game with payoffs $V_i(x_i, \mathbf{x}_{-i})$ if and only if for all x_i and x_i' and all $\mathbf{x}_{-i}$

$$\varphi(x_i, \mathbf{x}_{-i}) - \varphi(x_i', \mathbf{x}_{-i}) = V_i(x_i, \mathbf{x}_{-i}) - V_i(x_i', \mathbf{x}_{-i}) \qquad \text{for all } i.$$

A potential function mirrors each agent's payoff function. Changing actions from x_i to x_i' increases the potential by exactly the same amount as it increases agent i's payoffs. Not all payoff functions $V_i(x_i, \mathbf{x}_{-i})$ allow for a potential function. Monderer and Shapley (1996) show for (continuous, twice-differentiable) payoffs V_i, there exists a potential function if and only if $\frac{\partial^2 V_i(\mathbf{x})}{\partial x_i \partial x_j} = \frac{\partial^2 V_j(\mathbf{x})}{\partial x_j \partial x_i}$ for all $i \neq j$. A key property is that the potential is preserved when restricting the domain. That is, for any possible choices x_i and x_i' in an agent's strategy space, the potential increases by exactly the same amount as agent i's payoffs.

5.5.2 Best Reply Equivalence and Potential for Quadratic Payoffs

Since all games in the class have the same best replies, we can analyze the equilibria for all games in the class by studying the equilibria for one game in the class. Any game with continuous actions and quadratic payoffs (5.9) has a potential function when $g_{ij} = g_{ji}$ since $\frac{\partial^2 \Pi_i}{\partial x_i \partial x_j} = -\delta g_{ij} = \frac{\partial^2 \Pi_j}{\partial x_j \partial x_i}$. We can then analyze the Nash equilibria for all the games

[19] Blume (1993) and Young (1998) introduced potential techniques to the study of discrete network games. Blume (1993) focuses on lattices, while Young (1998) looks at 2×2 coordination games played on networks. Bramoullé, Kranton, and D'amours (2014) first apply potential techniques to the study of network games with continuous actions.

in the class using the potential function for quadratic payoffs (5.10):

$$\varphi(x_i, \mathbf{x}_{-i}) = \sum_i (x_i^0 x_i - \frac{1}{2}x_i^2) - \frac{1}{2}\delta \sum_i \sum_j g_{ij} x_i x_j$$

or, in matrix form,

$$\varphi(\mathbf{x}) = (\mathbf{x}^0)^T \mathbf{x} - \frac{1}{2}\mathbf{x}^T(\mathbf{I} + \delta \mathbf{G})\mathbf{x}.$$

Since the potential property holds for constrained actions, we can analyze all games with best replies (5.6), (5.7), and (5.8) with this potential function.

5.6 Equilibria in Constrained Continuous Action Games

In this section we analyze the equilibria in games where agents' actions are continuous, but constrained. We begin with preliminary results relating the maxima of the potential function to the set of Nash and asymptotically stable equilibria.

5.6.1 The Potential Function, Nash Equilibria, and Asymptotically Stable Equilibria

For games with continuous actions, an action vector $\mathbf{x}$ is a Nash equilibrium if and only if it solves the first-order conditions for maximizing the potential on X (Bramoullé, Kranton, and D'Amours 2014). To see this, consider maximizing the potential function:

$$\max_{\mathbf{x}} \varphi(\mathbf{x}; \delta, \mathbf{G}) \text{ s.t. } x_i \in X_i \text{ for all } i, \tag{P}$$

where $X_i = \mathbb{R}$, $X_i = [0, \infty)$ or $X_i = [0, L]$. The first-order conditions for this problem mimic each agent i's individual best reply.[20] Each agent chooses his action in the game *as if* he wants to maximize the potential, given other agents' actions. Thus we have:

Lemma 1 *With continuous actions, a profile* $\mathbf{x}$ *is a Nash equilibrium if and only if* $\mathbf{x}$ *satisfies the first-order conditions of problem (P).*

The potential function can also be used to identify asymptotically stable equilibria. Mathematically, it plays the role of a Lyapunov function for the system of differential equations. Say that a local maximum is *strict* if it is the only maximum in some open neighborhood. Starting from an equilibrium which is a strict local maximum of the potential and modifying actions slightly, individual adjustments will lead back to the

[20] $\varphi(x_i, \mathbf{x}_{-i})$ is strictly concave in each x_i, so for any $\mathbf{x}_{-i}$ a single x_i satisfies the ith Kuhn-Tucker condition.

equilibrium. Locally, there is no way to increase the potential, which reflects all agents' best replies. In contrast, if the equilibrium is not a strict local maximum, but, say, a saddle point, then modifying agents' actions slightly, there will be a direction in which the potential is increasing and individual reactions lead away from the equilibrium. Formally:

Lemma 2 *With continuous actions, a profile* $\mathbf{x}$ *is a stable Nash equilibrium if and only if* $\mathbf{x}$ *is a strict local maximum of* φ *over* X.

With these tools in hand, we will attack the analysis of the constrained continuous action games, where $X_i = [0, \infty)$ for all i or $X_i = [0, L]$ for all i.

5.6.2 Existence of Nash Equilibria

For games with a finite upper bound, existence of a Nash equilibrium is guaranteed. First, by standard results, for $X_i = [0, L]$ the strategy space is compact and convex. Since the best reply (5.7) is continuous, existence follows from Brouwer's fixed point theorem. This argument holds also for any directed network. Alternatively, since the potential function φ is continuous, it has a global maximum over X, and by Lemma 1 this maximum is a Nash equilibrium.

With no upper bound, a Nash equilibrium may fail to exist. When $X_i = [0, \infty)$, existence depends on whether actions are strategic substitutes or complements and on the extent of network effects. In a game of pure complements, if $|\delta|\lambda_{\max}(\mathbf{G}) < 1$, there exists a Nash equilibrium that is equivalent to the unconstrained equilibrium $\mathbf{x}^*$. For $|\delta|\lambda_{\max}(\mathbf{G}) > 1$, there is no Nash equilibrium with positive actions. Social interactions feed back into each other and diverge to infinity. With pure substitutes, on the other hand, existence is guaranteed. An agent will never choose an action that is greater than his autarkic optimum (i.e., $f_i(\mathbf{x}_{-i}) \leq x_i^0$). We can then assume without loss of generality that actions are bounded from above by $L = \max_i x_i^0$ and existence follows. All these arguments extend to directed networks. In general, with a mix of complements and substitutes, Bramoullé, Kranton, and D'amours (2014) show that if $|\lambda_{\min}(\delta\mathbf{G})| < 1$, a Nash equilibrium exists.[21]

The literature lacks existence results for larger payoff impacts. We conjecture that existence holds if the substitutes somehow dominate the complements in the strategic mix.

5.6.3 Unique versus Multiple Equilibria

Uniqueness is naturally related to the curvature of the potential function, as shown in Lemma 1. In particular, when φ is strictly concave, the first-order conditions of problem

[21] Note that $|\lambda_{\min}(\delta\mathbf{G})| = |\delta|\,\lambda_{\max}(\mathbf{G})$ under pure complements.

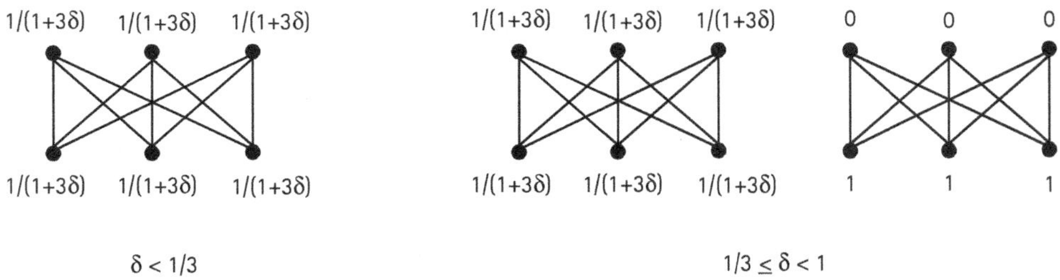

FIGURE 5.3 Unique vs. multiple equilibria in a regular graph.

(P) have at most one solution. Thus, for $X_i = [0, \infty)$, or $X_i = [0, L]$, there is a unique equilibrium when $|\lambda_{\min}(\delta\mathbf{G})| < 1$, since $\nabla^2\varphi = -(\mathbf{I} + \delta\mathbf{G})$, and the potential is strictly concave when $\mathbf{I} + \delta\mathbf{G}$ is positive definite (Bramoullé, Kranton, and D'amours, 2014):

Proposition 1 *There is a unique Nash equilibrium if* $|\lambda_{\min}(\delta\mathbf{G})| < 1$.

Proposition 1 provides the best known condition valid for all continuous action games in the class. Researchers have derived stronger results for specific cases. Belhaj, Bramoullé, and Deroian (2014) show that the equilibrium is unique for any game with pure complements. For pure substitutes and homogeneous agents, Proposition 1's condition is necessary and sufficient for regular graphs (Bramoullé, Kranton, and D'amours 2014). When $\delta > 0$, Proposition 1's condition becomes $|\lambda_{\min}(\mathbf{G})| < 1/\delta$. The lowest eigenvalue—a negative number—gives a measure of overall substitutabilities in the network. When it is small in magnitude, the magnitude of the ups and downs in the network is smaller, and there is only one equilibrium.

Figure 5.3 illustrates in a complete bipartite graph for six agents. The lowest eigenvalue for this network is -3. Hence for $\delta < 1/3$, there is a unique Nash equilibrium where all agents play $1/(1 + 3\delta)$. For higher δ, there are three equilibria. One of these additional equilibria is illustrated, involving all agents on one side of the network playing action 0 and agents on the other side playing 1. The third equilibrium has the same pattern but with the play of the sides reversed.

The special case of local public goods with perfect substitutes and homogeneous agents ($\delta = 1$, $g_{ij} \in \{0, 1\}, x_i^0 = x^0$) yields precise structural results (Bramoullé and Kranton 2007). Every maximal independent set of the graph yields a Nash equilibrium. Agents inside the set choose x^0 and all agents outside the set choose 0.[22] In any connected graph, there are multiple equilibria and the number of equilibria can grow exponentially with the number of agents.

[22] An *independent set* of agents is a set such that no agent in the set is linked. A *maximal independent set* is an independent set that is not a subset of any other independent set. A maximal independent set has the property that all agents outside the set are linked to at least one agent in the set.

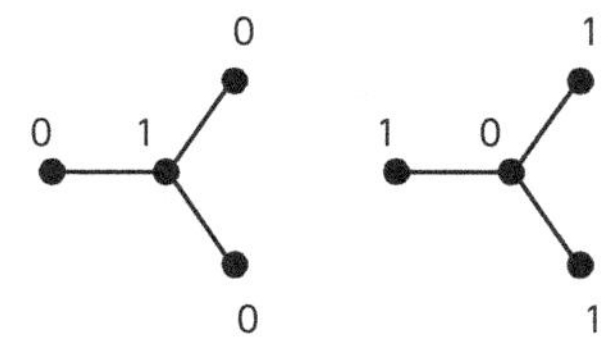

FIGURE 5.4 Maximal independent sets and equilibria in the star for $\delta = 1$.

Figure 5.4 shows the equilibria in a star for $\delta = 1$. The agent in the center constitutes a maximal independent set, and the agents in the periphery constitute a maximal independent set.

Our general knowledge of how unique versus multiple equilibria depend on parameters and the network is still very fragmented. We conjecture that multiplicity tends to be higher when $|\delta|$ is greater, when there are more substitutes in the strategic mix, and when $|\lambda_{\min}(\mathbf{G})|$ is greater.

5.6.4 Stability and the Lowest Eigenvalue

In light of the large possible number of equilibria, stability is a natural refinement. From Lemma 2, we can show that a stable equilibrium exists for any $\mathbf{G}$ and almost any δ. Moreover, stability is then related to the local curvature of the potential.

Consider a Nash equilibrium $\mathbf{x}$ with unconstrained agents U and all other agents strictly constrained. Now perturb agents' actions slightly by adding a vector $\varepsilon = (\varepsilon_1, ..., \varepsilon_n)$ such that $\mathbf{x} + \varepsilon \in X$. When $|\lambda_{\min}(\delta \mathbf{G}_U)| < 1$ the potential function is strictly concave in $\mathbf{x}_U$, and the best replies converge back to $\mathbf{x}$. When $|\lambda_{\min}(\delta \mathbf{G}_U)| > 1$, the potential function is not concave in $\mathbf{x}_U$. Some small perturbation ε can lead to large changes in best replies, and the equilibrium is not stable.

Proposition 2 *A Nash equilibrium with unconstrained agents U and all other agents strictly constrained is stable if and only if* $|\lambda_{\min}(\delta \mathbf{G}_U)| < 1$.

The lowest eigenvalue is key to the set of Nash and stable equilibria. There are only a few results that relate this eigenvalue to a network's structure.[23] Intuitively, the lowest eigenvalue—a negative number—gives the extent of substitutabilities in the network. Overall, $|\lambda_{\min}(\mathbf{G})|$ tends to be larger when the network is more "two-sided," so that agents can be subdivided into two sets with few links within the sets but many links between them. Loosely speaking, an action then reverberates between the two sides. For n agents, the graph with the highest $|\lambda_{\min}(\mathbf{G})|$ is the complete bipartite network with as equal sides as possible (that is, agents are divided into as equal size sets as possible and all agents in each set are linked to all agents in the other set, with no links within the sets).

[23] See summary in Bramoullé, Kranton, and D'Amours (2014).

FIGURE 5.5 Stable equilibria and the lowest eigenvalue.

Figure 5.5 illustrates Proposition 2 in two graphs that highlight the importance of the lowest eigenvalue. Each network contains six agents, nine links, and three links per agent. The network on the left is a complete bipartite graph, with lowest eigenvalue of -3. The network on the right has lowest eigenvalue of -2. Consider in each network perturbing the play of agent 1 in the lower left corner. In the bipartite graph, this perturbation directly impacts the three agents on the other side of the network (agents 4, 5, and 6) who must adjust, leading to adjustments on the other side, and so on. In the prism graph in the right panel, the perturbation directly impacts three agents (2, 3, and 6), two of whom are linked to each other (2 and 3). The play of these two agents jointly adjusts to the perturbation, dampening its effect. In the bipartite graph, the interior Nash equilibrium is stable for $\delta \leq 1/3$; for the prism graph, the equilibrium is stable for greater impact parameters $\delta \leq 1/2$.

Overall, among the multiple Nash equilibria, stable equilibria tend to contain more constrained agents. If $|\lambda_{\min}(\delta\mathbf{G})| > 1$, any stable equilibrium involves at least one constrained agent. More generally, the number of constrained agents in stable equilibria tends to increase when $|\delta|$ and $|\lambda_{\min}(\mathbf{G})|$ increase. In addition, with pure substitutes stable equilibria involve the largest sets of constrained agents among all Nash equilibria.[24]

At this point, however, we know little about the selection power of stability. Exploratory simulations show that a large proportion of Nash equilibria is typically unstable. We do not yet know how this proportion depends on the structure of the netwok. The shape of stable equilibria and the selection power of stability deserves to be studied more systematically.

5.6.5 Individual Network Position and Equilibrium Action

The general knowledge of how individual network positions affect equilibrium actions is still spotty. While alignment with Bonacich centralities is preserved in some circumstances, this relationship does not hold generally.[25]

[24] See Proposition 6 in Bramoullé, Kranton, and D'amours (2014).

[25] See the example in Section IV.A. in Belhaj, Bramoullé, and Deroian (2014).

In the following discussion, we consider homogeneous agents ($x_i^0 = x^0$ for all i) and study how the possible nested structure of neighborhoods affects play. Whether an agent i plays more or less than agent j depends on whether actions are pure substitutes or complements. In particular, suppose that i's neighborhood is nested in j's, so that j's neighbors are a superset of i's neighbors: $\forall k \neq i,j,\ g_{ik} \leq g_{jk}$. Then under pure complements, $x_i \leq x_j$ in the unique equilibrium.[26] An agent with more neighbors plays a higher action. In contrast, under pure substitutes and if $\forall i,j, \delta g_{ij} < 1$, then $x_i \geq x_j$ in any Nash equilibrium. An agent with more neighbors now plays a lower action. The usefulness of these results, of course, depends on the structure of the network and the extent to which agents are nested.

Recent theoretical work has drawn attention to a specific class of networks—nested split graphs—where any two agents' neighborhoods are always nested.[27] On nested split graphs, action is then weakly increasing in an agent's degree (i.e., number of neighbors), or centrality, under pure complements and weakly decreasing in degree under pure substitutes. Future research could possibly extend these results to more complex structures and interactions.

5.6.6 Interdependence

When actions are unconstrained, all agents are interdependent but constraints on actions can break this pattern. Strictly constrained agents do not change their actions in response to small changes in neighbors' actions and hence break the chain reaction from a possibly distant exogenous shock. Depending on the agents' positions in the network, this "dam" effect can break interdependence and leave parts of the network without effects on other parts.

Consider first pure complements ($\delta < 0$). In $\mathbf{x}^*$ all direct and indirect effects are aligned, and $\partial x_i/\partial x_j^0 > 0$ for any i and j who are path-connected. With constraints on actions, the direction of alignment is unchanged,[28] but x_i may now be unaffected by x_j^0, and x_i is only weakly increasing in x_j^0 for any $|\delta|$. Since agents who reach the upper bound do not transmit positive shocks, the right derivative $(\partial x_i/\partial x_j^0)^+ > 0$ if and only if i and j are connected by a path of unconstrained agents.[29] In that case,

$$(\partial x_i/\partial x_j^0)^+ = \left[(\mathbf{I}+\delta\mathbf{G}_U)^{-1}\right]_{ij} = \sum_{k=0}^{+\infty}(-\delta)^k[\mathbf{G}_U^k]_{ij}, \tag{5.12}$$

[26] See Proposition 4 of Belhaj, Bramoullé, and Deroian (2014), which shows that this property holds in any Nash equilibrium for a broad class of network games with non-linear best-replies.

[27] Konig, Tessone, and Zenou (2104) and Baetz (2014) show that nested split graphs emerge as outcomes of natural network formation processes. Belhaj, Bervoets, and Deroian (2014) show that they solve network design problems. We refer to these papers for precise definitions and further discussions of these networks' properties.

[28] See Corollary 4 of Belhaj, Bramoullé, and Deroian (2014).

[29] See Proposition 5 in Belhaj, Bramoullé, and Deroian (2014). The left and right derivatives of x_i with respect to x_j^0 may differ. The argument extends to left-derivatives.

FIGURE 5.6 Measures of interdependence: Complements with unconstrained vs. constrained agents.

and the magnitude of the impact is equal to a weighted sum of the number of these interior paths. As $\mathbf{x}^0$ increases, more agents reach the upper bound and interdependence decreases on both margins. Fewer agents affect each other, and all the positive pairwise impacts have lower magnitude.

Figure 5.6 illustrates consequences of constraints to the interdependence of agents—in the communities graph under complementarities. The left panel contains the communities graph shown previously, with the partial derivatives of the impact of play from an impact to the agent on the far left in the unconstrained equilibrium $\mathbf{x}^*$. The right panel shows the magnitudes of the partial derivatives for interdependence in the equilibrium when agents are constrained to play in the range $[0,6]$. The two agents connecting the communities are strictly constrained, playing 6, and all other agents play 5. A small change in x^0 of the agent to the far left has no impact on the agent in his community with links to the other community. This agent then blocks the impact from reaching further in the network, and the agents in the two communities are not interdependent.

Substitutes involve a number of complexitites, but the basic idea holds. Consider pure substitutes ($\delta > 0$) and a stable equilibrium $\mathbf{x}$ with unconstrained agents U and strictly constrained agents.[30] Now, a necessary condition for $(\partial x_i/\partial x_j^0)^+ \neq 0$ is that i and j are connected by a path of unconstrained agents, and this condition is generically sufficient. Equation (5.12) holds.

5.6.7 Network Comparative Statics—Adding Links

When studying network comparative statics, a natural first step is to look at the addition or strengthening of links. That is, consider the equilibria for a network G and compare them to the equilibria for G′ where $\forall i,j,\ g_{ij} \leq g'_{ij}$. An increase in g_{ij} is in many ways similar to a simultaneous shock to x_i^0 and x_j^0. Hence we can use insights from the study of interdependence to determine the effect of the new or stronger link.

As with interdependence, the effect depends on substitutes vs. complements. Under pure complements, the action of every agent in G′ is greater than or equal to the action

[30] These conditions ensure the existence of stable equilibrium with same set of unconstrained agents following a small enough increase in $\mathbf{x}^0$.

in G.[31] The action of an agent k is affected by g_{ij} if and only if there is a path of unconstrained agents connecting k with i or j. Under pure substitutes, the comparative statics are more complex. We do not know how to sign the effects at the individual level and equilibrium multiplicity further aggravates the issue.

The potential function, however, gives some traction on the problem, at least for aggregate outcomes. When $\delta > 0$, the potential is higher in $\mathbf{G}$ for any vector of actions: $\varphi(\mathbf{x}, \mathbf{G}') \leq \varphi(\mathbf{x}, \mathbf{G})\ \forall \mathbf{x}$. In addition, $\varphi(\mathbf{x}, \mathbf{G}) = \frac{1}{2}(\mathbf{x}^0)^T\mathbf{x}$ for any equilibrium $\mathbf{x}$. Thus the largest $\sum_i x_i^0 x_i$ in equilibrium decreases weakly following an expansion of the network.[32] In this sense, the direct and indirect negative impacts of the new link dominate the indirect positive effects.

5.7 Binary Action / Threshold Games

This section studies binary action games and shows how the common framework advanced in this chapter can make progress on questions of existence, uniqueness, and stability of equilibria. A full-fledged analysis is a ripe topic for future research. For ease of exposition in this section, we set the two actions to $a = -1$ and $b = 1$, so $X = \{-1, 1\}^n$.

5.7.1 Existence

While existence of a pure strategy equilibrium is not guaranteed a priori, we find existence follows naturally from the potential formulation in Section 5.5.1. Since the potential property is preserved on a constrained domain, the maxima of the potential function within the constrained space are Nash equilibria. Thus we can state the following new result:

Proposition 3 *In any network where $g_{ij} = g_{ji}$ and agents play a game with best replies (5.8), there exists a Nash equilibrium in pure strategies.*

The existence of a pure strategy Nash equilibrium does not extend to directed networks, except in the special case of pure complements. When agents always desire to take the same actions as their neighbors, existence of a pure strategy equilibrium is implied by standard results of the theory of supermodular games. However, for directed networks and a strategic mix there is no guarantee, as in Jackson's (2008, p. 271) "fashion" game where some agents desire to differentiate from neighbors, and there is no pure strategy equilibrium.

[31] See Corollary 4 of Belhaj, Bramoullé, and Deroian (2014).

[32] See Bramoullé, Kranton, and D'amours (2014). Theorem 2 in Ballester, Calvó-Armengol, and Zenou (2006) is a special case.

5.7.2 Unique versus Multiple Equilibria

For binary action games, strict concavity of the potential function does not guarantee a unique equilibrium. With the constraints on the action space, there can be more than one vector that maximizes a strictly concave potential subject to the constraints.

We can construct, however, a sufficient condition for a unique equilibrium that depends on the network structure. When each agent has a dominant strategy, there is a unique equilibrium, and whether each agent has a dominant strategy, in turn, depends on the impact parameter δ and each agent's degree. For pure complements ($\delta < 0$), an agent with $x_i^0 > 0$ strictly prefers to play 1 no matter what his neighbors choose if and only if $|\delta| k_i < x_i^0$. Similarly, playing -1 is strictly dominant for an agent with $x_i^0 < 0$ if and only if $x_i^0 < -|\delta| k_i$. An agent with $x_i^0 = 0$ is a priori indifferent and hence does not have a strictly dominant strategy. Similar conditions hold for pure substitutes. We thus have:

Proposition 4 *Consider pure substitutes or pure complements. A binary action game has a unique Nash equilibrium in dominant strategies if and only if*

$$|\delta| < \min_i \left(\frac{|x_i^0|}{k_i} \right).$$

Uniqueness in binary games then generally depends on the heterogeneity of agents and the distribution of idiosyncratic preferences.

This uniqueness condition bears a similarity to the conditions for continuous action games in that there is a unique equilibrium in binary action games when payoff impacts are small enough. With continuous actions, a small enough $|\delta|$ guarantees dampened adjustments to others' play. With binary actions, a small enough $|\delta|$ guarantees that idiosyncratic preferences dominate social interactions altogether, and agents never adjust to their neighbors.

While small payoff impacts guarantee unique equilibria, large impacts generally lead to multiple equilibria. Consider $|\delta| \geq \max_i \left(\frac{|x_i^0|}{k_i} \right)$. Under pure complements, all agents choosing -1 and all agents choosing 1 are both Nash equilibria. Social interactions completely swamp any idiosyncratic preferences, and full coordination occurs if and only if $|\delta| \geq \max_i \left(\frac{|x_i^0|}{k_i} \right)$. Under pure substitutes, equilibria involve agents playing different actions. But full anticoordination is impossible as soon as the network has a triangle; two connected agents then must play the same action. It is not possible for three agents to play different actions when only two actions are available.

For particular binary games with large payoff impacts, Nash equilibria have an intuitive graph-theoretic characterization. An action profile is a Nash equilibrium of the best-shot game if and only if the set of contributors is a maximal independent set of the graph (see Section 5.6). Each contributor is connected to agents who free-ride on his contribution.[33] This result implies that in the best shot game any connected network has multiple equilibria and, moreover, the number of equilibria may increase exponentially with n (Bramoullé and Kranton, 2007).

5.7.3 Stability

To refine the set of Nash equilibria, we can invoke a notion of stability. As noted above, asymptotic stability—defined by the system of differential equations (5.3)—does not apply to discrete action spaces. For binary choice games, we use asynchronous best-reply dynamics to study stability. For asynchronous best-reply dynamics subject to log-linear trembles, profiles that globally maximize the potential are the stochastically stable outcomes for all potential games (Blume 1993; Young 1998).[34] This stability notion depends on the specific payoffs of each game.[35]

In what follows, we provide a first discussion of stability in a binary game where agents have quadratic payoffs (5.10). Let $g_{ij} \in \{0, 1\}$. Simple computations show that the potential function is then

$$\varphi(\mathbf{x}) = \sum_{x_i=b} x_i^0 - \sum_{x_i=a} x_i^0 - \delta(n_{aa} + n_{bb} - n_{ab}) - \frac{1}{2}n$$

where n_{aa} is number links between a-players and similarly for n_{ab} and n_{bb}.

Consider pure complements ($\delta < 0$). Since $n_{aa} + n_{bb} + n_{ab} = |\mathrm{G}|$ (i.e., the number of links in the graph), the potential is

$$\varphi(\mathbf{x}) = \sum_{x_i=b} x_i^0 - \sum_{x_i=a} x_i^0 + 2|\delta|(n_{aa} + n_{bb}) - |\delta||\mathrm{G}| - \frac{1}{2}n.$$

This potential combines two forces. On one hand, φ is greater when individuals play the action for which they have some intrinsic preference: b for $x_i^0 > 0$ and a for $x_i^0 < 0$. On the other hand, φ is greater when there are more links between agents playing the same action. These two forces can be aligned. When all individuals intrinsically prefer the same action, full coordination on this action is the unique stable equilibrium. In general, however, a stable equilibrium can involve coordination on different actions in different parts of the network.

[33] The relation between Nash equilibria and maximal independent sets also appears when agents play games of anti-coordination on the network and one action has a much higher relative payoff than the other, see Bramoullé (2007).

[34] In the literature, researchers have analyzed stochastic stability in coordination games (Blume 1993; Young 1998, Jackson and Watts 2002) anti-coordination games (Bramoullé 2007) and the best-shot game (Boncinelli and Pin 2012).

[35] Hence, while all games with best replies (5.8) have the same Nash equilibria, the stochastically stable equilibrium sets could diverge.

Next consider strategic substitutes ($\delta > 0$). The potential is then

$$\varphi(\mathbf{x}) = \sum_{x_i=b} x_i^0 - \sum_{x_i=a} x_i^0 + 2\delta n_{ab} - \delta|\mathbf{G}| - \frac{1}{2}n.$$

Here φ is greater when there are more links between agents playing different actions. The two forces are aligned only in the special case when no link connects two agents who prefer the same action, in which case the network is bipartite. The profile where every agent plays his preferred action is then the unique stable equilibrium.

5.7.4 Interdependence

With binary actions, in equilibrium agents are not typically indifferent between the two actions, and small changes in individual parameters do not lead to a change in play. We then consider larger changes, and show changes in individual parameters only have impact on own play and play of others in critical configurations where what we call *switching cascades* can occur.

To illustrate, consider a pure complements game where $g_{ij} \in \{0,1\}$ and $\delta = -1$. Consider one of the extremal equilibria—either an equilibrium where most agents play 1 or one where most agents play -1. Consider an initial change from a situation where $x_j^0 \ll 0$ and j plays -1 to the situation where $x_j^0 \gg 0$ and j plays 1. When does this change in j's preferences and action affect the play of an agent i? A clear necessary condition is a path of agents playing -1 connecting i to j in the initial equilibrium. As with bounded agents in Section 5.6, agents playing 1 cannot transmit positive shocks. If there is an agent playing 1 on all paths between i and j, x_i is unaffected by the change in x_j^0.

Unlike in Section 5.6, however, this condition is typically not sufficient. Agents playing -1 might not change their actions if some of their neighbors switch to 1. Changing actions depends on idiosyncratic preferences and on the number of switching neighbors. Overall, we observe that interdependence displays a non-monotonic pattern. When $\mathbf{x}_{-j}^0$ is low, agents have a strong preference for playing -1 and an increase on x_j^0 does not propagate. When $\mathbf{x}_{-j}^0$ is high, an increase in x_j^0 also has no impact because agents playing 1 block the transmission of shocks. The increase in x_j^0 eventually affects x_i only when $\mathbf{x}_{-j}^0$ takes some critical intermediate value. In future research, it would be interesting to understand more deeply how shocks propagate in binary action games.

5.8 Econometrics of Social Interactions

In this section, we connect the above theory to the empirical analysis of social interactions. The connections provide econometric models a precise game-theoretic microfoundation and set the stage for estimation of equilibria. Social scientists have long been trying to assess the importance of social interactions for outcomes as diverse as academic performance, welfare participation, smoking, obesity, and delinquent behavior.[36] In a typical regression, researchers try to estimate the impact of peers' outcomes $\sum_j g_{ij}x_j$ on individual outcome x_i. Because individual and peers' outcomes are determined at the same time, regressions define a set of simultaneous equations. This econometric system of equations is formally equivalent to the system of equations characterizing Nash equilibria with best replies (5.5), (5.6), (5.7), or (5.8).

Simultaneity raises two main econometric challenges: multiplicity and endogeneity. First, the econometric system may have multiple solutions.[37] Second, the variable $\sum_j g_{ij}x_j$ on the right hand side of the regressions is endogenous. To address these challenges, applied researchers must determine the *reduced form* of the system of simultaneous equations. That is, they must understand how outcomes $\mathbf{x}$ depend on parameters, observables, and unobservables. Formally, determining the reduced-form is equivalent to solving for the Nash equilibria of a network game.[38]

When the outcome is continuous and unbounded, a standard econometric model of peer effects is: $x_i = x_i^0 + \delta\sum_j g_{ij}x_j + \varepsilon_i$ where δ is a key parameter to be estimated, usually called the "endogenous peer effect," ε_i is an error term, and x_i^0 depends on individual and peers' covariates. We know from Section 5.4 that this system generically has a unique solution. The reduced-form is then given by $\mathbf{x} = (\mathbf{I} - \delta\mathbf{G})^{-1}(\mathbf{x}^0 + \varepsilon)$, and this equation provides the basis of many empirical analysis of social and spatial effects.[39]

However, most outcomes of interest (academic performance, etc.) are naturally bounded. These bounds are neglected in the previous approach and, in fact, in most studies of peer effects, which can yield biased estimates. A truncated version of the previous model,

$$x_i = \min(\max(0, x_i^0 + \delta\sum_j g_{ij}x_j + \varepsilon_i), L),$$

is a way to incorporate bounds on continuous actions. As for binary outcomes, a direct extension of classical discrete choice models is:

$$\hat{x}_i = x_i^0 + \delta\sum_j g_{ij}x_j + \varepsilon_i$$

with $x_i = 1$ if $\hat{x}_i \geq 0$ and $x_i = -1$ otherwise (Koreman and Soetevent 2007). These correspond to best reply (5.8).

[36] See the chapters by Vincent Boucher and Bernard Fortin, by Sinan Aral, by Emily Breza, and by Lori Beaman in this handbook.

[37] A related issue is that the system may not have any solution.

[38] In the literature, researchers have also considered games of incomplete information when agents do not know the outcomes and error terms of others (Brocke and Durlauf 2001; Lee, Li, and Lin 2014; Blume et al. 2014). Interestingly, the information structure has little impact on the analysis of continuous unbounded outcomes but deeply modifies the econometrics of binary actions.

[39] See, for example, Case (1991), Bramoullé, Djebbari, and Fortin (2009), Lee (2007), and Anselin's (2000) reviews.

The techniques and results presented in this chapter can then be combined with classical methods to estimate models with multiple equilibria. The researcher could, for instance, assume that all equilibria are equally likely (Koreman and Soetevent 2007); consider a flexible selection mechanism (Bajari, Hong and Ryan, 2010); build a likelihood from some evolutionary process (Nakajima 2007); or derive informative bounds from dominance relations (Tamer 2003).

5.9 Conclusion

This chapter presents a formal framework and technical tools to analyze a broad class of network games. These games all share the same underlying incentives—linear best replies with successive constraints on agents' choices. The chapter makes new connections between continuous action games and binary action games that share the same basic structure. We conclude here by reiterating future research directions and by connecting the analysis in this chapter to studies of games outside our class.

While much progress has been made on the network features that yield unique and stable equilibria, many interesting issues are still little understood. The following are some areas for future research: (i) existence of equilibria when network effects are large, (ii) the relationship between network structure and equilibrium multiplicity, (iii) comparative statics on the network in the presence of substitutes, (iv) interdependence in binary action games, and (v) the implications of our framework for the analysis of network games with discrete action spaces.

The research discussed below considers successively further departures from the assumptions that characterize the class of games covered by our framework.

Directed networks. In general, the analysis does not cover directed networks ($g_{ij} \neq g_{ji}$) since an exact potential function does not exist. The analysis does extend, however, in at least three cases. First, a game has a "weighted" potential when there are scalars α_i such that $\forall i,j, \alpha_i g_{ij} = \alpha_j g_{ji}$, and most results hold.[40] In particular, it is easy to extend the model to situations with undirected links and individual-specific δ_i. Second, when network effects are small, we can apply the theory of concave games developed by Rosen (1965) and obtain the following generalization of Proposition 1. For any network **G**, the games with continuous actions have a unique Nash equilibrium if $|\lambda_{\min}(\delta(\mathbf{G}+\mathbf{G}^T)/2)| < 1$. Third, Belhaj and Deroian (2013) identify a balance condition under which the comparative statics on aggregate actions presented in Section 5.6 generalize to directed networks.

Non-linear best replies. In our framework, underlying best replies are linear and constraints add complexity. Researchers have started to analyze games with broader forms of non-linearities. Allouch (2014) studies the private provision of local public goods, combining the frameworks of Bergstrom, Blume and Varian (1986) and

[40] See Monderer and Shapley (1996) and Section VI.B in Bramoullé, Kranton and D'amours (2014).

Bramoullé and Kranton (2007). With pure substitutes ($\delta > 0$), Proposition 1 and the key role of the lowest eigenvalue extend to a wide class of non-linear best replies. The analysis does not consider, however, large network effects.

Researchers have also applied results from supermodular games to network games with pure complements. Belhaj and Deroian (2010) consider payoffs with indirect network effects and some specific symmetric networks. They show that action is aligned with Bonacich centrality in the highest and lowest equilibria, but not on intermediate equilibria. Belhaj, Bramoullé, and Deroian (2014) analyze network games with continuous, bounded actions, pure complements, and non-linear best replies. They derive a novel uniqueness condition and study interdependence.

Multidimensional strategies. In the games mentioned thus far, players choose a single number. In some contexts, players' actions are naturally multidimensional. A firm could choose both quality and quantity of a good, for example. Individuals can adopt different technologies to interact with different people. Bourlès and Bramoullé (2014) advance a model of altruism in networks, where players care about the utility of their neighbors and can transfer money to each other. An individual strategy specifies a profile of transfers. In recent research on conflict and networks, agents may also allocate different levels of resources to conflicts with different neighbors.[41] More generally, multidimensional strategies emerge when players can play different actions with different neighbors. Little research has been conducted to date on such games.

Incomplete information. The papers discussed so far analyze games of complete information, where payoffs and the network structure are common knowledge. In the games studied in this chapter, however, the best replies and convergence to a Nash equilibrium do not require that agents know the whole network or what all agents play. Agents respond simply to the play of their neighbors. In some contexts, however, these assumptions may be inappropriate. Agents may not have complete local information, or changes in the network may prevent convergence. Agents may then face residual uncertainty on others' connections when taking their actions. Actions would then depend on players' beliefs about their own and neighbors' positions and actions. Galeotti et al. (2010) study such network games of incomplete information.[42]

Network formation. In the games studied in this chapter, networks are fixed. In reality, networks evolve, possibly in a way that can be influenced by actions. Researchers have analyzed the joint determination of actions and links for many of the network games discussed in this chapter, as elaborated in Fernando Vega-Redondo's chapter in this volume. This work includes: coordination games (Jackson and Watts 2002; Goyal and Vega-Redondo 2005); anti-coordination games (Bramoullé et al. 2004); public goods in networks (Galeotti and Goyal 2010); and games with quadratic payoffs and continuous actions (Cabrales, Calvó-Armengol, and Zenou 2011). A broad conclusion of this literature is that endogenizing the network can lead to fewer possible outcomes, as equilibrium networks tend to have specific shapes. Thus far researchers generally

[41] See Franke and Ozturke (2009), Huremovic (2014), and Sanjeev Goyal's chapter in this volume.

[42] The literature on these games is reviewed in Jackson and Zenou (2014).

assume that the payoffs from the actions constitute the only incentive for network formation. In reality, people form friendships or other links for a variety of reasons, and there could be possible multiple costs and benefits from making and breaking links. Future research could engage these more challenging but potentially fruitful avenues.

References

Allouch, Nizar (2015). "On the private provision of public goods on networks." *Journal of Economic Theory* 157, 527–552.

Anselin, Luc (2010). "Thirty years of spatial econometrics." *Papers in Regional Science* 89(1), 3–25.

Bajari, Patrick, Han Hong, and Stephen P. Ryan (2010). "Identification and estimation of a discrete game of complete information." *Econometrica* 78(5), 1529–1568.

Ballester, Coralio, Antoni Calvó-Armengol, and Yves Zenou (2006). "Who's who in networks. Wanted: the key player." *Econometrica* 74(5), 1403–1417.

Ballester, Coralio, Antoni Calvó-Armengol, and Yves Zenou (2010). "Delinquent networks." *Journal of the European Economic Association* 8(1), 34–61.

Belhaj, Mohamed, and Frédéric Deroïan (2010). "Endogenous effort in communication networks under strategic complementarity." *International Journal of Game Theory* 39(3), 391–408.

Belhaj, Mohamed, Yann Bramoullé, and Frédéric Deroïan (2014). "Network games under strategic complementarities." *Games and Economic Behavior* 88, 310–319.

Bergstrom, Theodore, Lawrence E. Blume, and Hal Varian (1986). "On the private provision of public goods." *Journal of Public Economics* 29(1), 25–49.

Blume, Lawrence E. (1993). "The statistical mechanics of strategic interaction." *Games and Economic Behavior* 5(3), 387–424.

Blume, Lawrence E. (1995). "The statistical mechanics of best-response strategy revision." *Games and Economic Behavior* 11(2), 111–145.

Blume, Lawrence E., William Brock, Steven Durlauf, and Raji Jayaraman (2015). "Linear social interaction models." *Journal of Political Economy* 123(2), 444–496.

Bonacich, Phillip (1987). "Power and centrality: A family of measures." *American Journal of Sociology* 92(5), 1170–1182.

Boncinelli, Leonardo and Paolo Pin (2012). "Stochastic stability in best shot network games." *Games and Economic Behavior* 75(2), 538–554.

Bourlès, Renaud and Yann Bramoullé (2014). "Altruism in networks." Working paper, Aix-Marseille School of Economics.

Bramoullé, Yann (2007). "Anti-coordination and social interactions," *Games and Economic Behavior* 58(1), 30–49.

Bramoullé, Yann, Habiba Djebbari, and Bernard Fortin (2009). "Identification of peer effects through social networks." *Journal of Econometrics* 150(1), 41–55.

Bramoullé, Yann, and Rachel Kranton (2007). "Public goods in networks." *Journal of Economic Theory* 135(1), 478–494.

Bramoullé, Yann, Dunia López-Pintado, Sanjeev Goyal, and Fernando Vega-Redondo (2004). "Network formation and anti-coordination games." *International Journal of Game Theory* 33(1), 1–19.

Bramoullé, Yann, Rachel Kranton, and Martin D'Amours (2011). "Strategic interaction and networks." Working paper, Duke University.

Bramoullé, Yann, Rachel Kranton, and Martin D'Amours (2014). "Strategic interaction and networks." *American Economic Review* 104(3), 898–930.

Brock, William A. and Steven N. Durlauf (2001). "Discrete choice with social interactions." *Review of Economic Studies* 68(2), 235–260.

Bulow, Jeremy I., John D. Geanakoplos, and Paul D. Klemperer (1985). "Multimarket oligopoly: Strategic substitutes and complements." *Journal of Political Economy* 9(3), 488–511.

Cabrales, Antonio, Antoni Calvó-Armengol, and Yves Zenou (2011). "Social interactions and spillovers." *Games and Economic Behavior* 72(2), 339–360.

Case, Anne C. (1991). "Spatial patterns in household demand." *Econometrica* 59(4), 953–965.

Ellison, Glenn (1993). "Learning, local interaction, and coordination." *Econometrica* 61(5), 1047–1071.

Fisher, Franklin M. (1961). "The stability of the cournot oligopoly solution: The effects of speeds of adjustment and increasing marginal costs." *Review of Economic Studies* 28(2), 125–135.

Franke, Jörg and Tahir Ozturk (2009). "Conflict networks." Ruhr Economic Papers No. 116.

Galeotti, Andrea and Sanjeev Goyal (2010). "The law of the few." *American Economic Review* 100(4), 1468–1492.

Galeotti, Andrea, Sanjeev Goyal, Matthew O. Jackson, Fernando Vega-Redondo, and Leeat Yariv (2010). "Network games." *Review of Economic Studies* 77(1), 218–244.

Goyal, Sanjeev, and Fernando Vega-Redondo (2005). "Network formation and social coordination." *Games and Economic Behavior* 50(2), 178–207.

Hirshleifer, Jack (1983). "From weakest-link to best-shot: The voluntary provision of public goods." *Public Choice* 41(3), 371–386.

Huremović, Kenan (2014). "Rent seeking and power hierarchies - a noncooperative model of network formation with antagonistic links." Working Paper, Aix Marseille School of Economics and GREQAM.

Jackson, Matthew O. (2008). *Social and Economic Networks*. Princeton, NJ: Princeton University Press.

Jackson, Matthew and Alison Watts (2002). "On the formation of interaction networks in social coordination games." *Games and Economic Behavior* 41(2), 265–291.

Jackson, Matthew O. and Yves Zenou (2014). "Games on networks." *Handbook of Game Theory with Economic Applications*, Vol. 4, Peyton Young and Shmuel Zamir, eds., Elsevier.

König, Michael, Claudio Tessone, and Yves Zenou (2014). "Nestedness in networks: A theoretical model and some applications." *Theoretical Economics* 9(3), 695–752.

Lee, Lung-fei (2007). "Identification and estimation of econometric models with group interactions, contextual factors and fixed effects." *Journal of Econometrics* 140(2), 333–374.

Lee, Lung-fei, Ji Li, and Xu Lin (2014). "Binary choice models with social network under heterogeneous rational expectations." *Review of Economics and Statistics* 96(3), 402–412.

Monderer, Dov and Lloyd S. Shapley (1996). "Potential games." *Games and Economic Behavior* 14(3), 124–143.

Morris, Stephen (2000). "Contagion." *Review of Economic Studies* 67(1), 57–78.

Nakajima, Ryo (2007). "Measuring peer effects on youth smoking behaviour." *Review of Economic Studies* 74(3), 897–935.

Soetevent, Adriaan R. and Peter Kooreman (2007). "A discrete-choice model with social interactions: With an application to high school teen behavior." *Journal of Applied Econometrics* 22(3), 599–624.

Tamer, Elie (2003). "Incomplete simultaneous discrete response model with multiple equilibria." *Review of Economic Studies* 70(1), 147–165.

Weibull Jörgen W. (1995). *Evolutionary Game Theory*. Cambridge, MA: MIT Press.

Young, P. (1998). *Individual Strategy and Social Structure*. Princeton, NJ: Princeton University Press.

CHAPTER 6

REPEATED GAMES AND NETWORKS

FRANCESCO NAVA

6.1 Introduction

In many strategic environments, interaction is local and segmented. Competing neighborhood stores serve different yet overlapping sets of customers; informal lending and insurance arrangements often have to be fulfilled by relatives and friends; the behavior of the residents of an apartment block affects their contiguous neighbors to a larger extent than neighbors in a different block; a nation's foreign or domestic policy typically generates larger externalities for neighboring nations than for remote ones. One classic case is the private provision of local public goods. In addition to local interaction, one notable feature of these environments is local monitoring: whereas participants are aware of their own neighbors' identities and actions, they are not necessarily aware of the identity and actions of their neighbors' neighbors. Within these strategic environments, it is of particular interest to study long-run interaction, when incentives can only be provided locally in a decentralized manner. The main objective of this literature is analyzing such interactions within a repeated game framework that differs from the standard one in that actions can only be observed locally.

Three main lines of research have been developed in such environments. The first, and most classical, develops Folk Theorems for games with local monitoring, and establishes that network structure is usually irrelevant for enforcing cooperation when the frequency of interaction is sufficiently high. The other two explicitly study the link between network structure and the equilibrium payoffs by focusing on environments in which discount rates are fixed. One strand analyzes how the monitoring structure affects the maximal level of equilibrium cooperation, and broadly finds that larger and/or better connected groups are more cooperative. The other evaluates how different communication protocols affect the set of equilibrium payoffs and the incentives to cooperate in environments with local monitoring.

The analysis of community enforcement was initially developed in the context of repeated games with random matching. Pioneering studies by Kandori (1992) and Ellison (1994) focused on environments with pairwise matching, and established how collective punishments could sustain efficient equilibrium outcomes when bilateral punishments would fail. Subsequent and related contributions on random matching games include Harrington (1995), Takahashi (2010), and Deb (2014).[1] Although the matching literature and the literature focusing on stable local interactions (and therefore on networks) share several methodological insights, there are significant differences both in the assumptions on feasible interactions and in the broad aims. Whereas most random matching games assume all players potentially interact, and thus exchange information about deviant behavior, all network games constrain interactions, monitoring, and information exchange to take place on a stable network, which represents the topology of relationships in a society. Whereas most random matching games (with a few exceptions, including Harrington 1995) focus on Folk Theorems and seldom on optimal punishments, the study of network games aims to establish a relationship between the underlying network structure and the equilibrium correspondence (or alternatively the most efficient equilibrium payoff).

The chapter begins by presenting relevant definitions in the context of a baseline environment with local monitoring and local interaction. It proceeds with a survey of Folk Theorems for network games in Section 6.3. Sections 6.4 and 6.5 discuss community enforcement at a given frequency of interaction. In particular, Section 6.4 surveys results on optimal punishments and network structure, while Section 6.5 presents results on communication. Section 6.6 hints at related applications (on reciprocity, informal insurance, and lending), at relevant omissions, and at possible extensions. Static games with local interactions and the related literature are discussed in a separate chapter of this book (Chapter 8).

6.2 A Baseline Setup

Environments considered in the literature invoke different assumptions about information, matching, and the availability of individual punishments. This section introduces a baseline environment which nests a large number of possible setups to discuss contributions more transparently in the following sections.

The Stage Game: Consider a game, the stage game, played by a set N of n players in which any player i can interact with a subset of players $N_i \subseteq N \backslash \{i\}$, which is called the *neighborhood* of player i. As customary, assume that $j \in N_i$ if and only if $i \in N_j$. This

[1] Harrington (1995) shows that relationships with low frequencies of interaction can be supported using relationships that interact more frequently. Takahashi (2010) shows that cooperation can be sustained in repeated Prisoners' Dilemmas if all that is observed are partners' past play. Deb (2014) offers a general Folk Theorem for anonymous random matching environments.

structure of interaction defines an undirected graph (N, G) in which $ij \in G$ if and only if $j \in N_i$. Refer to G as the *interaction network*. In the stage game players interact with a possibly random subset of their neighbors. In particular, for any undirected subgraph $\bar{G} \subseteq G$, let $f(\bar{G}|G)$ denote the probability that the realized network of interactions is $\bar{G}$. Let $\bar{N}_i \subseteq N_i$ denote the realized neighborhood of i in this subgraph. Two extreme cases are generally considered in the literature. In the first case $f(G|G) = 1$, which I refer to as *local interaction*, while in the second $f(\bar{G}|G) > 0$ implies that $|\bar{N}_i| \leq 1$, which I refer to as *pairwise interaction*. The former scenario captures environments in which players interact with all of their neighbors in every period, while the latter environments in which interactions take place only between pairs of players. Refer to f as the *matching technology*.

Assumptions on information vary across setups, but consistently require that players know their neighborhood, N_i, their realized neighborhood, $\bar{N}_i$, and the matching technology, f. When players are privately informed about their neighborhood, their beliefs regarding the interaction network, conditional upon observing their neighborhood, are derived from a common prior distribution over the set of interaction networks. Beliefs regarding the realized interaction network are then constructed by simply applying Bayes' rule.

The action set of player i is denoted by A_i. Given a subset M of players, let A_M denote $\times_{j\in M} A_j$ and a_M an element of A_M. Also, let $-i$ denote the set $N\backslash\{i\}$. The stage game payoffs are common knowledge. The payoff of any player i depends only on actions chosen in his realized neighborhood, and is denoted by $v_i(a_i, a_{\bar{N}_i}|\bar{N}_i)$. As a convention, the payoff equals zero when $\bar{N}_i$ is empty. Payoffs are *separable*, if for any player $i \in N$ the stage game payoff satisfies

$$v_i(a_i, a_{\bar{N}_i}|\bar{N}_i) = \sum_{j\in\bar{N}_i} u_{ij}(a_i, a_j),$$

where $u_{ij}(a_i, a_j)$ is the payoff of player i from the relationship $ij \in G$.

The stage game is *separable* if: (a) payoffs are separable; (b) action sets have the product structure, $A_i = \times_{j\in N_i} A_{ij}$ for any $i \in N$; and (c) for any action profile $a_N \in A_N$, the stage game payoff on any link $ij \in G$ satisfies

$$u_{ij}(a_i, a_j) = x_{ij}(a_{ij}, a_{ji}),$$

for some map $x_{ij} : A_{ij} \times A_{ji} \to \mathbb{R}$. In a separable game, players choose actions that are specific to each interaction, and payoffs in an interaction depend only on actions chosen in that specific interaction. Any pairwise interaction game can be represented as a separable game, if the identity of players is known to their realized partners. If so, action sets have the product structure, as players can tailor behavior to every opponent. Thus, non-anonymous random matching games are separable. Anonymous random matching games, instead, are not separable, since action sets do not have the product structure as players cannot choose a different action for every realized interaction.

The stage game is *binary-symmetric* if: (i) payoffs are separable; (ii) action sets are binary, $A_i = \{C, D\}$ for any $i \in N$; and (iii) payoffs are symmetric, for any link $ij \in G$,

$$u_{ij}(a_i, a_j) = \eta_{ij} u(a_i, a_j),$$

for some map $u : \{C, D\}^2 \to \mathbb{R}$ and a some scalar $\eta_{ij} \in \mathbb{R}_+$. In such games, players must choose the same action in every interaction and cannot discriminate across neighbors. For convenience, refer to action C as cooperation and action D as defection. Results on binary-symmetric games are generally developed for stage games in which: (iv) the payoff $u(a_i, a_j)$ of player i in an interaction with j satisfies

$i \backslash j$	C	D
C	1	$-l$;
D	$1+g$	0

(v) mutual cooperation is efficient, $g - l < 1$; and (vi) defection is a best response when the opponent cooperates $g > 0$. The first assumption restricts the class of binary games by imposing a common payoff for mutual defection across relationships; the second uniquely pins down an efficient action profile; while the third rules out the trivial case in which mutual cooperation is an equilibrium of the stage game. Naturally, if $l > 0$, the stage game has a unique Bayes Nash equilibrium in which all players play D, and all pairwise interactions amount to a Prisoners' Dilemma. If instead $l < 0$, the stage game always possesses a mixed strategy Bayes Nash equilibrium, and all pairwise interactions amount to an anti-coordination game.[2]

Local interaction games with separable and symmetric payoffs capture environments in which behavior cannot be targeted to individual neighbors, while separable games capture environments in which players can make decisions contingent on the identity of their realized neighbor. For instance, decentralized competition between sellers, when prices set are independent of identity of buyers, fits in the class of local interaction games with separable payoffs; whereas non-anonymous negotiations between traders in an spatial economy fit in the class of separable games.

Repetition and Local Monitoring: The players play the infinite repetition of the stage game. The interaction network, G, is realized prior to the beginning of play and remains fixed throughout the game. Realized interactions, $\bar{G}(t)$, however, are drawn independently every period, t, from a distribution over the set of subnetworks of G.[3]

Monitoring is *local* implying that a player observes only the past play in his realized neighborhood. Local monitoring is a key assumption in the networks approach to community enforcement as it implies that realized interactions are not anonymous.

[2] When $l < 0$, pure strategy equilibria also exist in some networks, as miscoordinating with neighbors can be best reply. In particular, if beliefs are concentrated on bipartite graphs (which have only cycles of even length; Bramoullé 2007), pure equilibria exist, since all players can successfully miscoordinate their action with all their neighbors.

[3] A subnetwork G' of a networks G is a subset of G. That is, $G' \subseteq G$.

This differs from classical random matching models requiring anonymity, as players may now develop relationship-specific reputations to enforce good behavior. Formally, when the stage game is not separable, a history h_i^t of length t for player i consists of a sequence

$$h_i^t = \left\{N_i, \bar{N}_i(0), \bar{a}_i(0), \bar{N}_i(1), ..., \bar{a}_i(t-1), \bar{N}_i(t)\right\}$$

that satisfies $N_i \subseteq N$, $\bar{N}_i(s) \subseteq N_i$, and $\bar{a}_i(s) \in \times_{j \in \bar{N}_i(s) \cup \{i\}} A_j$ for any value of s. When the stage game is separable, however, players monitor only the neighbor-specific actions played in their realized interactions and therefore $\bar{a}_i(s) \in \times_{j \in \bar{N}_i(s)} [A_{ij} \times A_{ji}]$ for any value of s. Denote by H_i^t the set of histories of length t for player i, and by H_i the corresponding set of possible histories, $H_i = \cup_{t=0}^{\infty} H_i^t$. A strategy for player i is a map that assigns to every history in H_i an action in A_i. A full history h^t of length t similarly consists of a sequence

$$h^t = \left\{G, \bar{G}(0), \bar{a}(0), \bar{G}(1), ..., \bar{a}(t-1), \bar{G}(t)\right\}$$

satisfying $G \subseteq \{ij | i, j \in N\}$, $\bar{G}(s) \subseteq G$, and $\bar{a}(s) \in A_N$. Denote by H^t the set of full histories of length t and by H the set of possible full histories $H = \cup_{t=0}^{\infty} H^t$.

Players discount the future by a common factor $\delta \leq 1$. To construct the payoffs in the infinitely repeated game, fix a player $i \in N$ and a history $h_i \in H_i$, and let h_i^t denote the subhistory of length $t > 0$ of h_i. Define

$$w_i^t\left(h_i^t\right) = \sum_{s=0}^{t-1} \frac{v_i(a(s)|\bar{N}_i(s))}{t}$$

to be the average payoff up to period t and $w_i\left(h^i\right) = \{w_i^t\left(h_i^t\right)\}_{t=1}^{\infty}$ to be the sequence of average payoffs. Repeated game payoffs conditional on h_i are defined as

$$\mathcal{V}_i(h_i) = \begin{cases} (1-\delta)\sum_{t=0}^{\infty} \delta^t v_i(a(s)|\bar{N}_i(s)) & \text{if} \quad \delta < 1 \\ \Lambda\left(w_i\left(h_i\right)\right) & \text{if} \quad \delta = 1 \end{cases}$$

where $\Lambda(\cdot)$ denotes a suitable limit operator, such as the limit inferior or the Banach-Mazur limit of a sequence.[4]

A full history h uniquely pins down the history of play in the dynamic game. An observed history h_i is associated uniquely with an information set $\mathcal{I}(h_i)$ for player i and vice versa. A system of beliefs defines, at each information set $\mathcal{I}(h_i)$ of player i, the conditional probability of each full history $h \in \mathcal{I}(h_i)$.

Departures: Although the baseline setup allows for much flexibility, it does not capture the full range of environments considered in the literature. Some studies model the

[4] If ℓ_∞ denotes the set of bounded sequences of real numbers, a Banach-Mazur limit is a linear functional $\Lambda : \ell_\infty \to \mathbb{R}$ such that: (i) $\Lambda(e) = 1$ if $e = \{1, 1, ...\}$; (ii) $\Lambda(x^1, x^2, ...) = \Lambda(x^2, x^3, ...)$ for any sequence $\{x^t\}_{t=0}^{\infty} \in \ell_\infty$ (Aliprantis and Border 2005). It can be shown that, for any sequence $\{x^t\}_{t=0}^{\infty} \in \ell_\infty$,

$$\liminf_{t\to\infty} x^t \leq \Lambda\left(\{x^t\}_{t=1}^{\infty}\right) \leq \limsup_{t\to\infty} x^t.$$

interaction network as a directed graph. Others allow for *global interaction*, while assuming that monitoring is local. If so, players may be affected by the action chosen by every other player in the game, but only observe the behavior of a subset of players. Other frameworks have considered *imperfect local monitoring* by having players only observe a noisy signal of their neighbors' actions. Finally, many setups have focused on communication, by adding to the stage game a communication stage, modelled in one of many possible ways.

6.3 Limiting Results and Network Irrelevance

A significant body of literature provides conditions on the interaction network for a Folk Theorem to apply. These studies generally establish in many environments that a Folk Theorem obtains under very weak conditions on the network structure, and thus yield limited insights about the optimal monitoring structure. A key concern in these papers is ensuring that players do not cooperate off the equilibrium path, as grim trigger strategies may provide such strong incentives to cooperate on the equilibrium path that players prefer to cooperate even after observing a deviation. Ellison (1994) resolves this problem by introducing either a public randomization device or a milder version of grim trigger strategies tailored to make players indifferent between cooperating and defecting on-path, and then noting that cooperation is more appealing on-path than off-path (since off-path at least one opponent is already defecting). The literature on local monitoring addresses similar concerns with related approaches, either by allowing some form of communication, or by constructing suitable strategies with mild punishments. Further complications arise, however, with local monitoring as, upon observing defections, players try to infer the spread of defection and the beliefs of other players about future play.

All of the limiting results presented in this section apply to stage games with local monitoring that are not separable, since a Folk Theorem would trivially obtain otherwise. Most results are developed for stage games with local interaction and in which the network structure is common knowledge. Thus for expositional ease, restrict attention to such scenarios unless specified otherwise.

Ben-Porath and Kahneman's (1996) seminal contribution considers games with public randomization and in which players can make public announcements about the past behavior of other players whom they observed. The analysis characterizes the minimal level of observability required to obtain efficient outcomes for arbitrary stage games. The main result establishes that, when the discount factor tends to one, the limit set of sequential equilibrium payoffs contains the set of individually rational payoffs, whenever every player is observed by at least two other players. For arbitrary stage game payoffs, two monitors are required to guarantee that inconsistent public announcements about past play can be sanctioned by the community. Results also establish that if payoffs are assessed by the limit inferior of the average payoff (that is if $\delta = 1$), every individually rational payoff is a sequential equilibrium payoff even when players are monitored by only one other player.

Renault and Tomala (1998) develop similar insights in a model with global interactions, local monitoring, no discounting, and no explicit communication. Their main finding establishes that a Nash Folk Theorem applies if and only if the monitoring network is 2-connected (that is, if there are two independent paths connecting any two players, or equivalently, if the subgraphs obtained by suppressing any one player are still connected). The result abstracts from sequential rationality, which considerably simplifies the problem as punishments need not be incentive compatible. Although explicit communication is ruled out, the no discounting assumption and the restriction to Nash equilibrium imply that players can use any finite number of future periods to privately communicate with neighbors at no cost. Tomala (2010) extends the analysis to partially known networks, in which players only know their neighbors and the number of players in the network, and derives a Nash Folk Theorem.

More recently, Laclau (2012) considers a local interaction setup analogous to Renault and Tomala (1998), while allowing for imperfect local monitoring and explicit communication between neighbors (private local cheap talk). Monitoring is imperfect, as players observe their payoff, but not the actions chosen in their neighborhood. Her main contribution identifies necessary and sufficient conditions on the network of interactions for a Nash Folk Theorem to hold when the payoff of every player is responsive to unilateral deviations (in that players monitor unilateral deviations in their neighborhood, despite local monitoring being imperfect). In a recent companion paper, Laclau (2014) extends conclusions to a model in which communication is global (players can communicate with all opponents), and can be either private or public. Contrary to Laclau (2012), where a Nash Folk Theorem is established, the analysis here applies to sequential equilibria of the infinitely repeated game with imperfect local monitoring. As before, payoffs are assumed to be sensitive to unilateral deviations. If so, a sequentially rational Folk Theorem holds provided that a joint pairwise identifiability condition regarding payoff functions is satisfied. The condition requires players to detect the identity of the deviating player, whenever they detect a unilateral deviation in their neighborhood. The analysis then shows that, when payoffs are sensitive to unilateral deviations, a necessary and sufficient condition on the network topology for the Folk Theorem to hold for all payoff functions is that no two players have the same set of neighbors (not counting each other). The main contribution of both papers consists in the analysis of imperfect local monitoring, which had been neglected by the earlier literature.

Three related studies, Xue (2004), Cho (2010), and Cho (2011), analyze cooperation in binary-symmetric Prisoners' Dilemma games. Even though it is not difficult to construct sequential equilibria supporting cooperation in these environments, the classical modification of a trigger strategy devised in Ellison (1994) to enforce a cooperative equilibrium has an undesirable feature. Namely, it is not stable to mistakes in that defections spread over the network, and cooperation is never recovered whenever an agent defects by mistake in the repeated game. The main aim of these three studies, thus, consists of constructing equilibria that sustain cooperation and revert to cooperation after any history of play. The classical solution to this complication involves bounding the length of the punishment phase. That is, if an agent observes his neighbor playing defection, then he punishes his neighbor by defecting for finitely many periods. Local monitoring, however, may cause discrepancies in beliefs between agents about a neighbor's future actions (that is, the expected date at which a player ends a defection phase may not be common knowledge in his neighborhood). If there is such a discrepancy at some history, then an agent whose neighbors have different beliefs may not be able to satisfy the expectations of all his neighbors in any period which in turn may cause an infinite repetition of defection phases and, thus, a failure of stability. Furthermore, bounded punishment strategies may not even constitute a sequential equilibrium in a general networked setting. In order to prove the existence of a cooperative and stable sequential equilibrium, such discrepancies of beliefs may be resolved through some form of coordination in punishments. To this end, Xue (2004) restricts attention to line-shaped networks, and shows that in such graphs cooperation is a stable equilibrium when players comply with specific bounded punishment strategies. Cho (2010) establishes a similar results for acyclical networks by allowing agents to communicate locally with their neighbors. In contrast to Laclau (2012), the focus is on sequential equilibria; while in contrast to Laclau (2014), communication is only local, and not public, and therefore players cannot easily coordinate their punishments. Both Xue (2004) and Cho (2010) exploit the acyclicity of the network structure to simplify the inference problem associated with contagion, as players expect punishments to dissipate at the periphery of the network. Borrowing an idea from Ellison (1994), Cho (2011) instead shows that a cooperative and stable sequential equilibrium exists for any possible monitoring structure if players have access to a public randomization device. If so, the inference problem is solved through coordinating behavior rather than by restricting the class of network structures.

Nava and Piccione (2014) study a broader class of binary-symmetric games which satisfy the additional requirements (iv)–(vi) described in Section 6.2. In contrast to the earlier results, but similarly to Tomala (2011), the study allows for uncertainty about the interaction network. In particular, to capture behavior in large markets the analysis postulates that players are privately informed of their neighborhood. Their main result establishes that, for sufficiently high discount rates and any prior beliefs with full support about the network structure, sequential equilibria exist in which efficient stage-game outcomes are played in every period. Standard results do not apply in this framework because bilateral enforcement may not be incentive compatible when punishments in one relationship affect outcomes in all the others. For instance, punishing a neighbor indefinitely with a grim trigger strategy is not viable if cooperation in other relationships is disrupted (see Figure 6.1), and mild trigger strategies such as in Ellison (1994) work only for particular specifications of payoffs (e.g., Prisoners'

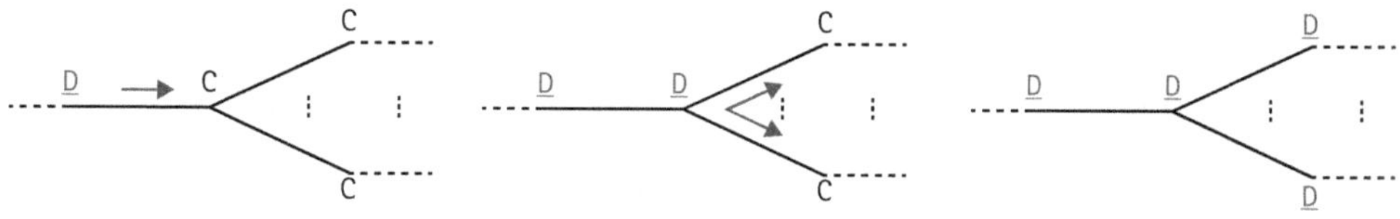

FIGURE 6.1 With trigger strategies, the central player prefers not to punish a single defection, as it would destroy cooperation in all his remaining relationships.

Dilemma). Equilibrium strategies supporting efficient outcomes are built so that players believe that cooperation will eventually resume, after any history of play.

The result is constructive, and exploits simple bounded-punishment strategies which are robust with respect to the players' priors about the monitoring structure. In particular, in the equilibria characterized only local information matters to determine players' behavior. Efficiency is supported by strategies that respond to defections with further defections. When the players' discount rate is smaller than one, the main difficulty in the construction of sequentially rational strategies that support efficiency is the preservation of short-run incentive compatibility after particular histories of play which involve several defections. When defections spread through a network, two complications arise. The first occurs when a player expects future defection coming from a particular direction. Suppose that somewhere in a cycle, for example, a defection has occurred and reaches a player from one direction. If this player does not respond, he may expect future defections from the opposite direction caused by players who are themselves responding to the original defection (see Figure 6.2). This player's short-term incentives then depend on the timing and on the number of future defections that he expects. In such cases, the verification of sequential rationality and the calculation of consistent beliefs can be extremely demanding. The analysis circumvents this difficulty by constructing consistent beliefs, which imply that a player never expects future defections to reach him (as unexpected behavior is always blamed on a neighbors' defection). Such beliefs are generated trivially when priors assign positive probability only to acyclic monitoring structures. More importantly, such beliefs can always be generated when priors have full support. The second complication arises when a player has failed to respond to a large number of defections. On the one hand, matching the number of defections of the opponent in the future may not be incentive compatible, say when this player is currently achieving efficient payoffs with a large number of different neighbors (as was the case with grim trigger strategies). The restriction that a player's action is common to all neighbors is of course the main source of complications here. On the other hand, not matching them may give rise to the circumstances outlined in the first type of complications, that is, this player may then expect future defections from a different direction. The former hurdle is circumvented by bounding the length of punishments, while the latter, as before, by constructing appropriate consistent beliefs.

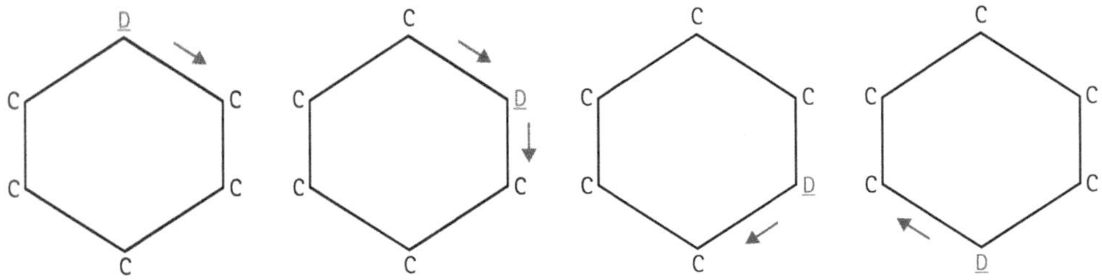

FIGURE 6.2 Cycling defections complicate incentives constraints, as beliefs about the timing of future defections become the key force driving behavior.

Some of these difficulties do not arise when players are patient (that is, if $\delta = 1$) as short-term incentives are irrelevant and punishments need not be bounded. Indeed, stronger results hold for the case of limit discounting in which payoffs are evaluated according to Banach-Mazur limit of the average payoff. If so, efficiency is resilient to histories of defections. In particular, there exists a sequential equilibrium such that, after any finite sequence of defections, paths eventually converge to the constant play of efficient actions in all neighborhoods in every future period. An essential part of the construction is that in any relationship in which defections have occurred the number of periods in which inefficient actions are played is "balanced": as the game unfolds from any history, both players will have played the inefficient action an equal number of times before resuming the efficient play. Remarkably, such balanced retaliations eventually extinguish themselves and always allow the resumption of cooperation throughout the network. Although the analysis is restricted to homogeneous discount rates and symmetric stage games with deterministic payoffs, the equilibria characterized are robust with respect to heterogeneity in payoffs and discount rates, and with respect to uncertainty in payoffs and population size, as long as the ordinal properties of the stage games are maintained across the players. These equilibria also persist as babbling equilibria in setups with communication. In addition, results extend to accommodate monitoring structures in which players interact with fewer players than they observe.

6.4 Fixed Discounting and Network Amplification

Much of the literature on community enforcement (discussed in the introduction and in Section 6.3) focuses on the case of sufficiently high discount factors and does not characterize efficient equilibria at fixed discount factors. A major concern in these papers was ensuring that players did not cooperate off the equilibrium path. The literature on repeated networked games with fixed discount factors abstracts from such a concern by analyzing the most cooperative equilibrium in games with continuous action sets. Such equilibria make players indifferent between cooperating and defecting on-path (as otherwise a player could be asked to cooperate more). By essentially the same argument as in Ellison (1994), this implies that players weakly prefer to defect off-path. Hence, a contribution of this literature is to show that grim trigger strategies provide the strongest possible incentives for cooperation on-path, not that they provide incentives for punishing off-path. The characterization of the most cooperative equilibrium has implications for the efficiency and stability of various network configurations which are the main objectives of this literature.

This approach was pioneered by several papers in public economics analyzing the effect of the size and structure of a group on the maximum equilibrium level of public good provision. Classical references, however, characterize maximum cooperation only for complete networks and public monitoring, and find few unambiguous relationships between group structure and maximum cooperation. Pecorino (1999) shows that public good provision is easier in large groups because a deviation causing everyone else to defect is more costly in large groups. Haag and Lagunoff (2007) consider a broader class of public goods games in a similar setup and characterizes the maximal average level of cooperation (MAC) over all stationary subgame perfect equilibrium paths. The MAC is shown to be increasing in monotone shifts of, and decreasing in mean preserving spreads of the distribution of discount factors. The latter suggests that more heterogeneous groups are less cooperative on average. Furthermore, in a class of Prisoners' Dilemma games, the MAC exhibits increasing returns to scale for a range of heterogeneous discount factors. That is, larger groups are more cooperative, on average, than smaller ones. By contrast, when the group has a common discount factor, the MAC is invariant to group size.

Haag and Lagunoff (2006) relax the public monitoring assumption and examine optimal network structure in a binary-symmetric Prisoners' Dilemma with local interactions and local monitoring, in which each individual's discount factor is randomly determined. A planner chooses a local interaction network before the discount factors are realized in order to maximize utilitarian welfare. A local trigger strategy equilibrium (LTSE) describes a sequential equilibrium in which each individual conditions his cooperation on the cooperation of at least a subset of neighbors. The main results restrict attention to the LTSE associated to the highest utilitarian welfare, and demonstrate a trade-off in the design problem between suboptimal punishments and social conflict. Potentially suboptimal punishments arise in designs with local interactions since monitoring is local. Owing to the heterogeneity of discount factors, however, greater social conflict may arise in more connected networks. When individuals' discount factors are known to the planner, the optimal network exhibits a cooperative core and an uncooperative fringe. Uncooperative players are impatient, and are connected to cooperative ones who are patient and tolerate their free riding so that social conflict is kept to a minimum. By contrast, when the planner knows only the ex-ante distribution over individual discount factors, in some cases the optimal design partitions individuals into isolated cliques, whereas in other cases incomplete graphs with small overlap are possible.

Two recent and related studies have addressed similar questions in the context of continuous action games with local monitoring, namely Wolitzky (2013) and Ali and Miller (2013). Both models feature smooth actions and payoffs so that, with grim trigger strategies, binding on-path incentive constraints imply slack off-path incentive constraints. Wolitzky (2013) studies cooperation in repeated networked games with a fixed and common discount factor. The setup displays local monitoring, while allowing for global interaction, and generalizes environments analyzed in Kandori (1993) and Ellison (1994). In particular, the analysis considers public goods games with continuous actions in which players choose of a level of cooperation (in that higher actions are privately costly but benefit everyone). Payoffs are separable, but depend on the action chosen by every other player in the game. In every period, a monitoring network is realized and players receive signals about the global structure of the realized network. Players perfectly observe the actions of their realized neighborhood, but observe nothing about any other player's action. A distinguishing feature of the environment analyzed is that in every period the monitoring network must be observed by players after actions are chosen. The assumption is substantial in the equilibrium construction, and results do not generally apply to alternative specifications in which uncertainty about the realized monitoring persists over time.[5]

The study characterizes the maximum level of cooperation that can be robustly sustained in Perfect Bayesian equilibrium (in that it can be sustained for any information that players may have about the realized monitoring network). The robustness criterion captures the perspective of an outside observer, who knows what information players have about each other's actions, but not what information players have about each other's information about actions, and who must make predictions that are robust to higher-order information. Determining the maximum level of cooperation for any specification of players' higher-order information appears intractable, as the strategies sustaining the maximum level of cooperation could in principle depend on players' private information in complicated ways. However, the main theoretical contribution establishes that the robust maximum level of cooperation is always sustained by simple grim trigger strategies, where each player cooperates at a fixed level unless he ever observes another player failing to cooperate at his prescribed level, in which case he stops cooperating forever. Grim trigger strategies also maximize cooperation when players have perfect knowledge of who observed whom in the past (as is the case when the monitoring network is fixed over time). Interestingly, it is when players have less information about the monitoring structure that more complicated strategies can do better than grim trigger. This is the case because the actions of different players are strategic complements when players know who observed whom in the past, as defecting makes every on-path history less likely when monitoring is local and strategies are

[5] The key role of the assumption is to ensure that stage game actions are one-dimensional (so that players simply choose a level of cooperation, rather than a map from the realized monitoring network to a level of cooperation).

grim triggers.[6] The strategic complementarity breaks down, however, when players can disagree about who has observed whom.

The analysis then compares different economies in terms of the maximal level of cooperation that can be achieved. Results are developed for two special cases: equal monitoring (when in expectation all players are monitored equally well); and fixed monitoring network (when the monitoring network is fixed over time). With equal monitoring, the effectiveness of a monitoring technology in supporting cooperation is completely determined by one simple statistic, its effective contagiousness, which captures the cumulative expected present discounted number of players who learn about a deviation. Naturally, higher levels of cooperation can be sustained if news about a deviation spreads throughout the network more quickly. Cooperation in the provision of pure public goods (when the marginal benefit of cooperating is independent of group size) is increasing in group size if the expected number of players who learn about a deviation is increasing in group size, while cooperation in the provision of divisible public goods (when the marginal benefit of cooperating is inversely proportional to group size) is increasing in group size if the expected fraction of players who learn about a deviation is increasing in group size. Hence, cooperation in the provision of pure public goods tends to be greater in larger groups, while cooperation in the provision of divisible public goods tends to be greater in smaller groups. In addition, there is a sense in which making monitoring more uncertain reduces cooperation. With a fixed monitoring network, instead, a novel notion of network centrality determines both which players cooperate more in a given network and which networks support more cooperation overall, thus linking the graph-theoretic property of centrality with the game-theoretic property of robust maximum cooperation. For example, adding links to the monitoring network necessarily increases all players' robust maximum cooperation, which formalizes the idea that individuals in better-connected groups cooperate more.

Ali and Miller (2013) analyze community enforcement in a pairwise interaction game in which the network is common knowledge. Their analysis compares interaction networks in terms of maximal level of cooperation in variable-stakes Prisoners' Dilemmas. Results establish that cliques are optimal network structures when players' equilibrium path behavior is stationary. Results are developed in the context of a continuous time model in which all players discount the future at a common fixed rate. Every link of the network is governed by an independent Poisson recognition process with a common recognition rate. Whenever a link is recognized, an instantaneous two-period interaction is played within the selected relationship. In the first subperiod, both players propose stakes at which they intend to interact, and the smallest of the two proposals determines the actual stakes in the relationship. In the second subperiod, players engage in a Prisoners' Dilemma. If both cooperate, each receives a payoff which coincides with the agreed stakes; if both defect, each receives a payoff equal to zero; whereas when one defects while the other cooperates, the cooperating player incurs a

[6] Actions are strategic complements if a player is willing to cooperate more at any on-path history whenever another player cooperates more at any on-path history.

cooperation-loss which may depend on the agreed stakes, while the defecting player receives a deviation-gain which may also depend on the agreed stakes.[7] The stage game is said to satisfy strategic complementarity whenever stakes exceed the difference between deviation gains and cooperation losses.[8] Monitoring is local. Thus, players ignore both the actions chosen in interactions to which they did not belong and the time at which these interactions took place. The analysis restricts attention to stationary strategies in which behavior is independent of the history of play at any equilibrium path. A stationary equilibrium in which players always cooperate for any possible equilibrium-path history is said to be a mutual effort equilibrium. Any stationary grim trigger strategy profile that prescribes stakes such that incentive constraints bind at every equilibrium path history is therefore a Perfect Bayesian equilibrium by the arguments above.

The main result establishes that any symmetric network[9] with degree d possesses a symmetric contagion equilibrium that Pareto dominates every distinct mutual effort equilibrium (and thus identifies the optimal stakes). The result also implies that no other stationary equilibrium has a higher value if the stage game satisfies strategic complementarity. The argument relies on a measure of network viscosity (which is minimal in the clique) that captures the incentives to comply with equilibrium strategies. This measure differs from the effective contagiousness in Wolitzky (2013) because in public goods games every player may punish a deviator upon receiving news of a defection, whereas in separable games only neighbors can effectively punish a deviation.

Results exploit the characterization of the optimal stakes to analyze how network structure affects aggregate welfare. Adding links has two roles in the model: it helps information diffusion through contagion; and it increases the number of interactions (as links are recognized at the same rate) and consequently the expected surplus when cooperating. The main welfare implication of the model is the optimality of cliques. In particular, for any network in which the maximal degree is no more than d, no player attains a mutual effort equilibrium payoff that exceeds his optimal equilibrium payoff in the symmetric network of degree d. Moreover, if the stage game satisfies strategic complementarity, then the value in every equilibrium is less than the optimal value in the symmetric network of degree d.

[7] The deviation-gain is assumed: to exceed stakes; to be zero when stakes are equal to zero; to be strictly increasing and strictly convex in stakes; to have a first derivative that is greater than 1 at zero and diverging to infinity as stakes diverge. The cooperation-loss is also assumed to be zero when stakes are equal to zero.

[8] When strategic complementarity holds, mutual cooperation is efficient in the stage game. But the assumption is stronger. Even with two players, efficiency of mutual cooperation does not ensure that optimal equilibria are mutual effort, which is why the stronger assumption is invoked.

[9] A permutation of the players $\pi : N \to N$ is a graph automorphism if $ij \in G$ implies $\pi(i)\pi(j) \in G$. A network G is symmetric if for any two links $ij, kl \in G$ there exists a graph automorphism π such that $\pi(i) = k$ and $\pi(j) = l$. In a symmetric network, all links are isomorphic to each other.

Results also extend such logic to a model in which players incur an additively separable cost of forming links which depends only on the number of neighbors they have. Costs are said to be concave if the average link cost weakly decreases with the number of links; while costs are convex otherwise. When costs are concave, there exists a symmetric Perfect Bayesian equilibrium on the complete network that yields to each player a payoff that is higher than his payoff in any mutual effort equilibrium of any other incomplete network. Moreover, when the game satisfies strategic complementarity, the claim holds for every equilibrium (not just for mutual effort equilibria). When costs are convex, the welfare maximizing network may no longer be the complete network, and the analysis applies only to regular networks. In such cases it is possible to find the clique size that maximizes the payoff of a player in the welfare maximizing equilibrium, and the associated optimal value. No mutual effort equilibrium on any regular network attains payoffs that exceed such value. Moreover, if the stage game satisfies strategic complementarity, no equilibrium on any regular network attains a higher value.

All the papers discussed provide novel and interesting insights linking interaction and monitoring networks to measures of aggregate welfare. These observations can in principle explain why community enforcement may lead to substantially different levels of cooperation across societies. The main limitation of these studies, however, is the restrictive class of games to which results apply, as results are generally developed for Prisoners' Dilemma type stage games possessing a mutual minmax Nash equilibrium. Generalizing techniques to arbitrary stage game does not seem straightforward, as the characterization of the equilibrium with the highest utilitarian welfare may become intractable.

6.5 Fixed Discounting and Communication

A separate strand of the literature analyzes how equilibrium outcomes are affected by the availability of different communication technologies. These studies include Lippert and Spagolo (2011), Wolitzky (2014), and Ali and Miller (2014).

Lippert and Spagnolo (2011) consider environments with local interaction and separable stage games. In particular, they focus on stage games in which every pair of players plays an asymmetric Prisoners' Dilemma in which the interaction network may be direct, but is necessarily common knowledge. In this setup, they first consider two benchmark cases: public monitoring (when each agent observes the full history of play); and local monitoring. The main focus, however, is a variant of the local monitoring model in which players have access to a fixed number of rounds of private cheap talk in every period of the game. The communication network coincides with the interaction network (as was implicitly the case in Renault and Tomala 1998; Cho 2010; and Laclau 2012).

With cheap talk, the possibility of transmitting soft information about privately observed defections to other agents may foster cooperation in games with fixed discount factors. Grim trigger strategies, which are optimal (in the sense of Abreu 1988) under public monitoring and which correspond to the contagion strategies studied in random matching games, are no longer optimal when information transmission is endogenous and players account for their incentives to communicate truthfully. When cooperation in the network is disciplined by such strategies, cheap talk is never used in equilibrium, as an agent reverts to non-cooperative play forever after observing a defection. This triggers a contagious process that eliminates all prospects of future cooperation in the network, thereby removing any motive for truthful communication. When forgiving strategies are used, instead, agents do have incentives to transmit information truthfully to avoid the collapse of cooperation as upon observing a defection, non-defecting agents continue cooperating and spread information on the deviation until only the initial deviator can be punished by a neighbor who benefits from such punishment. As information transmission within the network speeds up punishment phases, forgiving equilibria strictly dominate contagious equilibria.

Another central finding of the analysis is that with asymmetric stage-games, interaction networks display a rather general end-network effect that occurs under any informational assumption. Network structures such as trees may not sustain cooperative behavior, as agents with only outgoing links cannot be sanctioned if they defect. This end-network effect is a special case of gatekeeping and characterizes those gatekeepers as key players to cooperation in the network. Circular networks overcome this problem, ensuring that all defections can be met with punishment and that networks of relations are sustainable in equilibrium. The results provide an intuitive explanation for the importance of "closure" and "density." When monitoring is local and agents play according to grim trigger strategies, the enforceability of cooperation in bilateral relationships may hinder global cooperation in the larger networks, as a pair may not be willing to sacrifice their bilateral relationship to be part of the multilateral punishment mechanism which could sustain cooperation in the larger network. This argument extends from bilateral relations to larger subnetworks and establishes why coalitional agreements may undermine global cooperation ones by softening third-party punishments. This problem, however, can be overcome by forgiving strategies.

Wolitzky (2014) analyzes games with fixed discounting under different communication protocols. His main contribution establishes a direct relationship between different communication technologies and the set of sequential equilibrium payoffs. Results apply to separable stage games with local interaction, in which monitoring is local and imperfect, and in which the interaction network is common knowledge. In particular, the actions chosen by the two players in a relationship determine a signal realization pinning down payoffs in that relationship. Signals are random variables that depend only on the actions chosen by the two players on a link (and are thus independent across relationships). Signals are locally public, but local monitoring is imperfect, as players observe only their action and the signal realizations in the interactions to which they belonged (but not necessarily the actions of their opponents). As the stage game is separable, the study aims at characterizing the community enforcement for a given discount factor. Results apply to stage games which possess a mutual minmax Nash equilibrium in every realized interaction.

The analysis first establishes that different communication protocols replicate any sequential equilibrium of a corresponding game with public information. The public information benchmark analyzed here is one in which all players observe the signal realizations on every link (but not necessarily the actions chosen by players other than themselves). In every period of the game, communication is modelled as an infinite number of rounds in which messages can be sent. Three communication technologies are considered: public cheap talk, private cheap talk, and tokens. The first result extends contributions in Ben-Porath and Kahneman (1996), and establishes that any equilibrium payoff of the game with public monitoring is also an equilibrium payoff of a corresponding game with local monitoring and public cheap talk. The second result builds on the contribution by Renault and Tomala (1998), and considers environments in which cheap talk is private and constrained to take place only on the interaction network (that is, when the interaction and communication networks coincide). The result establishes that any public monitoring equilibrium payoff is also an equilibrium payoff of a game with local monitoring and private cheap talk if and only if the network is 2-connected. The main departures from Renault and Tomala (1998) are: (a) that 2-connectedness is not only sufficient, but also necessary (in that for any network that is not 2-connected there exists a game in which private cheap talk cannot replicate public monitoring); and (b) that 2-connectedness is sufficient for replication even when the frequency of interaction is low. A final replication result considers environments with private cheap talk in which tokens can be exchanged in every relationship at each communication round. The main difference between cheap talk and tokens is that players must own tokens before transferring them. Results establish that public monitoring outcomes can always be replicated as sequential equilibria with tokens. Although the equilibrium construction relies both on tokens and private cheap talk, the same conclusions would hold if cheap talk were ruled out, since infinitesimal amounts of valueless tokens could be used to communicate. Message spaces and monetary endowments need not to be tailored to the specific game provided that a spanning tree exists in which all non-leaf players have a positive token endowment.

The final contribution presents sufficient conditions for tokens to expand the set of equilibrium payoffs compared both to games without communication, and to games with private cheap talk. Sufficient conditions require: (a) the network to possess a subtree; (b) every game played by two linked players to have a product structure; (c) the set of public information equilibria to include the convex hull of the locally public equilibria of the game with private information. These conditions simplify in many common environments, and only require the existence of a small subtree in which a strategy with tokens expands the equilibrium set. The essentiality of tokens then follows since tokens expand the equilibrium payoff hull in the entire game when they do so in a subtree (as the remaining players can always comply with a strategy with private cheap talk in which tokens play no role).

The analysis of tokens builds on and is closely related to the literature on microfoundations of money. One of the most important themes in that literature asks when letting individuals exchange inherently valueless tokens can expand the equilibrium payoff sets in dynamic decentralized economies; for instance, Kocherlakota (1998, 2002). Results here carry out a similar exercise in the context of a more general setting.

Ali and Miller (2014) analyze the same environments discussed in their 2013 paper (in Section 6.4) while allowing for pre-play communication. In particular, before selecting stakes, partners may communicate to their neighbors information about the behavior of other players. The analysis studies both evidentiary communication (when players can conceal information, but cannot falsify it), and cheap talk. The analysis focuses on ostracism strategies in which players target punishments toward defecting players while cooperating with those they believe to be cooperative. To understand the impact of strategic communication, the analysis first characterizes two classical benchmarks. The first is bilateral enforcement, which identifies equilibria that abstract from community enforcement or communication (which in this setup amounts to bilateral grim trigger strategies played independently in each relationship). The second benchmark is mechanical communication, which characterizes settings in which players are constrained to reveal all their information truthfully. Permanent ostracism is an equilibrium with mechanical communication, since defectors must reveal themselves as such in all their future interactions. As permanent ostracism employs the harshest feasible punishment against defectors, it supports at least as much cooperation as any other equilibrium, and it coincides with the most cooperative equilibrium of a model with public monitoring.

When communication is strategic, one may conjecture that, while defecting players have a strong incentive to conceal their own misdeeds, cooperating players should have aligned interests in revealing and punishing the guilty. The main result establishes that this intuition is wrong. If defecting players are permanently ostracized, then their victims have a strong incentive to conceal such defections and to defect on other cooperating players. This strategic motive implies that permanent ostracism cannot be optimal with strategic communication and that the players are no better off than under bilateral enforcement. In other words, truthful communication is incentive compatible with permanent ostracism only if community enforcement is redundant. This stark negative conclusion applies to every network, even when communication is evidentiary. In fact, consider a permanent ostracism equilibrium and a relationship between two neighbors. Suppose by contradiction that they cooperate at stakes that would not be attainable under bilateral enforcement. Each player's incentives to cooperate must then be driven by the threat of punishments from others. Now consider a history at which one of them knows that everyone, except the two of them, has defected and should be ostracized. Because all the other players are defecting, this player's only incentive to cooperate arises from his continuation play with the one cooperative neighbor he has left, just as under bilateral enforcement. Thus, he must strictly prefer to conceal his information and to defect at the equilibrium stakes, rather than telling the truth and reducing their stakes to keep on cooperating.

This result is most pronounced in Prisoners' Dilemmas, but analogues apply to general separable stage games. In any symmetric permanent ostracism equilibrium, each player's equilibrium payoff in a relationship is bounded above by the highest payoff attainable in a bilateral enforcement equilibrium in that relationship. Asymmetric permanent ostracism equilibria allow for more flexibility, but a bound on payoffs, arising from bilateral enforcement equilibria, still applies regardless of the network structure. Thus, the incentives to conceal information generally constrain the surplus that can be attained through permanent ostracism.

The negative theoretical conclusion on permanent ostracism contrasts with the prevalence of ostracism in communities and markets. Observed community enforcement norms, however, often involve forgiveness, in that players are only ostracized temporarily. The analysis provides a rationale for such norms by showing how forgiveness may encourage truthful communication between cooperative victims. In particular, when ostracism is temporary and players are forgiven at random times, innocent players communicate truthfully and cooperate with each other at levels beyond those attainable under bilateral enforcement (if players are sufficiently patient or society is sufficiently large). Temporary punishments may thus facilitate community enforcement by maintaining social collateral that fosters communication and cooperation among non-defecting players in the wake of defections.

The results on communication and ostracism should be contrasted with community enforcement schemes without information transmission, such as contagion equilibria introduced for anonymous random matching environments by Kandori (1992) and Ellison (1994), and applied to social networks by Wolitzky (2013), Ali and Miller (2013), and others. Contagion offers a useful benchmark for attainable payoffs in the absence of institutions or communication, but it also represents a fragile form of collective reputation, in that a single defection destroys a player's trust in the entire community. Ostracism, by contrast, reflects the principle that players ought to trust those partners who have never defected to their knowledge, while punishing those who have done so. Thus, with ostracism, reputations are entirely at the individual level. Hybrid community enforcement norms can be envisioned in which cooperative players communicate truthfully to other cooperators while ostracizing those who have defected in the past so long as they know of no more than d defecting players, and defect on all their partners otherwise. Such equilibria improve upon permanent ostracism, but average stakes are bounded by contagion with $n - d$ players.

Results in Ali and Miller (2014) rely on several modeling assumptions and innovations. Players interact at random privately observed times, which contrasts with classical repeated games in which all players are known to have interacted in every period. This generates non-trivial incentives at the communication stage, as players may now conceal an interaction from their partners. Incentives would differ if the timing of interactions were public. Unraveling would compel a player to reveal all details of his past interactions, since a partner could rationally consider his failure to disclose as evidence of a deviation. If so, strategic communication would be as effective as permanent ostracism with mechanical communication. However, equilibria would be fragile, and even the slightest chance of interactions happening at privately observed times would again undermine any incentive for truthful communication.

The variable stakes model allows for a tight comparison of equilibria at a fixed discount rate and offers more scope for cooperation. Prisoners' Dilemma games with fixed stakes partly obscure incentives to ostracize by limiting the extent to which players can tailor their actions to the environment. In fact, if the stakes in each relationship were fixed, permanent ostracism would do no better than bilateral enforcement (as players, who are unwilling to cooperate under bilateral enforcement, would be unwilling to cooperate when only two cooperating players remain). In contrast, variable stakes enable partners to adjust the terms of their relationship based on their mutual history (for instance, by reducing their stakes once some players have been ostracized), shifting focus from technological constraints to the incentives for truthful communication.

6.6 Comments: Applications and Omissions

Applications: The use of implicit social sanctions to deter misconduct has been widely documented in economics (Milgrom, North, and Weingast 1990; Greif 1993), political science (Ostrom 1990; Fearon and Laitin 1996), sociology (Coleman 1990; Raub and Weesie 1990), and law (Bernstein 1992). Some of these studies have stressed the importance of community cohesion for attaining socially desirable outcomes in trust-based transactions, for example, Coleman (1990), Greif (1993), McMillan (1995), Fearon and Laitin (1996), Uzzi (1996), Dixit (2006). Coleman seminal's contribution identifies a notion of social capital, and relates such notion to the underlying social architecture. In Coleman's findings, the enforcement of cooperation is more effective in networks with high closure and cohesion, as cohesion facilitates the implementation of social sanctions, thereby increasing welfare. Other studies highlight the importance of information dissemination within the community for the effectiveness of such community-based sanctions. Greif (2006) finds that contract enforcement between medieval Maghribi traders is effective only when a close-knit community disseminates information so to align its members' incentives to comply with the community-based sanctions against deviant behavior.

Coleman's notion of social capital has motivated many of the more applied theoretical contributions in this field. For instance, Vega-Redondo (2006) considers a novel approach to network formation in the context of a repeated binary-symmetric Prisoners' Dilemma with random payoffs. The social network specifies not only the local interaction structure, but also the diffusion of information about past play, and the availability of new cooperation opportunities. Search plays an important role in this environment, as agents always look for new partners when relationship-specific payoffs are volatile. In this context, the analysis develops a notion of social capital and shows how the social network adapts to changes in the environment. Network effects are important in enhancing cooperation; and the social network endogenously adapts by displaying more cohesiveness whenever the environment deteriorates. Conclusions are obtained by numerical simulations and supported by approximate mean-field analysis.

More recently, Balmaceda and Escobar (2014) build on results from Haag and Lagunoff (2006, 2007), discussed in Section 6.4, to show that cohesive communities (in which players are partitioned into isolated cliques) emerge as welfare-maximizing network structures. Cohesive communities generate local common knowledge, which allows players to coordinate their punishments, and, as a result, yield high equilibrium payoffs. Results provide an additional theoretical rationale for Coleman's link between cohesion and social capital, but apply only to environments in which monitoring is local, while interactions are centralized (in that all community members interact with a single player who knows the full history of play). The analysis also establishes that optimal networks are minimally connected, when players monitor every other community member in their component of the social network. If so, as in Burt (1992, 2001), bridging structural holes in the monitoring network becomes the sole consideration identifying the optimal social network (as cohesion within a component is imposed by assumption).

Other recent studies have theoretically analyzed and empirically documented the impact of network structure on different kinds of cooperation, such as favor exchange (Möbius 2001; Hauser and Hopenhayn 2004; Karlan et al. 2009; Jackson, Rodriguez-Barraquer, and Tan 2011) and risk-sharing (Ambrus, Möbius, and Szeidl 2010; Bramoullé and Kranton 2007; Bloch, Genicot, and Ray 2008). These studies are survey and discussed in Chapter 28 of this handbook. Although much empirical work remains to be done, empirical findings hint at different measures of centrality as determinants of cooperation within social interactions. For example, Karlan et al. (2009) find that indirect network connections between individuals in Peruvian shantytowns support lending and borrowing, consistent with findings showing that more central players cooperate more. More subtly, Jackson, Rodriguez-Barraquer, and Tan (2011) find that favor-exchange networks in rural India exhibit high support (the property that linked players share at least one common neighbor).

Endogenizing Networks: General results on network formation are discussed in several chapters of this handbook (Chapters 5–7). Most studies on repeated interactions have focused on optimal network design, rather than network formation, as in a repeated setup many well-documented network formation games generate large multiplicity of equilibrium networks (often including efficient networks). To see this, consider a pairwise linking process in which players simultaneously propose the partnerships they wish to engage in, and in which a partnership forms if and only if both players propose it. Consider a Prisoners' Dilemma game, in which the formed network is common knowledge. It is straightforward to see that any network G can arise in an equilibrium of this game if it yields an individually rational net-payoff to each player, via the following strategy profile: if network G arises, then players follow the prescribed equilibrium, but if any other network forms then each player perpetually defects. This simple punishment deters players from deviating in the network formation stage. A similar logic applies to more complex games in which the network may not be common knowledge since any link remains local common knowledge among the two neighbors.

Separating Monitoring from Interaction: Most studies analyze environments in which the monitoring network and the network of interactions coincide (as was the case in the baseline setup presented in Section 6.2). However, conclusions generally carry over to the case in which players monitor more individuals than they interact with (as payoffs in any interaction can always be set to zero). Models with local monitoring and global interaction have only been analyzed in a limited number of studies which include Renault and Tomala (1998), Laclau (2012), and Wolitzky (2013).

Omissions: Some notable contributions to the literature have been omitted from the main discussion to streamline exposition. Ahn (1997) and Ahn and Souminen (2001) are precursors to several subsequent, but more general, contributions. They analyze cooperation in the context of binary-symmetric seller-buyer games with local monitoring and cheap talk, and present somewhat strong conditions for efficient outcomes to obtain. Kinateder (2008) considers a particular Prisoners' Dilemma game with global interaction, local monitoring, and in which players can truthfully communicate information to neighbors over time. The Folk Theorem extends to this setup, although the set of sequential equilibria and the corresponding payoff set may be reduced for discount factors strictly below 1. If players are allowed to communicate strategically, truthful communication arises endogenously only under additional assumptions. An additional implication of his analysis is that, when the discount factor is below 1, the viability of cooperation depends on the network's diameter, but not on its clustering coefficient. Mihm, Toth, and Lang (2009) consider strategic interaction in separable stage games with local monitoring. Their main contributions establish why strategic interdependencies between relationships on a network may facilitate efficient outcomes, and derive necessary and sufficient conditions to characterize the efficient equilibria of the network game in terms of the architecture of the underlying network.

Large Bipartite Networks: More recently, two studies have considered a novel and interesting approach to analyzing repeated networked games with a large number of players, namely Fainmesser and Goldberg (2011), and Fainmesser (2012). Fainmesser and Goldberg (2011) analyze repeated games in large bipartite networks with local monitoring and incomplete information about the network structure (players are informed of their neighbors and of several additional characteristics about the underlying graph). The model characterizes networks in which each agent cooperates in some equilibrium with every client to whom he is connected. To this end, the analysis establishes that in the proposed game: (a) the incentives of an agent to cooperate depend only on her beliefs with respect to her local neighborhood (a subnetwork whose size is independent of the size of the entire network); and (b) when an agent observes the network structure only partially, his incentives to cooperate can be calculated as if the network was a random tree with him at its root. The characterization sheds light on the welfare costs of relying only on repeated interactions for sustaining cooperation, and on how to mitigate such costs. Fainmesser (2012) builds on this analysis by considering buyer-seller games in large bipartite networks, in which sellers have the option to cheat their buyers, and buyers decide whether to repurchase from different sellers. While endowing sellers with incomplete knowledge of the network, the analysis derives conditions that determine whether a network is consistent with cooperation between every buyer and seller that are connected. Three network features reduce the minimal discount factor sufficient for cooperation: moderate and balanced competition, sparseness, and segregation.

References

Abreu, Dilip J. (1988). "On the theory of infinitely repeated games with discounting." *Econometrica* 56(2), 383–396.

Ahn Iltae (1997). "Three essays on repeated games without perfect information." PhD Thesis, University of Pennsylvania.

Ahn Iltae and Suominen Matti (2001). "Word-of-mouth communication and community enforcement." *International Economic Review* 42(2), 399–415.

Ali, Nageeb S. and Miller David A. (2013). "Enforcing cooperation in networked societies." Mimeo.

Ali, Nageeb S. and Miller David A. (2014). "Ostracism." Mimeo.

Aliprantis, Charalambos D. and Border Kim C. (2005). *Infinite Dimensional Analysis*. Springer.

Balmaceda Felipe and Escobar Juan F. (2014). "Trust in cohesive communities." Mimeo.

Ben-Porath Elchanan and Kahneman Michael (1996). "Communication in repeated games with private monitoring." *Journal of Economic Theory* 70(2), 281–297.

Bernstein Lisa (1992). "Opting out of the legal system: Extralegal contractual relations in the diamond industry." *Journal of Legal Studies* 21(1), 115–157.

Bloch Francis, Genicot Garance, and Ray Debraj (2008). "Informal insurance in social networks." *Journal of Economic Theory* 143, 36–58.

Bramoulle Yann (2007). "Anti-coordination and social interactions." *Games and Economic Behavior* 58(1), 30–49.

Bramoulle Yann and Kranton Rachel (2007). "Risk-sharing networks." *Journal of Economic Behavior and Organization* 64, 275–294.

Burt, Roland S. (1992). *Structural Holes: The Social Structure of Competition*. Cambridge, MA: Harvard University Press.

Burt, Roland S. (2001). "Structural holes versus network closure as social capital." *Social Capital: Theory and Research* Chapter 2, first issue, 31–56, https://books.google.co.uk/books?hl=en&lr=&id=u_KTkBHY_kgC&oi=fnd&pg=PR7& dq=Burt+Structural+Holes+versus+Network+Closure+as+Social+Capital,+Social+Capital:+Theory+and+Research,+31-56,+2001.&ots=PLXeYmfxMO&sig=W2shYlshCP_1ftFoYhBepv9sdBc#v=onepage&q&f=false.

Cho Myeonghwan (2011). "Public randomization in the repeated prisoner's dilemma game with local interaction." *Economic Letters* 112(3), 280–282.

Cho Myeonghwan (2010). "Cooperation in the Prisoner's Dilemma game with local interaction and local communication." Mimeo.

Coleman, James S. (1990). *Foundations of Social Theory*. Cambridge, MA: Belknap Press.

Deb Joyee (2014). "Cooperation and community responsibility: A Folk Theorem for random matching games with names." Mimeo.

Dixit Avinash (2006). *Lawlessness and Economics. Alternative Modes of Governance*. Oxford: Oxford University Press.

Ellison Glenn (1994). "Cooperation in the Prisoner's Dilemma with anonymous random matching." *Review of Economic Studies* 61(3), 567–588.

Fainmesser, Itay P. (2012). "Community structure and market outcomes: A repeated games in networks approach." *American Economic Journal: Microeconomics* 4(1), 32–69.

Fainmesser, Itay P. and Goldberg David A. (2012). "Cooperation in partly observable networked markets." Mimeo.

Fearon, James D. and Laitin David D. (1996). "Explaining interethnic cooperation." *American Political Science Review* 90(4), 715–735.

Greif Avner (1993). "Contract enforceability and economic institutions in early trade: The Maghribi Traders' coalition." *American Economic Review* 83(3), 525–548.

Greif Avner (2006). *Institutions and the Path to the Modern Economy: Lessons from Medieval Trade*. Cambridge, UK: Cambridge University Press.

Haag Matthew and Lagunoff Roger (2006). "Social norms local interaction, and neighborhood planning." *International Economic Review* 47(1), 265–296.

Haag Matthew and Lagunoff Roger (2007). "On the size and structure of group cooperation." *Journal of Economic Theory*, Elsevier, 135(1), 68–89.

Harrington, Joseph E. (1995). "Cooperation in a one-shot Prisoners' Dilemma." *Games and Economic Behavior* 8(2), 364–377.

Hopenhayn, Hugo A. and Hauser Christine (2004). "Trading favors: Optimal exchange and forgiveness." Meeting Papers 125, *Society for Economic Dynamics*.

Jackson, Matthew O., Rodriguez-Barraquer Tomas and Tan Xu (2012). "Social capital and social quilts: Network patterns of favor exchange." *American Economic Review* 102(5), 1857–1897.

Kandori Michihiro (1992). "Social norms and community enforcement." *Review of Economic Studies* 59(1), 63–80.

Karlan Dean, Mobius Markus, Rosenblat Tanya and Szeidl Adam (2009). "Trust and social collateral." *Quarterly Journal of Economics* 124, 1307–1361.

Kinateder Markus (2009). "Repeated games played on a network." Mimeo.

Kocherlakota, Narayana R. (1998). "Money is memory." *Journal of Economic Theory*, Elsevier, 81(2), 232–251.

Kocherlakota, Narayana R. (2002). "The two-money theorem." *International Economic Review* 43, 333–346.

Laclau Marie (2012). "A Folk Theorem for repeated games played on a network." *Games and Economic Behavior* 76(2), 711–737.

Laclau Marie (2014). "Communication in repeated network games with imperfect monitoring." *Games and Economic Behavior* 87(C), 136–160.

Lippert Steffen and Spagnolo Giancarlo (2011). "Networks of relations and word-of-mouth communication." *Games and Economic Behavior* 72(1), 202–217.

McMillan John (1995) "Reorganizing vertical supply relationships." *Trends in Business Organization*, Tübingen: Mohr, ISBN 3161463919, 203–222.

Mihm Maximilian, Toth Russell and Lang Corey (2009). "What goes around comes around: A theory of indirect reciprocity in networks." Mimeo.

Milgrom, Paul R., North Douglas C. and Weingast Barry R. (1990). "The role of institutions in the revival of trade: The law merchant, private judges, and the champagne fairs." *Economics and Politics*, 2(1), 1–23.

Mobius Markus (2001). "Trading favors." Mimeo.

Nava Francesco and Piccione Michele (2014). "Efficiency in repeated games with local interaction and uncertain local monitoring." *Theoretical Economics* 9(1), 279–312.

Ostrom Elinor (1990). *Governing the Commons: The Evolution of Institutions for Collective Action*, Cambridge, UK: Cambridge University Press.

Pecorino Paul (1999). "The effect of group size on public good provision in a repeated game setting." *Journal of Public Economics* 72, 121–134.

Raub Werner and Weesie Jeroen (1990). "Reputation and efficiency in social interactions: An example of network effects." *American Journal of Sociology* 96(3), 626–654.

Renault Jerome and Tomala Tristan (1998). "Repeated proximity games." *International Journal of Game Theory* 27(4), 539–559.

Takahashi Satoru (2010). "Community enforcement when players observe past partners' play." *Journal of Economic Theory* 145(1), 42–62.

Tomala Tristan (2010). "Fault reporting in partially known networks and Folk Theorems." *Operations Research* 59(3), 754–763.

Uzzi Brian (1996). "The sources and consequences of embeddedness for the economic performance of organizations: The network effect." *American Sociological Review* 61(4), 674–698.

Vega-Redondo Fernando (2006). "Building social capital in a changing world." *Journal of Economic Dynamics and Control* 30(11), 2305–2338.

Wolitzky Alexander (2013). "Cooperation with network monitoring." *Review of Economic Studies* 80(1), 395–427.

Wolitzky Alexander (2015). "Communication with tokens in repeated games on networks." *Theoretical Economics* 10(1), 67–101.

Xue Jun (2004). "Essays on cooperation, coordination, and conformity." PhD Thesis, Pennsylvania State University.

CHAPTER 7

STOCHASTIC NETWORK FORMATION AND HOMOPHILY

PAOLO PIN AND BRIAN W. ROGERS

7.1 Introduction

Many aspects of our lives are affected by social interactions with others. Often, the way these relationships operate is influenced by interactions between others. For example, an opinion communicated by one person to another is likely to have been influenced by the outcomes of previous conversations between different pairs of individuals. That is, a given relationship typically does not live in isolation, but rather is embedded in a social structure consisting of many relationships. It is now well-recognized that, accordingly, if one wants to understand many kinds of social phenomena, it is imperative to study the *networks* of these social relationships, rather than restricting the level of the analysis to any particular relationship. It thus becomes important to understand the structure of social networks, how and why they form, and how their properties relate to the interactions, behaviors, and processes of diffusion occurring through them.

So it is not surprising that researchers have employed a vast range of methodologies towards these ends. One characteristic along which to broadly categorize studies of network formation is in terms of those based primarily on strategic motivations as opposed to those based primarily on random events. Naturally, real networks are formed through a combination of randomness and self-motivated behaviors. This observation is reflected in modeling choices and, indeed, there are numerous analyses that, to varying extents, incorporate both random and strategic considerations into modeling link formation processes. Nevertheless, the literature can largely be sorted according to whether strategic considerations or, instead, random events, constitute the main guiding force behind which links are formed and which links are not formed. We here take up the latter category, and aim to survey and discuss the literature that studies network formation by viewing random events as the essential ingredients. See in this handbook Vannetelbosch and Mauleon for a discussion of strategic-based models, and Vega-Redondo for a survey of the literature that explicitly takes into account the coevolution of networks and behaviors.

We posit that random network models form the foundation of modern network theory. The literature has demonstrated that random network models are far superior to purely strategic models in terms of generating networks with properties that are consistent with field data on the structure of large social networks. This conclusion has several derivative implications. First, a number of random network models offer tractable frameworks within which to study incentives, and where it is possible to draw conclusions about how network structure impacts behavior. Such an approach can either allow strategic considerations to be brought into the analysis of link formation or, quite distinctly, it can allow the researcher to study, for example, diffusion processes or opinion formation from a strategic perspective in a network context. This kind of work is particularly important since, without modeling the incentives of agents in the network, it is not possible to discuss, for example, which network structures perform better than others from a social perspective, nor is it possible to provide policy-oriented conclusions.

Second, empirical work on network formation can be based only on random network models. As data becomes increasingly available along with the computing power to analyze it, empirical work will continue the trend of becoming increasingly important and useful. It is worth nothing that random network models constitute the benchmark with respect to which network properties are often described. That is, recognizing that real networks are not purely random, once we understand the macro features of purely random networks, we can begin to understand their discrepancies from real networks as attributable to non-random behavior. Always, for example, when we refer to *small world networks*, to *skewed degree distribution* or to *high clustering* in a specific network, we make an implicit comparison between the particular network we refer to and the expected outcome of some random network generation process.

By definition, the work we survey models the connections that form between agents as probabilistic events. The simplest such models, which serve as a baseline for the rest of the literature, view these events as independent and identically distributed across pairs of agents. Implicit in such an approach, of course, is that agents have no observable characteristics that are relevant to the formation of the network through affecting the probabilities with which different agents connect to each other. A more descriptive or realistic approach would have to accommodate the possibility of various kinds of correlations in linking outcomes. In essence, the likelihood with which an analyst views a connection between a given pair of agents to be formed is influenced by what is known about the characteristics of the agents, and possibly also by the presence or absence of other links. Agents' characteristics will influence link outcomes if, for example, there are complementarities or substitutability of the characteristics in the context of the relationship.

In this regard, the single most important force relating to agents' characteristics is *homophily*, which is the tendency of agents with similar characteristics to link with each other. Because homophily is so robustly observed across many contexts and dimensions, models that incorporate (the possibility of) homophily are central to the study of random networks in general, and often times it is the presence of homophily that allows a model to capture real-world properties that are missing from baseline models. Homophily has been documented by sociologists along a multitude of dimensions in self-reported friendship networks, and it is present in nearly all social networks if one examines the relevant characteristics. It has also been shown that homophily plays an important role in governing the outcomes of many important network-based phenomena, through studying models of naïve learning (Golub and Jackson 2012a), population games (Dalmazzo et al. 2014), strategic network formation games (Jackson and Rogers 2005; De Martí and Zenou 2013), and also random network processes (Currarini et al. 2009, 2010; Bramoullé et al. 2012 and structural models for empirical estimation that we discuss in Section 7.4.2). With modern communication technologies, the costs of maintaining geographically separated connections have been dramatically reduced, allowing for different patterns of social interactions. In particular, homophily is increasing along a number of prominent dimensions (see, e.g., Rosenblat and Mobius 2004).

Often it is difficult to disentangle from real data the role of homophily from that of peer effects and contagion (on this see, among others, Ozgür and Bisin 2013; Mele 2014; Goldsmith-Pinkham and Imbens 2013—and Chandrasekhar in this book). Actually, the following econometric problem is a main open issue: what are the conditions on the data for which this identification is possible (Angrist 2014; Shalizi and Thomas 2011)? We will see that most current answers to this question are indeed based on specific random networks models.

Our presentation below is delineated by a classification of models according to how they treat population dynamics. In particular, we first distinguish network formation models that are "one shot" from those that are dynamic and, in the latter case, according to whether the number of nodes varies over time, and how. In particular, and also following the chronological order of how these models came into the economics literature, Section 7.2 illustrates one-shot models with a fixed number of nodes, Section 7.3 presents growing random network models, and Section 7.4 discusses dynamical models in which the number of nodes is fixed or approaches a steady state. In Section 7.5 we discuss the economic literature on homophily and its deep relation with the one on random networks. Finally, Section 7.6 concludes.[1]

[1] We remark that many of our topics have received coverage in the books of Vega-Redondo (2007) and Jackson (2008b). Relative to those contributions, we try here to convey a unified view that includes also the more recent developments, especially in the fields of network games and empirical estimation models.

7.2 ONE-SHOT MODELS WITH A FIXED POPULATION

There are $2^{n(n-1)}$ directed unweighted networks between n nodes, and $2^{\frac{n(n-1)}{2}}$ undirected ones. A one-shot network formation model is a probability distribution P_n among this finite (but large) number of networks.[2] We distinguish in this section between those models that consider each link as an independent event, those that impose some of the network characteristics, and those that assume a probability distribution over the characteristics, assuming a correlation in the probability of link occurrence.

7.2.1 Independent Link Formation: A Baseline

The Erdös and Rényi (1960) model is at the foundation of modern network theory. It is a simple model in which, starting from n nodes, each of the $\frac{n(n-1)}{2}$ potential undirected links is randomly formed according to an i.i.d. probability $p(n)$. This model presents remarkable *threshold properties*: that is, topological characteristics that arise almost surely if and only if $p(n)$ is above a certain threshold as n grows without bound. Informally, we say that a network property $\mathcal{A}$ is *monotonic* if, given that it holds a.s. for a certain $p(n)$, it also holds for any $\hat{p}(n) > p(n)$, that is, it continues to hold when adding links to the network. So, for example, the property of being path-connected, or having at least one cycle, are monotonic properties, while the property of there existing isolated nodes is not.[3] Then we define as $\mathcal{P}_{n,p(n)}(\mathcal{A})$ the probability that the outcome of the i.i.d. generation process on n nodes, with probability $p(n)$, satisfies property $\mathcal{A}$. The authors show that for each monotonic property $\mathcal{A}$ there exists a discrete function $t(n) \in [0, 1]$, which is also a probability depending on n, such that if $p(n)$ grows asymptotically larger than $t(n)$ then at the limit $n \to \infty$ we have $\mathcal{P}_{n,p} \to 1$, if instead $p(n)$ grows asymptotically smaller than $t(n)$ then at the limit $n \to \infty$ we have $\mathcal{P}_{n,p} \to 0$ (while there may be an interior outcome when $\lim_{n\to\infty} \frac{p(n)}{t(n)}$ is a positive constant).

To provide some examples, $t(n) = 1/n$ is the threshold above which the network has almost surely at least one cycle and a unique giant component, while $t(n) = \log(n)/n$ is the threshold above which the network is almost surely connected.[4,5] Note that a specific monotonic property that we consider can also have clear economic implications: so, for example, Elliott et al. (2014), in a model on financial contagion, show that contagion on random networks display the same threshold properties.

[2] The recent paper of Acemoglu et al. (2014) uses exactly this general definition.

[3] Path-connected means that between every pair of nodes there exists a sequence of links connecting them. A cycle is a path that terminates at the same node where it originates.

[4] A component is a maximal set of connected nodes. A giant component is a component that contains a non-trivial proportion of nodes as n grows.

[5] All this is treated extensively, with more examples, also in Bollobás (1981, 1998) and in Jackson (2008b).

7.2.2 Random Graphs of a Given Degree Distribution

A generalization on the Erdös and Rényi (1960) model is given when some of the characteristics of the network are fixed. In this sense $P_n(g)$ is positive only if g satisfies those characteristics. In principle it is possible to fix many of the characteristics, such as the clustering coefficient or the diameter.[6] As an example Bianconi (2008) studies "network ensembles with the same degree distribution, the same degree correlations and the same community structure of any given real network" (on this see also Bianconi et al. 2008, and the discussion on exponential random graph models in Section 7.2.3). There is also a category of models based on an exogenous differentiation between the nodes, which determines heterogeneous probabilities of linking (we discuss this in Section 7.5.1).

However, most of the literature, and particularly in theoretical economics, has focused on networks with a predetermined degree distribution. These are called *configuration models* (Bender and Canfield 1978; Molloy and Reed 1995, 1998; Chung and Lu 2002): given a set of nodes n, and an n-dimensional vector $\mathbf{d} \in \{1,2,\ldots,n-1\}^n$ for the degree distribution, we attribute positive probability only to those networks with that degree distribution. If the network that we want is directed, then there is a natural way to define the probabilities: we simply assign to node i a uniformly random subset of size d_i from the other nodes. This provides, with the help of combinatorics, an unambiguous way to compute P_n that will determine i.i.d. uniform expectations for each node on her neighbors in the realized networks.

If instead we attribute positive probabilities only to undirected networks then we accept only those outcomes where node i sends a link to node j if and only if node j reciprocates. Recent works like Karrer and Newman (2011) show that in this case things can be complicated: should we attribute uniform probabilities to each of the admissible network outcomes, or should we instead try to maintain uniform expectations for each node on her neighbors in the realized networks?[7]

The configuration model has found extensive application in a recent literature on Bayesian network games, especially since Galeotti et al. (2010). In this literature it is assumed that agents (nodes in the network) know the degree distribution, their own degree and, as a consequence, the probability distribution of their neighbors.[8] In this

[6] There are different ways to measure clustering, but the essential idea is to measure the frequency with which two nodes that have a common neighbor are connected. The distance between a pair of nodes is the number of links in the shortest path connecting them. A graph's diameter is the maximum distance between a pair of nodes. The degree of a node is the number of links it has. A degree distribution describes the frequencies of different degrees in the population. The chapter by Chandrasekhar defines these concepts, and others that we discuss, more formally.

[7] It is possible to show with a very simple example that, especially when n is low, the two considerations are incompatible: there is only one undirected network with $n = 4$ and $\mathbf{d} = (1,2,2,3)$, and in this network node 1, who has degree 1, can only be matched with node 4, who has degree 3.

[8] Papers that have anticipated some of the theoretical results are Jackson and Yariv (2007), Sundararajan (2007), López-Pintado (2006, 2008), and Galeotti and Vega-Redondo (2011). See also Bramoulle and Kranton in this book.

framework, one can consider the case of no correlation in the degrees of linked nodes, or it could be the case instead that, for example, because of *assortative degree correlation* (see also Section 7.3 on *assortativity*), nodes with high degree expect to meet other high degree nodes with higher probability than in the case of no correlation.[9] Then every agent chooses, before the network is realized, an action to be played in a given game on the realized network, computing in the Bayesian way the expected action profile of her neighbors.[10]

7.2.3 Incorporating Other Network Statistics

It is often the case in real world social networks that if node i has two *friends* j and k, then it is likely that also j and k are *friends* together. This property is known as *clustering*; it was first considered and defined in the literature on complex networks by Watts and Strogatz (1998) (and alternative definitions are in Newman 2003) and was then empirically measured by Ravasz and Barabási (2003) and Vázquez (2003). However, the models that we discuss below, dealing with this issue, date back to the 1970s and 1980s.

The main idea is that a network exhibits clustering if it shows a considerable amount of closed triangles (which would be unlikely to appear in the Erdös and Rényi 1960 model). So, even if the classical intuition of why clustering arises is dynamical (and we will discuss in Section 7.3 models reflecting such an intuition), one could imagine a one-shot network formation model in which triangles are assigned with some specific probability. This is what is essentially done in the class of *Exponential Random Graph Models* (ERGM, also called p^* models—see Frank and Strauss 1986; Strauss 1986; Wasserman and Pattison 1996; Anderson et al. 1999; Park and Newman 2004, 2005, see also Besag 1974 for a similar concept in the field of spatial econometrics). Since Section 13.4 in the chapter by Chandrasekhar in this handbook presents a detailed treatment of ERGMs, we refer the reader there for details, and we keep our discussion here brief.

Formally, the analyst specifies a set of simple statistics that are to be satisfied by the network g, such as the number of links $L(g)$, the number of closed triangles $T(g)$, and possibly a number of more complicated structures such as loops of a certain length or

[9] On the cases with degree correlation, see the discussion in Galeotti et al. (2010) and the general framework proposed in Feri and Pin (2015).

[10] In this literature, applications to games of vaccination against the risk of contagion are in Galeotti and Rogers (2013, 2015) and in Goyal and Vigier (2015); an application to public goods is in López-Pintado (2013); applications to peer effects and influence are in López-Pintado (2012) and Jackson and Lopez-Pintado (2013). Also, some market models are built on this framework: Nermuth et al. (2013) provide an application to a price competition between firms, in a bipartite network of firms and consumers; Fainmesser and Galeotti consider instead a game of purchasing choice in a network of consumers with positive complementary peer effects for the consumption of a specific good. Note finally that in support of this approach, Charness et al. (2014) provides experimental evidence of the results from Galeotti et al. (2010).

clusters of k nodes, obtaining, overall, ℓ structures of interest, denoted by $(\alpha_1(g), \alpha_2(g), \ldots, \alpha_\ell(g))$. Now, the probability of a network g that satisfies exactly the ℓ-dimensional vector α of natural numbers, one for each of the statistics, will be given by a probability

$$\begin{aligned} P_n(g) &\propto \exp\left(\beta_1\alpha_1(g) + \beta_2\alpha_2(g) + \cdots + \beta_\ell\alpha_\ell(g)\right) \\ &= \frac{\exp\left(\beta_1\alpha_1(g) + \beta_2\alpha_2(g) + \cdots + \beta_\ell\alpha_\ell(g)\right)}{\sum_{g' \in \mathcal{G}} \exp\left(\beta_1\alpha_1(g') + \beta_2\alpha_2(g') + \cdots + \beta_\ell\alpha_\ell(g')\right)} \\ &= \exp\left(\beta_1\alpha_1(g) + \beta_2\alpha_2(g) + \cdots + \beta_\ell\alpha_\ell(g) - \kappa\right), \end{aligned} \tag{7.1}$$

where κ is a common constant for the model.

An important issue is that calibrating an ERGM model on real data becomes computationally impossible, due to the need to estimate the value of the constant or, equivalently, from equation (7.1), of the quantity

$$\sum_{g' \in \mathcal{G}} \exp\left(\beta_1\alpha_1(g') + \beta_2\alpha_2(g') + \cdots + \beta_\ell\alpha_\ell(g')\right).$$

All available approximation techniques (see Bhamidi et al. 2008; Chatterjee and Diaconis 2013) rely on the assumption of *independent links*, which is against the spirit of the model.[11]

Chandrasekhar and Jackson (2014) show that these techniques prove themselves to be inaccurate when tested against simulations. The reason is, as they show, that if the parameters of the ERGM are all non-negative (with exclusion of the parameter attached to the number of links), then asymptotically the ERGM is indistinguishable from an Erdös and Rényi (1960) graph (or a mixture of them). For this reason, Chandrasekhar and Jackson (2014) propose *Statistical Exponential Random Graph Models*, in which, instead of computing exactly equation (7.1), one assumes sparsity of the network and approximates for each ℓ-dimensional vector α the log-likelihood $K(\alpha)$ of obtaining one of the networks satisfying α from the random network formation process. Then, the problem reduces to estimating

$$P_n\left(g|\alpha_i(g') = \alpha_i \; \forall \, i \in \{1, \ldots, \ell\}\right) = \frac{K(\alpha)\exp(\beta \cdot \alpha)}{\sum_{\alpha'} K(\alpha')\exp(\beta \cdot \alpha')}, \tag{7.2}$$

where it is now possible to compute the sum in the denominator. Chandrasekhar and Jackson (2014) successfully calibrate this method to real social network data for 75 Indian villages with a population each in the order of hundreds.

[11] He and Zheng (2013) try to solve the variational problem of Chatterjee and Diaconis (2013) with an approximation similar to mean field.

7.2.4 A Simple Approach to Capture Basic Properties

Granovetter (1973) is a pioneering paper in sociology, which after having analyzed job contact networks through interviews, proposed the *strength of weak ties* idea. Essentially, since the people we meet more often are also those that share most of our daily experience, and are exposed to correlated sources of information, it is through the people we meet occasionally (with whom we have the weak ties) that we obtain the most valuable information. Also, the weak ties make it possible to decrease the overall average distance of a social network, bridging together communities that would otherwise be disconnected, or at least very far apart.[12]

With this concept in mind, Watts and Strogatz (1998) proposed a model where, starting from a fixed lattice (a ring, or a two-dimensional grid) that establishes naturally an Euclidean distance between the nodes, some links are rewired randomly, and this rewiring makes the distances that they bridge in the underlying structure arbitrarily large. Historically, this was the first step in the field of *complex networks*, that we will discuss in Section 7.3. It is somewhat surprising that the Watts and Strogatz (1998) model has not been used more in the theoretical economic literature. One potentially interesting application that could prove useful in future research is to the idea of "searchability" of a network. For example, if one needs to find a particular input for production from a node in the network, one way to do this is to ask one's neighbors, who can ask their neighbors, and so forth. We have a less than complete understanding of which networks are more easily searchable, and so perhaps more efficient under varying conditions.[13] One could also consider labor models of job search based on this framework, especially given that Kleinberg (2000, 2004, 2006) derives analytical results to study decentralized search and information diffusion in the context of this model.

7.3 Network Formation with a Growing Population

In the late 1990s, increased availability of data allowed researchers to study the empirics of *complex networks* at a greater level of detail across a range of applications, including friendship networks, professional networks such as coauthorship, and more recently using social media and other online data such as blogs. This allowed an analysis of some key aspects of the macroscopic structure of large social networks for the first time. As it turns out, many networks share a number of common empirical features. A good

[12] This is reminiscent of the work of Burt (2004), who observes how the nodes who have these long-range weak ties are also the key nodes in the diffusion and aggregation of new ideas.

[13] We discuss in Section 7.5.1 another way to tell the weak ties story that has been adopted in some theoretical economic papers, where the underlying distance is given by a segregation of the nodes in communities, given by homophily.

survey on this is Newman (2003). We will briefly discuss three of the most important empirical regularities.

The first relates to the distribution of connectivity in the network across nodes. Specifically, starting from the analysis of the worldwide web (see, e.g., Albert et al. 1999), and then generalized to other complex networks (by Barabasi and Albert 1999), it was shown that the degree distribution exhibits heavy tails. That is, there are many nodes with relatively few connections and, perhaps more importantly, there are also many nodes that are highly connected. The observation of heavy-tailed degree distributions has sometimes been strengthened to claim that networks exhibit a power-law degree distribution, at least in the upper tail. Intuitively, if one is interested in understanding processes that operate across a network, the degree distribution is of central importance. For example, if a disease is spreading through a population, the outcome will be heavily influenced by the prevalence of high degree nodes, since they are both more likely to be infected and then more likely to spread the disease to many other nodes.

The second regularity is high clustering, as discussed in Section 7.2.3. That is, two given nodes who have a common neighbor are more likely to be connected. Networks with high clustering thus have strong correlations in link patterns, and tend to have highly interconnected subsets of individuals. The level of clustering is also very important for how certain processes operate on a network. If one is interested in how information, for example, about job openings spreads, a highly clustered network will generate a lot of repeated information.

The third regularity is positive assortativity, which is to say that highly connected nodes are more likely to have highly connected neighbors than poorly connected nodes (see, e.g., Newman 2002).

The fairly robust presence of these properties begs for a common explanation of why they should arise across a wide range of settings. A rough intuition is the following. As a network evolves, there is naturally some level of heterogeneity across nodes, arising perhaps partly through chance, and this results in differential numbers of connections across nodes. Then, as new connections are formed, it is easier or more likely to come into contact with the nodes that already have more connections. This could be because of their initial popularity (even if it is by chance), or more mechanically if there is some process of searching for nodes along existing links, then nodes with more links are more likely to be found. The key implication is that any initial differences in connectivity will tend to be exacerbated through the process of forming additional links. Thus, eventually one expects there to be a significant number of very highly connected nodes, since more connections lead to a fast rate of acquiring even more connections. At the same time, one expects there to be many poorly connected nodes, as having few connections means that new connections are harder to come by.

If there is a search process that is responsible for the formation of new connections, then that process will also tend to create clustering. One can imagine meeting the friends of current friends and connecting to them, as in Jackson and Rogers (2007), or copying the connections of one's neighbors, as in Vázquez (2003). Either of those processes directly induces patterns in which one's friends tend to be friends of each other, precisely because the common friend generated the meeting in the first place.

Such a search process operates over time, and nodes accumulate new connections gradually. Thus, a node's degree is typically correlated with its age: it is older nodes that have had more time to accumulate many connections. So if one focuses on the most highly connected nodes, they tend to be older nodes and, as such, tend to be connected to other old nodes, as those are the ones who were available to receive connections earlier on in the evolution of the network.

What this intuition suggests is that models that explicitly account for network growth, especially those based on an explicit search process, have the potential to offer a parsimonious explanation for some of the most robust empirical regularities of real networks. Our focus in this section is on understanding the essential properties of such models.

7.3.1 Uniform Link Formation: A Baseline

Jackson (2008b) presents a simple model of a growing network that we summarize briefly here, as it provides a useful benchmark and introduces some of the techniques used in analyzing such models.

Time is discrete, and at time 0 there are m nodes arranged in a fully connected network. At time t a new node enters (let us call this node also t, without ambiguity) and casts with uniform probabilities m links to some of the $t-1+m$ nodes that are currently present in the system.[14] In this model, if we aggregate out- and in-degrees, the expected degree of node i at time t is

$$m+\frac{m}{i+1}+\frac{m}{i+2}+\cdots+\frac{m}{t}\simeq m\left(1+\int_i^t 1/\tau\; d\tau\right)=m\left(1+\log\left(\frac{t}{i}\right)\right). \tag{7.3}$$

Nodes that have expected degree less than d at time t are those such that

$$m(1+log(t/i))<d \;\Rightarrow\; i>t\,e^{-\frac{d-m}{m}}. \tag{7.4}$$

From here, it is straightforward to see that the process thus generates an exponential degree distribution. In a sense, this model can be seen as the analogue of the Erdos-Renyi model for the setting of a growing network, in that all nodes form their sets of links independently and with uniform attachment probabilities. Notice, in particular, that the probability of acquiring new links from entering nodes is independent of a node's current degree. As this model is therefore lacking a "rich get richer" aspect of link accumulation, it is not surprising that it generates a thin-tailed

[14] It is essentially this rule of *attachment*, that here is uniform, that characterizes the different growing network models. For a general framework for growing networks see the model proposed by Lobel and Sadler (2015a,b).

degree distribution. Moreover, the model does not generate high clustering. Notice that as the network grows, it becomes increasingly sparse. In asking whether two given nodes are connected, conditioning on the event that they have a common friend does not change the answer, and so clustering coefficients tend to zero as the population grows. Notice, however, that assortativity does occur. The highest degree nodes tend to be the oldest nodes, who tend to be connected to each other. One takeaway is that assortativity is perhaps one of the most basic consequences of a wide range of growing network models.

Before going on, let us examine the implicit assumptions that we have used to obtain the expression (7.4), which delivers the exponential degree distribution. We have adopted a continuous time approximation to derive a closed-form expression in equation (7.3). Also, we have taken expected outcomes to compute the degree distribution in (7.4), not considering any other higher-order moment of the distribution around this first-order approximation. This simplifying approach is what is called *mean field* approximation, and we will use it also below to consider the more complicated models. The technique is extremely helpful because the underlying link formation model, while simple at the individual level, generates a highly path-dependent stochastic process that is difficult to analyze directly. Some research has derived analytical results without having to invoke a mean field approximation, but more often the literature has proceeded by testing, through simulations, that the outcomes predicted by the mean field approach provide a reasonably good description of the real process.

7.3.2 Preferential Attachment

We now turn our attention to models that do a better job of capturing the empirical regularities discussed above. In particular, the main point of this subsection is that by adjusting the link probabilities away from uniform attachment in a fairly natural way, one can readily generate a heavy-tailed degree distribution. This idea originated in Price (1965, 1976), who built on Simon (1955), but he did not see the generality of his model. The mathematics underlying the approach first appeared in Yule (1925). However, the idea re-entered the literature more recently in Barabasi and Albert (1999), at which point the power of the idea became more widely recognized.

The essential idea is to specify the linking process such that a node's probability of attracting a new link at any given moment of time is proportional to its current degree at that time. Let us denote node i's degree at time t by $k_i(t)$. Adjusting the model above to account for preferential link attachment and applying the mean field approximation, we have

$$\frac{dk_i(t)}{dt} = m\left(\frac{k_i(t)}{2mt}\right) = \frac{k_i(t)}{2t},$$

since each of the m links formed at time t reaches agent i with probability equal to agent i's degree divided by the sum of all degrees in the network. Starting from the initial condition $k_i(i) = m$, we reach the solution

$$k_i(t) = m\sqrt{t/i}.$$

Thus, the nodes i that have degree $k_i(t) \leq d$ at time t, for some given d are such that

$$m\sqrt{t/i} < d \Rightarrow i/t > \left(\frac{m}{d}\right)^2.$$

This produces a degree distribution given by $F(d) = 1 - \left(\frac{m}{d}\right)^2$, which is a power-law distribution and, most notably, has a heavy-tailed (scale-free) degree distribution.

Notice that the preferential attachment model retains the assortativity from the first growing model, and generates a power-law degree distribution through a constant rate of entry of new nodes that form links with probabilities in proportion to existing nodes' current degrees.

However, at least when the model is interpreted such that link formation is independent across nodes, preferential attachment on its own does not predict high clustering. A second observation is that, while the power-law distribution exhibits heavy tails, it is a very precise prediction, and frequently does not match well observed distributions, except possibly in the upper tail.[15] We now describe a model that is motivated by these two obervations.

7.3.3 Network Search: Combining Uniform and Preferential Meetings

We will present a simplified version of the network growth model of Jackson and Rogers (2007). The model is based on the idea that some connections are formed largely due to random or idiosyncratic forces, while other connections are found through existing links. That is, one of the main ways that nodes find each other is by meeting them through their existing neighbors.

More specifically, we shall assume that each entering node first forms a given number, m_R, of connections to other nodes found uniformly at random, exactly as in the model of Section 7.3.1. Then, from these partners, the node meets a given number m_N of additional nodes who are connected to his original partners. The idea is that these network-based meetings capture a process of meeting "friends of friends."

There is a related idea of vertex copying in the physics literature first studied by Vázquez (2003). In this approach, an entering node first finds an existing node at random, and then makes additional links to all of that node's partners. While it was not fully appreciated at the time, vertex copying produces many of the same insights as the friends of friends model.

[15] Note also that it is practically impossible for all but the largest data sets to actually exhibit a power law distribution (on this see, e.g., Virkar et al. 2014).

To study the friends of friends model, we will now account for the orientation of links, distinguishing inlinks from outlinks. Each node forms $m = m_R + m_N$ outlinks at birth (and never forms additional outlinks). Inlinks, on the other hand, are accumulated from entering nodes over time. Applying the mean-field analysis as before, the growth of node i's indegree, now denoted $k_i(t)$, is governed by

$$\frac{dk_i(t)}{dt} = \frac{m_R}{t} + m_N\left(\frac{k_i(t)}{tm}\right), \tag{7.5}$$

where the first term accounts for the possibility to be found at random, and the second term accounts for the possibility that one of i's in-neighbors is found at random, and then i is found though the network. Using the initial condition that $k_i(i) = 0$, one reaches an indegree distribution described by

$$F(d) = 1 - \left(\frac{rm}{d+rm}\right)^{1+r},$$

where $r = m_R/m_N$ is the ratio of the number of links formed at random versus through the network.

This degree distribution is interesting, since it predicts a power-law in the upper tail, but a somewhat thinner lower tail, with a shape depending on the value of the parameters. To see this, note that for large degree d, that is, the upper tail, the distribution is approximately $\bar{F}(d) = 1 - (rm/d)^{1+r}$. On the other hand, for lower degrees, the distribution is not well approximated by a power-law.

As it turns out, more careful empirical analysis often shows that power-law-like degree distributions are mostly found only for the highest degree nodes, and so this class of degree distributions is fairly well-suited to capturing this feature. Notice finally that the friends of friends meetings (or similarly the vertex copying model) directly generates high clustering: whenever two nodes meet through the network, it is precisely because they have a friend in common.

7.4 Dynamic Network Formation

Here we consider a class of dynamic models where the number of nodes is not necessarily growing, and may even be constant. Most of the modeling with this approach comes from the economic literature, because of necessities that we will go through case by case. Such models have been analyzed only very recently in the other disciplines.[16]

[16] The first work from economists considered situations in which not only the network was endogenous, but also some action played by nodes in the network: first papers are Jackson and Watts (2002) and Goyal and Vega-Redondo (2005). This approach was then broadened in more complex

7.4.1 Search and Matching

One way to model network formation and, if desired, at the same time behavior mediated by that network is to base meetings on a search model. This idea has its roots in the seminal work of Diamond (1982a,b), where the focus was on labor markets rather than networks, per se, but the general methodology is well suited to a more network-centric analysis.

This approach can incorporate both random and strategic network formation elements. This is probably why it appears only in the economic literature. We focus here primarily on the random components; see Vannetelbosch and Mauleon in this book for a specific treatment of strategic factors.

Currarini et al. (2009) apply such an approach to studying link formation in the presence of types, focusing on the role that type-based preferences play in equilibrium. Let us summarize how the model works. Agents enter the system over time and search for partners, whom they meet randomly. A given agent stays active in the system until the marginal benefit of additional search falls below its marginal cost. This random meeting process may exhibit biases, that we discuss in Section 7.5.

Immorlica et al. (2013) adopt a new approach also based on a matching and search framework. In their model, agents are born into the system over time and initiate a given number of connections to others found uniformly at random. At each date, agents choose to cooperate or defect and play a Prisoners' Dilemma with their current partners. Then, agents can sever any set of connections that they wish. Some agents are randomly chosen to die, and then at the next date, any agent with unused connections re-initiates new meetings in the subsequent period. Immorlica et al. (2013) study the extent to which cooperation can be supported in equilibrium. Under favorable parameters, full cooperation can be supported. Otherwise, if there is any cooperation, it takes the form of a specific fraction of the population cooperating, while others defect, so that the two behaviors co-exist. Even though, in their model, agents are anonymous, and are not threatened by a contagion of defecting behavior, cooperation can be supported because of a network effect: in equilibrium, cooperating agents are able to accumulate many profitable relationships, whereas defectors live in relative isolation, losing all of their links after every period.

Pin and Rogers (2015) apply a similar approach to studying immigration policy. In their model, agents enter either as immigrants or as the offspring of citizens. The main question is to understand how a policy of punishing defecting immigrants impacts equilibrium outcomes, when individuals are motivated by the outcomes of playing the

settings in papers outside of economics in, for example, Marsili et al. (2004), Ehrhardt et al. (2006a,b), Holme and Newman (2006), Gross and Blasius (2008), and König and Tessone (2011). These results are recently coming back into the economic literature with Fosco et al. (2010) and König et al. (2014). With a more applied view, Lee and Fong (2013) have a market model with a fixed number of farsighted agents, while Goldsmith-Pinkham and Imbens (2013) use a two-period model with endogenous network and actions for empirical estimation. Models in which agents make choices in an evolving network are treated extensively by Vega-Redondo in this book.

Prisoners' Dilemma in a society where connections are made and lost according to the model of Immorlica et al. (2013). The main finding is that, while increasing punishment increases the level of cooperation in society, it also has the consequence of increasing the rate of defection among citizens. The intuition is that there is a natural level of cooperation in society dictated by the model's parameters. When the cost of defecting is increased for immigrants through punishment, there will be a partial substitution into defecting behavior by citizens.

7.4.2 Structural Models

One of the most useful notions of equilibrium for strategic network formation games is *pairwise stability* (see Vannetelbosch and Mauleon in this book): in a nutshell, a network is pairwise stable if, (*i*) when two nodes are linked together, neither of them prefer to delete the link; and (*ii*) if they are not linked together, it is not the case that both of them prefer to create the link. One interpretation of this notion is through a matching pool mechanism: in society, people are randomly and sequentially matched in couples, and every couple can reconsider their link, and create it with mutual agreement, if absent, or delete it if present, if at least one of them wishes to do so. This implies a Markov process between all possible network configurations, where pairwise stable equilibria are absorbing states (as discussed in Jackson and Watts 2002).

One can introduce some noise into this system, in the form of possible errors made by the agents, or in the form of shocks on their utility functions, so that the Markov process becomes ergodic, that is, with no more absorbing states. In this case it is possible to compute ergodic probabilities of each network, as if they were the probabilities of a one-shot network formation model, as those discussed in Section 7.2. If errors are small, pairwise stable networks would still be likely outcomes, but every possible network will have some positive probability to be in place.

There are at least three recent papers that adopt this approach as a structural model for empirical estimation: Goldsmith-Pinkham and Imbens (2013), Mele (2014), and Graham (2014). They will be covered also by Chandrasekhar in this book, and as they are related to homophily we will come back to their results in Section 7.5. Here we just present the main idea, abstracting from the specific models.

Suppose that there is an undirected[17] network g_t between n nodes at every moment t in discrete time, and that at time t a matching technology provides positive probability to any pair (i,j) of nodes (but just one pair per period) to consider their mutual link. Node i gets a utility $U_i(g_t \cup \{ij\})$ from the network with that link in place (which could be g_t itself, if originally $ij \in g_t$), and a utility $U_i(g_t \backslash \{ij\})$ from the network without that link in place (which could also be g_t, in the complementary case). We call $\Delta_{t,j} U_i = U_i(g_t \cup \{ij\}) - U_i(g_t \backslash \{ij\})$, and the same holds for node j. Then, $g_{t+1} = g_t \cup \{ij\}$ if and

[17] Mele (2014) actually considers a directed network.

only if both

$$\Delta_{t,j}U_i + \eta_i > 0 \quad \text{and} \quad \Delta_{t,i}U_j + \eta_j > 0,$$

where η_i and η_j are i.i.d. random shocks. Otherwise, $g_{t+1} = g_t \backslash \{ij\}$.

This Markov process is ergodic, and when the shocks follow a Weibull distribution (as in the assumptions of the standard logit model) the problem of estimating ergodic probabilities is very similar to the one discussed in Section 7.2.3 for ERGM models. In this case, only for some limiting cases, or when a *potential function* exists for the utilities, is it possible to numerically compute these probabilities.

7.5 Homophily

When observing a social network from the real world, and even considering our own perceptions of social networks surrounding us, we always find evidence that similar people are more likely to be connected together than if they were different along some dimension: this applies to cultural classification, age, gender, and so on.

Formally, suppose that the population of a society is heterogeneous, which we model by proposing a partition into a finite number of types, which could denote a classification like age groups, or self-reported racial groups. Each type θ has a frequency in the population of $w_\theta \in [0,1]$, so that $\sum_{\theta'} w_{\theta'} = 1$. Now we check the friends of some agent i, where i belongs to type θ. In this set of friends, those that are also of type θ are a proportion H_i of the neighborhood of i that we call the *homophily* of agent i. In the same way we can define the homophily of a whole group θ considering the aggregate measure

$$H_\theta \equiv \frac{\text{average \# of friends of } \theta\text{'s members, which are also in } \theta}{\text{average \# of friends of } \theta\text{'s members}} .$$

The problem is that H_θ is not a relative measure; therefore we normalize it with respect to w_θ through *imbreeding homophiliy*:[18]

$$h_\theta \equiv \frac{H_\theta - w_\theta}{1 - w_\theta} ,$$

which is positive only when there is positive bias in favor of own type, and reaches its maximum at 1 when all members of θ have only friends from θ.

So, we say that a type θ is homophilous if it has $H_\theta > w_\theta$, or equivalently $h_\theta > 0$, and we can also compare more or less homophilous types. This, however, does not help us in understanding the causes of this observed segregation: is it due to the choices of the individuals in θ, to the choices of the rest of the population, or to biases in the meeting opportunities, for example, because a type is correlated with the occupation

[18] This is also known as the Coleman (1958) index.

of individuals? Also, the biases in meetings could result as an effect: as Montgomery (1991, 1992) has shown, if job places are allocated through informal contacts, as in the Granovetter (1973) weak ties story, then homophilous behavior results in occupational segregation. Note that we consider the types as exogenous, and we refer to homophily as the tendency of similar people to connect together. However, types could be endogenous, and diffusion processes (see Lamberson, Chapter 18 in this book) could cause people that are connected together to become similar. When both homophily and diffusion are present, the identification of causalities for observed segregation becomes an extremely hard task, as discussed in Chandrasekhar in this book.

The classical economic theory was based on the assumption of a *representative agent*, abstracting from heterogeneity. However, a preliminary attempt to explain the observed homophily in residential segregation with an elegant and general model is in Schelling (1971).[19] Separately, from the pioneering work of Becker (1973), matching theory has focused attention on *assortative matching*, thereby emphasizing one of the sources of observed homophily.[20] Below we discuss the recent contributions based on homophily and the implications of homophily in various economic models. Then we end with a discussion of the relationships between the models, which are also based on some randomness in the network formation process, and the empirical estimation of real network data sets.[21]

7.5.1 Static Link Formation

The first implication of homophily is that, as soon as there is some heterogeneity in the characteristics of the nodes, not every pair of nodes will have the same probability to be linked together. The most intuitive way to apply this is generalizing the Erdös and Rényi (1960) model in a way that assigns to each pair (i,j) a linking probability $p(i,j)$ that depends on the characteristics of the nodes. In a simplified setting, the characteristic of a node is just a type, as in the definition of homophily above, the linking probability depends on the types of the two nodes, and it is typically higher when the two types coincide. An intuition that comes out is that a few rare links will connect parts of the population that are different and otherwise largely disconnected, and this also reminds one of the concepts of *weak ties* and *structural holes* that we introduced in Section 7.2.4

[19] Schelling (1971) is based on a grid structure with cells that resemble some of the topological properties of networks, referring to neighbors as those cells that are adjacent in the grid. Agents have a preference for having neighbors of their same type, and can jump to another random location (in a network setting, they would reshuffle all their links) if unsatisfied. The main result is that even mild preferences for own type can result in dramatic segregation.

[20] Note that assortative matching, which refers to an outcome in which types (skill, beauty, productivity, etc.) of paired agents are positively correlated, is quite distinct from the network-based notion of assortativity, which refers instead to a correlation between the degrees of linked nodes.

[21] Jackson and Rogers (2005), De Martí and Zenou (2013), and Boucher (2015) consider network formation games that are deterministic. We do not cover them in this chapter.

(there the underlying distance between nodes was given by an exogenous Euclidean topology, while here it is given by the exogenous characterization). Jackson (2008a,b) surveys properties of such models, and in Section 7.5.3 we discuss Golub and Jackson (2012a), who use a benchmark case for their analysis.

In general, one could think that the characteristics of an individual are not just a single label, but that it is better described by a vector of (potentially continuous) variables: age, education, skills, political opinion, rate of self-identification with a particular culture, and so on. This representation may not be completely exogenous (as discussed above), but one can still treat it as such in the context of a model. In this way every node i is characterized a vector $a_i \in \mathbb{R}^m$, where m is the number of characteristics that we consider, and we can think of i as a point in an m-dimensional space that can be endowed with many possible metrics. In this setting, Gauer and Landwehr (2014) and Iijima and Kamada (2014) are two recent theoretical papers that assume that the probability of linking $p(i,j)$, of two nodes i and j, depends in a monotonic decreasing way on their distance in this m-dimensional space. The two approaches differ in the distance metric they consider: Gauer and Landwehr (2014) consider the Euclidean distance when $m = 1$, while Iijima and Kamada (2014) adopt the *k'th norms*, in which the distance between two points in the type space is the k'th smallest distance among the m dimension-wise distances between them. The latter choice allows the expected realized network to obtain a higher level of clustering and maintain a small diameter.

7.5.2 Linking Based on Friendship

We will discuss two friendship models that allow for an interesting analysis of homophily. The first is based on a matching model and the second is based on a growing network formation model.

Let us return first to the analysis of Currarini et al. (2009, 2010), which we anticipated in Section 7.4.1 as a model of search and matching. In their model each individual is endowed with a type that comes from a discrete set $\theta \in \Theta$ (we continue to denote the proportion of type-θ agents in the population by w_θ), and the combination of random meeting and preference-based biases over types have some interesting consequences. If a type has a high representation in the matching pool, and if members of that type place high value on meeting others of the same type, then the expected benefit of search will remain high for this type, and so they will have a greater number of total connections. Building on this intuition, and assuming types prefer meeting others of the same type, if a given type constitutes a large share of the population, then they will optimally search more, and so their representation in the matching pool will be even higher than w_θ, leading to an even more attractive calculation for their continued search. Thus, the two sources of biases are inherently intertwined and, in this way, can reinforce each other.

Their model allows one to understand a number of empirical regularities of segregation patterns in friendship networks as a consequence of homophily-related biases. The first empirical regularity is that individuals in larger groups tend to form more total connections. Formally, the equilibrium number of connections should be increasing in w_θ. It is important to note that this prediction is not universally born out in the literature. In the present model, a necessary condition for agents of different types to prefer different numbers of total connections is that they have type-based preferences. Thus, in a given application, if all biases are opportunity-based, rather than preference-based, one would not expect such a pattern in the data.

The second empirical regularity relates to *inbreeding homophily* h_θ, which captures the excess proportion of same-type friends over the type's representation in the population, normalized to have a maximum value of unity. The empirical regularity is that h_θ is inverse-U shaped, showing the greatest levels of homophily for groups of intermediate sizes, but generally positive even for the smallest and largest groups. In the model, this prediction relies on a combination of preference-based and opportunity-based same-type biases. The opportunity-based bias is important because it allows all groups, including the small groups, to meet same-types at relatively high rates.[22]

We now turn attention to the analysis of Bramoullé et al. (2012), who extend the model from Jackson and Rogers (2007) that we discussed in Section 7.3.3 to incorporate types and homophily-based biases. Agents are again assigned a type from a finite set. Random meetings are biased in the following way. Let $p(\theta,\theta')$ denote the probability that an entering agent of type θ meets an agent of type θ' in the random meeting process. In general, $p(\theta,\theta')$ can differ arbitrarily from the type frequencies, allowing for biases. Then, conditional on the realization of p, an agent from the appropriate type is drawn uniformly at random.[23] Letting $p(\theta)$ be the probability of being born of type θ (the w_θ discussed above), we let $P_j^t(\theta_t,\theta_j)$ denote the probability that a node born in period j of type θ_j receives a link from a node of type θ_t born at time $t > j$, the following expression is a mean–field approximation of the overall linking probability, that generalizes equation (7.5):

$$P_j^{t+1}(\theta,\theta_j) = m_R\frac{p(\theta)p(\theta,\theta_j)}{tp(\theta_j)} + m_N\sum_{\theta'\in\Theta} p(\theta)p(\theta,\theta')\frac{\sum_{\lambda=j}^{t} P_j^{\lambda}(\theta',\theta_j)}{tp(\theta')}\frac{1}{m}, \tag{7.6}$$

where m, m_R and m_n are those described in Section 7.3.3. This multi–dimensional set of $|\Theta| \times |\Theta|$ differential equations can be expressed in a more compact way as

$$\mathbf{P}_j^{t+1} = \frac{m_R}{t}\mathbf{B} + \frac{m_S}{mt}\mathbf{B}\sum_{\lambda=j}^{t}\mathbf{P}_j^{\lambda}, \tag{7.7}$$

[22] While Currarini et al. (2009) assume exogenous meeting biases, a micro–foundation for this is provided by Currarini and Vega-Redondo (2013). Currarini and Vega-Redondo (2013) assume that agents have a trade–off in making new connections between homophily and diversity. This results in different searching strategies which in turn endogenously biases their matching pools.

[23] The search-based meetings are allowed to be similarly biased, although much of the analysis assumes neutrality for these meetings.

where

$$B(\theta,\theta') \equiv p(\theta)\frac{p(\theta,\theta')}{p(\theta')}. \tag{7.8}$$

Following the mean field approach, we can transform equation (7.7) into a continuous differential equation in matrix form, which has a unique solution.

This model generates a number of testable predictions. One finding of interest is that, for quite different reasons, inbreeding homophily exhibits the same inverse-U shape as a function of group size as in the Currarini et al. (2009) analysis. A main object of interest in the Bramoullé et al. (2012) analysis concerns *integration*, that is, the tendency of an agent's set of friends to become less and less biased over time. In their linking process, the friends of friends meetings create a channel through which an agent meets a less-biased set of nodes compared to direct search of her immediate neighborhood. Since as an agent ages, she tends to be found more and more through the friends of friends channel, compared to through random search, her neighborhood begins to reflect a diminished bias over time. In particular, when search-based meetings are unbiased, the authors show that in the long run, every agent's local neighborhood converges to the type frequencies in the population. Nonetheless, the network at large can still be heavily biased, as the neighborhoods of younger agents will generally be same-type biased.

Using very different contexts, Bramoullé et al. (2012) and Bramoullé and Rogers (2010) present empirical evidence supporting these theoretical predicitons. Bramoullé et al. (2012) study citations between published papers in physics, where subfields (which are well-defined in physics) correspond to the type of the paper. Bramoullé and Rogers (2010) use the Add Health data, where gender defines the type of an individual and connections are self-reported friendships (see also note 25 for more information on the data). In both cases, nodes with larger neighborhoods have more diverse connections, in the sense of a type distribution that is less homophilic and closer to the population frequencies.

These findings are actually quite reminiscent of the work of Chaney (2014) in the context of international trade. Indeed, the main mechanism at work is similar in the two cases. In the setup of Chaney (2014), the friends of friends channel translates into the finding that the trading partners of one's trading partners are located further away in space, and he finds an analogue to long-run integration. We refer the reader to Chaney in this book for more discussion.

One takeaway from these papers is that a combination of type-based biases, along with their equilibrium consequences, is quite capable of capturing a broad range of empirically relevant homophilous patterns. While the two models are quite different from each other, both of them impose a combination of biases that have interesting interactions that generate the main results. Whether a search model or a network growth model is more appropriate for a given application, or whether one wants to explicitly model preferences, depends on the details and goals of a given study.

7.5.3 Implications of Homophily

While the previous subsections have discussed some of the ways in which homophily can arise, we now turn our attention to understanding some of the various consequences of homophilous connection patterns. That is: what is the importance of homophily in social networks?

In many important applications, the interactions between agents, the structure of which is captured through the network, are influenced by the types of the agents. Agents with similar views or backgrounds may communicate at lower cost, for example. Or it could be that agents with similar opinions are more likely to talk to each other, and this will clearly have an influence on how opinions evolve over time. The spread of epidemics will be strongly influenced by how segregated different groups in the population are from each other.

Taking up the first idea, it is possible to study opinion evolution under simple rules of updating.[24] What will the presence of homophily imply about eventual opinions in a population? Golub and Jackson (2012a,b) study these questions and derive some interesting results. Consider a process in which agents update their opinions by taking weighted averages of the opinions of their neighbors. Such a model was first proposed by French (1956) and DeGroot (1974), and is a natural way to capture a naïve learning dynamic or a myopic best reply under a simple utility specification. The key insight here is that homophily is tied to segregation, in that highly homophilous societies tend to have groups that live in relative isolation from each other, due to the relative sparsity of connections between groups. That segregation, in turn, has important implications for opinion dynamics. Under general conditions (essentially that the network is path-connected and aperiodic), opinions converge to a consensus in the long run. But an important object of interest is the speed of that convergence. Golub and Jackson (2012a,b) demonstrate that more homophilous societies converge much more slowly.

Coming now to the idea of contagion in networks, Golub and Jackson (2012a,b) show also that on an exogenous network homophily does not play a role in the contagion process, as long as the diameter of a network is kept fixed. However, in an economic process of contagion in which agents react endogenously to the risk of being infected, Galeotti and Rogers (2013) present a simple model that shows how the level of homophily impacts the contagion of an infectious disease. In their work, there are two populations who interact with each other, with a parameter β that controls the probability that any given interaction is same-type. Thus, for $\beta > 1/2$, the connections are homophilous, while for $\beta < 1/2$ most connections are across groups as in, for example, the case of sexual contacts and gender. Agents strategically choose whether or not to immunize at a cost. Agents who remain vulnerable anticipate outcomes according to the steady-state of a standard SIS diffusion process (see, e.g., Bailey et al. 1975). Under

[24] See Golub and Sadler in this book for a deeper discussion and for the distinction between Bayesian and naïve learning models.

these dynamics, agents become infected through their connections to other infected agents, and they recover at a given exogenous rate.

In equilibrium, agents balance the cost of immunization against the cost of time spent infected. To illustrate the importance of homophily, consider starting from a world in which the two groups are completely separate, that is, $\beta = 1$ and there are no cross-group connections. Suppose that the cost of immunization is slightly lower (relative to the benefits of being healthy) in, say, group A, so that group A immunizes at a slightly higher rate. Now consider the effects of gradually mixing the two groups, so that the world is homophilous, but not perfectly so. The essence of the result is that equilibrium outcomes in the two groups diverge, so that the consequence of mild homophily is to magnify the slight underlying difference between the groups. As individuals from group A come into contact with those in group B, they are exposed to slightly more infection, since group B has lower immunization. This causes group A to adjust upwards their already higher immunization rate. Conversely, group B benefits from the lower exposure due to interacting with group A, so that they have less incentive to immunize. As the two groups interact more and more, eventually group B stops immunizing all together.

One remark is that this is a very simple model, relying on a basic diffusion process and a simple homophily structure. A promising avenue of further research is to explore various directions for modeling strategic decisions that depend on type-dependent contact patterns.

Another domain where network structure and, in particular, homophily will have important effects is in market settings including, for example, product adoption. In a world in which consumers communicate with each other about product experiences, there are a range of pricing and competition questions to be addressed, so as to understand differences in such environments compared to cases in which consumers have complete information ex ante. Two papers that study demand and pricing effects are Campbell (2013) and Chuhay (2014). We think this is an area that will gain increasing attention.

7.5.4 Empirical Estimation of Homophilous Biases

As pointed out in the beginning of this section, even when we are sure that the classification into types is exogeneous, it is very difficult to empirically disentangle the sources of observed segregation. Schelling (1971) already demonstrated that this can depend on the choices of one or another group, or possibly many groups, and on the meeting constraints that people from different groups have, because, for example, some variable, for example, workplace or hobbies is correlated with type, and this bias in opportunities can itself recursively be caused by all of the former factors. From the point of view of the econometrician, homophilous biases can be included in one–shot homophilous models (Section 7.5.1), matching pool models (Section 7.4.1), structural models (Section 7.4.2), or in some of the mean field results from growing network models (Section 7.3). In all cases the problem is to estimate the level of this bias for each group.

It is possible to analyze data where simultaneously more similar environments are observed, for example, many high schools where different ethnic groups or races are differently represented, as in the Add Health data set.[25] In this case the units of observation for the analysis are the average behavior of members of one type in each school, and it is clear that many environments are needed to obtain statistical significance. Under these conditions, matching pool models are particularly suited, as they allow the analyst to also specify a bias in the meeting opportunities, as has been done by Currarini et al. (2010), who find a great heterogeneity both in opportunities and in choices between the different ethnic groups.[26] Both sources of bias are significant in the data, and they also differ significantly across races. For example, Asians and Blacks have a much stronger chance-based bias than Whites. Estimated preference biases range from valuing inter-race connections at 90% of the level of intra-race connections for Asians, to 55% for Blacks.

When instead the researcher has a single snapshot of the network, or data from only a limited number of points in time (as panel data on the same network, or only a few separated environments), then the neighborhood of each single node must be used as a unit of observation. The most natural theoretical framework to test seems to be one-shot network formation models, and the benchmark models, where the probability of linking $p(i,j)$ depends only on the types of i and j, have been applied to empirical estimation by Newman (2004) and Copic et al. (2009) (in this case they are sometimes called *block models*). However, these applications have proved themselves not to be robust, and require Monte Carlo methods to estimate the confidence levels. Another approach is to use structural models, as in Mele (2014), Goldsmith-Pinkham and Imbens (2013), and Graham (2014). This approach provides more stable outcomes, but has the drawback of being computationally difficult, so that a modern computer is able to process only a network whose size is on the order of some hundreds of nodes.[27]

7.6 Conclusion

Thus far, the most successful models of network formation are based primarily on random events. Analyses without an explicit random component produce severely

[25] The National Longitudinal Survey of Adolescent Health (commonly referred to as Add Health) is a program project that started in 1994 collecting data from a representative sample of almost 100 U.S. high schools. It was designed by J. Richard Udry, Peter S. Bearman, and Kathleen Mullan Harris.

[26] Another possibility is to specify an underlying random network of opportunities on which agents establish links with some preferential bias, as in Franz et al. (2010).

[27] Here we have highlighted only the models underlying the empirical approach on detecting homophily in social networks. More details, and in particular the description of the methods, are in Chandrasekhar in this book.

stylized network structures. As a result, our current theoretical toolkit for studying social networks in economics has benefited greatly from graph theory and combinatorial techniques in mathematics. This approach has allowed us to learn a great deal about the connections between the micro-level stochastic processes of link formation and the large-scale structural characteristics they produce in a network. Leveraging observational and empirical work on the structure of social networks to identify and calibrate models that best fit macroscopic network features, we can then make inferences about the linking processes by which those networks form.

The tradeoff inherent in this approach is that one need not model explicitly behavior as arising from incentives or objectives of the agents involved. This omission largely precludes addressing welfare and policy questions. For this reason, much of the random network formation literature has come from disciplines outside economics. We think an area of particular interest to economists, and to those seeking to understand social networks, will be bringing together what is known about random network formation with analyses that bring economic behavior to the forefront. For example, van der Leij and Kovarik (2012) show how the "friends of friends" meeting process can originate from optimizing behavior of the agents under particular constraints.

More generally, it is important to have a more complete understanding of the decisions that go into meeting and linking with other agents. Some aspects of these events are random even from the perspective of the agents themselves. Other events may be dictated by particular circumstances that are essentially deterministic to the agents, but are based on variables that are not observed by the analysts, and so appear to be random. Thus, a complete picture must be informed by (i) an understanding of the properties of random formation processes, (ii) a set of theoretical frameworks with which to model the incentives of agents and understand their optimal behavior, and (iii) empirical work that identifies the relevant characteristics of agents and their environments, to best understand their decisions, and (iv) finally structural work to estimate the resulting models.

Acknowledgments

We gratefully acknowledge help from Ben Golub, Dunia López-Pintado, Angelo Mele, Yves Zenou, and participants at the 10th International Winter School on Inequality and Social Welfare Theory in Canazei.

References

Acemoglu, D., A. Malekian, and A. Ozdaglar (2014). "Network security and contagion." Mimeo.

Albert, R., H. Jeong, and A.-L. Barabási (1999). "Internet: Diameter of the world-wide web." *Nature* 401(6749), 130–131.

Anderson, C. J., S. Wasserman, and B. Crouch (1999). "A p^* primer: Logit models for social networks." *Social Networks* 21(1), 37–66.

Angrist, J. D. (2014). "The perils of peer effects." *Labour Economics* 30, 98–108.

Bailey, N. T. et al. (1975). *The Mathematical Theory of Infectious Diseases and its Applications*. Charles Griffin & Company Ltd, 5a Crendon Street, High Wycombe, Bucks HP13 6LE.

Barabasi, A.-L. and R. Albert (1999). "Emergence of scaling in random networks". *Science* 286, 509–512.

Becker, G. S. (1973). "A theory of marriage: Part i." *The Journal of Political Economy* 81(4), 813–846.

Bender, E. A. and E. R. Canfield (1978). "The asymptotic number of labeled graphs with given degree sequences." *Journal of Combinatorial Theory, Series A* 24(3), 296–307.

Besag, J. (1974). "Spatial interaction and the statistical analysis of lattice systems." *Journal of the Royal Statistical Society. Series B (Methodological)*, 192–236.

Bhamidi, S., G. Bresler, and A. Sly (2008). "Mixing time of exponential random graphs." In *Foundations of Computer Science, 2008. FOCS'08. 49th Annual IEEE Symposium*, pp. 803–812. IEEE.

Bianconi, G. (2008). "The entropy of randomized network ensembles." *EPL (Europhysics Letters)* 81(2), 28005.

Bianconi, G., A. C. Coolen, and C. J. P. Vicente (2008). "Entropies of complex networks with hierarchically constrained topologies." *Physical Review E* 78(1), 016114.

Bollobás, B. (1981). "The diameter of random graphs." *Transactions of the American Mathematical Society* 267(1), 41–52.

Bollobás, B. (1998). *Random Graphs*. Springer.

Boucher, V. (2015). "Structural homophily." *International Economic Review* 56(1), 235–264.

Bramoullé, Y., S. Currarini, M. O. Jackson, P. Pin, and B. W. Rogers (2012). "Homophily and long-run integration in social networks." *Journal of Economic Theory* 147(5), 1754–1786.

Bramoulle, Y. and R. Kranton (2015). "Network games." *This book: Chapter 8*.

Bramoullé, Y. and B. W. Rogers (2010). "Diversity and popularity in social networks." Mimeo.

Burt, R. S. (2004). "Structural holes and good ideas." *American Journal of Sociology* 110(2), 349–399.

Campbell, A. (2013). "Word-of-mouth communication and percolation in social networks." *The American Economic Review* 103(6), 2466–2498.

Chandrasekhar, A. (2015). "Econometrics of network formation." *This book: Chapter 13*.

Chandrasekhar, A. and M. O. Jackson (2014). "Tractable and consistent random graph models." Mimeo.

Chaney, T. (2014). "The network structure of international trade." *American Economic Review* 104(11), 3600–3634.

Chaney, T. (2015). "Networks in international trade." *This book: Chapter 28*.

Charness, G., F. Feri, M. A. Meléndez-Jiménez, and M. Sutter (2014). "Experimental games on networks: Underpinnings of behavior and equilibrium selection." *Econometrica* 82(5), 1615–1670.

Chatterjee, S. and P. Diaconis (2013). "Estimating and understanding exponential random graph models." *The Annals of Statistics* 41(5), 2428–2461.

Chuhay, R. (2014). "Strategic diffusion of information in social networks with homophily." Working paper.

Chung, F. and L. Lu (2002). "The average distances in random graphs with given expected degrees." *Proceedings of the National Academy of Sciences* 99(25), 15879–15882.

Coleman, J. S. (1958). "Relational analysis: The study of social organizations with survey methods." *Human Organization* 17(4), 28–36.

Copic, J., M. O. Jackson, and A. Kirman (2009). "Identifying community structures from network data via maximum likelihood methods." *The BE Journal of Theoretical Economics* 9(1). http://www.degruyter.com/view/j/bejte.2009.9.1/bejte.2009.9.1.1523/bejte.2009.9.1.1523.xml?format=INT.

Currarini, S., M. O. Jackson, and P. Pin (2009). "An economic model of friendship: Homophily, minorities, and segregation." *Econometrica* 77(4), 1003–1045.

Currarini, S., M. O. Jackson, and P. Pin (2010). "Identifying the roles of race-based choice and chance in high school friendship network formation." *Proceedings of the National Academy of Sciences* 107(11), 4857–4861.

Currarini, S. and F. Vega-Redondo (2013). "A simple model of homophily in social networks." Mimeo.

Dalmazzo, A., P. Pin, and D. Scalise (2014). "Communities and social inefficiency with heterogeneous groups." *Journal of Economic Dynamics and Control,* forthcoming.

De Martí, J. and Y. Zenou (2013). "Ethnic identity and social distance in friendship formation." Mimeo.

DeGroot, M. H. (1974). "Reaching a consensus." *Journal of the American Statistical Association* 69(345), 118–121.

Diamond, P. A. (1982a). "Aggregate demand management in search equilibrium." *The Journal of Political Economy* 90, 881–894.

Diamond, P. A. (1982b). "Wage determination and efficiency in search equilibrium." *The Review of Economic Studies* 49(2), 217–227.

Ehrhardt, G., M. Marsili, and F. Vega-Redondo (2006a). "Diffusion and growth in an evolving network." *International Journal of Game Theory* 34(3), 383–397.

Ehrhardt, G. C., M. Marsili, and F. Vega-Redondo (2006b). "Phenomenological models of socioeconomic network dynamics." *Physical Review E* 74(3), 036106.

Elliott, M., B. Golub, and M. O. Jackson (2014). "Financial networks and contagion." *American Economic Review* 104(10), 3115–3153.

Erdös, P. and A. Rényi (1960). "On the evolution of random graphs." *Publications of the Mathematical Institute of the Hungarian Academy of Science* 5, 17–61.

Fainmesser, I. P. and A. Galeotti (2015). "Pricing network effects." *Review of Economic Studies.*

Feri, F. and P. Pin (2015). "The effect of externalities aggregation on network games outcomes." Mimeo.

Fosco, C., F. Vega-Redondo, and M. Marsili (2010). "Peer effects and peer avoidance: The diffusion of behavior in coevolving networks." *Journal of the European Economic Association* 8(1), 169–202.

Frank, O. and D. Strauss (1986). "Markov graphs." *Journal of the American Statistical Association* 81(395), 832–842.

Franz, S., M. Marsili, and P. Pin (2010). "Observed choices and underlying opportunities." *Science and Culture* 76(9–10), 471–476.

French Jr, J. R. (1956). "A formal theory of social power." *Psychological Review* 63(3), 181.

Galeotti, A., S. Goyal, M. O. Jackson, F. Vega-Redondo, and L. Yariv (2010). "Network games." *The Review of Economic Studies* 77(1), 218–244.

Galeotti, A. and B. W. Rogers (2013). "Strategic immunization and group structure." *American Economic Journal: Microeconomics* 5(2), 1–32.

Galeotti, A. and B. W. Rogers (2015). "Diffusion and protection across a random graph." *Network Science* 3(3) (September 2015), 361–376.

Galeotti, A. and F. Vega-Redondo (2011). "Complex networks and local externalities: A strategic approach." *International Journal of Economic Theory* 7(1), 77–92.

Gauer, F. and J. Landwehr (2014). "Continuous homophily and clustering in random networks." *Institute of Mathematical Economics Working Paper* (515).

Goldsmith-Pinkham, P. and G. W. Imbens (2013). "Social networks and the identification of peer effects." *Journal of Business & Economic Statistics* 31(3), 253–264.

Golub, B. and M. O. Jackson (2012a). "How homophily affects the speed of learning and best-response dynamics." *The Quarterly Journal of Economics* 127(3), 1287–1338.

Golub, B. and M. O. Jackson (2012b). "Network structure and the speed of learning: Measuring homophily based on its consequences." *Annales d'Economie et de Statistique* 107/108. http://www.jstor.org/stable/23646571.

Golub, B. and E. Sadler (2015). "Learning in social networks." *This book: Chapter 19.*

Goyal, S. and F. Vega-Redondo (2005). "Network formation and social coordination." *Games and Economic Behavior* 50(2), 178–207.

Goyal, S. and A. Vigier (2015). "Interaction, protection and epidemics." *Journal of Public Economics* 125, 64–69.

Graham, B. S. (2014). "An empirical model of network formation: Detecting homophily when agents are heterogenous." Mimeo.

Granovetter, M. S. (1973). "The strength of weak ties." *American Journal of Sociology* 78, 1360–1380. http://www.jstor.org/stable/2776392?seq=1#page_scan_tab_contents

Gross, T. and B. Blasius (2008). "Adaptive coevolutionary networks: A review." *Journal of The Royal Society Interface* 5(20), 259–271.

He, R. and T. Zheng (2013). "Estimation of exponential random graph models for large social networks via graph limits." In *Proceedings of the 2013 IEEE/ACM International Conference on Advances in Social Networks Analysis and Mining*, pp. 248–255. ACM.

Holme, P. and M. E. Newman (2006). "Nonequilibrium phase transition in the coevolution of networks and opinions." *Physical Review E* 74(5), 056108.

Iijima, R. and Y. Kamada (2014). "Social distance and network structures." Mimeo.

Immorlica, N., B. Lucier, and B. Rogers (2013). "Cooperation in anonymous dynamic social networks." Mimeo.

Jackson, M. O. (2008a). "Average distance, diameter, and clustering in social networks with homophily." In *Internet and Network Economics*, pp. 4–11. Springer. http://link.springer.com/book/10.1007/978-3-540-92185-1.

Jackson, M. O. (2008b). *Social and Economic Networks.* Princeton, NJ: Princeton University Press.

Jackson, M. O. and D. Lopez-Pintado (2013). "Diffusion and contagion in networks with heterogeneous agents and homophily." *Network Science* 1(1), 49–67.

Jackson, M. O. and B. W. Rogers (2005). "The economics of small worlds." *Journal of the European Economic Association* 3(2-3), 617–627.

Jackson, M. O. and B. W. Rogers (2007). "Meeting strangers and friends of friends: How random are social networks?" *The American Economic Review*, 890–915.

Jackson, M. O. and A. Watts (2002). "The evolution of social and economic networks". *Journal of Economic Theory* 106(2), 265–296.

Jackson, M. O. and L. Yariv (2007). "Diffusion of behavior and equilibrium properties in network games." *The American Economic Review* 97, 92–98.

Karrer, B. and M. E. Newman (2011). "Stochastic blockmodels and community structure in networks." *Physical Review E* 83(1), 016107.

Kleinberg, J. (2000). "Navigation in a small world." *Nature* 406(6798), 845–845.

Kleinberg, J. (2004). "The small-world phenomenon and decentralized search." *SiAM News* 37(3), 1–2.

Kleinberg, J. (2006). "Complex networks and decentralized search algorithms." In *Proceedings of the International Congress of Mathematicians (ICM)*, Volume 3, pp. 1019–1044.

König, M. D. and C. J. Tessone (2011). "Network evolution based on centrality." *Physical Review E* 84(5), 056108.

König, M. D., C. J. Tessone, and Y. Zenou (2014). "Nestedness in networks: A theoretical model and some applications." *Theoretical Economics* 9(3), 695–752.

Lamberson, P. J. (2015). "Diffusion in networks." *This book: Chapter 18.*

Lee, R. S. and K. Fong (2013). "Markov-perfect network formation an applied framework for bilateral oligopoly and bargaining in buyer-seller networks." Working paper.

van der Leij, M. and J. Kovarik (2012). "Risk aversion and social networks." Technical report, Instituto Valenciano de Investigaciones Económicas, SA (Ivie).

Lobel, I. and E. Sadler (2015a). "Information diffusion in networks through social learning." *Theoretical Economics* 10, 807–851. https://econtheory.org/ojs/index.php/te/article/viewFile/20150807/13864/409.

Lobel, I. and E. Sadler (2015b). "Preferences, homophily, and social learning." Technical report.

López-Pintado, D. (2006). "Contagion and coordination in random networks." *International Journal of Game Theory* 34(3), 371–381.

López-Pintado, D. (2008). "The spread of free-riding behavior in a social network." *Eastern Economic Journal* 34(4), 464–479.

López-Pintado, D. (2012). "Influence networks." *Games and Economic Behavior* 75(2), 776–787.

López-Pintado, D. (2013). "Public goods in directed networks." *Economics Letters* 121(2), 160–162.

Marsili, M., F. Vega-Redondo, and F. Slanina (2004). "The rise and fall of a networked society: A formal model." *Proceedings of the National Academy of Sciences of the United States of America* 101(6), 1439–1442.

Mele, A. (2014). "A structural model of segregation in social networks." Mimeo.

Molloy, M. and B. Reed (1995). "A critical point for random graphs with a given degree sequence." *Random Structures & Algorithms* 6(2-3), 161–180.

Molloy, M. and B. Reed (1998). "The size of the giant component of a random graph with a given degree sequence." *Combinatorics, Probability and Computing* 7(3), 295–305.

Montgomery, J. D. (1991). "Social networks and labor-market outcomes: Toward an economic analysis." *The American Economic Review* 81, 1408–1418. http://www.jstor.org/stable/2006929?seq=1#page_scan_tab_contents

Montgomery, J. D. (1992). "Job search and network composition: Implications of the strength-of-weak-ties hypothesis." *American Sociological Review* 57, 586–596. http://www.jstor.org/stable/2095914?seq=1#page_scan_tab_contents

Nermuth, M., G. Pasini, P. Pin, and S. Weidenholzer (2013). "The informational divide." *Games and Economic Behavior* 78, 21–30.

Newman, M. E. (2002). "Assortative mixing in networks." *Physical Review Letters* 89(20), 208701.

Newman, M. E. (2003). "The structure and function of complex networks." *SIAM Review* 45(2), 167–256.

Newman, M. E. (2004). "Detecting community structure in networks." *The European Physical Journal B-Condensed Matter and Complex Systems* 38(2), 321–330.

Ozgür, O. and A. Bisin (2013). "Dynamic linear economies with social interactions." Mimeo.

Park, J. and M. Newman (2005). "Solution for the properties of a clustered network." *Physical Review E* 72(2), 026136.

Park, J. and M. E. Newman (2004). "Solution of the two-star model of a network." *Physical Review E* 70(6), 66–146.

Pin, P. and B. W. Rogers (2015). "Cooperation, punishment and immigration." *Journal of Economic Theory*, forthcoming.

Price, D. d. S. (1965). "Networks of scientific papers." *Science* 169, 510–515.

Price, D. d. S. (1976). "A general theory of bibliometric and other cumulative advantage processes." *Journal of the American Society for Information Science* 27(5), 292–306.

Ravasz, E. and A.-L. Barabási (2003). "Hierarchical organization in complex networks." *Physical Review E* 67(2), 026112.

Rosenblat, T. S. and M. M. Mobius (2004). "Getting closer or drifting apart?" *The Quarterly Journal of Economics* 119(3), 971–1009.

Schelling, T. C. (1971). "Dynamic models of segregation." *Journal of Mathematical Sociology* 1(2), 143–186.

Shalizi, C. R. and A. C. Thomas (2011). "Homophily and contagion are generically confounded in observational social network studies." *Sociological Methods & Research* 40(2), 211–239.

Simon, H. A. (1955). "On a class of skew distribution functions." *Biometrika* 42, 425–440. http://www.jstor.org/stable/2333389.

Strauss, D. (1986). "On a general class of models for interaction." *SIAM Review* 28(4), 513–527.

Sundararajan, A. (2007). "Local network effects and complex network structure." *The BE Journal of Theoretical Economics* 7(1).

Vannetelbosch, V. and A. Mauleon (2015). "Network formation games." *This book: Chapter 8.*

Vázquez, A. (2003). "Growing network with local rules: Preferential attachment, clustering hierarchy, and degree correlations." *Physical Review E* 67(5), 056104.

Vega-Redondo, F. (2007). *Complex social networks.* Cambridge, MA: Cambridge University Press.

Vega Redondo, F. (2015). "Links and actions in interplay." *This book: Chapter* 7.

Virkar, Y., A. Clauset, et al. (2014). "Power-law distributions in binned empirical data." *The Annals of Applied Statistics* 8(1), 89–119.

Wasserman, S. and P. Pattison (1996). "Logit models and logistic regressions for social networks: I. An introduction to Markov graphs and p." *Psychometrika* 61(3), 401–425.

Watts, D. J. and S. H. Strogatz (1998). "Collective dynamics of 'small-world' networks." *Nature* 393(6684), 440–442.

Yule, G. (1925). "The growth of population and the factors which control it." *Journal of the Royal Statistical Society* 88(1), 1–58. http://www.jstor.org/stable/2341575.

CHAPTER 8

NETWORK FORMATION GAMES

ANA MAULEON AND VINCENT VANNETELBOSCH

8.1 Introduction

The organization of agents into networks plays an important role in the determination of the outcome of many social and economic interactions. Networks of relationships help determine the careers that people choose, the jobs they obtain, the products they buy, and how they vote. Such networks also matter for the trade of goods and services, the provision of insurance in developing countries, R&D collaborations among firms, and trade agreements between countries. In many economic networks, the network is not exogenous, but agents decide about the links they want to build. A central question is predicting the networks that agents will form.

Jackson and Wolinsky (1996) propose the notion of pairwise stability to predict the networks that one might expect to emerge in the long run. A network is pairwise stable if no agent benefits from deleting a link and no two agents benefit from adding a link between them. It suffices to check whether two agents have incentives or not to add a link between them, and whether a single agent has incentives or not to delete one of her links. Mutual consent is only required for adding a link. Pairwise stability is a very important tool in network analysis. It is simple and tractable. In some applications it is powerful enough for ruling out most network architectures. But, in other applications, pairwise stability does not yield precise predictions. There are situations where an agent has no incentive to delete one of her links but would benefit from deleting simultaneously more links. Myerson's (1991) linking game allows for such a possibility and models the network formation as a noncooperative game where agents simultaneously choose the links they want to form. A link between two agents is formed if and only if both agents wish to build the link. A network is pairwise Nash stable if it corresponds to a Nash equilibrium of Myerson's linking game and any pair of agents have no incentive to form a link that does not exist in the network. Pairwise Nash stability is a refinement of pairwise stability. However, one shortcoming of pairwise

(Nash) stability is the lack of farsightedness. Agents do not anticipate that other agents may react to their changes. For instance, farsighted agents might not add a link that appears valuable to them as this might induce the formation of other links, ultimately leading to lower payoffs for them.

Another central question about network formation is whether the networks formed by the agents are efficient from an overall societal perspective. Moreover, as the relevance of social and economic networks has been recognized, there are more and more policies that provide incentives to build links. The effectiveness of such policies is highly dependent on our understanding of how such networks form.

Some of the literature on network formation has been motivated by concrete empirical evidence on the properties that networks have, and has explored the circumstances under which networks will or will not exhibit those properties (see Goyal and Joshi 2006a, among others). In this chapter we rather focus on solution concepts (Section 8.2), and we illustrate the bites that these solution concepts have on economic applications (Section 8.3). So, we first propose some myopic and farsighted definitions for modeling network formation. We then investigate in three models of economic networks whether the networks formed by farsighted agents are efficient and different from those formed by myopic agents. There are alternative methodologies to analyze network formation, like stochastic network formation models (see Chapter 7 by Pin and Rogers in this handbook). More recently, the study of network formation has been combined with games on networks (see Chapter 9 by Vega-Redondo in this handbook).

We focus on situations where the formation of a link requires the consent of both agents. An alternative approach was developed by Bala and Goyal (2000), who propose a connections model of network formation where each agent unilaterally decides the links she wants to form and the costs of link formation are incurred only by the agent who initiates the link, while the benefits accrue to both agents linked. Bala and Goyal (2000) find that the only strict Nash networks are the center-sponsored star (where one agent forms all the links) and the empty network. Galeotti, Goyal, and Kamphorst (2006) introduce heterogeneous agents in the connections model, and they show that strict Nash equilibrium networks exhibit high centrality and short average distances between agents even in presence of considerable heterogeneity. Hojman and Szeidl (2008) assume that benefits from connections exhibit decreasing returns and decay with network distance. They find that the unique equilibrium network is a periphery-sponsored star (where a single agent—the center—maintains no links and all other agents maintain one link to the center). We refer to Goyal (2007) for an extensive analysis of one-sided link formation models where each agent can unilaterally form links with any subset of the other players.

8.2 Network Formation: Solution Concepts

Let $N = \{1, ..., n\}$ be the finite set of players who are connected in some network relationship. The network relationships are reciprocal and the network is thus modeled as a nondirected graph. A network g is a list of players who are linked to each other. We write $ij \in g$ to indicate that i and j are linked in the network g. Let g^N be the set of all subsets of N with cardinality 2, so g^N is the complete network. The set of all possible networks on N is denoted by $\mathcal{G}$ and consists of all subsets of g^N. The network obtained by adding link ij to an existing network g is denoted $g + ij$, and the network obtained by cutting link ij from an existing network g is denoted $g - ij$. For any network g, we denote by $N(g) = \{i \mid \exists j \text{ such that } ij \in g\}$ the set of players who have at least one link in the network g. A path in a network g between i and j is a sequence of players $i_1, \ldots, i_K$ such that $i_k i_{k+1} \in g$ for each $k \in \{1, \ldots, K-1\}$ with $i_1 = 1$ and $i_K = j$. A non-empty network $h \subseteq g$ is a component of g, if for all $i \in N(h)$ and $j \in N(h) \setminus \{i\}$, there exists a path in h connecting i and j, and for any $i \in N(h)$ and $j \in N(g)$, $ij \in g$ implies $ij \in h$.[1] We denote by $C(g)$ the set of components of g. A component h of g is minimally connected if h has $\#N(h) - 1$ links (i.e., every pair of players in the component are connected by exactly one path). The partition of N induced by g is denoted by $\Pi(g)$, where $S \in \Pi(g)$ if and only if either there exists $h \in C(g)$ such that $S = N(h)$ or there exists $i \notin N(g)$ such that $S = \{i\}$.

A network utility function (or payoff function) is a mapping $u : \mathcal{G} \to \mathbb{R}^N$ that assigns to each network g a utility $u_i(g)$ for each player $i \in N$. A network $g \in \mathcal{G}$ is strongly efficient relative to u if it maximizes $\sum_{i \in N} u_i(g)$. A network $g \in \mathcal{G}$ Pareto dominates a network $g' \in \mathcal{G}$ relative to u if $u_i(g) \geq u_i(g')$ for all $i \in N$, with strict inequality for at least one $i \in N$. A network $g \in \mathcal{G}$ is Pareto efficient relative to u if it is not Pareto dominated, and a network $g \in \mathcal{G}$ is Pareto dominant if it Pareto dominates any other network. In Figure 8.1 we provide an example of the networks that could be formed by three players and their utilities. For instance, g_4 is a star network where player 1 gets 5 as utility and players 2 and 3 obtain 1 as utility. We will use this example to illustrate the solution concepts.

8.2.1 Pairwise Stability and Closed Cycles

A simple way to analyze the networks that one might expect to emerge in the long run is to examine a sort of equilibrium requirement that players not benefit from altering the structure of the network. A weak version of such condition is the pairwise stability notion defined by Jackson and Wolinsky (1996). A network is pairwise stable if no player benefits from severing one of their links and no other two players benefit from adding a link between them, with one benefiting strictly and the other at least weakly.

Definition 1 (Jackson and Wolinsky 1996). A network g is pairwise stable with respect to u if and only if (i) for all $ij \in g$, $u_i(g) \geq u_i(g - ij)$ and $u_j(g) \geq u_j(g - ij)$, and (ii) for all $ij \notin g$, if $u_i(g) < u_i(g + ij)$ then $u_j(g) > u_j(g + ij)$.

[1] We use the notation $\subseteq$ for weak inclusion and $\subsetneq$ for strict inclusion, and # refers to the notion of cardinality.

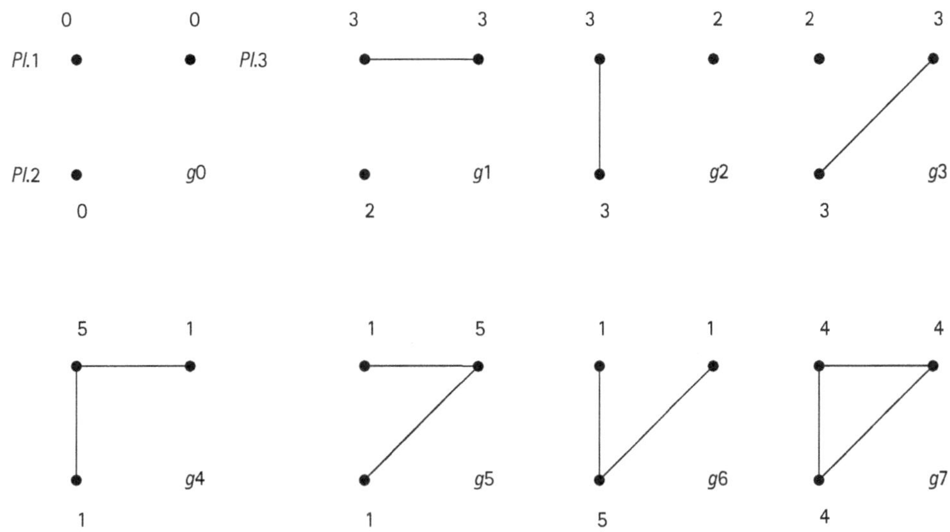

FIGURE 8.1 The networks that can be formed among three players with their utilities.

Two networks g and g' are adjacent if they differ by one link. That is, g' is adjacent to g if $g' = g + ij$ or $g' = g - ij$ for some ij. A network g' defeats g if either $g' = g - ij$ with $u_i(g') > u_i(g)$ or $u_j(g') > u_j(g)$, or if $g' = g + ij$ with $u_i(g') \geq u_i(g)$ and $u_j(g') \geq u_j(g)$ with at least one inequality holding strictly. Hence, a network is pairwise stable if and only if it is not defeated by another (necessarily adjacent) network. We say that the network utility function u exhibits no indifference if for any g and g' that are adjacent either g defeats g' or g' defeats g. In the 3-player example of Figure 8.1, both the partial networks g_1, g_2, and g_3 and the complete network g_7 are pairwise stable. Nobody has an incentive to delete one of her links in the complete network. She would end up worse off in the star network after having cut one of her two links. In any partial network, the player who has no link does not want to link with another player to form a star network. Moreover, both players who are linked have no incentives to cut this link to move to the empty network. The empty network g_0 is not pairwise stable because two players have incentives to link to each other and the star networks g_4, g_5, and g_6 are not pairwise stable since the peripheral players have incentives to add the missing link to form the complete network.

Pairwise stable networks do not always exist. Jackson and Watts (2002) introduce the notion of improving paths. An improving path is a sequence of networks that can emerge when players form or sever links based on the improvement the resulting network offers relative to the current network. If a link is added, then the two players involved must both prefer the resulting network to the current network, with at least one of the two strictly preferring the resulting network. If a link is deleted, then it must be that at least one of the two players involved in the link strictly prefers the resulting network. Formally, an improving path from a network g to a network $g' \neq g$

is a finite sequence of networks $g_1,\ldots,g_K$ with $g_1 = g$ and $g_K = g'$ such that for any $k \in \{1,\ldots,K-1\}$ either (i) $g_{k+1} = g_k - ij$ for some ij such that $u_i(g_{k+1}) > u_i(g_k)$ or $u_j(g_{k+1}) > u_j(g_k)$, or (ii) $g_{k+1} = g_k + ij$ for some ij such that $u_i(g_{k+1}) > u_i(g_k)$ and $u_j(g_{k+1}) \geq u_j(g_k)$. If there exists an improving path from g to g', then we write $g \mapsto g'$. For a given network g, let $M(g) = \{g' \in \mathcal{G} \mid g \mapsto g'\}$ be the set of networks that can be reached by an improving path from g. Notice that g is pairwise stable if and only if $M(g) = \emptyset$.

Improving paths emanating from any network lead either to some pairwise stable network or to some closed cycle. Jackson and Watts (2002) define the notion of a closed cycle. A set of networks C is a cycle if for any $g \in C$ and $g' \in C$, we have $g \in M(g')$. A cycle C is a closed cycle if no network in C lies on an improving path leading to a network that is not in C.

Proposition 1 (Jackson and Watts 2002). *For every* $g \in \mathcal{G}$, *either* g *is pairwise stable or there is a closed cycle* C *such that* $C \subseteq M(g)$.

Jackson and Watts (2001) provide a condition on the network utility function that rules out cycles.[2] The network utility function u is exact pairwise monotonic if g' defeats g if and only if $\sum_{i\in N} u_i(g') > \sum_{i\in N} u_i(g)$ and g' is adjacent to g. Exact pairwise monotonicity implies that strongly efficient networks are pairwise stable. Jackson and Watts (2001) show that, if u is exactly pairwise monotonic, then there are no cycles.[3]

8.2.2 Pairwise Nash Stability

An alternative way to model network formation is Myerson's (1991) linking game where players choose simultaneously the links they wish to form and where the formation of a link requires the consent of both players. A strategy of player $i \in N$ is a vector $\sigma_i = (\sigma_{i1},\ldots,\sigma_{ii-1},\sigma_{ii+1},\ldots,\sigma_{in})$ where $\sigma_{ij} \in \{0,1\}$ for each $j \in N \setminus \{i\}$. If $\sigma_{ij} = 1$, player i wishes to form a link with player j. Given the strategy profile $\sigma = (\sigma_1,\ldots,\sigma_n)$, the network $g(\sigma)$ is formed where $ij \in g(\sigma)$ if and only if $\sigma_{ij} = 1$ and $\sigma_{ji} = 1$.

Definition 2 (Belleflamme and Bloch 2004; Goyal and Joshi 2006a). A strategy profile σ is a pairwise Nash equilibrium of Myerson's linking game if and only if, for each player i, each strategy $\sigma'_i \neq \sigma_i$, $u_i(g(\sigma)) \geq u_i(g(\sigma'_i,\sigma_{-i}))$ and there does not exist a pair of players i and j such that $u_i(g(\sigma)+ij) \geq u_i(g(\sigma))$, $u_j(g(\sigma)+ij) \geq u_j(g(\sigma))$ with strict inequality for one of the two players.

[2] Jackson and Watts (2002) and Tercieux and Vannetelbosch (2006) propose random dynamic models of network formation to select from the set of pairwise stable networks.

[3] The connections model of Jackson and Wolinsky (1996) satisfies exact pairwise monotonicity for certain values of the parameters.

A network g is pairwise Nash stable with respect to a network utility function u if there exists a pairwise Nash equilibrium σ such that $g = g(\sigma)$.[4] Pairwise Nash stability is a refinement of pairwise stability. Pairwise Nash stability requires that a network is immune both to the formation of a new link by any two players and to the deletion of any number of links by any player.[5] In the three-player example of Figure 8.1, both the partial networks g_1, g_2, and g_3 and the complete network g_7 are pairwise Nash stable since these networks are pairwise stable and no player has incentives to cut more than one link at a time. If the players who have no link in the partial networks g_1, g_2, and g_3 would get a utility of 5 instead of 2, then the complete network g_7 would be pairwise stable but not pairwise Nash stable. However, g_1, g_2, and g_3 would remain pairwise stable and pairwise Nash stable. Calvo-Armengol and Ilkilic (2009) provide conditions on the network link marginal utilities such that the sets of pairwise and pairwise Nash stable networks coincide. Let $\alpha \geq 0$. The network utility function u is α-submodular in own current links on $G \subseteq \mathcal{G}$ if and only if $u_i(g) - u_i(g - l) \geq \alpha \sum_{ij \in l}(u_i(g) - u_i(g - ij))$ for all $g \in G$, $i \in N$ and $l \subseteq \{jk \in g \mid j = i \text{ or } k = i\}$. This condition applies only to marginal utilities from existing links, and it imposes that the marginal benefits from a group l of links already in the network are higher than the sum of marginal benefits of each single link in l scaled by α. Under the condition that utilities are α-submodular on the set of pairwise stable networks for some $\alpha \geq 0$, if a player does not gain from cutting any single link, then she does not gain from deleting more links simultaneously.

Proposition 2 (Calvo-Armengol and Ilkilic 2009). *The set of pairwise stable networks coincides with the set of pairwise Nash stable networks if and only if u is α-submodular on the set of pairwise stable networks, for some $\alpha \geq 0$.*

Calvo-Armengol and Ilkilic (2009) show that the condition of α-submodularity holds in the connections and co-author models of Jackson and Wolinsky (1996) and in the information transmission model of Calvo-Armengol (2004).[6]

Goyal and Vega-Redondo (2007) strengthen the concept of pairwise Nash stability by proposing the notion of bilateral equilibrium that allows for simultaneous deletion and addition of a link. A network can be supported in a bilateral equilibrium of Myerson's linking game if it is pairwise Nash stable and no pair of players benefit, at least one

[4] A strategy profile σ is a Nash equilibrium of Myerson's linking game if and only if, for each player i, each strategy $\sigma_i' \neq \sigma_i$, $u_i(g(\sigma)) \geq u_i(g(\sigma_i', \sigma_{-i}))$. A network g is Nash stable with respect to a network utility function u if there exists a Nash equilibrium σ such that $g = g(\sigma)$. But the concept of Nash stability is too weak for analyzing network formation when the formation of a link needs the approval of both players. For instance, the empty network is always Nash stable regardless u. De Sinopoli and Pimienta (2010) show that all Nash equilibria are regular when players incur a positive cost to propose links.

[5] Gilles and Sarangi (2010) extend Myerson's linking game to include additive link formation costs: if player i attempts to form a link with player j (i.e., $\sigma_{ij} = 1$), then player i incurs a cost $c_{ij} \geq 0$ regardless of σ_{ji}. See also Slikker and van den Nouweland (2001); Gilles, Chakrabarti, and Sarangi (2012).

[6] Goyal and Joshi (2006a) explore conditions on the network utility function under which pairwise Nash stable networks are or are not asymmetric. Hellman (2013) studies how externalities between links affect the existence and uniqueness of pairwise stable networks.

of them strictly, from simultaneously deleting some of their links and adding the link between them. Denote $\sigma_{-i,j} = (\sigma_1,...,\sigma_{i-1},\sigma_{i+1},...,\sigma_{j-1},\sigma_{j+1},...,\sigma_n)$ the strategy profile σ less the strategies of players i and j.

Definition 3 (Goyal and Vega-Redondo 2007). A strategy profile σ is a bilateral equilibrium of Myerson's linking game if and only if, for each player i, each strategy $\sigma_i' \neq \sigma_i$, $u_i(g(\sigma)) \geq u_i(g(\sigma_i',\sigma_{-i}))$ and there does not exist a pair of players i and j and a pair (σ_i',σ_j') such that $u_i(g(\sigma_i',\sigma_j',\sigma_{-i,j})) \geq u_i(g(\sigma))$, $u_j(g(\sigma_i',\sigma_j',\sigma_{-i,j})) \geq u_j(g(\sigma))$ with strict inequality for one of the two players.

In the 3-player example of Figure 8.1, the partial networks g_1, g_2, and g_3 cannot be supported in a bilateral equilibrium of Myerson's linking game. For instance, g_1 is not a bilateral equilibrium outcome because from $\sigma_1 = (\sigma_{12},\sigma_{13}) = (0,1)$, $\sigma_2 = (\sigma_{21},\sigma_{23}) = (0,0)$, $\sigma_3 = (\sigma_{31},\sigma_{32}) = (1,0)$, players 1 and 2 can now deviate to $\sigma_1' = (\sigma_{12}',\sigma_{13}') = (1,0)$, $\sigma_2' = (\sigma_{21}',\sigma_{23}') = (1,0)$, $\sigma_3 = (\sigma_{31},\sigma_{32}) = (1,0)$, and player 2 is strictly better off at σ' while player 1 is indifferent between σ and σ'. However, the complete network g_7 is a bilateral equilibrium outcome since it is pairwise Nash stable and no pair of players have incentives to cut some of their links.

8.2.3 Farsighted Stability

Pairwise stability and pairwise Nash stability are myopic definitions. Players are not farsighted in the sense that they do not anticipate how others might react to their actions. For instance, adding or severing one link might lead to subsequent addition or deletion of another link. If players have very good information about how others might react to changes in the network, then these are things we want to allow for in the definition of the stability concept. For instance, a network could be stable because players might not add a link that appears valuable to them given the current network, as that might in turn lead to the formation of other links and ultimately lower the payoffs of the original players. The notion of farsighted improving paths captures the farsightedness of the players.

A farsighted improving path is a sequence of networks that can emerge when players add or delete links based on the improvement the end network offers relative to the current network. Each network in the sequence differs by one link from the previous one. If a link is added, then the two players involved must both prefer the end network to the current network, with at least one of the two strictly preferring the end network. If a link is deleted, then it must be that at least one of the two players involved in the link strictly prefers the end network.

Definition 4 (Jackson 2008; Herings, Mauleon, and Vannetelbosch 2009). A farsighted improving path from a network g to a network $g' \neq g$ is a finite sequence of networks $g_1,\ldots,g_K$ with $g_1 = g$ and $g_K = g'$ such that for any $k \in \{1,\ldots,K-1\}$ either: (i) $g_{k+1} =$

$g_k - ij$ for some ij such that $u_i(g_K) > u_i(g_k)$ or $u_j(g_K) > u_j(g_k)$, or (ii) $g_{k+1} = g_k + ij$ for some ij such that $u_i(g_K) > u_i(g_k)$ and $u_j(g_K) \geq u_j(g_k)$.

If there exists a farsighted improving path from g to g', then we write $g \to g'$. For a given network g, let $F(g) = \{g' \in \mathcal{G} \mid g \to g'\}$ be the set of networks that can be reached by a farsighted improving path from g. Notice that $g \to g'$ means that g' is the endpoint of at least one farsighted improving path from g. In the 3-player example of Figure 8.1, we have $F(g_0) = \{g_1, g_2, g_3, g_7\}$, $F(g_1) = \{g_2, g_3, g_7\}$, $F(g_2) = \{g_1, g_3, g_7\}$, $F(g_3) = \{g_1, g_2, g_7\}$, $F(g_4) = \{g_1, g_2, g_3, g_7\}$, $F(g_5) = \{g_1, g_2, g_3, g_7\}$, $F(g_6) = \{g_1, g_2, g_3, g_7\}$, and $F(g_7) = \emptyset$. The computations of farsighted improving path are not always obvious. For instance, the only way to go from g_1 to g_2 is via g_4 (and $g_2 \in F(g_1)$). Indeed, players 1 and 2 make a link to go from g_1 to the intermediate network g_4 in the anticipation that player 3 subsequently cuts her link with player 1. At the same time it holds that $g_4 \notin F(g_1)$. To go from g_1 to the terminal network g_4 is worse for player 2 and player 3. The only thing player 1 can do at g_1 is to delete her link with player 3, which leads to g_0. This is not helpful for player 1, since once at g_0 she can only form a link with player 2 (or 3) to go to g_2 (or g_1), but player 2 (or 3) will never form the missing link to go to g_4.

Jackson (2008) defines a network to be farsightedly pairwise stable if there is no farsighted improving path emanating from it. That is, g is farsightedly pairwise stable if $F(g) = \emptyset$. This concept refines the set of pairwise stable networks, and so may fail to exist in economic networks. Another drawback of the definition is that it does not require that a farsighted improving path ends at a network that is stable itself. Hence, Herings, Mauleon, and Vannetelbosch (2009) propose the concept of pairwise farsightedly stable set.[7]

A set of networks G is pairwise farsightedly stable if three conditions are satisfied. First, all possible pairwise deviations from any network $g \in G$ to a network outside G are deterred by a credible threat of ending worse off. Second, there exists a farsighted improving path from any network outside the set leading to some network in the set (external stability condition). Third, there is no proper subset of G satisfying the first two conditions.

Definition 5 (Herings, Mauleon, and Vannetelbosch 2009). A set of networks $G \subseteq \mathcal{G}$ is pairwise farsightedly stable if

(i) $\forall\, g \in G$,

(ia) $\forall\, ij \notin g$ such that $g + ij \notin G$, $\exists g' \in F(g + ij) \cap G$ such that $(u_i(g'), u_j(g')) = (u_i(g), u_j(g))$ or $u_i(g') < u_i(g)$ or $u_j(g') < u_j(g)$,

(ib) $\forall\, ij \in g$ such that $g - ij \notin G$, $\exists g', g'' \in F(g - ij) \cap G$ such that $u_i(g') \leq u_i(g)$ and $u_j(g'') \leq u_j(g)$.

[7] An alternative approach to model network formation is simply to model it explicitly as a non-cooperative extensive form game. But, the equilibrium outcome is usually quite sensitive to the exact network formation process. See Aumann and Myerson (1988), among others.

(ii) $\forall g' \in \mathcal{G} \setminus G$, $F(g') \cap G \neq \emptyset$.
(iii) $\nexists\, G' \subsetneq G$ such that G' satisfies conditions (ia), (ib), and (ii).

Condition (ia) captures that adding a link ij to a network $g \in G$ that leads to a network outside of G, is deterred by the threat of ending in g'. Here g' is such that there is a farsighted improving path from $g + ij$ to g'. Moreover, g' belongs to G, which makes g' a credible threat. Condition (ib) is a similar requirement, but then for the case where a link is deleted. Condition (ii) requires that from any network outside G there is a farsighted improving path leading to some network in G. Condition (iii) is a minimality condition motivated by the fact that the set $\mathcal{G}$ (trivially) satisfies the first two conditions. In the 3-player example, $\{g_7\}$ is pairwise farsightedly stable. Since $g_7 \in \bigcap_{g \in \mathcal{G} \setminus \{g_7\}} F(g)$, condition (ii) of the definition is satisfied. In addition, condition (i) is also satisfied, since any deviation from g_7 may lead back to g_7. The set $\{g_7\}$ is clearly minimal, so condition (iii) is satisfied too. Since $F(g_7) = \emptyset$, condition (ii) implies that g_7 belongs to any pairwise farsightedly stable set. Using condition (iii) it follows that $\{g_7\}$ is the only pairwise farsightedly stable set.

Proposition 3 (Herings, Mauleon, and Vannetelbosch 2009). *A pairwise farsightedly stable set of networks exists.*

Herings, Mauleon, and Vannetelbosch (2009) provide easy to verify conditions for a set G to be pairwise farsightedly stable.

Proposition 4 (Herings, Mauleon, and Vannetelbosch 2009).

(i) *If for every $g' \in \mathcal{G} \setminus G$ we have $F(g') \cap G \neq \emptyset$ and for every $g \in G$, $F(g) \cap G = \emptyset$, then G is a pairwise farsightedly stable set.*
(ii) *The set $\{g\}$ is a pairwise farsightedly stable set if and only if for every $g' \in \mathcal{G} \setminus \{g\}$ we have $g \in F(g')$.*

Notice that the minimality condition implies that if $\{g\}$ is a pairwise farsightedly stable set, then g does not belong to any other pairwise farsightedly stable set. But there may be pairwise farsightedly stable sets not containing g.

The next proposition provides a full characterization for unique pairwise farsightedly stable sets.

Proposition 5 (Herings, Mauleon, and Vannetelbosch 2009).

(i) *The set G is the unique pairwise farsightedly stable set if and only if $G = \{g \in \mathcal{G} \mid F(g) = \emptyset\}$ and for every $g' \in \mathcal{G} \setminus G$, $F(g') \cap G \neq \emptyset$.*
(ii) *The set $\{g\}$ is the unique pairwise farsightedly stable set if and only if for every $g' \in \mathcal{G} \setminus \{g\}$ we have $g \in F(g')$ and $F(g) = \emptyset$.*

Thus, if G is the unique pairwise farsightedly stable set and the network g belongs to G, then $F(g) = \emptyset$, which implies that g is pairwise stable. So, pairwise farsighted stability is a refinement of pairwise stability when there is a unique pairwise farsightedly stable set.[8]

In Section 8.3 we show that propositions 4 and 5 turn out to be helpful and powerful for characterizing pairwise farsightedly stable sets in economic networks. If for every $g' \in \mathcal{G} \setminus \{g\}$ we have $g \in F(g')$, then $\{g\}$ is a pairwise farsightedly stable set. If, moreover, $F(g) = \emptyset$, then $\{g\}$ is the unique pairwise farsightedly stable set. If, on the other hand, $F(g) \neq \emptyset$, then there exists another pairwise farsightedly stable set.[9]

Dutta, Ghosal, and Ray (2005) propose an alternative approach to model network formation among farsighted players. They develop a model of dynamic network formation where players are farsighted and evaluate the formation of links in terms of its consequences on the entire discounted stream of payoffs. Dutta, Ghosal, and Ray (2005) provide an example where there is a network g that strictly Pareto dominates all other networks, but which is not reached in equilibrium.[10] However, $\{g\}$ is the unique pairwise farsightedly stable set.

8.2.4 Coalitions, Side Payments, and Bargaining

Pairwise (farsighted) stability only considers deviations by at most a pair of players at a time. It might be that some coalition of players could all be made better off by some complicated reorganization of their links, which is not accounted for under pairwise stability. Dutta and Mutuswami (1997) and Jackson and van den Nouweland (2005) define the notion of strong stability where a strongly stable network is a network which is stable against changes in links by any coalition of players. A network g' is said to be obtainable from g via deviations by coalition $S \subseteq N$ if (i) any new links that are added can only be between players belonging to S and (ii) at least one player of any deleted link is member of S.

Definition 6 (Jackson and van den Nouweland 2005). A network g is strongly stable if for any $S \subseteq N$, g' that is obtainable from g via deviations by S, and $i \in S$ such that $u_i(g') > u_i(g)$, there exists $j \in S$ such that $u_j(g') < u_j(g)$.

[8] If G is the unique pairwise farsightedly stable set, then G is the set of farsightedly pairwise stable networks. In addition, if g is a farsightedly pairwise stable network then it belongs to all pairwise farsightedly stable set of networks.

[9] Incorporating the notion of farsighted improving paths into the original definition of the vNM stable set (von Neumann and Morgenstern 1944) leads to the vNM pairwise farsightedly stable set. From Proposition 4, it follows that a vNM pairwise farsightedly stable set is a pairwise farsightedly stable set. However, vNM pairwise farsightedly stable sets do not always exist. Another farsighted concept is the largest pairwise consistent set (Chwe 1994; Page, Wooders, and Kamat 2005). But, it often fails to eliminate implausible pairwise stable networks.

[10] Page and Wooders (2009) propose a model of network formation whose primitives consist of a feasible set of networks, player preferences, rules of network formation, and a dominance relation on feasible networks. Rules may range from noncooperative, where players may only act unilaterally, to cooperative, where coalitions of players may act in concert.

Strong stability is a refinement of pairwise stability. Jackson and van den Nouweland (2005) provide some conditions on the network utility function so that the set of strongly efficient networks coincides with the set of strongly stable networks. The network utility function u is top convex if some strongly efficient network maximizes the per-capita sum of utilities among players. Let $\rho(u,S) = \max_{g \subseteq g^S} \sum_{i \in S} u_i(g)/\#S$. The network utility function u is top convex if $\rho(u,N) \geq \rho(u,S)$ for all $S \subseteq N$. Suppose that u is such that (i) players belonging to the same component get the same utility, and (ii) there are no externalities across components (i.e., payoffs of players belonging to a component in a given network do not depend on the structure of other components). It turns out that, under the conditions they impose, the notion of strong stability eliminates the inefficient pairwise stable networks. Moreover, Grandjean, Mauleon, and Vannetelbosch (2011) show that the set of strongly efficient networks is the unique pairwise farsightedly stable set if and only if u is top convex. So, pairwise farsighted stability selects the pairwise stable networks that are immune to deviations by coalitions if and only if u is top convex.

Proposition 6 (Jackson and van den Nouweland 2005; Grandjean, Mauleon, and Vannetelbosch 2011). *Take any u such that* (i) $u_i(g) = u_j(g)$ *for all* $i, j \in S \in \Pi(g)$ *and* (ii) $u_i(g) = u_i(h)$ *with* $h \in C(g)$ *and* $i \in N(h)$.

(a) *The set of strongly efficient networks is the set of strongly stable networks if and only if u is top convex.*
(b) *The set of strongly efficient networks is the unique pairwise farsightedly stable set if and only if u is top convex.*

Strong stability makes sense in situations where players have substantial information about the overall structure and potential payoffs and can coordinate their actions.[11]

There are a number of papers that look at the endogenous determination of payoffs together with network formation. Currarini and Morelli (2000) develop a sequential network formation game, where players propose links and demand payoffs. Following an exogenously given order, each player proposes in turn the links she wants to form and she demands some payoff. Once all proposals are made, links are formed if both players involved in the link proposed it and the demands of the players are compatible. The payoffs in Currarini and Morelli (2000) are endogenously generated but are highly asymmetric and sensitive to the order in which players make proposals.[12] Bloch and Jackson (2006, 2007) investigate the role played by transfers payments in

[11] The definition of strong stability of Dutta and Mutuswami (1997) considers a deviation to be valid only if all members of a deviating coalition are strictly better off, while the definition of Jackson and van den Nouweland (2005) is slightly stronger by allowing for a deviation to be valid if some members are strictly better off and others are weakly better off. Under the weaker definition of Dutta and Mutuswami, a network is strongly stable if it corresponds to a strong Nash equilibrium of Myerson's linking game.

[12] Goyal and Vega-Redondo (2007) develop a model where players form links with others to create surplus and where rents are split among buyers, sellers and intermediaries.

the formation of networks. They study whether different forms of transfers (direct transfers, indirect transfers, or contingent transfers) can solve the conflict between stability and strong efficiency when there are network externalities that usually lead to the emergence of inefficient networks when transfers are not feasible. They find that indirect transfers together with contingent transfers are needed to guarantee that strongly efficient networks form. Indirect transfers enable players to take care of positive externalities by subsidizing the formation of links by other players; while contingent transfers enable players to overcome negative externalities by preventing the formation of links.[13]

8.3 Some Models of Economic Networks

We now investigate in some models of economic networks whether the pairwise farsightedly stable sets of networks coincide or not with the set of pairwise (Nash) stable networks and the set of strongly efficient networks. We mostly focus on three models of network formation: (1) networks of R&D collaborations, (2) networks of free trade agreements, and (3) criminal networks. In (1), myopia predicts that firms partition themselves into two nearly symmetric coalitions while farsightedness leads to two asymmetric coalitions, with the largest one comprising roughly three-quarters of the total number of firms. Both myopia and farsightedness do sustain networks that do not maximize the social welfare nor the sum of the profits. In (2), the global free trade network is the unique strongly efficient network, and both myopia and farsightedness can sustain it. However, myopia and farsightedness do not impede the emergence of some inefficient networks. In (3), farsightedness reduces the conflict between stability and efficiency by destabilizing some strongly inefficient networks and sustaining one of the strongly efficient networks, namely the complete network.

Thus, depending on the application, myopia and farsightedness may lead to divergent predictions, and farsightedness can help to support the emergence of efficient networks. An interesting question is to find properties on the network utility function so that either myopia and farsightedness lead to the same conclusion, or farsightedness solves the conflict with efficiency while myopia does not.[14] Whenever some of the properties are not satisfied, it is important to understand whether the agents are myopic or farsighted in order to use the appropriate stability concept. Knowing which

[13] Navarro (2014) studies a model of dynamic network formation with farsighted players, similar to the Dutta, Ghosal, and Ray (2005) model, but when side payments can be made between connected players.

[14] In many-to-one matching problems with substitutable preferences, Mauleon, Vannetelbosch, and Vergote (2011) show that, contrary to the vNM (myopically) stable sets (Ehlers 2007), vNM farsightedly stable sets cannot include matchings that are not in the core. For one-to-one matching problems, Mauleon, Molis, Vannetelbosch, and Vergote (2014) provide conditions on preference profiles such that farsightedness coincides with myopia.

networks are likely to be formed and to evolve can help to shape more effective policy recommendations.

8.3.1 R&D Networks

We consider Goyal and Joshi's (2003) two-stage game in a setting with n competing firms that produce some homogenous good. In the first stage, firms decide the bilateral R&D collaborations they are going to establish in order to maximize their respective profits. R&D collaborations reduce marginal costs of production. Given a network g, the marginal cost for firm i is given by $c_i(g) = c_0 - (1 + \sum_{j \neq i} \delta^{t(ij)-1})$ where c_0 is the initial marginal cost, $\delta \in (0, 1]$ and $t(ij)$ is the number of links in the shortest path between i and j (setting $t(ij) = \infty$ if there is no path between i and j). Each firm benefits both from her own R&D (reducing her marginal cost by 1) and from the R&D done by the firms she is connected to (reducing her marginal cost by $\sum_{j \neq i} \delta^{t(ij)-1}$). Let $N_i^k(g) = \{j \mid t(ij) = k\}$ be the set of firms that are connected to firm i by a path of at least k links. Then, $c_i(g) = c_0 - 1 - \sum_{k=1}^{n-1} \#N_i^k(g)\delta^{k-1}$. We focus as in Mauleon, Sempere-Monerris, and Vannetelbosch (2014) on the case $\delta = 1$ where each firm fully benefits from the research done by the firms she is connected to.[15] There are small but positive costs to forming links, $\gamma > 0$.

In the second stage, firms compete in quantities in the oligopolistic market, taking as given the costs of production. Let $p = a - \sum_{i \in N} q_i$ with $a > 0$ be the linear inverse demand function. Thus, firm i's profits in a R&D network g is given by

$$u_i(g) = (q_i(g))^2 - d_i(g)\gamma$$

where the equilibrium output is $q_i(g) = (a - c_0 + (n+1)\#S(i) - \sum_{S \in \Pi(g)} (\#S)^2)/(n + 1)$, $d_i(g)$ is the number of links of firm i and $S(i)$ is the group of firms to which firm i is connected.

Lemma 1 (Mauleon, Sempere-Monerris, and Vannetelbosch 2014). *For R&D networks,*

- **(i)** *any network g where all components are not minimally connected is not pairwise stable;*
- **(ii)** *any minimally connected network g is not pairwise stable;*
- **(iii)** *any network g where* $\#\Pi(g) > 2$ *and all components are minimally connected is not pairwise stable.*

Since each firm benefits perfectly from the research done by the firms she is connected to ($\delta = 1$) and there is an infinitesimal small cost for forming links ($\gamma > 0$), any R&D network where a component is not minimally connected cannot be pairwise

[15] In Goyal and Joshi's (2003) original model, each firm benefits only from her own R&D and from the R&D done by the firms she is directly linked to. In Goyal and Moraga-Gonzalez (2001), firms even benefit, although imperfectly, from the R&D done by firms to whom they are not connected, while in Mauleon, Sempere-Monerris, and Vannetelbosch (2008), benefits decrease with the distance.

stable. Mauleon, Sempere-Monerris, and Vannetelbosch (2014) also show that any minimally connected network is not pairwise stable because any firm who has more than one link in the R&D network increases her profits from cutting a link with a firm who has only one link. In addition, any R&D network consisting of at least three components and where all components are minimally connected is not pairwise stable because two minimally connected components of cardinality smaller than $(n+1)/2$ have incentives to add a link between them to form one component.

Thus, the only candidates for being pairwise stable are R&D networks consisting of two minimally connected components. In fact, Mauleon, Sempere-Monerris, and Vannetelbosch (2014) find that a R&D network g is pairwise stable if and only if g consists of two minimally connected components with the cardinality of the largest component equal to $\text{int}((n+3)/2)$ for n even and to $(n+1)/2$ for n odd (where $\text{int}(\cdot)$ denotes the integer part).[16]

Proposition 7 (Mauleon, Sempere-Monerris, and Vannetelbosch 2014). *For R&D networks, a network g is pairwise stable if and only if $C(g)=(h_1,h_2)$, h_1 and h_2 are minimally connected, $N(h_1)\cup N(h_2)=N$, and $\#N(h_1)=\text{int}((n+3)/2)$ if n even and $\#N(h_1)=(n+1)/2$ if n odd.*

Pairwise Nash stability coincides with pairwise stability since no firm has an incentive to delete more than one link at the pairwise stable R&D networks.[17] The R&D networks that maximize social welfare include any minimally connected network, and the strongly efficient R&D networks consist of one minimally connected component of cardinality greater than $3n/4$ with the remaining firms having no links.[18] Hence, we observe a conflict between pairwise stability and efficiency since pairwise stability leads to the emergence of R&D networks that split the firms into two coalitions of nearly equal size.[19]

[16] Firms in the largest component have incentives to delete one link isolating one firm until the cardinality of the component is equal to $\text{int}((n+3)/2)+1$ if n even or to $(n+3)/2$ if n odd. So, only any network g such that $C(g)=(h_1,h_2)$, h_1 and h_2 are minimally connected, $N(h_1)\cup N(h_2)=N$, and $\#N(h_1)=\text{int}((n+3)/2)$ if n even and $\#N(h_1)=(n+1)/2$ if n odd, is a candidate for being pairwise stable. Since firms in the smallest component h_2 have no incentives to delete links, firms in the largest component h_1 have no incentives to delete one link isolating more than one firm, and two firms $i\in N(h_1)$ and $j\in N(h_2)$ do not have incentives to form the link ij, this R&D network is pairwise stable.

[17] The pairwise (Nash) stable R&D networks cannot be supported in a bilateral equilibrium of Myerson's linking game because there is a pair of firms who benefit from simultaneously deleting some of their links and adding the link between them. For instance, one firm belonging to the smallest component cuts all her links and adds a new link to some firm belonging to the largest component.

[18] Yi (1998) shows that a network consisting of one minimally connected component h^* with the remaining firms organized as singletons maximizes industry profits, where $\#N(h^*)$ $(\leq n)$ is the solution to $4(a-c_0)+3(n+1)^2(\#N(h^*)+1)-4(n+2)(\#N(h^*)^2+(n-\#N(h^*)))=0$.

[19] Goyal and Joshi (2003) find that, in case each firm only benefits from R&D done by firms she is directly linked to, the complete network maximizes the social welfare and is the unique pairwise stable network when linking costs are small.

Once firms are farsighted, Mauleon, Sempere-Monerris, and Vannetelbosch (2014) show that the set of all networks g consisting of two minimally connected components h_1 and h_2 such that $\#N(h_1) = \text{int}((3n+1)/4)$ and $N(h_1) \cup N(h_2) = N$ is a pairwise farsightedly stable set of networks.[20]

Proposition 8 (Mauleon, Sempere-Monerris, and Vannetelbosch 2014). *The set* $G^* = \{g \mid C(g) = (h_1, h_2)$, h_1 *and* h_2 *are minimally connected,* $\#N(h_1) = \text{int}((3n+1)/4)$ *and* $N(h_1) \cup N(h_2) = N\}$ *is a pairwise farsightedly stable set.*

First, profitable deviations from any network $g^* \in G^*$ are deterred. For instance, for $n > 5$, the only profitable deviations from g^* to $g^* - ij$ are such that j is isolated in $g^* - ij$. But such deviations are deterred because there is a farsighted improving path from $g^* - ij$ to some $g' \in G^*$ and the initial deviator, firm i, is worse off at g'. At each step of a farsighted improving path, one firm belonging to the largest component in g^* becomes isolated, and this firm then links to the smallest component in g^* looking forward to the end network g' where the smallest component in g^* becomes now in g' the largest component with size $\text{int}((3n+1)/4)$. Firm i, who initially deleted the link ij, ends in the smallest component and so is worse off in g'. Second, there is a farsighted improving path from any network g consisting of minimally connected components to some $g^* \in G^*$.[21] For instance, in case g has only components of cardinality smaller than $\text{int}((3n+1)/4)$, first at each step of a farsighted improving path from g, two firms belonging to the two largest components form a link until we reach a network g' with one component of cardinality greater or equal than $\text{int}((3n+1)/4)$ and some smaller components. From g', at each step, two firms belonging to the two smallest components form a link until we reach a network g'' consisting of only two components (with one of them having cardinality larger than $\text{int}((3n+1)/4)$). Next, at each step, a firm of the largest component is isolated, and then this firm is linked to one firm of the smaller component, and so forth until we reach g^*.[22] Finally, since $g \notin F(g')$ for all $g, g' \in G^*$, any proper subset violates condition **(ii)** in Definition 5. Hence, this set G^* is also a vNM farsightedly stable set.

However, the set of all pairwise stable networks is neither a pairwise farsightedly stable set of networks nor a vNM farsightedly stable set. Indeed, the set of all pairwise stable networks violates the external stability condition (i.e., condition **(ii)** in Definition 5).

[20] Roketskiy (2012) studies collaboration between farsighted firms competing in a tournament and finds that von Neumann-Morgenstern farsightedly stable sets of networks consist of two asymmetric mutually disconnected complete components. See also Grandjean and Vergote (2014).

[21] If $g^* \in F(g)$ then $g^* \in F(g')$ for any $g' \supset g$ with $\Pi(g') = \Pi(g)$ because profits only depend on the cardinality of the components and forming links is costly.

[22] Firms who have linked the largest components along the sequence as well as firms who have isolated some other firms from the largest component once there were only two components prefer the end network g^* to the network of the sequence from which they moved since they belong to the component of cardinality $\text{int}((3n+1)/4)$ in g^*.

Remember that strongly efficient R&D networks consist of one minimally connected component of cardinality greater than $3n/4$ with the remaining firms having no links. Hence, we obtain that efficient R&D networks may not emerge in the long run even if firms are farsighted. Thus, in case of quantity competition and homogenous goods, farsightedness does not solve the conflict between stability and efficiency. In fact, farsightedly stable R&D networks lead to a collaboration architecture close to the equilibrium structure of Bloch's (1995) sequential game for forming research associations where firms partition themselves into two asymmetric associations, with the largest one comprising roughly three-quarters of industry members.[23] But, the network approach differs from the group formation approach in the decision making for establishing R&D collaborations. Mutual consent is needed for forming a new link between two firms, whereas the consent of all members of the association is required when a firm joins the association.

Finally, in case of price competition and homogenous goods, all networks would give zero profits for all firms. Since forming links is costly, farsighted stability would then only support the empty network, which is also the unique pairwise stable network.

8.3.2 Free Trade Networks

We consider the Goyal and Joshi (2006b) and Zhang, Xue, and Zu (2013) three-stage game in a setting with n countries. Each country has one firm producing some homogeneous good that can be sold in the domestic market and in each foreign market. A firm's ability to sell in foreign markets, however, depends on the level of import tariffs set by the foreign countries. In the first stage, countries decide the free-trade agreements (or links) they are going to establish in order to maximize their respective social welfare. The collection of bilateral links between the countries defines a network of free trade agreements. In the second stage, if two countries have negotiated a free trade agreement, then each country offers the other a tariff-free access to her domestic market. Otherwise, each country imposes an optimal non-zero tariff on the imports from the other country in order to maximize her social welfare. In the third stage, each firm chooses how much to produce for her domestic market and how much to export to each foreign country, taking as given the output decisions of the other firms, the settled tariffs, and the network of free trade agreements. Firms compete in quantities in each country's market and markets are of equal size. In country i's market, firms face an inverse linear demand, $p_i = a - q_i$, where $a > 0$, q_i^j is the export level of firm j to country i, and $q_i = \sum_j q_i^j$ is the aggregate quantity in country i. Firms have a constant and identical marginal cost $c > 0$, where $a > c$.

The social welfare of country i is defined as the sum of consumer surplus, firm's profits and tariff revenues. Given a free trade network g, Zhang, Xue, and Zu (2013) shows that

[23] Bloch (2005) provide a survey on group and network formation in industrial organization.

the social welfare of country i at equilibrium is given by

$$u_i(g) = \frac{\left[d_i(g)(2n+1) - (n-4)\right]}{2\left[d_i(g)(2n+5) - (n-2)\right]}(a-c)^2 + \sum_{j\in N_i(g)} \left[\frac{2(d_j(g)+1)}{d_j(g)(2n+5)-(n-2)}\right]^2 (a-c)^2 + \sum_{k\in N\setminus N_i(g),\, k\neq i} \left[\frac{(2d_k(g)-1)}{d_k(g)(2n+5)-(n-2)}\right]^2 (a-c)^2,$$

where $N_i(g)$ is the set of countries with whom country i has negotiated a free trade agreement and $d_i(g)$ is the number of links or free trade agreements of country i. Thus, country i's welfare is a function of the number of links of country i, $d_i(g)$, the number of links of countries linked to country i, $d_j(g)$ with $j \in N_i(g)$, and the number of links of countries without link to country i, $d_k(g)$ with $k \in N \setminus N_i(g)$. Given g, global welfare is defined as $\sum_{i\in N} u_i(g)$. A free trade network g is strongly efficient if $\sum_{i\in N} u_i(g) \geq \sum_{i\in N} u_i(g')$ for all $g' \neq g$. Goyal and Joshi (2006b) find that the complete network g^N is strongly efficient if tariffs are exogenously given. In case tariffs are endogenously determined, Zhang, Xue, and Zu (2013) show that the complete global free trade network is again the unique strongly efficient network.[24]

Proposition 9 (Goyal and Joshi 2006b). *The complete global free trade network g^N is pairwise stable.*

Goyal and Joshi (2006b) show that the complete global free trade network g^N is pairwise stable, implying that global free trade, if reached, will prevail. However, the global free trade network is not the unique pairwise stable network. Moreover, the global free trade network may not be pairwise Nash stable. For instance, Zhang, Xue, and Zu (2013) show that, in the case of 10 countries, the global free trade network is not pairwise Nash stable because a single country has incentives to simultaneously delete all her links. Furusawa and Konishi (2007) study the trading network generated by countries that trade a numeraire good and a continuum of differentiated goods. They find that, when all countries are symmetric, the global free trade network in which every pair of countries sign a free trade agreement is pairwise stable, and it is the unique pairwise stable network if goods are not highly substitutable. However, if countries are asymmetric in the market size, the global free trade network may not be attained.[25]

[24] Given g, let $\widehat{u}_i(g)$ be the welfare generated by country i. It consists of three parts: consumer surplus of country i, producer surplus provided by firm i to her country and her linked countries, and tariff revenues provided by firm i to her unlinked countries. Zhang, Xue, and Zu (2013) show that, in any g, $\sum_{i\in N} u_i(g) = \sum_{i\in N} \widehat{u}_i(g)$ and $\widehat{u}_i(g)$ is an increasing function of $d_i(g)$. Hence, $\widehat{u}_i(g)$ is maximized when g is the complete network g^N, and so it is the unique strongly efficient network.

[25] Mauleon, Song, and Vannetelbosch (2010) find that the asymmetry consisting of having unionized and non-unionized countries, could also impede the formation of the global free trade network.

Goyal and Joshi (2006b) leave open the question whether a sequence of bilateral free trade negotiations can lead to global free trade from the empty network or any preexisting network.[26] Zhang, Xue, and Zu (2013) answer this question in case of farsighted countries.

Proposition 10 (Zhang, Xue, and Zu 2013). *The complete global free trade network* $\{g^N\}$ *is a pairwise farsightedly stable set.*

Thus, the complete network constitutes a pairwise farsightedly stable set and, therefore, starting from any other free trade network, there is a farsightedly improving path leading to the complete network. In particular, there is a farsighted improving path from the empty network to the complete network involving only link addition. One could interpret this result in the sense that there is a suitable sequence of bilateral free trade agreements that constitutes building blocks towards global free trade. However, starting from an arbitrary free trade network, a farsighted improving path to the complete network may involve link deletion. When this is the case, some bilateral free trade agreements are stumbling blocs that should be eliminated to facilitate the convergence to global free trade. If, for some reasons, such bilateral free trade agreements cannot be eliminated, countries can get trapped in a partially connected free trade network.

Proposition 11 (Zhang, Xue, and Zu 2013). *The complete global free trade network* $\{g^N\}$ *may not be the unique farsightedly stable set.*

Since $F(g^N) \neq \emptyset$, from Proposition 5, there can be other pairwise farsightedly stable sets of networks that do not contain the complete network g^N. Therefore, common expectation of the eventual emergence of the complete network and suitable sequence of bilateral free trade agreements are essential in achieving global free trade.

8.3.3 Criminal Networks

We consider a simplified version of Calvo-Armengol and Zenou's (2004) model where criminals compete with each other in criminal activities but benefit from being friends with other criminals by learning and acquiring know-how on the crime business.[27] Players are referred to as criminals. If two players are connected, then they are part of

[26] Mauleon, Song, and Vannetelbosch (2010) show that, in presence of unionized and non-unionized countries, starting from the network in which no country has signed a free trade agreement, all sequences of networks due to continuously profitable deviations do not lead (in most cases) to the global free trade network, even when global free trade is pairwise stable.

[27] Calvo-Armengol and Zenou (2004) mostly focus on the case where the criminal network is exogenously given. In their original model, players choose first either to participate to the labor market or to be involved in criminal activities. In case they become criminals, they decide how much effort they devote to delinquent behavior.

the same criminal group. Each group S of criminals has a positive probability $p_S(g)$ of winning the loot $B > 0$. We assume that the bigger the criminal group, the higher its probability of getting the loot. Criminals learn from others belonging to the same group how to be more efficient in criminal activities,[28] and the probability of winning the loot is given by $p_S(g) = \#S/n$. The network also determines how the loot is shared among the criminals in the group. Let $S(i) \in \Pi(g)$ be the criminal group criminal i belongs to and $c_i(g) = \max_{j \in S(i)} d_j(g)$ be the maximum degree in the criminal group $S(i)$, where $d_j(g)$ is the number of links (or degree) of criminal j. A criminal i who belongs to $S(i) \in \Pi(g)$ expects a share $\alpha_i(g)$ of the loot given by

$$\alpha_i(g) = \begin{cases} 1/(\#\{j \in S(i) \mid d_j(g) = c_j(g)\}) & \text{if } d_i(g) = c_i(g) \\ 0 & \text{otherwise.} \end{cases}$$

That is, within each criminal group, the criminal who has the highest number of links obtains the loot. If two or more criminals have the highest number of links, then they share the loot equally among them. Criminals can be caught with some positive probability. In case criminal i is caught, i's rewards are punished at a rate $\phi > 0$ with $\phi < n/(n-1)$. We assume that the higher the number of links criminal i has, the lower i's probability of being caught. The probability of being caught is simply given by $q_i(g) = (n - 1 - d_i(g))/n$. Then, criminal i's payoff is given by

$$\begin{aligned} u_i(g)) &= p_{S(i)}(g)\alpha_i(g)(1 - q_i(g)\phi)B, \\ &= \begin{cases} \frac{\#S}{n} \frac{1}{\#\{j \in S(i) | d_j(g) = c_j(g)\}} (1 - \frac{n-1-d_i(g)}{n}\phi)B & \text{if } d_i(g) = c_i(g) \\ 0 & \text{otherwise.} \end{cases} \end{aligned}$$

There are many networks that are pairwise stable. Calvo-Armengol and Zenou (2004) find that the complete network is pairwise stable and strongly efficient. In addition, it can be easily verified that any network consisting of complete components, where no two components have the same degree, is pairwise stable. However, such pairwise stable networks are not strongly efficient.

Proposition 12 (Herings, Mauleon, and Vannetelbosch 2014). *For criminal networks, each network g such that $g = \bigcup_{S \in \Pi(g)} g^S$ and $\#S \neq \#S'$ $\forall S, S' \in \Pi(g)$ is pairwise stable.*

In addition, any network consisting of complete components, where no two components have the same degree, is pairwise Nash stable since no criminal has an incentive to delete more than one link.[29] The set of strongly efficient networks consists

[28] Patacchini and Zenou (2008) provide evidence that peer effects and the structure of social interactions matter strongly in explaining criminal or delinquent behavior.

[29] However, such a network may not be supported in a bilateral equilibrium of Myerson's linking game. For instance, take g such that $g = g^{S_1} \cup g^{S_2}$ where $S_1, S_2 \in \Pi(g)$, $S_1 \cup S_2 = N$ and $\#S_1 = \#S_2 + 1$. Then, both criminals $i \in S_1$ and $j \in S_2$ benefit from simultaneously adding the link ij between them and deleting one link of i with some other criminal in S_1 to form g'. Indeed, $u_i(g) = (1 - (n - \#S_1)\phi/n)B/n < u_i(g') = (1 - (n - \#S_1)\phi/n)B/\#S_1$ and $u_j(g) = (1 - (n - \#S_2)\phi/n)B/n < u_j(g') = u_i(g')$.

of all networks with a single component where at least one player has $n-1$ links. Some of them are not pairwise stable. Indeed, pairwise stable and strongly efficient networks consist of a single component where $d_i(g) = n-1$ for some i and either $d_j(g) \leq n-3$ or $d_j(g) = n-1$ for any $j \neq i$. In any such network, adding one link for criminals with $d_j(g) \leq n-3$ amounts to receiving more know-how on the crime business, but this additional know-how is not enough to have a chance to win the competition for the loot. So, when myopic criminals decide with whom to establish links, it is not excluded that their decisions lead to some strongly inefficient network.

Herings, Mauleon, and Vannetelbosch (2009) show that from any criminal network there is a farsighted improving path ending in the complete network. Hence, using Proposition 4 we obtain the following proposition.

Proposition 13 (Herings, Mauleon, and Vannetelbosch 2009). *For criminal networks, the set* $\{g^N\}$ *is a pairwise farsightedly stable set.*

From any network g' consisting of complete components, where no two components have the same size, farsighted criminals may add links anticipating that their moves ultimately lead to the complete network. Moreover, from g^N there is no farsighted improving path leading to any g'. Hence, the set $\{g'\}$ cannot be a pairwise farsightedly stable set and farsightedness helps in reducing the conflict between stability and efficiency.

8.3.4 Other Models

There are other economic models that have been analyzed under both the myopic and the farsighted perspective.[30] Grandjean, Mauleon, and Vannetelbosch (2011) find that, in the Jackson and Wolinsky (1996) connections model, farsightedness does not eliminate the conflict between stability and strong efficiency that may occur when costs are intermediate. However, farsightedness helps to reduce the conflict when costs are large enough. In the Kranton and Minehart (2001) model of buyer-seller networks, Grandjean, Mauleon, and Vannetelbosch (2011) show that pairwise farsighted stability may sustain the strongly efficient network while pairwise stability only sustains networks that are strongly inefficient or even Pareto dominated. Finally, Grandjean (2014) investigates the Bramoullé and Kranton (2007) model of risk-sharing networks in developing countries. When the cost for forming links is small, farsighted players form strongly efficient networks while myopic players do not.

[30] We do not intend to cover all economic applications of network formation. We only mention those that, up to now, have been studied adopting both perspectives.

8.4 Conclusion

The outcomes of real-life network formation are affected by the degree of farsightedness of the players. Consider the case where the worth of link creation turns non-negative after some threshold in the connectedness of the network is reached, both for the players and on aggregate, but the players' benefits are negative below this threshold. If network externalities take this form, myopic players can be stuck in insufficiently dense networks. Farsighted players may take care of this problem and achieve the efficient network. If players have a limited degree of farsightedness, their ability to pass the threshold will depend on the number of reactions to their moves they can foresee from the starting network.

Strategic models of network formation have been evaluated in the experimental laboratory.[31] Kirchsteiger, Mantovani, Mauleon, and Vannetelbosch (2013) design a simple network formation experiment to test between pairwise stability and farsighted stability, but find evidence against both of them. Their experimental evidence suggests that subjects are consistent with an intermediate rule of behavior, which can be interpreted as a form of limited farsightedness. Herings, Mauleon, and Vannetelbosch (2014) propose an intermediate concept, namely level-K farsighted stability, that can be used to study the influence of the degree of farsightedness on network stability. Level-K farsighted stability is a tractable concept with myopic and full farsighted behavior as extreme cases.

References

Aumann, R. and R. Myerson (1988). "Endogenous formation of links between players and coalitions: An application of the Shapley value." In *The Shapley value*, A. Roth, ed., Cambridge, UK: Cambridge University Press.

Bala, V. and S. Goyal (2000). "A non-cooperative model of network formation." *Econometrica* 68, 1181–1229.

Belleflamme, P. and F. Bloch (2004). "Market sharing agreements and collusive networks." *International Economic Review* 45, 387–411.

Berninghaus, S. K., K.-M. Ehrhart, and M. Ott (2012). "Forward-looking behavior in Hawk-Dove games in endogenous networks: Experimental evidence." *Games and Economic Behavior* 75, 35–52.

Bloch, F. (1995). "Endogenous structures of association in oligopolies." *Rand Journal of Economics* 26, 537–556.

[31] There is a growing literature on experiments with network formation. See Corbae and Duffy (2008), Goeree, Riedl, and Ule (2009), Berninghaus, Ehrhart, and Ott (2012), and Falk and Kosfeld (2012), among others.

Bloch, F. (2005). "Group and network formation in industrial organization: A survey." In *Group Formation in Economics: Networks, Clubs and Coalitions*. G. Demange and M. Wooders, eds., Cambridge, UK: Cambridge University Press.

Bloch, F. and M. O. Jackson (2006). "Definitions of equilibrium in network formation games." *International Journal of Game Theory* 34, 305–318.

Bloch, F. and M. O. Jackson (2007). "The formation of networks with transfers among players." *Journal of Economic Theory* 133, 83–110.

Bramoullé, Y. and R. Kranton (2007). "Risk-sharing networks." *Journal of Economic Behavior & Organization* 64, 275–294.

Calvó-Armengol, A. (2004). "Job contact networks." *Journal of Economic Theory* 115, 191–206.

Calvó-Armengol, A. and R. Ilkilic (2009). "Pairwise stability and Nash equilibria in network formation." *International Journal of Game Theory* 38, 51–79.

Calvó-Armengol, A. and Y. Zenou (2004). "Social networks and crime decisions: The role of social structure in facilitating delinquent behavior." *International Economic Review* 45, 939–958.

Chwe, M. S. (1994). "Farsighted coalitional stability." *Journal of Economic Theory* 63, 299–325.

Corbae, D. and J. Duffy (2008). "Experiments with network formation." *Games and Economic Behavior* 64, 81–120.

Currarini, S. and M. Morelli (2000). "Network formation with sequential demands." *Review of Economic Design* 5, 229–249.

De Sinopoli, F. and C. Pimienta (2010). "Costly network formation and regular equilibria." *Games and Economic Behavior* 69, 492–497.

Dutta, B., S. Ghosal, and D. Ray (2005). "Farsighted network formation." *Journal of Economic Theory* 122, 143–164.

Dutta, B. and S. Mutuswami (1997). "Stable networks." *Journal of Economic Theory* 76, 322–344.

Ehlers, L. (2007). "von Neumann-Morgenstern stable sets in matching problems." *Journal of Economic Theory* 134, 537–547.

Falk, A. and M. Kosfeld (2012). "It's all about connections: Evidence on network formation." *Review of Network Economics* 11(3), Article 2.

Furusawa T. and H. Konishi (2007). "Free trade networks." *Journal of International Economics* 72, 310–335.

Galeotti, A., S. Goyal, and J. Kamphorst (2006). "Network formation with heterogeneous players." *Games and Economic Behavior* 54, 353–372.

Gilles, R. P., S. Chakrabarti, and S. Sarangi (2012). "Nash equilibria in network formation games under consent." *Mathematical Social Sciences* 64, 152–158.

Gilles, R. P., and S. Sarangi (2010). "Network formation under mutual consent and costly communication." *Mathematical Social Sciences* 60, 181–185.

Goeree, J. K., A. Riedl, and A. Ule (2009). "In search of stars: Network formation among heterogeneous agents." *Games and Economic Behavior* 67, 445–466.

Goyal, S. (2007). *Connections: An Introduction to the Economics of Networks*. Princeton, NJ: Princeton University Press.

Goyal, S. and S. Joshi (2003). "Networks of collaboration in oligopoly." *Games and Economic Behavior* 43, 57–85.

Goyal, S. and S. Joshi (2006a). "Unequal connections." *International Journal of Game Theory* 34, 319–349.

Goyal, S. and S. Joshi (2006b). "Bilateralism and free trade." *International Economic Review* 47, 749–778.

Goyal, S. and J. L. Moraga-Gonzalez (2001). "R&D networks." *RAND Journal of Economics* 32, 686–707.

Goyal, S. and F. Vega-Redondo (2007). "Structural holes in social networks." *Journal of Economic Theory* 137, 460–492.

Grandjean, G. (2014). "Risk-sharing networks and farsighted stability." *Review of Economic Design* 18, 191–218.

Grandjean, G., A. Mauleon, and V. Vannetelbosch (2011). "Connections among farsighted agents." *Journal of Public Economic Theory* 13, 935–955.

Grandjean, G. and W. Vergote (2014). "Network formation among rivals." CEREC Discussion Paper 2014-9. Université Saint-Louis, Brussels, Belgium.

Hellmann, T. (2013). "On the existence and uniqueness of pairwise stable networks." *International Journal of Game Theory* 42, 211–237.

Herings, P. J. J., A. Mauleon, and V. Vannetelbosch (2009). "Farsightedly stable networks." *Games and Economic Behavior* 67, 526–541.

Herings, P. J. J., A. Mauleon, and V. Vannetelbosch (2014). "Stability of networks under level-K farsightedness." CORE Discussion Paper 2014-32. Université catholique de Louvain, Belgium.

Hojman, D. A. and A. Szeidl (2008). "Core and periphery in networks." *Journal of Economic Theory* 139, 295–309.

Jackson, M. O. (2008). *Social and Economic Networks.* Princeton, NJ: Princeton University Press.

Jackson, M. O. and A. Watts (2001). "The existence of pairwise stable networks." *Seoul Journal of Economics* 14, 299–321.

Jackson, M. O. and A. Watts (2002). "The evolution of social and economic networks." *Journal of Economic Theory* 106, 265–295.

Jackson, M. O. and A. Wolinsky (1996). "A strategic model of social and economic networks." *Journal of Economic Theory* 71, 44–74.

Jackson, M. O. and A. van den Nouweland (2005). "Strongly stable networks." *Games and Economic Behavior* 51, 420–444.

Kirchsteiger, G., M. Mantovani, A. Mauleon, and V. Vannetelbosch (2013). "Limited farsightedness in network formation." CORE Discussion Paper 2013-33. Université catholique de Louvain, Belgium.

Kranton, R. E. and D. F. Minehart (2001). "A theory of buyer-seller networks." *American Economic Review* 91, 485–508.

Mauleon, A., E. Molis, V. Vannetelbosch, and W. Vergote (2014). "Dominance invariant one-to-one matching problems." *International Journal of Game Theory* 43, 925–943.

Mauleon, A., J. J. Sempere-Monerris, and V. Vannetelbosch (2008). "Networks of knowledge among unionized firms." *Canadian Journal of Economics* 41, 971–997.

Mauleon, A., J. J. Sempere-Monerris, and V. Vannetelbosch (2014). "Farsighted R&D networks." *Economics Letters* 125, 340–342.

Mauleon, A., H. Song, and V. Vannetelbosch (2010). "Networks of free trade agreements among heterogeneous countries." *Journal of Public Economic Theory* 12, 471–500.

Mauleon, A., V. Vannetelbosch, and W. Vergote (2011). "von Neumann-Morgenstern farsightedly stable sets in two-sided matching." *Theoretical Economics* 6, 499–521.

Myerson, R. (1991). *Game Theory: Analysis of Conflict.* Cambridge, MA: Harvard University Press.

Navarro, N. (2014). "Expected fair allocation in farsighted network formation." *Social Choice and Welfare* 43, 287–308.

Page, F. H., Jr. and M. Wooders (2009). "Strategic basins of attraction, the path dominance core, and network formation games." *Games and Economic Behavior* 66, 462–487.

Page, F. H., Jr., M. Wooders, and S. Kamat (2005). "Networks and farsighted stability." *Journal of Economic Theory* 120, 257–269.

Patacchini, E. and Y. Zenou (2008). "The strength of weak ties in crime." *European Economic Review* 52, 209–236.

Roketskiy, N. (2012). "Competition and networks of collaboration." Mimeo, New York University.

Slikker, M. and A. van den Nouweland (2001). "A one-stage model of link formation." *Games and Economic Behavior* 34, 153–175.

Tercieux, O. and V. Vannetelbosch (2006). "A characterization of stochastically stable networks." *International Journal of Game Theory* 34, 351–369.

von Neumann, J. and O. Morgenstern (1944). *Theory of Games and Economic Behavior.* Princeton, NJ: Princeton University Press.

Yi, S-S. (1998). "Industry profit-maximizing joint-venture structure in a linear Cournot oligopoly." *Economics Letters* 58, 361–366.

Zhang, J., L. Xue, and L. Zu (2013). "Farsighted free trade networks." *International Journal of Game Theory* 42, 375–398.

CHAPTER 9

LINKS AND ACTIONS IN INTERPLAY

FERNANDO VEGA-REDONDO

9.1 Introduction

In social contexts, agents not only choose their actions but often also have substantial control over whom they interact with when choosing those actions.[1] Therefore, in the end, if every agent behaves in this fashion, there must be a joint determination of both choice dimensions (i.e., what, in common jargon, is described as a *co-evolution of links and actions*). In the study of many phenomena of interest, such an integrated consideration of the link-action interplay is key to understand the problem at hand. In particular, if we abstract from it, either the model may fail to produce sharp predictions or, if it does, those predictions could be quite misleading.

In this chapter, I cannot review exhaustively the wide range of theoretical research that bears on the issue. Thus what I will do is illustrate some of the main ideas involved through a collection of paradigmatic applications. Specifically, I will start by focusing on three key but abstract problems: *coordination, cooperation, and intermediation*. Then, somewhat more concretely, I will turn to briefly summarizing how network co-evolution shapes the outcome for other important phenomena such as bargaining, local public goods, learning, and conflict. For all of these cases, I shall argue that a modeling approach that jointly accounts for the adjustment of links and actions provides important insights on the problem at hand that would be largely missed if they were separately considered.

[1] See Chapters 5 and 8 in this volume, the first coauthored by Yann Bramoullé and Rachel Kranton, and the second by Ana Mauleón and Vincent Vannetelbosch. Whereas the former focuses on how agents play on given networks, the latter considers the issue of how the network itself is shaped by agents' decisions.

9.2 Coordination

Coordination is certainly one of the key aspects of social interaction. The easiest way to model it is through a simple bilateral game with two actions, A and B, and a payoff table of the following form:

i \ j	A	B
A	a, a	c, d
B	d, c	b, b

(9.1)

where i and j are generic players connected by the network and a, b, c, d are payoffs satisfying $a > c$ and $b > d$. Thus the bilateral game has two pure strategy equilibria, and let us suppose that $a > b$, so that the one given by (A,A) defines the unique Pareto efficient configuration. Suppose first that the social network is given by an undirected graph $\Gamma = \{N,L\}$, where $N = \{1,2,\ldots,n\}$ is the set of players and L the set of links of the form ij (equivalently, ji) for any pair $\{i,j\}$ of connected players.

To make our point in a particularly transparent manner, let us suppose that the network Γ is a ring and every player i is connected to $i-1$ and $i+1$ (with $n+1$ identified with 1 and 0 with n). An important assumption is that every player must choose the same action, A or B, when playing the game with each of her neighbors in Γ. This is the only case in which the network could have any effect on the pattern of play. Then, it is clear that at least two action profiles define Nash equilibria of the population-wide (network-based) game where every player chooses her action independently. These are the two homogeneous configurations: $s_A = \{A,A,\ldots,A\}$ and $s_B = \{B,B,\ldots,B\}$. In fact, if we rule out the knife-edge case where $a+c = d+b$, they are the only two equilibria.

This begs the question of whether these two action profiles are equally robust in the face of some dynamic process of adjustment, say the classical best-response dynamics. In the absence of any noise or perturbation, the answer must be positive: essentially, both appear equally "solid" (i.e., locally stable for those dynamics). This is why the early evolutionary literature concerned with tackling the resulting equilibrium-selection issue introduced the possibility that, over time, the agents may be subject to what is often labeled as "mutations" (i.e., rare and random changes in their current action).[2] In this context, the focus then turned to identify those states that are called stochastically stable (i.e., those that have positive weight in the invariant distribution of the process

[2] The seminal papers are Kandori, Mailath, and Rob (1993) and Young (1993), which studied the problem under the assumption that interaction is global (i.e., every player confronts everyone else). In a subsequent paper, Ellison (1993) analyzed the case considered here, with interaction being local and players placed along a ring.

when the probability of mutation becomes arbitrarily small). It was then found that the so-called risk-dominant action is selected in the long run—or, more precisely, the only *stochastically stable state* is the one where every agent chooses that action. How does one define such an action? It is the one that yields the higher expected payoff when facing an equal probability that the other player chooses any of the two actions, A or B. If, for example, $b + d > a + c$, action B qualifies as risk-dominant, while action A may continue being the Pareto-efficient action. The intuition for the selection of the risk-dominant action is widely understood by now, so we do not elaborate on it here. Suffice to say that risk dominance makes the transition from s_A to s_B so much easier in terms of the number of mutations required that, when the probability of mutation is small,[3] the process spends "most of the time" at, or in the vicinity, of s_B.

Suppose now that the network is not fixed but can change over time, either by agents creating new links or/and destroying existing ones, in addition to their ability to adjust actions. Jackson and Watts (2002) and Goyal and Vega-Redondo (2005) have studied this problem in different frameworks—the former in a two-sided network-formation context, while the latter considers a one-sided one. Here I summarize the approach pursued by Jackson and Watts. Their framework is essentially that described for a fixed network but with the *added* possibility that agents can also form or/and destroy existing links when randomly given the opportunity. Links are taken to be costly, with a constant creation/maintenance cost for every prevailing link, which is split equally between the two agents involved. Being a two-sided mechanism, the assumption is that any given link under consideration remains in place (when existing) or it is formed (when not present) if, and only if, its net benefits are positive for both players connected by it. Then, the possibility of mutation (now affecting both actions and links) is again added to the dynamics, and the question is posed as to what is the long-run outcome where the system spends most of the time when the mutation probability is very small (i.e., infinitesimal). The main conclusion is that, in this context, an efficient action carries a positive long-run weight, even if it is risk-dominated, provided that the linking cost lies at an intermediate level (higher than the off-diagonal payoffs, c and d, but still lower than the highest payoff a).

What is the intuition for this conclusion? It is based on the following simple insight. If the costs are as described, any configuration with two completely connected and homogeneous components where every player chooses the same action (A in one and B in the other) is stationary in the absence of mutations. But then, through drift, the relative sizes of the two components can change in either direction and at the same rate (in terms of single mutations). The only case in which a single mutation will not work to produce a persistent transition is when the whole population is in a single component, all choosing the same action, A or B. In this case, *two* mutations are required to escape the situation, since this is what is needed to create a non-trivial

[3] This conclusion presumes that all mutations are infinitesimals of the same order. However, in some cases there may be reasons why this is not a good assumption, and deviating from it alters significantly the selection results. For an early discussion of this issue, see Bergin and Lipman (1996).

component consisting of at least two agents. But then, given that the same number of mutations are needed to transit away from an homogeneous population in A or B, these two configurations should obtain in the long run with positive frequency, even as the mutation rate becomes infinitesimally small.

The former analysis has the following limitations. On the one hand, by relying crucially on low-probability mutation, its analysis, in effect, must be viewed as having a very (ultra-)long-run character and therefore questionable practical relevance. On the other hand, it does not shed light on what may be the topology, or the relative sizes of the components. But when we think of social coordination, those relative sizes are clearly an important aspect of the problem. However, for low mutation, it is easy to see that the prediction of the model by Jackson and Watts (2002)—also of that by Goyal and Vega-Redondo (2005)—is that most of the time the whole population will be in a *single and completely connected homogeneous component.*

To arrive at a less drastic conclusion, it seems necessary to relax the assumption that the perturbation component of the model operates at an infinitesimal level throughout. This was the route pursued by Ehrhardt, Marsili, and Vega-Redondo (2008) in a model where, to focus the discussion, the coordination game is assumed to have $a = b > 0$ and $c = d = 0$—hence it is a pure and action-symmetric coordination game. Over time (modeled continuously), agents form links when they meet (at random) some other agent with whom they can obtain positive payoffs. Their actions, on the other hand, are chosen as a (weak) best response to the current state. Finally, the "perturbation" required to make the analysis interesting is conceived in a minimalist manner as *link volatility*. That is, it is assumed that all existing links vanish at a constant rate. Combining all these features, an ergodic process is obtained whose unique invariant distribution can be characterized. In particular, the long-run states (i.e., those that have positive weight in the invariant distribution) are found to display the following features. If the population is large, there is a threshold value $\hat{\eta}$ for the volatility rate η such that

1. if $\eta > \hat{\eta}$, in every long-run state the population is fragmented into "maximum miscoordination" (i.e., half of the population chooses each action) and the underlying network is sparsely connected (no giant component exists);
2. if $\eta < \hat{\eta}$, in every long-run state one action attracts a majority of the population and the underlying network displays a giant component, whose fractional size increases as volatility decreases.

Thus the magnitude of volatility has a major effect not only on the extent of coordination but also on the corresponding density of connections prevailing in the society. This conclusion contrasts with that obtained in the aforementioned models of Jackson and Watts (2002) and Goyal and Vega-Redondo (2005), where coordination and connectivity are at a maximum in every long-run state. This is not the case in the model of Ehrhardt, Marsili, and Vega-Redondo (2008) because, since noise (i.e., volatility) operates at a non-infinitesimal level, full connectivity is no longer maintained at long-run states, and hence it interplays with the extent of action coordination in richer ways.

9.2.1 Summary of Additional Literature

The co-evolution of links and actions in coordination games has been studied from a number of other different perspectives. Hojman and Szdeil (2006) consider a one-sided model of network formation similar to that of Goyal and Vega-Redondo (2005) but assume that the partners of any given player are not just her direct neighbors but indirect ones as well (i.e., everyone in her component). They find that equilibria are selected by a criterion that embodies some trade-off between efficiency and risk dominance. Staudigl (2011) posits a model where the underlying coordination game is a partnership (in our notation, $d = c$) and both links and actions are adjusted according to a logit formulation. The dynamics also includes a process of volatility as described before, by which links are destroyed at a constant rate. The paper derives the invariant distribution of the process and shows that the efficient action is selected in the long run when the noise displayed by the logit formulation becomes arbitrarily small.

Another specific scenario where actions and links co-evolve under strategic complementarities is the one studied by König, Tessone, and Zenou (2014), which is based on the linear-quadratic model introduced by Ballester, Calvó-Armengol, and Zenou (2006). They show that the model leads to networks called "nested split graphs," which are networks that exhibit a strict hierarchical structure. More general games displaying strategic complementarities have been studied by Hiller (2012) and Baetz (2014). They model the network-formation problem as a simultaneous game where agents decide jointly on actions and links, the key difference between them being that the former assumes that the value function is convex while the latter postulates that it is concave. Baetz also restricts to one-sided network formation. They both show that some interesting structures arise at equilibria, such as core-periphery networks in the first paper and stratified networks in the second.

Finally, we refer to those papers that have addressed essentially the same issue (the co-evolution of conventions and connections in coordination games), but in a model where agents, rather than choosing their neighbors directly, do it indirectly by selecting a "location." More specifically, the key assumption is that each agent interacts with all those placed at the same location. In general, the main insight gained from these models (see, e.g., Ely 2002 or Bhaskar and Vega-Redondo 2004)—is that allowing for such a possibility of migration tends to favor the rise of efficient conventions. This approach is reminiscent of the celebrated work of Schelling (1972, 1978), who also explored the dynamic implications of individual self-motivated moves on some measure of social welfare (e.g., ethnic integration). There, however, individual mobility can lead to undesired consequences (segregation), even if individuals do not display a marked preference for it. Recently, a quite versatile model studying such interplay between individual incentives, local peer effects, and social welfare has been proposed by Badev (2013), who has applied it to the study of teenage smoking.

9.3 Cooperation

Cooperation is another key phenomenon where the interplay of network-based interaction and link adjustment may play a key role. In principle, however, it is possible to conceive "behavioral" setups where such an interplay is not needed for cooperation to arise. That is, local interaction alone can still support cooperation in some cases, provided agents behave in a somewhat less than fully rational manner. A good example is provided by the well-known model of Eshel, Samuelson, and Shaked (1998), which we outline next.

In their model agents are arranged on a ring and interact with their two neighbors. If they behave altruistically (say, by producing a local public good), they benefit their neighbors by one unit but have to pay a cost of $C > 0$ (net of the gross benefits they might derive from the public good). This obviously makes behaving egoistically (i.e., not providing the public good) a dominant strategy. Hence no truly rational agent will ever provide the public good, and the prediction in that case is that full egoism must prevail throughout the population.

Assume instead that agents do not guide their actions by direct payoff maximization but simply adjust their behavior by mimicking the action of whoever achieves the highest average payoff in their neighborhood. Then, Eshel et al. show that some altruism can be supported as a stable configuration. To see this, first note that, of course, an all-altruism situation will remain stable if the alternative egoist behavior is not present at all. (This simply follows from the fact that imitation by itself cannot bring up any new behavior.) The problem, however, is that such a homogeneous cooperative state is extremely fragile and any deviation from it will lead to a large change away from altruism. Thus suppose instead that the initial state is chosen at random. What is then the likely outcome? It is not difficult to see that, if altruism is not too costly (specifically, if $C \leq 1/2$), the final outcome of an imitation-based adjustment process as described is very likely to lead to a configuration where at least 60% of the agents are altruists. In fact, the same happens as well if, as we contemplated in our discussion of coordination games, the imitation dynamics is perturbed by some small "mutation"—it turns out that at least 3/5 of the population must be altruist in any stochastically stable state.

The problem with this apparently promising conclusion is that it is quite fragile. For, as shown by Mengel (2009), even if agents enjoy a "radius of information" that is just slightly longer than the range at which they interact with others (for example, they interact with their immediate neighbors but observe instead the payoffs obtained by their first- and second-order neighbors), full defection is the only stochastically stable state. This happens as well if, instead of the ring, we allow for some irregular networks. Such an ambiguous state of affairs is indeed underscored by recent experimental evidence, which shows that cooperation on *fixed networks* generally remains low and is hardly affected by the type of local interaction considered. For example, Grujić et al. (2010) and Gracia-Lázaro et al. (2012) have conducted controlled Prisoner's Dilemma experiments on large networks, varying their size and degree of heterogeneity.

Comparing these treatment scenarios with a control setup in which the network is "reshuffled" at every round (and hence the network is largely irrelevant), they find no significant differences in the amount of cooperation displayed by the treatment and control groups for the different cases.

Given the theoretical and empirical considerations outlined above, the question arises as to what mechanism can be effective in supporting cooperation in a network context. And again, the natural option that is now considered is that an endogenous co-evolving network could prove an effective way to overcome the aforementioned limitations. In fact, experimental evidence (see, e.g., the interesting paper by Wang, Suri, and Watts 2012)[4] does provide strong support for this suggestion. Next, I discuss two papers (Fosco and Mengel 2011 and Vega-Redondo 2006) that study, in two very different theoretical frameworks, the significant contribution that co-evolving networks may lend to the rise of cooperation.

The paper by Fosco and Mengel (2011) considers what is probably the simplest *co-evolutionary counterpart* of the model proposed by Eshel, Samuelson, and Shaked (1998). It can be seen, therefore, as a natural way of assessing whether network endogeneity can indeed remedy some of the aforementioned problems. The model involves a stochastic dynamic process through which agents play a Prisoner's Dilemma with their current neighbors and can also adjust their links over time. Its main novelty revolves around the contrast between two different ways in which local effects can be conceived and formalized. A first one is in terms of what is called the *radius of interaction*, which specifies the network range at which any given agent interacts (i.e., plays the game) with others and obtains corresponding payoffs. The second one is in terms of the *radius of information*, which indicates how far into the network an agent can access payoff-relevant information. (This information includes not only the payoffs others obtain and the actions they choose, but also who are the available new partners.) Then, given how payoffs and information are jointly determined by these two radii, the postulated adjustment involves a natural imitation dynamics, both for actions and links.

The methodology used by Fosco and Mengel to analyze their model is a standard one: the outlined dynamics is perturbed with some small noise/mutation and the focus is on the stochastically stable states—those states that are visited with significant frequency even for infinitesimal noise. The main conclusion is that all stochastically stable states are polymorphic—hence they display *some* amount of cooperation—but the degree of cooperation achieved and the details of its implementation crucially depends on both the radius of information and that of interaction. For example, if the former is longer than the latter, the population is split into two components (the cooperative and the

[4] Their experimental setup involved groups of 24 individuals who were involved in an iterated Prisoner's Dilemma lasting 12 rounds. At every such round, they had to choose whether to be cooperators and defectors and also could update their links (only a certain fraction, which was a design parameter). The extent of cooperation was substantially increased by link flexibility, and the effect became stronger the higher the fraction of links available for revision, even if the cost of a cooperator meeting a defector grew higher.

uncooperative one), which means that, in this case, cooperation is maintained by having the cooperators being shielded from the exploitation of opportunistic defectors.

If, instead of allowing for some bounded rationality (e.g., imitation, as above) one insists on sticking to the traditional full-rationality paradigm, repeated play has been the classical equilibrium approach in which the game-theoretic literature has understood the rise of altruistic behavior in social contexts. That is, if the same set of agents are facing each other over time, the (credible) threat of future punishment can sometimes be enough of a deterrence against opportunistic behavior. But if interaction is purely bilateral, the possibilities of supporting altruism in this manner may be quite limited due to a number of factors (e.g., highly impatient agents or too-tempting rewards for opportunistic behavior). And when this happens, it is conceivable that embedding the bilateral interaction in a larger social network can remedy the problem by enlarging significantly the scope of punishment and reward. But this then begs the question of whether the given ("initial") network will display the features required for such a state of affairs—or, if not, whether there are endogenous forces at play that may lead the network on that direction.

The model studied by Vega-Redondo (2006) aims at shedding light on this issue. In it agents are involved in a *separate* Repeated Prisoner's Dilemma with each of their neighbors on an evolving social network. The focus is on grim-trigger (subgame-perfect) equilibria, which embody the threat of reversion to indefinite defection with any neighbor who is found to have violated the cooperative "social norm." Two different social norms (or, equivalently, types of equilibria) are considered: the network-based and the network-free ones.

(a) Under the *network-based norm*, agents are supposed to punish (in equilibrium) not only those partners who have defected in their own bilateral interaction but also those who have done it with third parties.[5] Thus, in this case, even though the actions played in different bilateral interactions are independently chosen by players, agents' behavior is *not strategically* independent across those interactions.
(b) Under the standard *network-free norm*, each bilateral repeated game is played in a strategically independent manner across connected pairs. Thus, the behavior in any given bilateral interaction only depends on what has happened in that same interaction.

The main objective of the chapter is to contrast the implications—both on the resulting network as well as on the corresponding ability to support cooperation—of the two previous social norms. This is done in terms of a co-evolutionary process of network and strategy change that includes the following two components, in operation at each point in time.

[5] This punishment, however, is only implemented with some delay since information about the defection on other agents is assumed to be channeled gradually though the network itself.

1. *Link creation*: Individuals explore the network and search for new linking opportunities; when some are found new links are created iff, given the prevailing network, cooperation can be strategically supported on them.
2. *Volatility*: The payoffs associated to existing links are redrawn with some given probability; once this is done, all links are reconsidered and only those that can still support cooperation are maintained.

The analysis combines mean-field techniques (i.e., the study of the deterministic dynamics given by the expected law of motion) as well as numerical simulations. Some of the conclusions are as expected. For example, it is found that volatility has a negative impact on the density of the networks that are sustainable in the long run (recall that each prevailing link must be a cooperative one). The social norm in place, however, has a very substantial effect on how the population confronts the detrimental consequences on increased volatility. While under the network-based norm, the impact is gradual and moderate, in the alternative network-free norm it is abrupt (i.e., of the threshold type) and much larger in magnitude. Such contrasting behavior can be understood as follows. Under the network-based norm, the links *endogenously reconfigure* themselves so that the social structure becomes more cohesive, and hence the population can support cooperation more effectively. Instead, under the network-free norm, no such reconfiguration occurs (nor is relevant, for that matter), which means that the population ends up being much less successful in coping with volatility.

9.3.1 Summary of Additional Literature

The literature that studies how alternative interaction structures, endogenous or not, bear on the rise of cooperation is so large and diverse that my account of it here will be unavoidably sketchy and very partial. I started this section by illustrating the fact that, in order for cooperation to be robustly sustained in a network setup, agents displaying simple (say, imitative) behavior must enjoy not only the flexibility of adjusting their actions but also exhibit some plasticity in their linking behavior. In a network setup, a further illustration of this point can be found in the work of Eguíluz et al. (2005), Pacheco, Traulsen, and Nowak (2006), Ule (2008), and Bilancini and Boncinelli (2009). A similar idea arises as well in network-free environments if agents can endogenously determine when to maintain a given partnership or return to a common matching pool. Fujiwara-Greve and Okuno-Fujiwara (2012), and Izquierdo, Izquierdo, and Vega-Redondo (2014) study evolutionary models of this sort, in which the evolutionary stable equilibria is found to support cooperation under some conditions. Still a different evolutionary mechanism of partner selection is given by group selection. In this case, evolution proceeding at different levels (within and across groups) can again support cooperative behavior if the discipline imposed by group selection is strong/fast enough.

Simple illustration of this mechanism can be found in Vega-Redondo (1996) and Traulsen and Nowak (2006).

Concerning repeated interaction, the role of embeddedness (or what Nowak 2006 has called indirect reciprocity) in supporting cooperation has been highlighted by the literature studying this phenomenon at the interface of the theories of games and networks.[6] An early instance is the paper by Raub and Weesie (1990), whereas recent contributions include those of Ali and Miller (2009), Lippert and Spagnolo (2011), Haag and Lagunoff (2011), Jackson, RodrÃ-guez-Barraquer, and Tan (2012), Fainmesser (2014), and Immorlica, Lucier, and Rogers (2014). The first three papers stress the role that network structure (exogenously postulated) has in the possibility of supporting cooperation at equilibrium. The latter two instead introduce network endogeneity into the analysis, although only concerning link destruction within some exogenously given network of ex-ante linking possibilities.

9.4 Intermediation

Often, the social network not only determines the pattern of *direct* interaction among agents but also specifies how these agents connect indirectly—say to communicate, collaborate, or compete. This is the phenomenon we may call intermediation. It reflects the natural idea that, even if agents are far from each other in the social network, they can rely on others lying along a path, joining them in order to establish a valuable connection. Then, as a natural follow-up question, the issue arises as to whether those agents facilitating such indirect connection will obtain (and possibly compete for) some share in the value/surplus thus generated. In line with the theme of this chapter, our emphasis here will be on how these considerations shape the incentives to form links (i.e., direct connections) and how these links feed back on the actual distribution of the surplus among all the agents involved.

The notion of intermediation can play an important role in many different contexts (e.g., trade brokerage, technological or scientific research, trust and "social collateral," job search and referral, etc). (see Chapter 32). Here I shall approach it from a general and abstract viewpoint, focusing my discussion on a model studied by Goyal and Vega-Redondo (2007) that proposes a stylized formulation of the problem that is not specifically tailored to any given scenario.

Given any fixed population of n ex-ante identical agents and some given social network, the starting assumption of the model is that any pair of agents i and j who can connect, *directly or indirectly,* can earn a unit of surplus. If their connection is direct (i.e., there is a link between them in the social network), they need no one else to earn this surplus, so any efficient and symmetric bargaining procedure should have them divide the "pie" equally. Instead, if they need "intermediaries" to connect indirectly, the

[6] For a discussion of how networks bear on strategic play in repeated games see Chapter 6 by Francesco Nava in this volume.

key issue is whether some of these agents are essential (i.e., they are in *every* path joining them). If so, they should be expected to also demand a share of the surplus. And since, in fact, they are as crucial as either i or j in generating the surplus, all of them should demand the same share, and only they (the truly essential individuals) will obtain a positive cut. Formally, this reasoning can be provided with microfoundations as the Kernel of a coalitional bargaining game involving the whole population.

Clearly, the preceding analysis allows for a wide range of payoff possibilities, depending on the underlying network. Just to illustrate this richness, consider the following two polar cases: a ring network and a star network. In the case of the ring, every two agents are of course connected, and the important point is that there are two possible and disjoint paths in every case. This means that no player is an essential intermediary and, therefore, every pair of agents in the population can not only generate a unit of surplus but divide it between the two of them alone. In the end, every player obtains the same total payoff, equal to $\frac{1}{2}(n-1)$.

Now contrast the former situation with the one where the social network is a star and, say, agent 1 is at the center of it. Clearly, on the one hand, agent 1 can generate a unit of surplus with each of the $n-1$ spoke agents without any intermediation required. But, being at the center, the central agent can also intermediate, in an essential manner, the connections between every pair $\{i,j\}$ of other agents in the population ($i \neq 1 \neq j$). Thus, for each of these $\frac{1}{2}(n-1)(n-2)$ cases, agent 1 will get 1/3 of the unit surplus generated. Overall, therefore, the payoff of agent 1 will be much higher than that of the other agents. In the terminology put forward by Burt (1992), agent 1 fills a "structural hole" in the social network and is able to extract hefty rents from that.

But then the important issue is whether such asymmetry in agents' network positions can be understood/rationalized as the outcome of a network-formation mechanism that reflects the strategic incentives of the agents involved. In other words, one would like to know which structures and corresponding outcomes should obtain when agents' linking decisions are taken in anticipation of the intermediation rents induced. For, in general, it is clear that in parallel to the incentives of agents to create (and fill in) structural holes, there are the opposing incentives to destroy essential intermediation through the creation of additional links.

To explore this tension, Goyal and Vega-Redondo (2007) propose a network-formation model where, as usual, any (two-sided) link can be created by bilateral consensus of the two agents to be connected while it is destroyed if any of the agents involved objects to its remaining in place.[7] On the other hand, links are assumed costly, so that only those links that provide a benefit above their cost are formed. The key two assumptions of the model are as follows. First, any two agents directly or indirectly connected can earn a unit surplus. Second, that surplus is equally divided among the two agents in question and any other agent who might be a crucial intermediary in establishing the indirect connection between the former two. Under these conditions, the main conclusion is

[7] Specifically, the concept used to model network formation is *Strict Bilateral Equilibrium*, which is a refinement of (strict) Nash equilibrium that requires the robustness of both individual and bilateral deviations.

very clear-cut: there is a certain threshold such that if the linking cost is below it, then the only equilibrium network is the complete one; instead, if the cost lies above the threshold, the only equilibrium architecture is a star.

The result for a low linking cost below the threshold is clear. For, if links are very cheap, it must be worthwhile establishing direct connections to every other agent and thus save on intermediation payments. But why is the star architecture the only strategically stable architecture for linking costs above the threshold? The reason is that, for any structure other than the star, at least one of the two following statements is true:

(a) some *individual agents* have at least weak incentives to *unilaterally* severe their links (i.e., they have no strict incentives to maintain all of them);
(b) there is some pair of agents who can divide the network into separate components (each of them lying in one of them) by destroying some of the links under their control.

To understand the essential point, suppose that the two alternative configurations to be considered are just a ring and a star. In both of them, if the linking costs are sizable and the population is large, agents have *strict individual incentives* to keep all their links.[8] The ring, however, allows for the possibility that any two agents at opposite sides of it can act in coordination and achieve the following outcome: first, they may break the ring into two separate (line) components, opening a "structural hole"; second (but simultaneously), they can establish a new link between themselves, close the aforementioned hole, and by so doing become central and crucial to many pairs of other agents. This is precisely the profitable bilateral deviation that renders the ring an unstable configuration. Obviously, the same argument does *not* apply to the star, which is stable in the face of any possible deviation, unilateral or bilateral. So, in sum, again we find that by endogenizing the prevailing network one obtains sharper predictions and important insights that hinge upon the action-link interplay embodied in the network-formation game.

9.4.1 Summary of Additional Literature

The general issue of intermediation has received significant attention in the recent literature, with a particular focus on bargaining and trade (see Chapter 27 by Condorelli and Galeotti and Chapter 28 by Thomas Chaney). In an abstract context, stylized models have been proposed by Gale and Kariv (2007), Blume et al. (2009), Manea (2013), Nava (2014), and Siedlarek (2014). These papers differ in their specific details

[8] Note that, in a ring, the elimination of one of the links would put the two agents who were involved in that link at the end of a line. Thus, in order to still connect with many others, a high number of crucial intermediaries would then be needed, and hence high intermediation rents should be paid. The ring, instead, provides the two-path redundancy that allows agents to save paying those rents.

of the trading protocol (e.g., whether agreements are bilateral or multilateral) but have in common two key features: the network is taken to be exogenously given and both the network as well as all the characteristics of the market participants (costs and valuations) are common knowledge. In contrast, a recent paper by Condorelli and Galeotti (2012) studies trade and intermediation in a context where agents have incomplete information on agents' values.

On the specific context of financial networks, the issue of intermediation has drawn much recent attention, in light of the widely held that it has an important bearing on the robustness of the financial system. Interesting examples of this growing literature are Gofman (2011), Babus (2012), Farboodi (2014), Fainmesser (2012),[9] and Glode and Opp (2014). Gofman focuses on the issue of how the (given) network of trading relationships impinges on the efficiency of the system, whereas Babus and Farboodi study a context where the network is endogenous and responds to the incentives of the agents involved (say, banks or other financial institutions). Both Babus and Farboodi show that agents' incentives lead to polarized hub-spoke networks, a prediction consistent with empirical evidence on financial markets. Finally, the papers by Fainmesser and Glode-Opp explore the role of intermediation in financial markets under two different important extensions: repeated interaction (Fainmesser) and asymmetric information (Glode and Opp). Both show that intermediation can mitigate (and even eliminate) inefficiencies—those due to adverse selection in the first case, and to opportunism in the second.

Finally, in line with the issue of structural holes discussed in abstract terms by Goyal and Vega-Redondo (2007), there are two other papers that have studied the problem in a similar vein: Buskens and van der Rijt (2008) and Kleinberg et al. (2008). Both focus as well on the implications on network formation but, in contrast with the model of Goyal and Vega-Redondo, assume that the payoff for an agent is defined locally (i.e., it depends only their own set of neighbors). In the first case, an agent's payoff is identified with the negative of a magnitude introduced by Burt (1992), which he called *network constraint*.[10] Instead, in the second paper, the gross payoff of an agent i is identified with both the number of neighbors he has as well as the number of neighbor pairs he can intermediate, each of the latter being associated a weight that decreases with the number of alternative two-step paths that compete with the one provided by i.[11] Both papers single out complete multipartite (or multilevel) networks as a prominent prediction of their models.

[9] This is an earlier version of the aforementioned Fainmesser (2014). In that earlier paper the focus is on financial markets.

[10] Intuitively, the network constraint experienced by an individual is high is for the pair of neighbors he can intermediate there are many alternative two-step paths that could perform a similar function.

[11] In the setup proposed by Kleinberg et al. (2008), net payoffs are obtained by subtracting from gross payoffs the cost of links, which are assumed one-sided (and thus paid by one party) but permit two-way flows of payoffs.

9.5 Additional Contexts

As advanced, now I briefly review other socio economic phenomena where an explicit account of the interplay between actions and links also delivers important insights.

9.5.1 Bargaining

The traditional paradigm used in economics to model a market economy has been that of Walrasian equilibrium (see Chapter 2 by Alan Kirman in this volume). The standard version of this approach presumes that prices of homogeneous goods are uniform throughout the economy and trade is *anonymously* conducted at those prices. But at least since the seminal work of Rubinstein and Wolinsky (1985)—see also Gale (1987) and Rubinstein and Wolinsky (1990)—economic theory has made a substantial effort to enrich such a description/model of a market system with the explicit introduction of so-called micro-structure. Much of this recent literature (see Chapter 26 in this volume by Mihai Manea) has adopted a network approach, with buyers and sellers assumed to bargain bilaterally over time with those partners from the other side of the market to whom they are connected in some underlying trading network.

When such a trading network is bipartite-complete (every buyer and every seller are connected to each other) and the discount factor converges to one (i.e., agents become arbitrarily patient), it is straightforward to see that there must be a unique limit price at which all transactions are conducted—that is, the outcome is "arbitrage-free." But, naturally, the most interesting setup is one where the network is *not* complete and, possibly, the agents may occupy asymmetric positions in the network and thus enjoy significantly different bargaining power. In this case, the key question arises as to whether the network in place displays enough connectivity (and a suitable topology) to ensure the "law of one price." This question was addressed by Manea (2011) in an homogeneous context, where all buyers and sellers respectively have the same valuations and costs. Note that, in this case, a single prevailing price is equivalent to a *uniform payoff* for each side of the market. Manea shows that such a price and payoff uniformity situation obtains under buyer and seller homogeneity if, and only if, the underlying trading network displays what is called a perfect matching (i.e., there is a collection of buyer-seller pairs (links) where every buyer and every seller is involved in exactly one such pair).[12]

Then, in line with our approach here, from the previous characterization we are led to the following additional question: Should one expect arbitrage-free networks to obtain if, endogenously, the links are shaped by agents' own (optimizing) decisions? The answer to this question, provided as well by Manea (see the online appendix

[12] In general, a matching—possibly non-perfect—is defined as a collection of links satisfying the restriction that every node is in involved it at most one of them (possibly none).

to the aforementioned paper), is again a sharp one: a network is *pairwise stable* (cf. Jackson and Wolinsky 1996 and Chapter 5) if, and only if, it is arbitrage-free (or nondiscriminatory[13]). Thus, in this sense, price/payoff uniformity is intimately associated to the possibility that the price and the underlying trading networks be endogenously co-determined. A conceptual limitation of this result is that freedom of arbitrage imposes too little structure on the trading network. For, in general, a wide range of arbitrage-free networks—including in *every case* the complete one—are pairwise-stable.

Polanski and Vega-Redondo (2014a) extend the previous analysis to a general context, with an arbitrary distribution of buyers' valuations and sellers' costs. In this context, price uniformity no longer entails payoff uniformity within each side of the market and the outcome (in particular, the payoffs) depends on a rich interplay between network topology and types (valuations and costs). They characterize those networks that induce an arbitrage-free outcome in terms of a condition that can be viewed as a market-based version of the well-known result of Graph Theory—the so-called Marriage Theorem by Hall (1935)—that characterizes the existence of perfect matchings in binary networks. But then, we can ask once more: How are matters affected if (under type heterogeneity) the trading network is endogenous?

Polanski and Vega-Redondo address the previous question but, in contrast with Manea (2011), they do it under the assumption that links are costly (infinitesimally so). This, in essence, implies that the unilateral and bilateral incentive conditions embodied by the notion of pairwise stability require a *strictly positive* gain of linking. Under these conditions, every pairwise stable network is still found to be arbitrage-free. However, the converse conclusion is no longer true because costly linking rules out those networks that include "strategically irrelevant" links—in particular, the complete network is always discarded. An important consequence is that pairwise stability introduces a specific network structure, which is generally nontrivial. It turns out, in particular, that pairwise-stable networks lead to inherent waste (allocation inefficiency) if the matching procedure is decentralized. (Roughly, decentralized matching does not allow the reliance on some global mechanism and prescribes that pairs of agents bargain and trade when they can profitably do so.) This illustrates once more the interesting insights that may arise when a network-based model of economic interaction—trade, in the present case—is subject to the discipline of incentive compatibility in the process of link creation and destruction.

Two additional recent papers in a similar vein are those by Elliot (2014) and Elliott and Nava (2014). The former considers a setup where, in a first stage, the agents have to decide on whether to devote resources to partner-specific investments that will allow them to bargain with those partners in a second stage. The first stage, therefore, can be essentially conceived as determining the trading network on which subsequent bargaining (modeled as the outcome of a cooperative matching procedure)

[13] Manea (2011) uses this terminology since, as explained, under buyer and seller homogeneity they imply a uniform payoff across buyers and sellers.

takes place in the second stage. Elliott finds that, depending of the protocol through which linking costs are divided between partners, inefficiencies of two polar sorts (under- or over-investment) can arise. On the other hand, the paper by Elliott and Nava (2014) models both partner selection and bargaining as part of a single non-cooperative dynamic game. Their analysis relies on an intuitive comparison of the "Rubinstein-type payoffs" that would arise in bilateral bargaining with those belonging to the Core of the economy. They show, specifically, that an efficient equilibrium exists if, and only if, the aforementioned Rubinstein payoffs lie in the Core. These two papers therefore provide an additional illustration of how a modeling approach where links and actions are endogenously determined can enhance the sharpness of the conclusions. In their case, it sheds light on the important issue of how the primitives of the model (e.g., valuation and costs) impinge on the efficiency of the outcome.

9.5.2 Local Public Goods

In Section 9.2, I discussed coordination games, which is a particularly stark example where the actions of players are strategic complements—that is, the incentives for choosing a particular action grow with the number of partners who choose that same action. A polar context is one where players' actions are strategic substitutes (see Chapter 8). In this case, as the number of partners who choose a certain action grows, it is instead the incentive to use the *alternative* strategies that becomes stronger. One example of this type of network games are anti-coordination games, which are studied, for example, by Bramoullé et al. (2004). They focus on how actions and links are jointly co-determined as part of an equilibrium in an overall simultaneous-move game, and characterize how the networks thus induced depend on the cost of forming links.

Another interesting case is provided by the study of local public goods, as studied for example by Bramoullé and Kranton (2007) in a context where agents are located on a *given* network. These authors suppose that agents have to choose a level of costly effort that generates positive spillovers on their immediate neighbors. A natural interpretation is that individual effort is devoted to gathering information, which is then freely shared with neighbors. Depending on the network structure, a wide multiplicity of *very different* equilibria exist. Thus, if agents are arranged in a star, one equilibrium has the center be the sole contributor of effort, while the opposite configuration where the peripheral agents contribute (and the center not) defines an equilibrium as well. Clearly, the former is much more cost-effective (i.e., efficient) than the latter. This begs the question of how matters would be affected if the network were allowed to be determined endogenously. The issue has been addressed by Galeotti and Goyal (2010), who pose a network-formation model where agents can choose (unilaterally and at an individually born cost) with whom to connect and hence enjoy spillovers from.[14] Their

[14] Another way to tackle the problem is to postulate specific mechanisms/criteria of equilibrium selection. This is the approach pursued by López-Pintado (2008), who studies classical (myopic)

main result is that any arbitrarily small asymmetry in the cost of effort induces a single robust equilibrium where the lowest-cost individual is the center of a star and all others establish connections to him. So, in this sense, allowing for a joint determination of action and links supports an efficient effort profile even if, as explained, the associated network configuration (a star) is one where a large inefficiency could in principle obtain at equilibrium if that same star network were to be considered exogenous.

9.5.3 Conflict

Conflict is present in many economic and social interactions, and indeed most games of theoretical and practical interest do have a "conflict" component (see Chapter 10 in this volume by Sanjeev Goyal and coauthors). As a paradigmatic representation of "pure" conflict situations, there are two leading formulations in the literature: the contest-function approach pioneered by Buchanam and Tullock (1962) and that going under the general term of Colonel Blotto games (see Roberson 2006, or Kovenock and Roberson 2012 for a recent survey). Somewhat surprisingly, however, the literature that approaches the problem from a network viewpoint is scarce.

There is some recent literature, discussed in Chapter 13, where a network defines the nature of the conflict by specifying the routes or nodes through which value is generated—and, correspondingly, where attack and defense should gravitate. Another approach is the one pursued by the recent paper of König et al. (2014), which relies on a *binary signed network* to represent the patterns of alliances (friends and enemies) in an all-out conflict set out to determine the shares in some fixed surplus. Just very few papers, however, do model a network of interrelated bilateral conflicts, with the important spillovers that should typically flow across them. A notable exception is the paper by Franke and Öztürk (2009), which explores a context where agents are involved in a set of bilateral conflicts, each one of them modeled as a contest game. The interconnection among different conflicts derives from the fact that, for every agent, the resources devoted to all conflicts in which he is involved induce a total cost that is given by a *convex* function of the aggregate amount. They consider different stylized contexts (star, regular, complete networks) and study how the intensity of conflict depends on the parameters of the model.

All of the papers mentioned above posit a fixed network, signed or not, which is exogenously given. In contrast, the recent papers by Hiller (2012) and Huremovic (2014) extend the analysis to contexts where the network itself is chosen by the agents as part of a network-formation game. In Hiller's model, agents have to choose with what other agents they want to have positive and negative relationships (i.e., their friends and foes). These one-sided decisions, in the end, determine a signed, binary, and directed

best-response dynamics through which a large population adjusts their behavior over time. Relying on a mean-field approach, she shows that a unique stable equilibrium exists, and the fraction of agents devoting effort decreases as the connectivity of the network rises or becomes more spread out.

network. The key feature of his model is that, for any individual agent, the strength he musters in each of the conflicts he has with his enemies increases with the number of his friends. Instead, Huremovic's approach involves two-sided links, with two agents becoming enemies if at least one of them decides so. Then, between all pairs of enemies, a corresponding collection of interrelated contest games are simultaneously played, in a setup quite similar to that considered by Franke and Öztürk (2009).

The notions of network stability are formally different in the aforementioned papers by Hiller and Huremovic, but conceptually they are quite similar. On the one hand, positive links are created and maintained as usually postulated in the literature (i.e., by bilateral consensus). Instead, the formation of negative (or conflict) links is carried out in a polar fashion: whereas such links can be created unilaterally (i.e., if just one of the agents involved decides to "fight"), their removal requires bilateral agreement (i.e., "peace treaties" must be signed by both parties). Interestingly, despite their very significant differences, the two models deliver quite parallel predictions and insights. Specifically, they show that at a strategically stable outcome, the population must be arranged as a complete k-partite network. Each of the k parts can be conceived as coalitions, with all agents in each of them being friends among themselves but in an all-out conflict with those from all other coalitions. This is an interesting conclusion in two respects. First, empirically, it is in line with evidence observed in real-world contexts so diverse as large online social media, or the pattern of country alliances in international conflict (see, e.g., Leskovec et al. 2010 and Antal et al. 2006). Second, theoretically, it is consistent with the predictions of Structural Balance Theory (see Heider 1946 and Cartwright and Harary 1956), which has been long used in sociology as a canonical framework to study signed social networks. In a sense, one can view the models of Hiller and Huremovic as providing alternative game-theoretic foundations of Structural Balance Theory.

9.5.4 Learning

The study of how the structure of social networks impinges on social learning has spanned a large literature, which is the subject of Chapter 12. Here, in order to illustrate the phenomenon at hand (i.e., the effect on learning of an endogenous action-link interplay), I choose a particularly simple model that goes back to the work of De Groot (1974) and has received fresh attention from recent work by Golub and Jackson (2010, 2012). In this model, the pattern of "influence" in a society is represented by a directed weighted network, with an agent i being influenced by another one j as captured by the corresponding weight a_{ij} in the adjacency matrix A defining the social network. The basic modeling assumption is that the action (or belief) displayed by any given agent at some point in time t is simply a convex combination of the actions displayed at $t-1$ by his neighbors (i.e., those who have an influence on him, generally including himself). The main implication is then that, if the network is connected (everyone is at least indirectly connected to everyone else), as the number of rounds grows unboundedly the system converges to a state of consensus where everyone in the population ends up sharing the same action/belief. Furthermore, the effect that the initial action of a player has on such a final outcome depends on some suitable measure of his network centrality.

The approach just summarized presumes that the social network is exogenously given, independently of the learning outcome. That is, the assumption is that there is no converse effect, from the learning outcome to the influence matrix. But, in fact, empirical evidence suggests that such a reciprocal relationship is strong in many real-world contexts. For example, it has been amply documented in the political arena, where an agent's ideology (or *a priori* position in the political spectrum) affects significantly the range of agents with whom he interacts (and hence the set of individuals who can influence him) (see, e.g., Adamic and Glance 2005; Boutyline and Willer 2013; or Colleoni et al. 2014). In quite different context, a similar phenomenon has also been shown to arise among Wikipedia editors, as they edit different articles and correspondingly change their editing concerns—see Crandall et al. (2008). The main feature that transpires from all this evidence is that agents tend to display significant homophily (i.e., a tendency to connect to those who share their actions or beliefs). This then must have important implications on the learning process. Intuitively, one expects that homophily tends to segment the interaction of the population into distinct communities and hence must also affect the extent to which the actions/beliefs of the different agents tend to converge—or, at the very least, the rate at which this happens.[15] And then, by having the forces at work feed on each other, the overall interplay of action and link dynamics may well lead to an exacerbation of social segmentation.

To study the problem, a model displaying an endogenous co-determination of influence and learning in the De Groot framework has been recently proposed by Polanski and Vega-Redondo (2014b). In contrast to the received approach, it is assumed that, for any given network of interagent influence, a large number of learning rounds are conducted so that:

(a) at the the start of each of them, agents receive individual signals (possibly correlated among them but stochastically independent in time) that determine their initial action/belief;
(b) social learning then proceeds for only a finite (possibly large) number of steps.

Because of (b), full convergence of actions is generally not achieved at the end of each round. Hence, in the absence of complete consensus, (a) implies that one can *non-trivially* define the degree of correlation among the actions displayed by the different agents across all the different rounds, each of these starting with a fresh pattern of signals. Then, given the induced pattern of correlations, an influence network is said to be in *equilibrium* if it satisfies the following homophily condition: for each

[15] Golub and Jackson (2012) study how homophily affects the rate at which the population converges to consensus in a connected network.

pair of individuals connected by the social network, their bilateral influence weight is proportional to their corresponding bilateral correlation.

The equilibrium analysis just outlined sheds light on the following important questions:

(i) Will the population be fragmented in separate influence components?
(ii) What is relationship between link weights, informational/influence redundancy, and segmentation?

The relevance of (i) is illustrated by the empirical evidence on action/belief segmentation discussed above. The interest of (ii), on the other hand, relates to the celebrated hypothesis of Granovetter's that strong ties are (informationally) weak. As explained in Easley and Kleinberg (2010, Chapter 3), the usual motivation for this hypothesis is based on the notion of "strong triadic closure," the postulate that when two agents are *strongly* connected to a common third agent, they themselves will tend to become connected as well. Instead, the model of Polanski and Vega-Redondo (2014b) relies on the converse idea that whenever two agents have common neighbors, these act as shared "anchors" that render their actions correlated and hence make their link strong. The two approaches highlight different mechanisms. Whereas the first has link strength impinge on link formation, in the second it is the topology of existing connections that determines the pattern of link strengths. One expects that, in the real world, both of these mechanisms should act in a complementary manner whenever link strengths and action choices evolve in full interplay.

9.6 Summing Up

In this chapter I have stressed the importance of modeling socioeconomic network phenomena in a theoretical setup where actions and links are in reciprocal interplay. By endogenously determining the prevailing network, this approach often succeeds as well in producing sharp predictions on agents' behavior. In a sense, formulated in such a general fashion, the previous point is a trivial one: extending the range of endogenous features can never harm the modeling effort, the only possible drawback being that it makes the analysis more involved. This is why I have chosen to illustrate its practical significance by discussing a set of concrete and paradigmatic contexts. Many of these contexts concern important problems that are discussed in more detail in other chapters of this volume. Specifically, I have focused on coordination, cooperation, and intermediation in Sections 9.2 through 9.4 while, more briefly, I have discussed bargaining, public goods, conflict, and learning in Section 9.5.

In all of those cases, we have seen that, allowing for a genuine co-determination of actions and links sheds new and sharper light on the phenomena at hand, by (a) singling out definite network structure where otherwise there would be little basis to select it, and (b) identifying specific behavior associated to that structure, in cases where a wide (equilibrium) multiplicity would be *a priori* possible. Obviously, the empirical relevance of the analysis has to be qualified by the fact that, in the real world, network formation is also subject to many factors other than just action choice (anticipated or effective). For example, individuals tend to obtain important economic information from friends, but these are often not chosen with such "instrumental" considerations in mind. However, the fact that so many important phenomena (e.g., learning, cooperation, or intermediation) do hinge upon the reciprocal feedback of link and action choice suggests that, even if carried out in a stylized manner, a proper study of such interplay can hardly be avoided.

References

Adamic, L. and N. Glance (2005). "The political blogosphere and the 2004 U.S. election: Divided they blog." *Proceedings of the 3rd International Workshop on Link Discovery*, 36–43.

Ali, S. N. and D. A. Miller (2009). "Enforcing cooperation in networked societies." Unpublished manuscript, University of California at San Diego.

Antal, T., P. Krapivsky, and S. Redner (2006). "Social balance on networks: The dynamics of friendship and enmity." *Physica D* 224, 130–136.

Babus, A. (2012). "Endogenous intermediation in over-the-counter markets." Working paper, Imperial College London.

Badev, A. I (2013). "Discrete games in endogenous networks: Theory and policy." U.S. Federal Reserve Board, mimeo.

Baetz, O. (2015). "Social activity and network formation." *Theoretical Economics* 10, 315—340.

Ballester, C., A. Calvó-Armengol, and Y. Zenou (2006). "Who is who in networks. Wanted: The key player." *Econometrica* 74, 1403–1417.

Bhaskar, V. and F. Vega-Redondo (2004). "Migration and the evolution of conventions." *Journal of Economic Behavior and Organization* 55, 397–418.

Bergin, J. and B. J. Lipman (1996). "Evolution with state-dependent mutations." *Econometrica* 64, 943–956.

Bilancini, E. and L. Boncinelli (2009). "The co-evolution of cooperation and defection under local interaction and endogenous network formation." *Journal of Economic Behavior and Organization* 70, 186–195.

Blume, L. E., D. Easley, J. Kleinberg, and E. Tardos (2009). "Trading networks with price-setting agents." *Games and Economic Behavior* 67, 36–50.

Boutyline, A. and R. Willer (2013). "The social structure of political echo chambers: Ideology and political homophily in online communication networks." Working paper, University of California at Berkeley and Stanford University.

Bramoullé, Y., S. Goyal, D. Lpez-Pintado, and F. Vega-Redondo (2004). "Network formation and anti-coordination games." *International Journal of Game Theory* 33, 1–19.

Bramoullé, Y. and R. Kranton (2007). "Public goods in networks." *Journal of Economic Theory* 135, 478–494.

Buchanan, J. M. and G. Tullock (1962). *The Calculus of Consent: Logical Foundations of Constitutional Democracy*. Ann Arbor: University of Michigan Press.

Burt, R. S. (1992). *Structural Holes: The Social Structure of Competition*. Cambridge, MA: Harvard University Press.

Buskens, V. and A. van der Rijt (2008). "Dynamics of networks if everyone strives for structural holes." *American Journal of Sociology* 114, 371–407.

Cartwright, D. and F. Harary (1956). "Structural balance: A generalization of Heider's theory." *Psychological Review* 63, 277.

Colleoni, E. A. Rozza, and A. Arvidsson (2014). "Echo chamber or public sphere? predicting political orientation and measuring political homophily in twitter using big data." *Journal of Communication* 64, 317–332.

Condorelli, D. and A. Galeotti (2012). "Bilateral trading in networks." mimeo, University of Essex.

Crandall, D., D. Cosley, D. Huttenlocher, J. Kleinberg, and S. Suri (2008). "Feedback effects between similarity and social influence in online communities." *Proceedings of the 14th ACM SIGKDD International Conference on Knowledge Discovery and Data Mining.*

DeGroot, M. H. (1974). "Reaching a consensus." *Journal of the American Statistical Association* 69, 118–121.

Easley, D. and J. Kleinberg (2010). *Networks, Crowds, and Markets: Reasoning about a Highly Connected World.* Cambridge: Cambridge University Press.

Eguíluz, V. M., M. Zimmermann, C. J. Cela-Conde, and M. San Miguel (2005). "Cooperation and the emergence of role differentiation in the dynamics of social networks." *American Journal of Sociology* 110, 977–1008.

Ehrhardt, G., M. Marsili, and F. Vega-Redondo (2008). "Networks emerging in a volatile world." European University Institute, Working Paper Series, no. 2008/08.

Elliott, M. (2014). "Inefficiencies in networked markets." forthcoming in *American Economic Journal: Microeconomics.*

Elliott, M. and F. Nava (2014). "Decentralized bargaining: Efficiency and the Core." mimeo, California Institute of Technology and London School of Economics.

Ellison, G. (1993). "Learning, local interaction, and coordination." *Econometrica* 61, 1047–1071.

Ely, J. C. (2002). "Local conventions." *Advances in Theoretical Economics* 2, Article 1.

Eshel, I., L. Samuelson, and A. Shaked (1998). "Altruists, egoists and hooligans in a local interaction model." *American Economic Review* 88, 157–179.

Fainmesser, I. P. (2012). "Intermediation and exclusive representation in financial networks." mimeo, Brown University.

Fainmesser, I. P. (2014). "Exclusive intermediation." mimeo, Johns Hopkins University.

Farboodi, M. (2014). "Intermediation and voluntary exposure to counterparty risk." Booth School of Business, University of Chicago.

Fosco, C. and F. Mengel (2011). "Cooperation through imitation and exclusion in networks." *Journal of Economic Dynamics and Control* 35, 641–658.

Franke, J. and Öztürk, T. (2009). "Conflict Networks." Ruhr Economic Papers, no. 16.

Fujiwara-Greve, T. and M. Okuno-Fujiwara (2012). "Behavioral diversity in voluntary separable repeated Prisoner's Dilemma." Working paper, http://ssrn.com/abstract=2005115.

Gale, D. (1987). "Limit theorems for markets with sequential bargaining." *Journal of Economic Theory* 43, 20–54.

Gale, D. M. and S. Kariv (2007). "Financial networks." *The American Economic Review* 97, 99–103.

Galeotti A. and S. Goyal (2010). "The law of the few." *The American Economic Review* 100, 1468–1492.

Glode, V. and C. Opp (2014). "Adverse selection and intermediation chains." mimeo, The Wharton School, University of Pennsylvania.

Golub, B. and M. O. Jackson (2010). "Naïve learning in social networks: Convergence, influence, and the wisdom of crowds." *American Economic Journal: Microeconomics* 2, 112–149.

Golub, B. and M. O. Jackson (2012). "How homophily affects the speed of learning and best response dynamics." *Quarterly Journal of Economics* 127, 1287–1338.

Goyal, S. and F. Vega-Redondo (2005). "Network formation and social coordination," *Games and Economic Behavior* 50, 178–207.

Goyal, S. and F. Vega-Redondo (2007). "Structural holes in social networks." *Journal of Economic Theory*, 137, 460–492.

Gracia-Lázaro, C., A. Ferrer, G. Ruiz, A. Tarancón, J. A. Cuesta, A. Sánchez, and Yamir Moreno (2012), "Heterogeneous networks do not promote cooperation when humans play a Prisoner's Dilemma." *Proceedings of the National Academy of Sciences of the USA* 109, 12922–12926.

Granovetter, M. (1973). "The strength of weak ties." *American Journal of Sociology* 78, 1360–1380.

Grujić, J., C. Fosco, L. Araujo, J. A. Cuesta, and A. Sánchez (2010). "Social experiments in the mesoscale: Humans playing a spatial Prisoner's Dilemma." PLoS ONE 5, e13749. doi:10.1371/journal.pone.0013749.

Haag, M. and R. Lagunoff (2006). "Social norms, local interaction, and neighborhood planning." *International Economic Review* 47, 265–296.

Hall, P. (1935). "On representatives of subsets." *Journal of the London Mathematical Society* 10, 26–30.

Heider, F. (1946). "Attitudes and cognitive organization." *The Journal of Psychology* 21, 107–112.

Hiller, T. (2012). "Peer effects in endogenous networks." University of Bristol, Bristol Economics Working Papers no. 12/633.

Hiller, T. (2012). "Friends and enemies: A model of signed network formation." Working paper TC 2012, University of Bristol.

Hojman, D. and A. Szeidl (2006). "Endogenous networks, social games and evolution." *Games and Economic Behavior* 55(1), 112–130.

Huremovic, K. (2014). "Rent seeking and power hierarchies: A noncooperative model of network formation with antagonistic links." Working paper, European University Institute, Florence. 55, 112–130.

Immorlica, N., B. Lucier, and B. W. Rogers (2014). "Cooperation in anonymous dynamics social networks." mimeo, Washington University in St. Louis.

Jackson, M. and A. Wolinsky (1996). "A strategic model of social and economic networks." *Journal of Economic Theory* 71, 44–74.

Jackson, M. O. and A. Watts (2002). "On the formation of interaction networks in social coordination games." *Games and Economic Behavior* 41, 265–291.

Jackson, M. O., T. Rodriguez-Barraquer, and X. Tan (2012). "Social capital and social quilts: Network patterns of favor exchange." *American Economic Review* 102, 1857–1897.

Kandori, M., G. Mailath, and R. Rob (1993). "Learning, mutation, and long-run equilibria in games." *Econometrica* 61, 29–56.

Kleinberg, J., S. Suri, E. Tardos, and T. Wexler (2008). "Strategic network formation with structural holes." *Proceedings of the 9th ACM Conference on Electronic Commerce.*

König, M., D. Rohner, M. Thoenig, and F. Zilibotti (2013). "Networks in conflict: Theory and evidence from the Great War of Africa." Working paper, Universities of Lausanne and Zurich.

König, M., C. J. Tessone, and Y. Zenou (2014). "Nestedness in networks: A theoretical model and some applications." *Theoretical Economics* 9, 695—752.

Kovenock, D. and B. Roberson (2012). "Conflicts with multiple battle fields." In *The Oxford Handbook of the Economics of Peace and Conflict*, M. Garfinkel and S. Skaperdas, eds. Oxford: Oxford University Press.

Leskovec, J., D. Huttenlocher, J. Kleinberg (2010). "Predicting positive and negative links in online social networks." *Proceedings of the 19th International World Wide Web Conference.*

Lippert, S. and G. Spagnolo (2011). "Networks of relations and word-of-mouth communication." *Games and Economic Behavior* 72, 202–217.

López-Pintado, D. (2008). "The spread of free-riding behavior in a social network." *Eastern Economic Journal* 34, 464–479.

Manea, M. (2011). "Bargaining in stationary networks." *American Economic Review* 101, 2042–2080 .

Manea, M. (2013). "Intermediation in networks." mimeo, MIT.

Mengel, F. (2009). "Conformism and cooperation in a local interaction model." *Journal of Evolutionary Economics* 19, 397–415.

Nava, F. (2015). "Efficiency in decentralized oligopolistic markets." *Journal of Economic Theory*. 157, 315–348.

Nowak, M. A. (2006). "Five rules for the evolution of cooperation." *Science* 314, 1560–1563.

Pacheco, J. M., A. Traulsen, and M. A. Nowak (2006). "Active linking in evolutionary games." *Journal of Theoretical Biology* 243, 437–443.

Polanski, A. and F. Vega-Redondo (2014a). "Bargaining and arbitrage in endogenous trading networks." Working paper, University of East Anglia and Bocconi University.

Polanski, A. and F. Vega-Redondo (2014b). "Homophily and influence: The strength of weak ties revisited." Working Paper, University of East Anglia and Bocconi University.

Raub, W. and J. Weesie (1990), "Reputation and efficiency in social interactions: An example of network effects." *American Journal of Sociology* 96, 626–654.

Roberson, B. (2006). "The Colonel Blotto Game." *Economic Theory* 29, 1–24.

Rubinstein, A. and A. Wolinsky (1985). "Equilibrium in a market with sequential bargaining." *Econometrica* 53, 295–328.

Rubinstein, A. and A. Wolinsky (1990). "Decentralized trading, strategic behaviour and the Walrasian outcome." *Review of Economic Studies* 57(1), 63–78.

Schelling, T. (1972). "Dynamic models of segregation." *Journal of Mathematical Sociology* 1, 143–186.

Schelling, T. (1978). *Micromotives and Macrobehavior*. New York: Norton.

Siedlarek, J. P. (2014). "Intermediation in Networks." mimeo, European University Institute.

Traulsen A. and M. A. Nowak (2006). "Evolution of cooperation by multilevel selection." *Procceedings of the National Academy of Sciences of U.S.A.*, 103, 10952–10955.

Ule, A. (2008). *Partner Choice and Cooperation in Networks: Theory and Experimental Evidence*. Berlin-Heidelberg: Springer Verlag.

Vega-Redondo, F. (1996). "Long-run cooperation in the one-shot Prisoner's Dilemma: A hierarchic evolutionary approach." *Biosystems* 37, 39–47.

Vega-Redondo, F. (2006). "Building up social capital in a changing world." *Journal of Economic Dynamics and Control* 30, 2305–2338.

Wang, J., S. Suri, and D. J. Watts (2012). "Cooperation and assortativity with dynamic partner updating." *Proceedings of the National Academy of Sciences U.S.A.* 109, 14363–14368.

Young, P. (1993). "The evolution of conventions." *Econometrica* 61, 29–56.

CHAPTER 10

CONFLICT AND NETWORKS

MARCIN DZIUBIŃSKI, SANJEEV GOYAL, AND
ADRIEN VIGIER

10.1 Introduction

Conflict remains a central element in human interaction. Networks—social, economic and infrastructure—are a defining feature of society. So it is natural that the two should intersect in a wide range of empirical contexts. This motivates the recent interest on conflict and networks. The aim of the paper is to provide a survey of this research.

We find it useful to start with specific empirical phenomena involving conflict and networks.

1. **Robustness of infrastructure networks:** Highways, aviation, shipping, pipelines, train systems, and telecommunication networks are central to a modern economy. These networks face a variety of threats ranging from natural disasters to human attacks. The latter may take a violent form (guerrilla attacks, attacks by an enemy country, and terrorism) or a nonviolent form (as in political protest that blocks transport services).[1] A network can be made robust to such threats through additional investments in equipment and in personnel. As networks are pervasive,

We thank the editors for very helpful comments on an earlier draft. We also thank Vessela Daskalova, Julien Gagnon, Michiel de Jong, and Anja Prummer for helpful discussions. Marcin Dziubiński acknowledges support from Homing Plus programme of the Foundation for Polish Science, via the project "Strategic Resilience of Networks." Sanjeev Goyal acknowledges support from a Keynes Fellowship and the Cambridge-INET Institute.

[1] The US Office of Infrastructure Protection says, "Our nation's critical infrastructure is crucial to the functioning of the American economy... (It) is increasingly connected and interdependent and protecting it and enhancing its resilience is an economic and national security imperative," Department of Homeland Security (2012). For an introduction to network based conflict, see Arquilla and Ronfeldt

the investments needed could be very large; this motivates the study of targeted defence. What are the "key" parts of the network that should be protected to ensure maximal functionality? Moreover, taking a longer-term view, how should networks be designed to enhance their robustness to threats?

2. **Cybersecurity:** As energy, communication, travel, and consumer interaction increasingly adopt digital networks, cybersecurity has emerged as a major priority. In the United States, this is a responsibility of the Department of Homeland Security (DHS). Its mission statement reads,"Our daily life, economic vitality, and national security depend on a stable, safe, and resilient cyberspace. We rely on this vast array of networks to communicate and travel, power our homes, run our economy, and provide government services."[2] At the heart of these developments is the question of how to design networks so that they are robust to attacks.
3. **Criminal networks:** Criminal activity, being illegal, it is especially difficult for participants to enforce formal contracts. Trust and networks of favor exchange are especially important in crime. This suggests that personal connections may be important for crime; however, the investigation and capture of one agent by police can expose connected others. What is the best way to organize criminal network?
4. **Civil wars and armed conflict:** Conflict takes place between countries or communities that are geographically contiguous (Caselli et al. 2014). Conflict between two entities, however, typically has spillovers on neighboring third parties, which is turn may travel through the network of relations. We wish to understand how the network structure shapes conflict and determines the winners and losers.
5. **Strategic alliances:** A common feature of civil unrest and international conflict is the salience of networks of alliances. For example, through the nineteenth century and the early part of the twentieth century, shifting strategic alliances were a salient feature of European politics.[3] Empirical research shows that violent international conflict was more common in the hundred years prior to 1950 as compared to the years after that. The stability of alliances exhibits a corresponding time line: alliances were much less stable in the period prior to 1950 than in the period since. Finally, we know that international trade has grown steadily since the 1950s (Jackson et al. 2014). Is there a systematic relation between these stylized facts?

Inspired by applications 1, 2, and 3, we start with a discussion of the design and defence of networks that face threats. As networks carry out a variety of functions, different aspects of networks generate value depending on the context. Similarly, threats come in different forms: in some cases, the threat is posed by an intelligent adversary

(2001) and Zhu and Levinson (2011); for news coverage of the effects of natural disasters and human attacks on infrastructure networks, see Eun (2010), Kliesen (1995), *India Today* (2011), and Luft (2005).

[2] In 2009, roughly 10 million computers were infected with malware designed to steal online credentials. The annual damage caused by malware is of the order of 9.3 billion Euros in Europe, while in the United States the annual costs of identity theft are estimated at 2.8 billion USD (Moore, Clayton, and Anderson 2009). One indicator of the economic magnitude of the problem is the valuation of security firms: Intel bought McAfee in 2010 for 7.68 billion USD (bbc.co.uk; August 19, 2010).

[3] The Triple Alliance between Germany, Austria-Hungary, and Italy and the Triple Entente involving Britain, France, and Russia played a key role in shaping World War I.

(such as the police or investigating agency, terrorists, or political protestors), while in others it comes from nature (in the form of floods and earthquakes). Similarly, the dynamics of the threat also vary. Viruses and worms spread through computer connections; contagion is an important aspect of these threats. On the other hand, an earthquake or a storm damages a specific port or an airport or a railway station. By varying these different dimensions of the problem we generate an ensemble of different scenarios. The key question here is: how should networks be designed and defended in the face of threats? Section 10.2 provides a survey of the existing research and concludes with the discussion of a number of open questions.[4]

Motivated by application 4, we then turn to the study of conflict between nodes located in a network. Section 10.3 takes up the case of conflict in fixed networks. Connections between nodes determine who is in conflict with whom. The nodes choose how much to invest in conflict and the conflicts yield prizes to winners. We study both static models and also the dynamics of resource accumulation through war and conquest. The section ends with a discussion of open problems.

The study of war naturally leads us to the study of alliances in conflict: in application 5, a salient feature of civil and international conflict is the existence of alliances among warring parties. As the example of World War I illustrates, these alliances can have a decisive influence on the shape of conflict. Section 10.4 starts with the study the nature of conflict under given alliance structures and then moves on to the formation and stability of alliances.

Section 10.5 contains concluding remarks.

10.2 Network Design and Defense

The examples in the introduction illustrate a range of empirical contexts where networks face threats. The key question in this field is how to design and defend networks against these threats. The research on this question is at an early stage. We provide a survey of this work and point to a number of interesting open problems.

While there are different aspects of networks that create value, in the literature to date much attention has centered on the setting where network connectivity is central to value. Thus network value is increasing and convex in its size (i.e., the number of nodes). The threat to the network is modeled as a game of conflict between a Designer (and the nodes in the network) and an Adversary. The Designer chooses a network. The Designer (or the nodes) and the adversary then allocate their resources across the network. There is conflict between the attack and defense resources. In the infrastructure example, attack or damage of a specific part of the network (a node or a link) compromises the

[4] The problem of network design and defence has been extensively studied in electrical engineering and computer science; for an overview of this work, see Alpcan and Başar (2011), Anderson (2001) and Roy et al. (2010). The economics literature surveyed below contributes to this field by developing a general framework that combines strategic interaction with a rich formulation of network value.

network by disrupting flows along paths. We study this disruption in terms of break down in connectivity of the network. In the cybersecurity example, the spread of worms and viruses through the network connections is central to the damage. We develop a model of contagion through networks. The literature has focused on zero-sum games. We shall follow the literature in this regard.

We start with the problem of contagion in networks. We first set up and solve the first best solution. There are two players, a Designer and an Adversary. The Designer chooses both the design of the network and the allocation of defence resources. The Adversary observes these choices and then attacks particular nodes of the network. We then move to a discussion of the game where the Designer creates the network, but the nodes in the network choose defense allocations. This is motivated by applications in cybersecurity where individual computer users generally choose their own security.

We then turns to infrastructure robustness. We will first discuss optimal design and defence. Finally, motivated by the interest in the robustness of infrastructure networks, we will study optimal defense of a given network. As networks are pervasive, the investments needed to protect them can be very large; this motivates the study of targeted defense. What are the "key" nodes to defend to maximize functionality of the network? We also study how networks affect the intensity of conflict, a question that will reappear in the subsequent sections, when we study conflict among nodes located in networks.

10.2.1 Connectivity and Network Value

We now introduce some terminology and notation. There is a set of nodes $N = \{1,\ldots,n\}$, $n \geq 2$. A link between two nodes i and j is represented by $g_{ij} \in \{0,1\}$: we set $g_{ij} = 1$ if there is a link between i and j, and $g_{ij} = 0$ otherwise. Links are undirected (i.e., $g_{ij} = g_{ji}$). The nodes and the links together define a network g.

A path between two nodes i and j in network g is a sequence of nodes $i_1,\ldots,i_k$ such that $g_{ii_1} = g_{i_1 i_2} = \ldots = g_{i_{k-1} i_k} = g_{i_k j} = 1$. Two nodes are said to be connected if a path exists between them. A component of the network g is a maximal set of nodes such that any two elements in it are connected. $\mathcal{C}(g)$ is the set of components of g and $C_i(g)$ is the component containing node i. We let $|C|$ indicate the cardinality (or size) of the component C. A maximum component of g is a component with maximal cardinality in $\mathcal{C}(g)$. A network with a single component is said to be connected.[5] A network g' on N' is a subnetwork of g if and only if $N' \subseteq N$, and $g'_{ij} = 1 \Rightarrow g_{ij} = 1$ and $i,j \in N'$. We let $\mathcal{G}(g)$ denote the set of all subnetworks of g.

[5] The complete network, or a clique, g^c, has $g_{ij} = 1$, for all pairs (i,j). The empty network, g^e, has $g_{ij} = 0$ for all pairs (i,j). A core-periphery network has two types of nodes, N_1 and N_2. Nodes in N_1 constitute the periphery and have a single link each, and this link is with a node in N_2; nodes in N_2 constitute the core and are fully linked with each other and with a subset of nodes in N_1. When the core contains a single node, we have a star network. For a general introduction to networks concepts and terminology, see Goyal (2007).

Following Myerson (1977), we assume that the value of a network is the sum of the value of the different components and that the value of any component is a function of its size only. Let the function $f : \mathbb{N} \rightarrow \mathbb{R}_{+}$ specify a value to component size. Our interest is in network generated value and so we assume increasing and convex returns to size of component.

Assumption A.1: *The value of network g is given by*

$$\Pi(g) = \sum_{C \in \mathcal{C}(g)} f(|C|). \tag{10.1}$$

where f is (strictly) increasing, (strictly) convex and $f(0) = 0$.

Increasing and convex network value functions arise naturally in the large literature on network externalities (see, e.g., Katz and Shapiro 1985, and Farrell and Saloner 1986). In that literature, the value to a consumer from buying a product is related to the number of other consumers who buy the same product (i.e., belong to the same network). In its simplest form this gives rise to the quadratic form $f(n) = n^2$. This functional form also arises in the communications model in the literature on network economics (see, e.g., Goyal 1993; Bala and Goyal 2000) and is consistent with Metcalfe's Law, concerning the nature of value in telecommunication networks.

On the other hand, suppose that subsets of nodes perform various tasks, each task being of equal value normalized to 1. A task is carried out if and only if the subset of nodes performing that task is connected. The value of the network is the total value of tasks performed. A component with m nodes thus generates value $2^m - 1$ (as there are exactly $2^m - 1$ tasks which m nodes can perform). This yields a network value that is exponential in the size of components; it is consistent with Reed's Law (Reed 2001) on value of networked systems.

Conflict and contagion: In this section we will study the optimal defense and design of networks that face contagious attacks. In their influential paper on computer security, Staniford et al. (2002) identify stealth worms and viruses as the main threats to security in computer networks. Using data from actual attacks, they argue that adversaries scan the network to explore its topology and the vulnerabilities of nodes, prior to attack. In the first instance, the objective is to deploy a worm on selected nodes in the network. Deployed worms then exploit communication between nodes to progressively take control of neighboring nodes in the network. The likelihood of capture of a node and the spread of the worm in a network depend on the strength of the worm, the topology of connections, and on vulnerabilities of individual nodes. These considerations motivate the following theoretical model, due to Goyal and Vigier (2014).

They consider a setting with two players: a Designer and an Adversary. The Designer moves first and chooses a network and an allocation of defense resources. The Adversary then allocates attack resources on nodes; if an attack succeeds then the Adversary decides how successful resources should navigate the network. The model has three important ingredients: the value of the network (summarized in assumption

A.1 above), the technology of conflict between defense and attack resources, and the spread of successful attack resources through the network.

They assume that the value of a network is increasing and convex in the number of interconnected nodes (assumption A.1 above). They model the conflict between defense and attack resources on a network node as a *Tullock contest*.[6] The contest defines the probability of a win for Designer and Adversary as a function of their respective resources. The resources of the loser of the contest are eliminated, the winner retains his resources. In case the Adversary wins a contest on a node, the winning attack resources can move and attack neighboring nodes. The dynamics of conflict continue as long as both defense and attack resources co-exist. The initial network design and the conflict dynamics yield a probability distribution on surviving nodes (i.e., nodes that have not been captured by the Adversary). The Designer and Adversary are engaged in a zero sum game; so, given a defended network, we consider the minimum payoff of the Designer given all possible attacks. An *optimal defended network* maximizes this (minimum) payoff.

We let $d \in \mathbb{N}$ (resp. $a \in \mathbb{N}$) denote the total resources of the Designer (resp. Adversary). A strategy for the Designer is a pair $(g, \mathbf{d})$, where g is a network defined on nodes in N and $\mathbf{d}$ is a vector specifying the defense resources allocated at each node such that $\sum d_i = n$. A strategy for the Adversary is a pair $(\mathbf{a}, \Delta)$. The vector $\mathbf{a}$ specifies the attack resources initially allocated at each node. The matrix $\delta = (\delta_{ij})_{i,j \in N}$, on the other hand, describes the spread of attack resources during the course of time.

Given a defended network $(g, \mathbf{d})$, let K denote the subset of protected nodes and O the subset of unprotected nodes. Further, for $i \in N$ let $O_i \subseteq O$ denote the subset of unprotected nodes that can be reached from i through some path such that each node on that path lies in O. The set O_i will sometimes be called the unprotected neighbourhood of i. Similarly, let $K_i \subseteq K$ denote the subset of protected nodes that can be reached from i through some path such that each node on that path lies in O.

Attack resources a_i and defense resources d_i located on a node i engage in a *contest* for control of the node. If $a_i + d_i > 0$, then following Tullock (1980):

$$\text{probability of successful attack} = \frac{a_i^\gamma}{a_i^\gamma + d_i^\gamma}, \tag{10.2}$$

where $\gamma > 0$. If a_i is 0, then the probability of successful attack is 0, irrespective of the value of d_i: a node is safe if it is not under attack.

We will provide an informal sketch of the dynamics; for details refer to Goyal and Vigier (2014). At the start, the Adversary captures all nodes that are attacked and unprotected. After that the Adversary captures O_i. He then reallocates a_i attack resources to an un-captured and protected node. The result below holds for a range

[6] Here we build on the rich literature on rent seeking and conflict, see Garfinkel and Skaperdas (2012), Tullock (1980), and Hirshleifer (1995).

of spread matrices. A defended network will be called optimal if it maximizes the minimum expected network value from all possible attacks.

The key to the analysis is whether or not a few nodes are "essential" to the network value given by assumption A.1. In the case where $f(x) = x^2$, as n grows, the impact of eliminating a few nodes vanishes. On the other hand, if $f(x) = 2^x - 1$, the impact does not vanish: $\lim_{n\to\infty}(n-a)^2/n^2 = 1$, whereas $\lim_{n\to\infty}(2^{n-a}-1)/(2^n-1) = 1/2^a < 1$. The methods of analysis for the two cases involve different arguments.

We will henceforth assume that the following limit exists, and define:

$$\ell = \lim_{n\to\infty} \frac{f(n-1)}{f(n)}.$$

A defended network $(g,\underline{d})$ is optimal if $\overline{\Pi}^e(g,\underline{d}) \geq \overline{\Pi}^e(g',\underline{d}')$ for all defended networks $(g',\underline{d}')$.

Given $\epsilon > 0$, a defended network $(g,\underline{d})$ is ϵ-optimal if $\overline{\Pi}^e(g,\underline{d}) \geq (1-\epsilon)\overline{\Pi}^e(g',\underline{d}')$ for all defended networks $(g',\underline{d}')$. A star network in which all defense resources are allocated to the central node is referred to as a center-protected star.

We are now ready to state the main result from Goyal and Vigier (2014).

Theorem 1. *Assume that (A.1) holds, $a/d \in \mathbb{N}$ and $n > a+1$. Let $\epsilon > 0$ and consider the class of connected networks. There exists n_0 such that, for all $n > n_0$:*

1. *If $\ell < 1$ the CP-star is uniquely optimal.*
2. *If $\ell = 1$ the CP-star is ϵ-optimal.*

We illustrate the general line of argument with an example, by comparing the expected network value achieved with a CP-star to the value achieved with a symmetric 2-hub network as illustrated in Figure 10.1. The defended network has $|K| = 2$ protected nodes, with one link between them, and each protected node has $n-2/2$ nodes in its unprotected neighbourhood. We assume that d is even and each protected node has $d/2$ defense units allocated to it. To simplify the exposition, we also assume $a = d$. The aim is again to find a way to attack this network and leave the Designer with expected network value less than $\frac{d}{d+a}f(n-a)$; this will show that the CP-star performs best in this case too.

Consider the following attack strategy, where the Adversary allocates 1 unit of resource to exactly $a/2$ nodes of the periphery of each protected node. There are four possible outcomes of the two contests on the hubs: either both hubs survive, both hubs are captured, or one hub survives and the other is captured. Given the equal resources engaged in contests, it follows that the first two outcomes each arise with probability 1/4. The two outcomes define terminal states of the dynamics, represented at the top and the bottom end of Figure 10.1. There is a probability 1/2 that one of the hubs survives and the other is captured. This is represented in the middle of the Figure 10.1. Capture of a hub triggers the capture of its respective peripheral nodes. All attack resources then target the surviving hub, inducing a second round of contests. With

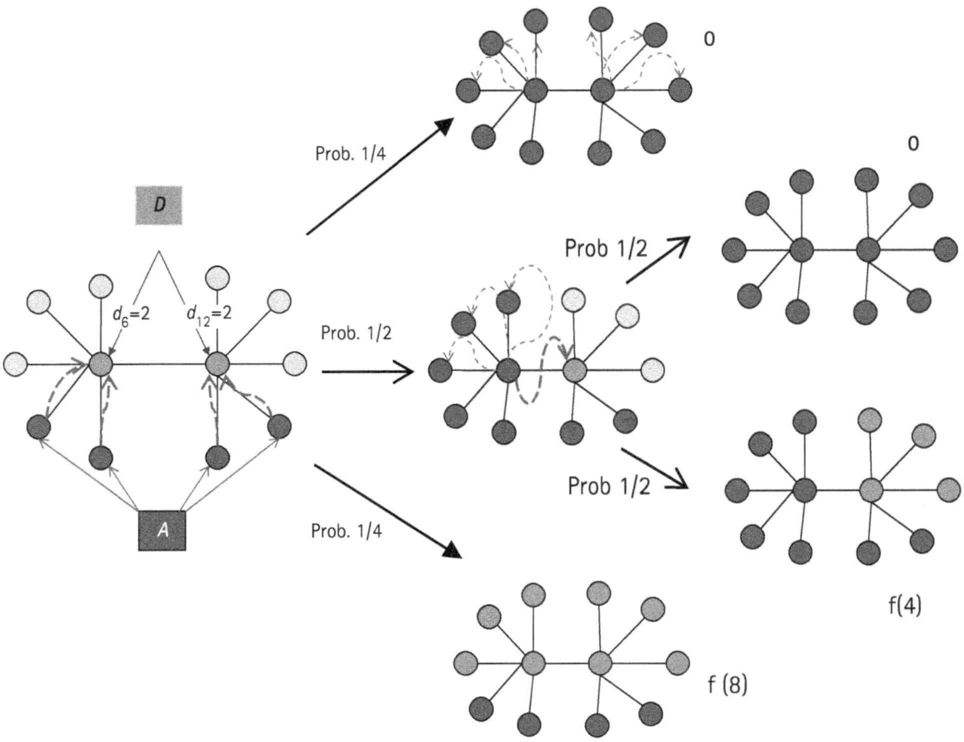

FIGURE 10.1 Mimic attack on two-hub network: $n = 12$, $a = d = 4$.

probability 1/2 the hub survives the attack, and with probability 1/2 it is captured. If the hub is captured, then this triggers the capture of the remaining peripheral nodes. This brings to an end the dynamics of conflict.

The probability density P on surviving nodes is: with probability 1/2 all nodes are captured, with probability 1/4 half the nodes survive, and with probability 1/4 all nodes survive. Observe that this distribution is first order stochastically dominated by the distribution P' such that with probability 1/4 all nodes are captured, with probability 1/2 half the nodes survive, and with probability 1/4 all nodes survive. But P' is in turn second order stochastically dominated by the distribution P'' in which all nodes are captured with probability 1/2, and all nodes survive with probability 1/2. Noting that P'' is the distribution facing the Designer if he chooses a CP-star finishes to show that the CP-star dominates the 2-hub network examined here, given that f is increasing and convex. Goyal and Vigier (2014) generalize these ideas to cover all connected networks and establish:

Theorem 1 is a powerful result. It holds for all payoff functions which satisfy (A.1): so the result does not depend on the curvature (i.e., the extent of convexity) of f. The result holds for all γ in the Tullock contest function: so the conclusion is robust with respect to the technology of conflict. The result holds for all resource configurations between the Designer and the Adversary such that $a/d \in \mathbb{N}$.

Empirical work on networks draws attention to the prominence of the hub-spoke network architecture (see, e.g., Goyal 2007; Newman 2010). In an influential paper, Albert et al. (2000) argue that these architectures are vulnerable to strategic attacks since potential adversaries can significantly reduce their functionality by removing only a few hub nodes. By contrast, the above analysis highlights the attractiveness of these architectures in a setting where defense resources are scarce and network value is convex.

Decentralized defense: Theorem 1 provides us a result on the optimal defense and design of a network facing an intelligent adversary threat. In the context of cybersecurity, investments in protection are typically made by individual nodes. Heterogeneities in the network structure create corresponding differences in individual incentives and in externalities. Thus Theorem 1 provides us a benchmark. We now turn to the question of how network design should address the variety of network externalities. In this context, the standard understanding of externalities is that individual returns to security may be lower than collective returns, due to the risks of contagion. However, in a setting where the Adversary chooses targets, there is an additional and novel consideration: investing in security diverts the attack to other nodes. This potentially negative externality brings a new set of considerations into play. We follow Cerdeiro et al. (2015) in this discussion.[7]

The Designer first chooses the network over the n nodes. Given this network, each of the n nodes (simultaneously) chooses whether to protect or not; protection carries a fixed cost. Finally, the Adversary chooses a node to attack. If the attacked node is protected, then all nodes survive the attack. If the attacked node is not protected, then this node and all nodes with a path to the attacked node through unprotected nodes are eliminated. Nodes are assumed to derive benefits from their connectivity: the payoff of a node is increasing in the size of its surviving component. A node's net payoffs are equal to its connectivity payoffs less the amount spent on protection. The Designer is utilitarian: he seeks to maximize the sum of nodes' payoffs. The Adversary is intelligent, purposefully choosing the attacked node so as to minimize connectivity-related payoffs.

We start with a study of the first best design and defense profile. We show that for low protection costs, all nodes should be protected and any connected network is optimal. For intermediate costs of protection, the Designer chooses a star network and protects its center only. The Adversary then eliminates a single spoke of the star. If protection costs are high, the Designer splits the network into equal size components and leaves all nodes unprotected. The Adversary eliminates one of these components.

This sets the stage for the decentralized problem. Observe that if defense is sufficiently expensive (so that no protection is first best), no protection is the unique equilibrium defense of any first best network. At the other extreme, if protection is sufficiently cheap (so that full protection is first best), there exist

[7] For a general survey of games played on networks, see chapter 5 by Bramoulle and Kranton in this volume.

networks that implement the first best in every equilibrium. Departures from first best welfare will therefore arise only for intermediate costs of protection; that is, when a center-protected star is optimal. The Designer cannot attain first best payoffs in equilibrium, as the only equilibria on star networks are those where either all or no node protects.

We now examine the optimal design problem in greater detail. When a center-protected star is first best but all nodes protect in equilibrium, protection decisions involve negative externalities and exhibit strategic complementarities. Nodes have incentives to protect and divert the Adversary's attack to other parts of the network. How can the Designer induce some nodes to be eliminated in equilibrium? Connected networks are not the best way to address the overprotection problem. When a connected network has an equilibrium achieving higher welfare than full protection, there always exists a disconnected network that welfare-dominates it. Thus, if the Designer is to avoid the overprotection problem, he must disconnect the network and sacrifice some nodes.

The analysis summarized so far assumes that individual coordinate on equilibria that achieve maximum equilibrium welfare. In general, however, some of these networks may feature multiple equilibria that achieve vastly different welfare levels. How can the Designer tackle potential coordination problems? To illustrate the issue, suppose that the costs of protection are such that maximum equilibrium welfare is achieved via full protection on a connected network. The network where nodes are arranged on a cycle has a full protection equilibrium. However, if the cost of protection outweighs the benefits of surviving in isolation, there is another equilibrium on this network where no node protects and the Adversary brings down the entire network. Cerdeiro et al. (2015) provide a necessary and a sufficient condition for a network to induce full protection in any equilibrium. Such networks are *sparse* in the following sense: they must feature a node that can block the Adversary's attack, thus saving a large part of the network.

The contribution of the paper lies at the intersection of economics and computer science literature. For an early contribution in the study of decentralized defense, see Kunreuther and Heal (2004). Aspnes et al. (2006) study security choices by nodes in a fixed network when nodes only care about their own survival, attack is random, and both protection as well as contagion are perfect. The focus is on computing the Nash equilibria of the game. They provide approximation algorithms for finding the equilibria. In a recent paper, Acemoglu et al. (2013) study the incentives for protection in a setting when both defense and contagion are imperfect.[8]

The relationship with Goyal and Vigier (2014) is worth discussing as they highlight the large effects of decentralized defense for optimal network design. In Goyal and Vigier (2014), the optimal design is a star network and optimal allocation of resources is exclusively on the central node. By contrast, when individual nodes choose security,

[8] There is also a very active research program in financial contagion (see, e.g., Blume et al. 2011; Acemoglu et al. 2015; Cabrales et al. 2010 and Elliot et al. 2014). For a survey of this issues, see Chapter 20 by Cabrales et al. in this volume.

the optimal design has to address problems of too much as well as too little protection. This best way to tackle overprotection is by disconnecting the network and sacrificing some nodes. Potential under-protection problems are addressed by creating equal components. Finally, coordination problems in security are mitigated through the creation of "sparse" networks that contain critical nodes.

10.2.1.1 Noncontagious Threats

In its strategy statement, the U.S. Office of Infrastructure Protection says, "Our nation's critical infrastructure is crucial to the functioning of the American economy... (It) is increasingly connected and interdependent and protecting it and enhancing its resilience is an economic and national security imperative," Department of Homeland Security (2012). In these contexts the primary cost of an attack is in terms of nodes (and links) that are eliminated and the consequent loss in the connectivity of the network. This motivates the study of networks in a setting with noncontagious attacks. In parallel with our discussion of contagious risk, we start with a study of optimal defense and design.

The presentation here draws on Dziubiński and Goyal (2013). There is a Designer and an Adversary. The Designer moves first and chooses a network and defense allocation. The Adversary moves next. Costs of attack are sunk; the Adversary can choose up to $k \leq n-2$ nodes to eliminate/remove. The costs of the Designer are linear: there is a cost $c_l > 0$ for every link, and a cost $c_d > 0$, for defending a node. Defense is perfectly reliable. Given network g, the set of defended nodes, Δ, the set of attacked nodes, $X \subseteq N$, the payoff to the Defender and Adversary, are respectively:

$$\Pi^{D}(g, \Delta, X; c_d, c_l) = \Phi(g - (X \setminus \Delta)) - c_d|\Delta| - c_l|g|$$
$$\Pi^{A}(g, \Delta, X) = -\Phi(g - (X \setminus \Delta)).$$

It is useful to consider the connectivity based value function; for the analysis of general value functions that satisfy assumption A.1, see Dziubiński and Goyal (2013). In this case the residual network has value 1 if it is connected and 0 otherwise. Now the equilibrium has a simple structure. One of the following three possible outcomes arises: one, the network is empty and there is no defense; two, there is no defense, the network involves redundant links; and three, the star network with protected center.

The exact levels of costs for each of the above outcomes can be derived by applying a result due to Harary (1962). Harary (1962) showed that a network that cannot be disconnected by removal of k nodes requires exactly $\lceil \frac{n(k+1)}{2} \rceil$ links. Moreover, any such network is regular of degree $(k+1)$, or almost regular, having one node of degree $(k+2)$ (if both n and k are odd). The set of these graphs is denoted by $\mathcal{M}(n,k)$.[9] Dziubiński and Goyal (2013) establish the following result.

[9] The set $\mathcal{M}(n,k)$ is not empty, as it includes *Harary graphs* defined by Harary to obtain the upper bound on the number of links.

Proposition 1. *Consider the Designer-Adversary game under connectivity based value function and suppose that $k \leq n-2$. In equilibrium*

1. *The Designer chooses network g and defense Δ:*
   - *If $c_l < 1/\left\lceil \frac{n(k+1)}{2} \right\rceil$ and $c_d > c_l\left(\left\lceil \frac{n(k-1)}{2} \right\rceil + 1\right)$, then $g \in \mathcal{M}(k,n)$ and $\Delta = \varnothing$.*
   - *If $c_l(n-1) + c_d < 1$ and $c_d < c_l\left(\left\lceil \frac{n(k-1)}{2} \right\rceil + 1\right)$, then g is a star and the central node is defended.*
   - *Otherwise g is empty and $\Delta = \varnothing$.*
2. *The Adversary chooses a separating cut for g and Δ, if it exists; if it does not exist then all cuts yield the same payoff.*

The proposition above illustrates the trade-off faced by the Designer. If costs of defense are high relative to the costs of linking, the Designer chooses a regular and dense network. On the other hand, when costs of defense are relatively low, the Designer chooses the star network and defends the hub node.

A comparison between Goyal and Vigier (2014) and Dziubiński and Goyal (2013) helps us to understand the role of contagion in the optimal defense and design. In the latter paper, when defense units are 0, the Designer defends the network by adding more links, and so the optimal network is $(k+1)$-connected. By contrast, in Goyal and Vigier (2014), when there is no defense, the Designer defends the network by separating it into distinct components. This is due to the implicit cost to linking introduced by the possibility of contagion.

Finally, in Dziubiński and Goyal (2014), the equilibrium will typically involve protection of multiple nodes. By contrast, Goyal and Vigier (2014) show that under a wide variety of circumstances, the Designer will assign all resources to the central node of a star.

The defense of a network: In some contexts, such as trains or roads or telecommunications, the network involves very large and time-consuming investments. So it is important to study the problem of defending a given network. The focus is on where to allocate resources to maintain the network in the face of threats that potentially damage or knock out nodes. The presentation draws on Dziubiński and Goyal (2014).

They consider a two-player sequential move game with a Defender and an Adversary. In the first stage, the Defender chooses an allocation of defense resources. In the second stage, given a defended network, the Adversary chooses the nodes to attack. Successfully attacked nodes (and their links) are removed from the network, yielding a residual network. The goal of the Defender is to maximize the value of the residual network, while the goal of the Adversary is to minimize this value.

Fix a network g on a set of nodes $N = \{1, \ldots, n\}$, where $n \geq 3$. A *defense* is a set of nodes $\Delta \subseteq N$. The set of attacked nodes $X \in\subseteq N$ chosen by the Adversary is called a *cut*. Removing a set of nodes $X \subseteq N$ from the network creates *residual network* $g - X$. It is assumed that the defense is perfect: a protected node cannot be removed by an attack, while any attacked unprotected node is removed with certainty. Given a defense Δ and a cut X, a set $Y = X \setminus \Delta$ will be removed from the network.

Defense resources are costly: the cost of defending a node is $c_d > 0$. Given network g, Defender's payoff from strategy $\Delta \subseteq N$, when faced with opponent strategy $X \subseteq N$, is

$$\Pi^D(\Delta, X; g, c_d) = \Phi(g - (X \setminus \Delta)) - c_d|\Delta|. \tag{10.3}$$

where $\Phi(\cdot)$ satisfies assumption (A.1).

Attack resources are costly: the cost of attacking a node is given by $c_a > 0$. Given defended network (g, Δ), payoff to the Adversary from strategy $X \subseteq N$ is

$$\Pi^A(\Delta, X; g, c_a) = -\Phi(g - (X \setminus \Delta)) - c_a|X|. \tag{10.4}$$

They study the (sub-game perfect) equilibrium of this game.

Dziubiński and Goyal (2014) show that the Adversary should target nodes that separate the network, while the Defender must protect nodes that block these separators (i.e., their *transversal*). They then study the relation between network architecture and the intensity of conflict (the sum of resources allocated to attack and defense) and the prospects of active conflict (when some nodes are defended while some are attacked). To get a sense of the issues, it is useful to begin with a simple example.

Example 1. *Defending a star network.*

Consider the star network with $n = 4$ and $\{a\}$ as central node (as in Figure 10.1). The value function is $f(x) = x^2$. As is standard, we solve the game by working backward. For every defended network (g, Δ) we characterize the optimal response of the Adversary. We then compare the payoffs to the Defender from different (g, δ) profiles and compute the optimal defense strategy. Equilibrium outcomes are summarized in Figure 10.2.

A number of points are worth noting.

1. Observe that removing node a disconnects the network; this node is a *separator*. Moreover, there is a threshold level of cost of attack (7) such that the Adversary either attacks a or does not attack at all when $c_a > 7$. Protecting this node is also central to network defense.
2. The intensity of conflict exhibits rich patterns: when costs of attack are very large there is no threat to the network and no need for defense. If the costs of attack are small, intensity of conflict hinges on the level of defense costs. When they are low, all nodes are protected and there is no attack (the costs of conflict are nc_d), if they are high then there is no defense but all nodes are eliminated (the costs of conflict are nc_a). For intermediate costs of attack and defense, both defense and attack are seen in equilibrium.
3. The size of the defense may be nonmonotonic in the cost of attack. Fix the cost of defense at $c_d = 3.5$. At a low cost of attack ($c_a < 1$) the Defender protects all nodes, in the range $c_a \in (1, 5)$ he protects 0 nodes, in the range $c_a \in (5, 13)$ he protects $\{a\}$, and then in the range $c_a > 13$, he stops all protection activity. Similarly, the size of the attack strategy may be nonmonotonic in the cost of attack.

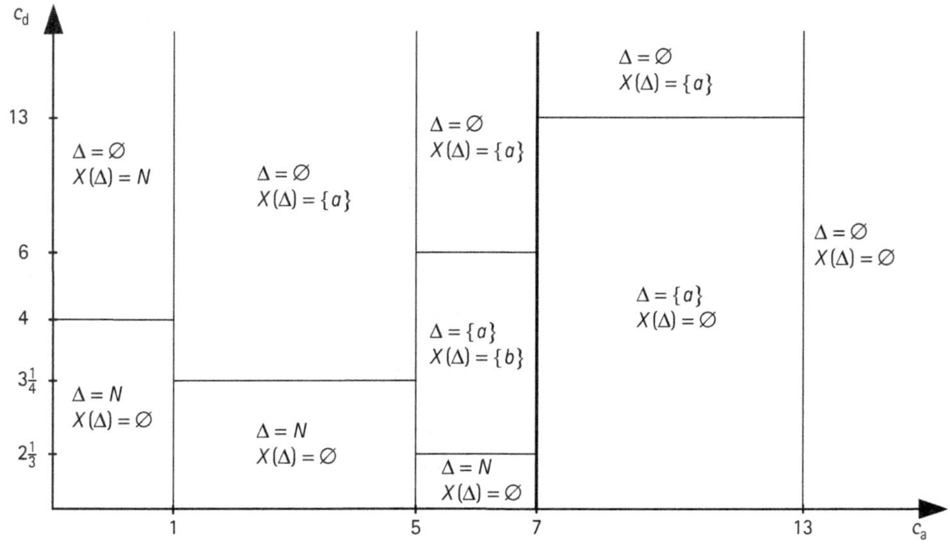

FIGURE 10.2 Equilibrium outcomes: star network ($n = 4$) and $f(x) = x^2$.

Turning now to the analysis for general networks, we note first that given the convexity in the value function of networks, disconnecting a network is especially damaging. A cut $X \subseteq N$ is a *separator* if $|\mathcal{C}(g)| < |\mathcal{C}(g - X)|$. However, a network will normally possess multiple separators and the Adversary should target the most effective ones. A separator $S \subseteq N$ is *essential* for network $g \in \mathcal{G}(N)$, if for every separator $S' \subsetneq S$, $|\mathcal{C}(g - S)| > |\mathcal{C}(g - S')|$. The set of all essential separators of a network g is denoted by $\mathcal{E}(g)$. Figures 10.3 and 10.4 illustrate essential separators in some well-known networks.

The second element is the level of costs. As illustrated by Example 1, the network defense problem can be divided into two parts, depending on the cost of attack. Given $x \in \mathbb{N}$, $\Delta f(x) = f(x+1) - f(x)$ is the marginal gain to a node in the value of a component of size x. Under Assumption A.1, $\Delta f(x)$ is strictly increasing. It is useful to separate two levels of costs: one, high costs with $c_a > \Delta f(n-1)$, and two, low costs with $c_a < \Delta f(n-1)$.

We present the case of high cost as it brings out some of the main insights in a straightforward way. Facing a high cost, the Adversary must disconnect the network (i.e., choose a separator or not attack the network at all). Clearly, the Adversary would never use an essential separator that yields a lower payoff than the empty cut. Given cost of attack c_a and network g, the set of individually rational separators is $\mathcal{E}(g, c_a) = \{X \in \mathcal{E}(g) : \Phi(g) - \Phi(g - X) \geq c_a |X|\}$.

We now turn to equilibrium strategies of the Designer. Again, it is instructive to start with the setting where cost of attack is high. An optimal strategy of the Defender should block a subset of individually rational essential separators in the most economical way. Given a family of sets of nodes, $\mathcal{F}$, and a set of nodes M, $\mathcal{D}(M, \mathcal{F}) = \{X \in \mathcal{F} : X \cap M \neq$

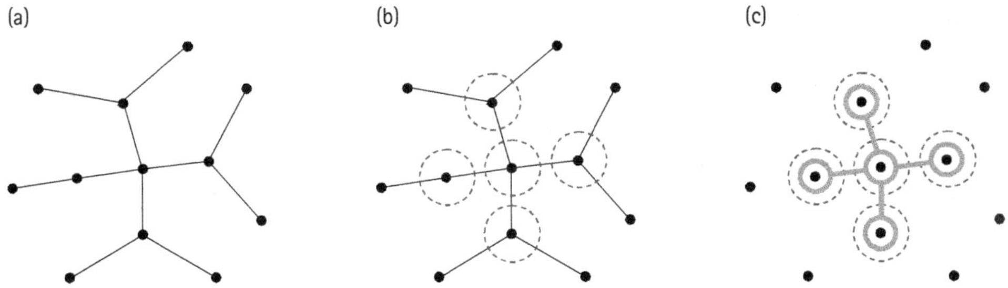

FIGURE 10.3 (a) Tree network, (b) essential separators, (c) minimum transversal.

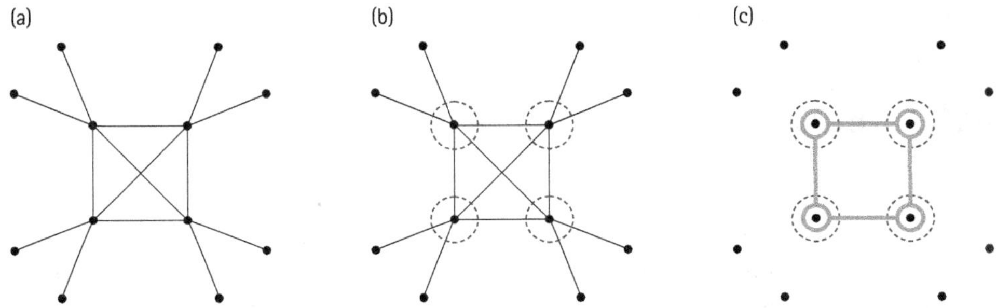

FIGURE 10.4 (a) Core-periphery network, (b) essential separators, (c) minimum transversal.

$\varnothing\}$ are the sets in $\mathcal{F}$ that are blocked (or *covered*) by M. The set M is called a *transversal* of $\mathcal{F}$, if $\mathcal{D}(M,\mathcal{F}) = \mathcal{F}$. The set of all transversals of $\mathcal{F}$ is denoted by $\mathcal{T}(\mathcal{F})$. Elements of $\mathcal{T}(\mathcal{F})$ with the smallest size are called *minimum* transversals of $\mathcal{F}$. Let $\tau(\mathcal{F})$ denote the *transversal number* of $\mathcal{F}$, i.e., the size of a minimum transversal of $\mathcal{F}$. Figures 10.3 and 10.4 illustrate the transversal in some well-known networks.

Dziubiński and Goyal (2014) develop the following result on optimal defense and attack.

Proposition 2. *Consider a connected network $g \in \mathcal{G}(N)$ and suppose $c_a > \Delta f(n-1)$. Let (Δ^*, X^*) be an equilibrium.*

- $|\Delta^*| \leq \tau(\mathcal{E}(g,c_a))$ *and Δ^* is a minimum transversal of $\mathcal{D}(\Delta^*,\mathcal{E}(g,c_a))$.*
- $X^*(\Delta) = \varnothing$, *if $\Delta \in \mathcal{T}(\mathcal{E}(g,c_a))$; $X^*(\Delta) \in \mathcal{E}(g,c_a)$ with $X^*(\Delta) \cap \Delta = \varnothing$, otherwise.*

Optimal defense is characterized in terms of minimal transversal of the appropriate hypergraph of separators (or defense covers all nodes). If cost of attack is such that elimination of single nodes is not worthwhile, optimal attack is bounded above by the transversal number of the graph. Optimal attack is either empty or targets essential separators.

Example 1 suggests that defense size is falling in defense costs and is nonmonotonic in attack cost. The attack size is nonmonotonic in both attack cost and defense cost.

Dziubiński and Goyal (2014) show that these patterns are true more generally. The authors then study the relation between the network architecture and the *intensity of conflict*: this is the sum of expenditures of defense and attack. Their analysis characterizes minimal intensity of conflict and the corresponding networks that sustain it. This allows them to show how network architecture matters for the intensity of conflict.

They then turn to the problem of defense when nodes makes security choices: they show that the equilibrium in this decentralized defense game can also be characterized in terms of transversals and separators of the underlying network. Second, they find that defense exhibits properties of strategic substitutes and a threshold public good. Three, they show that the welfare gap between decentralized equilibrium and first best outcomes is unbounded: interestingly, individual choice may lead to too little and to too much protection, relative to the choice of a single (centralized) Defender.

10.2.2 The Design of Criminal Networks

Illegal organizations, like other institutions, rely on cooperation and coordination among their members. As legal enforcement of formal contracts is problematic, such organizations are especially reliant on trust among members. Information sharing on identity and personal information may be an important factor in building internal trust and cohesion, but it can leave the organization vulnerable to 'serial' exposure. How does this trade-off affect the design of a criminal organization? In an early paper, Baccara and Bar-Isaac (2008) study this question.

The first point to note is that there is significant heterogeneity across the information structures of different criminal/illegal organizations. On the one hand, there is the view that these organizations have a centralized information and enforcement structure. In the mafia, there is the so-called cupola that holds large amounts of information about the organization itself and carries out the enforcement needed for the organization to function. These crucial agents are shielded from the authorities since they are typically not directly involved in criminal activities. On the other hand, recent studies on modern terrorism suggest a decentralized organization characterized by the presence of independent "cells." These cells consist of agents who know each other and enforce each others' actions but who have a very vague idea of how the organization looks outside the cell boundaries. Thus, even if authorities detect a cell, it is difficult to expand the detection further. This structure is similar to other organizations observed in history, including the anarchist and the revolutionary organizations in the late nineteenth century in Europe.

These empirical observations motivate a model with the following ingredients. Individuals are engaged in an infinitely repeated multi-person Prisoner's Dilemma, augmented with the possibility of additional punishment, which can help encourage "good" behavior. The additional punishment of a player requires personal information about this specific person; this information makes the person vulnerable. Examples of this kind of information include identity, whereabouts, or some incriminating evidence about a person. They explore the trade-off between the enhancement in internal cohesion derived by exchanging internal information and the increase in vulnerability to detection that this exchange implies.

The model has $N = \{1,2\ldots,n\}$, $n \geq 2$ agents. Agents are engaged in an infinitely repeated prisoner's dilemma. A Designer attempts to sustain cooperation among the n agents. Links are directed, and if i is linked to j then he can inflict an *additional* punishment on j in case the latter fails to cooperate. This implies that the returns from connectedness are positive for the first link, but zero afterwards (as an agent cannot be punished more than once). Observe that in the absence of an Adversary, a network made up of paired nodes is the optimal architecture.

The external authority, or the Adversary, attempts to inflict the most disruption possible on the network by allocating attack resources across the nodes. Contagion follows the direction of the link and is assumed to be unidirectional. The timing of the game is such that the Adversary moves first. The Designer observes the allocation of attack resources, and chooses links between the nodes.

Note that if nodes are sufficiently patient then cooperation can be sustained even without links between nodes. The empty network is in that case the optimal organization. If nodes are impatient, on the other hand, then even adding links between nodes will not suffice to induce cooperation. Again, the empty network is in that case the optimal organization. The case of interest is therefore that of intermediary values of the nodes' impatience. Let us now turn to this case.

Generally speaking, the agent's probability of getting detected directly depends on the resources allocated on him and on his activity status. Baccara and Bar-Isaac (2008) study two polar cases-one where detection probability is independent of activity in the organization, and one where detection is only possible if the agent is cooperating with the criminal organization. They characterize the optimal information structure within the organization. In the independent detection case, they find that if the probabilities of detection are sufficiently similar, either it is optimal to create no information links or the optimal structure consists of "binary cells" (pairs of agents with information about each other but with no information links with other members of the organization). Given this characterization, they then consider the optimal budget allocation for the Adversary. They show that there are circumstances in which allocating the budget symmetrically induces the organization to exchange no information. In these cases, a symmetric allocation is optimal. However, sometimes a symmetric allocation induces the agents to form a binary cell structure. Baccara and Bar-Isaac (2008) show that in this case, the authority optimizes by not investigating one of the agents at all while investigating the others equally.

In the latter cooperation-based detection case, since each agent's probability of detection is a function of the level of cooperation within the organization, an optimal information structure may require lower levels of cooperation from some of the agents to shield them from detection. Even though agents are ex ante symmetric, they show that the optimal information structure can be asymmetric, resembling a hierarchy with an agent who acts as an information hub, does not cooperate at all, and thus remains undetected. If each individual agent's contribution to the organization is sufficiently high, the optimal organization can also be a binary cell structure. Moreover, the optimal strategy of the external agent is different under cooperation-based detection. For example, devoting considerable resources to scrutinizing a single agent makes that agent relatively likely to be detected whether linked or not under the agent-based cooperation model, thus making it cheap for the organization to link the agent and induce him to cooperate. In contrast, under cooperation-based detection, it is costly to make such a scrutinized agent cooperate (and thereby increase considerably the probability that he is detected).

The driving forces in these two detection approaches are thus very different. In the independent detection model, the Adversary chooses a strategy that makes it unappealing for agents to be vulnerable. In the cooperation-detection model, however, the external authority's strategy of targeting someone makes it less attractive to have him cooperate.

We now briefly relate the findings of Goyal and Vigier (2014) and Baccara and Bar-Isaac (2008). In both papers there is a trade-off between connections and vulnerability. However, the models differ along a number of dimensions, and these differences serve to highlight the rich theoretical possibilities in this literature. In Goyal and Vigier (2014) the gains from large-scale connectivity are key; by contrast, in Baccara and Bar-Isaac (2008) the size of the network plays no essential role in defining network value.[10] Two, the former paper studies conflict between defense and attack; by contrast, there are no defense resources in the latter. Three, the Designer moves first in the former model, while the Adversary moves first in the latter model. Four, links are undirected in the former, while they are directed in the latter. These differences are substantive and taken together lead to very different insights.

10.2.3 Open Questions

The design and defense of networks that face threats is an important practical problem. Networks perform a variety of functions, and this gives rise to different potential sources of network value. The papers we have surveyed approach the network value question in different ways, but connectivity and component size have been a prominent feature of many papers. The discussion of Baccara and Bar-Isaac (2008) highlights the key role of the network value function in shaping an answer to the design question. As research in this field matures, we believe that closer attention to the source of network value would be important. In this section, we started with the first best scenario: where the design and defense are both controlled by a single player. In applications, individual nodes often have control of these variables. Our discussion of Cerdeiro et al. (2015) suggests

[10] This is best seen by comparing optimal networks in the absence of an Adversary in the two settings: in Baccara and Bar-Isaac (2008), linked pairs of players is an optimal criminal organization. By contrast, in Goyal and Vigier (2014) any optimal network must be connected.

that optimal networks with decentralized choice may be very different from first best networks. In important contexts such as epidemiology and cybersecurity, individuals choose links in addition to security. In future work, it would be important to study the impact of these choices.[11] Finally, we would like to comment on the nature of defense. Following on the early work of Aspnes et al. (2006) and others, most of the recent work surveyed in the section has assumed that defense is perfect. This is a natural first step, but clearly it is a strong assumption. The dynamics of contests on networks remains a poorly understood problem.

10.3 Resources, Conflict, and Networks

In economics and in biology, we think of agents and organisms as seeking to expand their influence and to capture territory. One possible avenue through which to obtain resources is to appropriate them through conflict. However, agents may face constraints on whom they can target for conflict. The extensive literature on wars shows that a significant majority of them take place among physically proximate entities (Caselli et al. 2014).The traditional models of conflict have focused on bilateral conflicts or on groups of countries in conflict (Garfinkel and Skaperdas 2012). As bilateral conflicts create spillovers to other conflicts and as the spillovers are mediated by the pattern of neighborhood relations, it is important to develop general models of conflict in networks. The literature surveyed below is a first step in this direction.

We start with a static conflict game on a network. The presentation draws on Franke and Öztürk (2009). There is a set $N = \{1, \dots, n\}$, where $n \geq 3$ of agents that are located in an undirected network g. The set of rivals of agent i is given by $N_i(g)$, and so agent i is engaged in $n_i(g) = |N_i(g)|$ conflicts. The outcome of each bilateral conflict is probabilistic and depends on the investment in conflict by the respective rivals. For concreteness, we shall suppose that the conflict technology is the linear Tullock contest function. In a conflict between i and j, and given investments e_{ij} and e_{ji} the probability of winning for agent i is given by

$$p_{ij}(e_{ij}, e_{ji}) = \frac{e_{ij}}{e_{ij} + e_{ji}}, \tag{10.5}$$

so long as $e_{ij} + e_{ji} > 0$. In case $e_{ij} + e_{ji} = 0$, the probability of either player winning is $1/2$. Each agent i chooses a investment for each of his conflict links $\mathbf{e}_\mathrm{i} = \left(e_{ij}\right)_{j \in N_i}$. The cost of investment in conflict is given by the function $c(e_i)$. For simplicity, we assume that $c(e_i) = (\sum_{j \in N_i} e_{ij})^2$. The reward from winning a conflict is V, while the cost of losing is $-V$. We may now write the payoffs of agent i in network g, with investment

[11] For a survey of the literature on co-evolution of networks and behavior, see (chapter 9 by Vega-Redondo in this volume.

profile $e = (e_1, \ldots, e_n)$, as

$$\pi_i(e) = V \sum_{j \in N_i} [p_{ij}(e_{ij}, e_{ji}) - p_{ji}(e)] - c(e_i). \tag{10.6}$$

The interest is in understanding how the network shapes conflict. We will focus on Nash equilibrium in conflict investments.

Define $e^*(g)$ to be an equilibrium for network g. Franke and Öztürk (2009) start by showing that there exists a unique equilibrium in this model and that it is interior. Define $E_i^*(g)$ to be the aggregate equilibrium investment by player i in network g. Equilibrium investments satisfy the following property: for each $i \in N$ and each $k \in N_i$,

$$V \frac{e_{ik}^*}{[e_{ik} + e_{ki}]^2} = E_i(g). \tag{10.7}$$

Let $e^*(g)$ be the equilibrium profile of investments in network g and let $E^*(g) = \sum_{i \in N} E_i^*(g)$. be an aggregate equilibrium investments or the conflict intensity.

Franke and Öztürk (2009) provide results on some well-known special classes of networks. We present their results on regular networks and the star network.

Proposition 3. *Conflict equilibrium exhibit the following properties.*

1. *Regular networks: Conflict intensity is increasing in degree, d, and in number of agents n. Conflict intensity is higher in g_1 than in g_2 if and only if $n_1\sqrt{d_1} > n_2\sqrt{d_2}$. Individual investment and expected payoff is decreasing in degree and does not depend on number of agents. Expected equilibrium payoff is negative for all agents.*
2. *Star network: Conflict intensity is increasing in the number of peripheral agents. For the center agent, link specific (aggregate) investment is decreasing (increasing) and the expected payoff is decreasing in number of peripheral agents. For the periphery agent, conflict investment is declining and payoffs are increasing.*

Franke and Öztürk (2009) offer us an interesting first look at conflict in networks. Their results illustrate how the cost function and network creates spillovers and how these spillovers in turn shape behavior by different agents in a network. As the authors note, the problem of characterizing conflict in general networks remains an open problem.

In a recent paper, Konig et al. (2014) study conflict in a network setting where links may be positive (as between allies) or negative (as between enemies). They provide a characterization of conflict investments as a function of the network: investments by allies are strategic substitutes, while investments by enemies are strategic complements. They then show that equilibrium investments are proportional to Bonacich centrality of the allies and enemies networks, respectively. They then apply these results to understanding the nature of conflict in the Congo.

The authors assume a linear Tullock contest function and the conflict is static. In real-world conflicts dynamics play an important role, as winners of current conflicts acquire more resources and power that they can use to subsequent conflicts. We examine the dynamics of conflict in the next section.

10.3.1 Dynamics of Conflict

We now take up the study of conflict as a dynamic process with resource accumulation for the winner and elimination of the losers. By way of motivation, consider the example of a kingdom or a country seeking to expand its territory by means of military conquest. This drive toward expansion will typically be geographically constrained as it is difficult to conduct military campaigns far away from established territory. Indeed, empirical research shows that the vast majority of conflicts are amongst physically neighboring entities. Moreover, once a particular territory has been conquered and integrated, new territories close to the conquered territory become accessible. In addition, the resources of the newly acquired territory can be used for further conquest. Historians use archaeological evidence to argue that the Roman empire used resources from occupied territories around the Mediterranean to supply legions during the invasion of Western Europe and Britain.

Conflict may also take nonmilitary forms, such as a company seeking to expand into new markets. It is easier for a company to expand into markets that are closely aligned to its current markets, either geographically or in terms of product lines. Once a foothold has been established in a new market, this generates resources for further expansion. In addition, markets closely aligned to the newly entered market become more accessible.

The presentation here draws on De Jong et al. (2014), who study a framework where a number of agents seek to capture resources through conflict. At the start, individuals have resources. Each individual is located at a node of a network. As time goes by, opportunities arise for individuals to "fight" and capture the resources of neighboring nodes. A conflict is modeled as a Tullock contest. The conflict yields a winner and a loser. The winner captures the resources of the loser and expands his influence in the network.

There are three ingredients in this framework: the initial resources of individuals, the network structure, and the technology of conflict (the parameters of the Tullock contest function). Their goal is to characterize the dynamics of conflict and the rise and fall of empires.

A set of players $P = \{1,2,\ldots n\}$ engage in conflict over a network with nodes $N = \{1,2,\ldots n\}$. Nodes have attached resources $r = (r_1, r_2, \ldots, r_n)$. The set of all links form a connected network g.

Each node is controlled by one of the players. Let $S_i^t \in P$ denote the player in control of node i in period t, with $t = 0,1,2,\ldots$ Each player starts with ownership of their "own" node, so that $S_i^0 = i$. Ownership of nodes may change during the game, depending on conflict outcomes. Let $G^t = \{g_{ij} \in g \mid S_i^t \neq S_j^t\}$. This is a sub-network of g containing all the links that connect nodes controlled by different players. Note that the value of G^0 is g. At the beginning of period t, a link $L^t \in G^t$ is selected with equal probability.

The players who control the nodes at two ends of the link each use the combined resources from their respective nodes. The winner of the contest gains control of the loser's nodes. Define $R_u^t = \sum_{S_k^t = u} r_k^t$, the total resources available to player u in period t. Let $g_{ij} = L^t$ be the selected link in G^t at round t. Then players S_i^t and S_j^t engage in a Tullock contest. Suppose, without loss of generality, that $S_i^t < S_j^t$. Then W_i^t is defined to be the binary variable describing the winner of the contest in period t, taking value 1 if player S_i^t wins and 0 if player S_j^t wins. The distribution of this variable is

$$P(W_i^t = 1) = \frac{(R_i^t)^\gamma}{(R_t^t)^\gamma + (R_j^t)^\gamma}$$

where $\gamma > 0$ is the parameter of the Tullock contest function.

The network is then updated depending on the value of W.

$$S_v^{t+1} = S_v^t \qquad \text{for } v \neq i,j$$

$$S_v^{t+1} = W^t S_i^t + (1 - W^t) S_j^t \qquad \text{for } v = i,j$$

The game ends when there is only player left, who is denoted the victor.

In this basic model, one player is eliminated in every period, so the dynamics must end in period $n-1$. Moreover, *For* $\gamma < +\infty$ *and* $r \in \mathbb{N}^n$, *the game ends in* $n-1$ *periods and the probability for any player to win the game is strictly positive.*

We now turn to how the probability of victory is affected by starting resources and network position. To make progress, De Jong et al. (2014) specialize the model and consider the case of linear Tullock function. Define $R = \sum_{i=1}^{n} r_i$. They establish:

Proposition 4. *Let* $\gamma = 1$. *Then the probability for player i to win the game is given by* $P(i) = r_i / R$.

The intuition behind this result is as follows. In order to win the game, a player must eventually capture all other resources. The player can do this in a single contest, fighting all other resources at the same time, in which case the probability of winning is as above. If the player engages in an intermediate contest, the increased probability of winning the final contest compensates exactly for the probability of losing the intermediate contest for γ equal to one. Network position influences the timing and frequency of contests, and therefore has no effect on win probability. For $\gamma < 1$, the Tullock contest function is everywhere concave in resources, and so resources gained in the intermediate contest do not compensate for the chance of losing the intermediate contest. If $\gamma > 1$, then the Tullock contest function is convex on part of its domain, so depending on resources, the resources gained from the intermediate contest may compensate for the chance of losing. In order to illustrate the effects of interaction between the technology of conflict and network structure, we consider a specific network, the star.

Proposition 5. *Let G be a star network with n peripheral nodes, and let resources be homogeneous among peripheral nodes, so that* $r = (r_c, r_p, \dots r_p)$. *Then the win probabilities for central and peripheral nodes are given as follows*

$$P_c(r_c, r_p, \gamma) = \prod_{i=0}^{n-1} \frac{(r_c + r_p)^\gamma}{(r_c + ir_p)^\gamma + r_p^\gamma} \tag{10.8}$$

$$P_p(r_c, r_p, \gamma) = \frac{1}{n}(1 - P_c(r_c, r_p, \gamma)). \tag{10.9}$$

It is important to explore the interaction between resources, network, and technology of conflict a bit more closely. To illustrate the richness of this relationship, it is worth looking at the limiting case when $\gamma \to +\infty$. When $\gamma \to +\infty$, the Tullock contest reduces to the all-pay auction, where the player with higher resources wins with probability 1. Consider a network with three nodes in a line, with resources r_1, r_2, and r_3. First, note that the player with the lowest resources has negligible probability of winning, regardless of network position. If any player has resources higher than the combined resources of the other two players, that player will win with probability close to 1, regardless of network position. So, the network position is irrelevant. Let $r_2 > r_1 = r_3$ and $r_2 < r_1 + r_3$. Then player 2 wins with probability 1, and the marginal effect of resources is 0 for all players. Now switch the location of player 2 and player 1. Then player 2 wins with probability $\frac{1}{2}$, and players 1 and 3 win with probability $\frac{1}{4}$. For players 1 and 3, adding any amount of resources increases win probability by a further $\frac{1}{4}$. To summarize, the network position can have either no effect or a very large effect, and a small amount of resources can have either no impact or a very large impact, depending on the network and current resource allocation.

These results suggest a very rich interaction between resources, networks, and the technology of conflict in shaping conflict dynamics. A general analysis of this problem remains an open problem.

The framework presented above raises a number of interesting further questions. The network is taken as given. In a recent paper, Huremovic (2014) studies conflict in an evolving network. We have assumed that nodes engage in conflict, but a natural question is whether they have an incentive to engage in conflict. If the technology of conflict is very equalizing (γ close to 0) then nodes would prefer to not fight, as there is no gain in terms of additional resources, while there is a cost in terms of positive probability of elimination. Another issue relates to alliances: so far we have assumed that nodes remain independent. But alliances are a salient feature of international as well as civil conflict. We take up the role of alliances in shaping conflict next.

10.4 Alliances, Networks, and Conflict

International and civil wars impose enormous direct costs on the parties involved and have large indirect costs on third parties. Empirical work shows that around 40%

of the wars with more than 1000 casualties involved more than two countries and large conflicts such as world wars and the Vietnam War involved alliances with many nation states. More generally, alliances have been a central element in violent conflict throughout history. These considerations motivate a study of alliance formation and how it shapes the intensity of conflict.[12]

We begin with a presentation of a recent paper by Jackson and Nei (2014).[13] They develop a model for the incentives of countries to attack each other, to form alliances, and to trade with each other. There is a set $N = \{1, \ldots, n\}$, where $n \geq 3$, of countries. Countries are linked through alliances, represented by a network of alliances g. If two countries are linked then they are allies. Let $g - i$ denote the network obtained by deleting all alliances that involve country i. Let $\mathcal{CL}(g)$ denote the set of cliques in network g. Each country $i \in N$ is endowed with a military strength $M_i \in \mathbb{R}_+$. For any subset of countries $C \subseteq N$, let $M(C) = \sum_{i \in C} M_i$ be their collective military strength. If there is a war between C_1 and C_2, with C_1 being the aggressor, then C_1 wins if $M(C_1) > \rho M(C_2)$. The parameter $\rho > 1$ reflects a relative advantage of being the Defender, and $\rho < 1$ reflects a relative advantage of being the aggressor.

The notion of "vulnerability" plays a key role in the analysis. A country i is vulnerable at a network g if there exists a country j and a coalition $C \subseteq N_j(g) \cup \{j\}$ such that $j \in C$, $i \notin C$ and $M(C) > \rho M(i \cup (N_i(g) \cap C^c))$, where C^c is the complement of C. In this case, country j is said to be a potential aggressor at a network g. Thus, no country is vulnerable at a network g if for any coalition C of a potential aggressor j and any target country $i \notin C$, the aggressors cannot successfully attack the country. At this point, it is being assumed that winning is desirable and that losing a war is undesirable.

Jackson and Nei (2014) introduce the notion of war-stable networks to study the incentives of countries to form coalitions to defeat and conquer other countries. Define $E_i k(g, C)$ as the net gains to country k if country i is conquered by coalition C (of which k is a member), when country i is conquered by coalition given by $C' = \{i\} \cup (N_i(g) \cap C^c)$. It is assumed that there is a cost to maintaining a link $c_{ij} > 0$ between any pair of countries i, j. These costs will be taken to be small relative to the spoils from a successful war.

With this notation in place, a network g is war stable if the following conditions are satisfied:

1. no country is vulnerable at g.
2. $\forall g_{j,k} \notin g$, no country is vulnerable at $g + jk$,
3. $\forall g_{jk} \in g$, both j and k are vulnerable at $g - g_{jk}$.

Suppose that countries are ordered as follows: $M_1 \geq M_2 \geq \cdots \geq M_n$. Jackson and Nei (2014) establish the following result:

Proposition 6. *Let $n \geq 3$. There are no nonempty war-stable networks. The empty network is war-stable if and only if $\rho M_n \geq M_1 + M_2$.*

[12] There is an important body of research on groups in conflict, for a survey, see Garfinkel and Skaperdas (2012). In this survey, due to space constraints, we do not cover conflict among groups.

[13] For an early paper on network formation with antagonistic links, see Hiller (2012).

The intuition behind this result is as follows: For no country to be vulnerable and for every alliance to be productive (in terms of condition 3) networks to be sparse. However, sparse networks are susceptible to condition 2: allies of a country can join forces and defeat it. This tension suggests rapidly shifting alliances and is reminiscent of the empirical patterns from the nineteenth century. Jackson and Nei (2014) report that during the nineteenth century and the first half of the twentieth century, roughly one-third of the alliances present at any time were dissolved within the next five years. By contrast, in the period from 1950 until 2000, this probability was around 0.05!

This sharp difference in the performance of alliances motivates a closer examination of other significant economic changes. Jackson and Nei (2014) focus on the changes in the size of international trade. They report that international trade has had two major periods of growth. The latter part of the nineteenth century and beginning of the twentieth saw a sharp rise in international trade. This rising trend was disrupted by the world wars. International trade picked up after the Second World War, recovering its pre–First World War level in the 1960s and then continuing to grow at an increasing rate thereafter. In particular, in 1850 international trade amounted to 5.1% of total world output, this share rose to 11.9% in 1913. It then remained below this level until the 1960s, picking up thereafter and reaching 25% in 2012. These changes lead Jackson and Nei (2014) to propose a richer model of alliances and wars that incorporates the role of international trade.

They propose that a country gets a payoff $u_i(g)$ from network g, reflecting gains from trade. The notion of vulnerability is now adapted to take into account this additional consideration. A country i is said to be *vulnerable despite trade* in a network g if there exists a country j and a coalition $C \subseteq N_j(g) \cup \{j\}$ such that $j \in C$, $i \notin C$ and (i) $M(C) > \rho M(i \cup (N_i(g) \cap C^c))$, and (ii) $u_k(g-i) + E_{ik}(g, C) \geq u_k(g)$, with some strict inequality.

The $u_k(g-i)$ reflects the trade implications of successful elimination of i: it is worth noting that a country k may stand to benefit or to lose from such a conquest. Taking this general effect into account, Jackson and Nei (2014) define network g to be *war and trade stable* if the following three conditions are met:

1. no country is vulnerable despite trade at g;
2. $\forall g_{jk} = 0$, if $u_j(g + g_{jk}) > u_i(g)$ then $u_k(g + g_{jk}) < u_k(g)$, and g_{jk} is not war-beneficial;
3. $\forall g_{jk} = 1$, either $u_j(g - g_{jk}) \leq u_j(g)$ or j is vulnerable despite trade at $g - g_{jk}$, and similarly for k.

In other words, a network of alliances is war and trade stable if no country is vulnerable despite trade, if no two countries can add an alliance that is mutually profitable (through economic or through war means), and either economic or war considerations prevent every country from severing any of its links.

For simplicity, suppose that

$$u_i(g) = f(d_i(g)) - cdi(g) \quad (10.10)$$

where $d_i(g)$ is the degree of i, f is concave, nondecreasing, and there is some $d \leq n-1$ such that $f(d) < cd$. Let $\bar{d}$ maximize $f(d)..cd$. In addition, let $E_{ij}(g;C) = E(d_i(g))/|C|$. Under these simplifying assumptions, Jackson and Nei (2014) establish the following existence result.

Proposition 7. *Consider the symmetric model with $d \geq 2$.*

- *If $E(d^*) \leq 2[f(d^*) - f(d^* - 1) - c]$, then any d-regular network (in any configuration), is war and trade stable network if $\rho \geq \frac{d^*+1}{d^*-1}$.*

The above proposition illustrates one route through which trade supports stable networks and thereby contains conflict. The condition provides sufficient gains from trade such that the potential spoils of a war are outweighed by the lost trade value: this in turns means that a country is never attacked by one of its own trading partners. Each country then has enough alliances to protect itself against attacks from outside, which allows a wide range of networks to be sustained.

To summarize, the discussion above develops a simple model of network formation that yields two interesting insights. The first insight is that in a pure conflict setting individual attempts to form alliances and attack opponents leads to shifting and unstable alliances: this instability may make peace hard to sustain. The second insight is that the presence of large gains from trade can sustain stable alliance structures where no country is vulnerable to attack by a coalition of enemies.

In this model conflict is implicit: countries do not allocate resources and wage war on other countries. Thus the impact of alliances on the incentives to allocate resources for conflict are not explicitly considered. This public good aspect to individual contributions to a coalitional conflict are potentially important. Investing in conflict within an alliance has public good properties: the study of alliance formation in a setting where countries choose investments in conflict remains an open problem.[14]

10.5 Concluding Remarks

In recent years, a new literature has begun to study of the relation between conflict and networks. This chapter provides a survey of this nascent literature.

In the first part, the focus was on settings where network connectivity is the key source of value. Motivated by cybersecurity and infrastructure applications, the aim was to study the design and the defense of networks under threat. The basic framework involves two players: an Adversary and a Designer/Defender. The Adversary uses resources to target nodes that maximize damage of the network, while the Designer/Defender uses defense resources and links to maximize the value of the residual network. We presented a number of results on optimal attack and defense

[14] For a survey on free riding and coalition formation among agents in conflict, see Bloch (2012).

targeting and on optimal design. The discussion reveals that the technology of conflict, the respective resources of the Adversary and Designer, and the network value function all play an important role in shaping conflict and in the design of the optimal network.

The second part of survey is motivated by the observation that most international conflict and civil unrest happens between physically proximate entities. Bilateral conflicts between two neighbors have spillovers on the neighbors of the neighbors. This motivates a study of conflict in networked environments. We started with a static model of conflicts and showed that aggregate conflict intensity and individual investments in conflict vary in interesting ways with the structure of the network. We then presented a model with a focus on the dynamics of conflict and conquest where winners captured the resources of the losers. Here we derived results on how the network and resources determine the winner for specific technologies of conflict and for specific networks.

In the third part, we moved to a study of alliance formation among competing nodes. The existing research provides us with insights into how gains from international trade are key to understanding the structure of alliances and the decline of international conflict in the last 50 years.

References

Acemoglu, D., A. Malekian, and A. Ozdaglar (2013). "Network security and contagion." NBER Working Paper 19174, *National Bureau Of Economic Research.*

Acemoglu, D., A. Ozdaglar, and A. Tahbaz-Salehi (2015). "Systemic risk and stability in financial networks." *American Economic Review* 105, 564–608.

Albert, R., H. Jeong, and A.-L. Barabási (2000). "Error and attack tolerance of complex networks." *Nature* 406(6794), 378–382.

Alpcan, T. and T. Başar (2011). *Network Security: A Decision and Game Theoretic Approach.* Cambridge, UK: Cambridge University Press.

Anderson. R. (2001). *Security Engineering: A Guide to Building Dependable Distributed Systems.* New York: John Wiley & Sons.

Arquilla, J. and D. Ronfeldt (2001). *Networks and Netwars: The Future of Terror, Crime, and Militancy.* Santa Monica, CA: Rand.

Aspnes, J., K. Chang, and A. Yampolskiy (2006). "Inoculation strategies for victims of viruses and the sum-of-squares partition problem." *Journal of Computer and System Sciences* 72(6), 1077–1093.

Baccara, M. and H. Bar-Isaac (2008). "How to organize crime?" *Review of Economic Studies* 75(4), 1039–1067.

Bala, V. and S. Goyal. (2000). "A noncooperative model of network formation." *Econometrica* 68(5), 1181–1230.

Bloch, F. (2012). "Endogenous formation of alliances in conflicts." In *The Oxford Handbook of the Economics of Peace and Conflict,* M. Garfinkel and S. Skaperdas, eds. Oxford: Oxford University Press.

Blume, L., D. Easley, K. J., K. R., and T. T. (2011). "Network formation in the presence of contagion risk." In *Proceedings of the 12th ACM Conference on Electronic Commerce.*

Cabrales, A., P. Gottardi, and F. Vega-Redondo (2010). "Risk sharing and contagion in networks," Working paper.

Caselli, F., M. Morelli, and D. Rohner (2014). "The geography of inter-state resource wars." Working paper, Columbia University.

Cerdeiro, D., M. Dziubiński, and S. Goyal (2015). "Contagion risk and network design." Cambridge-INET Institute 2015-04.

De Jong, M., A. Ghiglino, and S. Goyal (2014). "Resources, conflict and empire." mimeo.

Department of Homeland Security (2012), *Office of Infrastructure Protection Strategic Plan: 2012-2016*. Washington, DC.

Dziubiński, M. and S. Goyal. (2013). "Network design and defence." *Games and Economic Behavior* 79(1), 30–43.

Dziubiński, M. and S. Goyal (2014). "How to defend a network." Cambridge-INET Working Paper 2014-01.

Elliot, M., B. Golub, and M. Jackson (2014). "Financial networks and contagion." *American Economic Review* 104, 3115–3153.

Eun. H. (2010). "Impact analysis of natural disasters on critical infrastructure, associated industries, and communities." PhD thesis, Purdue University, West Lafayette, IN.

Farrell, J. and G. Saloner (1986). "Installed base and compatibility: Innovation, product preannouncements, and predation." *American Economic Review* 76, 940–955.

Franke, J. and T. Öztürk (2009). Conflict networks. Ruhr Economic Papers 116, University of Dortmund.

Garfinkel, M. and S. Skaperdas (2012). *The Oxford Handbook of the Economics of Peace and Conflict.* Oxford University Press.

Goyal, S. (1993). "Sustainable communication networks." Tinbergen Institute Discussion Paper TI 93–250, Rotterdam-Amsterdam. 1993.

Goyal S. (2007). *Connections: An Introduction to the Economics of Networks*. Princeton, NJ: Princeton University Press.

Goyal, S. and A. Vigier (2014). "Attack, defence, and contagion in networks." *Review of Economic Studies* 81(4), 1518–1542.

Harary. F. (1962). "The maximum connectivity of a graph." *Proceedings of the National Academy of Science* 48(7), 1142–1146.

Hiller, T. (2012). "Friends and enemies: A model of signed network formation." Working paper, Bristol University.

Huremovic, K. (2014). "Rent seeking and power hierarchies: A noncooperative model of network formation with antagonistic links." Nota di Lavoro 45.2014, Fondazione Eni Enrico Mattei, Milan, Italy.

India Today (2011). "Political agitations affect railway service." March 26.

Jackson, M. and S. Nei (2014). "Networks of military alliances, wars, and international trade." mimeo.

Katz, M. and C. Shapiro (1985). "Network externalities, competition and compatibility." *American Economic Review* 75(3), 424–440.

Kliesen, K., (1995). "The economics of natural disasters." *The Regional Economist.*

Konig, M., M. Rohner, D. Thoenig, and Zilibotti, F. (2014). "Networks in conflict: Theory and evidence from the great war of Africa." mimeo.

Kunreuther, H. and G. Heal (2004). "Interdependent security." *The Journal of Risk and Uncertainty* 26(3), 231–249.

Luft, G. (2005). "Pipeline sabotage is terrorists weapon of choice." *Energy Security* March 28.

Myerson, R. B. (1977). "Graphs and cooperation in games." *Mathematics of Operations Research* 2, 225–229.

Newman, M. (2010). *Networks: An Introduction*. New York: Oxford University Press.

Roy, S., C. Ellis, S. Shiva, D. Dasgupta, V. Shandilya, and Q. Wu (2010). "A survey of game theory as applied to network security." In *2010 43rd Hawaii International Conference on System Sciences* 1–10.

Staniford, S., V. Paxson, and N. Weaver (2002). "How to own the internet in your spare time." In *Proceedings of the 11th USENIX Security Symposium*, 149–167, Berkeley, CA. USENIX Association.

Tullock, G. (1980). *Efficient Rent Seeking*, 97–112. College Station, TX: Texas A&M University Press.

Zhu, S. and D. Levinson (2011). "Disruptions to transportation networks: A review." Working paper, University of Minnesota.

CHAPTER 11

KEY PLAYERS

YVES ZENOU

11.1 Introduction

Social networks are important in several facets of our lives. For example, the decision of an agent whether or not to buy a new product, attend a meeting, commit a crime, or find a job is often influenced by the choices of her friends and acquaintances.[1] Social networks have recently gained popularity with the advent of sites such as MySpace, Friendster, Orkut, Twitter, Facebook, and so on. The number of users participating in these networks is large and still growing. Therefore, we need a clear understanding of the functioning of these networks. One crucial aspect of social networks is key players, who are those elements in the network that are considered important, in regard to some criteria. For example, identifying key players is one of the goals in an online interaction media such as blog posts (Sathik and Rasheed 2009).

The aim of this article is to survey the research on key players in the economics of networks. First, we quickly review the key-player literature in sociology, which mostly provides centrality measures that are not microfounded. Then, we expose the economic literature by examining the theoretical models of the key player. Most of these models are within the framework of network games with strategic complementarities where the utility of the action of a player increases with the marginal increase of the actions of her neighbors or links. We expose the canonical network model of games with strategic complementarities where, at the Nash equilibrium, the action (or effort) of a player is proportional to her position in the network, as measured by her Katz-Bonacich centrality, a well-known centrality measure in sociology. For example,

[1] For overviews of the network literature, see Goyal (2007), Jackson (2008, 2011, 2014), Ioannides (2012), Jackson, Rogers, and Zenou (2015), and Zenou (2015).

in the context of crime, the effort (i.e., the number of crimes) each delinquent exerts will be proportional to her position in the network, as measured by her Katz-Bonacich centrality. Then, we can determine who is the key player, which is the agent who once removed from the network generates the highest decrease in total activity. For example, in the context of crime, the key player is the criminal who, once removed, reduced total crime the most. Importantly, we show that the key player need not be the player exerting the highest effort. We are then able to extend this framework to more complex situations where, for example, competition between agents is introduced or where imperfect information on the strength of interaction is included. We also develop another framework for the key player when the focus is on the diffusion of information about a product or behaviors. In that case, the key players are the first nodes in the network that need to be targeted so that the diffusion of a product is the largest possible.

More generally, which measure of centrality is appropriate to predict behavior depends on context. In models with behavioral complementarities or where individuals may spread information, measures of an individual's centrality need to incorporate aspects of the network beyond the number of friends of an individual (i.e., degree centrality). When there are complementarities in behaviors such as in crime or education, the Katz-Bonacich centrality and the key-player intercentrality measures seem to be the right measure describing the activity of each agent. If the objective of the planner is such that the diffusion of information spreads the most, then it is not clear that we should first approach the agent with the highest Katz-Bonacich or intercentrality in a network. It may be that the agent with the highest diffusion centrality[2] would be the right node to target.

In the second part of this article, we review the empirical evidence on the key-player policies based on the theoretical models developed in the first part. A natural application is the one on criminal networks. Using both data on juvenile crime in the United States and adult crime in Sweden, we show how key-player policies outperform other reasonable policies such as targeting the most prolific criminals. We then look at R&D networks where key firms are defined such that, if they exit from the market and thus from the network of collaborations, the reduction in total welfare will be the highest. Interestingly, General Motors, which was bailed out by president-elected Barack Obama in 2008, was ranked highest among the key firms in the same year. We also study education and how the centrality of a group of students affects own educational outcome and financial networks where the key players are the banks that should be bailed out. Finally, we examine the diffusion of a microfinance program in India and show that targeting the individuals with the highest diffusion centrality will have the highest impact on the adoption of this microfinance program in the village.

[2] A concept introduced by Banerjee et al. (2013) and defined below.

11.2 Key Players: Theoretical Considerations

11.2.1 The Sociological Approach

The problem of identifying key players in a network is an old one, at least in the sociological literature. Indeed, one of the focuses of this literature is to propose different measures of network centralities and assert the descriptive and/or prescriptive suitability of each of these measures to different situations. Borgatti (2003, 2006) was among the first researchers to really investigate the issue of key players, which is based on explicitly measuring the contribution of a set of agents to the cohesion of a network. The basic strategy is to take any aggregate network property, such as density or maximum flow, and derive a centrality measure by deleting nodes and measuring the change in the network property. Measures derived in this way have been called "vitality measures" (see Koschützki et al. 2005, for a review).

To be more precise, Borgatti (2003, 2006) identifies two key-player problems. First, he puts forward the "Key Player Problem/Negative" (KPP-Neg), which is defined in terms of the extent to which the network depends on its key players to maintain its cohesiveness. It is a "negative" problem because it measures the amount of *reduction* in cohesiveness of the network that would occur if the nodes were not present. Borgatti gives examples of such problems. In public health, a key-player problem is whenever the planner needs to select a subset of population members to immunize or quarantine in order to optimally contain an epidemic. In the military or criminal justice context, the key-player problem arises when a planner needs to select a small number of players in a criminal network to be neutralized (e.g., by arresting, exposing, or discrediting) in order to maximally disrupt the network's ability to mount coordinated action.

The second key player problem identified by Borgatti (2003, 2006) is called the "Key Player Problem/Positive" (KPP-Pos) where the planner is looking for a set of network nodes that are optimally positioned to quickly *diffuse* information, attitudes, behaviors, or goods. In a public health context, a health agency will need to select a small set of population members to use as seeds for the diffusion of practices or attitudes that promote health, such as using bleach to clean needles in a population of drug addicts. In a military or criminal justice context, the planner needs to select an efficient set of agents to surveil, to turn into something else (double-agents for example), or feed misinformation.

To identify key players, one uses centrality or the position of nodes in networks. There are numerous different ways of quantifying centrality, each having its distinct importance and logic. A basic one is simply counting the number of connections of a given node, or its degree. But there are also many richer definitions that keep track of how close a given node is to others, on average, or whether a given node is a critical connector on paths between other nodes, or whether a given node is well connected to other important nodes. Let us give a more precise definition of the most prominent centrality measures.

Degree centrality simply measures the number of links of each agent and captures a direct measure of popularity. *Betweenness centrality* of a given agent is equal to the number of shortest paths between all pairs of agents that pass through the given agent. In other words, an agent is central if she lies on several shortest paths among other pairs of agents. Betweenness centrality thus captures the importance as an intermediary. Such central agents have control over the flow of information in the network, which is related to the notion of *structural holes* developed by Burt (1992), who postulates that social capital is created by a network where people can broker connections between otherwise disconnected segments of the network. *Closeness* and *decay centrality* are measures of how close an agent is to all other agents in the network. The most central agents can quickly interact with all others because they are close to all others. These measures of centrality capture how easily an individual reaches others (i.e., how informed a given individual is in the context of information flows). *Eigenvector* centrality is a measure of the influence of an agent in a network. It assigns relative scores to all agents in the network based on the concept that connections to high-scoring agents contribute more to the score of the agent in question than equal connections to low-scoring agents. It thus captures indirect reach so that being well-connected to well-connected other agents makes you more central. Google's PageRank is a variant of the eigenvector centrality measure. Finally, *Katz-Bonacich centrality* (due to Katz 1953 and Bonacich 1987) takes all possible paths in a network (not only the shortest ones) but puts a lower weight on nodes that are further away from the agent.[3] As a result, Katz-Bonacich centrality captures the influence on friends and their friends. If there are strong network externalities, it can be shown that Katz-Bonacich centrality becomes proportional to eigenvector centrality (see Wasserman and Faust 1994, Chap. 5.2).[4] Therefore, these different measures fundamentally capture different aspects of centrality.[5]

11.2.2 The Economic Approach

The standard centrality measures described above only consider network structure and do not take additional information into account. In economics, we first study the behavior of agents and their interactions and then deduce which measure of centrality will explain these behaviors. For this purpose, we would now like to explicitly model the *behavior* of individuals since it will help us define a key-player centrality measure. We will first focus on the "Key Player Problem/Negative" defined above and will have

[3] For a mathematical definition of the Katz-Bonacich centrality, see formula 11.4.

[4] See Dequiedt and Zenou (2014), who propose an axiomatic approach to derive the degree, eigenvector, and Katz-Bonacich centralities.

[5] For a review of the different existing centrality measures, see Wasserman and Faust (1994) and Jackson (2008).

criminal networks in mind. In that case, the aim of the planner is to identify the key player in a (connected) network, which is the individual (criminal) to be removed from the network so that total crime is minimized. What is crucial here is that when a criminal is removed from the network, the remaining criminals will optimally react by adjusting their criminal effort. As a result, we first need to define the way in which criminals exert their crime effort and examine how the key-player policy affects their criminal activities. We will then focus on "Key Player Problem/Positive" and study the role of key players in diffusion. In that case, the planner wants to target a set of network nodes that are optimally positioned to quickly *diffuse* information, attitudes, behaviors, or goods.

11.2.2.1 Games with Strategic Complementarities

Although games on networks can take many forms, there are two prominent and broadly encompassing classes of games.[6] The distinction between these types of games relates to whether a given player's relative payoff to taking an action is *increasing* or *decreasing* in the set of neighbors who take this action. The first class of games on networks, of which coordination games are the canonical example, are games of *strategic complements*. In games of strategic complements, an increase in the actions of other players leads a given player's higher actions to have relatively higher payoffs as compared to that player's lower actions (Ballester et al. 2006, 2010). Games of *strategic substitutes* are such that the opposite is true: an increase in other players' actions leads to relatively lower payoffs to higher actions of a given player (Bramoullé and Kranton 2007; Bramoullé, Kranton, and D'Amours 2014).

For criminal behaviors, it seems relatively natural to consider games with strategic complementarities, since the higher my friends' criminal efforts, the higher is my marginal utility of exerting criminal effort. Indeed, there is no formal way of learning to become a criminal, no proper "school" providing an organized transmission of the objective skills needed to undertake successful criminal activities. Given this lack of formal institutional arrangement, the most natural and efficient way to learn to become a criminal is through the interaction with other criminals. Delinquents learn from other criminals belonging to the same network how to commit crime in a more efficient way by sharing the know-how about the "technology" of crime. Another way of understanding why strategic complementarities are important in crime is that the perception that one's peers will or will not disapprove can exert a stronger influence in committing crime. Indeed, if delinquency is seen as a badge of honor in a population (Wilson and Herrnstein 1985; Kahan 1997; Silverman 2004), then the delinquents who see others committing crimes infer that their peers value law-breaking. They are then more likely to break the law themselves, which leads other juveniles to draw the same inference and engage in the same behavior. In this respect, violence and crime can become status-enhancing.

[6] For a complete overview of the literature on games on networks, see Jackson and Zenou (2015) and Chapter 5 in this handbook by Bramoullé and Kranton.

Following Calvó-Armengol and Zenou (2004) and Ballester, Calvó-Armengol, and Zenou (2006, 2010), we would like to examine a simple network model with strategic complementarities in crime effort.[7] For this purpose, consider a game where $N = \{1,\ldots,n\}$ is a finite set of agents in network **g**. We represent these social connections by a graph **g**, where $g_{ij} = 1$ if agent i is connected to agent j and $g_{ij} = 0$ otherwise. Links are taken to be reciprocal, so that $g_{ij} = g_{ji}$.[8] By convention, $g_{ii} = 0$. We denote by **G** the $n \times n$ adjacency matrix with entry g_{ij}, which keeps track of all direct connections. In criminal activities, agents i and j share their knowledge about delinquent activities if and only if $g_{ij} = 1$. Each agent i decides how much effort to exert on crime, denoted by $y_i \in \mathbb{R}_+$. The utility of each agent i providing effort y_i in network **g** is given by:

$$u_i(\mathbf{y},\mathbf{g}) = \alpha_i y_i - \frac{1}{2}y_i^2 + \phi_1 \sum_{j=1}^{n} g_{ij} y_i y_j \tag{11.1}$$

where $\phi_1 > 0$ is the intensity of interactions and **y** is an n-dimensional vector of crime efforts. This utility has two parts. An individual part, $\alpha_i y_i - \frac{1}{2}y_i^2$ where the marginal benefits of providing criminal effort y_i are given by $\alpha_i y_i$ and increasing in own effort y_i, where α_i denotes the *exogenous heterogeneity* of agent i that captures the *observable* characteristics of individual i (e.g., sex, race, age, parental education). The second part of the utility function, $\phi_1 \sum_{j=1}^{n} g_{ij} y_i y_j$, corresponds to the *local-aggregate* effect of peers since each agent i is affected by the sum of efforts of the agents for which she has a direct connection. The higher are these number of active connections, the higher is the marginal utility of providing her own effort. This is a game with strategic complementarities since

$$\frac{\partial^2 u_i(\mathbf{y},\mathbf{g})}{\partial y_i \partial y_j} = \phi_1 g_{ij} \geq 0.$$

In equilibrium, each agent maximizes her utility (11.1) and the best-reply function, for each $i = 1,\ldots,n$, is given by:

$$y_i = \alpha_i + \phi_1 \sum_{j=1}^{n} g_{ij} y_j. \tag{11.2}$$

Denote by $\mu_1(\mathbf{G})$ the largest eigenvalue of network **g** and by $\boldsymbol{\alpha}$ the non-negative n-dimensional vector corresponding to α_i. Denote also by $\mathbf{I}_n$ the $(n \times n)$ identity matrix, $\mathbf{1}_n$ the n-dimensional vector of ones and $\mathbf{M}(\mathbf{g},\phi_1) \equiv (\mathbf{I}_n - \phi_1 \mathbf{G})^{-1}$. We have the following result:[9]

[7] The same model or an extension of it can be used for education, R&D, trade, etc. See below.

[8] This is only for the sake of the exposition. All the results go through with a directed and weighted network.

[9] Throughout this chapter bold-face, lower-case letters refer to vectors while bold-face, capital letters refer to matrices.

Proposition 1. *If $\phi_1 \mu_1(\mathbf{G}) < 1$, the network game with payoffs (11.1) has a unique Nash equilibrium in pure strategies given by:*

$$\mathbf{y}^* \equiv \mathbf{y}^*(\mathbf{g}) = \mathbf{b}_{\alpha}(\mathbf{g}, \phi_1), \tag{11.3}$$

where $\mathbf{b}_{\alpha}(\mathbf{g}, \phi_1)$ is the weighted Katz-Bonacich centrality defined as:

$$\mathbf{b}_{\alpha}(\mathbf{g}, \phi_1) = (\mathbf{I}_n - \phi_1 \mathbf{G})^{-1} \boldsymbol{\alpha} = \mathbf{M}(\mathbf{g}, \phi_1)\boldsymbol{\alpha} = \sum_{k=0}^{\infty} \phi_1^k \mathbf{G}^k \boldsymbol{\alpha}. \tag{11.4}$$

When there is no ex ante heterogeneity (i.e., $\alpha = \mathbf{1}$), the Katz-Bonacich centrality of agent i counts the total number of paths (not just the shortest paths) in $\mathbf{g}$ starting from i, weighted by a decay factor that decreases with the length of these paths. This is captured by the fact that the matrix $\mathbf{G}^k$ keeps track of the indirect connections in the network (i.e., $g_{ij}^{[k]} \geq 0$ measures the number of paths of length $k \geq 1$ in $\mathbf{g}$ from i to j). When there is individual heterogeneity (i.e., $\alpha \neq \mathbf{1}$), then paths have different weights depending on where they arrive. In particular, each path is also weighted by the sum of the αs to which it corresponds. Proposition 1 shows that more central agents in the network will exert more effort. This is intuitively related to the equilibrium behavior, as the paths capture all possible feedbacks. In our case, the decay factor depends on how others' effort enters into the payoff of own effort. It is then straightforward to show that, for each individual i, the equilibrium utility is:

$$u_i(\mathbf{y}^*, \mathbf{g}) = \frac{1}{2}\left[b_{\alpha_i}(\mathbf{g}, \phi_1)\right]^2 \tag{11.5}$$

so that the equilibrium utility of each criminal is proportional to the square of her Katz-Bonacich centrality.

11.2.2.2 The Key-Player Policy

We here focus on the "Key Player Problem/Negative" (KPP-Neg) where the key-player policy aims at removing the player (criminal) who reduces total activity (crime) in a network the most. The removal of the key player can have large effects on crime because of feedback effects or "social multipliers" (see, in particular, Kleiman 2009; Glaeser, Sacerdote, and Scheinkman 1996; Verdier and Zenou 2004). That is, as the fraction of individuals participating in a criminal behavior increases, the impact on others is multiplied through social networks. Thus, criminal behaviors can be magnified, and interventions can become more effective.[10]

[10] Contrary to the key player approach developed in this section, which is a *noncooperative game theoretical approach*, Lindelauf et al. (2013) have proposed to tackle the key-player problem in *cooperative game theoretical approach*, which is particularly well suited for organized crime and terrorist organizations. In particular, Lindelauf et al. (2013) use the Shapley value as a measure of importance in cooperative games that are specifically designed to reflect the context of the terrorist organization at hand. The advantage of this approach is that both the structure of the terrorist network, which usually reflects

The benchmark key-player policy. Formally, consider the previous model and denote by $Y^*(\mathbf{g}) = \sum_{i=1}^{n} y_i^*$ the total equilibrium level of crime in network $\mathbf{g}$, where y_i^* is the Nash equilibrium effort given by (11.3). Denote also by $\mathbf{g}^{[-i]}$ the network $\mathbf{g}$ without individual i. Then, in order to determine the key player, the planner will solve the following problem:

$$\max\{Y^*(\mathbf{g}) - Y^*(\mathbf{g}^{[-i]}) \mid i = 1,...,n\}.$$

When the original delinquency network $\mathbf{g}$ is fixed, this is equivalent to:

$$\min\{Y^*(\mathbf{g}^{[-i]}) \mid i = 1,...,n\}. \tag{11.6}$$

Definition 1. *Assume that $\phi\mu_1(\mathbf{g}) < 1$. The intercentrality or key-player centrality measure $d_i(\mathbf{g},\phi_1)$ is defined as follows:*

$$d_i(\mathbf{g},\phi_1) = \frac{b_{\alpha_i}(\mathbf{g},\phi_1)b_{1_i}(\mathbf{g},\phi_1)}{m_{ii}} \tag{11.7}$$

Ballester et al. (2006, 2010) have showed the following result:

Proposition 2. *A player i^* is the key player that solves (11.6) if and only if i^* is a delinquent with the highest intercentrality in $\mathbf{g}$, that is, $d_{i^*}(\mathbf{g},\phi_1) \geq d_i(\mathbf{g},\phi_1)$, for all $i = 1,...,n$.*

The intercentrality measure (11.7) of delinquent i is the sum of i's centrality measures in $\mathbf{g}$, and i's contribution to the centrality measure of every other delinquent $j \neq i$ also in $\mathbf{g}$. It accounts both for one's exposure to the rest of the group and for one's contribution to every other exposure. This means that the key player i^* in network $\mathbf{g}$ is given by $i^* = \arg\max_i d_i(\mathbf{g},\phi_1)$, where

$$d_i(\mathbf{g},\phi_1) = Y^*(\mathbf{g}) - Y^*\left(\mathbf{g}^{[-i]}\right). \tag{11.8}$$

Let us now discuss different extensions of this benchmark model and their implication for the key-player policy.[11]

a communication and interaction structure, and non-network features (i.e., individual based parameters such as financial means or bomb-building skills), can be taken into account. See also Husslage et al. (2014). Observe that, as in the sociological approach, there is no microfoundation for the use of their proposed centrality measures.

[11] Sommer (2014) further develops the notion of key player defined in (11.7). He states that the formula (11.7) implies that when someone is removed from the network, she does not participate in any criminal activity at all. Sommer (2014) proposes another policy where the removed criminal can still exert crime but without influencing anybody in the network. She is just isolated from all other players of the game.

Imperfect information. Consider the same utility function as in (11.1) but assume that the players do not know the exact value of the synergy parameter ϕ_1 (the state of the world), which is *common* to all agents but only *partially known* by the agents. Each agent i receives a private (independent) signal s_i about the state of the world, which allows her to update her beliefs about ϕ_1. There are M different states of the world so that ϕ_1 can take M different values: $\phi_1 \in \{\phi_{11}, \ldots, \phi_{1M}\}$. There are T different signals so that agents can be of T different types, which we denote by $\tau = 1, \ldots, T$. Individual i, who receives signal $s_i = \tau$, computes the following conditional expected utility:

$$\mathbb{E}[u_i(\mathbf{y},\mathbf{g}) | \{s_i = \tau\}] = \alpha_i \mathbb{E}[y_i | \{s_i = \tau\}] - \frac{1}{2}\mathbb{E}[y_i^2 | \{s_i = \tau\}] + \sum_{j=1}^{n} g_{ij} y_i \mathbb{E}[\phi_1 y_i y_j | \{s_i = \tau\}]$$

$$= \alpha_i y_i(\tau) - \frac{1}{2} y_i(\tau)^2 + \sum_{j=1}^{n} g_{ij} y_i(\tau) \mathbb{E}[\phi_1 y_j | \{s_i = \tau\}]$$

where $\mathbb{E}[.]$ is the expectation operator. De Marti and Zenou (2014, 2015) show the existence and uniqueness of the Nash equilibrium in effort and characterize it as a function of a weighted combination of the Katz-Bonacich centrality of the players and the information matrix, which keeps tracks of all the information received by the agent about the states of the world. The authors then analyze the key-player policy where the planner plays first. They assume that the planner has a prior (which is unknown to the agents) that may be different to the one shared by the agents because the authority may have superior information. The planner needs to solve the key player problem, that is $\max\{\mathbb{E}\left[Y^*(\mathbf{g})\right] - \mathbb{E}\left[Y^*(\mathbf{g}^{[-i]}\right]) \mid i = 1, \ldots, n\}$, which is the difference in aggregate activity according to her prior. They show that the formula of the key player is different to (11.7) and that it is, in fact, a convex combination between different intercentrality measures where the weights are the different parameters related to the information structure, in particular, the priors of the agents and the planner and the posteriors of the agents.

Network formation. In the key-player formula (11.7), it is assumed that when someone is removed, the remaining players in the network adjust their optimal criminal efforts but cannot form new links. This makes sense in the short run but it is possible that, in the long run, there is a "rewiring" of the network so that, following the removal of a criminal, the remaining players form new links. Liu et al. (2012) propose a simple dynamic formation model that incorporates this effect. More precisely, each period of time is divided into two subperiods. In the first period, a player is randomly selected from the network and must decide *myopically* whether she wants to form a link and with whom. In the second period, all players connected in the network play the effort game where their equilibrium utility is given by (11.5). What is interesting here is that when a person wants to form a link, she anticipates the rewiring of the network and thus, the new equilibrium efforts or, equivalently, the new Katz-Bonacich centralities of all individuals in the network (see (11.3)). The chosen person will then form a link with the individual that increases her utility the most. If there is no cost of forming new links and if there is some noise in the link-formation process, König, Tessone, and Zenou (2014) show that the steady-state equilibrium network will be a *nested split graph*, a well-known network in the graph-theory literature (Mahadev and Peled 1995), which also starts to be acknowledged in the economics literature (Belhaj, Bervoets, and Deroïan 2015; Hiller 2014; Lagerås and Seim 2015).[12]

Liu et al. (2012) develop this dynamic model but impose a cost of maintaining links, which is specific to each individual. Thus, the steady-state equilibrium network is no longer a nested-split graph. However, the Markov process for which the state variable is the network itself is still well defined, even though it is not ergodic. In this framework, an equilibrium is reached when there is an absorbing state. The planner can then decide who the key player is by first removing an individual in the network and then letting the other individuals play the dynamic network-formation game described above. The person who reduces total crime the most in equilibrium is the key player. The latter is not necessarily the key player in the static model defined in (11.7).

Congestion and competition effects. Another interesting extension is to generalize the utility function (11.1) to include competition effects so that:

$$u_i(\mathbf{y}, \mathbf{g}) = \alpha_i y_i - \frac{1}{2} y_i^2 + \phi_1 \sum_{j=1}^{n} g_{ij} y_i y_j - \rho \sum_{j=1}^{n} c_{ij} y_i y_j \tag{11.9}$$

where ρ is the degree of competition between agents and c_{ij} is the ijth cell of the matrix **C**, which keeps track of who is in competition with whom. For example, in the case of delinquent networks, criminals benefit from other criminals who are linked to them (strategic complementarities) but are also in competition with all criminals in the neighborhood where they operate, which leads to congestion effects. Thus, the utility of i will be lower the more competition there is in the same neighborhood since there will be less to steal. In that case, the matrix **C** will keep track of who resides or operates in the same neighborhood with whom. For example, consider a star network where individual 1 is the star. Assume that there are two neighborhoods, $\mathcal{N}_1$ and $\mathcal{N}_2$, and that criminals 1 and 2 commit their crime in the same neighborhood $\mathcal{N}_1$ (for example, they sell drugs) while criminal 3 operates alone in neighborhood $\mathcal{N}_2$. Then, the adjacency matrix **G** and the competition matrix **C** can be written as:

$$\mathbf{G} = \begin{pmatrix} 0 & 1 & 1 \\ 1 & 0 & 0 \\ 1 & 0 & 0 \end{pmatrix}, \quad \mathbf{C} = \begin{pmatrix} 0 & 1 & 0 \\ 1 & 0 & 0 \\ 0 & 0 & 0 \end{pmatrix}. \tag{11.10}$$

[12] A graph is nested split if agents can be ordered so that $g_{kl} = 1 \Rightarrow g_{ij} = 1$ whenever $i \leq k$ and $j \leq l$. This means, in particular, that if agent i has a lower degree than agent j, then she has less neighbors in the sense of set inclusion.

This model with utility (11.9) was originally proposed by Ballester, Calvó-Armengol, and Zenou (2006) and Calvó-Armengol, Patacchini, and Zenou (2009) and generalized by König, Liu, and Zenou (2014) for any matrix **G** and **C**. Let $\overline{\alpha} = \max\{\alpha_i \mid i \in N\}$ and $\underline{\alpha} = \min\{\alpha_i \mid i \in N\}$, with $\overline{\alpha} > \underline{\alpha} > 0$. If $\phi_1 \mu_1(\mathbf{G}) + n\left(\frac{\rho}{1-\rho}\right)\left(\frac{\overline{\alpha}}{\underline{\alpha}} - 1\right) < 1$, then there is a unique interior Nash equilibrium defined by:

$$y_i^* = \alpha_i + \phi_1 \sum_{j=1}^{n} g_{ij} y_j^* - \rho \sum_{j=1}^{n} c_{ij} y_i^* y_j^*. \tag{11.11}$$

Then, we can define the key player as in (11.7) but where $b_{\alpha_i}(\mathbf{g}, \phi_1)$, $b_{1_i}(\mathbf{g}, \phi_1)$ and m_{ii} are now defined with respect to both **G** and **C** and *not* only with respect to **G** as in (11.7).

Occupational choice before effort decisions. So far, we have assumed that the delinquency network was given. In some cases, though, delinquents may have opportunities outside the delinquency network. For instance, petty delinquents may consider entering the labor market and giving up delinquent activities. Here, we expand the model and endogenize the delinquency network by allowing delinquents to take a binary decision on whether to stay in the delinquency network or to drop out of it. Ballester, Calvó-Amengol, and Zenou (2010) show that there always exists a subgame-perfect Nash equilibrium of this two-stage game but it is usually not unique. Then, they study the key player policy. Initially, the planner must choose a player to remove from the network. Then, players play the two-stage delinquency game. In this context, there is an added difficulty to the planner's decision. The removal of a player from the network affects the rest of the players' decisions to become active delinquents. This fact should be taken into account by the planner in order to attain an equilibrium with minimum total delinquency. Denote the fixed wage earned in the labor market by ω. Then, in order to reduce delinquency, the planner should consider two policies: the one that provides a higher ω and the key-player policy. These policies are *complementary* from the point of view of their effects on total delinquency, although we are aware that they may be substitutes if we impose a budget-restricted planner who had to implement costly policies.[13]

Local average versus local aggregate network model. The utility in (11.1) corresponds to the *local-aggregate* model because what matters for peer effects is the sum of the activity of direct friends. There is another model introduced by Patacchini and Zenou (2012) where the utility is equal to:

$$u_i(\mathbf{y}, \mathbf{g}) = a_i y_i - \frac{1}{2} y_i^2 - \frac{1}{2} \lambda \left(y_i - \sum_{j=1}^{n} g_{ij}^* y_j \right)^2 \tag{11.12}$$

[13] See also Chen, Zenou, and Zhou (2015) for a model where agents provide two activities (for example, crime and education) and where the key player is determined. Chen, Zenou, and Zhou (2015) show that, by not taking into account the two activities, the planner may target the wrong key player.

where $g_{ij}^* = g_{ij}/g_i$. This model is called the *local-average model* because what matters for each individual i is the difference between her effort y_i and the average effort of her neighbors $\sum_{j=1}^n g_{ij}^* y_j$. Each individual i wants to minimize the social distance between herself and her reference group, where λ is the parameter describing the taste for conformity. Indeed, the individual loses utility $\frac{1}{2}\lambda\left(y_i - \sum_{j=1}^n g_{ij}^* y_j\right)^2$ from failing to conform to others. It is easily shown that there exists a unique interior Nash equilibrium given by:

$$y_i^* = \alpha_i + \phi_{12} \sum_{j=1}^n g_{ij}^* y_j \tag{11.13}$$

where $\alpha_i = a_i/(1+\lambda)$ and $\phi_{12} = \lambda/(1+\lambda)$. Interestingly, when one considers the key-player policy in the local-average model, then one can see that, in general, when the agents are ex ante homogeneous, which agent to remove from the network does not matter in terms of the aggregate effort level reduction, unless the agent holds a very special position in the network such that removing this agent generates isolated nodes in the network (Liu et al. 2014). If, on the contrary, the agents are ex ante heterogenous in terms of α_i, the key-player problem for the local-average network game and the general network game with utility (11.12) does not have an analytical solution. The difficulty is due to the fact that the row-normalized adjacency matrix of network $\mathbf{g}^{[-i]}$ is, in general, not a submatrix of the row-normalized adjacency matrix of network $\mathbf{g}$ (since $\mathbf{G}^{[-i]*}$ is not a submatrix of $\mathbf{G}^*$). Yet one can still determine the key player numerically using its definition given by (11.8), if the unknown parameters can be estimated in the best-response function (11.13).

Liu, Patacchini, and Zenou (2014) and Liu et al. (2014) extend this model to incorporate both local-average and local-agreggate effects in the utility function. This is defined as the *hybrid model* and is given by:

$$u_i(\mathbf{y}, \mathbf{g}) = a_i y_i - \frac{1}{2} y_i^2 + \lambda_1 \sum_{j=1}^n g_{ij} y_i y_j - \frac{1}{2}\lambda_2 \left(y_i - \sum_{j=1}^n g_{ij}^* y_j\right)^2. \tag{11.14}$$

The best-reply function for each agent i is then given by:

$$y_i^* = \alpha_i + \phi_1 \sum_{j=1}^n g_{ij} y_j + \phi_2 \sum_{j=1}^n g_{ij}^* y_j \tag{11.15}$$

where $\alpha_i = a_i/(1+\lambda_2)$, $\phi_1 = \lambda_1/(1+\lambda_2)$ and $\phi_2 = \lambda_2/(1+\lambda_2)$. Let $g^{\max} = \max_i g_i$ denote the highest degree in network $\mathbf{g}$. Then, if $g^{\max}\phi_1 + \phi_2 < 1$, the network game with utility (11.14) has a unique Nash equilibrium in pure strategies given by (11.15). We can once more calculate the key player in this extended model by using the formula (11.8).[14]

[14] Other extensions of the key-player policy have been considered. First, researchers have also defined "group" players so that the planner can remove more than one key player at a time (see, in particular,

11.2.2.3 Key Players: Maximizing Diffusion

The model of Section 11.2.2 states that when there are *strategic complementarities* in efforts between linked individuals as, for example, in crime, R&D, and financial activities, then it makes sense to use the methodology of Section 11.2.2 to identify the key players. However, in other activities, when strategic complementarities are less important, then other methodologies could be applied. This is particularly true in a diffusion process where the planner wants to target specific individuals (key players) to diffuse a technology or specific programs or to eradicate a virus or a sickness. In Section 11.2.1, this was referred to as the "Key Player Problem/Positive" (KPP-Pos) where the planner is looking for a set of network nodes that are optimally positioned to quickly *diffuse* information, attitudes, behaviors, or goods.

The canonical models of diffusion (the Susceptible-Infected-Recovered or SIR model and the Susceptible-Infected-Susceptible or SIS model) constitute a good starting point for this type of issue. The diffusion process works as follows. For example, in the SIS model, individuals exist in one of two states, S (Susceptible to be infected) and I (Infected), and transition between these states is analyzed over time. Transitions from I to S occur at a given recovery rate, but transitions from S to I occur at an endogenous rate that, in particular, depends on the expected number of individuals in state I that will be met in a given period. The SIS model and its variants have been proposed as useful in understanding a wide variety of processes including such diverse applications as behaviors, information diffusion, learning dynamics, myopic best response and imitation dynamics.[15]

Galeotti and Rogers (2013) extend the SIS model to analyze the spread of a harmful state (for example, human infection of various communicable diseases that spread through social contacts or tobacco use or an electronic virus on a computer network) through a population divided into two groups, labeled A and B. Individuals in the two groups are identical in every respect but the label. As in the SIS model, in each period, an individual interacts with k other individuals. A proportion of the interactions is with individuals from the same group while the remaining interactions are with individuals from the other group.

Their first set of results concerns the problem of a central planner attempting to eradicate the harmful state (infected state). The planner might be a governmental agency deciding how many vaccines to produce for a communicable disease and how

Ortiz-Arroyo, 2010, and Ballester, Calvó-Armengol, and Zenou 2010). Ballester, Calvó-Armengol, and Zenou (2010) show that the key group problem is NP-hard from the combinatorial perspective. However, they show that the error of approximation of using a greedy algorithm (where, in each step, the key-player formula (11.7) is used to remove a player) compared to directly solving the key group problem is at most 36.79%. Second, König, Liu and Zenou (2014) propose a slightly different definition of the key player, which is now the agent who, once removed, reduces *total welfare* the most (and not *total activity* as in (11.7)). Criminal networks are a natural application but R&D or financial networks could also be another application of the key-player policy.

[15] For an overview on diffusion and networks, see Jackson and Yariv (2011) and Chapter 18 by Lamberson in this handbook.

to allocate them across the population. Alternatively, the social planner could be a governmental agency that aims at eliminating smoking in schools via an educational program, and it has to decide how large the program should be and which students it should target or require to participate. How should the planner's policy depend on the structure of interactions among potential smokers? What information does the government need in order to determine the optimal size of the program (level of immunity) and how to efficiently distribute the resources to the population?

Galeotti and Rogers (2013) show that a central planner who aims for eradication optimally either divides the resources equally across groups, or concentrates entirely on one group, depending on whether there is positive or negative assortativity, respectively. To be more precise, where there are *positive assortative interactions*, the optimal allocation involves splitting the resources equally between the two groups and the minimum budget needed to eradicate the harmful state is independent of the exact degree of positive assortativity. In contrast, when there are *negative assortative interactions*, the optimal allocation is to first exclusively focus resources on one group until it is fully immunized and then immunize part of the other group, if required. Furthermore, under negative assortative interactions, the minimum budget necessary to achieve eradication is decreasing in the intensity of assortativity, reaching its lowest level when the network of interactions is bipartite.

This analysis, though interesting, focuses more on groups and less on targeting individuals. A recurring theme in the popular discussion as well as the academic literature on social networks has been the simple idea that it would be better to focus efforts—send information, coupons, or free samples—on individuals who are influential. Galeotti and Goyal (2009) identify conditions for the content of interaction under which optimal strategy targets more or less connected individuals. In particular, they find that, in the word of mouth application, it is optimal to target individuals who get information from a few others (marginalized consumers). By contrast, in the proportional adoption externalities application, it is optimal to seed the most connected individuals (as they are unlikely to adopt via social influence). Thus, the optimality of targeting highly connected nodes very much depends on the content of social interaction.

Another recent paper by Banerjee et al. (2013) addresses the issue of diffusion of a micro-finance program in India. Their key question is: How do the network positions of the first individuals in a village to receive information about a new product affect its eventual diffusion? To answer this question, Banerjee et al. (2013) develop a model of information diffusion through a social network that discriminates between *information passing* (individuals must be aware of the product before they can adopt it, and they can learn from their friends) and *endorsement* (the decisions of informed individuals to adopt the product might be influenced by their friends' decisions). To be more precise, at each point in time t, a node (household) i can be in two possible *information* states: either she is informed so that $s_{it} = 1$ or she is uninformed so that $s_{it} = 0$. Moreover, she can be in two possible *participation* states: $m_{it} = 1$ if he/she participates and $m_{it} = 0$ if not. Note that, by definition, if $m_{it} = 1$ then $s_{it} = 1$, as one cannot participate without being informed. Let I_t be the set of newly informed nodes at time t (i.e., $I_t = \{s_{it} = 1, s_{it-1} = 0\}$). Define I^t to be the historical stock. The authors develop an algorithm that works as follows. At the beginning of the period ($t = 0$), the initial set of nodes (i.e., the *leaders*) is informed so that $s_{i0} = 1$, $\forall i \in L$ and $s_{i0} = 0$ if $i \notin L$, where $I_0 = \{i \in N : i \text{ is a leader}\}$. This is the first stage where an initial set of households is informed (*injection points*). Then, those newly informed agents decide whether or not to *participate* based on their characteristics and the participation decisions of their neighbors. To be more precise, let p_{it} denote the probability that an individual who was just informed about microfinance decides to participate, where p_{it} is a function of the individual's characteristics X_i (which can account for homophily based on observables) and peer decisions. The authors assume a logistic function so that:

$$\log\left(\frac{p_{it}}{1-p_{it}}\right) = X_i'\beta + \lambda F_{it} \tag{11.16}$$

where F_{it} is a fraction whose denominator is the number of i's neighbors who informed i about the program and whose numerator is the number of these individuals who participate in microfinance. If we set λ to 0, then we have a pure information model without any endorsement effects, for each $i \in I_0$. In the case of pure endorsement effects, for the initial period, we can fix $F_{i0} = 0$. Next, each $i \in I^0$ transmits to $j \in N_i$ with probability $m_{i1}\phi_1^P + (1 - m_{i1})\phi_1^N$ so that ϕ_1^P and ϕ_1^N are, respectively, the probability that households that have been informed in previous periods pass information to each of their neighbors, independently, if they are participants (P) and if they are not (N). This is independent across i and j. Let I_1 be the set of j's neighbors who are informed via this process who were not members of I^0, and let $I(j)$ be the set of i's who informed j. Then, we iterate this process at time t so that newly informed agents are now I_t and have to decide whether to participate using the same rule as in (11.16) but for a different set of neighbors F_{it}. The process stops after T periods of information passing. If $\phi_1^N = 0$, so that only participating households pass information, and $T = +\infty$, this is a variant of the standard Susceptible, Infectious, Recovered (SIR) model described above. By allowing it to only operate for T periods, the authors can study what happens in finite time (since after enough rounds, everyone would be informed). They define that the key player(s) are the agents who diffuse information the best. As we will see below, they propose some centrality measures to define the key players in this context.

Finally, there is an interesting literature on *viral marketing*, which looks at how firms can use network information to better sell their products. Viral marketing may take the form of video clips, interactive Flash games, advergames, ebooks, brandable software, images, text messages, email messages, or web pages. The most commonly utilized transmission vehicles for viral messages include: pass-along based, incentive based, trendy based, and undercover based. The ultimate goal of marketers interested in creating successful viral marketing programs is to create viral messages that appeal to individuals with high *social networking potential* and that have a high probability of being presented and spread by these individuals and their competitors in their communications with others in a short period of time. In other words, one key aspect of viral marketing is to find the key consumers who have the most influence on other consumers. Indeed, in viral marketing, a company tries to use word-of-mouth effects to market a product with a limited advertising budget, relying on the fact that early adopters may convince friends and colleagues to use the product, thus creating a large wave of adoptions. While word-of-mouth effects have a history in the area of marketing that long predates the Internet, viral marketing has become a particularly powerful force in online domains, given the ease at which information spreads, and the rich data on customer behavior that can be used to facilitate the process.[16]

To illustrate the key-player idea in terms of the viral marketing framework, let us consider the algorithm problem posed by Domingos and Richardson (2001) to identify the "influential" sets of nodes in a network. Suppose that a firm is trying to market a new product, and wants to take advantage of word-of-mouth effects. One strategy would be as follows: this firm can collect data on the social network interactions among potential customers, chooses a set S of initial adopters, and markets the product directly to them. Then, assuming that they adopt the product, the firm will rely on their influence to generate a large cascade of adoptions, without having to rely on any further direct promotion of the product. Here, we focus on the following algorithmic problem: how does the firm choose the set S? There is a natural *influence function* $f(.)$ defined as follows: for a set S of nodes, $f(S)$ is the expected number of active nodes at the end of the process, assuming that S is the set of nodes that are initially active. From the marketer's point of view, $f(S)$ is the expected number of total sales if they get S to be the set of initial adopters. Now, given a budget k, how large can we make $f(S)$ if we are allowed to choose a set S of k initial adopters? In other words, we wish to maximize $f(S)$ over all sets S of size k. This turns out to be a hard computational problem since it is $NP-$hard to find the optimal set S. Kleinberg (2008) proposes to identify broad subclasses of the models that are not susceptible to strong inapproximability results, and for which good approximation results can be obtained. He assumes that $f(S)$ is a submodular function, which is a type of diminishing returns property: the benefit of adding elements decreases as the set to which they are being added grows.[17]

While the previous approach focused on *influence maximization*, Hartline, Mirrokni, and Sundararajan (2008) study *revenue maximization*. In their model, a buyer's decision to buy an item is influenced by the set of other buyers that own the item and the price at which the item is offered. They identify a family of strategies called influence-and-exploit strategies that is based on the following idea: Initially influence the population by giving the item for free to a carefully chosen set of buyers. Then, extract revenue from the remaining buyers using a "greedy" pricing strategy.[18]

[16] For an overview of the literature on viral marketing, see Kleinberg (2007) and Chapter 29 by Bloch and Chapter 30 by Mayzlin (2015) in this handbook.

[17] For more details, see Kempe, Kleinberg, and Tardos (2003, 2005).

[18] See also Campbell (2013), who develops a model based on random graphs for understanding the ways in which social learning, via word-of-mouth between friends, affects demand and, hence, the optimal pricing and advertising strategies of a firm.

11.3 Key Players: Empirical Results

There are very few empirical papers on key players in networks and most of them test the key-player model developed in Section 11.2.2.2. We will also give some empirical results on the empirical test of the model developed in Section 11.2.2.3.

11.3.1 Key Player: Econometric Methodology

There are two steps to empirically test the key-player policy of Section 11.2.2.2. First, one needs to have a credible estimate of the social multiplier ϕ_1 by estimating equation (11.2). Then, one must calculate the formula of the key player given in (11.7), using the estimated values of equation (11.2). Liu et al. (2012) were the first to propose an econometric methodology to test this key-player policy. In the real world, there is often more than one (connected) network. Denote by $r = 1, \ldots, \bar{r}$, the network where $\bar{r}$ is the total number of networks. In the empirical literature, it is relatively standard to define α_i not only as her own observable characteristics but also as the observable *average* characteristics of individual i's neighbors (*contextual effects*) (Manski 1993). As a result, α_i can be written as:

$$\alpha_i = \sum_{m=1}^{M} \beta_m x_i^m + \frac{1}{g_i} \sum_{m=1}^{M} \sum_{j=1}^{n} g_{ij} x_j^m \gamma_m \tag{11.17}$$

where $g_i = \sum_{j=1}^{n} g_{ij}$ is the number of direct links (the degree) of individual i, x_i^m is a set of M variables accounting for observable characteristics in individual characteristics of individual i, and β_m, γ_m are parameters. Using (11.17), equation (11.2) can be written as:

$$y_{i,r} = \phi_1 \sum_{j=1}^{n_r} g_{ij,r} y_{j,r} + \sum_{m=1}^{M} \beta_m x_{i,r}^m + \frac{1}{g_{i,r}} \sum_{m=1}^{M} \sum_{j=1}^{n_r} g_{ij,r} x_{j,r}^m \gamma_m + \eta_r + \varepsilon_{i,r}. \tag{11.18}$$

This corresponds to the best-reply function of our model (and we know that there exists a unique Nash equilibrium in efforts if $\phi_1 < 1/\mu_1(\mathbf{G})$). There are at least three different econometric problems that need to be addressed in order to have a credible estimation of ϕ_1. First, there is the *reflection problem* (Manski 1993), which is due to the difficulty in separating the *endogenous peer effect* (captured by ϕ_1) from the *contextual effect* (captured by the γ_m). Second, there is the *common-shock* problem, which is due to the fact that all individuals belonging to the same network r are affected by a common shock (for example, the teacher quality of a class) that affects their outcomes $y_{i,r}$. These two problems are usually solved by using the structure of the network (for example, one can use friends of friends' characteristics as an instrument for friends' actions) and network fixed effects η_r (Bramoullé, Djebbari, and Fortin 2009). Finally, there is the *correlated effect* so that individuals in the same network may behave similarly because they have similar unobserved individual characteristics. This issue is more difficult to address and researchers have either been explicitly modeling the network formation process (see, e.g., Mele 2013; Goldsmith-Pinkham and Imbens 2013; Chandrasekhar and Jackson 2014; Badev 2014; Del Bello, Patacchini, and Zenou 2015), or using instrumental variables (Bifulco, Fletcher, and Ross 2011; Patacchini and Zenou 2014) or natural experiments where agents are randomly allocated to the network (Carrell, Sacerdote, and West 2013; Algan et al. 2015; Hahn et al. 2015; Lindquist, Sauermann, and Zenou 2015). We refer to the literature surveys provided by Blume et al. (2011), Advani and Malde (2014), Jackson (2014), Jackson, Rogers, and Zenou (2015), Graham (2015), Topa and Zenou (2015) and Chapter 12 by Fortin and Boucher in this handbook for a detailed treatment of these econometric issues.

11.3.2 Key Players in Criminal Networks

"Key players" type of policies have been implemented in different countries. It is indeed estimated that only a few offenders are responsible for a very large proportion of all crimes committed. An example of such a policy is Operation Ceasefire in the United States. In this type of operation, police beefed up patrols in the area, attempting to locate gang members who had outstanding arrest warrants or had violated probation or parole regulations. Gang members who had violated public housing rules, failed to pay child support, or were similarly vulnerable were also subjected to stringent enforcement (Tita et al. 2003). This latter policy combines a strong law enforcement response with a "pulling levers" deterrence effort aimed at chronic gang offenders. The key to the success is to use a "lever pulling" approach, which is a crime deterrence strategy that attempts to prevent violent behavior by using a *targeted individual* or a *group's vulnerability* to law enforcement as a means of gaining their compliance. Operation Ceasefire was first launched in Boston, and youth homicide fell by two-thirds after the Ceasefire strategy was put in place in 1996 (Kennedy 1998). It was then implemented in Los Angeles in 2000 and there was also a considerable decrease in crime.

There have been similar policies in the UK with a large-scale policy intervention—the Street Crime Initiative (SCI)—that was introduced in England and Wales in 2002. This policy allocated additional resources to some police force areas to *specifically target street crime*, whereas other forces did not receive any additional funding. Machin and Marie (2011) show that robberies fell significantly in SCI police force areas relative to non-SCI force areas after the initiative was introduced. Moreover, the policy seems to have been a cost-effective one, even after extensively testing for possible displacement or diffusion effects on other crimes and into adjacent areas. Overall, they reach the conclusion that increased police resources can be used to generate falls in crime, at least in the context of the SCI program they study.

Machin, Marie, and Priks (2014) investigate the impact of the targeting of the *most prolific offenders* on area crime rates. They empirically study reforms in England and Wales (between 2000 and 2004) and Sweden (introduced in 2012) where the police has shifted the focus from general policing towards prolific offenders. For this purpose, the police has constructed lists with names of prolific offenders and has allocated resources to closely monitoring these individuals (for example, by visiting offenders' homes). The reforms were cost-neutral (i.e., the government provided no extra funding). Using a difference-in-difference approach, they show that the policy was successful by reducing burglaries by approximately 5–10% in England, Wales, and Sweden.

However, these policies are implemented based on intuitive judgements on whom to target rather than using a network approach. In particular, most policies have been targeting the most active or the most prolific criminals. We would now like to expose the empirical test of the key-player policy exposed in Section 11.2.2 and show how it outperforms more "intuitive" policies such as targeting the most active criminals.

Liu et al. (2012) were the first to test the key-player policy using the Add Health data[19] for the *local-aggregate* model where the utility is given by (11.1). First, they estimate equation (11.18) and obtain an estimated value of ϕ_1 equal to 0.0457. Then, using this estimated value, they can calculate the key player for each network using the intercentrality measure (11.8). They find that the key player is *not* necessarily the most active criminal in the network. They also find that it is *not* straightforward to determine which delinquent should be removed from a network by only observing his or her criminal activities or position in the network. Compared to other criminals, the key players are less likely to be female, are less religious, belong to families whose parents are less educated, and have the perception of being more socially excluded. They also feel that their parents care less about them, are more likely to come from single-parent families and have more troubles getting along with their teachers. Finally, Liu et al. (2012) show that the key-player policy outperforms other reasonable policies like targeting the most active criminals in the network. They calculate the average crime reduction for all 145 networks when a key-player policy, a random-target policy, and a policy that removes the most active criminal are implemented. For example, for all networks of size 4, the average crime reduction is 29.94% on average when the key-player policy is implemented, 23.86% for the random-target policy, and around 25% when the most-active criminal policy is implemented.

Liu et al. (2014) test the key-player policy also with Add Health data but for the general model with utility (11.14). We may first ask which model best matches the data at hand: the local-aggregate model (utility given by (11.1)) or the local-average model (utility given by (11.12))? Liu, Patacchini, and Zenou (2014) provide a test (the

[19] The National Longitudinal Survey of Adolescent Health (Add Health) has been designed to study the impact of the social environment on adolescents' behavior in the United States by collecting data on students in grades 7–12 from a nationally representative sample of roughly 130 private and public schools in the years 1994–95. The most interesting aspect of the Add Health data is the information on friendships that is based on actual friend nominations and helps us create the friendship network.

J test) that determines which model is more adequate for the data at hand. Liu et al. (2014) perform such a test and show that both models match the delinquency data of the Add Health data well. As a result, they use the general model where the utility of each agent is given by (11.14). They also find that the key player is not necessarily the most *active* student in delinquent activities. Only in 19 networks out of the 103 networks in the sample is the key player the most active student in delinquent activities. They also look at crime reduction. They find that, for a network with four students (which corresponds to the median network in their sample), the percentage reductions in aggregate delinquency by removing the key player, the most active delinquent, and a random delinquent are 70.19%, 53.13%, and 50.71%, respectively. Hence, targeting key players is more effective than targeting the most active delinquent in reducing total delinquency.

Lindquist and Zenou (2014) also test the key-player policy but with a different data set. They look at individuals in Sweden who are aged above 16 and who have been suspected of at least one crime. For this purpose, they have access to the official police register of all individuals who are suspected of committing a crime in Sweden. In this register, the police keeps records of who is suspected of committing a crime with whom. In this context, a (criminal) link exists between two individuals if they are suspected of committing a crime together (and are then convicted). Both the convictions data and the suspects data include crime type, crime date, and the sanction received. One advantage of this data set over the Add Health one is that links are not self-reported and are thus less subject to measurement errors. Another advantage is that information on links is available at each moment in time over a period of 20 years. As a result, they can add individual lagged crime as one of the individual level control variables.

They find an estimate of ϕ_1 of 0.167. This means that having only one friend who committed a crime increases an individual's own crime by 20%. If we consider the case of four individuals (their smallest network), then individual crime will increase by 100% compared to the case when the individual is committing a crime by herself. Lindquist and Zenou (2014) then consider two periods of three years each (2000 to 2002 and 2003 to 2005). The Period 1 data set includes 15,230 co-offenders who are suspected of committing (on average) 5.91 crimes each and who are distributed over 1,192 separate networks. The Period 2 data set includes 15,143 co-offenders who are suspected of committing (on average) 5.92 crimes each and who are distributed over 1,185 networks. Their data also include 3,881 individuals who are members of a network with four or more individuals in *both* periods. They show that 23% of all key players are not the most active criminal in their own networks; 23% do not have the highest eigenvector centrality; and 20% do not have the highest betweenness centrality.

Because they have two periods of time, Lindquist and Zenou (2014) can test the prediction of crime reduction following the key-player policy against the true outcome observed in Period 2 data. They thus look at the relative effect of removing the key player in those cases in which the key player is no longer part of the active network. For this purpose, they create an indicator variable for each person indicating whether or not they have died during the relevant time period and whether they have been placed in prison. Their results indicate that, in the real world, the key-player policy outperforms the random player policy by 9.58%. The key-player policy also outperforms the policy of removing the most active player by 3.16% and the policy of removing the player with the highest eigenvector and betweenness centrality by 8.12% and 2.09%, respectively.

11.3.3 Key Players in R&D Networks

R&D partnerships have become a widespread phenomenon characterizing technological dynamics, especially in industries with a rapid technological development such as, for instance, the pharmaceutical, chemical, and computer industries (see, e.g., Hagedoorn 2002; Powell et al. 2005). In those industries, firms have become more specialized in specific domains of a technology and they tend to combine their knowledge with the knowledge of other firms that are specialized in different technological domains (Powell et al. 1996; Weitzman 1998). The increasing importance of R&D collaborations has spurred research for theoretical models studying these relationships, and for empirical tests of these models.

We would like to determine the key players or, more exactly, the key firms in R&D networks. In other words, we would like to determine which firms are crucial for an industry in the sense that, if they exit the market (i.e., go bankrupt), the cost in terms of total activity or welfare for the remaining firms and consumers will be the highest possible. Following König, Liu, and Zenou (2014), consider the above model where the utility (or profit) function of each firm i is similar to (11.9). They estimate equation (11.11) and obtain values for both ϕ_1 and ρ with the predicted signs so that there are positive spillover effects of R&D collaborations, captured by $\phi_1 > 0$ and negative competition effects, captured by $-\rho < 0$. Using the MERIT-CATI data set, they then determined the key firms over a period of more than 40 years. König, Liu, and Zenou (2014) show that the key firms are usually not those with the largest number of R&D collaborations (degree), the largest number of patents, nor the highest eigenvector, betweenness or closeness centrality and, more importantly, not the firm with the highest market share in its sector. Interestingly, General Motors, which was bailed out in 2009 by President-elect Obama, was among the key firms. They show that, if General Motors had been removed from the market in 1990, then total welfare would have been reduced by 8.37% while total output would have been decreased by 2.14%.

11.3.4 Key Players in Education

The influence of peers on education outcomes has been widely studied both in economics and sociology (Sacerdote 2011). There are, however, fewer papers studying the network effects on education (see, in particular, Calvó-Armengol, Patacchini, and Zenou 2009; Bifulco, Fletcher, and Ross 2011; Patacchini, Rainone, and

Zenou 2014). Using a model similar to that of Section 11.2.2.1 and the Add Health data, Calvó-Armengol, Patacchini, and Zenou (2009) were able to show that the Katz-Bonacich centrality of a student is a key determinant of his/her grade. Here, we would like to go further by examining the importance of different centralities, including the Katz-Bonacich and the key-player centrality defined in (11.8), on educational outcomes.

In Section 11.3.1, we have seen that one of the most difficult empirical challenges in the estimation of network effects was the endogeneity of the adjacency matrix **G** or, equivalently, the fact that network formation is endogenous and may be due to correlation in unobservables between the individuals involved in the link formation. One way of addressing this issue is to consider a random experiment that exogenously assigns links to individuals. This is what is performed by Hahn et al. (2015), who propose to test the importance of centrality measures in educational achievement. They implement a field experiment in Bangladesh, whose design involved randomly grouping students within-classroom among grade-four students in rural primary schools. These experiments were conducted in 80 schools in two districts (Khulna and Satkhira) in Bangladesh.

Let us now describe the experiment. In June 2013, the authors conducted a survey of all students in the 80 schools by asking them to name up to 10 closest friends (who defined the network, as in the Add Health data), starting from the most to the least close friends. They also conducted a separate household survey, which contained questions on parent education, parent age, parent occupation, and other household characteristics and asked each student to perform a *baseline math test* to measure her ability, which will help the authors balance groups by average ability. In July 2013, one month later, groups were randomly allocated. Each group contains four students and any student has an equal chance of being in one group or the other since groups are generated by a random-number generator. To implement random grouping that has a relatively similar mean across groups, they rank students by their baseline test score. Hahn et al. (2015) then randomly select a student from each quartile to form a group of size four. Then, newly (randomly) formed groups were asked to solve a *general knowledge test*, which took place in July 2013 and which is performed *collectively* by each group of four. Each group was given a *math assignment* to be completed *collectively* by the end of the week. Finally, after each group had handed over its math assignment, each individual had to take an *individual math test*. Prizes were given to the most successful students.

Hahn et al. (2015) want to investigate the following question: *If, by chance, a student ends up in a group with a high average centrality, does this positively impact on his/her test score as compared to someone who finds him/herself in a group with a lower average centrality?* They consider six measures of centrality: the degree, closeness, betweenness, eigenvector, Katz-Bonacich, and key-player (or intercentrality) centrality. The authors calculate the average centrality of each group of four (not including *i*'s centrality) and compare them across the classroom (there are usually 40 students per classroom and thus 10 groups of 4) and across the 80 schools where groups were randomized. They find that the higher is the average centrality of a group to which a student *i* is allocated, the higher is his/her grade. This is true for both the *general knowledge test* (a homework performed by the group) and the *individual math test* (performed individually). This result is also true for *any* centrality measure with different magnitudes of the effects. Since these centrality measures are correlated, the authors then test one measure against the other. Interestingly, they show that for both the general knowledge test and the individual math test, the (average) Katz-Bonacich centrality as well as the key-player centrality have the most significant impact on educational outcomes. They also test the role of leadership in a group on own grade outcomes by looking at the impact of the individual with the highest centrality in the group on own outcome. They find that, for both tests, it is the Katz-Bonacich centrality that performs best. These results indicate that the composition of a group in terms of centrality is of importance for both individual and group outcomes and that Katz-Bonacich and key-player centralities are key for these outcomes.

11.3.5 Key Players in Financial Networks

Since the onset of the financial crisis in 2007, the discourse regarding bank safety has shifted strongly from the riskiness of financial institutions as individual firms to concerns about systemic risk. As the crisis evolved, so did the public debate, with concerns about systemic risk evolving from too-big-to-fail (TBTF) considerations to too-interconnected-to-fail (TITF) ones. As a result, it seems quite natural to consider the issue of key players or key banks in financial networks. Systemic risk is usually defined as the risk of default of a large portion of the financial system, which depends on the network of financial exposures among institutions.

There are several recent papers that model financial networks using network theory (see Acemoglu, Ozdaglar, and Tahbaz-Salehi 2015; Elliott, Golub, and Jackson 2014; Cohen-Cole, Patacchini, and Zenou 2015 just to mention a few) but do not analyze the targeting of banks in order to reduce the risk of contagion.[20] Demange (2014) has proposed a *threat index*, which measures the decrease in payment within the banking system following a reduction in net worth at one institution. The latter is related to the intercentrality measure of the key player. Battiston et al. (2012) propose another index called the *DebtRank*, which is a novel measure of systemic impact inspired by feedback-centrality that recursively takes into account the impact of the distress of an initial node across the whole network. Battiston et al. (2012) apply their methodology to analyze a data set on the USD 1.2 trillion FED emergency loans program to global financial institutions during 2008–2010. They find that a group of 22 institutions, which received most of the funds, forms a strongly connected graph where each of the nodes becomes systemically important at the peak of the crisis.

[20] See Chapter 21 by Acemoglu, Ozdaglar, and Tahbaz-Salehi in this handbook for an overview of the literature on systemic risk and networks and Chapter 20 by Cabrales, Gale, and Gottardi for an overview of the literature on financial contagion.

Finally, Denbee et al. (2014) slightly modify the benchmark model of Section 11.2.2 to model banks' liquidity holding decisions as a simultaneous game on an interbank borrowing network. They consider a network between banks, which is the sterling unsecured overnight interbank money market. This is where banks lend central bank reserves to each other, unsecured, for repayment the following day. The strength of the link between any two banks in their network is measured using the fraction of borrowing by one bank from the other. Hence, their network is weighted and directional. As well as relying on their own liquidity buffers, banks can also rely on their borrowing relationship within the network to meet unexpected liquidity shocks. Using daily data from January 2006 to September 2010, the authors estimate model (11.18) and find evidence for a substantial, and time varying, network risk. In the period before the Lehman crisis, the network is cohesive and liquidity holding decisions are complementary, and there is a large network liquidity multiplier. During the 2007–08 crisis, the network becomes less clustered and liquidity holding less dependent on the network. After the crisis, during Quantitative Easing, the network liquidity multiplier becomes negative, implying a lower network potential for generating liquidity.

They also identify *risk key players*; that is, the banks that contribute the most to the aggregate liquidity risk through these three periods. They show that the risk that key players took during these periods varies a great deal. They also find that the key players in the network are not necessarily the largest borrowers. In fact, during the credit boom, large lenders and borrowers are equally likely to be key players. This set of findings is of policy relevance, and gives guidance on how to effectively inject liquidity, to reduce the network risk, if the government decides to intervene.

11.3.6 Key Players and Diffusion in Networks

Banerjee et al. (2013) structurally estimate the model developed in Section 11.2.2.3 to the diffusion of microfinance loans in India, in a setting where the set of potentially first-informed individuals is known. They show that the *communication centrality* of the injection points is a strong predictor of eventual participation in microfinance and should therefore provide guidance to anyone trying to spread the news about microfinance in similar villages. They define communication centrality as follows. For each leader (injection points), they compute a score. This score is the fraction of households that would eventually participate if this household were the only one initially informed. To compute this fraction, they simulate the model with information passing and participation decisions being governed by the estimated values of ϕ_1^N, ϕ_1^P, and β. Banerjee et al. (2013) call this score the communication centrality of a node. However, the communication centrality cannot be computed without the estimations of ϕ_1^N, ϕ_1^P, and β, which could be very different if we were interested in the diffusion of other products or even microfinance in a very different context. As a result, the authors propose an approximation of communication centrality, called *diffusion centrality*, which is highly correlated with communication centrality, but requires considerably less data. In particular, it does not rely on estimating the diffusion model. The diffusion centrality of a node i in a network with an adjacency matrix G, passing probability $\phi_1^N = \phi_1^P = \phi_1$, and iterations T, as the ith entry of the following vector:

$$\delta(\mathbf{g}, \phi_1, T) = \left[\sum_{t=1}^{T} (\phi_1 \mathbf{G})^t\right] \mathbf{1} \tag{11.19}$$

Interestingly, the diffusion centrality of a node i becomes proportional to either Katz-Bonacich centrality or eigenvector centrality when $T \to +\infty$, depending on whether ϕ_1 is smaller than the inverse of the largest eigenvalue of the adjacency matrix or exceeds it (see Proposition 1). In the intermediate region of T, the measure differs from existing measures. As for the Katz-Bonacich centrality defined in (11.4) and the intercentrality or key-player centrality defined in (11.7), any method for computing a measure of diffusion centrality relies on the estimation of ϕ_1 or the choice of an appropriate value for ϕ_1. Extreme values of ϕ_1 either lead to no diffusion or to complete diffusion and thus do not distinguish nodes. Barnerjee et al. (2013) choose a prominent intermediate value of ϕ_1: the inverse of the largest eigenvalue of the adjacency matrix, $\mu_1(\mathbf{G})$. This is the critical value of ϕ_1 for which the entries of $(\phi_1 \mathbf{G})^T$ tend to 0 as T grows if $\phi_1 < 1/\mu_1(\mathbf{G})$ and some entries diverge if $\phi_1 > \mu_1(\mathbf{G})$.

Another interesting related application of the key-player policy (i.e., targeting injection points) is by Banerjee et al. (2014), where they try to identify the key player (eigenvector centrality, diffusion centrality) without knowing the whole network by asking people about the central players. Indeed, in many instances, learning who is central in a social network has the potential of being difficult and costly. Even for members of the community, knowing the structure of the network beyond their immediate friends is far from straightforward. As a result, the authors would like to answer the following question: Can we identify the members of a community who are best-placed to diffuse information simply by asking a random sample of individuals? Using the same data on 35 Indian villages as in Banerjee et al. (2013), they show that, by tracking sources of gossip, one can identify those who are most central in a network according to the "diffusion centrality" defined in (11.19). In particular, the authors find that respondents *accurately* nominate those who are diffusion central (not just those with many friends). In other words, they find that individuals in a network are able to identify central individuals within their community without knowing anything about the structure of the network.

11.4 Concluding Remarks

We provide an overview of the literature on key players in networks. There are different measures in the literature, which fundamentally capture different aspects of centrality and are therefore related to different behaviors. We have seen that both eigenvector and

Katz-Bonacich centralities are crucial in explaining educational outcomes and diffusion processes. This is also true in other contexts. For example, Acemoglu et al. (2012) focus on production networks and show that the sector with the highest *influence vectors*, a measure closely related to the Katz-Bonacich centrality, is the one with the highest share in total output and is more important to cause aggregate fluctuations. In other words, sectors that take more "central" positions in the network representation of the economy play a more important role in determining aggregate output. Studying the direct and spillover effects of local state capacity using the network of Colombian municipalities, Acemoglu, Garcia-Jimeno, and Robinson (2015) find that Katz-Bonacich centrality, betweenness centrality, and local clustering are strong predictors of where more state capacity should be directed.

In this chapter, we introduce a new centrality measure, intercentrality, which determines the key player, who is the agent targeted by the planner so that, once removed, she generates the highest level of reduction in total activity. This centrality measure is different than those proposed in the literature since it has a normative aspect. We believe that this key-player policy based on this new measure makes sense when there are strategic complementarities in actions between the different agents in a network such as, for example, in crime, education, R&D collaborations, interbank loans, and so on. We also consider another notion of key player based on targeting agents that are optimally positioned to quickly diffuse information, attitudes, behaviors, or goods. Then, we examine the empirical tests of the key-player policies for criminal networks, education, R&D networks, financial networks, and diffusion of microfinance. We show that implementing such a policy outperforms other standard policies, such as targeting the most active agents in a network.

We believe that key-player policies are crucial when resources are limited and multiplier effects are important. We have seen how these policies can be applied to different aspects of the economy, such as crime or education, but we believe that they have broader implications. For example, König et al. (2014) study the key-player policies for wars. They look at the recent civil war in the Democratic Republic of Congo (DRC), which involves many groups and a complex network of alliances and rivalries. They examine how the removal of each group involved in the conflict would reduce the conflict intensity. They show that, while large groups are, on average, more crucial than small ones, the relationship is not one-to-one. The hypothetical removal of some relatively small players turns out to have large effects on the containment of the DRC conflict. We hope that more empirical studies will be implemented in the future, showing the relevance of key players in many other aspects of the economy.

REFERENCES

Acemoglu, D., V. M. Carvalho, A. Ozdaglar, and A. Tahbaz-Salehi (2012). "The network origins of aggregate fluctuations." *Econometrica* 80, 1977–2016.

Acemoglu, D., A. Ozdaglar, and A. Tahbaz-Salehi (2015). "Systemic risk and stability in financial networks." *American Economic Review* 105, 564–608.

Acemoglu, D., C. Garcia-Jimeno, and J. A. Robinson (2015). "State capacity and economic development: A network approach." *American Economic Review* 105, 2364–2409.

Advani, A. and B. Malde (2014). "Empirical methods for networks data: Social effects, network formation and measurement error." Unpublished manuscript, University College London.

Algan, Y., Q.-A. Do, A. Le Chapelain, and Y. Zenou (2015). "How do social networks shape our values? A natural experiment among future French politicians." Unpublished manuscript, Sciences Po, Paris.

Badev, A. (2014). "Discrete games in endogenous networks: Theory and policy." Unpublished manuscript, Federal Reserve Board, Washington D.C.

Ballester, C., A. Calvó-Armengol, and Y. Zenou (2006). "Who's who in networks. Wanted: The key player." *Econometrica* 74, 1403–1417.

Ballester, C., A. Calvó-Armengol, and Y. Zenou (2010). "Delinquent networks." *Journal of the European Economic Association* 8, 34–61.

Ballester, C. and Y. Zenou (2014). "Key player policies when contextual effects matter." *Journal of Mathematical Sociology* 38, 233–248.

Banerjee, A., A. G. Chandrasekhar, E. Duflo, and M. O. Jackson (2013). "The diffusion of microfinance." *Science* 341, 6144.

Banerjee, A., A. G. Chandrasekhar, E. Duflo, and M. O. Jackson (2014). "Gossip: Identifying central people in social networks." NBER Working Paper No. 20422.

Battiston, S., M. Puliga, R. Kaushik, P. Tasca, and G. Caldarelli (2012). "DebtRank: Too central to fail? Financial networks, the FED and systemic risk." *Scientific Reports* 2, 541.

Belhaj, M., S. Bervoets, and F. Deroïan (2015). "Efficient networks in games with local complementarities." *Theoretical Economics*, forthcoming.

Bifulco, R., J. M. Fletcher, and S. L. Ross (2011). "The effect of classmate characteristics on post-secondary outcomes: Evidence from the Add Health." *American Economic Journal: Economic Policy* 3, 25–53.

Blume, L. E., W. A. Brock, S. N. Durlauf, and Y. M. Ioannides (2011). "Identification of social interactions." In *Handbook of Social Economics*, J. Benhabib, A. Bisin, and M.O. Jackson eds. Amsterdam: Elsevier Science.

Bonacich, P. (1987). "Power and centrality: A family of measures." *American Journal of Sociology* 92, 1170–1182.

Borgatti, S. P. (2003). "The key player problem." In *Dynamic Social Network Modeling and Analysis: Workshop Summary and Papers*, R. Breiger, K. Carley, and P. Pattison eds., 241–252. New York: National Academy of Sciences Press.

Borgatti, S. P. (2006). "Identifying sets of key players in a network." *Computational, Mathematical and Organizational Theory* 12, 21–34.

Bramoullé, Y., H. Djebbari, and B. Fortin (2009). "Identification of peer effects through social networks." *Journal of Econometrics* 150, 41–55.

Bramoullé, Y. and R. Kranton (2007). "Public goods in networks." *Journal of Economic Theory* 135, 478–494.

Bramoullé, Y., R. Kranton, and M. D'Amours (2014). "Strategic interaction and networks." *American Economic Review* 104, 898–930.

Burt, R. S. (1992), *Structural Holes: The Social Structure of Competition*. Cambridge, MA: Harvard University Press.

Calvó-Armengol, A., E. Patacchini, and Y. Zenou (2009). "Peer effects and social networks in education." *Review of Economic Studies* 76, 1239–1267.

Calvó-Armengol, A. and Y. Zenou (2004). "Social networks and crime decisions. The role of social structure in facilitating delinquent behavior." *International Economic Review* 45, 939–958.
Campbell, A. (2013). "Word-of-mouth communication and percolation in social networks." *American Economic Review* 103, 2466–2498.
Carrell, S. E., B. I. Sacerdote, and J. E. West (2013). "From natural variation to optimal policy? The importance of endogenous peer group formation." *Econometrica* 81, 855–882.
Chandrasekhar, A. and M. O. Jackson (2014). "Tractable and consistent random graph models." NBER Working Paper No. 20276.
Chen, Y.-J., Y. Zenou, and J. Zhou (2015). "Multiple activities for socially connected criminals." CEPR Discussion Paper No. 10709.
Cohen-Cole, E., E. Patacchini, and Y. Zenou (2015). "Static and dynamic networks in interbank markets." *Network Science* 3, 98–123.
Del Bello, C. L., E. Patacchini, and Y. Zenou (2015). "Neighborhood effects in education." IZA Discussion Paper No. 8956.
Dequiedt, V. and Y. Zenou (2014). "Local and consistent centrality measures in networks." CEPR Discussion Paper No. 10031.
Demange, G. (2014). "Contagion in financial networks: A threat index." Unpublished manuscript, Paris School of Economics.
Denbee, E., C. Julliard, Y. Li, and K. Yuan (2014). "Network risk and key players: A structural analysis of interbank liquidity." Unpublished manuscript, London School of Economics and Political Science.
de Marti, J. and Y. Zenou (2014). "Networks games under incomplete information." CEPR Discussion Paper No. 10290.
de Marti, J. and Y. Zenou (2015). "Networks games under incomplete information." *Journal of Mathematical Economics* 61, 221–240.
Domingos, P. and M. Richardson (2001). "Mining the network value of customers." *Proceedings of the 7th International Conference on Knowledge Discovery and Data Mining*, pp. 57–66.
Elliott, M. L., B. Golub, and M. O. Jackson (2014). "Financial networks and contagion." *American Economic Review* 104, 3115–3153.
Galeotti, A. and S. Goyal (2009). "Influencing the influencers: A theory of strategic diffusion." *Rand Journal of Economics* 5, 1–32.
Galeotti, A. and B. Rogers (2013). "Strategic immunization and group structure." *American Economic Journal: Microeconomics* 40, 509–532.
Glaeser, E. L., B. Sacerdote, and J. Scheinkman (1996). "Crime and social interactions." *Quarterly Journal of Economics* 111, 508–548.
Goldsmith-Pinkham, P. and G. W. Imbens (2013). "Social networks and the identification of peer effects." *Journal of Business and Economic Statistics* 31, 253–264.
Goyal, S. (2007), *Connections: An Introduction to the Economics of Networks.* Princeton, NJ: Princeton University Press.
Graham, B. S. (2015). "Methods of identification in social networks." *Annual Review of Economics* 7, forthcoming.
Hagedoorn, J. (2002). "Inter-firm R&D partnerships: An overview of major trends and patterns since 1960." *Research Policy* 31, 477–492.

Hahn, Y., A. Islam, E. Patacchini, and Y. Zenou (2015). "Teams, organization and education outcomes: Evidence from a field experiment in Bangladesh." CEPR Discussion Paper No. 10631.

Hartline, J., V. S. Mirrokni, and M. Sundararajan (2008). "Optimal marketing strategies over social networks." International World Wide Web Conference Committee (IW3C2), Beijing, China.

Hiller, T. (2014). "Peer effects in endogenous networks." Unpublished manuscript, University of Bristol.

Husslage, B., P. Borm, T. Burg, H. Hamers, and R. Lindelauf (2014). "Ranking terrorists in networks: A sensitivity analysis of Al Qaeda's 9/11 attack." CentER Discussion Paper No. 2014-028, Tilburg University.

Ioannides, Y. M. (2012), *From Neighborhoods to Nations: The Economics of Social Interactions*. Princeton, NJ: Princeton University Press.

Jackson M. O. (2008), *Social and Economic Networks*. Princeton, NJ: Princeton University Press.

Jackson, M. O. (2011). "An overview of social networks and economic applications." In *Handbook of Social Economics Volume 1A*, J. Benhabib, A. Bisin, and M. O. Jackson, eds., 511–579. Amsterdam: Elsevier Science.

Jackson, M. O. (2014). "Networks in the understanding of economic behaviors." *Journal of Economic Perspectives* 28, 3–22.

Jackson, M. O., B. W. Rogers, and Y. Zenou (2015). "The impact of social networks on economic behavior." CEPR Discussion Paper 10406.

Jackson, M. O. and L. Yariv (2015). "Diffusion, strategic interaction, and social structure." In *Handbook of Social Economics Volume 1A*, J. Benhabib, A. Bisin, and M. O. Jackson, eds., 645–678. Amsterdam: Elsevier Science.

Jackson, M. O. and Y. Zenou (2015). "Games on networks." In *Handbook of Game Theory, Vol. 4*, P. Young and S. Zamir, eds., 91–157. Amsterdam: Elsevier.

Kahan, D. M. (1997). "Social influence, social meaning, and deterrence." *Virginia Law Review* 83, 349–395.

Katz, L. (1953). "A new status index derived from sociometric analysis." *Psychometrika* 18, 39–43.

Kempe, D., J. Kleinberg, and É. Tardos (2003). "Maximizing the spread of influence in a social network." *Proceedings of the 9th International Conference on Knowledge Discovery and Data Mining*, pp. 137–146.

Kempe, D., J. Kleinberg, and É. Tardos (2005). "Influential nodes in a diffusion model for social networks." *Proceedings of the 32nd International Colloque on Automata, Languages and Programming*, pp. 1127–1138.

Kennedy, D. M. (1998). "Pulling levers: Getting deterrence right." *National Institute of Justice Journal* 236, 2–8.

Kleiman, M. A. (2009), *When Brute Force Fails. How to Have Less Crime and Less Punishment*. Princeton, NJ: Princeton University Press.

Kleinberg, J. (2007). "Cascading behavior in networks: Algorithmic and economic issues." In *Algorithmic Game Theory*, N. Nisan, T. Roughgarden, E. Tardos, and V. Vazirani, eds., 613–632. Cambridge: Cambridge University Press.

König, M. D., C. Tessone, and Y. Zenou (2014). "Nestedness in networks: A theoretical model and some applications." *Theoretical Economics* 9, 695–752.

König, M. D., X. Liu, and Y. Zenou (2014). "R&D networks: Theory, empirics and policy implications." CEPR Discussion Paper No. 9872.

König, M. D., D. Rohmer, M. Thoenig, and F. Zilibotti (2015). "Networks in conflict: Theory and evidence from the great war of Africa." Unpublished manuscript, University of Zurich.

Koschützki, D., K. A. Lehmann, L. Peeters, S. Richter, D. Tenfelde-Podehl, and O. Zlotowski (2005). "Centrality indices." In Brandes, U. and T. Erlebach (Eds.), *Network Analysis: Methodological Foundations*, Lecture Notes in Computer Science No. 3418, New York: Springer-Verlag, pp. 16–61.

Lagerås, A. and D. Seim (2015). "Strategic complementarities, network games and endogenous network formation." *International Journal of Game Theory*, forthcoming.

Lindelauf, R., H. Hamers, and B. Husslage (2013). "Cooperative game theoretic centrality analysis of terrorist networks: The cases of Jemaah Islamiyah and Al Qaeda." *European Journal of Operational Research* 229, 230–238.

Lindquist, M. J., J. Sauermann, and Y. Zenou (2015). "Network effects on worker productivity." CEPR Discussion Paper No. 10928.

Lindquist, M. J. and Y. Zenou (2014). "Key players in co-offending networks." CEPR Discussion Paper No. 9889.

Liu, X., E. Patacchini, and Y. Zenou (2014). "Endogenous peer effects: Local aggregate or local average?" *Journal of Economic Behavior and Organization* 103, 39–59.

Liu, X., E. Patacchini, Y. Zenou, and L-F. Lee (2012). "Criminal networks: Who is the key player?" CEPR Discussion Paper No. 8772.

Liu, X., E. Patacchini, Y. Zenou, and L-F. Lee (2014). "Who is the key player? A network analysis of juvenile delinquency." Unpublished manuscript, Stockholm University.

Mahadev, N. V. R. and U. N. Peled (1995), *Threshold Graphs and Related Topics*. Amsterdam: North Holland.

Machin, S. and O. Marie (2011). "Crime and police resources: The street crime initiative." *Journal of the European Economic Association* 9, 678–701.

Machin, S., O. Marie, and M. Priks (2014). "Targeting prolific offenders to reduce crime: Theory and evidence from two European experiments." Unpublished manuscript, Stockholm University.

Manski, C. F. (1993). "Identification of endogenous effects: The reflection problem." *Review of Economic Studies* 60, 531–542.

Mele, A. (2013). "A structural model of segregation in social networks." Unpublished manuscript, Johns Hopkins University, Carey Business School.

Ortiz-Arroyo, D. (2010). "Discovering sets of key players in social networks." In *Computational Social Network Analysis*, A. Abraham, A.-E. Hassanien, and V. Snášel, eds., 27–47. London: Springer Verlag.

Patacchini, E., E. Rainone, and Y. Zenou (2014). "Heterogeneous peer effects in education." CEPR Discussion Paper No. 9804.

Patacchini, E. and Y. Zenou (2012). "Juvenile delinquency and conformism." *Journal of Law, Economics, and Organization* 28, 1–31.

Patacchini, E. and Y. Zenou (2014). "Social networks and parental behavior in the intergenerational transmission of religion." Unpublished manuscript, Stockholm University.

Powell, W. W., K. W. Koput, and L. Smith-Doerr (1996). "Interorganizational collaboration and the locus of innovation: Networks of learning in biotechnology." *Administrative Science Quarterly* 41, 116–145.

Powell, W. W., D. R. White, K. W. Koput, and J. Owen-Smith (2005). "Network dynamics and field evolution: The growth of interorganizational collaboration in the life sciences." *American Journal of Sociology* 110, 1132–1205.

Sacerdote, B. (2011). "Peer effects in education: How might they work, how big are they and how much do we know thus far?" In *Handbook of Economics of Education, Vol. 3*, E. A. Hanushek, S. Machin, and L. Woessmann, eds., 249–277. Amsterdam: Elsevier Science.

Sathik, M. and A. Rasheed (2009). "A centrality approach to identify sets of key players in an online weblog." *International Journal of Recent Trends in Engineering* 2, 85–87.

Silverman, D. (2004). "Street crime and street culture." *International Economic Review* 45, 761–786.

Sommer, M. (2014). "Centrality with vertex idiosyncracy: Comparative statics and vertex-weighted key-player analysis." Unpublished manuscript, Stockholm University.

Tita, G. K., J. Riley, G. Ridgeway, C. Grammich, A. Abrahamse, and P. W. Greenwood (2003), *Reducing Gun Violence: Results from an Intervention in East Los Angeles*. Santa Monica, CA: RAND Corporation.

Topa, G. and Y. Zenou (2015). "Neighborhood and network effects." In *Handbook of Regional and Urban Economics, Vol. 5A*, G. Duranton, V. Henderson, and W. Strange, eds., 561–624. Amsterdam: Elsevier.

Verdier, T. and Y. Zenou (2004). "Racial beliefs, location and the causes of crime." *International Economic Review* 45, 731–760.

Wilson, J. Q. and R. J. Herrnstein (1985), *Crime and Human Nature*. New York: Simon and Schuster.

Wasserman, S. and K. Faust (1994), *Social Network Analysis. Methods and Applications*. Cambridge: Cambridge University Press.

Weitzman, M. L. (1998). "Recombinant growth." *Quarterly Journal of Economics* 113, 331–360.

Zenou, Y. (2015). "Networks in economics." In *International Encyclopedia of Social and Behavioral Sciences, 2nd Edition*, J. D. Wright, ed., 572-581. Amsterdam: Elsevier.

PART IV

EMPIRICS AND EXPERIMENTS

CHAPTER 12

SOME CHALLENGES IN THE EMPIRICS OF THE EFFECTS OF NETWORKS

VINCENT BOUCHER AND BERNARD FORTIN

12.1 INTRODUCTION

Aristotle, the Greek philosopher, writes, "Man is by nature a social animal... He who lives without society is either a beast or God" (*Politics*, Book I, Part II). In economics, the importance of social interactions outside the market is now well recognized. Individuals share information, learn from others, and influence each other in many contexts. This point is particularly important for policy evaluation. To estimate the overall effect of a social intervention, it is crucial to account not only for its direct impact on the treated but also its indirect impact on their peers (Manski 2013). Recently, a growing literature has attempted to investigate social interactions at the theoretical, empirical, and econometric levels using the *social network* approach.

At the theoretical level, social network theory has developed as a rigorous language to analyze social interactions (Jackson 2011). This theoretical framework explores in particular (1) how social networks influence outcomes, and (2) how this in turn affects network formation (Jackson 2010).

At the empirical level, while the literature on the formation of social networks is limited but burgeoning (Fafchamps and Gubert 2007; Comola 2008; Mayer and Puller 2008; Christakis et al. 2010; Mele 2013; Boucher 2015),[1] applications on the effects of social networks on outcomes is vast and expanding at a rapid pace. To name a few, we can mention Bertrand et al. (2000) on welfare participation, Cassar (2007) on coordination and cooperation, Patacchini and Zenou (2008) on criminal activity, Trogdon et al. (2008) on obesity, Karlan et al. (2009) on risk-sharing, and

[1] See also the survey of Chandrasekhar in this handbook (Chapter 13).

Calvó-Armengol et al. (2009) on education. Moreover, some recent papers attempt to simultaneously evaluate network formation and network effects (Conti et al. 2012; Hsieh and Lee 2015; Goldsmith-Pinkham and Imbens 2013; Badev 2013; Boucher 2014a).

At the econometric level, a rich literature has emerged focusing on issues raised by identification of peer effects in econometric models of social networks (Manski 1993; Bramoullé et al. 2009; Lee et al. 2010; Blume et al. 2015). Also, issues of estimation have been addressed either using experimental data and most often based on a reduced form approach (Sacerdote 2001; Barrera-Osorio et al. 2011) or using non-experimental data (Lin 2010; Card and Giuliano 2013). In the latter case, a number of studies have been inspired by the spatial econometric literature (Bramoullé et al. 2009; Lee et al. 2010; Lin 2010). Distribution-free methods such as generalized two-stage least squares and GMM, or approaches involving more structure such as (quasi-)maximum likelihood with or without autoregressive disturbances have been proposed.

As pointed out by Blume et al. (2015), the theoretical, empirical, and econometric literature on social networks lack integration. Regarding the integration between theoretical and empirical social network approaches, the so-called *identifiability problem* (Chiappori and Ekeland 2009) is rarely addressed. For instance, assume a noncooperative (Nash) game over networks in which each individual maximizes his utility, given his characteristics and the behaviors and characteristics of his reference group. In this case, the identifiability problem is whether the individual best response (or reaction) functions can be uniquely rationalized by a microeconomic theoretical model based on fundamentals, such as the utility parameters characterizing private and social preferences.[2] As we will show, a positive answer to this question may help the researcher to simulate overall (direct plus indirect) impact of hypothetical alternative policy reforms, such as a higher taxes on tobacco or an increase in penalties on crime, as long as he has full information on the best-response functions. This issue is related to the measurement of the so-called *social multiplier*.

Regarding the integration of empirical and econometric approaches, a crucial issue deals with the possibility of recovering the (stochastic) best-response functions from some *a priori* knowledge of the data-generating process, that is, the *identification problem*. Assume a standard linear-in-means model where the outcome of each individual depends linearly on his own characteristics, on the mean outcome of his reference group (*endogenous peer effect*), and on its mean characteristics (*contextual peer effects*). Assume also that the outcome is continuous.[3] Identifying the best-response functions involves specific econometric problems, such as (1) the *reflection problem* (as one's own behavior depends on peer behavior, which itself simultaneously depends

[2] See also Manski et al. (2000), where he discusses the importance of the *source* of social interactions for policy making: preferences, constraints or expectations.

[3] A discrete outcome generally helps to identify the model but may be the source of multiple equilibria (Brock and Durlauf 2007).

on one's own behavior), (2) the presence of *correlated effects* (e.g., due to endogeneity of network formation or network-invariant unobservables), and (3) *incomplete information* (e.g., due to partial knowledge of networks' architecture and individual outcomes). When individuals interact in groups[4] and the group sizes are the same, Manski (1993) shows that the reflection problem makes the best response model unidentified.[5] However, Bramoullé et al. (2009) shows that there is no reflection problem when individuals do not interact through groups, which is most likely to be the case. Therefore, two basic issues remain as related to the identification problem: the endogeneity of network formation[6] and problems of information. Problems of information happen either because the game played by the individuals is a game of incomplete information (Blume et al. 2015; Yang and Lee 2014), or because the econometrician has only partial knowledge of the network (Chandrasekhar and Lewis 2011) or of the characteristics and behaviors of the individuals (Liu et al. 2013a; Boucher et al. 2014). One important issue related with the latter point is whether the use of proxy variables (e.g., Cooley Fruehwirth 2013), such as hours of work or body mass index as proxies for work effort or healthy life habits, may allow us to identify peer effects in the structural econometric model.

In this chapter, we focus on some issues associated with the identifiability and identification problems in the social network approach. In particular, we put the accent on *four* aspects that we think are crucial for *policy analysis* and on which the chapter makes a number of contributions.

To give a broad perspective on the relevant literature, we first provide a discussion of recent empirical contributions of research on the effects of social networks on outcomes with a focus on their implications for policy-making. We insist, in particular, on the importance of the *social multiplier* and *key player* concepts to understand the role of the network structure in evaluating the impact of policy interventions on aggregate outcomes.

Second, we show that the presence or absence of a social multiplier crucially depends on the source of social interactions (complementarity vs. pure conformity). We prove that the social multiplier is generally not identified even when the peer effects can be recovered (identifiability problem). However, we also show that information on isolated individuals within the network generally allows us to identify this effect.

Third, we extend our analysis to the case where the structure of a network, as summarized by the *social interaction matrix*,[7] is not fixed but stochastic. In this context, two scenarios can be obtained depending on whether this matrix is exogenous—that is, not correlated with the random term, or endogenous. When the social interaction

[4] This means that the population is partitioned into groups, and that individuals are affected by all others in their group and by none outside of it.

[5] Manski implicitly assumes that the size of groups is infinite, which is akin to assuming that they are the same.

[6] The presence of network-invariant unobservables can be treated by introducing network fixed effects into the model (Bramoullé et al. 2009; Lee et al. 2010).

[7] Its mathematical definition is given in Section 12.3.1.

matrix is exogenous, it may depend on some individual variables. In that case, we show that it is important to account for the effect of a policy shock on the matrix when evaluating the impact of a policy. Also, we discuss the ability to test for the endogeneity of the network structure (identification problem).

Finally, we focus on issues of identification of the best-response functions when the econometrician does not observe the true interaction variable but only a proxy (identification problem). We insist on the importance of developing a microeconomic model based on sound and credible theoretical underpinnings to determine the parameters of the best-response functions that are identified.

This chapter mostly focuses on the standard linear-in-means model. Most papers on social interactions have considered the econometric framework since it is naturally related to the standard simultaneous linear model. Nonetheless, the essence of most of our results and discussions extends to other linear models such as in Blume et al. (2015), Boucher (2014a), and Liu et al. (2014), as well as to nonlinear models of social interactions.

The remainder of the chapter is organized as follows. Section 12.2 provides a survey of the recent contributions on the effects of networks on outcomes with an emphasis on their policy implications. Section 12.3 discusses the concept of social multipliers and presents our basic results as related to its identifiability. Section 12.4 discusses various types of network structures. We introduce the notion of stochastic network structure and discuss the cases where the network is exogenous or endogenous. Section 12.5 analyzes identification issues when the right interaction variable is unobservable and a proxy variable is used. Section 12.6 concludes.

12.2 What Has Been Learned About the Effects of Social Networks on Policy Outcomes?

One important interest of using the social network framework is for policy analysis purposes. This section considers a number of concepts with applications illustrating various types of peer effects and other externalities that can be fruitfully investigated in a network setting.

One of the central concepts in social networks related to policy analysis involves *social multipliers* (Glaeser et al. 2003). It arises when a change in a *common* (policy) shock exerts both a *direct effect* on individual action, and an *indirect effect* through social influence. The social multiplier can be defined as the ratio of the total (direct plus indirect) effect to the direct effect, and therefore exceeds 1 (around a stable equilibrium) under the assumption of positive spillovers or complementarity within the network.[8]

[8] This result is formally proven in Section 12.3.1.

The relevance of social multipliers in policy analysis stems from the fact that the impact of a (small) intervention may be strongly amplified at the aggregate level when social interactions across individuals are large.

Using social multipliers, policy-makers can obtain information on the impacts of shocks on various networks and involving different types of externalities, such as academic performance at school (Graham 2008), labor participation of mothers (Maurin and Moschion 2009), tax evasion among small businesses and professionals (Galbiati and Zanella 2012), and fast food consumption among teenagers in schools (Fortin and Yazbeck 2015). Importantly, in many papers, the social multiplier in a linear-in-means model is computed as $1/(1-\delta)$,[9] where δ is the endogenous peer effect; that is, the impact of the individual reference group's mean outcome on his/her own outcome. However, we show in Section 12.3.1 that this equality does not always hold and that the social multiplier depends on the mechanisms of social interactions (e.g., complementarity vs. conformity). We also show that those mechanisms may be hard to identify in the linear-in-means model even if peer effects are identified. However, identification may be possible when there are isolated individuals in the network (see Section 12.3.2).

While social multipliers apply to a common policy shock, it may be the case that the policy intervention directly affects only a subset of the social network. This point was discussed, among others, in Moffitt (2001) and empirically analyzed in Babcock and Hartman (2010) and Dieye et al. (2014), using a treatment evaluation approach. For instance, let us suppose that a fraction of students are given free textbooks to learn the material presented in class. This intervention will not only influence their academic performance but may also indirectly affect the performance of those who did not receive a textbook, through social interactions. This suggests that when one's treatment status influences the actions of others, experimental data are not generally sufficient to identify a causal treatment impact. A social network setting can be useful to provide an adequate analysis of this type of intervention (Dieye et al. 2014).[10]

Related to the previous point, one may ask which individuals in the class should be given a textbook to maximize the overall impact of the intervention on academic performance (for a given budget). This normative issue has been called the key-player problem in the social network literature (Borgatti 2006). At the econometric level, it has been addressed in a number of papers (e.g., Ballester et al. 2006; Liu et al. 2014) that have been surveyed by Zenou in this handbook (Chapter 11). In Section 12.3.3 we provide a simple approach that shows how to determine the key player in a linear-in-means model.

[9] There is a natural analogy between this formula and the standard Keynesian multiplier.

[10] Moffitt (2001) points out that this intervention also helps with the identification in a linear-in-means model when individuals interact in groups. The reason is that the policy introduces an exclusion restriction in the model since the students who do not receive a textbook are not directly affected by the intervention.

Up to now, it has been assumed that the shock does not influence the structure of the network. In contrast, Comola and Prina (2013) propose an approach in which the social network can be influenced by the exogenous shock. They apply it to experimental data collected to evaluate the impact of a new savings technology in Nepal. Their analysis accounts for the change in the network of financial exchanges between all villagers before and after the shock. According to their results, a non-negligible share of the peer effect in savings comes from the change in the structure of the network. In Section 13.4.1, we provide an analysis of peer effects when the network architecture is exogenous, but stochastic (i.e., it may be affected by changes in exogenous variables).

One of the most difficult problems arising when using network data for policy analyses involves the structure of the network. It is likely to be stochastic endogenous, at least when no experimental data are available.[11] For instance, individual preferences within a friendship network are likely to be similar. Thus, one should expect to observe the presence of homophily on *unobserved* individual characteristics—that is, the tendency of friends to link with similar other individuals. This source of spurious correlation between friends' outcomes may generate a serious identification problem and produce potentially large biases in the estimators of the structural econometric model.

Recently, some researchers (Conti et al. 2012; Hsieh and Lee 2015; Goldsmith-Pinkham and Imbens 2013; Badev 2013; Boucher 2014a) have made attempts to develop econometric models allowing for the joint estimation of network formation and network interactions. However, empirical results using Add Health data and focusing on outcomes such as smoking, sleep behavior and academic performance among students at the secondary level in the United States do not seem to detect much difference in peer effects when networks are assumed exogenous and when they are allowed to be endogenous. Among the many reasons which could explain this is the fact that the Add Health database includes a very large number of observable characteristics that can be introduced in the econometric model (Liu et al. 2013a). Another explanation is that, although statistically significant, the explanatory power of the individual characteristics on the probability that two individuals are friends is extremely small (Boucher 2014a). However, as noted by Badev (2013), even with this absence of bias on the estimated parameters, the endogeneous nature of the network structure may still strongly affect the impact of policy shocks.[12] In Section 12.4.1, we show that this argument holds even when the network is exogeneous. In Section 12.4.3, we provide a discussion on how to test for network endogeneity in a particular class of models (latent space models) where the probability of a relation between individuals depends on their positions in a partially observed "social space" (Hoff et al. 2002).

[11] Using experimental data also raises its own problems such as external validity and the use of an incomplete network. See Blume et al. (2011) for a critical analysis of identification problems when using experimental data.

[12] Badev (2013) finds that neglecting the endogeneity of the friendship network leads to a downward bias of 10% to 15% on the predicted impact of public policies.

As stressed by Manski (1993), one key issue faced by a researcher who studies social networks is, who interacts with whom? Except for cases when data from highly controlled field or lab experiments are available (e.g., Mas and Moretti 2009 and Beugnot et al. 2014, in experiments on work effort), problems such as incomplete information on the structure of networks or partial observability of outcomes are ubiquitous. Davezies et al. (2009) has provided a simple solution to the latter issue when individuals interact through a group. Boucher et al. (2014) has implemented their approach to estimate peer effects in academic achievements in Canadian secondary schools. The former issue is much more severe when individuals interact through more complex social networks, and network architecture is only partially known. Chandrasekhar and Lewis (2011) have developed a generalized instrumental approach that yields consistent estimators of the best-response functions when the researcher observes only a sample of the network. Liu et al. (2013a) study the special case where the network is fully observed, but the outcome variable is only observed for a subset of the individuals.

A related important observability issue arises when some crucial variables of the model of social interactions are not available. For instance, in the literature on the social epidemic of obesity (Christakis and Fowler 2007; Trogdon et al. 2008; Cohen-Cole and Fletcher 2008), the variable of an individual's effort to maintain a normal weight is usually not available. Therefore, researchers often regress the individual's BMI on his peers' average BMI and other variables to estimate peer effects. Does this approach allow us to identify true peer effects? In Section 12.5, we show that the answer to this question crucially depends on the nature of the microeconomic model incorporating both the technical relationship between the individual's weight and his effort, and the individual's utility function.

Up to now, the discussion has focused on models with linear social interactions. However, as stressed by a number of researchers (Sacerdote 2011), the standard linear model may hide large heterogeneity in the effects experienced by different types of individuals. For instance, in the education sector, Sacerdote (2011) reports that when one uses nonlinear models, one prevalent finding is larger peer effects in which high ability students benefit from the presence of other high ability students. Also, when the outcome of interest is discrete (e.g., smoking, marijuana use, criminality among teenagers), the network model becomes intrinsically nonlinear. This raises the problem of multiple equilibria (Brock and Durlauf 2001; Card and Giuliano 2013; Lee et al. 2014; Badev 2013).

12.3 Social Multiplier in Networks

As discussed in the previous section, one major policy implication of the peer effect literature is the potential presence of a social multiplier. In this section, we discuss the researcher's ability to identify the presence (or not) of a social multiplier from the structural econometric model.

12.3.1 A Negative Identification Result

This subsection provides a negative result regarding the identification of the primitive preferences parameters (the fundamentals of the microeconomic model). We assume that the individual strategy within a network obeys a linear structure as it is commonly assumed in the literature and that no individual is isolated.[13] We show that it is not possible to recover the social multiplier even when peer effects are identified.[14]

To see why, let us focus on a simple static model inspired by Bramoullé et al. (2009) and Blume et al. (2015). Assume workers with size n are interacting through a network within a firm and are producing an output. The $n \times n$ network adjacency matrix $\mathbf{A}$ is such that $a_{ij} = 1$ if the worker i is influenced by the worker j and $a_{ij} = 0$, otherwise. We assume the links do not differ in strength. Worker i's reference group with size n_i is the set of other workers by which he is influenced. For the remainder of this subsection, we assume that the social interaction matrix $\mathbf{G}$ represents a row-normalization of the adjacency matrix $\mathbf{A}$ such that rows of $\mathbf{G}$ sum to 1. This implies that there are *no isolated workers*. Therefore $g_{ij} = 1/n_i$ if i is influenced by j, and $g_{ij} = 0$ otherwise. Suppose that $\mathbf{G}$ is a *non-stochastic* (or fixed) and known social interaction matrix.

Consider the following quadratic utility functions:[15]

$$u_i(\mathbf{y}, c_0) = (c_0 + \boldsymbol{x}_i\boldsymbol{\beta} + \varepsilon_i)y_i - \frac{y_i^2}{2} + \alpha y_i \mathbf{g}_i\mathbf{y} \tag{12.1}$$

$$v_i(\mathbf{y}, c_0) = (c_0 + \boldsymbol{x}_i\boldsymbol{\beta} + \varepsilon_i)y_i - \frac{y_i^2}{2} - \frac{\lambda}{2}\left(y_i - \mathbf{g}_i\mathbf{y}\right)^2 \tag{12.2}$$

where $\boldsymbol{x}_i$ is the vector of individual i's observable characteristics and ε_i is a random term representing his unobservable characteristics. One assumes that $\boldsymbol{x}_i$ are strictly exogenous, and that $\alpha \in [0, 1)$ and $\lambda \geq 0$.[16] Both utility functions are separable into two components: a private and a social sub-utility. The first two expressions on the right-hand side of these equations are the same and correspond to the private sub-utility. The third expression, which describes the social sub-utility, features social interactions and differs across equations.

[13] This assumption is relaxed in the next section.

[14] Blume et al. (2015) provides a similar negative result on the identification of preferences but does not discuss its link with the social multiplier.

[15] To simplify the discussion, we ignore contextual peer effects. Including the latter would not change the basic argument.

[16] Those conditions on α and λ imply that the set of best-response functions is a contraction, which implies the existence of a unique Nash equilibrium of the game.

Preferences given by these utility functions come from two alternative sources of social interactions: *complementary* and *pure conformity*. A worker whose preferences are given by u_i could be selecting his production, y_i, taking into account the benefit received per unit produced, $c_0 + \boldsymbol{x}_i\boldsymbol{\beta} + \varepsilon_i$, the cost of producing, $y_i^2/2$, and the synergy created by working with his coworkers, $\alpha y_i \mathbf{g}_i\mathbf{y}$.[17] The latter expression, which reflects the social sub-utility of u_i, means that an increase in the average coworker's production, $\mathbf{g}_i\mathbf{y}$, positively influences his marginal productivity (*complementary*).[18] Alternatively, a worker whose preferences are given by v_i could be also choosing his production taking into account the benefit received from unit produced, $c_0 + \boldsymbol{x}_i\boldsymbol{\beta} + \varepsilon_i$, and the cost of producing, $y_i^2/2$. However, in this case there is no complementarity. Rather, his social sub-utility is positively affected by the degree to which he conforms with his peers' level of production, $-\frac{\lambda}{2}\left(y_i - \mathbf{g}_i\mathbf{y}\right)^2$. This refers to *pure conformity* due, for instance, to social norms or emulation between workers.[19]

Note that even if the utilities u_i and v_i represent different sources of social interactions, they lead to *observationally equivalent* sets of reduced form models:[20]

$$\mathbf{y}^u(c_0) = (\mathbf{I} - \alpha\mathbf{G})^{-1}(c_0\mathbf{1} + \boldsymbol{X\beta} + \boldsymbol{\varepsilon}) \tag{12.3}$$

$$\mathbf{y}^v(c_0) = (\mathbf{I} - \tilde{\alpha}\mathbf{G})^{-1}(\tilde{c}_0\mathbf{1} + \boldsymbol{X}\tilde{\boldsymbol{\beta}} + \tilde{\boldsymbol{\varepsilon}}) \tag{12.4}$$

where $\tilde{\alpha} = \lambda/(1+\lambda)$, $\tilde{\boldsymbol{\beta}} = \boldsymbol{\beta}/(1+\lambda)$, $\tilde{c}_0 = c_0/(1+\lambda)$ and $\tilde{\boldsymbol{\varepsilon}} = \boldsymbol{\varepsilon}/(1+\lambda)$ and where $\mathbf{1}$ is the vector of 1s of dimension n. This is important as only the model with complementarity features a social multiplier!

In other words, the presence (or not) of a social multiplier cannot be identified from the agents' decisions. Formally:

Proposition 1 *Let $c_1 = c_0 + c$, then $\Delta\mathbf{y}^u = \frac{c}{1-\alpha}\mathbf{1}$ and $\Delta\mathbf{y}^v = c\mathbf{1}$,*

with $\Delta\mathbf{y}^u = \mathbf{y}^u(c_1) - \mathbf{y}^u(c_0)$ and $\Delta\mathbf{y}^v = \mathbf{y}^v(c_1) - \mathbf{y}^v(c_0)$.

Suppose that the change (Δ) from c_0 to c_1 is a policy shock represented by the introduction of a wage subsidy. Consider first a worker whose preferences are represented by u_i. Without social interactions ($\alpha = 0$), this leads to a change in the production of any worker by c. With social interactions, however, a worker also

[17] Calvó-Armengol et al. (2009) took this approach in modeling peer effects in academic achievement.

[18] Note that *substitutability* in social interactions (with $\alpha < 0$), where the average peer's production reduces a worker's utility, is also possible as long as $|\alpha| < 1$.

[19] As is well-known, the literature on game theory associates the expression *strategic complementarity* or *supermodularity* to the case where an increase in a player's outcome increases the marginal utility (or payoff) of the other player's outcome. Here it is easy to show that both complementarity (as we define it) and pure conformity generate strategic complementarity.

[20] These *restricted* reduced forms of equations correspond to the (unique) Nash equilibrium implied by the set of best-response functions for the two models.

indirectly benefits from the other workers' change in pay so the overall impact of the change is larger: $|c|/(1-\alpha) > |c|$. In this case, the social multiplier is $1/(1-\alpha)$, which is strictly greater than 1.

Now consider a worker whose preferences are represented by v_i. In that case, there is no social multiplier (it is equal to 1)—that is, the impact of the wage subsidy on his production is not amplified by his peers' mean production. The basic intuition is that the subsidy induces the same direct effect (i.e., without social interactions) on a worker's production and on that of his peers. Therefore, the *gap* between these variables, or $(y_i - \mathbf{g}_i\mathbf{y})^2$, is not influenced by the policy shock. As a result, in this pure conformity model, social interactions between individuals will generate no indirect effect.

Note that it is possible that a worker's social utility has both complementary and pure conformity features. For instance, a worker may be affected both by the synergy in productivity from his co-workers and by his incentives to imitate their behavior.[21] In that case, the corresponding social multiplier will be equal to $\frac{1}{1-\alpha}$ and is independent of the parameter λ reflecting pure conformity.

In this situation, the best-response functions, obtained from maximizing each worker's utility with respect to his own production, *given* the others' production, are given by:

$$\mathbf{y} = \tilde{c}_0\mathbf{1} + \mathbf{X}\tilde{\boldsymbol{\beta}} + \delta\mathbf{G}\mathbf{y} + \tilde{\boldsymbol{\varepsilon}} \tag{12.5}$$

where $\delta = \frac{\alpha+\lambda}{1+\lambda}$. In (12.5), the parameters $\tilde{\boldsymbol{\beta}}$ capture the individual effects and δ captures the endogenous peer effect. When individuals do not interact through groups, the parameters of the linear-in-means model (12.5) are identified.[22] However, the parameters α and λ defining δ are not identified. Again, this shows that the social multiplier as well as the sources of social interactions are not identified.

Our analysis also shows the importance of providing the micro foundations (e.g., in terms of preferences, technology and decision process) of a social interactions model to fully address the identifiability problem, as was carefully done for instance by Blume et al. (2015). Thus, focusing the analysis only on the linear-in-means structural model does not allow to derive any identifiability result on the social multiplier.[23]

[21] In that case, the worker's utility function is given by: $(c_0 + \mathbf{x}_i\boldsymbol{\beta} + \varepsilon_i)y_i - \frac{y_i^2}{2} + \alpha y_i\mathbf{g}_i\mathbf{y} - \frac{\lambda}{2}\left(y_i - \mathbf{g}_i\mathbf{y}\right)^2$.

[22] The parameters are also identified when individuals interact through groups and there are many groups with different sizes (Davezies et al. 2009).

[23] This point was also made in Liu et al. (2013b) where they present a model featuring both complementarity and pure conformism. They provide a test for distinguishing between what they call the local aggregate (complementarity) and the local average (pure conformism). Their test relies on the fact that their local aggregate effect is defined on the non-normalised adjacency matrix. Our point is stronger: restricting the analysis to row-normalized matrices, the econometric model does not allow to identify the source of the interaction (pure conformity or complementarity).

12.3.2 Room for Identification when Some Individuals Are Isolated

The argument developed up to now depends on one important assumption: no individual is isolated. This implicitly follows from the assumption that the social matrix interaction **G** is row-normalized. This assumption is usually made to obtain identification results in the presence of correlated effects due to fixed network unobservables (Bramoullé et al. 2009; Lee et al. 2010; Kwok 2013). It is usually assumed that removing those individuals from the database can be done without any great loss of generality because, by definition, they do not affect the correlation structure of the network game. However, we will see that those isolated individuals provide identifiability conditions for the presence of a social multiplier as well as the source of social interactions.

Suppose now that the firm is also composed of isolated workers (referred to by $\check{ }$). Since their preferences do not include any social sub-utility component, their corresponding utilities are given by:

$$\check{u}_i(y_i, c_0) = (c_0 + \boldsymbol{x}_i\boldsymbol{\beta} + \varepsilon_i)y_i - \frac{y_i^2}{2} \tag{12.6}$$

$$\check{v}_i(y_i, c_0) = (c_0 + \boldsymbol{x}_i\boldsymbol{\beta} + \varepsilon_i)y_i - \frac{y_i^2}{2} \tag{12.7}$$

For those workers, both complementary and pure conformity models lead to the *same* best-response functions:

$$\check{\mathrm{y}} = c_0\mathbf{1} + \boldsymbol{X}\boldsymbol{\beta} + \boldsymbol{\varepsilon} \tag{12.8}$$

where $\boldsymbol{\beta}$ and c_0 are identified when $\boldsymbol{X}$ is full rank and can be estimated by a simple OLS.

We argue that isolated workers allow to test for the presence of a social multiplier. To see why, consider (12.5), where both sources of social interactions are potentially present, as well as (12.8) for isolated individuals. Since $\boldsymbol{\beta}$ is identified from (12.8) and $\tilde{\boldsymbol{\beta}}$ is identified from (12.5), λ is identified.[24] Also, since δ is identified from (12.5), the identification of λ is sufficient for the identification of α.[25] Then, all the parameters are identified, which is sufficient to test for the presence of a social multiplier.[26]

This argument, however, relies on the researcher's ability to distinguish between isolated and non-isolated individuals, which is strongly dependent on the observation of the exact network structure. Specifically, if some links are unobserved, one may incorrectly infer that an individual is isolated. In that case, the estimated parameters are likely to be biased and researchers should proceed to robustness checks—for instance, using simulations.

[24] Recall that $\tilde{\boldsymbol{\beta}} = \boldsymbol{\beta}/(1+\lambda)$.

[25] Recall that $\delta = (\alpha + \lambda)/(1+\lambda)$.

[26] Recall that the social multiplier is equal to $\frac{1}{1-\alpha}$.

Here an important remark is in order. In some cases for example, when the network structure is endogenous (see Section 12.4.2)—one may want to estimate the model with the full network matrix. Let $\mathbf{G}$ be such a matrix, where rows of isolated individuals sum to 0, while the other rows sum to 1.[27] Consider the following model:

$$\mathrm{y} = c_0 \mathbf{1} + \boldsymbol{X\beta} + \alpha \mathrm{G}\mathrm{y} + \boldsymbol{\varepsilon}$$

With this specification, and since $\mathbf{G}$ includes isolated and non-isolated individuals, the model *imposes* that the individual effects are the same for every individual. This implies that the economy *has* to feature a social multiplier. The model is then incompatible with preferences such as v_i and the social interaction parameter should not (unless explicit preferences are specified) be interpreted as an effect of conformism.[28]

12.3.3 Social Multiplier or Key Player?

In the previous sections, we have defined the social multiplier as the impact of a *common* shock (i.e., the wage subsidy). However, even when social interactions do not lead to the presence of a social multiplier, the policy-maker may want to know which individuals to target (e.g., Ballester et al. 2006; Graham 2014). Suppose that the policy-maker wants to affect only a subset $I \subset N$ of individuals. Which ones should he target in order to maximize the overall impact? Which criminals, once put in jail, generate the highest possible reduction in aggregate crime levels (Liu et al. 2014)? In this section, we briefly describe the identity of those key-players.[29]

Consider the following model where $\mathbf{G}$ is row-normalized:

$$\mathrm{y} = \tau \left(\mathbf{I} - \delta \mathrm{G}\right)^{-1} \left(c_0 \mathbf{1} + \boldsymbol{X\beta} + \boldsymbol{\varepsilon}\right)$$

where $\delta \in (0,1)$ and $\tau = \frac{1}{1+\lambda}$, with $\lambda \geq 0$. Remark that this specification includes the models of complementarity ($\lambda = 0$ and $\delta = \alpha$) and pure conformity ($\lambda > 0$ and $\delta = \frac{\lambda}{1+\lambda}$) as special cases. The overall impact of a shock c on all individuals $i \in I$ is given by:

$$\tau c \mathbf{1}' \left(\mathbf{I} - \delta \mathrm{G}\right)^{-1} \mathbf{e}_I$$

where $\mathbf{e}_I$ is a vector taking 1 at the i'th position for all $i \in I$ and 0 elsewhere. Letting $\mathbf{b} = \left(\mathbf{I} - \delta \mathrm{G}'\right)^{-1} \mathbf{1}$, this is equivalent to $\tau c \mathbf{b}' \mathbf{e}_I = \tau c \sum_{i \in I} b_i$. The vector $\mathbf{b}$ is a well-known measure of centrality called the PageRank (Brin and Page 1998) and is equal to the Katz-Bonacich centrality for the transpose of $\mathbf{G}$ (Calvó-Armengol et al. 2009). Then, the policy-maker should target the most central individuals. Note that the identity of

[27] This definition of the interaction structure is used, among others, by Bramoullé et al. (2009).

[28] When the interaction matrix G is not row-normalized, however, the full model may not be incompatible with conformism. See, for example, Boucher (2014a) where the interaction matrix is given by the Laplacian matrix of the network.

[29] For a complete survey, see Chapter 11 of this handbook by Zenou.

the key-players strongly depends on δ (**b** is a function of δ), but *not* on τ and c, which only affect the magnitude of the shock. To know the identity of the key players, the source of social interaction (complementarity or pure conformity) is then irrelevant, as long as δ is precisely estimated.

12.4 Types of Network Structures

Networks are convenient to represent the structure of interactions between individuals. Interactions of different natures lead to different types of network structures. Despite the important differences between those types (see below), the literature rarely distinguishes between them. In this section, we distinguish between three types of network structures: non-stochastic network structures, stochastic exogenous network structures, and stochastic endogenous network structures. We also discuss the problem of identification of the structural econometric model in these three cases.

For the remainder of this section, we assume that the interaction matrix **G** represents a row-normalization of the adjacency matrix **A** such that rows for non-isolated individuals sum to 1 while the other rows are 0. To clarify the exposition, we assume that there are no endogenous effects and that the structural econometric model is given by:

$$y = X\boldsymbol{\beta} + GX\boldsymbol{\gamma} + \boldsymbol{\varepsilon} \tag{12.9}$$

where $\varepsilon_i \sim_{iid} N(0, \sigma)$.

12.4.1 Exogenous or Non-stochastic?

Let's assume that the network structure **G** is non-stochastic. Prominent economic examples of such network structures include group interactions (Lee 2007; Boucher et al. 2014), provided that groups are non-stochastic, as well as panel data models. As discussed in Bramoullé et al. (2009), panel data models represent a special case of network interactions when individual i at time t has only one peer: himself at time $t-1$. Econometric methods based on non-stochastic network structures include for instance Lee et al. (2010).[30] When **G** is non-stochastic, parameters in (12.9) are identified as long as matrix X is of full rank and $\mathbb{E}(\boldsymbol{\varepsilon}|\mathbf{X}) = \mathbf{0}$. Moreover, they can be estimated by OLS.

Now, assume that **G** is stochastic; that is, the observed network **G** is the realization of a random variable. Assume also that **G** is exogenous in the sense that $\mathbb{E}(\boldsymbol{\varepsilon}|\mathbf{X}, \mathbf{G}) = \mathbf{0}$. For

[30] Note that despite the fact that Lee et al. (2010) considers non-stochastic network structures, the method has been applied in practice to cases where the network is explicitly stochastic.

example, $\mathbf{G}$ could represent the (row-)normalization of the adjacency matrix $\mathbf{A}$, defined as follows:

$$a_{ij} = \mathbf{1}(r(\mathbf{x}_i, \mathbf{x}_j) + \nu_{ij} > 0)$$

for some function r and where $\mathbf{1}(B)$ is an indicator function where $\mathbf{1}(B) = 1$ if B is true and $\mathbf{1}(B) = 0$ if B is false, and where $\nu_{ij} \sim_{iid} N(0,1)$. In that example, we immediately see that $\mathbb{E}(\boldsymbol{\varepsilon}|\mathbf{X},\mathbf{G}) = \mathbf{0}$, so the parameters in (12.9) can again be estimated by OLS.

Therefore, the fact that the network is (stochastic and) exogenous or non-stochastic has the same consequence for the identification and estimation of the model. However, the impact for the interpretation, and related policy recommendations, of the model are *not* the same. For example, suppose that (12.9) represents workers' productivity, $\mathbf{y}$, as a function of the workers' network $\mathbf{G}$ within a firm, and their individual characteristics $\mathbf{X}$, which may include, for instance, wages. Now, suppose that the government introduces a wage subsidy (i.e., a shock on $\mathbf{X}$). That policy may also influence the structure of the workers' network within the firm. This must be taken into account in the evaluation of the impact of the policy on the workers' productivity. In general, for exogenous networks, one cannot predict the impact of a policy shock without imposing assumptions on the process generating the network.

This might seem like a simple remark. However, very few models of peer effects include a description of network formation process. When they do, the emphasis is placed on whether or not the network is endogenous (i.e., whether or not $\mathbb{E}(\boldsymbol{\varepsilon}|\mathbf{X},\mathbf{G}) = \mathbf{0}$ holds). Our point is different: for policy's sake, any model of social interactions based on a stochastic network structure should include a description of the network formation process. This does not mean that researchers should always provide an explicit network formation process, but they should at least discuss the expected impact of their proposed policy shocks on the structure of the interaction network.

The distinction between stochastic and non-stochastic network structures is crucial in order to describe the impact of a policy shock. In the next section, we discuss the third type of network structure, which is when the network is stochastic and endogenous (i.e., when $\mathbb{E}(\boldsymbol{\varepsilon}|\mathbf{X},\mathbf{G}) \neq \mathbf{0}$). Specifically, we discuss our ability to test for the endogeneity of the network structure.

12.4.2 Testing for Network Endogeneity in Latent Space Models

Assume that the model in (12.9) can be decomposed as follows:

$$\mathrm{y} = X\boldsymbol{\beta} + GX\boldsymbol{\gamma} + \underbrace{\rho\boldsymbol{\nu} + \boldsymbol{\xi}}_{\varepsilon} \tag{12.10}$$

where ξ is an exogenous error term. Assume also that the network structure is stochastic and that $\mathbf{G}$ is the (row-)normalization of the adjacency matrix $\mathbf{A}$ defined as follows:

$$a_{ij} = \mathbf{1}\{\kappa - \phi|x_i - x_j| - \mu|\nu_i - \nu_j| + \eta_{ij} > 0)\} \tag{12.11}$$

where η_{ij} is an exogenous error term. The network formation processes characterized by (12.11), where the likelihood of a link between i and j can be represented as a function of the agents' individual characteristics, are called *latent space models* in the statistical literature (Hoff et al. 2002) and have been recently used in economics by Goldsmith-Pinkham and Imbens (2013) and Graham (2014). When ϕ and μ are greater than zero, this indicates the presence of homophily—that is, the closer two individuals are in terms of their observable and unobservable characteristics, the higher is the probability that they are linked.

Here, the network structure is endogenous due to the fact that the econometrician does not observe the relevant variable ν, which enters both the outcome equation (12.10) and the network formation equation (12.11). Note that this is different from the endogeneity that would result from self-selection on the outcome variable y, as in Badev (2013) and Boucher (2014a). Unless $\rho = 0$ or $\mu = 0$, the outcome structural equation is unidentified due to the endogeneity created by ν. This makes it impossible to obtain consistent estimates of its parameters. Goldsmith-Pinkham and Imbens (2013) discusses the possibility of testing for this particular problem. The idea is the following.

Suppose that $\rho \neq 0$ and $\mu > 0$. Let $\hat{a}_{ij}$ be the predicted value of a_{ij} assuming that $\mu = 0$ (that is, for a consistent estimator of κ and ϕ, conditional on $\mu = 0$). Similarly, let $\hat{\varepsilon}_{ij} = |\hat{\varepsilon}_i - \hat{\varepsilon}_j|$ where $\hat{\varepsilon}_i$ is the residual from $y_i - \hat{y}_i$, again assuming $\rho = 0$.

Now, consider the set of pairs such that $a_{ij} = 0$ (no link). Suppose that, for some pair of individuals (i,j), $\hat{a}_{ij}$ is "high" compared with the predicted value for the other linkages. This suggests that the link between i and j was more likely to be created than the other links (but is not linked in the data). Then, if $\mu > 0$, this implies that the value of $\hat{\varepsilon}_{ij}$ should be relatively large. Put differently, if the predicted value of the link is large, but no link is created, it implies that the unobserved shock on the link is strongly negative. If the model features homophily, this implies that the distance in the unobservables is large (e.g., strong difference in preferences).

To our knowledge, no formal test of endogeneity exists which fully exploits the information in $\hat{a}_{ij}$ and $\hat{\varepsilon}_{ij}$. In practice, researchers have used the fact that if the network is endogenous, "low-values" of $\hat{a}_{ij}$ should be correlated with $\hat{\varepsilon}_{ij}$ (Liu et al. 2013a). However, as discussed above, this is far from exploiting all the information provided by $\hat{a}_{ij}$ and $\hat{\varepsilon}_{ij}$. Also, if $\rho = 0$ or $\mu = 0$, the errors should not only be uncorrelated, but *independent*. Nonetheless, the same intuition holds. If $\rho = 0$ or $\mu = 0$, the residuals $\hat{\varepsilon}_{ij}$ from (12.10) should not provide any information on the probability that a link is created (conditional on the observables).

To explore further the empirical content of the model, we present numerical simulations. We show results for simple correlation tests, as well as for the Hilbert-Schmidt

Independence Criterion tests.[31] Note also that this "test" is only valid when the network formation process is given by (12.11) (i.e., when the utility of the network is separable across links; Bramoullé and Fortin 2010). However, we suspect that the same principle could be applied to more general models of network formation such as in Hsieh and Lee (2015), Mele (2013), or Chandrasekhar and Jackson (2013).

12.4.2.1 Simulations

We provide an example using Monte Carlo simulations for the model in (12.10), assuming that:

$$a_{ij} = \mathbf{1}\{\kappa - \phi|x_i - x_j| - \mu|\tilde{\nu}_i - \tilde{\nu}_j| + \eta_{ij} > 0)\} \tag{12.12}$$

We use two specifications of the model. An "exogenous" version where ν and $\tilde{\nu}$ are two *independent and identically distributed* variables, and an "endogenous" version where $\nu = \tilde{\nu}$ (i.e., the same variable as in (12.10)).[32]

We let $x, \nu, \tilde{\nu}, \varepsilon, \eta \sim N(0,1)$. Parameters are fixed such that $\phi = 0.5, \beta = 2, \gamma = 10, \kappa = 4$, $\mu = 6$, and $\rho = 20$. Parameters are chosen so as to facilitate the visual representation. We simulate 100 groups of sizes drawn from a *uniform* $(10, 50)$. The total population is comprised of 2,944 individuals and 49,304 within-group pairs.

We first regress the predicted value of the link (i.e., $\hat{a}_{ij}$) on the distance in the residuals of the outcome equation (i.e., $\hat{\varepsilon}_{ij}$), and on the distance between the residuals when individuals are linked (i.e., $\hat{\varepsilon}_{ij} * a_{ij}$) and on the status (linked or not) of the pair (i.e., a_{ij}). If the network structure is exogenous, the only predictor of $\hat{a}_{ij}$ should be a_{ij}. Looking at Table 12.1 (endogenous version), we see that we reject the null hypothesis that the network is exogenous since we measure a significant correlation between $\hat{\varepsilon}_{ij}$ and $\hat{a}_{ij}$. This result is confirmed by looking at Table 12.2 (exogenous version), where no correlation is measured.

The endogeneity created by the omitted variable ν can also be seen by looking at the conditional joint densities in Figures 12.1 through 12.4. If the network structure is exogenous, the realized network, (i.e., the set of pairs of individuals that are actually linked or not) should not affect the joint distribution of $\hat{\varepsilon}_{ij}$ and $\hat{a}_{ij}$. When the network is endogenous, we see a large difference between the joint density for linked pairs (Figure 12.1), and the joint density of unlinked pairs (Figure 12.2). When the network is exogenous, we do not (and should not; see Figures 12.3 and 12.4). This visual analysis is confirmed by the Hilbert-Schmidt independence test in Table 12.3.

[31] The Hilbert-Schmidt Independence Criterion is a kernel-based independence test. It tests whether the (Hilbert-Schmidt) norm of the cross-covariance operator is equal to zero. For generic kernels, the Hilbert-Schmidt Independence Criterion between two random variables is equal to zero if and only if the variables are independent. See Gretton et al. (2005) for details. We computed the Hilbert-Schmidt Independence Criterion using Arthur Gretton's (2007) *hsicTestGamma.m* function. See http://people.kyb.tuebingen.mpg.de/arthur/indep.htm.

[32] This strategy is used so that the distribution of the unobservables in (12.10) and (12.12) will be the same, which facilitates the visual representation (see Figures 12.1 through 12.4).

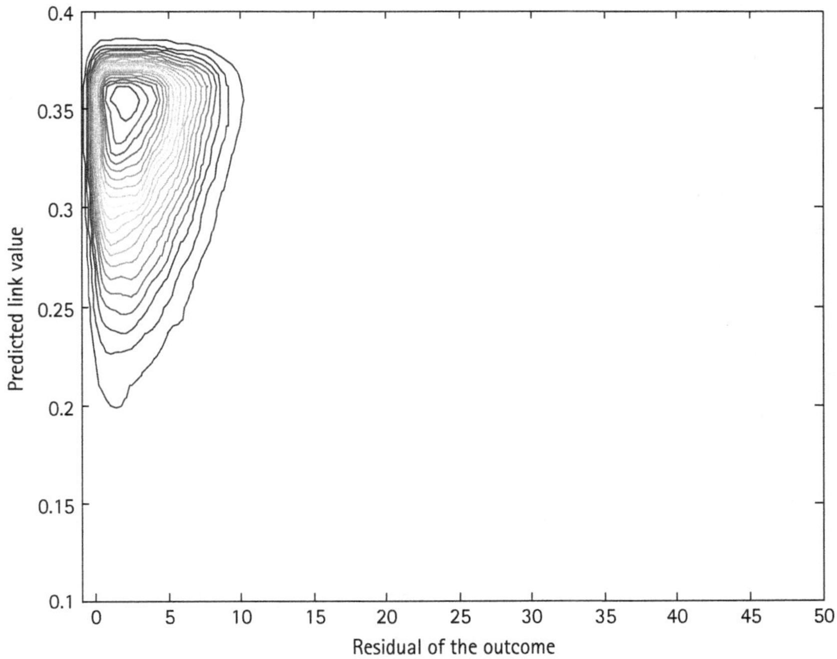

FIGURE 12.1 Kernel density estimates, endogenous model, linked.

Table 12.1 Regression on predicted value of link ($\hat{a}_{ij}$): endogenous version

Variable	Coefficient	(Std. err.)
$\hat{\varepsilon}_{ij}$	0.301**	(0.028)
$\hat{\varepsilon}_{ij} * a_{ij}$	1.898**	(0.145)
a_{ij}	5.228**	(0.736)
Intercept	310.639**	(0.471)

Joint test for $[\hat{\varepsilon}_{ij}] = [\hat{\varepsilon}_{ij} * a_{ij}] = 0$: $F(2.52385) = 177.45$

Since there exists no formal test exploiting all the empirical content of the model, we suggest that, in practice, researchers should proceed to a visual inspection of the conditional joint densities (as in Figures 12.1 and 12.2) and that they perform non-parametric independence tests such as the Hilbert-Schmidt Independence Criterion. Note that the visual inspection of the joint densities may provide more information as it gives an idea of the importance of the endogeneity. If joint densities are relatively similar, we can expect the bias due to the endogeneity to be small, even if evidence of endogeneity is found.

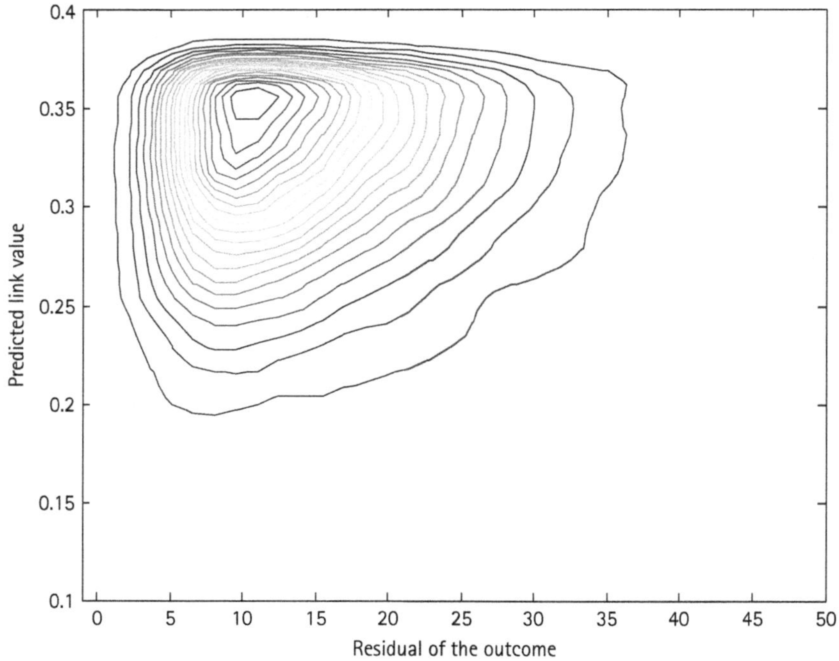

FIGURE 12.2 Kernel density estimates, endogenous model, no link.

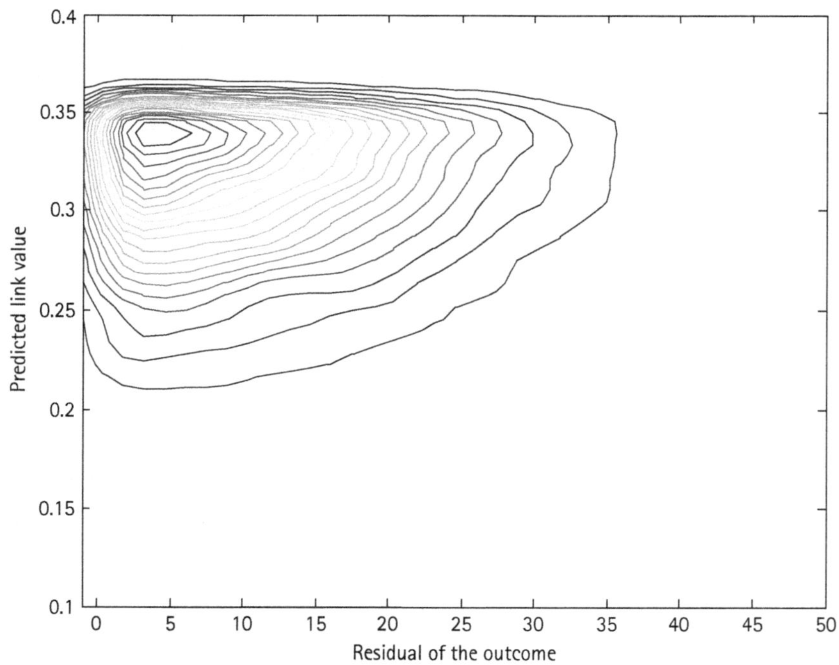

FIGURE 12.3 Kernel density estimates, exogenous model, linked.

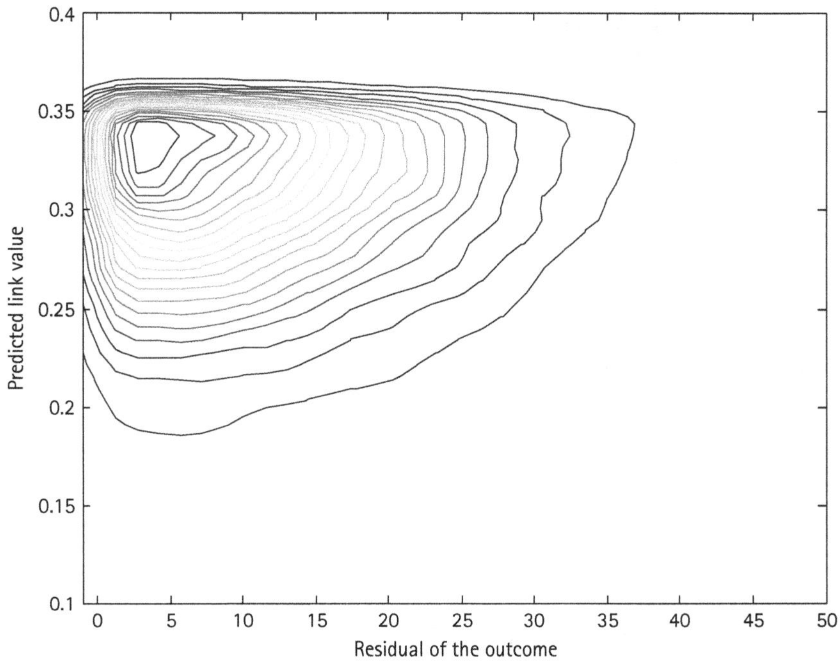

FIGURE 12.4 Kernel density estimates, exogenous model, no link.

Table 12.2 Regression on predicted value of link ($\hat{a}_{ij}$): exogenous version

Variable	Coefficient	(Std. Err.)
$\hat{\varepsilon}_{ij}$	0.035	(0.024)
$\hat{\varepsilon}_{ij} * a_{ij}$	-0.028	(0.042)
a_{ij}	7.187**	(0.617)
Intercept	301.958**	(0.342)

Joint test for $[\hat{\varepsilon}_{ij}] = [\hat{\varepsilon}_{ij} * a_{ij}] = 0$: $F(2, 49300) = 1.15$

Table 12.3 Hilbert-Schmidt Independence Criterion

Test	Stat.	Thresh(0.025)	$\phi_c = 0.05$
Exogenous ($a_{ij} = 0$)	0.3237	0.7434	0.6732
Exogenous ($a_{ij} = 1$)	0.3918	0.7367	0.6636
Endogenous ($a_{ij} = 0$)	1.2885	0.7925	0.7391
Endogenous ($a_{ij} = 1$)	1.3692	0.6557	0.5823

Note: For computational reasons, tests are performed on a 10% subsample.

12.5 Using Proxy Variables

As in many economic applications, the researcher does not always observe the "true" variable and must sometimes use a proxy. In the peer effects literature, prominent examples are the use of student achievement as a proxy for effort, or time studying (Sacerdote 2011), and body mass index (BMI) as a proxy for effort, or healthy life habits (Yakusheva et al. 2014). In this section, we use the example of BMI and present two model specifications. Since BMI is not directly transmitted from an individual to another nor can it be directly chosen by the individuals, peer effects will take source in the unobserved individual effort. As we will see, this distinction is important as it deeply affects the causal interpretation of peer effects.

We assume that individual's i's BMI (denoted by y_i) is linked to his observable characteristics, $\mathbf{x}_i$, his effort, e_i, and a random term, ν_i, in a linear way. The set of equations are given by:

$$\mathrm{y} = \mathbf{X}\boldsymbol{\psi} + \varrho \mathrm{e} + \boldsymbol{\nu} \tag{12.13}$$

where $\varrho < 0$. We also assume that preferences are as follows:

$$u_i(\mathrm{e},\mathrm{y}) = -(\mathbf{x}_i'\boldsymbol{\beta} + \varepsilon_i)y_i - \frac{e_i^2}{2} + \alpha e_i \mathbf{g}_i \mathbf{e} \tag{12.14}$$

The individual suffers a (linear) cost of being overweight (provided that $\mathbf{x}_i'\boldsymbol{\beta} + \varepsilon_i$ is positive) as well as a (quadratic) cost of making an effort. However, assuming complementarity, the cost of effort is lower if the individual's peers are also exerting effort ($\alpha > 0$).

Replacing (12.13) in (12.14) and solving for the first order conditions leads to: $\mathbf{e} = -\varrho\mathbf{X}\boldsymbol{\beta} + \alpha\mathbf{G}\mathbf{e} - \varrho\boldsymbol{\varepsilon}$. Note that this model only features endogenous interactions. Again, using (12.13), we can rewrite the model as:

$$\mathrm{y} = \mathbf{X}\boldsymbol{\theta} + \alpha\mathrm{G}\mathrm{y} - \alpha\mathbf{G}\mathbf{X}\boldsymbol{\psi} - \boldsymbol{\eta} \tag{12.15}$$

where $\boldsymbol{\theta} = [\boldsymbol{\psi} - \varrho^2\boldsymbol{\beta}]$ and $\boldsymbol{\eta} = \boldsymbol{\nu} - \alpha\mathrm{G}\nu - \varrho^2\boldsymbol{\varepsilon}$.

This model is then a special case of Liu and Lee (2010) so all the parameters are identified, with the exception of ϱ and $\boldsymbol{\beta}$.[33] This reflects the loss of information due to the fact that the econometrician does not observe the effort.

Remark that (12.15) results from a model with an *endogenous* peer effect on the effort only. There is *no* direct contribution of the peers' characteristics on the optimal level of effort (that is, no contextual effects). This is fundamental for the economic interpretation of the composite parameter $\alpha\boldsymbol{\psi}$. It is usually interpreted as a peer

[33] This is true for identification. However, their Cochrane-Orcutt type transformations cannot be directly applied, due to the shape of the errors in (12.15).

effect coefficient capturing the direct impact of the individuals' characteristics on the outcome. Here, however, it is simply a residual of the "true" social interaction parameter α, and not a second source of social interaction. The same reasoning holds for models of conformism.

12.5.1 Exogenous Effect

Consider the following alternative model:

$$u_i(\mathbf{e},\mathbf{y}) = -(\mathbf{x}_i'\boldsymbol{\beta} + \varepsilon_i)y_i - \frac{e_i^2}{2} - \frac{\lambda}{2}(y_i - \mathbf{g}_i\mathbf{y})^2 \tag{12.16}$$

Again, the individual suffers a cost of being overweight as well as a cost of making an effort. The individual also suffers a conformity cost on physical appearance. By substituting (12.13) in (12.16) and solving for the optimal choice of effort, we have:

$$\mathbf{e} = \mathbf{X}\boldsymbol{\zeta} + \tilde{\lambda}\mathbf{G}\mathbf{y} + \boldsymbol{\eta}$$

where $\boldsymbol{\zeta} = \frac{-\varrho}{1+\lambda\varrho^2}[\boldsymbol{\beta} + \lambda\boldsymbol{\psi}]$, $\tilde{\lambda} = \frac{\lambda\varrho}{1+\lambda\varrho^2}$, and $\boldsymbol{\eta} = \frac{-\varrho}{1+\lambda\varrho^2}[\boldsymbol{\varepsilon} + \lambda\boldsymbol{\nu}]$. We see that the optimal effort exerted by the individuals is a function of their peers' appearance. Rewriting the model in terms of the individual's BMI, we have:

$$\mathbf{y} = \mathbf{X}\breve{\boldsymbol{\zeta}} + \breve{\lambda}\mathbf{G}\mathbf{y} + \breve{\boldsymbol{\eta}}$$

where $\breve{\boldsymbol{\zeta}} = \frac{-1}{1+\lambda\varrho^2}[\varrho^2\boldsymbol{\beta} - \boldsymbol{\psi}]$, $\breve{\lambda} = \frac{\lambda\varrho^2}{1+\lambda\varrho^2}$, and $\breve{\boldsymbol{\eta}} = \frac{-1}{1+\lambda\varrho^2}[\varrho^2\boldsymbol{\varepsilon} - \boldsymbol{\nu}]$. Here, remark that the main parameter of interest, λ, is not identified. Identification results for linear models imply that $\frac{\varrho^2\boldsymbol{\beta}+\boldsymbol{\psi}}{1+\lambda\varrho^2}$ and $\frac{\lambda\varrho^2}{1+\lambda\varrho^2}$ are identified, but it is not possible to recover λ from those two expressions. This is important for the interpretation of the model in terms of causal impact of social interactions. The correlation between y_i and $\mathbf{g}_i\mathbf{y}$ (given by $\frac{\lambda\varrho^2}{1+\lambda\varrho^2}$) *does not* have an interpretation in terms of peer effects as it also captures the impact of the individual's effort on his BMI.[34] This point is related to Angrist's criticism (Angrist 2014), where the author is highly sceptical about the causal interpretation of the "peer effect" parameter $\breve{\lambda}$. Angrist's argument (see in particular Section 12.6) partly relies on the fact that one can always build a microeconomic model, free of any peer interaction, that will lead to a positive $\breve{\lambda}$, casting doubts on the causal interpretation of this parameter in terms of peer effects. Here, we argue that models of social interactions accounting for unobserved effort choices (as was carefully done by Cooley Fruehwirth 2013, for instance), brings credibility to the causal interpretation of the "social interaction" parameters.

[34] Note that testing $\breve{\boldsymbol{\lambda}} = 0$ allows to test for the *presence* of peer effects as $\breve{\boldsymbol{\lambda}}$ is proportional to the peer effect λ (assuming $\varrho < 0$). However, $\breve{\boldsymbol{\lambda}}$ does not provide an estimate of the size of the peer effect.

12.6 Conclusion

In this chapter, we first present a survey of recent developments and challenges in the empirics of social networks as they are related to policy analysis. We then discuss the concept of social multipliers (Glaeser et al. 2003). We show that in the standard linear-in-means model, the social multiplier is generally not identified except if we impose structure on the mechanisms of the social interactions (e.g., complementarity vs. pure conformity). We show that the presence of isolated individuals in the network helps to identify the social multiplier. However, we prove that estimating the linear-in-means model while including isolated individuals in the network will lead to biased estimators of peer effects when at least part the source of social interactions comes from conformity. We also provide a brief analysis of the key player (Borgatti 2006; Ballester et al. 2006) and how one can identify him/her within a network.

We then discuss three types of social networks: non-stochastic (or fixed), stochastic exogenous, and stochastic endogenous networks. We show that assuming that the network is non-stochastic or stochastic and exogenous has the same implication for the identification and the estimation of the model. However, the impact for the interpretation and policy recommendations flowing from the model are not the same. The reason is that when the network is stochastic and exogenous, an exogenous policy intervention may affect the structure of the network. This must be taken into account in the evaluation of the effect of the policy. As far as endogenous networks are concerned, we suggest performing a Hilbert-Schmidt Independence Test to check the exogeneity in the network formation.

Finally, we analyze situations where some choice variables of the model are unobservable to the researcher (e.g., healthy life habits), although proxy variables such as body mass index are observable. We show that, in this case, the issue of identifiability of the parameters crucially depends on the specification of the fundamentals of the model (preferences, technology, decision processes).

One basic message of our chapter is that in network econometric studies, more care should be taken in developing a microeconomic model based on sound theoretical underpinnings to determine the fundamentals that are identified. This theoretical *identifiability problem* (Chiappori and Ekeland 2009) is as important as the standard statistical *identification problem* (Manski 1993; Bramoullé et al. 2009; Blume et al. 2015) in order to verify whether estimated peer effects can be interpreted as true *causal effects*.

Further research is needed to determine the conditions under which one can recover the fundamentals of the social network microeconomic model from its corresponding structural econometric model. Indeed, knowledge of the fundamentals is required in order to allow the researcher to simulate the impact of hypothetical reforms on individual outcomes within the network. Moreover, our chapter shows the crucial importance of advancing research on testing for network endogeneity.

References

Babcock, P. and J. Hartman (2010). "Networks and workouts: Treatment size and status specific peer effects in a randomized field experiment." National Bureau of Economic Research Working Paper Series (16581).

Badev, A. (2013). "Discrete games in endogenous networks: Theory and policy." Mimeo.

Ballester, C., A. Calvó-Armengol, and Y. Zenou (2006). "Who's who in networks. Wanted: The key player." *Econometrica* 74(5), 1403–1417.

Barrera-Osorio, F., M. Bertrand, L. L. Linden, and F. Perez-Calle (2011). "Improving the design of conditional transfer programs: Evidence from a randomized education experiment in Colombia." *American Economic Journal: Applied Economics* 3(2), 167–195.

Bertrand, M., E. F. P. Luttmer, and S. Mullainathan (2000). "Network effects and welfare cultures." *The Quarterly Journal of Economics* 115(3), 1019–1055.

Beugnot, J., B. Fortin, G. Lacroix, and M.-C. Villeval (2014). "Social networks and peer effects at work." Mimeo.

Blume, L., W. Brock, S. Durlauf, and R. Jayaraman (2015). "Linear social interactions models." *Journal of Political Economy* 123(2), 444–496.

Blume, L. E., W. A. Brock, S. N. Durlauf, and Y. M. Ioannides (2011). *Identification of Social Interactions*. Vol. 1B of *Handbook of Social Economics*. North Holland, pp. 853–964.

Borgatti, S. P., (2006). "Identifying sets of key players in a social network." *Computational & Mathematical Organization Theory*, 12(1), pp. 21–34.

Boucher, V. 2014a. "Conformism and self-selection in social networks," Mimeo.

Boucher, V. (2015). "Structural homophily." *International Economic Review* 56(1), 235–264.

Boucher, V., Y. Bramoullé, H. Djebbari, and B. Fortin (2014). "Do peers affect student achievement? Evidence from Canada using group size variation." *Journal of Applied Econometrics* 29(1), 91–109.

Bramoullé, Y., H. Djebbari, and B. Fortin (2009). "Identification of peer effects through social networks." *Journal of Econometrics* 150(1), 41–55.

Bramoullé, Y. and B. Fortin (2010). "Social networks: Econometrics." *The New Palgrave Dictionary of Economics*, ed. S. N. Durlauf, S. N. and L. E. Blume. Online edition. Palgrave Macmillan.

Brin, S. and L. Page (1998). "The anatomy of a large-scale hypertextual web search engine." *Computer Networks and ISDN Systems* 30(1–7), 107–117.

Brock, W. A. and S. N. Durlauf (2001). "Discrete choice with social interactions." *The Review of Economic Studies* 68(2), 235–260.

Brock, W. A. and S. N. Durlauf (2007). "Identification of binary choice models with social interactions." *Journal of Econometrics* 140(1), 52–57.

Calvó-Armengol, A., E. Patacchini, and Y. Zenou (2009). "Peer effects and social networks in education." *The Review of Economic Studies* 76(4), 1239–1267.

Card, D. and L. Giuliano (2013). "Peer effects and multiple equilibria in the risky behavior of friends." *Review of Economics and Statistics* 95(4), 1130–1149.

Cassar, A. (2007). "Coordination and cooperation in local, random and small world networks: Experimental evidence." *Games and Economic Behavior* 58(2), 209–230.

Chandrasekhar, A. and M. Jackson (2013). "Tractable and consistent random graph models." mimeo.

Chandrasekhar, A. G. and R. Lewis (2011). "Econometrics of sampled networks." mimeo.

Chiappori, P.-A. and I. Ekeland (2009). "The microeconomics of efficient group behavior: Identification1." *Econometrica* 77(3), 763–799.

Christakis, N. A. and J. H. Fowler July (2007). "The spread of obesity in a large social network over 32 years." *The New England Journal of Medicine* 357(4), 370–379.

Christakis, N. A., J. H. Fowler, G. W. Imbens, and K. Kalyanaraman (2010). "An empirical model for strategic network formation," mimeo.

Cohen-Cole, E. and J. M. Fletcher (2008). "Is obesity contagious? Social networks vs. environmental factors in the obesity epidemic." *Journal of Health Economics* 27(5), 1382–1387.

Comola, M. (2008). "The network structure of informal arrangements: Evidence from rural tanzania." mimeo.

Comola, M. and S. Prina (2013). "Do interventions change the network? A dynamic peer effect model accounting for network changes." mimeo, Paris School of Economics.

Conti, G., A. Galeotti, G. Mueller, and S. Pudney, October (2012). "Popularity." Working Paper 18475, National Bureau of Economic Research.

Cooley Fruehwirth, J. C. (2013). "Identifying peer achievement spillovers: Implications for desegregation and the achievement gap." *Quantitative Economics* 4(1), 85–124.

Davezies, L., X. d'Haultfoeuille, and D. Fougère (2009). "Identification of peer effects using group size." *Econometrics Journal* 12, 397–413.

Dieye, R., H. Djebbari, and F. Barrera-Osorio (2014). "Accounting for peer effects in treatment response." mimeo.

Fafchamps, M. and F. Gubert (2007). "Risk sharing and network formation." *American Economic Review* 97(2), 75–79.

Fortin, B. and M. Yazbeck (2015). "Peer effects, fast food consumption and adolescent weight gain." *Journal of Health Economics* 42, 125–138.

Galbiati, R. and G. Zanella (2012). "The tax evasion social multiplier: Evidence from Italy." *Journal of Public Economics* 96(56), 485–494.

Glaeser, E. L., J. A. Scheinkman, and B. I. Sacerdote (2003). "The social multiplier." *Journal of the European Economic Association* 1(2/3), 345–353.

Goldsmith-Pinkham, P. and G. W. Imbens (2013). "Social networks and the identification of peer effects." *Journal of Business and Economic Statistics* 31(3), 253–264.

Graham, B. S. (2008). "Identifying social interactions through conditional variance restrictions." *Econometrica* 76, 643–660.

Graham, B. S. (2014). "Methods of identification in social networks," mimeo.

Gretton, A., O. Bousquet, A. Smola, and B. Schölkopf (2005). "Measuring statistical dependence with hilbert-schmidt norms." In: *Algorithmic Learning Theory*, S. Jain, H. U. Simon, and E. Tomita, eds., ALT 2005, LNAI 3734, pp. 63–77. Berlin Heidelberg: Springer-Verlag.

Hoff, P. D., A. E. Raftery, and M. S. Handcock (2002). "Latent space approaches to social network analysis." *Journal of the American Statistical Association* 97(460), 1090–1098.

Hsieh, C.-S. and L.-F. Lee (2015). "A social interactions model with endogenous friendship formation and selectivity." *Journal of Applied Econometrics.* Forthcoming

Jackson, M. O. (2010). *Social and Economic Networks.* Princeton, NJ: Princeton University Press.

Jackson, M. O. (2011). "An overview of social networks and economic applications." *The Handbook of Social Economics*, ed. Jess Benhabib, Alberto Bisin, and Matthew O. Jackson. North Holland: Elsevier Press.

Karlan, D., M. Möbius, T. Rosenblat, and A. Szeidl (2009). "Trust and social collateral." *The Quarterly Journal of Economics* 124(3), 1307–1361.

Kwok, H. H. (2013). "Identication problems in linear social interaction models: A general analysis based on matrix spectral decompositions," mimeo.

Lee, L. (2007). "Identification and estimation of econometric models with group interactions, contextual factors and fixed effects." *Journal of Econometrics* 140(2), 333–374.

Lee, L.-f., J. Li, and X. Lin (2014). "Binary choice models with social network under heterogeneous rational expectations." *Review of Economics and Statistics* 96(3), 402–417.

Lee, L.-f., X. Liu, and X. Lin (2010). "Specification and estimation of social interaction models with network structures." *The Econometrics Journal* 13(2), 145–176.

Lin, X. (2010). "Identifying peer effects in student academic achievement by spatial autoregressive models with group unobservables." *Journal of Labor Economics* 28(4), 825–860.

Liu, X. and L.-f. Lee (2010). "Gmm estimation of social interaction models with centrality." *Journal of Econometrics* 159(1), 99–115.

Liu, X., E. Patacchini, and E. Rainone (2013a). "The allocation of time in sleep: A social network model with sampled data." Mimeo.

Liu, X., E. Patacchini, and Y. Zenou (2013b). "Peer effects: Social multiplier or social norms?" Mimeo.

Liu, X., E. Patacchini, Y. Zenou, and L.-F. Lee (2014). "Criminal networks: Who is the key player?" mimeo.

Manski, C. (2013). "Identification of treatment response with social interactions." *Econometrics Journal* 16(1), S1–S23.

Manski, C. F. (1993). "Identification of endogenous social effects: The reflection problem." *The Review of Economic Studies* 60(3), 531–542.

Manski, C. F. (2000). "Economic analysis of social interactions." *Journal of Economic Perspectives* 14(3), 115–136.

Mas, A. and E. Moretti (2009). "Peers at work." *The American Economic Review* 99(1), 112–145.

Maurin, E. and J. Moschion (2009). "The social multiplier and labor market participation of mothers." *American Economic Journal: Applied Economics* 1(1), 251–272.

Mayer, A. and S. L. Puller (2008). "The old boy (and girl) network: Social network formation on university campuses." *Journal of Public Economics* 92(12), 329–347.

Mele, A. (2013). "A structural model of segregation in social networks." mimeo.

Moffitt, R. A. (2001). *Policy Interventions, Low-Level Equilibria, and Social Interactions.* MIT Press, 45–82.

Patacchini, E. and Y. Zenou (2008). "The strength of weak ties in crime." *European Economic Review* 52(2), 209–236.

Sacerdote, B. (2001). "Peer effects with random assignment: Results for Dartmouth roommates." *Quarterly Journal of Economics* 116(2), 681–704.

Sacerdote, B. (2011). "Peer effects in education: How might they work, how big are they and how much do we know thus far?" Vol. 3 of *Handbook of the Economics of Education.* Elsevier, Amsterdam: North Holland. Ch. 4, pp. 249–277.

Trogdon, J. G., J. Nonnemaker, and J. Pais (2008). "Peer effects in adolescent overweight." *Journal of Health Economics* 27(5), 1388–1399.

Yakusheva, O., K. A. Kapinos, and D. Eisenberg (2014). "Estimating heterogeneous and hierarchical peer effects on body weight using roommate assignments as a natural experiment." *Journal of Human Resources* 49(1), 234–261.

Yang, C. and L. Lee (2014). "Social interactions under incomplete information with heterogeneous expectations." mimeo.

APPENDIX

Proof 1 (of Proposition 1). Using the fact that $(\mathbf{1}+\alpha\mathbf{G})^{-1}=\sum_{k=0}^{\infty}\alpha^k\mathbf{G}^k$, we have: $\mathbf{y}^u(c_1)=\mathbf{y}^u(c_0)+c\sum_{k=0}^{\infty}\alpha^k\mathbf{G}^k\mathbf{1}$ and $\mathbf{y}^v(c_1)=\mathbf{y}^v(c_0)+\frac{c}{1+\lambda}\sum_{k=0}^{\infty}\tilde{\alpha}^k\mathbf{G}^k\mathbf{1}$. Since $\mathbf{1}$ is an eigenvector of $\mathbf{G}$ associated with the eigenvalue 1, this leads to $\mathbf{y}^u(c_1)=\mathbf{y}^u(c_0)+c\sum_{k=0}^{\infty}\alpha^k\mathbf{1}=\mathbf{y}^u(c_0)+\frac{c}{1-\alpha}\mathbf{1}$ and $\mathbf{y}^v(c_1)=\mathbf{y}^v(c_0)+\frac{c}{1+\lambda}\sum_{k=0}^{\infty}\tilde{\alpha}^k\mathbf{1}=\mathbf{y}^v(c_0)+c\mathbf{1}$, or $\Delta\mathbf{y}^u=\frac{c}{1-\alpha}\mathbf{1}$ and $\Delta\mathbf{y}^v=c\mathbf{1}$.

CHAPTER 13

ECONOMETRICS OF NETWORK FORMATION

ARUN G. CHANDRASEKHAR

13.1 Introduction

A growing empirical literature examines the role that networks play in a variety of economic phenomena. This research can be crudely partitioned into two branches. The first holds networks fixed and studies economic processes operating on networks (e.g., social learning, labor market search, peer effects in education). This is the subject of several other chapters in this handbook (e.g., "Some challenges in the empirics of the effects of networks," Boucher and Fortin, Chapter 12 in this volume; "Field experiments, social networks, and development," Breza, Chapter 16 in this volume; "Networked experiments: a review of methods and innovations," Aral, Chapter 15 in this volume). The second branch considers how and why economic networks form. To date, most work in this space has been theoretical, with very limited observational or experimental work on network formation. One major impediment to the progress of this research is the dearth of practicable empirical models of network formation.

This chapter will focus on the econometric issues that arise when an empirical researcher seeks to model network formation. A key goal of an empirical analysis of network formation is to estimate how and why relationships form. What governs the overall structure of the economy? Are there structural regularities that tell us something about the economic incentives underlying the interaction? The researcher may want to

I thank Emily Breza, Gabriel Carroll, Aureo de Paula, Bryan Graham, Han Hong, Itay Fainmesser, Paul Goldsmith-Pinkham, Matthew Jackson, Cynthia Kinnan, Michael Leung, Elena Manresa, Mounu Prem, Xiao Yu Wang, and Juan Pablo Xandri for helpful discussions.

estimate parameters driving network formation in order to (i) test economic hypotheses and (ii) conduct counterfactual analysis.[1]

A major difficulty is that the researcher typically has a data set consisting of a single network observed in a single period (e.g., social relationships among students in a single university, financial relationships between households in a single village). One reason that this is often the case is due to the prohibitive cost in collecting high-quality network data (see, e.g., Banerjee, Chandrasekhar, Duflo, and Jackson 2013). Thus, the relevant thought experiment is the following: given a single network of n agents, are the parameters driving the joint distribution of agents' linking decisions, their characteristics and possibly the economic environment (including observables, unobservables, etc.) consistently estimable as $n \to \infty$? This estimation problem involves a number of challenges, described below, including but not limited to the degree of correlation between linking decisions, concerns about multiple equilibria, and missing data. The intersection of such concerns arise in the network setting.

To highlight the core difficulty, that one large network may encode a tremendous amount of information or a very limited amount of information, observe the following. Suppose the researcher knows each pair of agents is linked with probability p, where p is the parameter to be estimated. We can imagine two extremes. At one extreme, any two pairs of agents are independent of each other. Then a single network effectively provides $\binom{n}{2}$ independent observations, so consistent estimation of p is trivial. At the other extreme, all pairs might be perfectly correlated (i.e., with probability p all agents are linked), and with probability $1-p$ no two agents are linked. Then it is clear that having a single network, no matter how large, is not enough to estimate p consistently.

Consider an economy with a set V of agents, whom we refer to as *nodes*. Assume there are $n = |V|$ nodes and each needs to decide whether or not to develop a relationship (e.g., a transaction, a friendship) with others. The resulting network of links, $g = (V,E)$, is called a *graph*, which consists of a set of nodes V and a set of edges E. If nodes share an edge, $ij \in E$. It is useful to represent this with an *adjacency matrix* $\mathbf{A} := \mathbf{A}(g)$, with

$$A_{ij} = \begin{cases} 1 & \text{if} \quad ij \in E \\ 0 & \quad\quad ij \notin E. \end{cases}$$

For simplicity, assume that if a link from i to j exists, then the link from j to i exists (networks are undirected), and a link either exists or it doesn't (the network is unweighted).[2]

We are interested in modeling the formation of g, in an economy where each agent may have some attributes and incentives to form relationships with other nodes.

Typically, the researcher has observations from a single network with many nodes (e.g., a university, a village). The researcher may also have a vector of covariates, $x = (x_1,\ldots,x_n)$. The goal is to use the data $(\mathbf{A},x)$ to estimate the parameters of network

[1] Another practical reason to develop network formation models is to be able to deal with missing data since data collection can be prohibitive.

[2] Most of what follows applies for directed and weighted graphs as well.

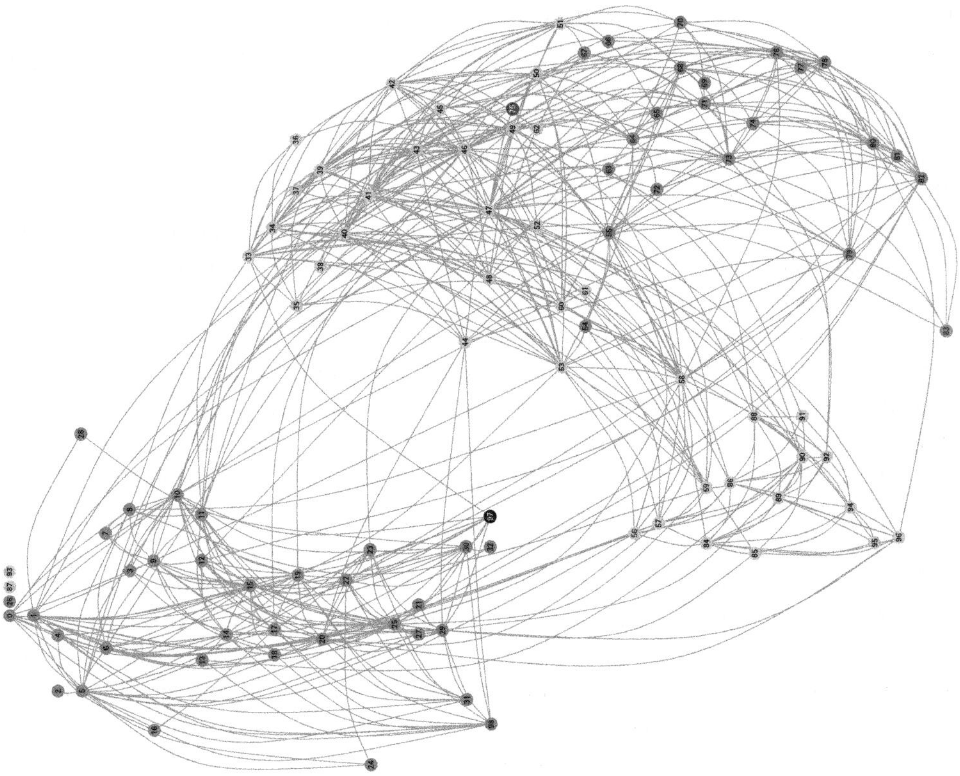

FIGURE 13.1 This figure, taken from Chandrasekhar, Kinnan, and Larreguy (2014) displays a village network from Karnataka, India. Nodes are households and are colored by caste. Edges denote whether members of households socialize with each other or engage in financial transactions.

formation from the observation of a single, large network. The challenge is that with a single network, there is potentially a lot of interdependency between the links. In sum, the researcher has a draw from

$$\mathrm{P}^n_\beta(\mathbf{A}|x),$$

which assigns likelihood to each of the possible $2^{\binom{n}{2}}$ graphs.

How much information is contained in such data for the econometrician? The crucial question is the extent to which the an event (e.g., i is linked to j but k is not linked to l), influences the likelihood of some other u and v being linked. It is useful to think of a vector of outcomes, the linking decisions for all $\binom{n}{2}$ pairs of nodes:

$$\mathrm{vec}(\mathbf{A}) = \left(\underbrace{1}_{\mathrm{A}_{1,2}}, \underbrace{0}_{\mathrm{A}_{1,3}}, \ldots, \underbrace{[\,?\,]}_{\mathrm{A}_{ij}}, \ldots, \underbrace{[\,?\,]}_{\mathrm{A}_{n-1,n}} \right).$$

Here we see the two extremes. If links are all independent, there are effectively $\binom{n}{2}$ observations in a given network.[3] However, if all links are completely correlated, the researcher actually has only one effective draw from the probability distribution, no matter the n.

In practice, the distributions we are trying to model may live somewhere in between. There is likely to be considerable interdependency among links, both for strategic reasons and due to (observable and unobservable) characteristics, but there may also be enough independence such that consistent estimation of the model's parameters is possible. This is very much like in the analysis of time-series or spatial data under standard assumptions: draws from adjacent periods or locations are highly correlated but draws that are far enough apart temporally are near-independent. A crucial distinction between the network case and the time-series or spatial case is that time-series or spatial contexts have natural embeddings into some metric space whereas in the network case there may not be such an embedding. Time has past, present, and future, and spatial data inherits, by construction, a natural geometry. For a network, there is usually no natural embedding of nodes into some space.

Our goal is to develop econometric models of network formation and have estimators that provide consistent estimates of the joint distribution, that are tractable, and that allow the researcher to conduct inference. Econometrically modeling network formation presents several challenges, some though not all unique to a networks setting. First, the econometric model must be tractable and have good asymptotic properties. Parameter estimates should be consistent, for instance, and estimating them should be feasible. The researcher faces a large network, which means that there are $2^{\binom{n}{2}}$ potential (undirected) networks. This can make it very difficult to conduct feasible estimation. Various techniques explored in the literature involve drawing a sample of networks over this immense, discrete space, or exploring the space of networks to determine bounds on parameter estimates. The enormity of the space can make it difficult to deal with.[4]

Second, the econometric model should reflect the structure of the network relevant to the economic problem. It is essential that the models used by the researcher reflect meaningful patterns in the data, which likely correspond to underlying economic forces. It is crucial, for instance, that network models used in applied work are consistent with stylized facts. A theoretical challenge here, as the reader will see, is that many natural network formation models are inconsistent with basic facts corresponding to empirical network data. In this chapter, we focus on the fact that economic networks (i) are sparse, and (ii) exhibit correlation in the location of links.

Third, since the decisions to link can be thought of as a discrete choice, endogeneity and the presence of unobserved heterogeneity present identification issues. For

[3] Estimation would be straightforward assuming away complications caused by unobserved heterogeneity, which we discuss below.

[4] Insights from empirical data can help here. While there are many possible networks and certain models are unable to be estimated in a computationally feasible model, conditioning on classes of models that replicate certain empirical features such as sparsity may reduce the problem significantly.

instance, network models, by their very nature, put outcome variables—whether or not certain links are built—on both the left- and right-hand side of equations. The event that $ij \in E$ can depend on the event that $kl \in E$ and vice versa. The structural equations may have multiple solutions, and therefore, issues surrounding multiple equilibria may need to be confronted (Bresnahan and Reiss 1991; Tamer 2003).

In this chapter, we will attempt to survey the literature and take stock of the approaches developed to deal with each of these issues. The models used in the literature often typically map to some preferences for agents over networks. A fairly general way to describe these models is

$$u_i(g,x,\epsilon;\beta),$$

where x is a vector of (observable and unobservable) agent attributes and ϵ is a vector of unobservable shocks. Let $g+ij$ be shorthand for the graph $(V,E\cup\{ij\})$ and similarly $g-ij$ be the shorthand for the graph $(V,E\setminus\{ij\})$. The basic decision individuals are making can be thought of as evaluating their marginal utility from maintaining or dropping a candidate link ij,

$$U_i(ij,g,x,\epsilon;\beta) := u_i(g+ij,x,\epsilon;\beta) - u_i(g-ij,x,\epsilon;\beta).$$

Therefore, the researcher is tasked with deciding how interdependent her model will be by making choices over specifications of preferences and disturbances. For instance, if

$$u_i(g,x,\epsilon;\beta) = \sum_j (\beta - \epsilon_{ij})A_{ij}$$

then it is easy to see that the decisions to form links ij and ik are independent, if shocks were independent.

On the other hand, preferences may be more complex. For instance, consider

$$u_i(g,x,\epsilon;\beta) = \sum_{j\neq i} (\beta_L - \epsilon_{ij})L_{ij} + \sum_{j<k\neq i} (\beta_T - \epsilon_{ijk})T_{ijk}$$

where L_{ij} denotes a link ij that is not part of a triangle ($A_{ij}=1$, $A_{ik}A_{jk}=0$ for all k) and T_{ijk} denotes a triangle (i,j,k mutually linked). Here, the agent receives different payoffs for being in different types of relationships. Having a direct friend may be less beneficial than having a friend who is also a mutual friend of another friend. This is related to a simple model of favor exchange, in which being part of a group of three generates a payoff, whereas having a link or two friends who themselves are unlinked may generate no payoff (Jackson, Barraquer, and Tan 2012).

Of course, there is nothing special in writing a model in which agents receive payoffs from links and from triangles. One could imagine writing a model based on utility from being in various subgraphs $g_\ell \subset g$ and payoff functions $v(g_\ell,x_\ell,\epsilon_\ell;\beta_\ell)$, which can depend on the covariates of those nodes involved in the subgraph $x_\ell = (x_{1,\ell},\ldots,x_{n,\ell})$:

$$u_i(g,x,\epsilon;\beta) = \sum_{i\in g_\ell} v(g_\ell,x_\ell,\epsilon_\ell;\beta_\ell).$$

How these preferences are specified, how an equilibrium is selected, and the nature of the errors all will influence to a large degree the ability of the model to both reflect realistic features of empirical data while also delivering estimators that are feasible and enable us to conduct inference.

This chapter makes a modest contribution to a growing set of review articles on network formation models across many disciplines (e.g., Albert and Barabasi 2002; Goldenberg, Zheng, Fienberg, and Airoldi 2010; Blume, Brock, Durlauf, and Ioannides 2010; Graham 2014b). Most recently, Graham (2014b) surveys methods of identification in social networks, paying particular attention to distinguishing between structural transitivity (correlation in links due to payoff incentives) versus homophily (correlation in links due to unobserved individual characteristics). My goal here is to acquaint readers, in a self-contained manner, with a number of network formation models used in the graph theory, statistics, sociology, and econometrics literatures, with a view to both how well they map to real-world data as well as the sorts of microfoundations they implicitly assume. I hope this serves as a toolbox and allows them to survey a wide set of modeling choices to see what might be useful for their specific aims.

The remainder of the chapter is organized as follows. In Section 13.2, I review some common network-based data sets used in the literature and document key stylized facts of empirical networks that are useful to bear in mind when thinking about modeling decisions.

Section 13.3 looks at models that are conditionally edge-independent. In these models the linking decisions only depend on agents' attributes and (independent) shocks, but a link elsewhere in the graph, conditional on these attributes and shocks, has no correlation with a considered link. The main concepts dealt with here pertain to heterogeneity. Agents may be members of unobserved groups and may even have unobserved heterogeneity that systematically influences their otherwise independent linking decisions. Such concerns, even without introducing interdependence in linking decisions, can generate challenging econometric problems.

Section 13.4 considers models that allow for higher order dependence between links.[5] In those models, agents' payoffs can depend on links elsewhere in the graph in a direct manner. This renders some models used in the literature entirely unusable, because estimators may not be feasible or consistent for the parameters they are aiming to recover. Models that succeed, of course, either limit the degree to which preferences across agents can be entangled by maintaining that payoffs can come from subgraphs that are not too large, or generate exogenous reasons why agents that are in some sense far enough away can never link to each other and therefore generate a form of independence that way.

Section 13.5 concludes and provides a discussion of the tradeoffs explored throughout the chapter.

[5] Surely models of unobserved heterogeneity and models of higher order dependence can be merged.

Table 13.1 Properties of various network data sets

Data Source	Average Degree	Density	Clustering	Average Path Length	Number of Networks	Average Number of Nodes
Add Health	3.361	0.042	0.411	-	14	80.36
China Villages	3.266	0.113	-	2.578	185	28.82
Harvard Dorm	10.830	0.019	0.181	3.260	1	569
Harvard Facebook-Based	7.917	0.003	0.174	4.571	1	2360
Karnataka Villages	17.378	0.089	0.303	2.337	75	213.37
Malawi Villages	6.189	0.061	0.234	3.089	21	133.48
Uganda Village	8.300	0.069	0.230	2.500	1	119

13.2 Stylized Facts

Let us first turn our attention to some data sets in order to establish stylized facts. I will focus on two themes:[6] (1) the density of links, and (2) the correlation in link presence.

Development economics and labor economics are two subfields that have made strong contributions to empirical network economics. Therefore, I focus on sets that make use of social network data from villages and schools. The data includes networks from the Add Health data set (the sample used in Badev 2013), Harvard social network data (used in Leider et al. 2009; Mobius et al. 2010), and several rural network data sets from the developing world (Cai et al. 2012; Banerjee et al. 2014; Ngatia 2011; De Weerdt and Dercon 2006). A link in the village data typically represents whether a member of one household has a social, financial, informational, or familial relationship with a member of another household. A link in the U.S. university/school data represents a friendship link. The results are presented in Table 13.1.

13.2.1 Density

The *density* of a graph is the share of possible links that are present:

$$density(g) := \frac{1}{\binom{n}{2}} \sum_{i<j} A_{ij}.$$

[6] The literature has focused on a number of other stylized facts about empirical networks. For example, networks have short paths and exhibit both positive correlation in degree and negative correlation between a node's degree and clustering. See Jackson (2008) or Jackson and Rogers (2007a) for further discussion.

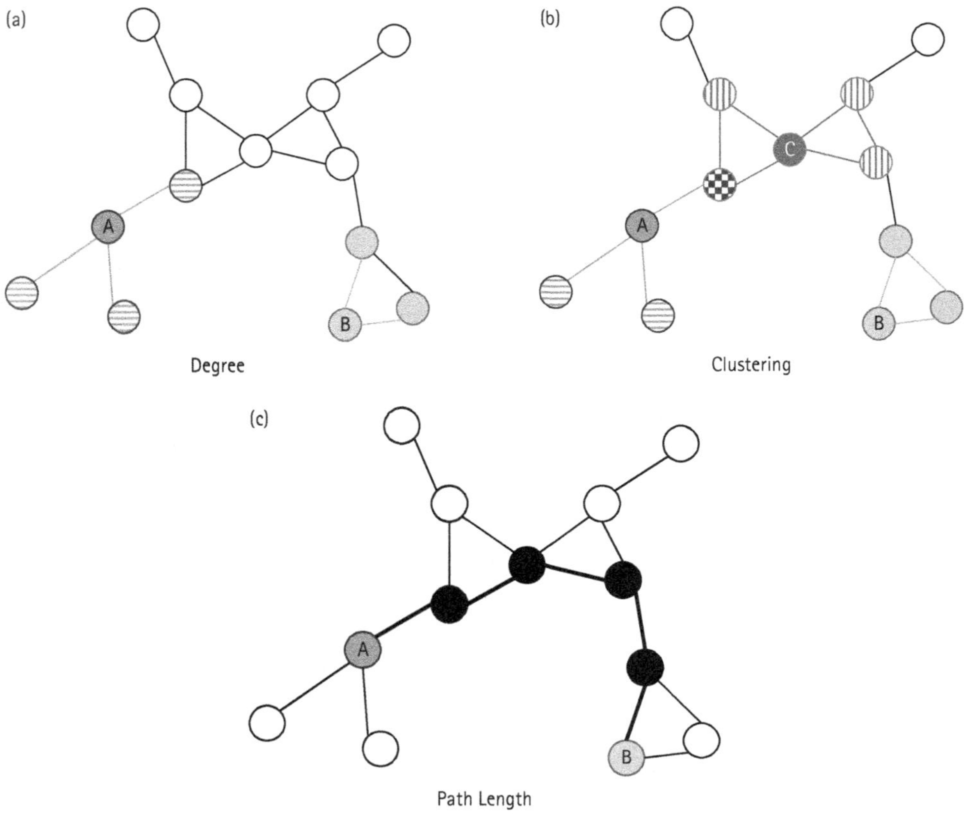

FIGURE 13.2 Illustrates network concepts. (A) Degrees are $d_A(g) = 3$ and $d_B(g) = 2$. (B) Clustering given by $c_A(g) = 0$, $c_B(g) = 1$, and $c_C(g) = 2/\binom{4}{2} = 1/3$. (C) Path length is $\gamma(A, B) = 5$ and $diam(g) = 6$.

It is useful to introduce some related concepts. The *degree* of a node is the number of neighbors it has and is the simplest measure of how central a node is in the network. A simple way to describe a network is to look at its *degree sequence*. This is the vector of degrees across all nodes:

$$d(g) = \sum_j A_{ij} = (d_1(g), \ldots, d_n(g))'.$$

Figure 13.2 illustrates this concept.

The *degree distribution* is the CDF of the degree sequence,

$$F_d(x) := \frac{|\{i : d_i \le x\}|}{n}.$$

It plays a central role in parsimonious models of random matching and diffusion of behavior (Jackson and Rogers 2007b; Jackson and Yariv 2011; Galeotti and Rogers

2013). Further, a large body of literature, mostly outside of economics, has paid considerable attention to the *shape* of the degree distribution. The researchers deal with questions as to whether there are "skinny tails" or "fat tails," as a means to examine whether there are particularly well-connected nodes in the network (e.g., the survey by Albert and Barabasi 2002).

Unsurprisingly, the mean of the degree distribution, the *average degree*, is a commonly used summary of how well-connected a network is

$$\bar{d}(g) := n^{-1} \sum_{i} \sum_{j} A_{ij}.$$

It is easy to see that in fact, the density is just a normalization of the average degree:

$$density(g) = \frac{1}{\binom{n}{2}} \sum_{i<j} A_{ij} = \frac{2}{n-1} \bar{d}(g).$$

A useful way to think about the density (or the expected degree) of a network is to do an asymptotic thought experiment. Imagine a sequence of distributions that generate networks, defined over a growing population of size n. Heuristically, we say that the sequence is *sparse* when the expected density converges to zero, since the graph is mostly empty. The usual case we imagine is one where the expected average degree is constant in n: individuals in large societies do not have many more links on average than in smaller societies. Notice that as long as $\mathrm{E}\bar{d} = o(n)$, under the density tends to zero.

On the other hand, we say that the sequence is *dense* if a typical node is linked to a number of nodes that is proportional to the number of potential partners in society. Dense graphs are those such that the expected average degree $\mathrm{E}\bar{d}^n \propto n$.

Throughout the literature, researchers vary as to what they mean when they say networks are sparse or dense. Sometimes sparsity refers to $\mathrm{E}\bar{d} = O(1)$, and sometimes sparsity refers to $\mathrm{E}\bar{d} = o(n)$. Correspondingly, sometimes density refers to $\mathrm{E}\bar{d} = \omega(1)$, and sometimes density refers to $\mathrm{E}\bar{d} = O(n)$. In this chapter I will be explicit about the rates when they are needed. Of course, these are all fictional constructions used for asymptotic thought experiments. After all, any realized network in the data is drawn from a finite sample.

Let us now turn to the empirical data. We can see from Table 13.1 that the average degree is low across all data sets (column 1). For instance, the Karnataka data has an average degree of 17 in villages that average 213 households. This is true also when we turn to the education data. The average degree in the Harvard dorm network is 10.8. Correspondingly, the density rates are rather low as well. Linking probabilities are in the 5% range, if not lower, and the highest probabilities are coming in the smallest networks. Therefore it seems reasonable that a researcher may want to use as a starting point assumptions that deliver sparse networks.

As we will discuss in later sections, sparsity can arise for a variety of natural reasons in economic settings: a lack of meeting opportunities, a budget constraint on time or resources, efficiency concerns, etc. It is important, therefore, that good network formation models respect this property of the data.

13.2.2 Clustering

The next concept concerns link correlation. If i has two neighbors j and k, what is the likelihood that j and k are linked? This is given by the *clustering coefficient*:

$$c(g) = n^{-1} \sum_i c_i(g), \; c_i(g) = \frac{\sum_{j<k} A_{ij} A_{ik} A_{jk}}{\binom{d_i(g)}{2}}.$$

A related measure looks at the global clustering coefficient. It addresses what share of all triples with ij linked and ik linked involve jk linked.

$$c^{global}(g) = \frac{\sum_{i<j<k} A_{ij} A_{jk} A_{ik}}{\sum_{i<j<k} 1\{A_{ij}A_{jk} + A_{ij}A_{ik} + A_{jk}A_{ik} > 0\}}.$$

Figure 13.2 computes clustering for three nodes, A, B, and C, which have a clustering of 0, 1, and 1/3 respectively.

Theoretical research has devoted considerable attention to addressing why agents may find it economically beneficial to have clustered links. First, there may be correlated tastes. Second, correlations in links could allow agents to sustain cooperation that they otherwise would not be able to. If i violates an agreement with j, then if they have a friend in common, k, then k could be enlisted to punish i as well (see Bloch et al. 2008; Karlan et al. 2009; Jackson et al. 2012).

Turning to the data, we see that empirical networks exhibit high clustering, in the range of 0.2 to 0.3. Note that if every link was drawn independently with a probability given by the link density (e.g., 0.07), then the implied clustering should be 0.07. The empirical clustering rates are 3 to 4 times as high. This suggests that the underlying economic processes generate incentives to correlate links much more than would be expected at random.

In many ways, it should be unsurprising that real-world networks have high clustering. After all, if there are incentives to link to a friend of a friend—perhaps it allows agents to better sustain cooperation—then we should see either three nodes all mutually linked or no linking at all.

13.3 Conditional Edge-Independence

Let us begin by looking at classes of models where each pair of individuals decide whether or not to link based only on their attributes and shocks. Conditional on these factors, the event that link ij forms is independent of the event that jk forms.

Under such an assumption, a simple model we could explore is one in which the marginal utility that i accrues by linking to j is given by

$$u_i(g+ij) - u_i(g-ij) = f(x_i, x_j) - \epsilon_{ij},$$

where x_i is a vector of individual attributes and ϵ_{ij} is a pair-level shock. Whether a link forms is given by

$$A_{ij} = 1\{f(x_i, x_j) - \epsilon_{ij} > 0\}. \tag{13.1}$$

For simplicity, let us take $f(\cdot,\cdot)$ to be symmetric, $\epsilon_{ij} = \epsilon_{ji}$, and ϵ_{ij} as drawn i.i.d. across the $\binom{n}{2}$ pairs with Type I extreme value. It follows that

$$\mathrm{P}(A_{ij}|x_i, x_j) = \frac{\exp f(x_i, x_j)}{1 + \exp f(x_i, x_j)}.$$

Under suitable assumptions on $f(\cdot,\cdot)$—for instance, the probability of a link increases in the similarity of i and j, $f(x_i, x_j) = -\beta|x_i - x_j|$—this allows for straightforward estimation of linking probabilities.[7]

In general, let us say $\epsilon_{ij} \sim \Lambda(\cdot)$, and so we have

$$\mathrm{P}(A_{ij}|x_i, x_j) = \Lambda(f(x_i, x_j)).$$

Such a model does not include individual-level unobserved heterogeneity, and this might be a serious limitation. After all, we think individuals who are altruistic may naturally be more likely to make friends, and this unobserved trait might systematically alter the distribution of links even conditional on observed characteristics.

To get at questions such as this, we could imagine adding in individual-level unobserved heterogeneity in the model. Consider extending the above to

$$u_i(g+ij) - u_i(g-ij) = f(x_i, x_j) + \nu_i + \nu_j - \epsilon_{ij},$$

where ν_i are unobserved fixed effects for each individual. Then the probability that a link ij forms is given by

$$p_{ij}(x_i, x_j) = \Lambda(f(x_i, x_j) + \nu_i + \nu_j).$$

The challenge for the researcher in this case is to not only estimate $f(\cdot,\cdot)$, but to do so either (a) in the presence of nuisance parameters ν_i, or (b) estimate ν_i as well in case they are of economic interest. They certainly will be if the researcher is interested in estimating counterfactual distributions of networks.

[7] One could simply do a logistic regression of A_{ij} on $|x_i - x_j|$.

13.3.1 Erdős-Rényi Model

To understand basic properties of conditional edge independent models, it is useful to remove covariates from the utility functions. It is easy to see that every link forms independently with common probability p, since

$$\mathrm{P}(A_{ij}=1)=\mathbf{1}\{\alpha>\epsilon_{ij}\}$$

for some constant α.

This gives us the Erdős-Rényi random graph model, which is easily the most studied graph formation model (Erdős and Rényi 1959). The Erdős-Rényi model has a number of special properties that correspond to specific restrictions on the resulting network. I focus on three implications here.

1. The expected number of links that a given node has is $(n-1)p$.
2. The probability that two neighbors of i, j, and k are linked is p.
3. The probability that i, j, and k are mutually linked is given by p^3.

While this is certainly not an exhaustive list of properties of Erdős-Rényi networks, it highlights the following tension that is relevant for empirical work. First, as noted above, empirical networks are often sparse. What this means is that the expected number of links that a node has is far smaller than n. When choosing how to model this, asymptotically, we typically think of a sequence $p=p_n$, which decreases in n. For example, if $p_n=d/n$, then for very large graphs, the expected degree is d irrespective of the size of the network.

Second, empirical networks are often clustered. Since clustering is essentially the share of times that jk forms given that ij and ik are links, notice that in this model, clustering has to be p_n.

Thus, if networks exhibit sparsity, so p_n is any sequence that tends to zero, then by definition there is no clustering in the limit. On the other hand, if networks exhibit non-vanishing clustering, there exists a lower bound $\underline{p}$ for the linking probability, bounded away from zero. In this case, the expected degree is at least $(n-1)\underline{p}$, which grows linearly in n. This implies that in networks of size 1000 versus 100, individuals will be linked to 10 times as many people.

It is worth making two observations about this model. First, even if the implied network formation parameter p_n is changing in n, the researcher may be able to estimate it well in the sense that (a) $\left|\widehat{p}/p_n-1\right|\xrightarrow{\mathrm{P}}0$ and (b) that, when appropriately normalized, the parameter estimate is asymptotically normally distributed

$$\sqrt{\frac{\binom{n}{2}}{p_n(1-p_n)}}(\widehat{p}-p_n)\rightsquigarrow\mathcal{N}(0,1),$$

under mild assumptions. The key assumption is that p_n does not tend to zero too fast, with the sharp threshold being that $p_n\cdot\binom{n}{2}\to\infty$.

Second, how shall we interpret p_n tending to zero when it comes to looking at microfoundations? One possible view is to think about a world where agents have (essentially) a selectivity threshold because they have a fixed budget of time. The population with which an agent interacts is indexed by ϵ_{ij}, which gives them the quality of the match. In larger populations, one can screen more selectively, and this would correspond to α_n being increasingly negative along the sequence, implying that in larger networks, higher shocks are required to justify a link. Another interpretation is that individuals have a fixed time budget and therefore can only ever meet a small subset of the population n, irrespective of n.

All remaining models in this section will add different layers of heterogeneity in terms of observables and unobservables. We will explore the degree to which the aforementioned trade-off between sparsity and clustering can or cannot be alleviated by these approaches.

13.3.2 Random Geometric Graphs

An alternative to the Erdős-Rényi models widely studied are random geometric graphs (Penrose 2003). Here nodes are stochastically placed in some embedding space and then link to each other if they are close enough in this space. For instance, if nodes are formed close enough in education-wealth space, then they link, otherwise they do not. The crucial point is that this places geometric restrictions, from the embedding space, on the topologies of the networks likely to arise. For nodes that are close enough, it may be impossible to have any subgraph between them except a clique.

A simplified version of the model is as follows. Let n nodes be assigned locations uniformly at random in $[0,1]^2$. A link exists between a pair if they are at most r_n apart:

$$A_{ij} = \mathbf{1}\{d(x_i, x_j) \leq r_n\}.$$

Having $r_n \downarrow 0$ corresponds to domain increasing asymptotics. This is depicted in Figure 13.3. One advantage of this perspective is that there are very well-understood properties that can exploit geometry. As such, the RGGs have well-understood subgraph counts with laws of large numbers and central limit theorems for the numbers of various subgraphs that should emerge (Penrose 2003).

Notice the formal relationship between a conditional edge independence model and the random geometric graph

$$\begin{aligned} A_{ij} &= \mathbf{1}\{\alpha - \beta\left|x_i - x_j\right| - \epsilon_{ij} > 0\} \\ &= \mathbf{1}\{\left|x_i\beta - x_j\beta\right| + \left|\epsilon_i - \epsilon_j\right| \leq \alpha\}, \text{ assuming } \epsilon_{ij} := -\left|\epsilon_i - \epsilon_j\right| \\ &= \mathbf{1}\{d((x_i\beta, \epsilon_i)', (x_j\beta, \epsilon_j)) \leq r\}, \text{ setting } r := \sqrt{\alpha}. \end{aligned}$$

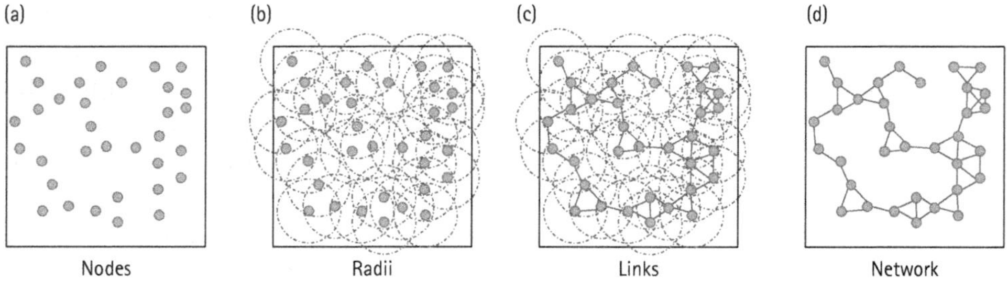

FIGURE 13.3 A random geometric graph formation process.

Under appropriate assumptions on the distributions of x and ϵ, the techniques used to study random geometric graphs and the properties of the resulting networks may be useful to help us understand how the distribution of observed and unobserved shocks may influence the globabl properties of the network in a meaningful way.

13.3.3 Observed Characteristics

The most commonly used approach to model correlation in link structure is to rely on observable node characteristics (Comola and Fafchamps 2013; Fafchamps and Gubert 2007ab). Since factors such as geographic location (as in trade models), education, ethnicity/caste, wealth, and so on are likely to influence whether agents form an economic relationship, it may very well be that the clustering in a network could be explained by such observable factors.

In this case, if there are covariates $x = (x_1, \ldots, x_n)'$ jointly drawn from a distribution $F_x(\cdot)$, one could consider

$$p(x_i, x_j) := \mathrm{P}(A_{ij}|x_i, x_j).$$

Usually this is a parametric model, but need not be. This leads to

$$\mathrm{P}_p(g|x) = \prod_{i<j} p(x_i, x_j)^{A_{ij}}(1 - p(x_i, x_j))^{1-A_{ij}}.$$

The expected graph would exhibit higher clustering when $i,j,$ and k are such that $p(x_i, x_j)$, $p(x_k, x_j)$ and $p(x_i, x_k)$ are all high.

The extent to which this is a good or a bad approach depends on the x's that the econometrician has at her disposal. Since, in principle, given that correlations in the x's are allowed, and having access to a large number of unobservable and observable x's could always represent the right probability distribution, what really matters is how much of the correlation is described by what the researcher sees. Chandrasekhar and Jackson (2014), for instance, look at data and show that including a large vector

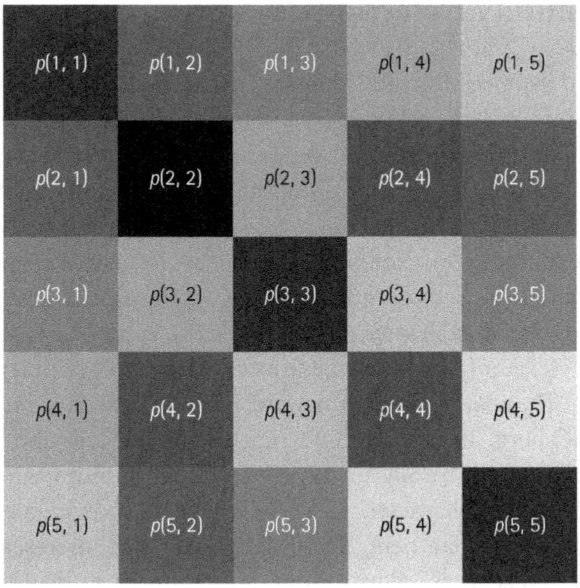

FIGURE 13.4 The adjacency matrix of a block model with $K = 5$, illustrated in the symmetric case. $p(k,l)$ is the linking probability for nodes from groups k and l. Darker shading indicates higher probability of linking.

of covariates of Indian village households (assets, caste, geography) is not enough to generate the clustering and linking patterns in the empirical data.

To get an intuition as to what could go wrong, let us look at *Block models* (Holland, Laskey, and Leinhardt 1983), which cover a specific parametrization of introducing covariates. Imagine that the only characteristic of an agent that matters to network formation is her group (e.g., caste, or caste $\times$ gender) and there are K communities. The adjacency matrix is formed as follows:

1. An individual belongs to group (or community) k with probability $\pi(k)$. The community is denoted x_i.
2. Conditional on group assignment, the probability that nodes i and j are linked depends on communities x_i, x_j: $p(x_i, x_j)$.

Figure 13.4 illustrates the model. Notice that within a group an Erdős-Rényi graph $p(x_i, x_i)$ forms. Across groups, there is a fixed linking probability given group identities as well. Therefore, the intuitive limitation of such a model, holding fixed the number of groups, is that we are simply looking at interconnected Erdős-Rényi graphs. This means that unless within a group networks can be thought of as dense, the clustering-sparsity trade-off will persist.

13.3.4 Community Detection

Now we turn to a more difficult problem, wherein individuals are members of communities/groups as before but these groups themselves are unobservable. The researcher's task is to (1) estimate the community structure and then (2) recover the linking probabilities. The group membership structure can be thought of as a collection of nuisance parameters.

There is a large literature on community detection in networks spanning computer science, sociology, statistics, and, more recently, economics. Surveying this literature is beyond the scope of this chapter, and I direct the reader to Clauset et al. (2004), Newman (2004), Jackson (2008), and Leskovec, Lang, and Mahoney (2010) for more in-depth reviews. The idea is straightforward, though, and has origins in sociology (Airoldi, Blei, Fienberg, and Xing 2008; Lazarsfeld, Henry, and Anderson 1968). Agents in a social network belong to some latent communities that may or may not coincide with observable characteristics. For instance, individuals in a village may share risk with others despite being from different castes, or researchers may form co-authorships across schools. A natural question is how to identify the underlying so-called communities that drive the observed network.

The logic of the identification of communities is simple but subtle. At a high level, we want to say that sets of nodes are likely to be in the same community if they have a lot of links within the set. At the same time, a natural requirement is that linking within the set should somehow exceed linking between members of different sets.

Much of the literature has focused on specifying objective functions to make precise what we mean by community and community detection. Because this involves assigning each node to a community, if there are K communities, this involves searching over a discrete space of size n^K. Therefore, the computer science literature has focused on developing simple algorithms to approximate solutions to the various objective functions that have been specified. These computational concerns are beyond the scope of this chapter.

Instead, I will discuss community detection from the perspective of statistics. After discussing a simple model of network formation based on unobserved communities, I will discuss the commonly studied Newman and Girvan objective function (frequently called NG Modularity) as well as an alternative, likelihood-based method by Bickel and Chen (2009), who present a nonparametric view of community detection models and relate them to another crucial question. They ask whether any reasonable probability distribution for a network can be covered by a model like the one specified below. The answer is, in some sense, "yes."

Specifically, as discussed by Bickel and Chen (2009) and studied extensively based on results for exchangeable arrays (Aldous 1981; Aldous 1985; Kallenberg 2006; Diaconis et al. 2008), under very mild assumptions—basically, on an unlabeled random network, the probability distribution of any realized network structure is the same across permutations—one can represent any probability distribution as

$$A_{ij} = f(\alpha, \xi_i, \xi_j, z_{ij})$$

where α, ξ_i, ξ_j, and z_{ij} are i.i.d. $U(0,1)$, $z_{ij} = z_{ji}$ and $f(\cdot)$ is a function symmetric in its second and third arguments.[8]

This provides theoretical motivation for an interesting avenue pursued by Bickel and Chen (2009) and Bickel, Chen, and Levina (2011). They look at parametric approximations to these nonparametric models. The basic idea is that every node is assigned to a community (or group) $1,\ldots,K_n$ independently of others. And then conditional on the vector of assignment to communities, edges are generated independently based on only group-identities. The researcher, of course, does not observe groups $x = (x_1,\ldots,x_n)$, and therefore is interested in estimating π, and has potential nuisance parameters x. The thought experiment is one where the number of communities grow as $n \to \infty$, though they do not fully develop these results in the present work.

The model implies that the probability of seeing a given graph is

$$\mathrm{P}_{\pi,p}(g) = \sum_x \left\{ \prod_{i<j} p(x_i,x_j)^{A_{ij}} (1 - p(x_i,x_j))^{1-A_{ij}} \right\} \prod_{i \in V} \pi(x_i).$$

There are $n + \binom{K_n}{2} + K_n$ parameters of the model: $x = (x_1,\ldots,x_n)$—the unobserved groups—as well as linking probabilities $p(m,l)$ for groups m,l and the group assignment probabilities, $\pi(\ell)$.

The main task, then, is to decide how to detect these groups. The first class of methods used in the literature come from computer science and focus on variations of minimum ratio cuts. The idea is simple. Given g, we can pick a collection of links into sets U and U' such that

$$\frac{\sum_{i\in U, j\in U'} A_{ij}}{|U| \cdot |U'|}$$

is minimized. The numerator is the number of links bridging members across the two groups, and the denominator is the product of the sizes of sets. This is intimately related to the conductance of the graph, which has also been used as an alternative criterion function. Nonetheless, actually solving for these cuts have long known to be NP-complete. Therefore, the literature has focused on approximation algorithms using, for instance, signs of eigenvectors associated with the adjacency matrix or related quantities and then applying repeated bisections. One limitation of this approach is that without modification, it does not adequately tell us about the group size or counts.

This observation motivates the second method, which is perhaps the most popular objective function used to partition the graph currently. The Girvan and Newman (2002) modularity is an objective function that picks community assignment to maximize the share of edges that link within a group minus the expected share that would link within a group given random location of links, holding fixed the community assignment of nodes.

[8] Note that $f(\cdot)$ is not unique and, further, often the models do carry labels.

It is useful to see their objective function, often called (NG)-modularity. First compute a matrix encoding the number of links between two blocks k and l

$$O_{k,l}(x|A) := \sum_{i<j} A_{ij}\mathbf{1}\{x_i = k, x_j = l\}.$$

Let $D_k := \sum_l O_{k,l}$ be the sum of degrees for those in community k. Then the objective function is given by

$$Q_{NG}(x|A) := \sum_{k=1}^{K} \frac{O_{kk}}{L} - \left(\frac{D_k}{L}\right).$$

What is conceptually wrong here? The objective function, in a sense, focuses on the diagonal of O only. Therefore it is giving up information by considering how often within group links stay within group versus the random allocation benchmark.

The approach taken by Bickel and Chen (2009) is to approach this from a likelihood perspective. Let n_k count the number of nodes in community k. Similarly, let

$$n_{a,b} := n_a n_b \text{ if } a \neq b,\ n_{a,b} := n_a(n_a - 1) \text{ if } a = b.$$

Then what is the likelihood under a known assignment of observing A? It is given by

$$Q_{LM}(x|A) := \frac{1}{2} \sum_{1 \leq a,b \leq K} \left\{ O_{a,b} \log p(a,b) + (n_{a,b} - O_{a,b}) \log(1 - p(a,b)) \right\}.$$

It must be the case that the likelihood is optimized by the $p(a,b)$ choices, and it is easy to see that this implies $p(a,b) = \frac{O_{a,b}}{n_{a,b}}$. Therefore,

$$Q_{LM}(x|A) := \frac{1}{2} \sum_{1 \leq a,b \leq K} n_{a,b} \left\{ \frac{O_{a,b}}{n_{a,b}} \log\left(\frac{O_{a,b}}{n_{a,b}}\right) + \left(1 - \frac{O_{a,b}}{n_{a,b}}\right) \log\left(1 - \frac{O_{a,b}}{n_{a,b}}\right) \right\}$$

gives us our (profile) likelihood.

Bickel and Chen (2009) study a general class of objective functions that nest LM and NG. Under an assumption that the underlying network formation process leads to dense enough graphs (the expected average degree grows faster than $\log n$) and that the objective function is uniquely maximized with smooth and bounded derivatives, they show that the assignments are correct with probability tending to 1 as $n \to \infty$.

Specifically, they show that for the LM objective and under some regularity conditions which NG sometimes meets,

$$\mathrm{P}(\widehat{x} = x) \to 1$$

and, further,

$$\sqrt{n}(\widehat{\pi} - \pi) \rightsquigarrow \mathcal{N}(0, \Sigma)$$

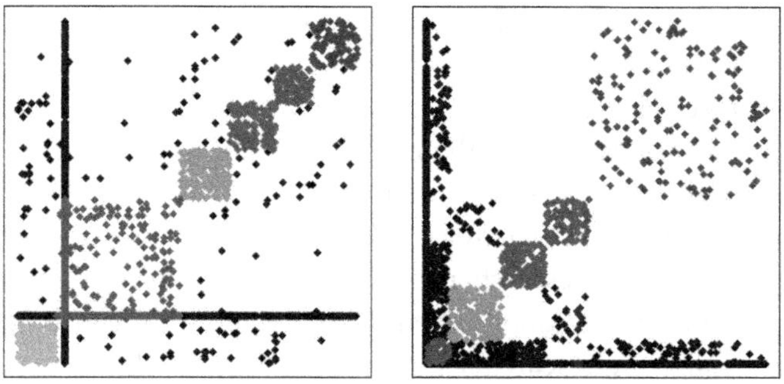

FIGURE 13.5 Figure 4 of Bickel and Chen (2009) reproduced. This presents the adjacency matrix, which consists of 621 individuals in a telephone communication network. The identified communities are of heterogenous sizes. Colors indicate within-group links and black indicates across-group links.

where Σ is a variance matrix which is unaffected by the estimation of x. They also show that p can be recovered with estimators that are consistent and asymptotically normally distributed.

What is particularly interesting is that the NG modularity can fail the identification/derivative conditions when, in fact, for every community there is higher linking probability within the community than the total sum of linking probabilities to any other community. The way that the inconsistency works is that merging two of the sparser communities can actually increase the modularity. One can see that this would be penalized by the likelihood approach. In that sense, the LM approach is a more robust objective function and, more generally, the statistical approach helps us see when and why other objective functions may fail.

Despite this, there are two main limitations to the entire enterprise. First, the computation is known to be challenging. As noted in Bickel and Chen (2009), it is NP-hard. While they and others report that certain algorithms work well, this inevitably makes implementation difficult. Second, the present theory is geared to studying dense graphs. As noted by the authors, it remains to be seen how valuable this approach could be for sparse networks, since classification will be imperfect.

13.3.5 Unobserved Node-level Heterogeneity

We now move to models where individual-specific unobserved heterogeneity is introduced, at the cost of additional parametric assumptions.

Individuals are often modeled to be heterogenous. They vary preferences and beliefs, and these differences can therefore systematically change one individual's decisions relative to another's in an otherwise identical circumstance. The models we have discussed so far have skirted this issue by making all heterogeneity observable or unobserved at a group level.

Nevertheless, in certain contexts it may very well be right to assume that individuals vary in their gregariousness, which the researcher will never observe, but gregarious individuals are simply more likely to build links.

Further, it could be the case that gregariousness is correlated with certain observable attributes, and therefore the researcher faces a standard problem that arises in settings with fixed effects: systematic unobserved heterogeneity that can be correlated with observables may cause the researcher to mis-estimate how the observables influence the outcomes.

Graham (2014a) studies precisely this problem. He considers a simple model where the probability of a link *ij* being formed conditional on a vector of covariates is given by

$$\mathrm{P}_{\beta,\nu}(A_{ij} = 1|x) = \frac{e^{x_{ij}'\beta+\nu_i+\nu_j}}{1+e^{x_{ij}'\beta+\nu_i+\nu_j}}. \tag{13.2}$$

Observe that every agent has a fixed effect ν_i that shifts the probability of a given link being formed. Since each agent is participating in $n-1$ linking decisions, we can think of each linking decision the way we would think about time in a panel data setting.

There are two goals that we might have in such a setting. First, we may want to understand how observables x_{ij} influence linking probabilities by estimating β. Since there are n individuals, this represents an n dimensional (nuisance) parameter. The natural difficulty in this problem is that the overall number of parameters in the model grows in n since there are $\dim(\beta) + n$ parameters. However, techniques developed by Chamberlain (1982) allow us to skirt the issue by using a conditional maximum likelihood estimator (CMLE).

A second goal might be to do counterfactuals. If we want to study how a shift in the distribution of x changes the overall resulting network structure, we must actually estimate ν_i. In this case, the researcher needs to use the fact that she observes individuals making $n-1$ linking decisions in order to estimate the n fixed effects. This presents a formidable and nonstandard statistical problem.

The solution to this problem lies in a literature in statistics going back to Holland and Leinhardt (1981). That literature ends with Chatterjee, Diaconis, and Sly (2010), who show how one can indeed consistently estimate all the ν_i in the above problem. I now discuss the basics of this statistical work before returning to Graham (2014a).

It is useful to address this problem from a different perspective. Let us begin with the observation that a key aspect of networks is its degree sequence: $d^n(g) = (d_1, \ldots, d_n)$. Therefore, a simple question to ask is whether it is possible to develop a model that, as the parameter varies over its domain, can capture all possible degree sequences. The answer is yes, and comes from a class of models introduced by Holland and Leinhardt (1981), but studied in greater detail by Chatterjee et al. (2010).

For Chatterjee et al. (2010), the first main result is to formalize the answer to this question. We consider a sequence $\{g_n\}$ of random graphs each chosen uniformly at random given d^n. This implies that the degree sequence is a sufficient statistic.

Additionally, let $f(\cdot)$ be any non-increasing function on $[0,1]$. Then there is a unique function $h:[0,1]\to\mathbb{R}$ such that

$$W(x,y) := \frac{\exp(h(x)+h(y))}{1+\exp(h(x)+h(y))},$$

$$f(x) = \int_0^1 W(x,y)dy,$$

and $\{g_n\}$ converges, in the sense described by Lovász and Szegedy (2006), almost surely to the limit represented by $W(x,y)$. This limit object is called a *graphon*.

A sequence of adjacency matrices can be placed on the unit square, and the limit is the case where we have each $x\in[0,1]$ corresponding to an agent, and therefore $W(x,y)$ encodes whether or not agents (x,y) are linked. Just as one would sum over every agent's friends to obtain a vector of degrees, one can integrate $W(x,y)$ over y to obtain the analogous limiting degree distribution, $f(\cdot)$.

What the result says is that if we consider a sequence of networks with a limit (in a specific technical sense that I will not cover here) given by $W(\cdot,\cdot)$, for any limit degree distribution, given by $f(\cdot)$, one can find a unique function $h(\cdot)$ that pins down $W(\cdot,\cdot)$ in the above.

Therefore, one may wonder whether we can compute the function $h(\cdot)$. Chatterjee et al. (2010) show that the answer is yes, and comes from the following, well-studied, model, the β-model.[9] Let $\nu=(\nu_1,\ldots,\nu_n)\in\mathbb{R}^n$ and set

$$p_{ij} = \frac{\exp(\nu_i+\nu_j)}{1+\exp(\nu_i+\nu_j)}.$$

Notice that this is exactly the model that Graham (2014a) later studies, only without covariates. Again, the technical challenge is that there are n parameters with only $\binom{n}{2}$ observations. This makes the problem somewhat nonstandard, and a major contribution of the Chatterjee et al. (2010) paper is to address this challenge and demonstrate that the parameters can be consistently estimated.

The two main results in Chatterjee et al. (2010) are:

1. *Consistency:* with probability $1-O\left(n^{-2}\right)$,

$$\max_{1\leq i\leq n} |\widehat{\nu}_i-\nu_i| \lesssim \sqrt{\frac{\log n}{n}} \to 0.$$

[9] Usually the model is written with parameter β, though we use ν to call attention to its economic interpretation as a fixed effect.

2. *No degree sequence is left out:* All possible expected degree sequences can be covered as ν varies over $\mathbb{R}^n$.

This implies that despite the challenge due to the rapidly growing number of parameters, they can consistently be estimated. Further, any limiting degree distribution can be modeled by the above for a suitable vector of parameters ν.

It is important to note that their result applies to dense networks and not to sparse networks. That means that the number of links that a typical node is a part of is on the order of n. To see why this matters, consider a simplified case where each agent had a constant order of friends. In this case, $f(x) = 0$ at every x in the statement of their first result. Next, we turn to the argument as to why one cannot recover the ν_i consistently in this case. A crucial piece of the argument is to show with probability going to 1 that every node's degree is close to its expected degree under the model. This requires the expected degree to grow at rate n, because otherwise, there is no growing information that allows us to precisely estimate every individual fixed effect. As noted by the authors, the arguments do not follow through for sparse networks precisely because there is not enough information in the system to estimate each parameter (for each node) tightly.

Returning to Graham (2014a), we add covariates to the problem and are interested in recovering both the fixed effects as well as the parameter driving the covariate effect. He shows the following:

1. $\widehat{\beta} \xrightarrow{P} \beta$,
2. with probability $1 - O(n^{-2})$, $\max_{1 \leq i \leq n} |\widehat{\nu_i} - \nu_i| \lesssim \sqrt{\frac{\log n}{n}}$,
3. and for I_0 the Fisher information for β,

$$n\left(\widehat{\beta} - \beta\right) \rightsquigarrow \mathcal{N}\left(\sqrt{2}B_0, 2I_0^{-1}\right).$$

Introducing these nuisance parameters generates a bias term in the asymptotic distribution, which then can be exactly characterized and therefore subtracted out.

13.3.6 Discussion

In this section, I have reviewed basic models where links are conditionally independent. The fundamental tension here, which we can see from the Erdős-Rényi model, is that if we model realistic levels of clustering, then the network must be dense with expected degree being on the order of n. However, this tends to be inconsistent with the sparsity of empirical networks.

While observable covariates can get us partway there, they are far from sufficient. It is possible, in theory, to model any distribution over networks as a function of two individual-specific latent variables and a pair-level latent variable. Estimating such a model directly from the data is infeasible, but a reasonable approach to get at this was developed by Bickel and Chen (2009), where individuals belong to communities and the number of communities grows asymptotically. Setting aside computational complexity issues, we find that a major limitation of their otherwise general result is that it depends on having a sufficiently dense network. Nonetheless, their results require that with probability approaching 1, the assignment to communities is exact. Perhaps requiring a less stringent assignment criterion would allow for more sparse network formation distributions to be modeled.

A different form of unobserved heterogeneity is studied by Chatterjee et al. (2010) and Graham (2014a), who emphasize the degree to which this is economically important since fixed effects may be correlated with observables, thereby influencing the researcher's estimates about the degree to which there is homophily in covariates. Again, however, the techniques required to be able to estimate the fixed effects lean heavily on the networks being dense, with an expected degree that is on the order of n.

The density requirements of the Bickel and Chen (2009) approach are less stringent than those required by the β-model. Perhaps this is not too surprising because, while Bickel and Chen (2009) are able to recover assignment to group correctly, they reduce the dimensionality of the driving probability distribution to just a collection of groups and probabilities of linking within and across groups. In that sense, there is more information per fundamental parameter in their model, as compared to the β-model case.

Many of the models studied in the next section relax the density requirements, by directly entangling agent payoffs with choices of their neighbors. In turn, links are correlated in a more direct manner without giving up on sparsity. On the other hand, these models raise more complicated problems when it comes to statistical inference, identification, and equilibrium selection.

13.4 Higher-Order Dependence

In this section, we look at network formation models that more directly allow for correlation in links. We will need to grapple with whether these models provide the researcher with more information, in the sense of admitting laws of large numbers and central limit theorems, as $n \to \infty$. Finally, point identification itself may become impossible without strong assumptions on preferences or equilibrium selection.

13.4.1 Exponential Random Graph Models

13.4.1.1 Dynamic, Myopic Best Response

There is an important relationship between strategic network formation models and potential functions, explored across a number of papers including Jackson and Watts (2001), Butts (2009), Mele (2013), and Badev (2013). We will use this as a starting point in our discussion of a class of models called exponential random graph models (ERGMs), which are among the most widely used models of network formation.

Recall the definition of a potential game. Consider a set of individuals with utility functions u and a strategy set Y. A game is a potential game if there exists a function $f(\cdot,\cdot)$ such that

$$u_i(y_i', y_{-i}) - u_i(y_i, y_{-i}) = f(y_i', y_{-i}) - f(y_i, y_{-i}).$$

for every player i and $y, y' \in Y$.

Butts (2009) and Mele (2013) study the steady state of simple models where nodes can dynamically revise their linking decisions, given the current network, but are not forward looking. If the utility function of the players admits a potential function, the resulting steady-state distribution corresponds to a specific distribution with a sufficient statistic given by the potential function.

That is, if players choose to revise their links via a logistic choice rule, and updating opportunities arise sequentially with every pair ij being selected with positive probability, then the sequence $\{g_t\}_{t\in\mathbb{N}}$ forms a Markov chain with equilibrium distribution:

$$P_{f(\cdot)}(g) = \frac{\exp(f(g))}{\sum_{g'} \exp(f(g'))}.$$

Thus, if agents are in a dynamic process where their payoffs are such that there is a potential function for the marginal utility of their links, then the stationary distribution of networks has the above form and is what the econometrician draws from.

Example 1 (links and triangles). *Imagine that an individual receives payoffs β_L for every link that she is part of and β_T for every triangle. In this case*

$$u_i(g) = \beta_L \cdot \sum_j A_{ij} + \beta_T \cdot \sum_j A_{ij} A_{jk} A_{ik}.$$

Define a function

$$f(g) := \beta_L \cdot \sum_i \sum_j A_{ij} + \beta_T \cdot \sum_i \sum_{j<k} A_{ij} A_{jk} A_{ik}.$$

It is clearly a potential function, since

$$u_i(g+ij) - u_i(g-ij) = \beta_L + \beta_T \cdot \sum_k A_{ik} A_{jk} = f(g+ij) - f(g-ij).$$

We now describe this for more general payoffs. The core argument is owed to Butts (2009) and Mele (2013). Chandrasekhar and Jackson (2014) extended the result to other subgraphs. Badev (2013) observes that adding an endogenous action to the mix does not change the fundamental result. He looks at a case where utility depends not only on links to others, but also on one's actions as well as one's neighbors' actions.

Assume that payoffs are additively separable in subgraphs that i is a member of:

$$u_i(g,x;\beta) = \sum_{g_\ell, i \in g_\ell} v(g_\ell, x_\ell; \beta_\ell),$$

where x_ℓ is a vector of covariates corresponding to the members of subgraph g_ℓ. Observe that this allows for an agent's utility to depend on "friends of friends" as desired in a number of models (Goldsmith-Pinkham and Imbens 2013; Mele 2013; Badev 2013) by including subgraphs of the form $g_\ell = \{ij, kj\}$. However, it also allows utility to depend on more complex subgraphs such as cliques.

The links and triangles example had no covariates, and so corresponded to

$$v(\{ij\};\beta) = \beta_L \text{ and } v(\{ijk\};\beta) = \beta_T.$$

Time is discrete, and in each period, at most one pair of individuals meet. Every ij ex ante have some positive probability of meeting this period, though the actual meeting probability can be history-dependent.

The stability concept used is pairwise stability with transfers:[10]

- $ij \in g$ implies that $u_i(g) + u_j(g) \geq u_i(g - ij) + u_j(g - ij)$, and
- $ij \notin g$ implies that $u_i(g) + u_j(g) \geq u_i(g + ij) + u_j(g + ij)$.

The utility function and the stability concept together imply the existence of a potential function under pairwise stability with transfers:

$$f(g) := \sum_{g_\ell \subset g} 2v(g_\ell, x_\ell; \beta_\ell).$$

One can check that for any g and $ij \in g$:

$$f(g) - f(g - ij) = \left(u_i(g) + u_j(g)\right) - \left(u_i(g - ij) + u_j(g - ij)\right).$$

Network formation occurs as follows. Agents can choose to form links in pairs. They choose to add a link whenever their *mutual* gain is positive. This is important, because this allows them to bargain and make side payments (either in cash or by exchange of favors) to add links whenever those links are mutually beneficial. Allowing for this may be important from theoretical first principles.

The dynamics are as follows. At period t the process is at some network g^t. Let g^t_{-ij} be the networks excluding link ij. In every period, there is some positive probability $p(ij, g^t_{-ij})$ of each given pair being recognized—where the two relevant agents meet. This is general in that the meeting process can depend on the present graph as well as characteristics of i and j.[11]

[10] This definition is from Bloch and Jackson (2006) and is related to the definition of pairwise stability allowing for side payments that appears in the conclusion of Jackson and Wolinsky (1996).

[11] This allows their meeting probability to depend on whether they have friends in common, how many friends in common they have, their covariates, other aspects of the network in general, etc.

In period t, the chosen i and j decide whether to add or delete the link, conditional on g_{-ij}^{t-1}. In the usual way by using extreme value disturbances, one can say that the probability that the link is added/kept is a logistic function of the mutual value of the link:

$$\frac{\exp\left(u_i(g_{-ij}^t + ij) + u_j(g_{-ij}^t + ij)\right)}{\exp\left(u_i(g_{-ij}^t + ij) + u_j(g_{-ij}^t + ij)\right) + \exp\left(u_i(g_{-ij}^t - ij) + u_j(g_{-ij}^t - ij)\right)}. \tag{13.3}$$

The dynamic process generates a sequence $\{g^t\}_{t\in\mathbb{N}}$ that is an aperiodic and irreducible Markov chain over the set of networks on n nodes. Therefore, there is a unique steady-state distribution. Moreover, it is a reversible Markov chain, and the unique steady-state distribution is given by

$$\mathrm{P}(g) = \frac{\exp(f(g))}{\sum_{g'} \exp(f(g'))} = \frac{\exp\left(\sum_{g_\ell \subset g} 2v(g_\ell, X_\ell)\right)}{\sum_{g'} \exp\left(\sum_{g_\ell \subset g'} 2v(g_\ell, X_\ell)\right)}.$$

Returning to our links and triangles example, the distribution is given by

$$\mathrm{P}_{\beta_L,\beta_T}(g) = \frac{\exp\left(\beta_L \cdot S_L(g) + \beta_T \cdot S_T(g)\right)}{\sum_{g'} \exp\left(\beta_L \cdot S_L(g') + \beta_T \cdot S_T(g')\right)}$$

where $S_L(g)$ is (twice) the total number of links in the network and $S_T(g)$ the triangles. The goal of the researcher would then be to estimate β_L, β_T.

In sum, what we see is that a simple, myopic best response dynamic where the payoffs admit a potential function corresponds to an invariant distribution of networks that has the above form. Specifically, this means that the network that the econometrician observes is drawn from a distribution where the sufficient statistic is given by the potential function itself. One way to think about this is that given $f(g)$, any graph g can be drawn uniformly at random.

13.4.1.2 ERGMs

A literature spanning several disciplines (computer science, economics, sociology, and statistics) has turned to a class of models called exponential random graph models (ERGMs) (see, e.g., Frank and Strauss 1986; Wasserman and Pattison 1996; Mele 2013).

The basic idea is that an ERGM specifies a vector of sufficient statistics, usually corresponding to counts of subgraphs in a network, and then formulates the probability as depending on the subgraph count with all networks with the same sufficient statistic value being drawn with equal probability. Above I described one approach to providing microfoundations for these models.

Let $S(g)$ denote a vector of characteristics of a newtork. For example,

$$S(g) = (S_L(g), S_T(g))' = \left(\sum_{i<j} A_{ij}, \sum_{i<j<k} A_{ij}A_{jk}A_{ki} \right)'$$

is a vector of the total number of links and the total number of triangles in the graph.

The ERGM specifies the probability of observing a network with a given vector of attributes S is given by[12]

$$\mathrm{P}_\beta(g) = \frac{\exp(\beta \cdot S(g))}{\sum_{g'} \exp(\beta \cdot S(g'))}. \tag{13.4}$$

ERGMs have become widely used for a number of reasons. First, they provide an intuitive formulation. They focus on aspects that researchers believe are important in network formation and that can encode rich types of interdependencies. Second, due to the Hammersley-Clifford theorem (1971), it turns out that any probability distribution over networks can be represented in this way. This result has very little content in practice, as it makes no claim about the complexity of the set of sufficient statistics. Trivially, any distribution can be captured by some (potentially very complicated) vector of sufficient statistics. Third, as described above, recent work provide utility-based microfoundations for such models (Butts 2009; Mele 2013).

It should be immediately obvious that an Erdős-Rényi random graph must have a sufficient statistic which is the number of edges in the network. This is because the binomial is in the exponential family, and the sufficient statistic is the number of successes (in this case, links).

Example 2 (Erdős-Rényi). *Let $S(g) = \sum_{i<j} A_{ij}$ denote the number of links in the graph. As we saw before, the likelihood of the Erdős-Rényi model is*

$$\mathrm{P}_p(g) = p^S (1-p)^{\binom{n}{2} - S},$$

which can be written as

$$\mathrm{P}_p(g) = \exp\left(\log \frac{p}{1-p} \cdot S(g) \right) \cdot (1-p)^{\binom{n}{2}}.$$

[12] Clearly, this formulation can depend on covariates, as well. So one can write

$$\mathrm{P}_\beta(g|x) = \frac{\exp(\beta \cdot S_x(g))}{\sum_{g'} \exp(\beta \cdot S_x(g'))},$$

where the vector of attributes can depend on x.

Setting $\beta := \log \frac{p}{1-p}$ *and observing* $(1-p)^{-\binom{n}{2}} = \sum_{g'} \binom{\binom{n}{2}}{S(g')} p^{S'}(1-p)^{\binom{n}{2}-S'}$, *yields*

$$\mathrm{P}_\beta(g) = \frac{\exp(\beta \cdot S(g))}{\sum_{g'} \exp(\beta \cdot S(g'))}.$$

Similarly, the β-model we know posits that given the degree distribution, the network is drawn uniformly at random. It is easy to see that $S(g) = d(g) = (d_1(g), \ldots, d_n(g))'$ in this case.

EXAMPLE 3 (β-model). *This is an ERGM with an n-dimensional parameter,* β, *with*

$$\mathrm{P}_\beta(g) = \frac{\exp(\beta \cdot S(g))}{\sum \exp(\beta \cdot S(g')} = \frac{\exp(\beta \cdot d(g))}{\prod_{i<j} 1 + \exp(\beta_i + \beta_j)}.$$

We have seen that the parameters in both of these models can be consistently estimated. However, this starts to become more of a challenge when $S(g)$ involves subgraphs with three or more nodes, such as triangles.

13.4.1.3 Estimation Problems

Let us assume the researcher believes that her data is best modeled by a distribution with a given known S. Generally, outside of models that reduce to link independence, there are two main problems with ERGMs in practice: (i) feasibility of estimation; and (ii) consistency of estimated parameter.

First, it can be very difficult to actually compute parameter estimates. The reason is the following. There are many possible networks, in fact $2^{\binom{n}{2}}$ undirected graphs. Notice that estimating the likelihood of a given network requires having an estimate of whether it is more or less likely than other possible networks. This means that either implicitly or explicitly, one needs an estimate of the denominator in (13.4). In simple models where links are conditionally independent, this is effectively trivial to do. But in the general case, this can be difficult.

Of course, computing this directly is entirely impossible. Therefore, the literature has moved to employing Markov Chain Monte Carlo (MCMC) sampling techniques (Snijders 2002; Handcock 2003). The basic idea is to construct a chain of networks that represents draws from the ERGM distribution. However, this approach also has its limitations. The developers of this method, as well as its practitioners, have long noted that there were degeneracy and convergence problems for many common specifications of ERGMs (usually those with links and triangles at a minimum).

Recent work by Bhamidi, Bresler, and Sly (2008) and Chatterjee and Diaconis (2013) has clarified exactly why these problems happen in the context of dense networks.[13] The MCMC procedure can only visit an infinitesimally small portion of the set of possible graphs. The crucial question, therefore, is whether a chain would actually mix

[13] For simulation-based evidence in the sparse case, see Chandrasekhar and Jackson (2014).

in polynomial time. These papers have shown that for broad classes of ERGMs, standard MCMC procedures will take exponential time to mix *unless the links in the network are approximately independent*. But if links are approximately independent then there is no real need for an ERGM formulation to begin with! There are more direct ways to represent and estimate such a model. It seems, then, that in cases where ERGMs are really needed they cannot be accurately estimated by such MCMC techniques.

The second problem concerns the consistency of parameter estimates. If we were indeed able to compute the maximum likelihood estimate, is it the case that along an asymptotic sequence of networks as $n \to \infty$ estimates become closer and closer to the truth? This is a crucial property for any frequentist estimator to have, especially given that in most data settings, the researcher only has access to a single network or a handful of networks. It may be the case, in fact, that having more nodes may not even increase information in the system. Shalizi and Rinaldo (2012) explore these issues. They show that for any exponential family distribution, projectivity is a sufficient condition for consistency. Then they show that any ERGM with a sufficient statistic that involves a triangle violates this condition.[14]

While ERGMs that have higher-order link dependencies are easily interpretable in a simple random utility framework, they lead to models that are difficult, if not impossible, to estimate and parameter estimates that are not necessarily consistent as $n \to \infty$. However, the aspiration of the microeconometric exercise is simple: preferences should incorporate payoffs from not simply friends but other links, which may depend upon local network structure. The next approach deals with modeling subgraphs directly.

13.4.2 Subgraph Generated Models (SUGMs)

One objective of the ERGM program was to model the correlation in links through probability models where the sufficient statistics were counts of subgraphs.

We now look at a related but more direct approach: build networks up from subgraphs as the fundamental units. This is the path taken in Bollobás, Janson, and Riordan (2011) and Chandrasekhar and Jackson (2015). A network is constructed from building up subgraphs of various types: links, triangles, other cliques, stars, and so on. These structures are layered upon each other yielding the network of interest. Figure 13.6 depicts the idea.

For exposition, let us consider the special case where networks are comprised of incentives to form links and incentives to form triangles. Define an *unsupported link* to be a link between two nodes that forms directly and not as part of larger subgraph,

[14] Chandrasekhar and Jackson (2014) approach this problem in a different way. They note that the reference distribution imposed by ERGMs is what generates the problem and point out that related classes of models with differing reference distributions can allow for consistency.

and the set of such links that form $\mathcal{LU}$. As before, a triangle is a set of three nodes that form together as a group, and the set of such triangles that form is denoted $\mathcal{T}$.

Node i forms a collection of unsupported links and triangles. These can intersect and overlap, and the resulting network that forms is given by

$$g = \{ij : ij \in \mathcal{LU} \text{ and/or } \exists k \text{ s.t. } ijk \in \mathcal{T}\}.$$

Two agents are connected if they are part of a link, or both are part of any triangle. Thus, g can be thought of as a projection of $\mathcal{LU}$ and $\mathcal{T}$.

If the researcher observed the sets $\mathcal{LU}$ and $\mathcal{T}$, then estimation would be straightforward under standard regularity conditions.[15] Often researchers only have binary information concerning whether two individuals in the society interact. For instance, we may have data indicating which pairs of agents are "friends" based on a survey, from observing that they are linked on social media, and so on.

From this perspective, the problem is that the formation of the subgraphs is not directly observed, and so must be inferred in order to estimate the parameters of interest. The observed network g is a projection of $\mathcal{LU}$ and $\mathcal{T}$. If

$$A_{ij}A_{jk}A_{ik} = 1,$$

is it the case that ijk formed as a triangle; ij, jk, and ik formed as links; ij and jk formed as links and ik formed as part of a different triangle ikm; and so on?

In sum, if the probability of links and triangles forming are given by some parameter vector, the researcher needs to use the observed graph g to infer how different subgraphs formed in order to estimate the parameter vector.

13.4.2.1 Utility Foundations

Before describing the statistical model and the estimation methods, it is useful to look at two (distinct) random utility foundations for subgraph generated models. In the first model, agents meet other agents in pairs, triples, or other subsets and then choose, via mutual consent, whether or not to establish links. The resulting vector of relationships that have formed comprise the observed graph g.

In the second model, agents are interested in investing in partnerships for production. Productive activities can be undertaken in pairs, in groups of three, and so on. Agents engage in costly search to find partners and other groups for production, and the equilibrium of this search process generates the observed graph g.

[15] In some instances, a researcher might have information on all the various groups a given individual is involved in: for example, in the case of a co-authorship network, the researcher may observe all the papers a researcher has written.

EXAMPLE 4 (Meetings). *The utility an agent i receives from being in a collection of unsupported links and a collection of triangles is given by*

$$u_i(\mathcal{LU},\mathcal{T},x) = \sum_{\{j:\ ij \in \mathcal{LU}\}} u_{i,L}(x_i,x_j) + \sum_{\{j<k:\ ijk \in \mathcal{T}\}} u_{i,T}(x_i,x_j,x_k).$$

The utility of an unsupported link can depend on

- *a selectivity factor:* $\beta_{L,0}$;
- *factors depending on both parties' characteristics* x_i,x_j: $h_L(x_i,x_j)$ *(for example,* $h_L(x_i,x_j) = |x_i - x_j|$*); and*
- *an idiosyncratic component:* $\epsilon_{i,ij}$.

The utility to i from forming an unsupported link ij is given by

$$u_{i,L}(x_i,x_j) = \beta_{L,0} + \beta_{L,1} h_L(x_i,x_j) - \epsilon_{i,ij}.$$

An agent consents to being in ij if and only if $u_{i,L}(ij,x) \geq 0$. *The unsupported link forms if both agents consent:*[16]

$$u_{i,L}(x_i,x_j) \geq 0 \text{ and } u_{j,L}(x_j,x_i) \geq 0.$$

Similarly, the utility of being in a trilateral relationship—a triangle—can depend on

- *a* selectivity factor: $\beta_{T,0}$;
- *factors depending on the parties' characteristics:* $h_T(x_i,x_j,x_k)$, *(for example,* $h_T(x_i,x_j,x_k) = \sum_{i<j} |x_i - x_j|/3$*); and*
- *an idiosyncratic component:* $\epsilon_{i,ijk}$.

The utility to i of being in a triangle $ijk = 1$ *is given by*

$$u_{i,T}(x_i,x_j,x_k) = \beta_{T,0} + \beta_{T,1} h_T(x_i,x_j,x_k) - \epsilon_{i,ijk}.$$

An agent consents to forming a triangle ijk if and only if $u_{i,T}(x_i,x_j,x_k) \geq 0$. *The triangle forms if all agents consent:*

$$u_{i,T}(x_i,x_j,x_k) \geq 0,\ u_{j,T}(x_j,x_i,x_k) \geq 0, \text{ and } u_{k,T}(x_k,x_i,x_j) \geq 0.$$

[16] Notice that as $\beta_{L,0}$ decreases, the agent becomes more selective, since holding the distribution of shocks and covariates fixed, the agent requires a higher draw of the idiosyncratic component to be happy with a link.

In sum, the utility of being in a society with the set $\mathcal{LU}$ unsupported links and the set of $\mathcal{T}$ triangles is

$$\begin{aligned} u_i(\mathcal{LU},\mathcal{T},x) &= \sum_{\{j:\, ij\in\mathcal{LU}\}} u_L(x_i,x_j) + \sum_{\{j<k:\, ijk\in\mathcal{T}\}} u_{i,T}(x_i,x_j,x_k) \\ &= \sum_{\{j:\, ij\in\mathcal{LU}\}} \left(\beta_{L,0}+\beta_{L,1}h_L(x_i,x_j)-\epsilon_{i,ij}\right) \\ &+ \sum_{\{j<k:\, ijk\in\mathcal{T}\}} \left(\beta_{T,0}+\beta_{T,1}h_T(x_i,x_j,x_k)-\epsilon_{i,ijk}\right). \end{aligned}$$

Individuals may not have the opportunity to meet all possible subgroups, so some subgraphs may be impossible to build. Let $\pi_L(x_i,x_j)$ denote the probability that two individuals i and j meet, and let $\pi_T(x_i,x_j,x_k)$ denote the probability that three individuals meet. Let F_L, F_T be the cumulative distribution functions for $\epsilon_{L,ij}, \epsilon_{T,ijk}$. Under independence

$$\begin{aligned} \mathrm{P}\left(ij\ \textit{forms as a link}|x_i,x_j\right) &= \mathrm{P}\left(ij\ \textit{meet}|x_i,x_j\right)\cdot \mathrm{P}\left(ij\ \textit{consent}|\ ij\ \textit{meet};x_i,x_j\right) \\ &= \pi_L(x_i,x_j)\cdot F_L\left(\beta_{L,0}+\beta_{L,1}h_L(x_i,x_j)\right)^2. \end{aligned}$$

and

$$\begin{aligned} \mathrm{P}\left(ijk\ \textit{forms as a triangle}|x_i,x_j,x_k\right) &= \mathrm{P}\left(ijk\ \textit{meet}|x_i,x_j,x_k\right)\cdot \mathrm{P}\left(ijk\ \textit{consent}|\ ijk\ \textit{meet};x_i,x_j,x_k\right) \\ &= \pi_T(x_i,x_j,x_k)\cdot F_T\left(\beta_{T,0}+\beta_{T,1}h_T(x_i,x_j,x_k)\right)^3. \end{aligned}$$

Therefore, under appropriate assumptions on functional forms, the frequency with which different subgraphs appear and how this frequency varies with member nodes' attributes, identifies the parameters of interest $(\beta_{L,0},\beta_{L,1},\beta_{T,0},\beta_{T,1})$.

While this example discussed models that implicitly or explicitly involved linking costs for various subgraphs in isolation, our next approach takes on an entirely different perspective in which there is an overall cost of forming relationships. This generates a trade-off, since individuals now need to optimally apply effort to maintain the right sort of structural relationships. The theory literature has studied models with search intensities (Currarini, Jackson, and Pin 2009 2010; Borgs, Chayes, Ding, and Lucier 2010; Golub and Livne 2010), and the next model develops an econometric analog.

EXAMPLE 5 (Search intensity). *An agent with attributes x_i can put in search effort to form a clique with m nodes, Cl_m, consisting of characteristic vector x. From this match, an agent obtains a utility*

$$u(x_i,m,x).$$

For simplicity, assume $u(x_i,m,x)=u_m$. Let

$$e(x_i,m,x)\in[0,\bar{e}_m]$$

denote the search effort. For now assume $e(x_i, m, x) = e_m$.

Whether or not the clique Cl_m *forms is stochastic. This probability depends on the vector of efforts of potential members,*

$$p(e_1, .., e_m) = p(e_1(x_1, m, x), \ldots, e_m(x_m, m, x)).$$

We can assume that the probability is non decreasing in each argument. For now set

$$p(e_1, .., e_m) = \sum_{j \in [1, \ldots, m]} e_j$$

where by assumption $\bar{e}_m$ *is such that* $m\bar{e} < 1$.[17]

The cost of effort is $c\left((e_i(x_i, m, x))_{m,x}, x_i\right)$, *which can depend on her characteristics and her search effort for all cliques of all attribute types. Here we have*

$$c\left((e_i(x_i, m, x))_{m,x}, x_i\right) = \sum_m \phi_m \frac{e_{i,m}^2}{2}.$$

The expected utility is

$$\sum_{m, Cl_m : i \in Cl_m} u(x_i, m, x) \cdot p((e_j(x_j, m, x))_{j \in Cl_m}) - c\left((e_i(x_i, m, x))_{m,x}, x_i\right).$$

Under our simplifying assumptions, expected utility is

$$\sum_{m, Cl_m : i \in Cl_m} u_m \cdot \left(\sum_{j \in [m]} e_{j,m} \right) - \sum_m \phi_m \frac{e_{i,m}^2}{2}.$$

It follows that $e_{i,m}^* = \frac{u_m}{\phi_m}$. *Normalizing utility by the cost parameter to define* $\beta_m := \frac{u_m}{\phi_m}$ *and observing that the frequency of a clique forming is given by*

$$p_m = p(e_1^*, \ldots, e_m^*) = m e_{i,m}^* = m\beta_m,$$

we see how to map from the frequency of a subgraph to the structural parameter.

Both examples presented here map two different interactions and random utility frameworks to a network wherein structural parameters are identified by the rate at which various subgraphs are present in the graph. We now turn to the problem of estimating subgraph frequency parameters which, under assumptions like those above, would allow a researcher to estimate utility parameters as well.

[17] This means that no agents can put in enough effort to guarantee a successful meeting.

13.4.2.2 *Subgraph Generation*

In SUGMs, subgraphs are directly generated by some process. Classic examples of this are Erdős-Rényi random graphs, where links are randomly generated, and the generalization of that model, stochastic-block models, in which links are formed with probabilities based on the nodes' attributes. The generalization to SUGMs is to allow richer subgraphs to form directly, and hence to allow for dependencies in link formation. The model can then be described by a list of probabilities, one for each type of subgraph, where subgraphs can be based on the subgraph shape as well as the nodes' characteristics.

Formally, there is a a finite number of different types of nonempty subgraphs, indexed by $\ell \in \{1,\ldots,k\}$, on which the model is based. Then a SUGM on n nodes is based on some list of k subgraph types: $(G_\ell)_{\ell\in\{1,\ldots,k\}}$. Each G_ℓ is a set of possible subgraphs on m_ℓ nodes, which are identical to each other up to the relabeling of nodes. Two examples for G_ℓ are (i) all triangles, and (ii) all stars, with one central node and four other nodes. The final ingredient is a parameter vector $\beta \in \mathcal{B}$, where $\mathcal{B}$ is a subset of some finite dimensional Euclidean space.

A network g is randomly formed as follows. Each of the possible subnetworks in G_ℓ is independently formed with probability

$$p_\ell(x(g_\ell),\beta_\ell)$$

for each $\ell \in \{1,\ldots,k\}$, where $x(g_\ell)$ is a set of covariates for the nodes in the candidate subgraph g_ℓ. The resulting network, g, is the union of all the links that appear in any of the generated subgraphs.

Figure 13.6 describes the process for a model with (p_L,p_T). Each of the $\binom{n}{3}$ potential triangles form with probability p_T, and each of the potential unsupported links form with probability p_L. Sometimes they overlap; in this case, an extra triangle forms relative to those that were truly generated, therefore eliminating an unsupported link.

13.4.2.3 *Estimation*

Before we describe estimation, it is useful to clarify the asymptotic thought experiment. As noted in Section 13.3, since the average degree (and other features) may not grow linearly in n and could even grow slowly in n, we must allow the linking probability to vary in n.

For simplicity, consider a sequence of SUGMs without covariates, where

$$p_\ell^n = F_\ell(\beta_\ell^n),$$

where $F_\ell(\cdot)$ is the uniform distribution on [0,1], β_ℓ^n is a parameter where subgraph ℓ forms if the random variable is less than β_ℓ^n. Typically $\beta_\ell^n \downarrow 0$ at appropriate rates.

Given g, the researcher seeks to estimate β_ℓ^n (equivalently p_ℓ^n). The quality of the estimates, whether the estimator is identified, consistent, and (when normalized appropriately) asymptotically normally distributed depends on the assumptions on the sequence $\beta^n = (\beta_1^n,\ldots,\beta_k^n)'$. We have already seen such a sequence and the asymptotic

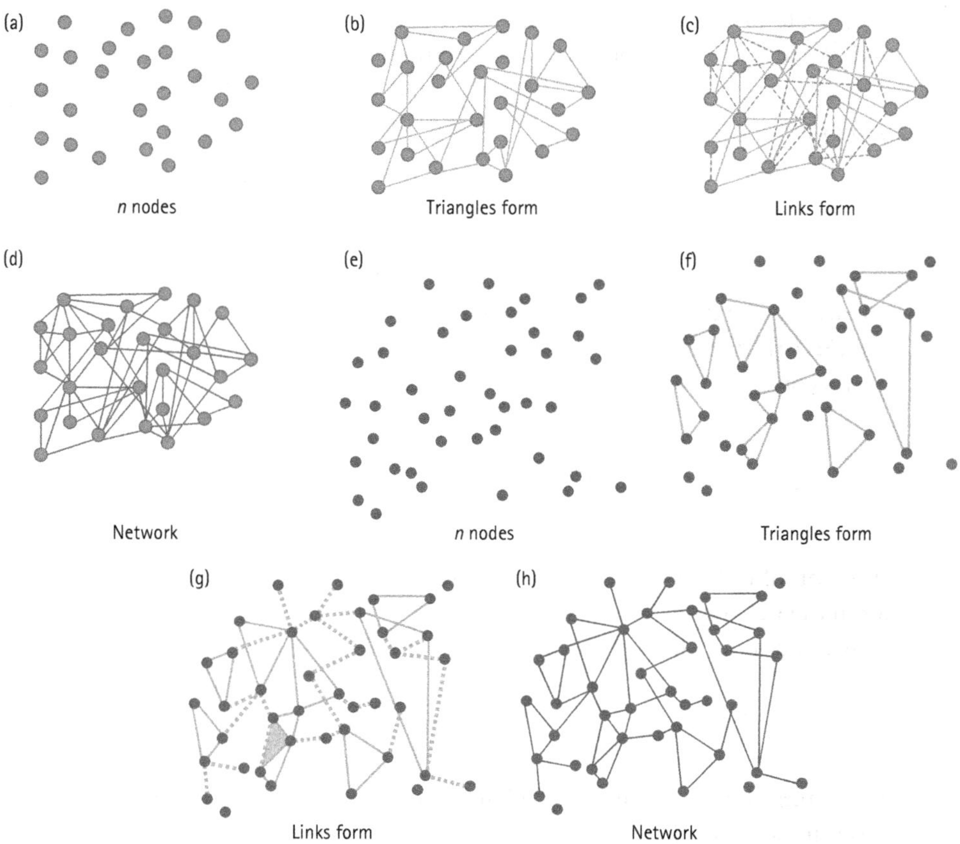

FIGURE 13.6 A figure depicting a SUGM process taken from Chandrasekhar and Jackson (2015). (A)–(D) shows the dense case and (E)–(H) shows the sparse case.

normality of the associated estimator when we studied the sparse Erdős-Rényi graph in equation (13.3.1). What we do now represents a generalization.

Incidental generation

What is the main difficulty that we face? As we can see in Figure 13.6, even though 20 triangles and 20 unsupported links form in the dense case, and though 9 triangles and 23 unsupported links form in the sparse case, we do not observe these sets $\mathcal{LU}$ and $\mathcal{T}$. Instead, we only observe the resulting graph g, which has 30 triangles and 4 unsupported links in the dense case and 10 triangles and 22 unsupported links in the sparse case. The generation of the extra triangles is called incidental generation.

There are two approaches to dealing with this problem. First, if the networks are sparse enough, then we can see that the error will be vanishing relative to the truly generated part, meaning that by directly counting subgraphs, one can recover the probability parameters of interest. Second, even if the networks are not sparse enough, because the rate of incidental generation varies as β varies, one can show how β is identified. We look at the general case first and then the sparse case.

General case

It is useful to define

$$S_\ell(g) := \frac{\sum_{i_1<\ldots<i_{m_\ell}} G_{i_1,\ldots,i_{m_\ell}}}{\binom{n}{m_\ell}}$$

as the share of subgraphs of type ℓ that exist in the graph. The vector $S(g) = (S_1(g),\ldots,S_k(g))'$ represents a set of moments that we will use in estimation.

When we consider our example of links and triangles, let

$$(S_L(g), S_T(g)) = \left(\sum_{i<j} A_{ij}/\binom{n}{2}, \sum_{i<j<k} A_{ij}A_{jk}A_{ik}/\binom{n}{3}\right)$$

be the fraction of links (whether isolated or in a triangle) and the fraction of possible triangles in the network g.

The researcher then estimates the minimizer of the GMM objective function.

$$\widehat{\beta} := \underset{\beta\in\mathcal{B}}{\operatorname{argmin}} \left[S(g) - \mathrm{E}_\beta S(g)\right]' W^n \left[S(g) - \mathrm{E}_\beta S(g)\right].$$

Under assumptions that allow for identification and limit the overall amount of correlation in the network, the parameters can be consistently estimated[18]

$$\delta(\widehat{\beta}, \beta^n) \xrightarrow{\mathrm{P}} 0.$$

The estimators, when appropriately normalized, are asymptotically normally distributed

$$D_n^{1/2}(\widehat{\beta} - \beta^n) \rightsquigarrow \mathcal{N}(0, I)$$

for some normalizing matrix sequence (D_n).

Let us highlight briefly the logic of identification and the asymptotic normality. The identification problem potentially comes from incidental generation: when one sees a triangle present, it could have been generated as a triangle, or as a mix of other triangles and links. The question is whether it is possible for a given vector of links and triangles to have two solutions (β_L, β_T) and (β_L', β_T').

The main identification result for the links and triangles case is that $(\beta_L, \beta_T) \neq (\beta_L', \beta_T')$ implies

$$\mathrm{E}_{\beta_L,\beta_T}\left[S_L(g), S_T(g)\right] \neq \mathrm{E}_{\beta_L',\beta_T'}\left[S_L(g), S_T(g)\right].$$

[18] $\delta(x,y) := \sum_k \frac{|x_k - y_k|}{\max(x_k, y_k)}$ is a distance metric.

To see this, consider the expectation of these moments where $q = q(\beta_L, \beta_T)$ is the unconditional probability of a link forming. We have

$$\mathrm{E}_{\beta_L,\beta_T}\left[S_L(g), S_T(g)\right] = \left[\beta_L + (1-\beta_L)(1-(1-\beta_T)^{n-2}), \beta_T + (1-\beta_T)(q-\beta_T)^3\right].$$

A link can be formed directly (with probability β_L) or by some triangle involving the two nodes in question (leaving $n-2$ choices for the third node). Meanwhile, a triangle forms directly (with probability β_T) or where the three links are formed by some combination of unsupported links and other triangles ($(q-\beta_T)^3$). This gives us two nonlinear equations and two unknowns with a unique solution under our assumption.

The intuition for identification is as follows. While the counts of triangles and links are being used, the key observation is that how triangles are glued together—whether they share nodes or share edges—tells us about what one parameter should be, holding fixed a budget of links. In that way, the arrangement of subgraphs relative to each other in a graph can give the researcher information about the degree of incidental generation versus direct generation.

Figure 13.7 provides simulation results to demonstrate how as sparsity declines, the direct estimation method is less suitable whereas the GMM approach works reliably.

Next, let us consider the logic as to why the parameter estimates are normally distributed. Standard arguments involve expanding the first-order condition around $(\widehat{\beta}-\beta^n)$. The key piece in the normality argument comes from showing that

$$(S_\ell - \mathrm{E}S_\ell)/\sigma_{n,\ell} \rightsquigarrow \mathcal{N}(0,1)$$

for $\sigma^2_{n,\ell} := \mathrm{var}(S_\ell)$.

The reason why asymptotic normality is not obvious here is because the summands in question can be correlated. It is useful to observe that in any SUGM that has a subgraph involving at least four nodes, any two links are always correlated. Consider the model of links, triangles, and four cliques (i, j, k, l all mutually connected). Then any g_{ij} and g_{kl} are surely correlated. Further, for any collection of *ijkl* and *mnop* of nodes, even if they have no nodes in common, the events $\{g_{ij}g_{ik}g_{il}g_{jk}g_{jl}g_{kl} = 1\}$ and $\{g_{mn}g_{mo}g_{mp}g_{no}g_{np}g_{op} = 1\}$ are correlated. In this case

$$(S_\ell - \mathrm{E}S_\ell)/\sigma_{n,\ell} = \sigma^{-1}_{n,\ell} \sum_{i<j<k<l} G_{i,j,k,l} - \mathrm{E}[G_{i,j,k,l}]$$

where summands $G_{i,j,k,l}$ and $G_{m,n,o,p}$ are correlated irrespective of the identities of the nodes in question.

On the other hand, the covariance between any two links is low. Moreover, when we look at subgraphs of four nodes, for instance, when there are few nodes in common between *ijkl* and *mnop*, the correlation in both subgraphs existing comes from a somewhat unlikely event: a four clique between *ijmn* and then some subgraphs forming to fill out the four-cliques *ijkl* and *mnop*.

If elements in an array can all be correlated with each other, if most of the correlation can be restricted to a small subset of the array, then Chandrasekhar and Jackson (2015)

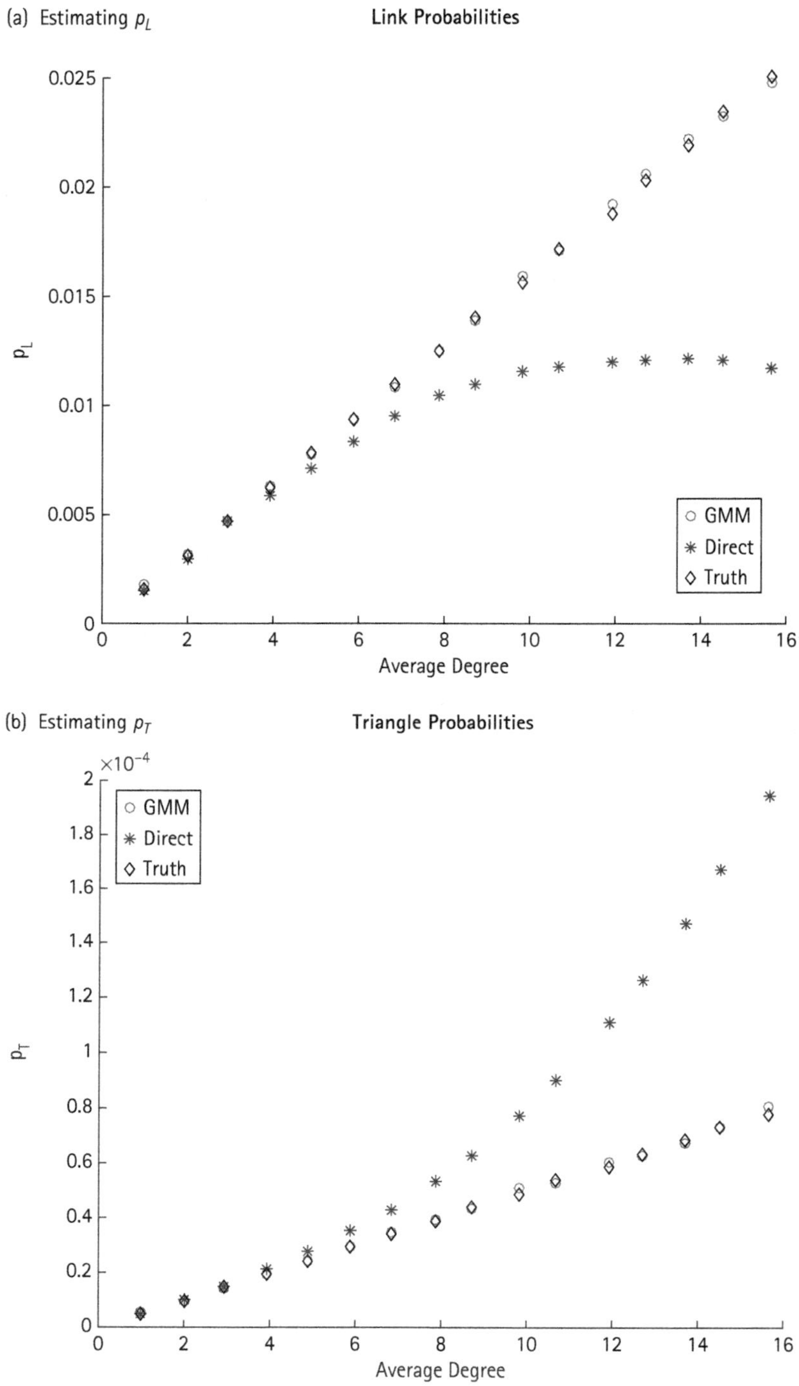

FIGURE 13.7 Parameter estimates as a function of density.

show that a central limit theorem applies. In the example described above there are many fewer four cliques that share nodes than those that are node-independent, allowing such a theorem to be applied. The result uses a key lemma in Stein (1986) that characterizes the distance of a random variable Y from a normal distribution (Z) in the Wasserstein metric by

$$d_W(Y,Z) \leq \sup_{\{f:||f||,||f''||\leq 2,||f'||\leq\sqrt{2\pi}\}} \left|E[f'(Y) - Yf(Y)]\right|.$$

Therefore, the goal is to show that the right-hand side tends to zero for $Y := (S_\ell - ES_\ell)/\sigma_n$. This strategy has been used widely, for instance by Bolthausen (1982), to prove central limit theorems for strongly mixing random fields, as well as by Baldi and Rinott (1989) to look at data that can be described by dependency graphs. In the latter example, every observation can be correlated with some (arbitrary) k other observations but is independent of the rest. Ross et al. (2011) provides an excellent survey of the techniques. The SUGM case is somewhat more complicated, because (i) in principle, all summands can be correlated with all other summands (therefore k can equal the number of observations), making one unable to rely on independence in the argument; and (ii) unlike in the time series or spatial case, one cannot rely on the idea that the summands can be embedded in some Euclidean space so that asymptotically each summand is infinitely far from most other summands. Ultimately, the normality result applies for SUGMs that admit considerable density as well as sparsity. For instance, in a links and triangles model, a node can have order $O(n^{1-\epsilon})$ expected degree for $\epsilon \in (0,1]$.

Sparse Case

While in the general treatment, we typically had $\beta_\ell^n \downarrow$ for parameters that did not correspond to direct links, the rate at which this was required was not particularly stringent. In particular, it allowed for a tremendous amount of incidental generation, and in fact, allowed for dense graphs (e.g., so long as less than half the links were present in expectation in the links and triangles model).

As we discussed previously, empirical network data is rather sparse. It turns out that in this case, our problem becomes much simpler. We can compute that it is exponentially more likely to have a directly formed subgraph than one formed as a combination (incidentally generated). In turn, this implies that the measurement error problem is small, so direct counting/regression approaches will work.

To make the required convergence rates concrete, in the links and triangles model we need

$$p_L = o(n^{-1/2}) \text{ and } p_T = o(n^{-3/2}).$$

This means that every node can participate in $o(\sqrt{n})$ unsupported links and $o(\sqrt{n})$ triangles, which as we know from the stylized facts, is more than sufficient to model empirical data.

To see how this would be estimated in practice, let us stay with the links and triangles SUGM with covariates where the probability of a triangle or a link can depend on covariates of the nodes:

$$p_T^n(x_T;\beta_T) = \frac{\exp(\beta_{0,T}^n + \beta_{1,T}(|x_i - x_j| + |x_j - x_k| + |x_i - x_k|))}{1+\exp(\beta_{0,T}^n + \beta_{1,T}(|x_i - x_j| + |x_j - x_k| + |x_i - x_k|))}, \text{ and}$$

$$p_L^n(x_L;\beta_L) = \frac{\exp(\beta_{0,L}^n + \beta_{1,L}(|x_i - x_j|))}{1+\exp(\beta_{0,L}^n + \beta_{1,L}(|x_i - x_j|))}.$$

Under an assumption that preserves sparsity—for instance, the constants $\beta_{0,\ell}^n \to -\infty$ at appropriate rates—the parameters β_ℓ can be estimated consistently with estimators that are asymptotically normally distributed. Further, the estimation procedure is rather direct. In the links and triangles example:

1. Consider all *ijk*, and conduct a logistic regression of whether a triangle exists on covariates of the triple. The coefficients are $\widehat{\beta}_T$.
2. Eliminate all nodes *ij* where *ij* was a member of some triangle. With the reduced data set, conduct a logistic regression of whether an edge exists on covariates of the pair. The coefficients are $\widehat{\beta}_L$.

13.4.3 Sequential Arrival and Preferential Attachment

A commonly studied class of random network models is *preferential attachment*. First introduced by Barabasi and Albert (1999), the model proceeds as follows.

Given a seed graph, in every period some nodes are born. These nodes link with some probability to previously existing nodes with probabilities proportional to the previously existing nodes' degrees. This generates a rich-get-richer phenomenon, where more popular nodes differentially become more likely to gain links. By tuning the number of nodes born each period, the number of edges born each period, and so on one can vary properties of the resulting distribution of graphs.

The main result of the literature is the ability to generate a power law tailed degree distribution. This is in contrast with (sparse) Erdős-Rényi graphs, which generate Poisson (thin) tails.

There are several difficulties in employing these models. Until recently, we knew little about whether parameter estimates were consistent. Nonetheless, practitioners made use of either a nonlinear least squares estimator an estimator by Hill (1975). A key contribution to this literature was made in Kolotilin (2013) and fills this gap. The paper shows that for a generalized class of preferential attachment models, the formation parameters can consistently estimated by a certain class of GMM estimators. At the same time, the nonlinear least squares and Hill estimators exhibit severe biases. However, the question as to whether these consistent GMM-based estimates are asymptotically normally distributed remains open. Kolotilin (2013) provides some simulations suggesting this may be the case, but there are no formal results in the literature.

In a seminal paper, Christakis et al. (2010) developed a sequential arrival model in this vein. In every period, one pair of agents have the opportunity to form a link and conditional on the opportunity the link is formed if both agents consent. As in the aforementioned models (e.g., Butts 2009; Mele 2013; and Badev 2013), agents are myopic and therefore engage in myopic best response.

A simplified version of the utility function they use is given by

$$u_i(g^{t-1} + ij) = \beta_0 + \beta_1 x_j - \beta_2 \left|x_i - x_j\right| + \beta_d d_j(g^{t-1}) + \beta_T \mathbf{1}\{\gamma_{g^{t-1}}(i,j) = 2\} + \epsilon_{ij}.$$

To make notation transparent $d_j(g^{t-1})$ is the degree in graph g^{t-1} of j whereas $\gamma_{g^{t-1}}(i,j)$ is the distance in the graph g^{t-1} from nodes i to j. Thus, individuals receive utility from being linked to more popular individuals and from completing a triangle.

A key difference in this model relative to those that led to stationary distributions (e.g., Butts 2009; Mele 2013; Badev 2013) is that their posited process is a one-time match process. The $\binom{n}{2}$ pairs meet in some unobserved sequence and therefore the econometrician has to integrate this sequence out. They use Bayesian MCMC methods to obtain draws from the posterior distribution of interest.

A related sequential arrival model in the literature is Koenig (2012). This paper studies a preferential attachment style network formation model. Agents arrive in a sequence and form links with preexisting nodes. The agents have utilities that correspond, essentially, to having incentive to link to the highest degree partner. In that way the model closely resembles the Barabasi and Albert (1999) model discussed before. The crucial departure from this literature is that he tinkers with the observation radius, meaning how much a newborn node i can see when making her decision. There are two regimes he considers. First, the agent at time t may see a large portion of the preexisting network (essentially all its predecessors and their links, similar to the usual Barabasi and Albert 1999 model). Second, the agent may only see a selected subset of nodes, their degrees, and their neighbors' degrees. By assumption the agent's payoff depends on her own degree and her neighbors' degrees and as such, and chooses to link with the highest degree nodes in her observation radius. The paper focuses on (i) how the observation radius affects the patterns of networks that emerge through this process and (ii) how the network patterns also vary with how much neighbors' degrees affect one's incentive to link with them (*noise* in linking decision). The author is able to generate patterns that both fit empirical data better than standard preferential attachment models which usually assume that a node born in period t sees the entire pre-existing network.

As the literature currently stand, neither of the models described here are known to have estimators that are consistent for large n. Results such as these are open questions.

13.4.4 Strong Homophily

There is a set of models in the literature that exploits a strong form of homophily to generate consistently estimable parametric utility functions. Instead of assuming additive separability over subgraph payoffs, they place less restriction on the local payoff structure.

However, the price paid for releasing the restrictiveness here is (i) more structure on which nodes are allowed, under the model, to link to which other nodes, (ii) intensive computational requirements, and (iii) partial identification. This approach has been taken in Goldsmith-Pinkham and Imbens (2013), Boucher and Mourifié (2012), and Leung (2014). Below, we review a simplified version of these models, using the notation closest to Leung (2014).

13.4.4.1 Overview

In this model, nodes have deterministic locations, $l_i \in \mathbb{R}^d$. Additionally, they have random characteristics x_i. Finally, there are pairwise shocks ϵ_{ij}. Let $w = \{(l_i, x_i, \epsilon_{ij})\}_{i,j\in V}$. As before, a general parametric model to write the utility of an equilibrium graph is

$$u_i(g, w; \beta).$$

As in previous sections, the literature focuses on pairwise stability. Define

$$U_{ij}^n(g, w, \beta) = u_i(g + ij, w, \beta) - u_i(g - ij, w, \beta).$$

This encodes the marginal value at the equilibrium graph of having link ij.

13.4.4.2 Estimation Requirements

In order to identify β, the econometrician will have some moment functions, $\psi_i^n(\beta)$, derived from the game. A crucial step, therefore, in conducting asymptotic inference is establishing a central limit theorem for the moments:

$$\frac{1}{\sqrt{n}} \sum_i \psi_i^n(\beta) \rightsquigarrow \mathcal{N}(0, \Sigma).$$

The main result is that if we assume that i and j are located far from each other, it is almost never worthwhile to form a direct link. Thus, one can show that the system $\{\psi_i^n : i \in V_n, n \in \mathbb{N}\}$ is an α-mixing random field.

To see what this means, consider two sets of nodes $V_A, V_B \subset V_n$. Let E_A and E_B be events in the σ-algebras generated by $\{\psi_i : i \in V_A\}$ and $\{\psi_i : i \in V_B\}$. Then

$$|\mathrm{P}(E_A \cap E_B) - \mathrm{P}(E_A)\mathrm{P}(E_B)|$$

looks at the probability of both events occurring versus the product of the probabilities that each occurs. If the events were independent, this is 0. Loosely, this is a measure of the degree of interdependence between random variables on V_A and V_B.

The α-mixing coefficient measures this. Formally,

$$\alpha_{k,l}(r) := \sup_{n\geq 1} \sup\{|\mathrm{P}(E_A \cap E_B) - \mathrm{P}(E_A)\mathrm{P}(E_B)| : E_j \in \mathcal{F}_j, j \in \{A,B\}, |V_A| \leq k,$$
$$|V_B| \leq l, d(V_A, V_B) \geq r\}$$

where $d(V_A, V_B)$ is the Hausdorff distance in l-space between the sets. If we look at all events taking place among nodes in V_A and V_B, with sizes k and l, if the sets are least r-distance apart, then the degree of interdependence over all events is at most $\alpha_{k,l}(r)$. A sufficient condition for a central limit theorem has this coefficient going to zero sufficiently fast as $n \to \infty$.

13.4.4.3 Driving Assumptions

The next step is to make assumptions on preferences such that the resulting distribution over links generated by strategic interaction will satisfy some mixing properties.

Assume that nodes arrive in communities. $C(i)$ is a set of nodes that are at most r_0 apart from each other. The distance between two nodes in "location" space, an admittedly clumsy term, is

$$d(i,j) := \left\| l_i - l_j \right\|.$$

Additionally, parametrize the marginal utility of having link ij as

$$U_{ij}(g, w; \beta) := U_{ij}(\underbrace{d(i,j)}_{\text{distance}}, \underbrace{z(g,x,\epsilon)}_{\text{endogenous}}, \underbrace{f(x,\epsilon)}_{\text{exogenous}}; \beta).$$

Note that $z = z(g,x,\epsilon)$ is a function that allows for endogenous effects of the network since it takes g as an argument. $f(x,\epsilon)$ encodes the direct effect of exogenous factors. In principle, endogenous effects allows for arbitrary correlation between links. To make progress, four key assumptions are made.[19]

A1. *Strong homophily:* Irrespective of the endogenous network effects on utility, it is typically not worthwhile to directly link to someone very far away:

$$\lim_{d(i,j)\to\infty} U_{ij}(d(i,j), \bar{z}, f(x,\epsilon); \bar{\beta}) < 0.$$

for $f(x,\epsilon) \leq \bar{f}$, and the probability of $f(x,\epsilon) > \bar{f}$ is sufficiently low.

If $f(x,\epsilon)$ is bounded, then A1. makes it impossible for certain nodes to link. Also observe that it is necessary to bound the endogenous effect. So very far nodes only link due to purely exogenous factors, and this is rare.

[19] For simplicity, let z and β be unidimensional and bounded by $\bar{z}, \bar{\beta}$ and let the marginal utility increase in these parameters.

A2. *Increasing domain:* All but finitely many nodes are within some minimum radius r_0 of any given node i:

$$\sup_{n\geq 1}\max_{i\in V_n}\left|j \in V_n : d(i,j) < r_0\right| < \infty.$$

This means that communities are small relative to the overall network.

A3. *Independent equilibrium selection/no coordination:*[20] For any given draw of the primitives x, ϵ, sets of nodes that aren't connected under any equilibrium independently form subnetworks.

Note that this is satisfied if equilbrium selection is degenerate. It is also satisfied under myopic best response dynamics.

13.4.4.4 Main Results

Let us sketch the proof of α-mixing from A1–A3. Consider an event

$$E := \{i \leftrightarrow j, d(C(i), C(j)) = r\},$$

that i and j are path connected in the graph when they are members of groups that satisfy $d(C(i), C(j)) = r$. We want to look at the probability of events $E_i := \{\psi_i \in H\}$ and the probability that $E_j = \{\psi_j \in H'\}$,

$$\left|P(E_i \cap E_j) - P(E_i)P(E_j)\right| \leq \underbrace{\left|P(E_i \cap E_j|E^c) - P(E_i|E^c)P(E_j|E^c)\right|}_{=0 \text{ by A3.}} P(E^c) + 3P(E).$$

If i and j are not path-connected, by A3, the events are independent. So we just need to know how likely it is that i and j are path connected in the graph if they are at least r apart.

$P(E)$ is a complicated event. There are many possibilities across many paths, potentially on the order of $n!$. A strong but sufficient condition provided in Leung (2014) illustrates that if groups of nodes are further and further apart along the sequence—that is, if there is drift—then one can drive this rate to zero. This brings us to the fourth assumption. I provide a simplified version here.

A4. *Diversity:* For every i and r, there exists a set of nodes S containing i such that any $k \in S$ satisfies $d(i,k) < r$ and $d(k,l) \geq 9\log r$ for $l \notin S$.[21]

This provides a sufficient condition to argue that

$$P(E) \leq \phi(r) \to 0 \text{ as } r \to \infty,$$

[20] Set of rationalizable networks is non-empty and the true data generating process is rationalized by the model and the selection mechanism.

[21] This applies when errors have exponential tails only.

which completes the α-mixing argument.

To see why A4 is important, consider bounding the probability that there exists some path between i and j:

$$\mathrm{P}(E) = \mathrm{P}\left(\underbrace{\bigcup_{\ell=1}^{n-2}}_{\#\text{ nodes in path}} \bigcup_{k_1,\ldots,k_\ell \in V_n} \left\{ A_{ik_1} = A_{k_1,k_2} = \cdots = A_{k_\ell,j} = 1 \right\} \right). \tag{13.5}$$

Typically, we would take a union bound, but in this case we would have more than $n!$ terms, which would not be useful at all. A4 allows us to turn a number of these terms off, permitting the union bound.

Assume $d(i,j) = r$. Then $i \leftrightarrow j$ implies that there exists k with $d(i,k) < r$ is directly connected to some node l such that $d(k,l) \geq 9 \log r$. This means that if two nodes r apart are connected, then some pair of nodes on that path must be $9 \log r$ apart. Formally,

$$\mathrm{P}(E) \leq \mathrm{P}\left(\bigcup_{\substack{k:d(i,k)<r, \\ l \in V_n, \\ d(k,l) \geq 9 \log r}} \left\{ A_{kl} = 1 \right\} \right). \tag{13.6}$$

This significantly reduces the number of events in the union. We can then apply the union bound:

$$\mathrm{P}\left(\bigcup_{\substack{k:d(i,k)<r, \\ l \in V_n, \\ d(k,l) \geq 9 \log r}} \left\{ A_{kl} = 1 \right\} \right) \leq (\#k : d(i,k) < r) \max_k \sum_{l=1}^{n} \mathrm{P}(A_{kl} = 1 \mid d(k,l) \geq 9 \log r).$$

Intuitively, as the sum ranges over nodes l increasingly distant from k, the summands on the RHS should be decaying to zero by homophily. Furthermore, as $9 \log r \to \infty$, the entire summation decays to zero. Some restrictions ensure that this decay is fast enough to overtake the growth of $(\#k : d(i,k) < r)$ as r increases. Then the RHS can be bounded by some $\phi(r) \to 0$ as $r \to \infty$, thus bounding $\mathrm{P}(i \leftrightarrow j)$.

In sum, exploiting strong homophily allows us to construct models of sparse networks and derive limit theorem by showing mixing properties. While the preferences can be left somewhat unrestricted locally (geographically), allowing for multiple equilibria and motivating partial identification approaches, by construction the interactions have to be localized. Figure 13.8 presents a schematic of the adjacency matrix implied if nodes were embedded in one dimension. Imagine that nodes were arranged geographically on a line. Nodes are more likely to establish links to their

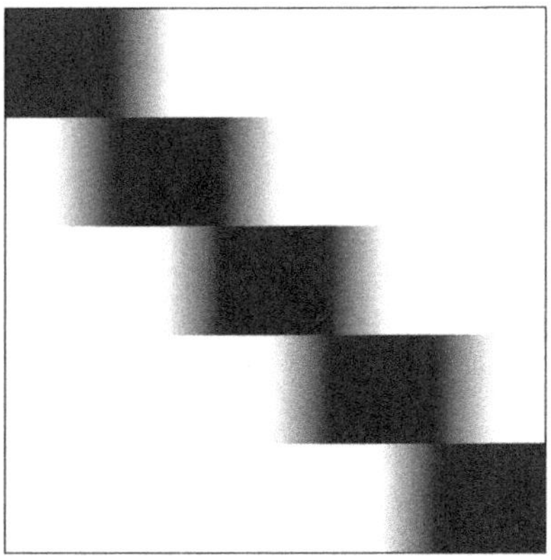

FIGURE 13.8 A schematic of the adjacency matrix of a network under strong homophily. Darker shades indicate greater likelihoods of link formation.

local, geographic neighbors, whereas links between distant nodes are unlikely. While the highly local link patterns are unrestricted, there are strong conditions placed on the off-diagonal terms beyond the main block. Just as in a random geometric graph, distant nodes simply cannot link, so too is the case here. Connectedness and other graphical properties of these models remain open for future research.

13.4.5 Network Types

de Paula, Richards-Shubik, and Tamer (2014) look at identifying preferences in network formation games. As in several preceding models, they assume that agents strategically form links under pairwise stability. The logic of the approach taken here is to consider local network structures, which they call "types," and to show how the relative proportions of various types map to preference parameters. Identification in this framework will be partial because there is a multiplicity of equilibria, so generally, there will be a set of parameters that are consistent with the model and data.

13.4.5.1 Driving Assumptions

A1. *Local payoffs:* Only links up to distance D are payoff relevant, and degree is bounded by L.

This allows control over the local structures that will become payoff relevant. Observe that it imposes a strong sparsity, that the expected average degree is $O(1)$, which may not be a binding restriction in the data (Section 13.2). On the other hand, this approach by construction will struggle to generate heavy-tailed degree distributions.

A2. *Preference shocks:* There is one preference shock $\epsilon_{il}(x)$ for each potential direct connection $l = 1, \ldots, L$ and characteristic $x \in \mathcal{X}$. Further, $supp(\mathcal{X})$ is finite, and the shock is independent of x and is drawn from a known distribution (up to unknown finite dimensional parameter).

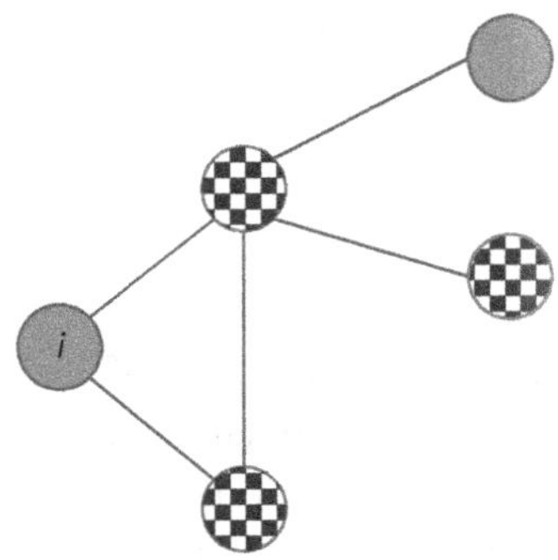

FIGURE 13.9 An example of a network type with $D = 2$ and $L > 2$. Characteristics are either blue (solid) or red (checkered) and $v_1 = Blue$.

A substantive restriction imposed here is on unobserved heterogeneity. Suppose agents j and k have the same observables. Then an agent i is different between using her lth link on j or k.

While the utility functions studied in the paper are general (required to satisfy A1 and A2), the leading example should be familiar to the reader (e.g., Christakis et al. 2010; Goldsmith-Pinkham and Imbens 2013; Mele 2013):

$$u_i(g,x) = \sum_j A_{ij}\left(f(x_i,x_j) + \epsilon_{ij}(x_j)\right) + \beta_T \sum_{j<k\neq i} A_{ik}A_{jk} + \beta_{FoF}\left(\left|\cup_{j\in N_i} N_j\right| - d_i(g)\right),$$

where $N_i = \{j : A_{ij} = 1\}$. To maintain A1, they impose the restriction that $\max_i d_i(g) \leq L$. Payoffs depend on direct friends and their traits, the number of triangles, and the number of friend-of-friends who aren't direct friends.

A3. *Value of indirect links close to value of closure:* The value of mutual friends is not large relative to the value of friends of friends: $\beta_T \leq \frac{L}{L-1}\beta_{FoF}$ and $\beta_{FoF} > 0$.

This assumption is not used in the general case, but relevant when studying the utility specified above. It imposes that the value of a triad is almost the same as a friend of friend. For instance, if the maximal degree is 11, then the utility from having a triangle can be at most 10% more than the utility of having a friend of friend. How reasonable this assumption is, of course, depends on context. It may be appropriate where closure is not important, and less relevant when closure is key (e.g., favor exchange or risk sharing without commitment).

Two novel concepts introduced in this paper are network types and preference classes. The *network type* describes the local network up to distance D from the reference node. The type is a pair $t = (A^{loc}, v)$ where A^{loc} is a local square adjacency matrix and v is a vector of characteristics whose length is the size of A^{loc}.

The nodes of A^{loc} consist of the reference node, its direct potential neighbors (L elements), its second-order neighbors ($L(L-1)$ elements), through its $D-1^{th}$ order neighbors ($L(L-1)^{D-1}$). Therefore, A^{loc} is a square of size $1 + L\sum_{d=1}^{D}(L-1)^{d-1}$ and v has this length. v_1 encodes i's characteristics, and v_k encodes the characteristics x_k of the kth node in A^{loc}. The set of network types is denoted $\mathcal{T}$.

A *preference class* is a subset $\mathcal{H} \subset \mathcal{T}$ of types defined as follows. Consider mapping characteristics and shocks $(x,\epsilon) \mapsto H(x,\epsilon) \in 2^{\mathcal{T}}$ to all types that satisfy

$$u\left(A^{loc}, v; \epsilon\right) \geq u\left(A^{loc}_{-j}, v; \epsilon\right)$$

for $j = 1, \ldots, L$. This means that it is the subset of types such that for some characteristic and shock vector, the reference agent wants to keep all her links in the local substructure. Essentially, preference classes are types that can be rationalized in the data for some collection of covariates and shocks.

The next two conditions are necessary for the observable distribution of network types to correspond to a pairwise stable network that could be generated under a given set of preference parameters and shocks. Let $\alpha_H(t)$ denote the share of nodes in preference class H that are of network type t. This is called an *allocation parameter.*

A4. *Existing links shouldn't be dropped:* All existing links are pairwise stable. For any type t and preference class H

$$t \notin H \implies \alpha_H(t) = 0.$$

Next, we define μ_{v_1} the share of nodes that have characteristic $x = v_1$. As all network types in a given preference class by construction have the same characteristic for the reference node, we can define the probability of an ϵ that would support all types in H

$$\mathrm{P}_{H|v_{1(t)}} := \mathrm{P}\left(\epsilon : H(x,\epsilon) = H | x = v_1\right).$$

A5. *No mutually beneficial links among far nodes:* For any two type pairs t, s where the nodes are at distance of at least $2D$, where adding a link between the nodes now generates two type pairs $\bar{t}, \bar{s}$:

$$\left(\mu_{v_1(t)} \sum_{\check{H} \in \mathcal{H}} \mathrm{P}_{\check{H}|v_1(t)} \alpha_{\check{H}}(t) \cdot \mathbf{1}\left\{\bar{t} \in \check{H}\right\}\right) \cdot \left(\mu_{v_1(s)} \sum_{\tilde{H} \in \mathcal{H}} \mathrm{P}_{\tilde{H}|v_1(s)} \alpha_{\tilde{H}}(s) \cdot \mathbf{1}\left\{\bar{s} \in \tilde{H}\right\}\right) = 0.$$

13.4.5.2 Main Results

The first main result is that given a distribution of preferences in the population, if there exists a pairwise stable network where the proportion of agents of type t is equal to π_t for every $t \in \mathcal{T}$, then there exists a vector of allocation parameters α satisfying assumptions 4 and 5 such that

$$\pi_t = \frac{1}{\mu} \sum_{H} \mu_{v_1(t)} \mathrm{P}_{H|v_{1(t)}} \alpha_H(t) \text{ for every } t \in \mathcal{T}.$$

The logic of the proof is as follows. For A4, we can prove the contrapositive: $\alpha_H(t) > 0 \implies t \in H$. This follows because under pairwise stability, all links are such that it is better for each participant to keep the link. This implies (by definition) that every node is in a network type that is within her preference class.

To prove A5 is satisfied, start with any pair of types t, s. Then consider the (new) pair of types that would exist if links were added between the two candidate nodes $(\bar{t}, \bar{s})$. The argument will show that the share of at least one of these new types is zero in the population.

They also show that under preference classes such as the one described above, A4 and A5 are necessary and sufficient for pairwise stable networks to exist. This argument relies on A3, which ensures that there is a balance between the value of a triangle and the value of a friend of friend.

The second main result concerns computation. The key observation is that A5 is stated as a product and, therefore, can be written as a quadratic function of allocation parameters. They have to define a Q such that A5 can be written as $\alpha' Q \alpha = 0$ at the optimal $\alpha_H(t)$. Q is defined in the following way: with elements such that both types t and s involved in $Q_{t,s}$ would prefer to link to each other if they were at length $2D$ apart.

The result is that given a structural parameter vector θ yielding $\mathrm{P}(\cdot)$, a network with type shares $\{\pi_x(t)\}$ satisfies A4 and A5 if and only if

$$\min_{\{\alpha_H(t):\, t \in H\}} \alpha' Q \alpha$$

such that

1. $\sum_{t \in H} \alpha_H(t) = 1, \forall H$
2. $\alpha_H(t) \geq 0, \forall t, H$
3. $\sum_H \mathrm{P}_{H|x} \alpha_H(t) = \pi_x(t), \forall t$

is equal to zero.

To see the argument, observe that A5 holds if and only if

$$\sum_{\bar{H} \in \mathcal{H}} \sum_{\tilde{H} \in \mathcal{H}} \alpha_{\bar{H}}(t) \cdot 1\{\bar{t} \in \bar{H}\} \alpha_{\tilde{H}}(s) \cdot 1\{\bar{s} \in \tilde{H}\} = 0.$$

Meanwhile, A4 is encoded in the constraint that allocation parameters are only over the preference classes.

13.4.6 Discussion

We have reviewed several classes of econometric models of network formation that allow us to identify (sets of) parameters and, at times, establish when the estimators are consistent with a known asymptotic distribution as $n \to \infty$. One common aspect of almost every model that we have looked at is that payoffs are localized in some way. Payoffs were explicitly limited to the structure of the subgraph(s) that a reference node is a member of, and different models took different stands on the degree to which, ex ante, certain nodes were allowed to link with others.

The different modeling choices involved trading off restrictions on what sort of adjacency matrices were admitted against the degree of structure placed on equilibrium selection. For instance, the models of strong homophily and network types bioth imposed very little on equilibrium selection. But this forced very specific network structures: a near-block diagonal structure in the case of strong homophily, and a strongly sparse structure with limited value on triadic closure in the case of network types. Subgraph-generated models allowed for more flexible modeling of network structure but required more separability and myopia in the microfoundations. The basic tension is that without limiting the degree of interdependence, through direct payoffs, strategic incentives, or otherwise, even a large network will not have enough information to either identify the formation parameters or estimate them consistently.

A role for experiments

Research on the econometrics of network formation has plenty to gain by interacting with empirical research. Experiments in the field, or even more in the lab, may teach us about what individuals take into account as they form networks (Burger and Buskens 2009; Kearns, Judd, and Vorobeychik 2012). How localized are individuals' preferences? How myopic are decisions? How separable are payoffs in subnetwork structure? Given the complexity of the modeling decisions and the sensitivity of the macrostructure of the resulting networks to these choices, it may be particularly important to take lessons from experimental economics as researchers develop further models.

The value of panel data

In this chapter, I have focused on estimating network formation when the researcher has access to an edge list from a single large network as well as a vector of covariates for each agent. Another strand of the literature has exploited panel data. I outline the ideas briefly in three seminal papers but a detailed discussion is beyond the scope of this chapter.

The first context in which panel data is useful is when the researcher wants to identify which agents influence others through the network when the network is unobserved. Manresa (2014) takes on this challenge by considering an environment in which agents' behavior depends on their neighbors' behavior, but the econometrician does not actually see the network.[22] A very simplified version of this framework is

$$y_{i,t} = \alpha + \beta \sum_j \frac{A_{ij}}{d_i} x_{j,t} + \delta x_{i,t} + \epsilon_{i,t}.$$

The econometrician observes $(y_{i,t}, x_{i,t})_{(i,t) \in V \times T}$, and the goal is to estimate (α, β, δ) as well as $\mathbf{A}$. If the underlying network is sparse, in the sense that the maximal degree is small relative to the number of time periods in the panel, the network structure can be recovered by using a pooled LASSO estimator (Tibshirani 1996; Meinshausen and

[22] Another issue that arises in the context of econometrics of network formation has to do with sampled networks (Chandrasekhar and Lewis 2013). In a panel setting, the Manresa (2014) results allow such problems to be circumvented. A discussion of the econometrics of sampled networks is beyond the scope of this chapter.

Bühlmann 2006; Yuan and Lin 2007; Meinshausen and Yu 2009).[23] The intuition is that the panel provides information as to when correlations in x's of two agents lead to correlation in outcomes and, provided there is enough sparsity in the system, one can sort through the $\binom{n}{2}$ potential links and identify which are present using LASSO approaches.

The second context in which panel data is useful is when the researcher has a second-stage equation which relates outcomes to neighbors' outcomes, as well as a first-stage equation that considers network formation. Both Goldsmith-Pinkham and Imbens (2013) and Graham (2014b) look at how unobserved heterogeneity can confound a system of equations an econometrician is interested in estimating. The intuition is that unobserved heterogeneity can make it so that certain agents are likely to interact and form a link. Moreover, the unobserved heterogeneity may directly influence outcomes, but by virtue of being unobserved, may bias the researcher's estimate of a peer effect. Goldsmith-Pinkham and Imbens (2013) develop a method to identify such endogeneity in a panel setting. Graham (2014b) shows that when a researcher has access to four periods of network data, by looking at how the network transitions, one can identify various parameters of network formation despite the unobserved heterogeneity.

13.5 Conclusion

In this chapter, I have surveyed econometric models of network formation that are estimated using data from a single large network observed for one period. My goal has been to acquaint readers with network formation models used across a variety of disciplines. I have tried to highlight how well various models reflect patterns in empirical data and the trade-offs they present when thinking about microfoundations.

Empirical network data tend to be very sparse and highly clustered. This represents a challenge to the modeler, because both sparsity and clustering imply that the network carries limited information. Since links are rare, this implies that there are fewer effective observations, which can limit the degree to which a researcher can model heterogeneity. Further, because links are correlated, the researcher could in principle have far less than $\binom{n}{2}$ independent units of information.

Models that treat links as conditionally edge-independent need to allow for correlation in links either through heterogeneity of nodes or through unobserved community assignment. The difficulty these models encounter is that they require a certain degree of network density in order to have enough information to identify the heterogeneity parameters of interest, but this may not be consistent with the sparsity of empirical data. Further, in many contexts, economic payoffs may not simply be bilateral, but instead depend on multilateral linking decisions. Meanwhile, models that more directly

[23] These results also have a connection to the literature on Gaussian graphical models.

model correlation in links through payoffs that depend on other subnetwork structures local to the node in question have to deal with a trade-off between how much structure they impose in terms of equilibrium selection and how much structure they impose on the shape of the network (e.g., through forcing a block diagonal structure or strongly limiting the value of triadic closure).

The challenges presented here require the researcher to become familiar with a broad set of tools. As we have seen, the tools used in modeling network formation can come from a variety of topics spread across several disciplines (e.g., statistics, computer science, sociology, statistical mechanics). Further, researchers working in this space may be able to draw inspiration from empirical work. For instance, experiments that vary aspects of the environment and observe the resulting patterns of interaction may provide some insight about what modeling assumptions are likely to be more relevant when developing models.

References

Airoldi, E. M., D. M. Blei, S. E. Fienberg, and E. P. Xing (2008). "Mixed membership stochastic blockmodels." *Journal of Machine Learning Research* 9, 1981–2014.

Albert, R. and A. Barabasi (2002). "Statistical mechanics of complex networks." *Reviews of Modern Physics* 74, 47–97.

Aldous, D. J. (1981). "Representations for partially exchangeable arrays of random variables." *Journal of Multivariate Analysis* 11, 581–598.

Aldous, D. J. (1985). *Exchangeability and Related Topics*. Springer.

Badev, A. (2013). "Discrete games in endogenous networks: Theory and policy." Mimeo, *University of Pennsylvania*.

Baldi, P. and Y. Rinott (1989). "On normal approximations of distributions in terms of dependency graphs." *The Annals of Probability* 1646–1650.

Banerjee, A., A. Chandrasekhar, E. Duflo, and M. Jackson (2013). "Diffusion of microfinance." *Science* 341, DOI: 10.1126/science.1236498, July 26, 2013.

Banerjee, A. V., A. Chandrasekhar, E. Duflo, and M. O. Jackson (2014). "Gossip: Identifying central individuals in a social network." Available at SSRN 2425379.

Barabasi, A. and R. Albert (1999). "Emergence of scaling in random networks." *Science* 286, 509.

Bhamidi, S., G. Bresler, and A. Sly (2008). "Mixing time of exponential random graphs." *Arxiv preprint arXiv:0812.2265*.

Bickel, P., A. Chen, and E. Levina (2011). "The method of moments and degree distributions for network models." *Annals of Statistics* 39, 2280–2301.

Bickel, P. J. and A. Chen (2009). "A nonparametric view of network models and Newman–Girvan and other modularities." *Proceedings of the National Academy of Sciences* 106, 21068–21073.

Bloch, F., G. Genicot, and D. Ray (2008). "Informal insurance in social networks." *Journal of Economic Theory* 143, 36–58.

Bloch, F. and M. Jackson (2006). "Definitions of equilibrium in network formation games." *International Journal of Game Theory* 34, 305–318.

Blume, L. E., W. A. Brock, S. N. Durlauf, and Y. M. Ioannides (2010). "Identification of social interactions."

Bollobás, B., S. Janson, and O. Riordan (2011). "Sparse random graphs with clustering." *Random Structures & Algorithms* 38, 269–323.
Bolthausen, E. (1982). "On the central limit theorem for stationary mixing random fields." *The Annals of Probability* 1047–1050.
Borgs, C., J. Chayes, J. Ding, and B. Lucier (2010). "The hitchhiker's guide to affiliation networks: A game-theoretic approach." *ArXiv:1008.1516v1.*
Boucher, V. and I. Mourifié (2012). "My friend far far away: Asymptotic properties of pairwise stable networks." *Available at SSRN*, 2170803.
Bresnahan, T. F. and P. C. Reiss (1991). "Entry and competition in concentrated markets." *Journal of Political Economy* 977–1009.
Burger, M. J. and V. Buskens (2009). "Social context and network formation: An experimental study." *Social Networks* 31, 63–75.
Butts, C. (2009). "Using potential games to parameterize ERG models." working paper, UCI.
Cai, J., A. De Janvry, and E. Sadoulet (2012). "Social networks and the decision to insure." *Available at SSRN 2161686.*
Chamberlain, G. (1982). "Multivariate regression models for panel data." *Journal of Econometrics* 18, 5–46.
Chandrasekhar, A. and R. Lewis (2013). "Econometrics of sampled networks," Stanford working paper.
Chandrasekhar, A. G. and M. Jackson (2014). "Tractable and consistent random graph models." SSRN Working Paper: http://ssrn.com/abstract=2150428.
Chandrasekhar, A. G. and M. Jackson (2015). "A network formation model based on subgraphs."
Chandrasekhar, A. G., C. Kinnan, and H. Larreguy (2014). "Social networks as contract enforcement: Evidence from a lab experiment in the field."
Chatterjee, S. and P. Diaconis (2013). "Estimating and understanding exponential random graph models." *The Annals of Statistics* 41, 2428–2461.
Chatterjee, S., P. Diaconis, and A. Sly (2010). "Random graphs with a given degree sequence." *Arxiv preprint arXiv:1005.1136.*
Christakis, N., J. Fowler, G. Imbens, and K. Kalyanaraman (2010). "An Empirical Model for Strategic Network Formation." NBER Working Paper.
Clauset, A., M. E. Newman, and C. Moore (2004). "Finding community structure in very large networks." *Physical Review E* 70, 066111.
Comola, M. and M. Fafchamps (2013). "Testing unilateral and bilateral link formation." *The Economic Journal.*
Currarini, S., M. Jackson, and P. Pin (2009). "An economic model of friendship: Homophily, minorities, and segregation," *Econometrica* 77, 1003–1045.
Currarini, S., M. Jackson, and P. Pin (2010). "Identifying the roles of race-based choice and chance in high school friendship network formation." *Proceedings of the National Academy of Sciences* 107, 4857–4861.
de Paula, A., S. Richards-Shubik, and E. Tamer (2014). "Identification of preferences in network formation games," Mimeo, Carnegie Mellon University.
De Weerdt, J. and S. Dercon (2006). "Risk-sharing networks and insurance against illness." *Journal of Development Economics* 81, 337–356.
Diaconis, P., S. Holmes, and S. Janson (2008). "Threshold graph limits and random threshold graphs." *Internet Mathematics* 5, 267–320.
Erdős, P. and A. Rényi (1959). "On random graphs," *Publ. Math. Debrecen* 6, 156.

Fafchamps, M. and F. Gubert (2007a). "Risk sharing and network formation." *The American Economic Review* 97, 75–79.

Fafchamps, M. and F. Gubert (2007b). "The formation of risk sharing networks." *Journal of Development Economics* 83, 326–350.

Frank, O. and D. Strauss (1986). "Markov graphs," *Journal of the American Statistical Association* 832–842.

Galeotti, A. and B. W. Rogers (2013). "Strategic immunization and group structure." *American Economic Journal: Microeconomics* 5, 1–32.

Girvan, M. and M. E. Newman (2002). "Community structure in social and biological networks." *Proceedings of the National Academy of Sciences* 99, 7821–7826.

Goldenberg, A., A. X. Zheng, S. E. Fienberg, and E. M. Airoldi (2010). "A survey of statistical network models." *Foundations and Trends in Machine Learning* 2, 129–233.

Goldsmith-Pinkham, P. and G. Imbens (2013). "Social networks and the identification of peer effects." *Journal of Business and Economic Statistics* 31:3, 253–264.

Golub, B. and Y. Livne (2010). "Strategic random networks: Why social networking technology matters." SSRN working paper.

Graham, B. S. (2014a). "An empirical model of network formation: Detecting homophily when agents are heterogenous."

Graham, B. S. (2014b). "Methods of Identification in Social Networks."

Hammersley, J. and P. Clifford (1971). "Markov fields on finite graphs and lattices."

Handcock, M. (2003). "Assessing degeneracy in statistical models of social networks."

Hill, B. M. (1975). "A simple general approach to inference about the tail of a distribution." *The Annals of Statistics* 3, 1163–1174.

Holland, P. W., K. B. Laskey, and S. Leinhardt (1983). "Stochastic blockmodels: First steps." *Social Networks* 5, 109–137.

Holland, P. W. and S. Leinhardt (1981). "An exponential family of probability distributions for directed graphs." *Journal of the American Statistical Association* 76, 33–50.

Jackson, M. (2008). *Social and Economic Networks*. Princeton, NJ: Princeton University Press.

Jackson, M., T. Barraquer, and X. Tan (2012). "Social capital and social quilts: Network patterns of favor exchange." *American Economic Review* 102, 1857–1897.

Jackson, M. and B. Rogers (2007a). "Meeting strangers and friends of friends: How random are social networks?" *The American Economic Review* 97, 890–915.

Jackson, M. and B. Rogers (2007b). "Relating network structure to diffusion properties through stochastic dominance," *The BE Journal of Theoretical Economics* 7, 1–13.

Jackson, M. and A. Watts (2001). "The existence of pairwise stable networks." *Seoul Journal of Economics* 14(3), 299–321.

Jackson, M. and A. Wolinsky (1996). "A strategic model of social and economic networks." *Journal of Economic Theory* 71, 44–74.

Jackson, M. and L. Yariv (2011). *Handbook of Social Economics.* San Diego: North Holland. "Diffusion, strategic interaction, and social structure."

Kallenberg, O. (2006). *Probabilistic Symmetries and Invariance Principles*. Springer Science & Business Media.

Karlan, D., M. Mobius, T. Rosenblat, and A. Szeidl (2009). "Trust and social collateral*." *Quarterly Journal of Economics* 124, 1307–1361.

Kearns, M., S. Judd, and Y. Vorobeychik (2012). "Behavioral experiments on a network formation game," in *Proceedings of the 13th ACM Conference on Electronic Commerce* ACM, 690–704.

Koenig, M. (2012). "The formation of networks with local spillovers and limited observability."
Kolotilin, A. (2013). "Estimation of a scale-free network formation model." *Available at SSRN.*
Lazarsfeld, P. F., N. W. Henry, and T. W. Anderson (1968). *Latent Structure Analysis.* Houghton Mifflin Boston.
Leider, S., M. Mobius, T. Rosenblat, and Q. Do (2009). "Directed altruism and enforced reciprocity in social networks*," *Quarterly Journal of Economics* 124, 1815–1851.
Leskovec, J., K. J. Lang, and M. Mahoney (2010). "Empirical comparison of algorithms for network community detection," in *Proceedings of the 19th International Conference on World Wide Web* ACM, 631–640.
Leung, M. (2014). "A random-field approach to inference in large models of network formation." Stanford working paper.
Lovász, L. and B. Szegedy (2006). "Limits of dense graph sequences." *Journal of Combinatorial Theory, Series B* 96, 933–957.
Manresa, E. (2014). "Estimating the structure of social interactions using panel data."
Meinshausen, N. and P. Bühlmann (2006). "High-dimensional graphs and variable selection with the lasso." *The Annals of Statistics* 1436–1462.
Meinshausen, N. and B. Yu (2009). "Lasso-type recovery of sparse representations for high-dimensional data." *Annals of Statistics* 37, 246–270.
Mele, A. (2013). "A structural model of segregation in social networks." Working paper.
Mobius, M., T. Phan, and A. Szeidl (2010). "Treasure hunt," Working paper, http://misrc.umn.edu/wise/papers/p1-11.pdf.
Newman, M. E. (2004). "Fast algorithm for detecting community structure in networks." *Physical Review E* 69, 066133.
Ngatia, M. (2011). "Social interactions and individual reproductive decisions." *Unpublished.* Working paper.
Penrose, M. (2003). *Random Geometric Graphs.* Oxford University Press.
Ross, N. et al. (2011). "Fundamentals of Stein's method." *Probability Surveys* 8, 210–293.
Shalizi, C. and A. Rinaldo (2012). "Consistency under sampling of exponential random graph models." *ArXiv 1111.3054v3.*
Snijders, T. (2002). "Markov chain Monte Carlo estimation of exponential random graph models." *Journal of Social Structure* 3, 240.
Stein, C. (1986). "Approximate computation of expectations," *Lecture Notes-Monograph Series* 7, i–164.
Tamer, E. (2003). "Incomplete simultaneous discrete response model with multiple equilibria." *The Review of Economic Studies* 70, 147–165.
Tibshirani, R. (1996). "Regression shrinkage and selection via the lasso." *Journal of the Royal Statistical Society. Series B (Methodological)* 267–288.
Wasserman, S. and P. Pattison (1996). "Logit models and logistic regressions for social networks: I. An introduction to Markov graphs andp." *Psychometrika* 61, 401–425.
Yuan, M. and Y. Lin (2007). "Model selection and estimation in the Gaussian graphical model." *Biometrika* 94, 19–35.

CHAPTER 14

SMALL-WORLD NETWORKS

DUNCAN J. WATTS

14.1 Introduction

The "small-world hypothesis" expresses the idea that every individual in a given population can reach every other via some "short" chain of intermediaries. Long an article of pop-cultural speculation (Guare 1990; Karinthy 1929), the hypothesis was first articulated formally by Pool and Kochen (1978), who asked: "How many pairs of persons in the population can be joined by a single acquaintance, how many by a chain of two persons, how many by a chain of three, etc.?" Pool and Kochen showed that when all individuals choose their acquaintances uniformly at random from the entire population, the number of pairs i,j connected by a path of length d increases exponentially fast in d. Thus under quite reasonable assumptions regarding the average number k of acquaintances per person, most pairs of individuals in even a very large population should be connected by paths only a few steps long. For example, for the population of the United States at the time—roughly 200 million—and assuming that individuals possessed an average of roughly 1,000 acquaintances, Pool and Kochen estimated that a randomly selected pair could be connected in only three or four steps.

This result is equivalent to a standard result from the theory of random graphs (Bollobas 1985; Erdos and Renyi 1959; Solomonoff and Rapoport 1951)—namely, that the average shortest path length separating two randomly chosen individuals should be proportional to $\log N$ (for $N >> k >> 1$), where N is the size of the population. (Strictly speaking, this result holds only when the network is connected, which is not guaranteed for small k; hence the second condition $k >> 1$.) As Pool and Kochen noted, however, social networks are far from random, but rather exhibit social structure in various manifestations and at multiple scales. Individuals, for example, are more likely to know each other when they share certain traits, such as geographical proximity, socioeconomic status or common interests—a tendency known as homophily (Kossinets and Watts 2009; Lazarsfeld and Merton 1954; McPherson, Smith-Lovin, and Cook 2001; Pin and Rogers 2015; Wimmer and Lewis 2010). Moreover, individuals are more likely to meet when they share mutual friends—a process known as triadic closure (Kossinets and Watts 2006; Rapoport 1957)—or when they participate in common activities or "foci" (Feld 1981; Kossinets and Watts 2006). Finally, social structure manifests itself at scales above that of individual decisions, as is the case with communities (Girvan and Newman 2002), organizations (McPherson and Smith-Lovin 1987), and geographical regions (Leskovec and Horvitz 2008; Ugander et al. 2011). For all these reasons, the more interesting and empirically relevant version of the small-world hypothesis is that path lengths are hypothesized to remain "short" (i.e. comparable to that of a random graph of the same size and density), even when the network in question is far from random. Unfortunately under these more general conditions the analytical methods that work well for the special case of random graphs fail badly, hence in the absence of modern digital computers Pool and Kochen were unable to proceed further in their theoretical analysis.

Inspired by Pool and Kochen's paper, however, social psychologist Stanley Milgram, along with his graduate student Jeffrey Travers, decided to test the small-world hypothesis *experimentally*, using what they called the "small-world method" (Milgram 1967; Travers and Milgram 1969):

1. Recruit one or more volunteers to be "targets" of the experiment.
2. For each target, recruit some large number of additional "senders."
3. Give each sender a description of the target sufficient to identify him or her uniquely (for example, name, address, occupation, and employer; wife's maiden name; and so on).
4. Instruct each sender to forward a message to the target in one of two ways:
    (a) If target is an acquaintance (that is, is known to sender on a first-name basis), forward directly to target.
    (b) If not, forward message, target information, and instructions to an acquaintance who is "closer" to target than the sender.
5. Repeat steps 3 and 4 with additional senders until chain either *completes* (reaches target) or *terminates* (is not forwarded for any other reason).[1]

Travers and Milgram implemented this protocol using a single target (a Boston stockbroker who was an acquaintance of Milgram's) and 296 senders: 100 from Boston; 100 blue-chip stockholders from Omaha, Nebraska; and 96 Omaha residents who were randomly selected from a list of people who had agreed to receive marketing literature. Famously, they found that the chains that reached the target were surprisingly short—approximately six steps on average, where unsurprisingly, chains initiated by stockholders in Nebraska were shorter than those from randomly selected Nebraskans, and chains that began in Boston were shorter still.

[1] In Milgram's version, subsequent senders also received a list of previous senders, in order to avoid cycles. In practice these cycles are extremely unlikely, and this requirement was dropped in subsequent small-world experiments.

These two seminal studies, one theoretical and the other experimental, are both often cited as providing support for the same claim: that individuals in even very large networks can be connected by "short" chains of intermediaries. In fact, however, they make subtly but importantly different claims, which in turn have different levels of generality and empirical support. The first claim, advanced by Pool and Kochen, holds only that for a randomly chosen pair of individuals there exists with high probability a short chain of intermediaries that connects them, where "short" is usually interpreted as proportional to the logarithm of the population size N (Watts and Strogatz 1998). Because this claim refers only to the structure, or *topology*, of the underlying social network, we refer to it as the *topological small-world hypothesis*. As first pointed out by Kleinberg (Kleinberg 2000a,b), however, the experimental results of Travers and Milgram appear to support a somewhat different and considerably stronger claim: that not only do these short chains exist, but that ordinary individuals can *locate* them in spite of having only local knowledge of the social network in question (Adamic et al. 2001; Adamic and Adar 2005; Kleinberg 2000a,b; Simsek and Jensen 2005; Watts, Dodds, and Newman 2002; West and Leskovec 2012a,b). Because this second, stronger claim implies the existence of some rule-following behavior on the part of the experiment's participants, we refer to it as the *algorithmic small-world hypothesis* (Goel, Muhamad, and Watts 2009; Kleinberg 2000b).

Distinguishing between the topological and algorithmic small-world hypotheses is important for three reasons. First, each hypothesis is relevant to different social processes (Watts 2003). The spread of a sexually transmitted disease along networks of sexual relations, for example, does not require that participants have any awareness of the disease, or intention to spread it; thus for an individual to be at risk of acquiring an infection, he or she need only be connected in the topological sense to existing infectives (Lamberson 2016). On the contrary, individuals attempting to "network"—in order to locate some resources like a new job (Beaman 2016; Granovetter 1995) or a service provider (Lee 1969)—must actively traverse chains of referrals, and thus must be connected in the algorithmic sense. Depending on the application of interest, therefore, either the topological or algorithmic distance between individuals may be more relevant—or possibly both together. Second, whereas the topological hypothesis has been shown to apply essentially universally, to networks of all kinds, the algorithmic hypothesis is mostly confined to social networks in which human agents make decisions about how to direct messages. And third, as I explain in more detail below, the two hypotheses are supported by very different amounts of empirical evidence.

14.2 The Topological Small-World Hypothesis

An important step in resolving the topological small-world hypothesis was made by Watts and Strogatz (1998), who analyzed a network model that incorporated

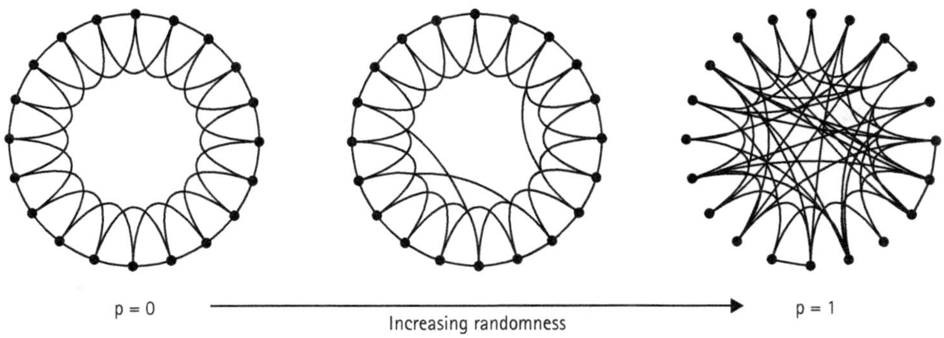

FIGURE 14.1 Schematic of Watts–Strogatz model of partly ordered, partly random networks, where $p =$ probability of random edge rewiring (from Watts and Strogatz 1998).

elements of both social structure and randomness. In the model, "social structure" was represented by a uniform one-dimensional lattice, where each node was connected to its k nearest neighbors on the lattice, and "randomness" was characterized by a tunable parameter p, which specified the probability that a link in the lattice would be *randomly rewired* (see Figure 14.1). Defining the *average path length* L of a network as $L = \langle d_{ij} \rangle$, where d_{ij} is the shortest path between each pair of vertices i,j, and the *clustering coefficient* C of a network as the average probability that two neighbors of a given node would themselves be neighbors, they showed: (a) when $p = 0$ (completely ordered) the network is "large" $\left(L(0) \sim N/2k\right)$ and "highly clustered" $(C(0) \approx 3/4)$; and (b) when $p = 1$ (completely random) it is "small" $\left(L(1) \sim \log N/\log k\right)$ and "poorly clustered" $\left(C(1) \sim k/N\right)$. Naively, one might suspect that path lengths will be short only when clustering is low, or equivalently, that when clustering is high path lengths will necessarily be long. In contrast with this intuition, however, and also Pool and Kochen's prediction, Watts and Strogatz found that their model exhibited a broad region of p values in which $C(p)$ is high relative to its random limit $C(1)$, yet $L(p)$ is roughly speaking as "small" as possible (see Figure 14.2). Noting that in a random graph of size N and average degree k, and assuming that $N >> k >> 1$, $L_r \sim \log N / \log k$ (Bollobas 1985; Erdos and Renyi 1959; Solomonoff and Rapoport 1951) and $C_r \sim k/N$, they proposed the following definition for a "small-world network":

Definition 1. A small-world network exhibits an average shortest path length$L \approx L_r \sim \log N/\log k$ and $C \gg C_r \sim k/N$.

Building on this insight, Walsh (1999) subsequently introduced a "small-world index" $\mu = \frac{C/C_r}{L/L_r} = \frac{CL_r}{C_r L}$ from which a small-world graph could be defined equivalently as:

Definition 2. A small-world network exhibits a small-world index $\mu >> 1$.

Numerous authors have since investigated the properties of this and other related models; see Newman (2000, 2003, 2010), Watts (2004), Jackson (2010), and Kleinberg

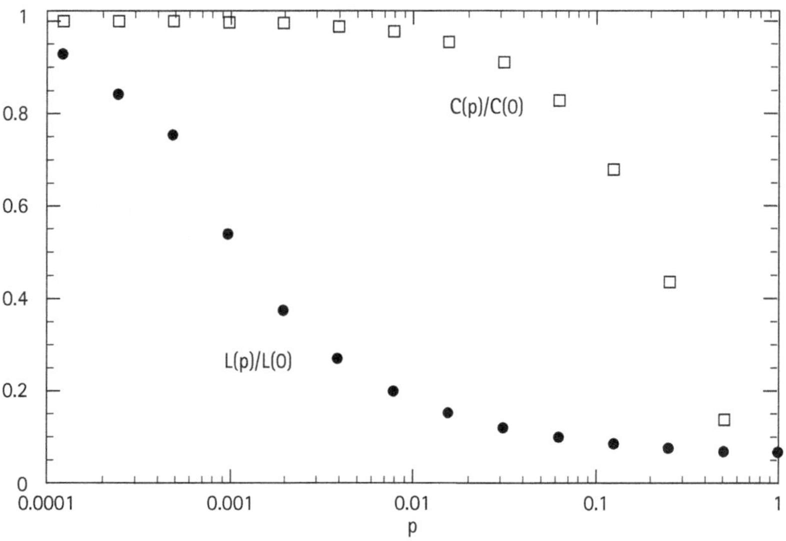

FIGURE 14.2 Normalized average path length $L(p)/L(0)$ and clustering coefficient $C(p)/C(0)$ versus p (from Watts and Strogatz 1998).

and Easley (2010) for reviews. Variations include (a) adding edges rather than rewiring them (Newman and Watts 1999); (b) choosing random edges according to non-uniform distributions (Marvel et al. 2013; Watts 1999b); (c) using substrates other than a 1D lattice (Bramoullé, Djebbari, and Fortin 2009; Jackson and Rogers 2005; Jin, Girvan, and Newman 2001; Watts 1999b), and (d) introducing degree-distributions of increasing variance (Amaral et al. 2000), including "scale-free" distributions (Barabasi and Albert 1999b), for which it can be shown that average path lengths scale as $\log \log N$ rather than $\log N$ (Chung and Lu 2002). Collectively, this literature has demonstrated that under a wide variety of assumptions and conditions, large-N networks exhibit a combination of high local clustering and short global path lengths, supporting Watts and Strogatz's original contention that small-world topologies should be widespread among real-world networks.

As Jackson and Rogers (2005) have pointed out, however, the majority of these models are "mechanical" in nature, meaning that the process by which links are arranged is specified exclusively in terms of network features (e.g., the probability of connecting as a function of nodal degree or network separation) that ignore possible underlying physical, psychological, or biological processes. One could argue that in abstracting away from domain-specific causal mechanisms, mechanical models have the advantage of generality—for example, any mechanism that mostly selects for local edges but also yields some arbitrarily small fraction of long-range links will generate small-world topology (Watts 2003). Nonetheless, it is also clearly of interest to investigate node-level generative mechanisms that are specific to particular substantive domains. In particular, a number of what Jackson and Rogers call economic models have been proposed that attempt to micro-found small-world features in the context of social networks. For example, Rosenblat and Mobius (2004) show that when agents situated on a circle prefer to form connections with partners who are similar to themselves but also pay a technology-dependent communication cost that increases with a potential partner's physical distance, then falling communication costs should (a) decrease average pairwise path-length in the resulting network, and (b) increase observed homophily as measured by the share of messages exchanged between similar types.[2] Carayol and Roux (2005) subsequently proposed a related model in which agents, again situated on a circle, maintain connections with one another only when the costs (which increase linearly with physical distance) are exceed by the benefits (which decrease exponentially with path length d_{ij}). When agents also have a preference for selecting geographically proximate partners—what Carayol and Roux call preferential meeting (as distinct from preferential attachment; Barabasi and Albert 1999a)—they find that small-world topologies emerged naturally. And finally, Jackson and Rogers (2005) proposed a variant of the Carayol and Roux model in which benefits again decrease exponentially with network distance between node, but where agents are now situated on initially disconnected "islands" (similar to the "caves" proposed by Watts 1999a), where links with inhabitants of the same island are less costly than with inhabitants of other islands. As with Carayol and Roux, Jackson and Rogers found that high clustering and short path lengths emerge naturally under reasonably broad conditions. Although motivated differently, therefore, economic models of network formation are largely consistent with the mechanical model literature in predicting that small-world topologies should be widespread.

14.2.1 Empirical Evidence

In addition to establishing theoretical conditions under which small-world topologies should arise, Watts and Strogatz also checked their prediction that a wide range of empirical networks would exhibit small-world structure by considering three network data sets to which they had access: the affiliation network of movie actors, the power transmission grid of the western United States, and the neural network of the nematode *C. elegans*. Strikingly they found that all three examples satisfied the small-world criteria encapsulated in Definition 1 (Watts and Strogatz 1998). Subsequently, hundreds of studies of large-scale networks have been conducted across a number of domains, including not only social networks (Golder, Wilkinson, and Huberman 2007;

[2] Rosenblat and Mobius also argue that the effects predicted by their model are consistent with changes observed in the co-authorship network of economists over the course of a decade from 1989–1999. Interestingly, Goyal et al. (2006) subsequently examined a similar data set, albeit over a longer time period (1970–2000) and also found that average path lengths had diminished, but offered a different explanation—that it was driven by the increasing interconnectivity of "stars," prominent authors with many coauthors.

Goyal, Van Der Leij, and Moraga-González 2006), but also biological (Sporns and Zwi 2004), technological (Sen et al. 2003), organizational (Kogut and Walker 2001), and virtual networks (Broder et al. 2000).

Three near-universal findings of these studies are that: (1) a majority of individuals in a given population are connected in a single "giant component"; (2) the typical shortest path length connecting pairs of nodes within the giant component is $O(\log N)$; and (3) the clustering coefficient is many orders of magnitude greater than would be expected for a random graph with the same N and k. In particular, two recent studies are of interest. First, Leskovec and Horvitz (2008) studied a network of 180M instant messenger users (where a link is defined as two users nominating each other as IM "buddies"), finding that users were separated by a mean of 6.6 steps and a median of 7 steps. Second, a more recent, and even larger, study of 721M Facebook users (Ugander et al. 2011) found largely confirmatory results: 99.91% of all individuals were contained within a single connected component; the average path length within that component was 4.74, close to $L_r \sim \log N / \log k = \log\left(721 \times 10^6\right) / \log 190 = 3.9$; and the clustering coefficient was $C \approx 0.14$, more than 500,000 times the expected clustering coefficient for an equivalent random graph $C_r \sim k/N = 190/\left(712 \times 10^6\right) = 0.00000026$.

Given the dramatic differences between the initial empirical examples identified by Watts and Strogatz and these more recent studies—differences spanning six orders of magnitude in size (from hundreds to hundreds of millions)—and given also the extraordinary diversity of domains over which network analyses have now been conducted, the consistency with which the basic small-world pattern of short path lengths combined with high clustering has been replicated is remarkable. Viewed as a whole, therefore, the combination of theoretical analyses and empirical findings provides overwhelming evidence that the topological small-world hypothesis as formalized in Definitions 1 and 2 is satisfied very generally in large-scale, real-world networks.

Interestingly, moreover, recent work by Marvel et al. (2013) suggests that small-world topologies, although very general, are neither inevitable nor universal, and hence the condition is not a trivial one. Starting from historical data about the spread of the Black Death—a pandemic of bubonic plague that devastated Europe in the fourteenth century—they first show that the observed wavelike pattern of diffusion can be obtained only if the contact probability decreases exponentially with physical distance; by contrast, if it decreases slower than exponential disease propagation is characterized by large numbers of satellite outbreaks with no distinct wave front. Next, they confirm an early speculation by Watts (1999b, p. 135) that for spatially embedded networks, small-world topology emerges only when the probability of a long-range (i.e., random) edge decrease slower than exponentially with distance. They conclude that the small-world phenomenon is not universal across time, but rather is distinct to the modern era, emerging sometime after the fourteenth century, possibly a consequence of long-range transportation technologies such as open ocean navigation and railways.

14.3 The Algorithmic Small-World Problem

As mentioned above, Travers and Milgram's results indicated not only that pairs of individuals in a large network were connected by short chains of intermediaries—the topological hypothesis—but also that at least some of these individuals could actually locate these short paths using only local information about the network structure. Pointing out this distinction, Kleinberg (2000a,b) proposed a more general class of partly ordered, partly random networks in which the spatial distribution of random links was permitted to vary according to the probability distribution $P_{ij} \propto r_{ij}^{-\gamma}$, where r_{ij} is the distance between nodes i and j measured on some underlying lattice of dimension D, and γ is a tunable parameter (see Figure 14.3 for details). With this network as a substrate, Kleinberg then defined a "local" message passing algorithm in which a given "sender" attempts to forward a message to a "target" individual where the sender is aware only of (a) its own location in the grid, (b) the location of each of its immediate neighbors, and (c) the location of the target. Consistent with Milgram's instructions to his subjects, each sender is assumed to forward the message deterministically to its neighbor who is closest to the target, at which point that individual then becomes the sender and the procedure repeats until the message reaches the target. For a given network parameterization γ, Kleinberg was able to compute the average time (i.e. number of steps) required for a randomly selected sender to locate a randomly selected target. He then proved that only when $\gamma = D$ would the network be "searchable" in the sense that they could be discovered by the message passing algorithm in time proportional to $(\log N)^{\alpha}$ (where $\alpha \geq 1$ is some constant). Interestingly, when $\gamma < D$, Kleinberg showed that although short paths existed, they could only be found by the algorithm in time proportional to $N^{1/D}$, whereas when $\gamma > D$ even the shortest paths were not sufficiently short. Motivated by this result, one could propose the following alternative definition of a small-world network in terms of its "searchability":

Definition 3. A searchable small-world network is one in which the length of the shortest search path is discoverable via a local algorithm$L_s \propto (\log N)^{\alpha}$, where $\alpha \geq 1$.

Subsequently, more realistic models of searchable small-world networks have been proposed, incorporating for example "social" distance (Adamic and Adar 2005; Kleinberg 2001; Lattanzi, Panconesi, and Sivakumar 2011; Watts, Dodds, and Newman 2002) as well as geographical distance (Liben-Nowell et al. 2005), and also heterogeneity of degree (Simsek and Jensen 2005). In particular, Watts et al. (2002) studied a variant of Kleinberg's local search algorithm on networks that were constructed out of multiple, cross-cutting "social dimensions," reflecting that in the social world individuals measure distance in multiple, possibly inconsistent ways—not only geographical proximity, for example, but also similarity of occupation, socioeconomic status, ethnic or racial identity, and so on (see Figure 14.4). Moreover, Watts et al. noted that although the

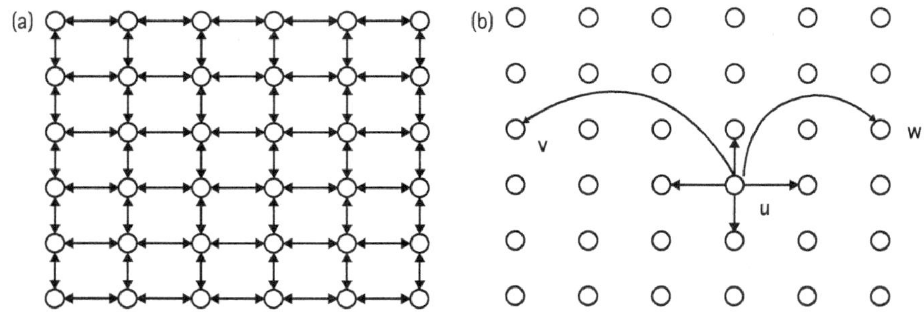

FIGURE 14.3 Kleinberg's small-world model for $D = 2$ (a) the underlying geometric grid used to determine social "distance"; (b) "local" edges are assigned automatically to neighboring nodes on the grid, while "global" edges are assigned randomly with probability proportional to $P_{ij} \propto r_{ij}^{-\gamma}$, where r_{ij} is the distance between nodes i and j on the grid (from Kleinberg 2000).

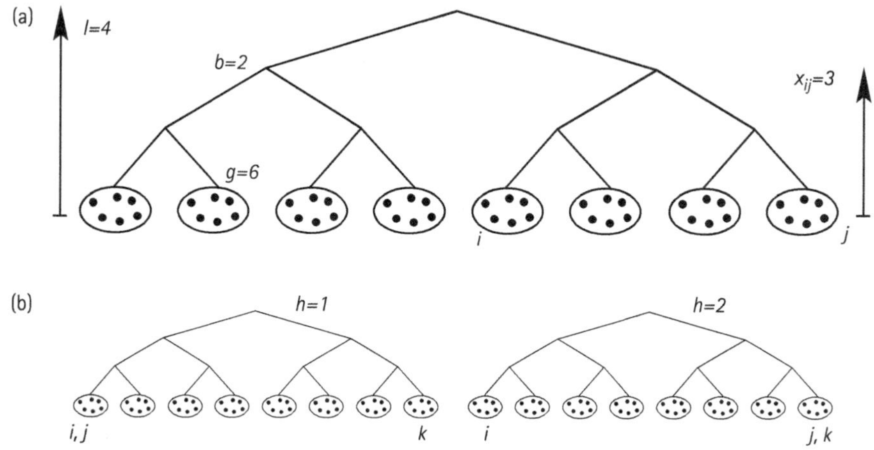

FIGURE 14.4 The generalized small-world model of Watts et al. (2002). (a) For any given social "dimension" individuals are allocated to small groups, which are then aggregated into groups of groups, and so on, ultimately inducing a distance metric x_{ij} on individuals i,j corresponding to the lowest common ancestor node in a hierarchy with branching ratio b. (b) Because the social world can be decomposed in this manner along more than one such dimension (e.g., geography, occupation, etc.) individuals i,j can construct correspondingly many distances between themselves (H in total), where the probability of being connected is proportional to $e^{-\alpha x_{ij}^h}$ in any dimension h.

observed path lengths in small-world experiments were indeed "short," it was not clear that they were short in the sense of Definition 3; that is, proportional to $(\log N)^{\alpha}$. For these reasons, and also because in the presence of chain attrition, the absolute length of chains is more relevant than its scaling behavior (i.e., if a chain doesn't complete, it doesn't matter whether or not it is as short as it could be; it is still too long), they proposed an alternative definition of a searchable small-world network:

Definition 4. Given a message failure probability p and minimal required chain completion rate r, a searchable small-world network exhibits an average search path length $L_s \leq \log r / \log(1-p)$.

Adopting this revised measure of searchability, and allowing both the number H of social dimensions to vary as well as the "homophily" tendency α of individuals to connect preferentially with similar others along any given dimension, Watts et al. showed that in contrast with Kleinberg's model, where searchability was demonstrable only for a unique parameter value $\gamma = D$, networks with cross-cutting social dimensions were searchable over a broad range of (α, H) parameter space. Interestingly, however, the most searchable networks tended to be for $H = 2, 3$, a result reminiscent of earlier "reverse small-world experiments" (Killworth and Bernard 1978, 1979), which found that experimental participants typically relied on two to three distance measures (most prominently geographical and occupational proximity) when targeting messages.

Finally, by assuming that message holders forward their message to an acquaintance with the minimum social distance to the target, Watts et al. showed that the model could account for a previously unexplained observation of Pool and Kochen's: that empirically estimated social distances tended to violate the triangle inequality. Specifically, for three individuals i, j, k in a social network, Pool and Kochen had noted that self-reported distances in social networks often exhibited the property $d(i,j) + d(j,k) < d(i,k)$, where $d(i, j)$ is individual i's self-reported distance to individual j. This result had puzzled previous researchers (e.g. Barnett 1989) as it clearly violates the triangle inequality, which in turn is an axiomatic property of distance. Watts et al. showed, however, that this violation follows naturally from the assumption that individuals assess "distance" along multiple dimensions (e.g. geography, occupation, etc.) simultaneously and choose the next recipient of a message by minimizing over all such dimensions. It is possible, in other words, for i to be close to j in one dimension (say geography) and j to be close to k in another dimension (say occupation), but for i to be distant from k in both dimensions.

14.3.1 Empirical Evidence

In contrast with the topological small-world hypothesis, empirical evidence in favor of the algorithmic small-world hypothesis—that individuals can locate these paths—is less than conclusive (Kleinfeld 2002; Watts 2003). As Kleinberg (2000a) noted in his initial formulation of the algorithmic small-world problem, the experiment conducted by Travers and Milgram (1969) clearly demonstrated that at least some people are able to construct short chains. But as Travers and Milgram themselves emphasized, it was also the case that most of the chains that started—roughly 80% in their case—never reached the target. Subsequent experiments (Dodds, Muhamad, and Watts 2003; Korte and Milgram 1970) found the same pattern: On the one hand, chains that reach their targets tend to be short in the topological small-world sense; but on the other hand, the rate of chain completion tends to be low. For example, in Korte and Milgram's (1970) follow-up study of white senders in Los Angeles attempting to reach one of 270 black targets in New York, the average length of completed chains was 7, but the completion rate was only 13%. And in the most recently repeated small-world experiment, conducted by Dodds et al. (2003), the pattern was even more striking: Completed chains were only 4 steps long, but only 0.4% of about 24,000 chains that started (a total of 384) reached their targets.

How should this mixed evidence be interpreted? The conventional interpretation, originally due to White (1970), is that message chains terminate randomly; that is for reasons that are unrelated to the attributes of the individual in question, the difficulty of the search, or the underlying structure of the network. Under this assumption, it is easily shown that the algorithmic hypothesis continues to hold, albeit with somewhat longer expected chain lengths (Dodds, Muhamad, and Watts 2003; White 1970). This "random attrition" hypothesis, however, can be challenged on two related but distinct grounds (Goel, Muhamad, and Watts 2009). First, it may simply be the case that some people are good at conducting social searches while others are not. Exceptionally determined or resourceful individuals may be able to construct short paths to distant targets, but most people lacking "social capital" (Coleman 1988) are effectively isolated from one another. A second objection, potentially compounding the first, is that chains terminate because most source-target pairs live in effectively separate populations that can only be reached via long or otherwise undiscoverable paths (Kleinfeld 2002). Thus, pairs of individuals who are already "close," in the sense of sharing social, geographical, and demographic attributes, can find each other but "distant" pairs cannot. Moreover, because many more pairs are distant than close, the bulk of message chains should not be expected to complete—as indeed is observed in small-world experiments.

In response to these criticisms, Goel at al. (2009) introduced a provably unbiased estimator of path length that (a) allowed for heterogeneous attrition, as a function of individual attributes (e.g., education, race, gender, age, etc.), and (b) allowed for precise computation of both the mean and the variance. Applying this estimator to data from two small-world experiments, Goel et al. found that the median "true" chain length (i.e., in a hypothetical experiment with zero attrition) could be estimated robustly to lie between six and seven steps, but also that the mean of this distribution was likely to be much larger. In other words, at least half of all chains that start would be expected to reach their targets within about six steps, consistent with the conventional interpretation of chain attrition, but at least some chains should be expected to be very long, consistent with the competing claim that some pairs of individuals are extremely unlikely to be able to locate each other.

Given that the variance in topological path lengths has been shown to be relatively small (Leskovec and Horvitz 2008), the possibility that search paths could vary so widely raises interesting and unresolved questions. If everyone is in some (i.e., topological) sense within just a few steps of everyone else, why is it that these short paths are discoverable in at least a substantial fraction of cases, but apparently not discoverable in at least some other potentially substantial fraction of cases? Could it be that some pairs of individuals are connected by a large number of distinct, hence redundant, short paths, whereas others are connected by only one or a few such paths, and that only in the case of extreme path redundancy is the error-prone process of social search able to work well? Or could it be instead that all individuals are connected via a comparable number of short paths, but that they vary in some important way in their strategy for choosing among the available alternatives? Or could it be finally, that individuals are similar both with respect to their topological connectivity and also their search strategy, but vary in their ability to mitigate the attrition of message chains—say because they are more or less persistent—and hence paths that in principle connect fail to do so in practice? Answers to these questions would require more sophisticated small-world style experiments in which, at a minimum, the underlying topological distance between senders and receivers were systematically varied. Such experiments would be useful, however, as the results could have potentially important implications for related questions of "social networking" and social capital, and would also settle more conclusively the empirical status of the algorithmic small-world hypothesis.

14.3.2 The Role of Hubs

An ancillary finding of Travers and Milgram that has subsequently attracted a great deal of attention (Gladwell 1999, 2000) was that a disproportionately large fraction of the chains—roughly 40%—that completed did so through one of only two individuals at the final step. Milgram labeled these individuals "sociometric stars," but subsequently they have been called variously stars, hubs, connectors, and funnels. Their purported role as essential elements of successful social search appears to be supported by recent theoretical models (Adamic et al. 2001; Adamic, Lukose, and Huberman 2003), which suggest that when otherwise random networks are characterized by extremely skewed degree distributions (Barabasi and Albert 1999b) the most connected individuals do indeed serve a critical role in directing decentralized searches. It is important to note, however, that hubs in these "scale-free" models are not simply more connected than ordinary nodes, but many orders of magnitude more connected. Hubs in these models, in other words, are more like hubs in airline networks than gregarious individuals in social networks. In particular, whereas airline hubs tend to be connected directly to a large fraction of total airports in a given region, and are also typically connected to most other hubs, in social networks even the most gregarious individuals are estimated to know only on the order of 1,000 others (Bernard et al. 1991; Killworth et al. 1990; Ugander et al. 2011), and thus are adjacent to a negligible fraction of the entire network. "Hubs" in social networks, therefore, cannot play the role that they do in truly scale-free networks. Furthermore, empirical work shows when choosing the next member of a message chain, senders focus mostly on geographical proximity to the target and occupational similarity. Only when all other cues have been exhausted do social searches seek out recipients based on number of acquaintances (Dodds,

Muhamad, and Watts 2003). Consistent with this finding, moreover, theoretical models find that networks can be searchable even when all individuals have exactly the same acquaintance volume (Watts, Dodds, and Newman 2002); hence, even though, as noted above, highly heterogeneous degree distributions do tend to shorten topological shortest path lengths (Chung and Lu 2002), it is unclear how important this feature is to small-world search.

14.4 Economic Implications

What are the implications of a given network exhibiting a small-world network topology? The short answer is: it depends. For example, the spread of "simple" contagious processes (Centola and Macy 2007), such as infectious diseases for which the probability of a susceptible individual being infected by an exposure to an infected individual is independent of previous exposures, is dominated by the presence of short global path lengths (Moore and Newman 2000; Newman and Watts 1999). Faced with emerging epidemics of infectious disease, therefore, disproportionally large gains may be realized by focusing on sources of random connections such as airline travel or needle sharing among strangers. By contrast, the spread of "complex" contagions—meaning processes for which an individual must be exposed to multiple "infected" neighbors before becoming infected himself—is determined more by dense local clustering (Centola 2010), in which case promoting or restricting spread would require very different types of interventions. Complicating matters further, Goel et al. (2012) have recently found that at least some forms of contagion can be characterized almost entirely by local spread, in which case diffusion may be largely independent of global metrics such as average shortest path length. Depending on the nature of the contagious process of interest, therefore, and also the objectives of the planner, different elements of small-world networks will be of interest.

As mentioned earlier, the algorithmic small-world hypothesis is far less general than the topological version, being restricted mostly to social and economic networks (Goel, Muhamad, and Watts 2009), although recent work (West and Leskovec 2012a,b) has proposed that it may also generate insights for human search in non-social networks, such as in networks of hyperlinked documents or distributed databases, especially in situations where centralized search is difficult. Even within the domain of social and economic networks, moreover, the economic implications of the algorithmic variant of the small-world phenomenon are likely to be different from the topological version, and may apply either to individuals within a network or to planners interested in the consequences of network structure for collective behavior. For individuals, obvious implications apply to a variety of real-world problem-solving activities, such as job search (Beaman 2016; Granovetter 1995), locating service providers (Lee 1969), or other types of search for specialized knowledge or resources (Kautz, Selman, and Shah 1997). At the collective level, meanwhile, insights regarding the conditions under which individuals can successful locate information or resources in a large network may have implications for organizational problem solving and hence performance (Bolton and Dewatripont 1994; Dodds, Watts, and Sabel 2003; Kogut 2012; Radner 1993). Although some intriguing speculations have been made along these lines, however, understanding precisely how particular features of small-world networks either constrain or facilitate individual and collective problem solving, as well as how problem-solving activities in turn impact social network structure, are largely open problems, both theoretically and empirically, and present outstanding opportunities for future work.

Acknowledgments

This chapter draws on two previous publications by the same author: Watts (2008) and Goel, Muhamad, and Watts (2009). The author is grateful to Steven Strogatz, Mark Newman, Peter Dodds, Jon Kleinberg, Roby Muhamad, and Sharad Goel for many fruitful collaborations and conversations.

References

Adamic, Lada A. and Eytan Adar (2005). "How to search a social networks." *Social Networks* 27(3), 187–203.

Adamic, Lada A., Rajan M. Lukose, Amit R. Puniyani, and Bernardo A. Huberman (2001). "Search in power-law networks." *Physical Review E* 6404(4), 046135.

Adamic, Lada A., Rajan M. Lukose, and Bernado A. Huberman (2003). "Local search in unstructured networks." In *Handbook of Graphs and Networks: From the Genome to the Internet*, Stefan and Schuster Heniz Georg Bornholdt, eds., 295–316. Weinheim, Germany: Wiley-VCH.

Amaral, Luis A. N., Antonio Scala, Marc Barthelemy, and H. Eugene Stanley (2000). "Classes of small-world networks." *Proceedings of the National Academy of Sciences of the United States of America* 97(21), 11149–11152.

Barabasi, Albert-Laszlo and Reka Albert (1999a). "Emergence of scaling in random networks." *Science* 286(5439), 509–512.

Barnett, George A. (1989). "Approaches to non-euclidean network analysis." In *The Small World*, Manfred Kochen, ed., 349–372. Norwood, NJ: Ablex.

Beaman, Lori (2016). "Social networks and the labor market." In *The Oxford Handbook on the Ecnomics of Networks*, Yann Bramoullé, Andrea Galeotti, and Brian Rogers, eds. Oxford, UK: Oxford University Press.

Bernard, H. Russell, Eugene C. Johnsen, Peter D. Killworth, and Scott Robinson (1991). "Estimating the size of an average personal network and of an event population: Some empirical results." *Social Science Research* 20, 109–121.

Bollobas, Bela (1985). *Random Graphs*. New York: Academic Press.

Bolton, Patrick and Mathias Dewatripont (1994). "The firm as a communication network." *Quarterly Journal of Economics* 109, 809–839.

Bramoullé, Yann, Habiba Djebbari, and Bernard Fortin (2009). "Identification of peer effects through social networks." *Journal of Econometrics* 150(1), 41–55.

Broder, Andrei, Ravi Kumar, Farzin Maghoul, Prabhakar Raghavan, Sridhar Rajagopalan, Raymie Stata, Andrew Tomkins, and Janet Wiener (2000). "Graph structure in the web." *Computer Networks* 33, 309–320.

Carayol, Nicolas and Pascale Roux (2005). *"Collective Innovation" in a Model of Network Formation with Preferential Meeting.* Berlin: Springer.

Centola, Damon (2010). "The spread of behavior in an online social network experiment." *Science* 329(5996), 1194.

Centola, Damon and Michael Macy (2007). "Complex contagions and the weakness of long ties." *American Journal of Sociology* 113(3), 702–734.

Chung, F. and L. Y. Lu (2002). "The average distances in random graphs with given expected degrees." *Proceedings of the National Academy of Sciences of the United States of America* 99(25), 15879–15882.

Coleman, James S. (1988). "Social capital in the creation of human capital." *American Journal of Sociology* 94, S95–S120.

Dodds, Peter Sheridan, Roby Muhamad, and Duncan J. Watts (2003). "An experimental study of search in global social networks." *Science* 301(5634), 827–829.

Dodds, Peter Sheridan, Duncan J. Watts, and Charles F. Sabel (2003). "Information exchange and the robustness of organizational networks." *PNAS* 100(21), 12516–12521.

Easley, David and Jon Kleinberg (2010). *Networks, Crowds, and Markets: Reasoning about a Highly Connected World.* Cambridge: Cambridge University Press.

Erdos, Paul and Alfred Renyi (1959). "On random graphs I." Publicationes Mathematicae Debrecen 6, 290–297.

Feld, Scott L. (1981). "The focused organization of social ties." *American Journal of Sociology* 86(5),1015–1035.

Girvan, Michelle and Mark E. J. Newman (20020. "Community structure in social and biological networks." *Proceedings of the National Academy of Sciences of the United States of America* 99(12), 7821–7826.

Gladwell, Malcolm (1999). "Six degrees of Lois Weisberg." *The New Yorker* 11, 52–63.

Gladwell, Malcolm (2000). *The Tipping Point: How Little Things Can Make a Big Difference.* New York: Little Brown.

Goel, Sharad, Roby Muhamad, and Duncan J. Watts (2009). "Social search in 'small-world' experiments." In *Proceedings of the 18th International Conference on World Wide Web.* Madrid, Spain: ACM.

Goel, Sharad, Duncan J. Watts, and Daniel G. Goldstein (2012). "The structure of online diffusion networks." In *Proceedings of the 13th ACM Conference on Electronic Commerce*: ACM, 623–638.

Golder, Scott, Dennis Wilkinson, and Bernardo A. Huberman (2007). "Rhythms of social interaction: Messaging within a massive online network." In *3rd International Conference on Communities and Technologies*, 41–66. London: Springer.

Goyal, Sanjeev, Marco J. Van Der Leij, and José Luis Moraga-González (2006). "Economics: An emerging small world." *Journal of Political Economy* 114(2), 403–412.

Granovetter, Mark S. (1995). *Getting a Job: A Study of Contacts and Careers.* Chicago: University of Chicago Press.

Guare, John (1990). *Six Degrees of Separation: A Play.* New York: Vintage.

Jackson, Matthew O. (2010). *Social and Economic networks.* Princeton, NJ: Princeton University Press.

Jackson, Matthew O. and Brian W. Rogers (2005). "The economics of small worlds." *Journal of the European Economic Association* 3(2–3), 617–627.

Jin, Emily M., Michelle Girvan, and Mark E. J. Newman (2001). "Structure of growing social networks." *Physical Review E* 6404(4), 046132.

Karinthy, Frigyes (1929). "Chains." In *Everything is Different.* Budapest: Atheneum Press.

Kautz, Henry, Bart Selman, and Mehul Shah (1997). "ReferralWeb: Combining social networks and collaborative filtering," *CACM* 40, 63–65.

Killworth, Peter D. and H. Russell Bernard (1978). "The reverse small world experiment." *Social Networks* 1, 159–192.

Killworth, Peter D. and H. Russell Bernard (1979). "A pseudomodel of the small world problem." *Social Forces* 58(2), 477–505.

Killworth, Peter D., Eugene C. Johnsen, H. Russell Bernard, Gene A. Shelley, and Christopher McCarty (1990). "Estimating the size of personal networks." *Social Networks* 12, 289–312.

Kleinberg, Jon (2001). "Small-world phenomena and the dynamics of information." In *Advances in Neural Information Processing Systems 14,* T. G. Dietterich, S. Becker, and Z. Ghahramani, eds. Cambridge, MA: MIT Press.

Kleinberg, Jon M. (2000a). "Navigation in a small world." *Nature* 406, 845.

Kleinberg, Jon M. (2000b). "The small-world phenomenon: An algorithmic perspective." In *Proceedings of the 32nd Annual ACM Symposium on Theory of Computing,* 163–170. New York: Association of Computing Machinery.

Kleinfeld, Judith S. (2002). "The small world problem." *Society* 39(2), 61–66.

Kogut, Bruce (2012). *The Small Worlds of Corporate Governance.* Cambridge, MA: MIT Press.

Kogut, Bruce and Gordon Walker (2001). "The small world of Germany and the durability of national networks." *American Sociological Review* 66(3), 317–335.

Korte, Charles and Stanley Milgram (1970). "Acquaintance linking between white and negro populations: Application of the small world problem." *Journal of Personality and Social Psychology* 15, 101–118.

Kossinets, Gueorgi and Duncan J. Watts (2006). "Empirical analysis of an evolving social network." *Science* 311(5757), 88–90.

Kossinets, Gueorgi and Duncan J. Watts (2009). "Origins of homophily in an evolving social network." *American Journal of Sociology* 115(2):405–450.

Lamberson, P. J. (2016). "Diffusion in networks." In *The Oxford Handbook on the Ecnomics of Networks,* Yann Bramoulle, Andrea Galeotti, and Brian Rogers, eds. Oxford, UK: Oxford University Press.

Lattanzi, Silvio, Alessandro Panconesi, and D. Sivakumar (2011). "Milgram-routing in social networks." In *Proceedings of the 20th International Conference on World Wide Web,* 725–734. ACM.

Lazarsfeld, Paul, and Robert Merton (1954). "Friendship as social process: A substantive and methodological analysis." In *Freedom and Control in Modern Society,* Monroe Berger, Theodore Abel, and Charles Page, eds., 18–66. New York: Van Nostrand.

Lee, Nancy H. (1969). *The Search for an Abortionist.* Chicago: University of Chicago Press.

Leskovec, Jure and Eric Horvitz (2008). "Planetary-scale views on a large instant-messaging network." In *17th International World Wide Web Conference,* 915–924. Beijing, China.

Liben-Nowell, David, Jasmine Novak, Ravi Kumar, Prabhakar Raghavan, and Andrew Tomkins (2005). "Geographic routing in social networks." *PNAS* 102(33), 11623–11628.

Marvel, Seth A., Travis Martin, Charles R. Doering, David Lusseau, and Mark E. J. Newman (2013). "The small-world effect is a modern phenomenon." *arXiv:1310.2636.*

McPherson, Miller J. and Lynn Smith-Lovin (1987). "Homophily in voluntary organizations: Status distance and the composition of face-to-face groups." *American Sociological Review* 52, 370–379.

McPherson, Miller, Lynn Smith-Lovin, and James M. Cook (2001). "Birds of a feather: Homophily in social networks." *Annual Review of Sociology* 27, 415–444.

Milgram, Stanley (1967). "The small world problem." *Psychology Today* 2, 60–67.

Moore, Cristopher and Mark E. J. Newman (2000). "Epidemics and percolation in small-world networks." *Physical Review E* 61(5), 5678–5682.

Newman, Mark E. J. (2000). "Models of the small world." *Journal of Statistical Physics* 101(3–4), 819–841.

Newman, Mark E. J. (2003). "The structure and function of complex networks." *SIAM Review* 45(2), 167–256.

Newman, Mark E. J. and Duncan J. Watts (1999). "Scaling and percolation in the small-world network model." *Physical Review E* 60(6), 7332–7342.

Newman, Mark (2010). *Networks: An Introduction.* Oxford, UK: Oxford University Press.

Pin, Paolo and Brian Rogers (2015). "Stochastic network formation and homophily." In *The Oxford Handbook on the Economics of Networks*, Yann Bramoulle, Andrea Galeotti, and Brian Rogers, eds. Oxford UK: Oxford University Press.

Pool, Ithiel de Sola and Manfred Kochen (1978). "Contacts and influence." *Social Networks* 1, 1–48.

Radner, Roy (1993). "The organization of decentralized information processing." *Econometrica* 61, 1109–1146.

Rapoport, Anatol (1957). "Contribution to the theory of random and biased nets." *Bulletin of Mathematical Biophysics* 19, 257–277.

Rosenblat, Tanya S. and Markus M. Mobius (2004). "Getting closer or drifting apart?" *Quarterly Journal of Economics* 119(3), 971–1009.

Sen, P., S. Dasgupta, A. Chatterjee, P. A. Sreeram, G. Mukherjee, and S. S. Manna (2003). "Small-world properties of the Indian railway network." *Physical Review E* 67(3), 036106.

Simsek, Ozgur and David Jensen (2005). "Decentralized search in networks using homophily and degree disparity." In *Proceedings of the 19th International Joint Conference on Artificial Intelligence (IJCAI 2005)*, 304–310. Edinburgh, Scotland: Morgan Kaufmann.

Solomonoff, Ray and Anatol Rapoport (1951). "Connectivity of random nets." *Bulletin of Mathematical Biophysics* 13, 107–117.

Sporns, O., and J. D. Zwi (2004). "The small world of the cerebral cortex." *Neuroinformatics*, 2(2), 145–162.

Travers, Jeffrey and Stanley Milgram (1969). "An experimental study of the small world problem." *Sociometry* 32, 425–443.

Ugander, Johan, Brian Karrer, Lars Backstrom, and Cameron Marlow (2011). "The anatomy of the Facebook social graph." *Arxiv preprint arXiv:1111.4503.*

Walsh, Toby (1999). "Search in a small world." In *Proceedings of the 16th International Joint Conference on Artificial Intelligence, Volume 2*, 1172–1177. Edinburgh, Scotland: Morgan Kaufmann.

Watts, Duncan J. (1999a). "Networks, dynamics, and the small world phenomenon." *American Journal of Sociology* 105, 493–527.

Watts, Duncan J. (1999b). *Small Worlds: The Dynamics of Networks Between Order and Randomness*. Princeton, NJ: Princeton University Press.

Watts, Duncan J. (2003). *Six Degrees: The Science of A Connected Age*. New York: W. W. Norton.

Watts, Duncan J. (2004). "The 'new' science of networks." *Annual Review of Sociology* 30, 243–270.

Watts, Duncan J. (2008). "Small-world networks." In *The New Palgrave Dictionary of Economics*, Steven N. Durlauf and Lawrence E. Blume, eds. Basingstoke: Palgrave Macmillan.

Watts, Duncan J., Peter S. Dodds, and Mark E. J. Newman (2002). "Identity and search in social networks." *Science* 296(5571), 1302–1305.

Watts, Duncan J. and Steven H. Strogatz (1998). "Collective dynamics of 'small-world' networks." *Nature* 393(6684), 440–442.

West, Robert and Jure Leskovec (2012a). "Automatic versus human navigation in information networks." In *Sixth International AAAI Conference on Weblogs and Social Media*. Dublin, Ireland: AAAI.

West, Robert and Jure Leskovec (2012b). "Human wayfinding in information networks." In *21st International World Wide Web Conference*. Lyon, France: Association of Computing Machinery.

White, Harrison C. (1970). "Search parameters for small world problem." *Social Forces* 49(2), 259–264.

Wimmer, Andreas and Kevin Lewis (2010). "Beyond and below racial homophily: ERG models of a friendship network documented on Facebook." *American Journal of Sociology* 116(2), 583–642.

CHAPTER 15

NETWORKED EXPERIMENTS

SINAN ARAL

15.1 Embracing Interdependence

Networked experiments have been conducted since at least the 1950s, when Bavalas (1950) and Leavitt (1951) manipulated communication networks to measure their effects on group performance. However, trends toward digitization and fine-grained data collection have recently expanded the scale, scope, and complexity of networked experiments. Early experiments typically operated on hundreds of student subjects, assembled in laboratories. Today, however, we can design, conduct, and analyze networked experiments much more rapidly; and not just on samples of hundreds of people, but rather on hundreds of millions of people at a time. As more and more social interactions, behaviors, decisions, opinions, and transactions are digitized and mediated by online platforms, our ability to quickly answer nuanced causal questions about the role of social behavior in population-level outcomes such as health, voting, political mobilization, consumer demand, information sharing, product rating, and opinion aggregation is becoming unprecedented.

The importance of this new line of inquiry in the social sciences is difficult to overstate. How we relate to one another, connect with one another, coordinate, cooperate, and conflict with one another, are arguably the most fundamental data-generating processes in the social sciences. They are, in some sense, what make the social sciences *social*. In recent years, networked studies have grown to represent larger fractions of total research in traditional social sciences, like economics and sociology, as well as in other important disciplines, such as computer science and physics, which are increasingly addressing social science questions. All of these disciplines are embracing interdependence, which has become a key assumption in the most influential theoretical and empirical work in these areas.

As many scientific disciplines are simultaneously exploring the implications of interdependence in our social world, networked experiments, which help us understand our interdependence more deeply, are becoming one of the most important tools of social science. The increasing scale and scope of modern networked experiments have created a new ability to engineer and randomize social settings to robustly estimate peer effects and the outcomes of social interactions, to explore the heterogeneity in these causal effects across subpopulations, and to unpack the nuanced behavioral mechanisms that underlie and explain these social effects. By understanding such causal behavioral mechanisms, how and why outcomes vary across individuals and how they change over time, we can develop more contextual, personalized and therefore more effective business and public policies. The ability to conduct large-scale networked experiments therefore portends a sea change in our scientific understanding of human relationships and behavior, and dramatic improvements in social policy as a result.

The precision and complexity of the experimental toolkit have also increased in recent years. As online platforms scaled to support hundreds of millions, even billions, of simultaneous users, and platform design became more open and precise, the ability to test complex dynamic hypotheses about social behavior expanded. For example, Facebook enables developers to customize application features for particular users, enabling feature and design randomization (e.g., Aral and Walker 2011a, 2012); Amazon Mechanical Turk enables the development of complex environments in which users can engage in precisely defined experimental micro-tasks (Mason and Watts 2012; Rand and Nowak 2011; Suri and Watts 2011); online labor markets and trading markets have been randomized to test market functions and market failure (Horton et al. 2010); and formal collaboration with platform developers and website administrators is enabling researchers to achieve even more comprehensive experimental control in large-scale in-vivo environments (Bakshy et al. 2012a,b; Muchnik, Aral, and Taylor 2013).

These digital tools are enabling a new era of experimental social science that has begun to reveal robust evidence of the nuanced causal determinants of human behavior. For example, recent large-scale digital experiments have revealed insights about product adoption and engagement (Aral and Walker 2011a, 2012; Bapna and Umyarov, forthcoming; Taylor, Bakshy, and Aral 2013), social commerce and advertising (Aral and Taylor 2011; Bakshy et al. 2012a; Tucker 2011), information sharing and diffusion (Bakshy et al. 2012b), herding behaviors in cultural markets (Muchnik et al. 2013; Salganik, Dodds, and Watts 2006; Tucker and Zhang 2011), health behaviors (Centola 2010, 2011), voting and political mobilization (Bond et al. 2012), performance in innovation contests (Boudreau and Lakhani 2011), the functioning of labor markets (Horton et al. 2010), coordination and cooperation (Fowler and Christakis 2010; Kearns et al. 2006; Mason and Watts 2012; Rand and Nowak 2011; Suri and Watts 2011), altruism and reciprocity (Bapna et al. 2011, Leider et al. 2009), and the growth and efficiency of two-sided matching markets (Tucker and Zhang 2010; Bapna et al. 2013). As a reference, I summarize the focus, context, and experimental procedures of recent large-scale networked experiments in Table 15.1. These studies have employed complex experimental designs that randomize social conditions at the system (e.g., Salganik et al. 2006), category (e.g., Tucker and Zhang 2010), item (e.g., Muchnik et al. 2013), group

Table 15.1 Examples of large-scale networked experiments

Study	Focus	Context	Experimental Procedure	Scale
		Product Diffusion and Engagement		
Aral and Walker (2014)	Impact of tie strength and embeddedness on social influence	Facebook app adoption	Randomize influence mediating messages at the dyad level	1.3M Users
Aral and Walker (2011a)	Viral features and social influence in product adoption	Facebook app adoption and use	Randomize viral features at the app level	1.5M Users
Aral and Walker (2012)	Social influence and susceptibility in product adoption	Facebook app adoption	Randomize influence mediating messages at the individual level	1.3M Experimental users; 12M observational users
Bapna and Umyarov (forthcoming)	Social influence in paid product adoption	Last.fm paid subscription adoption	Randomize gifted paid subscriptions at the individual level	40K Experimental users; 1.2M observational users
Taylor et al. (2013)	Downstream selection effects in online sharing behavior	Facebook offers	Randomize sharing mechanisms at platform level	1.2M Users
Hinz et al. (2011)	Seeding strategies to promote diffusion	Redeemable tokens; URLs to a viral video	Randomized initial seed recipients	120 Users/28 experiments; 1,380 students
		Social Commerce and Advertising		
Bakshy et al. (2012a)	Social advertising	Facebook social advertising	Randomize social signals in advertising	23M Users; 148K ads; 101M user-ad pairs
Tucker (2011)	Social advertising	Unnamed nonprofit	Randomize social ad text and targeting	630 Ads; 13K impressions
Aral and Taylor (2011)	Incentivized peer referral marketing	Online floral delivery site	Randomize incentive structure: selfish, generous, and fair	637 Users

Table 15.1 (*Continued*)

Study	Focus	Context	Experimental Procedure	Scale
		Information Sharing and Diffusion		
Bakshy et al. (2012b)	Social influence in information diffusion	Facebook news feed	Randomize influence mediating messages at the individual level	253M Users; 76M URLs; 1.2B user-URL pairs
		Herd Behaviors in Ratings and Cultural Markets		
Salganik et al. (2006)	Herd behavior in cultural markets	Artificial music lab	Randomize popularity information	14K Users
Tucker and Zhang (2011)	Popularity information and choice	Wedding service vendor website	Randomize availability of popularity information	3 Vendor categories; 90K click-throughs
Muchnik et al. (2013)	Social influence bias in online ratings	Ratings in an unnamed social news aggregation website	Randomize first rating at the comment level	116K Users; 101K comments; 10M user-comment impression pairs
Taylor et al. (2014)	Impact of identity on rating social content	Ratings in an unnamed social news aggregation website	Randomize availability of identity information	350K Comments; 17M impressions; 875K commenter-viewer pairs
		Health Behaviors		
Centola (2010)	Social influence in the spread of health behaviors	Artificial Health network	Randomize alters/ Network structure	1500 Users
Centola (2011)	Homophily and influence in the spread of health behaviors	Artificial health network	Randomize alters / network structure	700 Users

		Voting and Political Mobilization		
Bond et al. (2012)	Voting	Facebook voter registration campaign	Randomize social signals in voting	61M Users
		Cooperation and Coordination		
Kearns et al. (2006)	Graph coloring	Laboratory	Randomize network topology	2 Experiments; 55 users
Fowler and Christakis (2010)	Social influence in cooperation	Laboratory experiments	Random partnering in cooperation games	240 Users
Rand and Nowak (2011)	Cooperation games	Amazon Mechanical Turk Laboratory	Randomize fixed and fluid network structure	785 Users; 40 sessions
Suri and Watts (2011)	Public goods game	Amazon Mechanical Turk Laboratory	Randomize network topology	113 Experiments; 24 users/experiment
Mason and Watts (2012)	Cooperation and exploration games	Amazon Mechanical Turk Laboratory	Randomize network topology	256 Experiments; 16 users/experiment
		Reciprocity and Altruism		
Leider et al. (2009)	Altruism, directed altruism	Facebook dictator and helping games	Randomize anonymity and repeated interaction	802 Students; 2,360 students
Bapna et al. (2011)	Reciprocity games	Facebook game	Randomize anonymity	190 Users; 77 games

Table 15.1 (*Continued*)

Study	Focus	Context	Experimental Procedure	Scale
		Innovation Contest Performance		
Boudreau and Lakhani (2011)	Innovation contests	NASA TopCoder contest	Randomize incentives and sorting on competitive regime	Multiple settings
		Labor Market Outcomes		
Horton et al. (2010)	Labor markets	ODesk online labor market	Three different randomizations	Multiple settings
		Two-Sided Markets and Matching		
Tucker and Zhang (2010)	Growing two-sided markets	B2B exchange market	Randomize display of number of buyers and/or sellers	15 Categories; 3,314 listings
Bapna et al. (2012)	Effect of anonymous weak signaling on date matching	Dating website	Randomize anonymous weak signaling feature	10K Experimental users; 100K observational users

(e.g., Suri and Watts 2011), and individual (e.g., Aral and Walker 2012) levels to reveal important insights about human behavior at population scale.

The scale of modern-day experimentation also enables new levels of analysis. Smaller-scale randomized experiments are typically only sufficiently powered to estimate average treatment effects—the average effect of a policy in a population. But experimental tests of micro-level policies among millions of people allow researchers to unpack the heterogeneity of treatment effects across different populations and to explore the behavioral mechanisms that explain the treatment effects (Aral 2011). For example, the experiment conducted by Aral and Walker (2012) estimates heterogeneity in the impact of influence mediating messages on different types of consumers; Bapna and Umyarov (forthcoming) explore heterogeneity of influence across the degree distribution of users, and Muchnik et al. (2013) dig deeply into whether opinion change or selective turnout creates the social influence bias they estimate to exist in online ratings. These insights enable policies tailored for particular groups of people. Furthermore, creating a deeper understanding of the data-generating processes that explain the average behavioral effects of policy interventions will prepare policy-makers to respond to changes in these processes and to know how their interventions create dynamic changes in behavior over time.

Yet while the increasing scale, scope, and complexity of networked experiments creates tangible opportunities for scientific advancement, they also simultaneously create significant new challenges. As networked experiments become larger and more complex, the likelihood of nontrivial statistical interference, generating non-ignorable violations of the stable unit treatment value assumption (SUTVA) (Rubin 1986, 1990), must be explicitly considered. Although digital technologies enable us to observe more micro-level human interaction, they may also obscure new constraints on sampling, reducing the marginal cost of reaching individuals online while increasing the chances that we will miss large swaths of society that interact offline. Meanwhile, the digital platforms across which we observe micro-level behaviors online are constantly changing, as the algorithms that drive human behavior are altered week to week and month to month. Such changes, and vast behavioral dependencies on the rules of engagement embedded in platform technologies, must be considered as we generalize the findings of networked experiments conducted on digital substrates to the broader human condition.

The purpose of this chapter is to present a concise and accessible review of advances in modern networked experimentation and to summarize the opportunities and challenges they create for understanding our social world. The chapter is divided into two sections that discuss the *design* and *analysis* of networked experiments, respectively. I review innovations in *networked experimental design* in Section 15.2, assessing the implications of the experimental setting, sampling, randomization procedures, and treatment assignment. I then discuss the *analysis of networked experiments* in Section 15.3, with particular emphasis on modeling treatment response assumptions, inference, and estimation, including recent approaches to interference and uncertainty in dependent data. I conclude by discussing important challenges facing researchers as

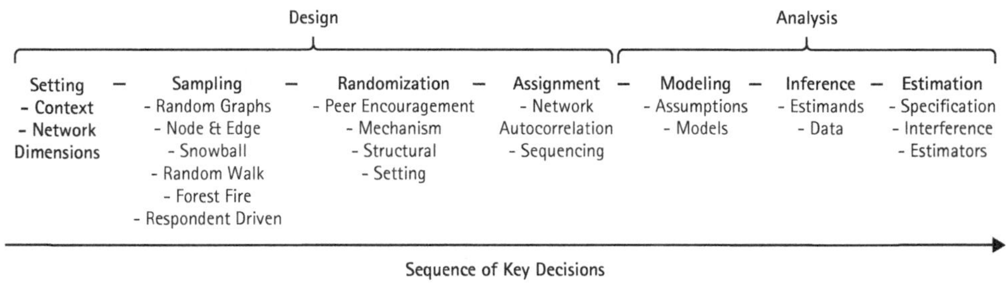

FIGURE 15.1 Key decisions in the design and analysis of networked experiments. The figure represents key decisions the researcher must make during the design and analysis of networked experiments. Though the figure presents these decisions sequentially, the process is, in a sense, iterative. Problems encountered during estimation (for example, interference), necessitate changes in one's approach to sampling, randomization, and modeling. Though there are feedback loops that connect the key decisions, this sequential representation helps frame the problems in a logical order.

we develop the future of networked experimentation, focusing on adaptive treatment assignment, novel randomization techniques, linking online treatments to offline responses and experimental validation of observational studies.

Figure 15.1 presents the sequence of key decisions in the design and analysis of networked experiments considered in Sections 15.2 and 15.3. In order to frame the discussion clearly, I use a particular paper, Aral and Walker (2011), to instantiate this sequence of key decisions. The choices made in Aral and Walker serve to concretely highlight the opportunities and challenges of networked experimentation. I have chosen this paper, to which readers may wish to refer directly while reading this chapter, not because it is the best example of networked experimentation (in fact it is an early example of large-scale, in vivo networked experimentation), but because I am personally aware of the opportunities and challenges encountered during the work.

The chapter provides a broad overview of networked experimental techniques, addressing key methodological concerns. It is intended as a review of the literature and a guide for researchers and practitioners interested in conducting experiments in networks. Given this aim, the chapter complements deeper treatments of narrower subtopics presented in this handbook in the chapters on laboratory experiments by Syngjoo Choi and Shachar Kariv, and on field experiments in developing countries by Emily Breza. The contribution of networked experimentation to our understanding of human social dynamics is likely to be large. I hope this overview, combined with the excellent contributions of my colleagues highlighted herein, provides a roadmap for extending networked experimentation in meaningful ways in the coming years.[1]

[1] This introductory section of this chapter relies heavily on a review of the literature I wrote with Dylan Walker in Aral and Walker (2014). Here I have expanded on and updated that review. Portions of the text as well as an earlier version of Figure 15.1 are reprinted by permission from Sinan Aral and Dylan Walker

15.2 Design of Networked Experiments

15.2.1 Setting

The settings in which networked experiments are conducted typically enable and constrain the inquiries researchers can pursue. One useful categorization divides experiments along two dimensions: the context in which the experiment takes place (in the field or in the lab) and the dimensions of networks that are examined (network structure, dyadic relationships, or individuals). Although in some sense network structure is always present, the focus of the inquiry, and thus the experimental manipulation, change simultaneously across these dimensions.

Context. Networked experiments typically take place either in a laboratory or in the field. The lab offers more experimental control but perhaps less realism and thus generalizability (depending on the research questions being asked), while the field offers less control but a more realistic setting in which to observe human behavior. There are also experiments that take place in "virtual labs," for instance in controlled laboratory environments created by researchers on the web (e.g., Salganik et al. 2006; Centola 2010, 2011; Mason and Suri 2012), as well as in "living labs," which are instrumented physical locations like domiciles, dwellings or communities in which inhabitants are digitally instrumented (e.g., Aharony et al. 2011), and in virtual worlds (Bainbridge 2007; Fiedler et al. 2011). Virtual labs, living labs, and virtual worlds have their own idiosyncrasies with regard to data collection, recruitment, sampling and so on, which are discussed in detail in the pioneering papers in these areas referenced here.

The trade-off between control and realism exists across lab settings and field settings even when the labs are virtual or the field is digitized. In a lab, the researcher can control many aspects of the environment that are difficult to control in the field. For example, complex network structure can be manipulated to study effects on information diffusion (Centola 2010), cooperation (Suri and Watts 2011) or coordination (Kearns et al. 2006). Participants are, in these cases, randomly inserted into precise, well-controlled and well-understood network structures to study how structure affects behavior. The information participants have access to can also be controlled, such that the interactions between complex network structures and information can be examined. Field studies sacrifice a measure of control for observation of more realistic networked behaviors. Subjects are observed and studied in their natural environment and through their typical modes of interaction with their real network connections. Field experiments allow us to study how real-life relationships affect our judgment and how true social influence affects our behavior or opinions. The appropriate setting depends on the

(2014). "Tie Strength, Embeddedness, and Social Influence: A Large-Scale Networked Experiment," *Management Science* 60(6), 1352–1370.

research question one is after: to understand how idealized behaviors change in response to complex changes in network structure, lab settings are quite powerful; to understand how behaviors evolve in real social structures, field experiments are perhaps more appropriate.

Aral and Walker (2011) conducted a field experiment of interactions and behaviors taking place on the live Facebook social network, to measure the effects of design features on product virality and to investigate the impact of network externalities on engagement and the likelihood of customer churn. The paper randomly enables viral product features, such as social notifications and the ability to invite friends, to test their effects on product diffusion, and measure how network externalities drive product use after adoption. I will return to this example periodically to illustrate various dimensions of networked experimentation.

Network Dimensions. Inquiry can focus on the role of structure, the functioning of dyadic relationships or the implications of changes in setting or technology on the networked social behaviors of individuals or groups. Choices made among these network dimensions frame experimental studies and the inferences and generalizations that can be drawn from them. Such choices also have implications for downstream choices regarding network sampling, randomization procedures, treatment assignment, modeling, measurement and inference. Analyses of the role of structure require sampling a set of individuals representative of the population. Analyses of dyadic peer influence may require samples of representative ego networks from larger graphs that maximize statistical precision and minimize interference between units. Analyses of changes in technology; for instance, the degree to which network externalities exist for a given technology, may require comprehensive samples of larger groups of interacting contacts to analyze the user experience under realistic network externality conditions, such as uniform access to the technology.[2]

For example, Aral and Walker (2011) are interested in estimating (i) dyadic peer influence conveyed through different interpersonal mechanisms (in their case, differences in peer influence conveyed by one to one personal invitations and one to many broadcast notifications), (ii) the overall impact of viral features on social contagion in product adoption processes, and (iii) the effect of network externalities on the persistence of behaviors inspired by social influence over time. Thus the work contemplates both dyadic and network-level dimensions of social contagion, which affect subsequent sampling, randomization, inference, and estimation decisions.[3]

[2] For a more comprehensive treatment of empirical investigations of network effects, I refer the reader to the excellent chapter on "Recent Developments in the Empirics of the Effects of Networks" by Bernard Fortin and Vincent Boucher in this handbook.

[3] It is important to note here that a networked experiment may also be interested in subsequent dynamic changes in network structure itself, a topic I do not cover in this chapter due to the dearth of prior research in this area. Although models of network dynamics consider link formation and disillusion, few, if any, experimental studies examine determinants of network evolution (for a notable exception, I refer the reader to work by Sharique Hasan, who has exploited natural experiments to study network evolution [Hasan and Bagde, forthcoming] and is building a living laboratory in India

15.2.2 Sampling

Given a setting, the network on which the experiment will be conducted must be generated or sampled from a graph or graph distribution. In cases where the network can be controlled, one may wish to initialize particular network structures based on theory. For example, a researcher may be interested in comparing coordination outcomes in cohesive or diverse networks (Suri and Watts 2011; Shore et al. 2014), or the propagation of simple or complex contagions across weak ties or wide bridges (Centola 2010). Alternatively, the researcher may generate synthetic graphs that display certain properties, such as power law degree distributions, particular path lengths, or clustering characteristics, using a model of network formation. Random graph models, which provide a family of network generation models that produce graphs with precisely solvable average properties (Newman 2002), trace their origins to Erdos and Reyni (1959, 1960, 1961), whose random graphs underpin today's modern graph generation procedures. Unfortunately, the original models also produced some unrealistic social network properties, like Poisson degree distributions and a lack of clustering. More modern generalized random graph models incorporate arbitrary non-Poisson degree distributions, clustering and more realistic power-law degree distributions (e.g., Aiello et al. 2001; Newman 2002).

In field experiments, naturally occurring network structure is typically taken as given, though not necessarily fixed, and randomization occurs through, for example, the perturbation of behaviors or information channels across the observed network. In this setting, individuals are sampled together with their peers from real graphs to study how changes in behavior affect social outcomes in the naturally occurring network. The goals of sampling can vary from estimating node or edge attributes, collecting subgraphs that are representative of local ego networks, or collecting a sample of network paths that reliably reproduces likely diffusion events. Several sampling methodologies can produce the network on which the experiment is conducted. For example, *node sampling* and *edge sampling*, which choose nodes and edges independently and uniformly at random from the population network, typically do not preserve graph properties, such as the degree distribution, density, connectivity, and clustering of the network (Stumpf et al. 2005; Lee et al. 2006).

Topology-based sampling is better equipped to preserve representative graph properties and structures. For example, *complete snowball sampling* (e.g., Aral et al. 2009), which uses breadth-first search from randomly selected seed nodes, can accurately preserve network properties within the snowball, but suffers from boundary bias as nodes at the edge of the snowball sample will be missing neighbors (Lee et al. 2006). Snowball sampling can also miss the connectivity of the global graph because, although clusters of representative subgraphs around random seed nodes are complete, the

to study network evolution through experimental means in a startup accelerator). Experimental studies of network evolution represent a fertile area for future research, one in which important insights are likely to be found.

connections between such subgraphs are typically under-sampled. Complete snowball sampling therefore best suits studies of micro-level social behavior (e.g., studies of average local peer effects or micro-level, interpersonal diffusion dynamics) rather than global diffusion or cascading behavior.

Random walk sampling, which begins by selecting a seed node uniformly at random and simulating a random walk along the network, more accurately captures global graph properties and community structure, but makes micro-level, dyadic social processes difficult to characterize. *Random jump sampling* mirrors the random walk, but jumps to any random node instead of returning to the seed node and therefore suffers from a lack of reliable reproduction of symmetric and/or complete local network structures. *Forest fire sampling* (Leskovec, Kleinberg, and Faloutsos 2005) conducts a partial-breadth first search, which samples some proportion of the edges of a randomly sampled "ambassador" node (and its corresponding dyadic node counterparts) with some probability, continuing the process until the "fire" burns itself out. These walk-based sampling methods replicate global graph properties and community structure more accurately than snowball sampling (because they are not constrained to local networks), but they do not sample complete local networks around seed nodes. *Respondent-driven sampling* (Heckathorne 1997; Salganik and Heckathorne 2004) relies on respondents to recruit their peers and a model of the recruitment process to weight the sample to compensate for nonrandom recruitment patterns. While such methods are well suited to recovering statistically valid samples of difficult-to-reach groups (e.g., IV drug users, homeless youth), they can also produce biased samples of local structure.

These sampling techniques make realistic testing of local social interactions more difficult to generalize because complete social environments are not observed. Structural determinants of social behavior are also difficult to analyze. For example, the debate about the roles and relative importance of wide bridges (Centola and Macy 2007) or thick bridges (Aral and Van Alstyne 2011) for the spread of complex contagions becomes more difficult to resolve in the absence of complete local structure. The point of this discussion is not to provide a comprehensive treatment of network sampling (e.g., Leskovec, Kleinberg, and Faloutsos 2005; Leskovec and Faloutsos 2006; Ahmed et al. 2014). Rather, the aim is to point out that sampling or graph generating procedures, which produce the graphs on which experiments are conducted, can enable and constrain the inferences that are possible as well as the degree to which those inferences are generalizable to larger populations or sets of graph structures.

In Aral and Walker (2011), we aimed to generalize our findings on social influence to users of Facebook and online social interactions more broadly. We therefore went to great lengths to ensure our experimental sample was representative of the Facebook population. We employed an ad-targeting agency to target advertisements about the application on which we conducted the experiment such that the adopting population was representative of the Facebook population. Distributions of individual characteristics of the resultant sample were then compared to published statistics about the overall Facebook population to ensure the sample was representative. After approximately 10,000 initial adopters were recruited, a complete one-step snowball sample was collected using the Facebook API to create a population of approximately 1.4 million friends of initial adopters. All edges originating from the initial adopters as well as ties between friends of initial adopters were sampled. In addition, any diffusion event was captured comprehensively, meaning if an invitation was sent from an initial adopter to a friend, who then adopted the application and invited a friend of a friend to adopt, and so on, every adoption in the chain prompted data collection about the adopter, their networks, and their characteristics to ensure complete observation of all diffusion events regardless of their depth. This example links the sampling strategy to the goals of the experiment. The representative sample of initial adopters allowed us to measure the impact of including viral invitation and broadcast notification features on the application's diffusion process, the dyadic peer influence conveyed through different interpersonal mechanisms and the effect of network externalities on the prolonged use of the application for a representative sample of Facebook users. Experimental results based on this sampling strategy can therefore be generalized to Facebook users as a whole.

15.2.3 Randomization

In some sense, the core of networked experimental design is the randomization procedure. Randomization creates the exogenous variation that identifies the causal effects of interest. If we aim to understand the impact of network structure on social behavior, exogenous variation should be introduced at the network level. If instead we aim to understand peer effects, exogenous variation should be introduced at the individual level. At least four categories of randomization procedures, aimed at different research questions, have gained traction in the last several years.

Peer encouragement designs randomly encourage particular behaviors in a set of nodes to analyze the effects of those behaviors on the nodes' peers. They enable studies of peer effects or social influence in a variety of settings and across a number of behaviors. For example, Bapna and Umyarov (forthcoming) gift premium subscriptions to users of a social music streaming website to estimate peer influence in the decision to adopt a premium service. Hinz et al. (2011) randomize initial seed recipients to test the impact of seeding strategies on the success of viral marketing campaigns in three different contexts. As with traditional encouragement designs, peer encouragement allows researchers to study social effects emanating from the adoption of products, services, or behaviors whose adoption cannot be guaranteed by policy instruments in real world settings.

Mechanisms designs change the mechanisms through which individuals interact. The focus could be on how different communication channels transmit information or the way behaviors spread in social networks through peer influence. For example, Bakshy et al. (2012b) randomize exposure to signals about friends' information sharing on Facebook. They find those who are exposed are significantly more likely to spread information. Aral and Walker (2011) randomize the channels of communication through which users can promote applications on Facebook and find active personalized messaging is more effective per message and generates more engagement, while passive broadcasts generate faster growth through more messages. Aral and Walker (2012, 2014) randomize the delivery of broadcast notifications to peers of selected Facebook users to estimate individual and dyadic correlates of social influence. Their randomization, at the ego-network level, also addresses known degree and homophily biases in independent random assignment designs (Thomas and Finegold 2013), a topic I cover in more detail in Section 15.3.

Structural designs randomize network structure, or individuals' positions within it, to understand the impact of structure on social behavior. For example, Kearns et al. (2006) study how network structure affects individuals coordinating to solve the graph coloring problem. By randomizing structure, they find that networks generated by preferential attachment make solving the coloring problem more difficult than do networks based on cyclical structures or "small worlds." Centola (2010) studies how network structure affects the spread of health behaviors in an online social network. By randomizing the degree of local clustering, he finds that diffusion is more likely across clustered-lattice networks than random networks. Mason and Watts (2012), Suri and Watts (2012), and Rand and Nowak (2011) all randomize network topology to study cooperation and public goods games played by Amazon Mechanical Turk users.

Setting designs change the settings in which social behavior takes place. Particular behaviors are not encouraged and interaction mechanisms are not necessarily altered. Instead, the environment in which the interaction takes place is altered and studied. For example, Aral and Taylor (2011) randomize the incentive structure in which social behaviors take place, to examine how incentives affect people's social behavior. Bapna et al. (forthcoming) and Taylor et al. (2014) randomize the presence of identifying information to study the effect of anonymity on social behavior in the context of online dating and online ratings respectively. Muchnik et al. (2013) study how the presence of popularity information affects the ratings of friends and enemies on each others' social content. Table 15.2 summarizes these four types of randomization procedure and their applications and gives examples of each.

Aral and Walker (2011) implemented a mechanism design experiment that randomly enabled and disabled certain features on a Facebook application to measure their impact on viral diffusion. As people downloaded the application, they are randomly assigned to one of three treatment groups: baseline, passive-broadcast, and active-personalized. Users assigned to the baseline treatment group received a version of the application in which both notifications and invitations were disabled. In the passive-broadcast treatment group (passive), only notifications were enabled. In the active-personalized treatment group (active), both notifications and invitations were enabled. There were no other differences between the baseline, passive, and active applications. In this sense, the randomization takes place at the user level and the environmental condition that is randomized is the state of the viral features that are enabled or disabled on the application.

Table 15.2 Common Networked Randomization Procedures

Design	Procedure	Application	Examples
Peer encouragement designs	Randomize encouragement of behavior in focal node; measurement of response in peers	Measurement of peer effects, social influence, or social contagion	Randomly encourage product adoption (Bapna and Umayurov [forthcoming]); randomly encourage thankful status updates on facebook (Eckles)
Mechanism designs	Randomize characteristics of channels of interaction between nodes; measure responses of ego and peers	Measurement of impact of interaction channels on diffusion, contagion, cooperation	Randomize viral features of a product (Aral and Walker 2011); randomize message delivery (Bakshy et al. 2012; Aral and Walker 2012)
Structural designs	Randomize properties of network structure; measure responses of ego and peers	Measurement of impact of network structure on diffusion, contagion, cooperation, team performance	Randomize network structure in behavior diffusion (Centola 2010); in cooperation games (Suri and Watts; Rand et al.)
Setting designs	Randomize environment in which social behavior takes place (e.g., availability of contextual information or incentive structures); measure responses of ego and peers	Measurement of social environment effects on interactions, transactions, diffusion, cooperation	Randomize incentives to propagate a product (Aral and Taylor 2011); randomize incentives and sorting (Boudreau and Lakhani 2011)

15.2.4 Treatment Assignment

Treatment assignments map nodes and/or edges to treatment conditions. Depending on the study design, assignments may be independent or correlated across nodes and edges in the network. Typical experimental treatment assignments independently assign units to treatment and control conditions. While this is certainly possible in network settings, treatment assignment with network autocorrelation may produce some desirable properties of the units chosen. Furthermore, in considering network autocorrelated assignments in which the assignment of some units is a function of the assignment of others, the sequencing of treatment assignments becomes critical.

Network Autocorrelation. Network autocorrelation in treatment assignment may be used to study group interventions; for example, in studies of inducement to peer pressure among sets of connected nodes (Mani et al. 2013) or in studies of the adoption of products and services that exhibit network externalities (Ugander et al. 2013). In the latter case, realistic parameter estimates of network effects may require assignment of entire network neighborhoods to encouragements to adopt a product because, in a world with universal access to the product, network externalities operate on clustered adoption with all peers of an adopter able to adopt the product. Blocked treatment assignment is also essential to design-based approaches to statistical interference (e.g., Ugander et al. 2013; Eckles et al. 2014), which I discuss in detail in Section 15.3. In these designs, clusters of treated nodes are optimally "surrounded" by peers with the same treatment assignment to reduce bias and error in estimation. In peer encouragement designs it may be useful to treat a set of seed nodes from a network and to block treatment of peers of those nodes (Airoldi et al. 2013). Such a setup can facilitate measurement of peer effects while minimizing interference. This example highlights another key design choice: the temporal sequencing of treatments.

Sequencing. The sequencing of treatment assignments is a critical aspect of randomization procedures that enable and constrain inference. Sequencing dilemmas are highlighted by viral marketing experiments, which utilize response driven sampling.[4] As networked experiments typically study cascading behaviors, products, or services, how to treat units sampled through the spread of social behavior is a fundamental decision. For example, Aral and Walker (2011) experimented with multiple "viral product design" strategies to estimate their impacts on peer to peer sharing of an online application. They randomly assigned subjects to different versions of the application with different viral messaging features and measured referrals as the different versions diffused through the Facebook network. An important choice in this context is how to assign sequentially referred users to treatments. If, for example, a user with version A refers a second user, should the referee be assigned to treatment A or

[4] While respondent-driven sampling asks respondents to name new respondents, I use the term response-driven sampling here to refer to new units sampled as a consequence of responding to a peer behavior induced by the experiment; for example, adopting a product through a peer referral designed by the experimenter.

randomly assigned to either A or B? As diffusion dynamics may depend on the consistency of peer experiences (e.g., due to network effects or consumer expectations), consistent treatment assignment—assigning all referees to the same treatment status as their referrers—enables estimation of the global treatment impact of implementing version A. Estimands can reflect what the world would look like if everyone received version A. However, consistent treatment assignment reduces the power of the experiment in estimating differences in micro-level peer effects—the effects of A or B on sharing—because fewer units are randomly assigned to treatment. Aral and Walker (2011) randomly assign response-driven samples, rather than maintaining consistent treatment assignment throughout the referral process, and discuss this trade-off in experimental design at some length. In essence, the decision maps directly to the goal of the research. As the goal of the research is to robustly estimate micro-level peer influence effects rather than global diffusion dynamics, random assignment of referred peers increases the statistical power of such estimation.

Sequential treatment assignment also plays a significant role in downstream inference. As I discuss in more detail in Section 15.3, two-stage clustered treatment assignment strategies can reduce selection bias due to latent homophily and the chances of selecting high-degree nodes in the second stage (Thomas and Finegold 2013). As peers of randomly selected users are likely to be of higher degree and similar in characteristics to their randomly selected peers, randomizing treatment assignment in the second stage can help alleviate these biases (Aral and Walker 2012). Sequential treatment assignment can also address other important statistical concerns, such as interference, that arise in interdependent data. As Airoldi et al. (2013) propose, sequential assignment can help alleviate such concerns by making treatment assignments a function of the previous treatment assignment of peers.

Finally, as we look toward the future of networked experimentation, adaptive sequential treatment assignment, in which treatment assignments are a function of prior treatment responses, could help us maximize precision, minimize interference, cost, and harm, or investigate treatment effect heterogeneity or particular hypotheses. For example, if the result of a particular hypothesis is becoming obvious, fewer units could be allocated toward testing that hypothesis as the experiment progresses. I discuss the potential of adaptive treatment assignment in Section 15.3.

15.3 Analysis of Networked Experiments

15.3.1 Modeling Treatment Response

The potential outcomes approach, formally the Rubin Causal Model or the Neyman-Rubin Causal Model, attempts to tackle the fundamental problem of causal inference: We wish to estimate differences in units' outcomes under different treatments, but are never able to observe a given unit under multiple treatments at the same time (Neyman 1923; Rubin 1974). As only one potential outcome can be observed for any individual, the differences between potential outcomes (causal effects) are impossible to observe, but they can be inferred under some basic assumptions, which I discuss in more detail in Section 15.3.3. A key assumption of the potential outcomes approach is that there is no interference between units, meaning "the observation on one unit [is] unaffected by the particular assignment of treatments to the other units." (Cox 1958: 19) If this stable unit treatment value assumption (SUTVA) (Rubin 1986, 1990) does not hold, then rather than having two potential outcomes under treatment and control, experimental units have many potential outcomes that depend on other units' treatments (Rosenbaum 2007).

Analysis of networked experiments therefore begins with a theoretical approach to treatment responses and assumptions about their interdependence.[5] There exists some data-generating process that translates treatment assignments and network structure into the observed outcomes of networked nodes. The assumptions we make about this data-generating process—whether it is assumed to be known or unknown, or in equilibrium or not—will guide our inference and our interpretations of experimental results.

For example, Manski (2013) provides precise theoretical assumptions about treatment response that vary the degree to which a unit's response depends on other units' treatment assignments. Manski's (2013) constant treatment response assumption considers a function that maps treatment assignment vectors to effective treatments such that a unit's response depends on its own assignment as well as the assignments of other units. If node responses are assumed to depend only on their own assignment in what Manski (2013) calls an *individualistic treatment response* assumption, the function simply maps units' own assignments to their outcomes. In this case, there is no interference and SUTVA holds. If, on the other hand, node responses depend on their own assignment and the assignments of a reference group with which they interact, then the function maps sub-vectors of treatments in the reference group to node outcomes. One example of such an assumption is the *neighborhood treatment response* assumption of Aronow and Samii (2012), which posits that effective treatments depend on the unit's assignment and the assignment of their immediate network neighbors. However, other assumptions about effective treatments are also possible. For example, when estimating the effectiveness of child vaccinations, a plausible assumption is that a child's illness is a function of their own vaccination status and the number or proportion of children vaccinated in their school, but not on the specific identities of those vaccinated. Manski (2013) refers to this as an *anonymous treatment response* assumption.

The "exposure mapping" Aronow and Samii (2012) present in their simulation of Add Health data models units' responses as a function of their own treatment and

[5] I return to a detailed treatment of recent approaches to interference in networked experiments in Section 15.3.3.

the number of their treated peers. Treatment responses may not only be a function of peer treatment assignments, but also of peer behavior. As Eckles et al. (2014) point out, under such assumptions, peer behavior may fully mediate the effects of peers' treatment assignments. Rather than comprehensively reviewing various treatment response assumptions, I present these as examples of the types of assumptions one can make. We can then use these assumptions to inform our models of treatment response, which we use to estimate and make inferences about networked behavior.

One of the most widely used models of treatment response is the *linear-in-means model*, which specifies a linear additive contribution of treatment assignment and peer effects to the mean treatment response, such that behavior is a function of a direct effect of treatment and the mean of peers' behaviors (which may mediate peers' treatment responses):

$$Y_i = \alpha + \beta X_i + \gamma \frac{\sum_{j=1,j=i}^{k} Y_j}{k_i} + \varepsilon_i \tag{15.1}$$

where β represents the direct effect of a unit's treatment assignment (*Xi*), and γ the indirect effects of the mean of peer behaviors (where *j* indexes *i*'s peers and *ki* represents i's degree). A similar model for binary outcomes is described in Brock and Durlauf (2001). The linear-in-means model has become a staple in the estimation of peer effects, in part because of its simplicity, generality, and applicability to a wide variety of social effects of interest in economics and sociology. Identification of the linear-in-means model has received a good deal of attention in the economics literature (e.g., Manski 1993; Graham and Hahn 2005; Lee 2007; Bramoulle et al. 2009) and has been used to estimate peer effects in a variety of settings. In more recent work, Dieye et al. (2014) propose a structural extension to the linear-in-means model that accounts for social network effects in treatment response and allows for a contextual effect of treatment alongside the endogenous peer effect.

However, the linear-in-means model is not without its limitations. For instance, in [1], γ captures the slope of the average peer effect of *i*'s neighbors, ignoring the possibility of dyadic peer effects that vary with the characteristics of *i* and *j*. Such a model is not well suited to situations in which peer effects are not simply linear in means. As Burke and Sass (2013) show for educational outcomes, average peer effects may be small while peer effects conditioned on the individual, the peer, or the dyad may be quite large. If high-ability students are influenced positively by other high-ability students but negatively by low-ability students, then a linear-in-means model will miss important heterogeneity in the data. When linear-in-means models of social effects in educational outcomes are compared to models with interactions by own achievement, large peer effects are detected by the conditional model but missed by the linear-in-means model (Burke and Sass 2013).[6] Similar extensions to the linear-in-means approach have been used in the study of peer influence in product adoption (Aral and Walker 2012), premium subscriptions (Bapna and Umyarov, forthcoming), online ratings (Muchnik

[6] Thanks to Michaela Kerrissey for referring me to Burke and Sass (2013).

et al. 2013), and in the role of identity in opinion formation (Taylor et al. 2014). It is also plausible that something other than the second moment of the distribution of peer behaviors is relevant to a unit's outcome. For example, there may be threshold effects (Granovetter 1978; Watts 2002), quadratic reinforcement effects (Centola and Macy 2007), effects from the variance of peer behaviors, or other more complex dependencies.

15.3.2 Inference

Estimands. In randomized experiments, the randomization distribution generated by the experimental design forms the basis of inference: one infers individuals' outcomes under treatments they were randomly denied from observations of individuals' outcomes under treatments they were randomly assigned. For simplicity I denote the randomization distribution of treatments under no intervention as η and under an arbitrary intervention as θ. Since any individual cannot be observed under multiple treatments, causal estimands are typically defined as averages of potential outcomes (Rosenbaum 2007; Hudgens and Halloran 2008).

Many different quantities of interest form the basis of relevant estimands in networked experiments. We may seek to understand individual-, group-, or population-level interventions and outcomes. In addition, we may be interested in experimental effects on the treated, peers of the treated, or the groups or populations in which treatment is studied. The estimands implied by these research goals vary and are being extended and re-conceptualized in some of the most recent literature. Precisely defining the experimental quantities of interest can help researchers draw bounded generalizations from experimental results. A large body of literature addresses the decomposition of treatment effects into direct, indirect and total effects across the units in an experiment (Halloran and Struchiner 1991, 1995). Using this decomposition, Hudgens and Halloran (2008) define several estimands useful for conceptualizing the quantities of interest in experimental studies with interdependence across groups. I describe how these estimands can be related to quantities of interest in networked experiments and illustrate their application in a peer encouragement design example. For simplicity, consider a basic peer encouragement design where treated nodes are given an encouragement to adopt a product and the researcher is interested in effects of this encouragement on product adoption outcomes (Y). In this context we can think about direct, indirect, and total effects as follows:

Direct causal effects, the effect of the treatment on the treated, can be formalized as the difference in individual average potential outcomes under treatment and no treatment conditions given the randomization distribution θ (i.e., in the case where an intervention is initiated): $\bar{D}_i(\theta) \equiv \bar{Y}_i(0;\theta) - \bar{Y}_i(1;\theta)$ where $\bar{D}_i(\theta)$ represents the average direct effect under the randomization distribution, θ, $\bar{Y}_i(0;\theta)$ represents the average potential outcome under control and $\bar{Y}_i(1;\theta)$ the average potential outcome under treatment. In the toy example, this represents the direct effect of the encouragement on product adoption outcomes among treated nodes in the network.

Indirect causal effects, the effect of the treatment on others, can generally be defined as the difference in individual average potential outcomes without treatment when a reference group[7] to which the individual is related in some way is treated and untreated: $\bar{I}_i(\theta,\eta) = \bar{Y}_i(\theta;\eta) - \bar{Y}_i(0;\theta)$. In our peer encouragement example this could represent the indirect effect of treatment on untreated peers of the treated, where a "peer" is a node with any arbitrary relationship to the treated node, including being connected, directly or indirectly, to the treated node in a social network or being a member of a group to which the treated node belongs (e.g., a classroom or a neighborhood).[8]

Total causal effects, the sum of the direct and indirect causal effects, can then be defined as the difference in individual average causal effects for an individual under treatment when an intervention is initiated (θ) and under no treatment when an intervention is not initiated (η): $\bar{T}_i(\theta,\eta) \equiv \bar{Y}_i(0;\eta) - \bar{Y}_i(1;\theta)$. In our example, the potential outcomes of interest are those of an encouraged node when the encouragement is introduced randomly to members of the node's reference group and those of a node that is not encouraged when no encouragement is introduced at all.

These estimands formalize the quantities of interest that form the basis for estimating the effects of networked social behavior. For example, in the peer encouragement example, we may be interested in estimating the micro-level peer influence effect of an encouragement to adopt a product on the peers of the encouraged. In other words, we may want to measure how peer influence affects the likelihood of adoption. Alternatively, we may be interested in the total effect on adoption in a reference group (e.g., a consumer population) of a program to encourage adoption by some members of the group. If peer spillovers exist, measuring the direct effect of treatment will underestimate the total effect of the intervention. Returning to our example, Aral and Walker (2011) are interested in the direct and indirect effects of viral features, but less so in isolating the total causal effect. The decision to randomly assign referee adopters to the baseline, active-personalized, and passive-broadcast conditions, rather than maintaining consistent treatment assignments throughout the diffusion process, enables robust estimation of direct and indirect effects and reflects a decision to forgo estimation of the total causal effects of the design decision. These trade-offs are inevitable in any networked experiment.

Data. It is worth mentioning briefly that the granularity of the data we can now collect through large networked experiments is changing the nature of what is estimable.

[7] A reference group is simply a group to which the individual belongs such that an intervention in that reference group could plausibly affect their potential outcomes. This could be a school, classroom, or social network to which the individual belongs.

[8] Indirect effects can exist as treatment effects on the untreated or reinforcement effects from treated nodes on other treated nodes. Assuming there is no complex interaction between treatments of distinct nodes, the indirect effects can be considered marginal effects holding own treatment status constant. Of course, interaction effects could exist and could subsequently be modeled.

Today's networked experiments, especially those conducted online, are collecting data on micro-level, individual behaviors and opinions in unprecedented detail and time stamped to the second. Our observations of human behavior have moved from static snapshots, to longitudinal panels, to continuous data streams (Domingos et al. 2000). Online platforms that record continuous streams of behavioral data are also themselves evolving over time. As a consequence, new machine learning methods are being devised to deal with key issues in the analysis of streaming data, including concept drift (Wang et al. 2003), clustering (Aggarwal et al. 2003), and querying (Babu and Widom 2001). Fine-grained data collection and new techniques for mining massive data streams are enabling new estimation strategies that incorporate subtle behaviors and time. As new data enable new models, future work should build theory around how granular, continuous time data can be incorporated into experimental analysis.

15.3.3 Estimation

Model Specification. The most straightforward approach to analyzing networked experiments is to estimate Average Treatment Effects (ATE): the difference in means between the treatment and control outcomes, as described in the estimands above. However, occasionally it may be fruitful to specify more sophisticated models that allow us to explore more of the heterogeneity in the data; for example, how treatment effects vary across subpopulations, what moderates the treatment effects, or the underlying behavioral processes that explain the outcomes of the experiment. Specifying a more sophisticated model allows more complex investigation, but at the cost of having to make more restrictive assumptions about treatment responses. For example, Aral and Walker (2011) incorporate time into their analysis by fitting hazard functions that estimate the rate at which individuals react to a treatment, rather than simply whether or not they react. They then validate their more sophisticated models with simple tests of differences in means that require fewer assumptions.

Interference. If the treatment response of one individual depends on the treatments of other individuals in the experiment, the treatments are said to "interfere" with one another. Interference is possible in a variety of settings including, for example, the effects of vaccination programs (Halloran and Struchiner 1991), housing vouchers (Sobel 2006), and educational interventions (Rosenthal and Jacobsen 1968). A substantial literature addresses contexts in which interference may exist between individuals in the same group, but not across groups (Sobel 2006; Rosenbaum 2007; Hudgens and Halloren 2008; Middleton and Aronow 2011; Tchetgen Tchetgen and VanderWeele 2012). This dependence structure, known as partial interference (Sobel 2006), is common in multi-level, group randomized, or cluster randomized studies, in which clusters represent well-defined groups, such as households (Tchetgen Tchetgen and VanderWeele 2012), neighborhoods (Ali et al. 2005), classrooms (Sobel 2006), or land plots (Kempton 1997).

Networked interference is in some sense a substantial departure from partial interference, as the pattern of interdependence is more complex and perhaps more uncertain in that it is not cleanly contained within well-defined groups. In another sense, however, recent approaches to interference in networks extend approaches to partial interference, modeling interference as a decreasing function of social distance and using the observed graph structure to define groups (or subgraphs) within which interference is more or less likely. Two sets of such strategies for addressing networked interference have emerged in recent years, which I refer to as *inference strategies* and *design strategies.*

Inference strategies correct for interference during analysis. For example, Aronow and Samii (2012) define exposure models that map assignment vectors and units' characteristics to exposure values or likelihoods of being subject to a given exposure. They then derive unbiased average treatment effect (ATE) estimators and variance estimators based on the assumptions of their exposure mapping. Airoldi et al. (2013) take a similar approach, deriving causal estimands based on the adjacency weighted matrix of exposure to treatment through social influence in a linear additive model. In Aronow and Samii's work, the basic approach is to model direct and indirect exposure and to base inference on the generalized probabilities of exposure implied by the model. For example, they consider an exposure model that assumes individuals' outcomes depend only on their own assignments and the assignments of their immediate neighbors. In their analysis of Add Health data, they assume a subject's outcomes depend only on their own assignment and a count of their treated neighbors. In both cases, each node's probability of being subject to a given exposure is precisely defined given a well defined experimental design and known network structure.

One important challenge with this approach is the difficulty of validating the exposure models. Aronow and Samii (2012: 19) acknowledge this criticism and make a convincing counterargument that "there is no escaping specification of exposure mappings for causal analysis." After all, analysis that does not specify a model of indirect exposure de facto embraces an assumption of no interference. However, it is possible to approach the problem by making less restrictive exposure assumptions and foregoing point identification to establish bounds on bias reduction. For example, Eckles et al. (2014) establish bounds on bias reduction achieved with estimands derived solely from successively more restrictive fractional neighborhood treatment response assumptions and an assumption of monotonicity in direct and indirect treatment responses (see their Theorem 2.2).[9] Using Manski's (2013) notion of effective treatments, they develop estimands that consider nodes as effectively treated if they satisfy successively more restrictive fractional neighborhood treatment response assumptions, where a node is considered treated if at least a fraction λ of its neighborhood is treated. For example, considering nodes as effectively treated if all of their neighbors are treated is more

[9] This is less restrictive than Aronow and Samii's (2014) generalized probability of exposure ($\pi_i(d_k)$) which precisely defines the probability that node i is subject to exposure d_k.

restrictive than including nodes as effectively treated if half of their peers are treated.[10] Assuming treatment responses are monotone, estimands with more restrictive assumptions on effectively treated nodes achieve bias reduction greater than or equal to estimands with less restrictive assumptions. Aral and Walker (2011) take a similar approach to interference, censoring peers any time after they have more than one treated neighbor, which defines a node as effectively treated at time t if and only if one of their peers was treated at time t.[11] These restrictive assumptions on effective treatment in experiments with independent assignment achieve bias reduction at the expense of increased variance (see Eckles et al. 2014, Figure 5). Though as I discuss below, when combined with design strategies like graph cluster randomization (Ugander et al. 2013), they can reduce bias and variance simultaneously under certain conditions.

Design strategies define treatment assignments over a network to minimize interference. More specifically, they use information about network structure to select treated nodes according to some interference minimizing protocol. Two types of design strategy take contrasting approaches to interference minimization (Walker and Muchnik 2014):

(1) Treatment clustering strategies randomize connected clusters of nodes, with high-edge density within clusters and low-edge density across clusters, into treatment and control groups to simulate global treatment and control conditions in which the entire network would either be treated or untreated. For example, Ugander et al. (2013) begin by defining an exposure condition called network exposure, in which a node is "networked exposed" if their treatment response under a given randomization distribution is the same as their treatment response under a randomization distribution in which the entire network is treated. They then investigate several conditions, analogous to fractional neighborhood treatments, which constitute network exposure. The main design element of the work is a "graph cluster randomization" procedure that partitions the graph into clusters across which treatments are randomized. The idea is to randomize treatment at the cluster level such that nodes are maximally surrounded by other nodes with the same treatment status. Although the protocol can admit any arbitrary clustering procedure and graph structure, Ugander et al. (2013) show that for any graph satisfying their restricted growth condition, specific graph cluster randomization protocols produce unbiased estimates with variance on the order of $1/n$ that is linear in the degrees of the graph. This is an important result because naïve clustering can produce estimators with variances that increase exponentially in the degrees. As Eckles et al. (2014) show, graph cluster randomization produces dramatic decreases in root mean squared error (RMSE) compared to independent assignment when peer effects are large (see their Figure 5). Graph cluster randomization is useful

[10] As Eckles et al. (2014) note, a fractional neighborhood treatment response assumption where $1 > \lambda > 0$ is less restrictive than Aronow and Samii's (2014) neighborhood treatment response assumption ($\lambda = 1$) and more restrictive than Manski's (2013) independent treatment response assumption ($\lambda = 0$).

[11] They then control for variation in degree by including a degree parameter in their model. A more comprehensive description of their approach to interference can be found in the paper's appendix (see Aral and Walker 2011).

when we want to estimate the effect of a global treatment. For example, if we are interested in changes in the degree to which local network externalities exist for a given technology, we may need to understand the effects of clustered adoption rather than of isolated adoption as the value to a new adopter is theoretically a function of the number of their peers who adopt the product.

One limitation to standard graph cluster randomization, which assigns all nodes in a cluster to the same treatment, is that it can become impossible to observe some nodes with a particular number of treated peers. This can complicate, for example, estimation of threshold effects in behavioral contagion studies because all nodes may not have a positive probability of being assigned to all threshold conditions (e.g., having 1, 2, 3, 4, or more treated peers). Eckles et al. (2014) therefore extend standard graph cluster randomization to allow for some nodes to be assigned to a different treatment than the rest of their cluster—a procedure they term "*hole punching*." Hole punching adds node-level randomness to the clustered treatment assignment by specifying independent switching variables that flip the assignment of some nodes in a cluster with a given probability. The result is that some fraction of the nodes in a cluster with a given cluster-level treatment assignment are assigned to the opposite condition, allowing for measurement of threshold effects in contagion or comparison of direct effects and peer effects.

(2) Treatment separating strategies assign treatments that maximize network distance between treated nodes in order to minimize interference between treatments. In contrast to Ugander et al. (2013), Coppock and Sircar (2013) develop a protocol that partitions the graph into non-interfering clusters according to assumptions about effective treatments described by what they call the "SUTVA degree" of a graph. The SUTVA degree corresponds to the maximum path length between two nodes over which interference is assumed to occur. A graph partition based on SUTVA degree produces a sample of nodes and edges that are relatively isolated from one another, enabling estimates of direct treatment and indirect exposure effects that minimize interference. Treatment separating strategies similar to Coppock and Sircar (2013) are not well suited to studying the effects of a global treatment or groups of commonly treated nodes, for instance in the case of estimating implications of changing network externality conditions. But they are well suited to studying micro-level peer effects. Airoldi et al. (2013) propose a different treatment separating randomization design called "insulated neighbors randomization" (INR), which assigns nodes to a non-exposure or k-level exposure status at random, such that the treatment assignments of the nodes already assigned are maintained and a percentage of shared neighbors between treated nodes are assigned to control. The purpose is to increase the causal information produced by the randomization at the expense of bias. Thomas and Finegold (2013) suggest a related block-randomized design in which a sample of nodes are selected to receive a treatment and a subset of their peers are randomly chosen to be denied access to (or awareness of) the treatment. They propose this design to deal with selection bias on degree and homophily between selected nodes and their peers, rather than interference per se. The peers of randomly selected nodes will be of higher degree and will exhibit observed and latent homophily with the selected

nodes. Block randomization can ameliorate this selection bias by randomizing at the level of the ego network. Aral and Walker (2012) achieve a similar goal through a mechanism randomization design that sends notifications of ego's behaviors only to randomly selected peers of the treated.

Eckles et al. (2014) observe larger bias and error reductions through design strategies than through inference strategies, though they admittedly explore specific design and inference strategies under a particular set of network and social effect conditions. Though the inference strategies they examine reduce bias, this bias reduction comes at a significant cost to precision. Combining design and inference approaches to networked interference also marginally improve both bias reduction and precision, though the effects on precision come mainly from assignment clustering rather than from neighborhood response assumptions or weighting.

Estimators (and their Variance). Since network exposure probabilities vary across nodes, estimators must adjust for different proportions of units in different exposure conditions. The Horvitz-Thompson (HT) estimator can produce unbiased estimates the differences in means of the ATE estimand in stratified samples, using inverse probability weighting to adjust for the varying proportions (Horvitz and Thompson 1952). However, heavy-tailed degree distributions in networks ensure high variability in the likelihood of network exposure, generating very large or small numbers of units with high weights, increasing the variance of the HT estimator, at times dramatically. Several adjustments have been suggested to reduce the variance of the HT estimator in the context of network experiments. The Hajek (1971) refinement, which uses the sum of the weights in the denominator, achieves efficiency gains at the cost of bias, though increases in bias are typically small relative to estimator's variability. Bootstrap methods have been evaluated to deal with a lack of coverage of standard difference in means estimators (Thomas and Finegold 2013) and the multiway dependence structure common in online settings (Bakshy and Eckles 2013). Ugander et al.'s (2013) graph cluster randomization further reduces the variance of both the HT and Hajek estimators by increasing the expected number of the individuals who are network exposed to a treatment or control condition and limiting the number of individuals outside of selected clusters who are network exposed (see Eckles et al. 2014, Figure 5).

15.4 The Future of Networked Experimentation

15.4.1 Adaptive Treatment Assignment

An area of innovation that may be relevant to the future of networked experimental design is the extension of sequential treatment assignment to more adaptive dynamic treatment assignment protocols. Such protocols might consider the sample characteristics or responses of the initially treated to inform downstream assignments to maximize precision, minimize interference or cost or harm, or to investigate treatment effect heterogeneity or a particular hypothesis. Sequential treatment assignment, in which a node's assignment is a function of network structure and prior assignments, has been proposed as an approach to minimize interference (Airoldi et al. 2013). However, one could imagine adaptive assignments that use response information obtained during early estimation to adjust the proportions or types of nodes assigned to a given treatment. Adaptive treatment assignment has been used in medical trials to expose fewer patients to avoidable harm than strategies with fixed assignments (Hu and Rosenberger 2006). Though these adaptive cases are difficult to optimize in situations in which the number of experimental units is large, heuristic approaches and near-optimal strategies have been shown to be valid in non-networked settings, even when the size of study population is unknown or unbounded (Press 2009).

Adaptive treatment assignment is essentially a bandit problem in which the researcher wishes to allocate treatments among experimental units, whose properties and responses are only partially known at the time of allocation, but which may become better understood as time passes during the experiment (Berry and Fristedt 1985; Berry 2006; Press 2009). Bandit problems exhibit the classic trade-off between the cost of gathering information and the benefit of exploiting information already gathered (e.g., exploration vs. exploitation). I can imagine a line of research that develops adaptive treatment assignment procedures in network contexts, deriving optimal solutions to relevant allocation problems, incorporating network information and inference into the approaches. Networked information and the regularity of assortativity and other networked characteristics may prove useful in cases analogous to contextual bandit problems because social structure provides so much information about the relationships between experimental units and their relative context.

15.4.2 Novel Randomization Techniques

The flexibility of modern application programming interfaces (APIs) is enabling new forms of randomization in networked experiments. As more comprehensive observability and new communication and messaging features imbue our online social networks with new functionality, new ways of creating exogenous variation at different network levels are constantly emerging. Several examples demonstrate the power of modern online environments to create useful variation with which to identify causal effects in social behavior.

Aral and Walker (2012) utilize the Facebook API to randomize the peer recipients of automated notifications of a node's behaviors. For example, when a user rated a movie or linked to a celebrity, packets of notifications informing their peers of their behaviors were automatically generated and randomly delivered to their Facebook friends. This randomization enables them to identify peer influence, susceptibility to influence, and individual and dyadic correlates of influence and susceptibility at the individual level. Bakshy et al. (2012b) use a similar design by directly manipulating Facebook's algorithms, randomly blocking information delivered to users' news feeds in order to identify causal information diffusion dynamics in networks. These types of micro-level randomization enable quite nuanced investigations of social behavior.

Bakshy et al. (2012b) and Aral and Walker (2012, 2014) examined correlates of social influence by randomizing messages between individuals in the network. However, these designs cannot estimate the causal effect of receiving a message from a particular individual with particular characteristics. Taylor et al. (2015) select ad exposures where viewers could potentially be exposed to social cues involving two different peers and then randomly choose which peer is shown, providing exogenous variation in the dyadic characteristics of friends displayed in the social cue.[12] This exogenous variation enables causal estimates of the influence of particular social cues (i.e., the age, gender, similarity, or tie strength associated with a particular peer) on behavior. These types of randomization will form the basis of personalized social advertising in the near future.

Finally, anonymity designs provide another novel randomization technique that can help identify subtle social effects. Bapna et al. (2014) enable randomly selected users of an online dating site to browse other users' profiles without leaving a digital signal of their interest (the default setting on the website is to provide viewing information to the user whose profile is being viewed). This enables causal estimates of the effects of anonymity on both browsing behaviors and the matches that are eventually formed. Interestingly, anonymous browsers are more likely to view other users who are of different races. Taylor et al. (2014) use a similar anonymity design to suppress the identity of users posting content to a social news aggregation website. They find strong identity effects on opinion formation as measured through ratings. These examples highlight the new designs that are possible today. In the future, nuanced randomization procedures such as these are likely to improve our understanding of subtle social effects at scale.

15.4.3 Linking Online Treatments to Offline Responses

Although digitization is enabling advances in networked experimentation, there remains a danger in relying too heavily on digital substrates to explore human behavior. Not only are digital samples biased toward those who are more active online, potentially missing large swaths of society, but limiting inquiry to digital behaviors constrains the theoretical reach of experimental work. It will therefore be important to devise strategies for linking online treatments to meaningful offline behaviors.

[12] This paper is based on Chapter 3 of Sean Taylor's PhD thesis (Taylor 2014).

For example, Bond et al. (2012) go to great lengths to validate the results of their Facebook voting experiment in public voting records. They find that users who received a social message were more likely to vote than users who received no message or an informational message with no social cue, raising doubts about the efficacy of information-only appeals to vote (Aral 2012). Similarly, in an ongoing collaboration with the Praekelt Foundation in South Africa, we are connecting data from experiments conducted on a mobile messaging platform with data on HIV testing conducting in physical clinics. We designed scratch cards with unique identifiers that are given to people who test in a facility and mention the program. They unlock their experimental incentive by entering the unique identifier on the card into their phone. When this identifier is entered, we can connect cell phone referral data to data on verified HIV tests. To maintain anonymity, we only record users' data as de-identified strings and never inquire about individual HIV test results, only whether a test was taken.

These examples provide evidence of the importance of linking online and offline behaviors in digital networked experiments. Validating the effects of digital treatments on physical behaviors or the digital measurement of physical behaviors will help establish and maintain the external validity of studies conducted on digital substrates that wish to generalize findings to a broader set of people and subjects.

15.4.4 Experimental Validation of Observational Methods

An important use of networked experiments is in benchmarking the results of observational studies against provably causal estimates. For example, Aral et al. (2009) adapted high dimensional propensity score matching (HDPSM) to dynamic network settings to identify peer influence in product adoption decisions. They analyzed a global instant messaging network of 27.4 million Yahoo users, combined with comprehensive daily product adoption data on a personalized news and weather application and users' longitudinal behavioral, demographic, and geographic characteristics. They found that previous methods overestimated peer influence by 300%–700%, and that homophily explained greater than 50% of the perceived behavioral contagion. Shalizi and Thomas (2012) subsequently argued that, theoretically speaking, homophily and contagion are "generically confounded." They claimed that latent unobserved (or unobservable) homophily will always thwart the identification of peer effects. Given the expense and difficulty of running experiments and the availability of observational data, it would be useful to develop reliable, nonexperimental causal inference strategies in networks. But how can we know how close our observational estimates are to provably causal estimates of social effects? How can we measure the error and bias of observational methods that potentially miss unobserved latent effects?

Eckles and Bakshy (2014) demonstrated one approach to the experimental validation of observational methods by conducting a "constructed observational study" evaluating the effectiveness of HDPSM in identifying peer effects in networks. Their study uses the results of a randomized experiment of peer influence in information diffusion among 35 million Facebook users (Bakshy et al. 2012b) to benchmark HDPSM methods applied to identify the same effects. In essence, they treat the results of the randomized experiment as a gold standard and measure how well, or how poorly, HDPSM does in identifying peer effects, compared to the experimental estimates, using the same data.

Three notable results emerge from their study: First, HDPSM methods can achieve a 70% bias reduction in estimates of peer effects compared to experimental benchmarks. This provides grounds for optimism in that observational techniques can effectively estimate peer effects and other causal effects in networks. Whether estimates that achieve 70% bias reduction are useful to policy-makers depends on the policy being evaluated. In many cases, a reliable estimate of the direction and order of the effects can help guide policy. In other cases, however, more reliable estimates may be necessary. Second, contextual variables used in the first-stage matching procedure are critical to bias reduction. Eckles and Bakshy (2014) find models that use a limited number of demographic features to create the matched sample only achieve approximately an 8% bias reduction. Contextual features relevant to the behavior being modeled improve matches and thus the reliability of matching estimates. Finally, the overestimates of social contagion by naïve methods found in Aral et al. (2009) and Eckles and Bakshy (2014) are quite similar. Both studies find naïve models overestimate social contagion by approximately 300%.[13] These benchmarks help us understand the errors produced by observational methods.

Constructed observational studies are one example of how networked experiments can be used to validate observational approaches to the identification of social effects. Using experimental results as a benchmark for true causal estimates can help us understand when and to what degree observational methods produce bias.[14]

15.5 Conclusion

Networked experiments are rapidly becoming one of the most precise tools available for studying social behavior. As more and more social behaviors are digitized and mediated by online platforms, our ability to quickly answer nuanced causal questions about the role of social behavior in population-level outcomes is becoming unprecedented. Interdependent behaviors potentially represent the most fundamental data-generating processes in the social sciences, as they are in some sense what make the social sciences social.

[13] Aral et al. (2009) find a 700% overestimate in the early weeks of the product's life cycle, immediately after the product launch. Their estimates then decrease to a consistent 300% overestimate.

[14] Of course, experiments are not necessarily a silver bullet. There exist many known potential biases in experimental settings as well. However, under certain assumptions and with robust execution, experiments provide the most reliable causal estimates of networked behavior.

Yet, while the increasing scale, scope, and complexity of networked experiments creates tangible opportunities for scientific advancement, they also simultaneously create significant new challenges. This chapter presented a review of advances in modern networked experimentation, summarizing the opportunities and challenges they create for understanding our social world. Fundamental innovations in networked experimental design are changing the implications of the experimental setting, sampling, randomization procedures, and approaches to treatment assignment. At the same time, the analysis of networked experiments is witnessing a rapid evolution, from modeling treatment response assumptions, measurement and inference to recent approaches to interference, and uncertainty in dependent data. These changes require networks researchers to be aware of the most modern approaches to experiments in interdependent data. At the same time, new challenges and opportunities are shaping the future of networked experimentation, particularly in adaptive treatment assignment strategies, novel randomization techniques, linking online treatments to offline responses and experimental validation of observational methods.

There is much work to do as we usher in this new era of networked experimentation and I cannot think of a more important set of social scientific endeavors. Scalable networked experimentation represents an application of the scientific method to questions about our social world, at unprecedented scale and precision. It behooves us to tackle such scientific challenges with vigor as we develop these new tools in the pursuit of human progress.

References

Aggarwal, C. C., Han, J., Wang, J., and Yu, P. S. (2003). "A framework for clustering evolving data streams." *Proceedings of the 29th International Conference on Very Large Data Bases* (29).

Aharony, N., Pan, W., Ip, C., Khayal, I., and Pentland, A. (2011). "Social fMRI: Investigating and shaping social mechanisms in the real world." *Pervasive and Mobile Computing* 7, 643–659.

Ahmed, N., Neville, J., and Kompella, R. (2014). "Network sampling: From static to streaming graphs." *ACM Transactions on Knowledge Discovery from Data*, 8.

Aiello, W., Chung, F., and Lu, L. (2001). "A random graph model for power law graphs." *Experimental Mathematics* 10(1), 53–66.

Airoldi, E., Toulis, P., Kao, E., and Rubin, D. B. (2013). "Estimation of causal peer influence effects." *Journal of Machine Learning Research* 28(3), 1489–1497.

Ali, M., Emch, M., von Seidlein, L., Yunus, M., Sack, D. A., Rao, M., Holmgren, J., and Clemens, J. D. (2005). "Herd immunity conferred by killed oral cholera vaccines in Bangladesh: A reanalysis." *Lancet* 366, 44–49.

Aral, S. (2011). "Commentary—identifying social influence: A comment on opinion leadership and social contagion in new product diffusion." *Marketing Science* 30(2), 217–223.

Aral, S. (2012). "Social science: Poked to vote." *Nature* 489(7415), 212–214.

Aral, S. and Walker, D. (2011a). "Creating social contagion through viral product design: A randomized trial of peer influence in networks." *Management Science* 57(9), 1623–1639.

Aral, S. and Walker, D. (2011b). "Identifying social influence in networks using randomized experiments." *IEEE Intelligent Systems* 26(5), 91–96.

Aral, S. and Walker, D. 2012. "Identifying influential and susceptible members of social networks," *Science* 337(6092), 337–341.

Aral, S. and Van Alstyne, M. W. (2011). "The diversity-bandwidth tradeoff." *American Journal of Sociology* 117(1), 90–171.

Aral, S. and Taylor, S. (2011). "Viral incentive systems: A randomized field experiment." Working Paper, Stern School of Business, NYU.

Aral, S., Muchnik, L., and Sundararajan, A. (2009). "Distinguishing influence-based contagion from homophily-driven diffusion in dynamic networks." *Proceedings of the National Academy of Sciences* 106(51), 21544–21549.

Aral S., Muchnik L., and Sundararajan, A. (2013). "Engineering social contagions: Optimal network seeding and incentive strategies." *Network Science* 1(2), 1–29.

Aronow, P. M. and Samii, C. (2012). "Estimating average causal effects under general interference." Summer Meeting of the Society for Political Methodology, University of North Carolina, Chapel Hill, July.

Bakshy, E. and Eckles, D. (2013). "Uncertainty in online experiments with dependent data: An evaluation of bootstrap methods." *Proceedings of the 19th ACM SIGKDD International Conference on Knowledge Discovery and Data Mining.*

Bakshy, E., Eckles, D., Yan, R., and Rosenn, I. (2012a). "Social influence in social advertising: Evidence from field experiments." *Proceedings of the 13th ACM Conference on Electronic Commerce*, 146–161. ACM, New York.

Bakshy, E., Rosenn, I., Marlow, C., and Adamic, L. (2012b). "The role of social networks in information diffusion." *Proceedings of 21st International Conference on World Wide Web*, 519–528. ACM, New York.

Bapna, R., Gupta, A., Rice, S., and Sundararajan, A. 2011. "Trust, reciprocity and the strength of social ties: An online social network based field experiment." Working paper, University of Minnesota, Minneapolis.

Bapna, R., Ramaprasad, J., Shmueli, G., and Umyarov, A. (2013). "One-way mirrors and weak-signaling in online dating: A randomized field experiment." Working paper, University of Minnesota, Minneapolis.

Bapna, R., Umyarov, A. Forthcoming. "Do your online friends make you pay? A randomized field experiment in an online music social network." Forthcoming in *Management Science.*

Bavelas, A. (1950). "Communication patterns in task-oriented groups." *The Journal of the Acoustical Society of America* 22(6), 725–730.

Babu, S. and Jennifer, W. (2001). "Continuous queries over data stream." *ACM Sigmod Record* 30(3), 109–120.

Berry, D. A. (2006). "Bayesian clinical trials." *National Review of Drug Discovery* 5, 27–36.

Berry, D. A. and Fristedt, B. (1985). *Bandit Problems: Sequential Allocation of Experiments.* London, UK: Chapman & Hall.

Bond, R. M., Fariss, C. J., Jones, J. J., Kramer, A. D. I., Marlow, C., Settle, J. E., and Fowler, J. H. (2012). "A 61-million-person experiment in social influence and political mobilization." *Nature* 489(7415), 295–298.

Boudreau, Kevin, J., and Lakhani, Karim R. (2011). "High incentives, sorting on skills—or just a taste for competition? Field experimental evidence from an algorithm design contest." No. 11-107. Harvard Business School Working Paper.

Bramoullé, Y., Habiba, D., and Fortin, B. (2009). "Identification of peer effects through social networks." *Journal of Econometrics* 150(1), 41–55.

Brock, W. A. and Durlauf, S. N. (2001). "Discrete choice with social interactions." *The Review of Economic Studies* 68(2), 235–260.

Burke, M. A. and Sass, T. R. (2013). "Classroom peer effects and student achievement." *Journal of Labor Economics* 31(1), 51–82.

Centola, D. (2010). "The spread of behavior in an online social network experiment." *Science* 329(5996), 1194–1197.

Centola, D. (2011). "An experimental study of homophily in the adoption of health behavior." *Science* 334(6060), 1269–1272.

Centola, D. and Macy, M. (2007). "Complex contagions and the weakness of long ties." *American Journal of Sociology* 113(3), 702–734.

Coppock, A. and Sircar, N. (2013). "Design of field experiments under unknown interference structures." Working paper.

Cox, D. R. (1958). *Planning of Experiments*. New York: Wiley-Interscience.

Dieye, R., Djebbari, H., and Barrera-Osorio, F. (2014). "Accounting for peer effects in treatment response." IZA Discussion Paper No. 8340, Institute for the Study of Labor, Bonn, Germany.

Domingos, P. and Geoff, H. (2000). "Mining high-speed data streams." *Proceedings of the 6th ACM SIGKDD International Conference on Knowledge Discovery and Data Mining.*

Eckles, D. and Bakshy, E. (2014). "Bias and high-dimensional adjustment in observational studies of peer effects." Working paper, Data Science Team, Facebook, Menlo Park, CA.

Eckles, D., Karrer, B., and Ugander, J. (2014). "Design and analysis of experiments in networks: Reducing bias from interference." Working paper, Data Science Team, Facebook, Menlo Park, CA.

Erdős, P. and Rényi, A. (1959). "On random graphs. I." *Publicationes Mathematicae* 6, 290–297.

Erdős, P. and Rényi, A. (1960). "On the Evolution of Random Graphs." *Publications of the Mathematical Institute of the Hungarian Academy of Sciences* 5, 17–61.

Erdős, P. and Rényi, A. (1961). "On the strength and connectedness of random graphs." *Acta Mathematica Academiae Scientiarum Hungarica* 12(1–2), 261–267.

Fowler, J. H. and Christakis, N. A. (2010). "Cooperative behavior cascades in human social networks," *Proceedings of the National Academy of Sciences USA* 107(12), 5334–5338.

Graham, B. S. and Hahn, J. (2005). "Identification and estimation of the linear-in-means model of social interactions." *Economics Letters* 88(1), 1–6.

Granovetter, M. (1978). "Threshold models of collective behavior." *American Journal of Sociology* 83(6), 1420–1443.

Hasan, S. and Bagde, S. Forthcoming. "Peers and network growth: Evidence from a natural experiment." *Management Science.*

Hajek, J. (1971). "Comment on 'An Essay on the Logical Foundations of Survey Sampling, Part One.'" in *The Foundations of Survey Sampling*, V. P. Godambe and D. A. Sprott, eds., 236. Holt, Rinehart, and Winston.

Halloran, M. E. and Struchiner, C. J. (1991). "Study designs for dependent happenings." *Epidemiology* 2, 331–338.

Halloran, M. E. and Struchiner, C. J. (1995). "Causal inference in infectious diseases." *Epidemiology*, 6, 142–151.

Heckathorn, D. D. (1997). "Respondent-driven sampling: A new approach to the study of hidden populations." *Social Problems* 44(2), 174–199.

Hinz, O., Skiera, B., Barrot, C., and Becker, J. U. (2011). "Seeding strategies for viral marketing: An empirical comparison." *Journal of Marketing* 75(6), 55–71.

Horton, J., Rand, D. G., and Zeckhauser, R. J. (2010). "The online laboratory: Conducting experiments in a real labor market." *Experimental Economics* 14(3), 399–425.

Horvitz, D. G. and Thompson, D. J. (1952). "A generalization of sampling without replacement from a finite universe." *Journal of the American Statistical Association* 47(260), 663–685.

Hu, F. and Rosenberger,W. F. (2006). *The Theory of Response-Adaptive Randomization in Clinical Trials*. Wiley: New York.

Hudgens, M. G. and Halloran, E. (2008). "Toward causal inference with interference." *Journal of the American Statistical Association* 103(482), 832–842.

Kearns, M., Suri, S., and Montfort, N. (2006). "An experimental study of the coloring problem on human subject networks." *Science* 313(5788), 824–827.

Kempton, R. A. (1997). "Interference between plots." In *Statistical Methods for Plant Variety Evaluation*, R. A. Kempton and P. N. Fox, eds., 101–116. London: Chapman & Hall.

Leavitt, H. J. (1951). "Some effects of certain communication patterns on group performance." *The Journal of Abnormal and Social Psychology* 46(1), 38.

Lee, L. (2007). "Identification and estimation of econometric models with group interactions, contextual factors and fixed effects." *Journal of Econometrics* 140(2), 333–374.

Lee, S., Kim, P., and Jeong, H. (2006). "Statistical properties of sampled networks," *Physical Review E* 73, 016102.

Leider, S., Möbius, M. M., Rosenblat, T., and Do, Q.-A. (2009). "Directed altruism and enforced reciprocity in social networks." *The Quarterly Journal of Economics* 124(4), 1815–1851.

Leskovec, J., Kleinberg, J., and Faloutsos, C. (2005). "Graphs over time: Densification laws, shrinking diamaters and possible explanations." In *Proceedings of the 11th ACM SIGKDD International Conference on Knowledge Discovery and Data Mining.*

Leskovec, J. and Faloutsos, C. (2006). "Sampling from large graphs," In *Proceedings of the 12th ACM SIGKDD International Conference on Knowledge Discovery and Data Mining*, 631–636.

Mani, A., Rahwan, I., and Pentland, A. (2013). "Inducing peer pressure to promote cooperation." *Scientific Reports* 3.

Manski, C. F. (1993). "Identification problems in the social sciences." *Sociological Methodology* 23.

Manski, C. F. 2013. "Identification of treatment response with social interactions." *The Econometrics Journal* 16(1): S1–S23.

Mason, W. and Suri, S. 2012. "Conducting behavioral research on Amazon's Mechanical Turk." *Behavioral Research* 44, 1–23.

Mason, W. and Watts, D. J. 2012. "Collective learning in networks." *Proceedings of the National Academy of Sciences* 109(3), 764–769.

Middleton, J. A. and Aronow, P. M. 2011. "Unbiased estimation of the average treatment effect in cluster-randomized experiments." Working paper, New York University, New York.

Muchnik, L., Aral, S., and Taylor, S. 2013. "Social influence bias: A randomized experiment." *Science* 341(6146), 647–651.

Newman, M. E. J. 2002. "Random graphs as models of networks." Working paper, Santa Fe Institute.

Neyman, J. 1923. "On the application of probability theory to agricultural experiments: Essay on principles, Section 9." Roczniki Nauk Roiniczych, tom X, 1–51 [in Polish]; reprinted in English with discussion in *Statistical Science* 5, 463–480.

Press, W. H. 2009. "Bandit solutions provide unified ethical models for randomized clinical trials and comparative effectiveness research." *Proceedings of the National Academy of Sciences* 106(52), 22387–22392.

Rand, D. G. and Nowak, M. A. 2011. "The evolution of antisocial punishment in optional public goods games." *Nature Communications. Article* 434.

Rosenbaum, P. R. 2007. "Interference between units in randomized experiments." *Journal of the American Statistical Association* 102, 477.

Rosenthal, R. and Jacobson, L. 1968. *Pygmalion in the Classroom: Teacher Expectation and Pupils' Intellectual Development*. New York: Holt, Rinehart and Winston.

Rubin, D. B. 1974. "Estimating causal effects of treatments in randomized and nonrandomized studie.," *Journal of Educational Psychology* 66, 688–701.

Rubin, D. B. 1986. "Which ifs have causal answers." *Journal of the American Statistical Association* 81(396), 961–962.

Rubin, D. B. 1990. "Comment: Neyman (1923) and causal inference in experiments and observational studies." *Statistical Science* 5(4), 472–480.

Salganik, M. J., Dodds, P. S., and Watts, D. J. 2006. "Experimental study of inequality and unpredictability in an artificial cultural market." *Science* 311(5762), 854–856.

Salganik, M. J. and Heckathorn, D. D. 2004. "Sampling and estimation in hidden populations using respondent-driven sampling." *Sociological Methodology* 34(1), 193–240.

Shalizi, C. R. and Thomas, A. C. 2011. "Homophily and contagion are generically confounded in observational social network studies." *Sociological Methods and Research* 40(2), 211–239.

Shore, J., Ethan, B., and Lazer, D. 2014. "Facts and figuring: An experimental investigation of network structure and performance in information and solution spaces." Harvard Business School Working Paper, No. 14-075.

Sobel, M. 2006. "What do randomized studies of housing mobility demonstrate? Causal inference in the face of interference." *Journal of the American Statistical Association* 101, 1398–1407.

Stumpf, M., Wiuf, C., and May, R. 2005. "Subnets of scale-free networks are not scale-free: Sampling properties of networks." *Proceedings of the National Academy of Sciences* 102(12), 4221–4224.

Suri, S. and Watts, D. J. 2011. "Cooperation and contagion in web-based, networked public goods experiments." *PLoS One* 6(3), e16836.

Taylor, S. J. 2014. "Social influence from online social signals: Three experiments." PhD thesis, NYU Stern School of Business.

Taylor, S., Bakshy, E., and Aral, S. 2013. "Selection effects in online sharing: Consequences for peer adoption." *14th ACM Conference on Electronic Commerce*, Philadelphia, PA.

Taylor, S., Muchnik, L., and Aral, S. 2014. "Identity and opinion: A randomized experiment." Working paper, MIT Sloan School of Management.

Taylor, S. J., Bakshy, E., Eckles, D., and Aral, S. 2014. "Asymmetric effects of personalized social cues: Evidence from field experiments." Working paper.

Tchetgen Tchetgen, E. J. and VanderWeele, T. J. 2012. "On causal inference in the presence of interference." *Statistical Methods in Medical Research* 21(1), 55–75.

Thomas, A. C. and Finegold, M. 2013. "Protocols for randomized experiments to identify network contagion." Working paper, CMU.

Tucker, C. 2011. "Social advertising." Working paper, MIT Sloan School of Management, Cambridge, MA.

Tucker, C. and Zhang, J. 2010. "Growing two-sided networks by advertising the user base: A field experiment." *Marketing Science* 29(5), 805–814.

Tucker, C. and Zhang, J. 2011. "How does popularity information affect choices? A field experiment." *Management Science* 57(5), 828–842.

Ugander, J., Karrer, B., Backstrom, L., and Kleinberg, J. 2013. "Graph cluster randomization: Network exposure to multiple universes." *19th ACM SIGKDD International Conference on Knowledge Discovery and Data Mining (KDD)*, 329–337. ACM, New York.

Walker, D. and Muchnik, L. 2014. "Design of randomized experiments in networks." *Proceedings of the IEEE* 102(12), 1940–1951.

Wang, H., Fan, W., Yu, P. S., and Han, J. 2003. "Mining concept-drifting data streams using ensemble classifiers" *Proceedings of the 9th ACM SIGKDD International Conference on Knowledge Discovery and Data Mining.*

Watts, D. J. 2002. "A simple model of global cascades on random networks." *Proceedings of the National Academy of Sciences* 99(9), 5766–5771.

CHAPTER 16

FIELD EXPERIMENTS, SOCIAL NETWORKS, AND DEVELOPMENT

EMILY BREZA

16.1 INTRODUCTION

The study of social networks has become an important part of the economics tool kit. Modeling agents as independent actors misses the countless ways in which actions are shaped by the traits, beliefs, and decisions of their peers. Similarly, in order to measure general equilibrium effects of a policy intervention, it is crucial to understand not only how a beneficiary is affected directly, but how an intervention spills over to others in the agent's community.

Measuring spillovers and peer effects is notoriously difficult, especially using observational data (Manski, 1993). Social relationships are often endogenous, and connected individuals are likely to have similar characteristics and experience common shocks. Thus, it should be no surprise that following the rise of experimental methods in microeconomics, many researchers have also begun to use experiments to tackle questions involving social networks and to revisit insights from foundational non-experimental papers.[1]

The way in which networks have been studied using experiments has evolved as the literature has matured. One of the simplest ways to measure spillovers is to use the spatial variation induced from randomization to measure geographical spillovers. Another common technique is to record baseline membership in groups such as kinship, caste, ethnicity, classroom, or borrowing groups and to again use variation

[1] Foundational non-experimental papers include Foster and Rosenzweig (1995), Conley and Udry (2010), McMillan and Woodruff (1999), Banerjee and Munshi (2004), Petersen and Rajan (1994), and Townsend (1994), among many others.

in treatment status or group-level treatment intensity to identify social effects. A more recent, but growing subset of the literature has begun to move away from mutually exclusive and symmetric group memberships towards incorporating more complex network structures and even global network (i.e., graph- or community-level) characteristics. Usage of network data allows researchers to ask three key questions. First, how does the embedding of individuals in a social network shape economic outcomes? Second, what is the effect of the community-level network structure on those same outcomes? Finally, how are networks themselves molded by external forces?

Much of the material for this chapter is drawn from developing countries. Arguably, development economics has led other subfields in the adoption of economic experiments. Part this trend is surely due to the feasibility and cost associated with field work in these settings. But, in addition, developing countries are likely to be places where social networks and spillovers play an especially important role. When formal institutions do not work very well, informal, network-based institutions often arise as substitutes. Further, other frictions such as barriers to technology adoption are potentially severe in developing countries, and network-based solutions may generate large welfare gains.

The current body of experimental economics research highlights the profound and diverse ways in which social structures shape economic decision-making. I begin in Section 16.2 with a discussion of the methodological challenges involved in conducting social network experiments along with recent innovations. Section 16.3 turns to one of the most well-studied topics, social learning and diffusion. Next, I present research measuring interactions of social network relationships with other-regarding preferences in Section 16.4, followed by work on peer monitoring and enforcement in Section 16.5. Both altruism and the scope for network-based contract enforcement are essential ingredients for risk-sharing, which is analyzed in Section 16.6. Section 16.7 explores arguably the newest experimental literature, which studies the formation and change of networks. A discussion of promising avenues for new work concludes the chapter.

16.2 Design and Implementation Issues

Much has been written to assist researchers in evaluating, designing, and implementing field experiments and can be applied directly to experiments with networks (Harrison and List 2004; and Duflo et al. 2007, for example). However, there are some issues that are distinct to network experiments, including elements of identification, experimental design, inference, network measurement, and network sampling. This discussion is related to other chapters in this volume, notably "Some Challenges in the Empirics of the Effects of Networks" by Boucher and Fortin and "Networked Experiments: A Review of Methods and Innovations" by Aral.

16.2.1 Identification

Before moving to more practical considerations, it is important to first discuss the fundamental problem of identification in the presence of spillovers.

Identification in randomized experiments is extremely simple when the stable unit treatment value assumption (SUTVA) of Rubin (1978) holds (i.e., when an individual is unaffected by the treatment status of others). For example, in an experiment with binary treatment assignment (i.e., treatment vs. control), randomization ensures that the potential outcomes of the members of each group are identical. Then the causal treatment effect is simply the difference in the mean outcomes between the groups.

Manski (2013) highlights the limits to identification when SUTVA no longer holds. Suppose now, in an experiment with simple binary treatments, that an individual may be affected either directly or indirectly by the treatment status of n members of that individual's community. Then, what Manski (2013) calls the *effective treatment* status of an individual no longer takes binary values (own treatment vs. control) but instead can take up to 2^n possible values (capturing the binary treatment status of each group member). If n is large relative to the number of participants in the experiment, there is no hope of identifying the treatment response for all possible treatment configurations without making stronger assumptions.

Given the intractability of this identification problem, in most cases researchers must place restrictions on the nature of the reference group and how spillovers are transmitted through a network. One common workhorse example is the linear-in-means peer effects model of Manski (1993). Bramoullé et al. (2009) extend the Manski (1993) model to the case of peer effects mediated by the social network.[2] Dieye et al. (2014) combine this model with experimental variation in cash transfers to study peer effects in schooling decisions in Colombia.[3]

It is important to keep in mind for the remainder of the chapter that in most analyses of spillovers in experiments, the researchers are making specific assumptions in the background about the nature of those spillovers. It is also important to note that one way to get around the identification problem of Manski (2013) is to treat each independent network or group as the unit of analysis (see Hudgens and Halloran 2008).[4]

16.2.2 Measurement

Defining the "true" social network is notoriously difficult and costly. One of the simplest network proxies to collect is geography, such as neighborhood indicators or spatial

[2] Also see Calvó-Armengol et al. (2009) and Lee et al. (2010), among others, for examples of model-based identification of social network spillovers.

[3] Note that in any estimation of the effect of the mean outcome of the peer group on an individual's own outcome, a type of exclusion restriction must be satisfied. That is, the experimental variation must affect the individual in only the ways specified by the model.

[4] In this spirit, Baird et al. (2014) present a two-stage experimental design combining within- and across-community randomization.

distances. However, geographical measures are only coarse proxies for bilateral ties and do not always map well to networks elicited through survey methods.[5] Another frequently used method is to employ common demographic characteristics (e.g., ethnicity, religion, caste, or kinship) or group memberships (e.g., churches, firms, social clubs, microfinance groups, classrooms) to proxy for social closeness. Again, this approach is limited, as it only allows for in- versus out-group comparisons.

Many network-based questions require more detailed network data, which in field studies is generally obtained from household surveys. In these surveys, researchers ask respondents to name individuals with whom they engage in a variety of activities such as risk-sharing, advice, or kinship.[6] Due to the difficulty of measuring networks, the recentness of the experimental networks literature, and setting-specific considerations, there is no consensus definition of what constitutes a link between two nodes. For example, some projects use actual historical interactions, others hypothetical interactions (i.e., financial transfers or advice). Many papers work with undirected graphs, while others instead take the directionality of interactions more seriously. Similarly, some research focuses on only one dimension of interaction, while others work with multigraphs over many dimensions of interactions.

Ideally, independent of the sampling strategy for the experiment itself, researchers would survey every member in each study network. This process often requires two steps, where an initial community-level census records the identities of all network nodes. A second survey then captures lists of each respondent's connections, matched to the census. Such surveys are often very costly, and are sometimes infeasible (for example, in urban areas, it is simply not possible to survey the entire graph), requiring researchers to sample subsets of nodes from the network.

16.2.3 Network Sampling

Measuring the network requires sampling sufficient numbers of nodes to generate accurate information about longer paths through the network. In general, small sampling rates will overestimate path lengths between arbitrary nodes and underestimate some node-specific measures such as degree. Chandrasekhar and Lewis (2011) show that sampling the network can cause nonclassical measurement error that systematically biases regressions that use network statistics as regressors. Even worse, the sign of the bias may depend on the underlying network.

[5] In their setting, Beaman et al. (2015) find geography-based centrality to be poorly correlated with survey-based measures of network centrality. Conley and Udry (2010) use differences between geographical and information neighbors for identification in their (nonexperimental) study of social learning.

[6] Other data sources have also been cleverly used for networks data. Leider et al. (2009) use online tasks to elicit networks of college students. Others have used mobile phone calling networks (Blumenstock et al. 2012), for example.

One approach that researchers have taken is to conduct full network surveys among only those individuals participating in one of the experimental treatments (i.e., individuals showing up to lab games). However, this method may miss a large fraction of potential links. If a fraction ψ is sampled, then information about approximately ψ^2 of the total links in a network is captured. If $\psi = 0.75$, then only 56% of potential links are recorded.[7] Chandrasekhar and Lewis (2011) highlight that first conducting a census of the community and then allowing sampled respondents to list connections with any other community members can dramatically improve coverage. Rates increase to approximately $1 - (1 - \psi)^2$. With $\psi = 0.75$, coverage under this method is 94%.[8] For lower levels of coverage due to sampling, Chandrasekhar and Lewis (2011) introduce a graphical reconstruction method for reducing bias.

There are some cases, often involving only direct social connections, where naive subsampling preserves the characteristics of interest. For example, measuring direct spillovers from a treated friend onto an untreated household only requires eliciting the direct connections between the treatment and spillover groups.[9] Banerjee et al. (2014a), who study dense urban communities where full networks surveys are impossible to implement, ask how access to credit changes a household's degree, which can be measured using survey responses from only treatment and control households.

However, taking only local measurements of the network places limitations on the types of questions that can be posed.[10] Subsampling considerations are particularly important when taking measurements of community-level (i.e., "global") network structures or when asking questions about indirect pathways of spillovers.[11]

16.2.4 Inference

One key problem with conducting inference along social networks is that outcomes of linked individuals are likely correlated. One common technique used in the broader experimental literature is to cluster standard errors to correct for within-group correlations. However, clustered standard errors require independence between members of different groups. It is an empirical regularity that real-world networks tend to contain one large connected component including the vast majority of nodes. For example, in the Banerjee et al. (2013) data, an average of 93% of households belong to the giant component in their villages.[12] Further, because clustered standard errors may suffer

[7] Related to Chandrasekhar and Lewis (2011), Manski (2013) suggests that this approach may be problematic, as treatment effects might flow through untreated individuals and then reflect back onto the treatment group.

[8] Reducing sampling rates to $\psi = 2/3$ only drops coverage to 89%.

[9] See Kremer and Miguel (2007) and Oster and Thornton (2012), for example.

[10] Measuring degree alone is likely to not be a sufficient proxy for more recursive or path-length dependent notions of centrality, for example.

[11] See Banerjee et al. (2013) or Alatas et al. (2014) for examples.

[12] Which is likely an understatement given the 50% sampling rate.

from small sample bias, the required number of distinct networks is large.[13] This is a second benefit, in addition to identification, from including many independent communities in an experimental design.

There are many types of research questions for which sampling a large number of independent graphs is not possible. Fafchamps and Gubert (2007) suggest using modified spatial standard errors, where spatial distance is replaced with network distance, and correlation between nodes decays with distance.

One standard component of the experimental design process that is tightly linked with inference is power calculations (Duflo et al. 2007). More work is needed by applied econometricians to construct analogous power calculations that incorporate network structure.

16.2.5 Other Implementation Considerations

In many experiments, researchers aim to measure the direct effects of an intervention on treated individuals and also the indirect effects of that treatment onto network neighbors or common group members. For these effects to be comparable, the potential outcomes of the treated and control groups need to be the same as the potential outcomes for the group receiving any spillovers. This creates design considerations when deciding how to define and sample the so-called "spillover" nodes.

The best method is to select the spillover group randomly from the same sample frame as the treatment group. This ensures that spillover nodes are exposed to treated versus control individuals with random intensity and that they have the same potential outcomes as the rest of the sample frame.

A second approach starts with defining the treatment and control groups, subsequently eliciting network connections from those individuals, and finally randomly sampling from that list of connections. Then spillovers can be measured by comparing the friends of treated individuals with friends of control individuals. Rather than collecting data on connections within the sample frame, this method relies instead on less intensive "snowball sampling." While the reduced form treatment effect on the friends is identified (subject to the discussion in Section 16.2.1), this group is likely to have different potential outcomes from the members of the sample frame themselves. One important difference is network centrality—individuals who are listed as friends by an arbitrary group of individuals are on average more central than that initial group of individuals. Thus, analyses using snowball sampling can overweight observations from more central nodes.

A different version of this problem also arises in experimental designs where the initially treated individuals are not randomly assigned at the individual level. An example is a village-level treatment where information about a preventative health

[13] Angrist and Pischke (2008) suggest a rule of thumb of 42. For smaller numbers, see Canay et al. (2014).

product is given to the head nurse in each treated village. One could measure spillovers onto the friends of the nurse, but those friends are not likely representative.

16.2.6 Endogeneity

One final consideration that is present in most empirical networks projects is the endogeneity of social connections. While this chapter describes many experiments that are overlaid on top of real-world networks, it is hard to escape the fundamental endogeneity of observed network structures. While some laboratory experiments are able to exogenously construct networks of different shapes (e.g., Choi et al. 2011) or to include individual-specific fixed effects (Chandrasekhar et al., 2014), this is generally not possible in many field settings. In Section 16.7, I discuss measurements of how networks change and evolve. This line of research is a key first step in solving the endogeneity problem.

16.3 Social Learning and Diffusion

Social learning is one of the most studied areas of the experimental social networks literature. This volume's chapter "Learning in Social Networks" by Golub and Sadler presents a thorough treatment of the theory. While most field experiments measure the joint information transmission and adoption decisions in a single outcome, laboratory experiments have the benefit of being able to closely control and measure the quality of signals, the content of information transmitted to and between subjects, and the network structure itself.

16.3.1 Testing Theories of Social Learning

Social learning is a core focus of Choi, Gallo, and Kariv's chapter, "Networks in the Laboratory." Here I discuss a small selection of papers in order to highlight the strengths of the laboratory for studying social learning. A dialogue between laboratory and field experiments is important to advance our understanding of how social learning works "in the wild."

Choi et al. (2012) and Choi et al. (2005) are key early studies that aimed to test the predictions of Bayesian learning presented in Gale and Kariv (2003). The authors experimentally formed networks of three individuals and randomly assigned private information to the nodes under different conditions. Participants took an action, which was observable to their network links each period. The authors find broad support for Bayesian theories, but they do note that over time, as the origin of information was harder to ascertain, the error rate increased.

Others have tested departures from the Bayesian model such as DeGroot learning, where as a rule-of-thumb, individuals naively average the beliefs of their peers. Chandrasekhar et al. (2012) formed seven-person networks of three varieties, chosen for the different learning patterns predicted under each model. Each node initially received a private signal, and every period entered a guess. Network neighbors where able to observe these guesses. At the network level, DeGroot performed better, explaining 76% of actions in contrast to 62% for Bayesian. DeGroot also performed better than a relaxation of the Bayesian model, where players were allowed to have incomplete information about the types of the other players. Grimm and Mengel (2014) randomly varied both the network structure and information about the network structure. They also find that DeGroot performed better than Bayesian. However, at odds with the DeGroot model, participants were able to partially account for the correlation in the beliefs of their neighbors when informed of the network structure. Also see Mueller-Frank and Neri (2013).

Möbius et al. (2015) test DeGroot learning models in a setting where individuals could communicate more than just their beliefs. They seeded signals in a real-life network of college students that they had measured in previous work. Students could update their guesses with the researchers at any time over the four-day study period, and were freely able to discuss the game with one another. The authors find that students were able to avoid double-counting information originating at a single source. This is consistent with participants communicating information to others along with the source of that information (tagging).

16.3.2 Social Learning and Technology Adoption

Studying social learning outside of the lab inevitably limits the questions that can be asked. In field applications, researchers generally take the network as given and only observe the final diffusion decision, rather than a path of intermediate beliefs. Is is also difficult to observe direct communication, both *who* spoke with whom and *what* was communicated. Further, it is often difficult to measure *a priori* beliefs about new technologies. Thus, results of diffusion experiments depend crucially on the network structure, the priors of the community, and the characteristics of the product that is being introduced.

One of the oldest empirical networks literatures has been the study of the adoption of agricultural technologies (such as fertilizer or high-yielding seeds) and basic or preventative health technologies (such as deworming drugs or bednets). Seminal work includes Foster and Rosenzweig (1995) and Conley and Udry (2010), and a recent wave of field experiments further explores this question.

16.3.2.1 *Health and Agricultural Technology Adoption with Peer Data*

Kremer and Miguel's (2007) study of the decision to purchase deworming drugs was one of the first field experiments in economics to measure adoption spillovers. The authors measure the treatment spillovers from a large randomized school-level deworming program in 75 Kenyan schools (Miguel and Kremer 2007). Three years after the program started, they collected social network information from a representative set of parents in their sample frame. Parents were asked to name up to five connections in each of several categories: friends, relatives, social contacts with children in primary school, and individuals with whom they spoke about health. The authors also collected information on the schools (and school treatment status) of the children of each of the reported links. They find that conditional on the total number of social connections, parents with two additional connections to treated schools were one standard deviation less likely to take up deworming drugs.

Miller and Mobarak (2014) study the decision to purchase improved cookstoves in 42 Bangladeshi villages and also find negative adoption spillovers. In the first round of their two-stage field experiment, they randomly offered new stove technologies at a randomized set of prices. In the second round, they returned to market the stoves to control group households and measured spillovers. Close social ties of round one participants were elicited in the original baseline survey. The authors instrument product purchase in round one with the random price variation and find that round two adoption decreased substantially with the number of friends who adopted in round one. They suggest that in their setting, it was both hard to learn about the benefits and that those benefits were not very large.

Oster and Thornton (2012) also analyze the diffusion of an unfamiliar product that was difficult to learn how to use. In their setting, menstrual cups were randomly introduced to adolescent girls attending four schools in Nepal. At baseline, the authors elicited links between each girl and the other students in the sample frame. They find that those girls in the treatment group who happened to have a higher share of friends who were also treated were more likely to use the technology. However, this effect was only present for the first six months, suggesting that information spillovers accelerated the adoption process.

Many other health adoption experiments that do not directly measure social network characteristics find strong evidence of (positive) spillovers based on geographical proximity to randomly treated households. Dupas (2014) finds that individuals were more likely to purchase insecticide-treated bednets if they had more neighbors who had previously received subsidies for their use. Banerjee et al. (2010) observe that villages bordering areas that received incentives for vaccination also had higher numbers of children receiving their full course of vaccines. Cohen et al. (2015) find that when neighbors were randomly exposed to rapid diagnostic tests (RDTs) for malaria, households increased their own demand for them. In an RCT in Morocco, Devoto et al. (2011) helped treated households obtain a private water tap. They find higher adoption rates among control households in areas that had higher densities of treated households. Finally, Godlonton and Thornton (2012) find that when individuals in rural Malawi were given incentives to take an HIV test, geographic neighbors were more likely to also learn their own HIV status (see also Ngatia 2012).

Taking these studies together, the results are decidedly mixed. Not surprisingly, the setting-specific benefits and ease of adoption appear to be of first-order importance. However, this raises a host of new questions including how the complexity of the information and uncertainty about the quality of the information affect both how people learn and the ultimate extent of learning.

16.3.2.2 Technology Adoption with Full Networks Data

Three relatively recent projects study diffusion using detailed network maps of the study communities.

Banerjee et al. (2013) trace out the adoption of microfinance and Cai et al. (2015) the adoption of crop insurance. These papers take very different, though complementary, approaches to studying social learning. Banerjee et al. (2013) measured take-up months after initial community members were informed about microfinance, allowing for many periods of information transmission along both short and long paths through the network. The authors are exactly interested in these longer paths. Cai et al. (2015) measured spillovers only days after seeding information, and therefore isolate direct, local spillovers from initially informed to uninformed households.

The Banerjee et al. (2013) project involved a detailed networks survey about 12 dimensions of relationships administered in 75 rural Indian villages. The authors partnered with a microfinance institution (MFI), which chose to enter 42 of these villages. Though the authors do not have randomized experimental variation in who was initially reformed, they argue that the fixed rule (based on demographic characteristics) used by the MFI gives quasi-exogenous variation in the network properties of those individuals who were first informed (injection points). They find that village-level adoption was higher when the injection points were more central in the network. Further the best-performing measure of centrality, which they call diffusion centrality, takes into account transmission through both short and long paths.[14,15] The authors also use a structural model to disentangle mechanisms behind the peer effect. They test whether individuals were differentially more likely to adopt if the links from whom they learned also chose to participate (endorsement effect). They do not find evidence of an endorsement effect, but find that individuals who adopted were more likely to pass information to others.

Cai et al. (2015) collected network data from rice farmers in 185 villages in China and invited a randomly selected group of farmers to intensive information sessions meant to encourage crop insurance take-up. A few days later, they returned and offered the insurance to a second group of farmers. The sessions did increase take-up and resulted in large spillovers to direct network connections. Because the second group was approached so soon after the information sessions, the spillovers likely isolate the effect of direct communications between the initially informed and their network

[14] For T periods of information diffusion, diffusion centrality measures the expected number of times all individuals hear about information originating at a common source.

[15] Degree, which is a local measure of centrality (i.e., number of links) does not perform as well.

neighbors. The spillover effect was half as large as the information session effect, and using pricing experiments, was measured to be equivalent to a 13% price reduction. In order to separate learning from an endorsement effect, they introduced randomized defaults into their experimental design, which increased the likelihood of purchase. Using these defaults as an instrument for take-up, they find small but imprecisely measured endorsement effects.

In a new paper studying the adoption of a productivity-enhancing agricultural technology, Beaman et al. (2015) mapped networks in 200 villages in Malawi and experimentally varied who in the network received training in the new technology. In one treatment arm, the authors selected who to initially inform based on simulations of threshold diffusion models, in which individuals need to learn about the technology through multiple network neighbors before adopting themselves. In their benchmark treatment, the initial seeds were selected by agricultural extension officers. The authors find that their model-driven approach increased adoption by 3–4 percentage points (on a base of 10% in the benchmark group) and reduced the chance of no individual in a village adopting.

When information about new technologies is scarce, these papers illustrate in three different settings how the success of diffusion and adoption is closely tied to the network properties of who is initially informed. From a policy perspective, they suggest that mis-allocation of information may be extremely costly, and that it is possible to use network properties to substantially improve outcomes. More work is needed to develop cost-effective tools to accomplish this.

16.3.2.3 Strategic Information Diffusion

The field experiments discussed thus far show a range of diffusion outcomes including negative, positive, and an absence of spillovers. Strategic motivations may be an additional mechanism hindering or aiding diffusion. Many types of goods such as jobs, popular concert tickets, or limited-quantity subsidies are rival. Conversely, some products exhibit complementarities in the number of end users such as technology platforms. It is natural to speculate that rivalries or complementarities should affect incentives to pass information and overall adoption rates.

Banerjee et al. (2012) study a setting of social learning about rival goods in 60 of the Banerjee et al. (2013) network villages. In each village, 10 randomly chosen households were notified about an opportunity to earn half a day's wage playing laboratory experiments. Invitees were also informed that they could invite others, but that participation had a strict cap of 24 subjects. When the researchers arrived to conduct the laboratory experiments, many individuals who were not directly invited were waiting to play. Indeed, individuals with a greater share of initially invited friends showed up. To capture rivalry in the decision of whom to invite, the authors estimate a diffusion model. They find that the patterns of participation were consistent with invited households only spreading information to peripheral (low degree) friends. Transmission rates to high-degree friends were very low, in comparison.

Better understanding the causes and implications of strategic passing is fertile ground for future experimental work.

16.3.3 Information Extraction

While many social learning applications investigate how information injected from outside flows through a network, the social network may also serve as an information reservoir. Many actors such as governments, employers, lenders, and NGOs may benefit from being able to tap into sources of information. A recent set of labor and credit market experiments investigate whether the network does in fact possess valuable information and if individuals can be provided with incentives for truthful revelation.

Pallais and Sands (2014) conducted job referral experiments in an online labor market. They asked previous employees to refer workers and compared the referred applicants to applicants who responded to an online job posting. On average, the referred workers had better observables and also performed better conditional on observables. The authors find evidence that the effect was driven by selection and that referral work quality was positively related to the work quality of the referrer.[16] A natural follow-up question is how much of this observed selection effect comes from selection in the network formation process (if more productive workers sort into relationships with other productive workers) versus information acquisition (referrers can better use the social network to learn about the quality of other workers). Further, what are the underlying incentives for individuals to refer some members of their social network and not others?

Beaman and Magruder (2012) tackle these questions in a laboratory-in-the-field experiment in Kolkata, India. They recruited participants to perform a set of tasks and then asked them to refer other highly skilled workers under experimentally varied incentive structures. Some participants received fixed payments for each referral, while others received performance-linked incentives.[17] First, they observe a high level of family referrals in the fixed incentive treatment and a large decrease when participants were instead given performance incentives. Second, they find that more productive participants were better able to refer other high-performing workers, but only did so under performance pay.[18] This suggests that for some individuals, there was both variation in the skill of their network connections and substantial (likely private) information about worker quality. However, incentives were required to induce information revelation.

[16] One alternative explanation is that more central workers tend to be better workers. Even if workers referred random friends, those friends would be more central on average than a randomly sampled worker.

[17] In order for the performance treatment to only measure selection and not induce monitoring, when the referrers brought their referrals to the lab, they were unexpectedly given the maximum payment before the referral started working.

[18] In contrast, performance pay did not increase the referral quality of the less-productive participants.

In a related study, Beaman et al. (2014) consider the distributive consequences of referrals. They also randomized between fixed and performance-based incentives, but in addition, requested (randomized) gender-specific referrals. In absence of a requested gender, men were much more likely to refer men. The authors argue that men were able to refer qualified women when incentivized, and they speculate that men likely either received higher social benefits from referring other men or paid lower search costs.

Bryan et al. (2015) investigate similar issues in the context of credit market referrals in South Africa. However, unlike Beaman and Magruder (2012) they find that holding incentives to monitor fixed,[19] providing incentives for borrowers to refer credit-worthy individuals did not result in better loan performance. This result is not consistent with individuals having superior information about the credit-worthiness of their network connections relative to the lender.[20] It may also be the case that demand for loans among the referrer's friend was low, limiting the scope for private information.

To understand which network structures are best-suited to facilitate information aggregation, Alatas et al. (2014) use network data from 631 Indonesian communities. In a prior study, Alatas et al. (2012) investigated how policy makers could most effectively identify the poor for subsidy delivery given that incomes and asset holdings could not easily be verified. They randomized between methods including proxy-means tests (PMTs) and community targeting. Alatas et al. (2014) investigate which networks produced the most accurate poverty rankings. This type of village-level analysis is only possible because of the very large number of independent networks measured in the original study. They present and estimate a quasi-Bayesian model of learning about other people's wealth levels when those levels are changing over time. They find that (as predicted by the model) networks of higher densities (i.e., higher maximal eigenvalue and higher frequency of links, fixing network size) and with degree distributions better suited for diffusion (i.e., first-order stochastic dominance) produced better information. They also show that villages that were expected to know more under the model versus those that were expected to know less performed better at allocating resources in community targeting versus proxy means.

A natural follow-up question is which members of the social network have the best information (holding incentives fixed)? While Alatas et al. (2014), Banerjee et al. (2013), and Banerjee et al. (2014c) investigate this question descriptively and provide suggestive evidence that central individuals are better informed, there is scope for more causal experimental evidence.

The experimental literature on social learning shows the profound ways in which social networks can matter for diffusing, storing, and revealing information. However, in many ways, the literature has barely scratched the surface.

[19] Incentives to monitor are discussed in Section 16.5

[20] The authors mention that at the time of the study, South Africa had a robust credit scoring system and that there may not have been much scope for additional soft information revelation. Unlike in Beaman and Magruder (2012), only borrowers who satisfied the lender's underwriting criteria were given a loan.

16.4 Other-Regarding Preferences and Social Networks

A large laboratory literature on other-regarding preferences demonstrates high levels of baseline altruism among anonymous pairs. Other research shows that the social context of laboratory experiments also affects payoffs. In an early experiment, Hoffman et al. (1996) find that transfers in the dictator game increased when dictators were placed in situations involving less social isolation. Bohnet and Frey (1999) adopted the clever experimental design of changing the anonymity of the dictator and recipient before transfer decisions were made. They find that transfers under two-way identification were almost twice as big as under double anonymity.

More recent work has studied the interactions of other-regarding behavior with real-life social networks. Goeree et al. (2010) explore transfers among 5th and 6th graders in a Pasadena girls school. In advance of dictator games, they elicited the full social network of the students.[21] They find that dictator transfers increased dramatically with inverse social distance, with strangers receiving $\frac{1}{6}$ share and direct links receiving $\frac{1}{2}$. Their results suggest that social proximity is important for determining how resources are shared. Branas-Garza et al. (2010) elicited networks and played similar games with undergraduates. They find that measures of centrality (betweenness and degree) were positively related to giving.

Close friends and central individuals may make larger transfers for many reasons. First, the social network may parametrize altruism. Namely, individuals who are directly linked may exhibit greater directed altruism toward one another. Second, the social network may be correlated with motives of reciprocity or social sanctions. Individuals with stronger social connections are more likely to interact in the future and accordingly may have stronger incentives to transfer more in dictator games. Third, more generous people may simply be more popular, leading to more observed links between generous individuals.

Leider et al. (2009) conducted a careful online experiment with Harvard students to shed light on these different mechanisms. By experimentally concealing or revealing the identities of partners in a dictator game (similar to Bohnet and Frey 1999), Leider et al. (2009) controlled whether the directed altruism or future interaction motive (or both or neither) was present. They focus on how social network relationships interacted with the treatment conditions. They find that when dictator transfers were efficient (i.e., transferred resources grew in size), dictators were more generous to direct links relative to random strangers. When the dictator observed the identity of the other party in a single-blind fashion (directed altruism), social closeness increased transfers by 52%. When the identities of the participants were fully non-anonymized (enforced reciprocity + directed altruism), direct links received 24% more than strangers.

Ligon and Schechter (2012) conducted a similar experiment in a very different empirical context, 15 villages in rural Paraguay. However, they find a very similar

[21] With 95% sampling.

decomposition of the altruism effect as Leider et al. (2009). They estimate an average effect relative to the anonymous game of 8% when only the transfer recipient could identify the dictator, 6% when only the dictator could identify the recipient, and 17% when the identities were completely non-anonymized. In comparison, the effect sizes in Leider et al. (2009) were 14%, 10%, and 40%, respectively. Ligon and Schechter (2012) also find that individuals with higher degree appeared to exhibit less directed altruism and more reciprocity.

Goeree et al. (2010), Leider et al. (2009), and Ligon and Schechter (2012) all examine dictator play between individuals with preexisting social relationships and find surprisingly consistent results. In their settings, it is possible to carefully measure the altruism effect and to separate directed altruism from reciprocity or social sanctions. However, it is much harder to understand the role selection or other endogenous factors may play. Leider et al. (2009) find that individuals with stronger levels of baseline altruism had friends who were also more altruistic on average. They raise the question of whether individuals with altruistic preferences sort into friendships with like-minded people, or whether social relationships themselves shape an individual's preferences. To explain their result, Leider et al. (2009) also reason that high-degree individuals may act more reciprocally because of their network position. It is also possible that those with an intrinsic preference for reciprocity select into positions of network centrality. Goeree et al. (2010) find that personal traits predict links (in a homophilic fashion), which in turn predict behavior in the dictator game.

These discussions suggest how important it is to understand the network formation process when interpreting experimental findings. The state of the existing experimental literature on network formation and change is discussed in Section 16.7. Similarly, more work is needed to understand how the social network shapes an individual's underlying preferences.

Another important primitive studied in the experimental economics literature is trust.[22] Glaeser et al. (2000) conducted a laboratory experiment with 196 Harvard undergraduates. Participants were paired (often with friends who accompanied them to the experiment) and then played two different trust games. Comparing across pairs of participants, the authors find that both trusting behavior and trustworthiness increased with social proximity. Binzel and Fehr (2013) find similar results in a laboratory experiment in a Cairo slum.[23] Also see Breza et al. (2014) (described below) and Barr (2003) for related evidence.

[22] The workhorse measure of trust is the game pioneered in Berg et al. (1995). Many have questioned whether the game really measures trust (see Schechter 2007 for a discussion). Such a debate is beyond the scope of this chapter. However, even if the Berg et al. (1995) game does not actually measure trust, it is still likely to measure the ability of a pair of individuals to successfully reach an efficient outcome in the absence of commitment. Karlan (2005), for example, shows that behavior in the game does predict loan repayment in Peru.

[23] The authors recruited pairs of friends and had subjects play trust and dictator games. In some treatments the friend pairs played together, while in others partners (who were likely to be strangers) were assigned. Using within-person variation, they find that trust was higher in the friend pairs.

As in the literature on altruism and social networks, most of these papers take the preexisting social network as fixed and vary with whom the trust games are played. Bearing in mind the same cautions about endogenous network formation and reverse causality, the lab-in-field literature suggests that social connections are likely to be quite important for sustaining trade and exchange.

16.5 Public Commitments, Peer Monitoring, and Enforcement

Examples of public commitments and peer monitoring are widespread in applications from labor to credit to going to the gym (i.e., Bandiera et al. 2010; Besley and Coate 1995; Babcock and Hartman 2010). Much theory is devoted to understanding how the network diffuses information and shares risk. However, some of these same network processes are also useful for peer monitoring and enforcement.

Breza et al. (2014) conducted non-anonymized trust games with third-party monitoring and punishment in 40 Indian villages and matched experimental outcomes with the Banerjee et al. (2013) networks data. They find evidence that when contracting parties could not call upon social proximity, peers could be used to improve outcomes. In their sample, 84% of the randomly assigned pairs in their experimental sessions were of social distance greater than two. If highly network central individuals were chosen to serve as the third party and were given a punishment technology, outcomes between these socially distant pairs were as good as those between socially close pairs without a third party. While network centrality was not randomly assigned, Breza et al. (2014) randomized both the identities of the players and the available contracting tools. They also find that no other demographic characteristics such as caste, elite status, or wealth could replicate the centrality effects. The result suggests that eigenvector centrality, which is usually thought to be important for diffusion, can be used in other capacities, namely enforcement.

Several field experiments have demonstrated the capacity for peer-based mechanisms to increase savings balances and loan repayment. In their field experiment studying different types of savings vehicles, Dupas and Robinson (2013b) find that social commitment increased savings and was more effective than their other individual-based treatments. Kast et al. (2012) further investigate this social commitment effect. Working with microfinance borrowing groups, they opened individual savings accounts for members, and some savers were randomly chosen to also receive monitoring by their microfinance groups. They find that those in the peer monitoring treatment saved almost twice as much as those in the control group. However, in a separate sample, they find that text-based SMS reminders were comparably effective and were not enhanced by the addition of a "Savings Buddy" who was chosen by the saver to monitor her performance.

Breza and Chandrasekhar (2015) ask whether such savings arrangements depend on the embedding of the savers and monitors in the social network. They conducted a field experiment in 60 villages in South India (also using the Banerjee et al. 2013 networks) with villagers who expressed a desire to save more. All savers were given a bundle of services including assistance with account opening, savings goal elicitation, and bi-weekly reminder visits by a surveyor. A randomly selected group of savers was also chosen to receive a monitor, who was updated about the saver's progress every two weeks. When savers were assigned to receive a randomly-chosen monitor, savings increased substantially (by 35% across all accounts) relative to the non-monitored group. The effects were even stronger when either the monitor had high eigenvector centrality or the saver and monitor were socially close. They also find suggestive evidence of a reputation-based mechanism, where monitors actively spread information about their savers to others. Sixty-three percent of monitors reported telling others about the saver's progress. Further, over a year later, other villagers were more likely to know if the saver exceeded her goal and to think that the saver was responsible if she had been randomly assigned to a more central monitor. When given the opportunity to choose their own monitors, savers picked monitors who were both closer and more central than average.

Credit is another financial product where some lenders have captured the power of peer enforcement. Many argue that peer monitoring is one of the reasons why microfinance is able to boast such high repayment rates. As discussed above, Bryan et al. (2015) conducted a clever field experiment to separately measure both peer screening and peer monitoring in credit markets. In their experiment, existing clients were asked to refer a friend to receive a loan and were randomly assigned into four different incentive treatments. They find very large monitoring effects, but no detectable screening effects. One could also speculate that the efficacy of such monitoring arrangements might depend on the network positions of the referrer and referred borrower.

These studies help to get inside of the black box of many informal financial products commonly observed in developing countries such as rotating savings and credit associations, self-help groups, and microfinance institutions, all of which incorporate groups of individuals from the same social network and rely on mechanisms that are likely to include mutual monitoring.

Another relevant application is public goods provision and the management of common pool resources.[24] However, there is surprisingly little experimental evidence on the topic. Carpenter et al. (2012) is one of the only laboratory experiments to consider how the graph structure when combined with social punishment can increase or limit the efficiency of public goods provision. Choi et al. (2011) consider the role networks can play in helping groups to coordinate on efficient levels of public goods. This area of study, especially in the field, represents a real research opportunity.

[24] A growing network theory literature (Bramoullé and Kranton 2007 and Elliott and Golub 2014 for example) exists on the topic as well as a large literature in political science that began with Ostrom (1992).

16.6 Risk-Sharing

Aside from social learning and diffusion, enabling the sharing of risk is one of the most important functions of social networks in economics. This feature of economic networks is well-theorized and has been the subject of many empirical, though often observational, studies (Kimball 1988; Coate and Ravallion 1993; Kocherlakota 2004; Ligon et al. 2000; Foster and Rosenzweig 2001; Udry 1994; and Townsend 1994). Because risk-sharing interacts with and encompasses many of the other mechanisms that have already been discussed, designing the "ideal" risk-sharing experiment is quite daunting. The ability for a decentralized group of individuals to share risk is shaped by many factors. Key components of risk-sharing include the sharing rule, the degree of commitment present, and opportunities to hide income or savings. Each of these is likely related to the social network through other-regarding preferences, peer enforcement, or information. Due to the complexity involved in studying risk-sharing, most of the existing experimental work comes from laboratory studies where it is much simpler to switch on and off mechanisms.

Barr and Genicot (2008) and Barr et al. (2012) analyze a multi-day lab experiment conducted in 14 villages in Zimbabwe. On the first day, individuals were invited to the lab to play a gamble choice game similar to Binswanger (1980). Participants were then invited back for a second day. They were allowed to form groups with other (non-anonymized) participants before they again played gamble choice games. Groups then shared their (risky) earnings equally under either external enforcement or limited commitment. In the limited commitment treatments, individuals could opt out of the group sharing after observing their own income realizations. Barr and Genicot (2008) show that full enforcement generated the largest groups and induced the most risk-taking by participants.[25] Barr et al. (2012) study the composition of the risk-sharing groups and find that participants tended to match by kinship and religious ties in the limited commitment treatments. However, matching by membership in a community-based organization only arose under full enforcement. They argue that only some types of ties were capable of meting out and withstanding the consequences of social sanctions.

Attanasio et al. (2012) conducted a similar exercise in 70 communities in Colombia and measured network connections between all participants in the lab. They find that close friends and relatives were more likely to form groups, and that they were also more likely to match assortatively on risk preferences. While only suggestive, these results highlight the scope for social proximity to facilitate risk-sharing under limited commitment.

Chandrasekhar et al. (2014) take up this idea in laboratory experiments in 34 of the villages from Banerjee et al. (2013). To better understand the role of social networks in sustaining self-insurance, they played insurance games with pairs of participants. They

[25] The increase in risk-taking may arise both from the full enforcement and the increase in group size.

exogenously varied both the social connections between the players and the contracting environment. They find that socially close pairs reached more cooperative outcomes than distant pairs, especially when full enforcement was switched off. They also find less cooperation when the pairs had unequal centralities, especially in the absence of full enforcement. Because each individual played several games, the authors could include individual-specific fixed effects and use within-participant variation for identification. This helps to rule out bias from person-specific unobservables that could be correlated with the individual's position in the network.

One application where it is easy to observe active risk-sharing groups is microfinance. Fischer (2013) ran laboratory experiments with microfinance groups in India to study the role of joint liability contracts. He highlights the opposing incentives for risk-taking induced by risk-pooling and peer monitoring. To separate these incentives, he exogenously imposed five different group financing structures. When group members could not monitor perfectly, joint liability (risk pooling) created significant incentives to free ride and to take more risk. However, under both monitoring and risk pooling (which approximates microfinance contracts) borrowing groups were not able to take risk, even when efficient. Giné et al. (2010) find similar evidence in laboratory games in Peru. In their setting, joint liability increased risk-taking, but less so when individuals were able to choose their own groups.

Because of the difficulties outlined above, field experiments with risk-sharing are quite rare. Existing field work has taken the approach of observing how individuals in a network with preexisting risk-sharing relationships react when given the chance to take up formal financial products.

Mobarak and Rosenzweig (2012) conducted an experiment in India to measure the effects of marketing formal rainfall insurance to households that were already engaged in informal risk-sharing. They randomized offers of insurance to members of different subcastes (which proxied for risk-sharing groups) and also randomized the placement of rain gauges (and thus the amount of basis risk). Using survey data, they show that ex ante, some castes were able to insure aggregate risk and others could insure idiosyncratic risk. The authors find that the ability to informally insure aggregate risk and the presence of basis risk each decreased the demand for formal insurance. However, the negative impacts of basis risk were mitigated when the network could only insure idiosyncratic risk, suggesting complementarities between the two. Mobarak and Rosenzweig (2012) provide important field evidence on the relationship between formal and informal insurance, which raises a new set of questions. For example, when an individual is able to take formal insurance, what are the follow-on effects on the former risk-sharing partners? Moreover, what determines the scope of insurance against idiosyncratic and aggregate risks in a network?

In most situations, informal risk-sharing generally improves welfare for risk averse agents. However, there is potentially a dark side. Using a lab-in-the-field experiment in rural Kenya, Jakiela and Ozier (2015) investigate the conventional wisdom that individuals are often unable to consolidate windfalls due to unsolicited demands from friends and family. Participants were given endowments of different sizes, which they then divided between risky and safe investment opportunities. In some treatments, the risky investment could be observed by other participants. The authors report that women often took costly actions to hide income from neighbors and family. Further, men and women both were willing to pay up to 15% of their earnings to keep their incomes fully hidden. They argue that their evidence is consistent with the conventional wisdom and that pressure to make transfers erodes efficiency. Dupas and Robinson (2013a) also suggest that unwanted demands from social contacts is one mechanism that may make access to private bank accounts so valuable to poor households. Jakiela and Ozier (2015) provide experimental evidence that social pressure can reduce efficiency ex post. However, the policy implications of this result depend crucially on why this inefficiency arises. Under many models of informal insurance with limited commitment, the ability for individuals to exert social pressure ex post is efficiency enhancing from an ex ante perspective, even if there are ex post incentives to hide income. More work is needed to resolve this important issue.

16.7 Network Formation and Change

As discussed above, most empirical networks projects, experiments included, take the social network structure as given and exogenously overlay a treatment such as access to information or credit. In most cases, authors must rely on conditioning on observables or structural assumptions to make the case that causality runs through the observed social network. Because the social network encodes patterns of interactions between individuals in real life, it is often extremely hard, if not impossible, to find sources of exogenous variation in network structure. However, a small recent literature has begun using field experiments to do just this.

In a nice experimental design, Fafchamps and Quinn (2012) study connections between judges of business idea competitions in Ethiopia, Tanzania, and Zambia. By randomly assigning the judges to committees that met over the three-day competitions, Fafchamps and Quinn (2012) sowed the seeds for new business linkages to be formed. A few months after the competition, co-committee members were more likely to remember and to speak with one another. They find suggestive evidence of peer effects that are positive for low-risk and low-cost actions such as opening a bank account, but negative for high-cost and high-risk actions such as launching new products.[26]

Feigenberg et al. (2013) is one of the first to document the effects of an intensive social intervention. The authors randomized the frequency of microfinance group meetings between weekly (status quo) and monthly during clients' first 10 months of borrowing. They find that 16 months after the experiment's end, individuals assigned to the weekly group met with fellow group members 37% more than pairs in the monthly group. They then tie these stronger social connections to evidence of improvements in risk-sharing;

[26] Also see Fafchamps and Quinn (2013).

participants reached more efficient group-level outcomes in a risky lottery experiment administered 16 months after the end of the intervention, and borrowers in the weekly group were substantially less likely to default on subsequent loans.

While Feigenberg et al. (2013) focus on changes to the strength of ties through microfinance, Banerjee et al. (2014a) analyze persistent changes in the number of ties caused by randomized exposure to microfinance. In 2012, the authors returned to 104 neighborhoods in Hyderabad, India that were involved in the Banerjee et al. (2015) randomized roll-out of microfinance. Half of the households were exposed to microfinance in 2006, six years earlier (treatment group), and the other half were exposed in 2008 (control group).[27] The authors first show that longer access to microfinance had persistent positive effects on business, credit, and durable consumption for the subset of households who had started a business before microfinance was first made available. Because the experiment was conducted in dense urban neighborhoods, a full network elicitation was not possible. Instead, respondents were asked to name all of the other households in their neighborhood with which they participated in each of eight different financial and nonfinancial activities and indicate whether each link was supported.[28] Treatment households without a preexisting business reported 8% fewer network connections than the control (mean degree 5.9), and the number of supported links decreased by 16% (control mean 2.7). The non-entrepreneurs who had longer access to microfinance were not more likely to report links with whom they had shared a microfinance group in the past. Total informal borrowing also decreased. However, for preexisting entrepreneurs, there was no detectable change in number of overall links or number of supported links. This subgroup also borrowed more from informal sources. This evidence suggest that microfinance and informal risk-sharing served as substitutes for those without a preexisting business (and with relatively low demand for credit) and complements for those who selected into entrepreneurship before microfinance was available.

Comola and Prina (2014) study how transfers between households changed in response to a randomized savings intervention. The authors collected data on actual financial transfers between all households in 19 villages in Nepal before and after the intervention.[29] Treated households sent larger transfers and increased the number of recipients relative to the control, and the effects were strongest for wealthier households with better financial access. The authors then use structural methods similar to Bramoullé et al. (2009) to decompose expenditure peer effects arising from the intervention. They are particularly interested in the channel of peer effects that results from the intervention's impact on the network structure.[30]

[27] Due to political reasons, access to microfinance ended in all study areas in 2010.

[28] In other words, whether there existed a third person who did that activity with both the respondent and the friend (Jackson et al. 2012).

[29] Prina (2015) describes the intervention and its direct effects.

[30] One limitation of the Comola and Prina (2014) study is that their measured networks are quite sparse with an average degree in the baseline data of 0.64. While the estimated causal impact of formal

Binzel et al. (2015) study the impacts of the roll-out of rural bank branches on village networks in India and find evidence of substitution between formal and informal borrowing. They measure decreases in the amount respondents reported being able to borrow from their village network (borrowing capacity) and find declines in amounts transferred to financial network links in non-anonymized dictator games. The dictator games also suggest that households shifted transfers away from financial links and towards social links.

Banerjee et al. (2014b) use their 75-village data set to study network change before and after microfinance. While again they do not have experimental variation, they use differences-in-differences to compare the network shapes of villages with or without access to microfinance, before and after the intervention. They find access to microfinance to be associated with lower link probabilities, lower levels of support, and less multiplexing (i.e., links with multiple dimensions of interaction). They further find that results are similar when looking at financial, social, and advice links.

To date, this line of inquiry has generally sought to understand how the shape of the network responds to external stimulus. The possibility of using exogenous network change to better understand causal links between network shape and other real outcomes is exciting. However, such an exercise would require that the underlying change to the network not be directly correlated with the outcome of interest (e.g., exclusion restriction). New developments in the modeling of network formation as discussed in "Econometrics of Network Formation" by Chandrasekhar in this volume will help advance this pursuit.

16.8 Conclusion and Open Issues

Experiments involving social networks both in the laboratory and the field have followed the broader growth of economic experiments. However, the addition of a network component requires new considerations ranging from identification to how to define and measure the network, how to measure spillovers, and how to conduct inference. Because of the recentness of this literature, there is still no firm consensus regarding how to approach these issues.

It is an exciting moment to reflect on the state of network experiments. This chapter reviews five interrelated strands of experimental networks research: diffusion and social learning, other-regarding preferences, peer monitoring and enforcement, risk-sharing, and network formation and change. In each of these areas, innovations in experimental techniques have permitted researchers to expand the scope of the study of peer effects from simple in- versus out-group spillover effects to incorporating more of the global network structure. Another recent trend in the literature has been to use experimental

savings accounts on realized transactions is identified given the experiment, the structural peer effects decomposition is likely quite sensitive to the way in which the network is specified.

variation to disentangle different network-based mechanisms underlying reduced form spillover effects.

Based on the large volume of research currently in progress, the next few years will mark an advance in the field's understanding of network-mediated phenomena. Further, as more experiments are completed, we will be able to better compare results across a range of different settings.

There are also many exciting directions for future research. One under-represented research area is the study of networks within and between firms. Information transmission is central for the spread of new productive technologies,[31] and understanding barriers to learning may help explain why productivity differences persist across firms. Supply chain networks may further have a credit or insurance component or may instead hasten the transmission of shocks (Cunat 2007; Garcia-Appendini and Montoriol-Garriga 2013; and Acemoglu et al. 2012). Pricing decisions may also evolve along networks (Gale and Kariv 2009), and there is scope for field research in this space.

Another application with great research promise is electronic payments and mobile networks. While observational studies have measured risk-sharing through mobile networks in response to shocks (Jack and Suri 2014 and Blumenstock et al. 2012), there has been little experimental work on the topic. Relatedly, Munshi (2003) and others have used non-experimental methods to demonstrate the importance of social networks in the migration decision and in migrant job search.[32] Revisiting these questions with new experimental network tools could also be fruitful.

Given the endogeneity concerns endemic to network field experiments, there may be high returns to pairing laboratory experiments with field experiments. While it is tremendously hard to randomly change the shape of real-world social networks, it is possible to randomly embed participants in communication networks in the lab. If researchers find suggestive evidence based on real-world network shapes, they can test those hypotheses in a controlled laboratory setting.

Finally, many of the applications discussed here have policy implications. For example, who should be given scarce information to maximize diffusion? Whose opinions should be elicited in order to target resources toward the poorest residents? How should firms organize themselves to minimize the effects of social comparisons? What is required to sustain informal insurance in underbanked communities? However, many of the implications are not quite ready for use by policy-makers, as they require detailed knowledge of network structure which is costly to collect. Therefore, advances in measurement techniques may help to move forward both the research and the applicability of its findings.

[31] Atkin et al. (2014) are diffusing new technologies to soccer ball firms in Pakistan and intend to measure spillovers. Menzel (2015) is measuring productivity spillovers across production lines in Bangladeshi factories.

[32] See "Community Networks and Migration" by Munshi in this volume. Also see the migration experiment by Bryan et al. (2014).

References

Acemoglu, D., V. M. Carvalho, A. Ozdaglar, and A. Tahbaz-Salehi (2012). "The network origins of aggregate fluctuations." *Econometrica* 80, 1977–2016.

Alatas, V., A. Banerjee, A. G. Chandrasekhar, R. Hanna, and B. A. Olken (2014). "Network structure and the aggregation of information: Theory and evidence from Indonesia." NBER Working Paper.

Alatas, V., A. Banerjee, and R. Hanna (2012). "Targeting the poor: Evidence from a field experiment in Indonesia." *American Economic Review* 102, 1206–1240.

Angrist, J. D. and J.-S. Pischke (2008). *Mostly harmless econometrics: An empiricist's companion*, Princeton University Press.

Atkin, D., A. Chaudhry, S. Chaudry, A. K. Khandelwal, and E. Verhoogen (2014). "Organizational barriers to technology adoption: Evidence from soccer-ball producers in Pakistan," Working paper.

Attanasio, O., A. Barr, J. C. Cardenas, G. Genicot, and C. Meghir (2012). "Risk pooling, risk preferences, and social network." *American Economic Journal: Applied Economics* 4, 134–167.

Babcock, P. and J. Hartman (2010). "Exercising in herds: treatment size and status specific peer effects in a randomized exercise intervention." Working paper.

Baird, S., A. Bohren, C. McIntosh, and B. Ozler (2014). "Designing experiments to measure spillover and threshold effects." Working Paper.

Bandiera, O., I. Barankay, and I. Rasul (2010). "Social incentives in the workplace," *The Review of Economic Studies* 77, 417–458.

Banerjee, A., E. Breza, A. G. Chandrasekhar, E. Duflo, and M. Jackson (2012). "Come play with me: Experimental evidence of information diffusion about rival goods," Work in Progress.

Banerjee, A., E. Breza, E. Duflo, and C. Kinnan (2014a). "Do credit constraints limit entrepreneurship? Heterogeneity in the returns to microfinance," Working paper.

Banerjee, A., A. Chandrasekhar, E. Duflo, and M. Jackson (2013). "The diffusion of microfinance." *Science* 341, 1236498.

Banerjee, A., A. Chandrasekhar, E. Duflo, and M. Jackson (2014b). "Network change," Working paper.

Banerjee, A., E. Duflo, R. Glennerster, and C. Kinnan (2015). "The miracle of microfinance? Evidence from a randomized evaluation." *American Economic Journal: Applied Economics* 7, 22–53.

Banerjee, A. and K. Munshi (2004). "How efficiently is capital allocated? Evidence from the knitted garment industry in Tirupur." *The Review of Economic Studies* 71, 19–42.

Banerjee, A. V., A. Chandrasekhar, E. Duflo, and M. O. Jackson (2014c). "Gossip: Identifying central individuals in a social network," Working paper.

Banerjee, A. V., E. Duflo, R. Glennerster, and D. Kothari (2010). "Improving immunisation coverage in rural India: Clustered randomised controlled evaluation of immunisation campaigns with and without incentives." *BMJ: British Medical Journal* 340, c2220.

Barr, A. (2003). "Trust and expected trustworthiness: Experimental evidence from Zimbabwean villages." *The Economic Journal* 113, 614–630.

Barr, A., M. Dekker, and M. Fafchamps (2012). "Who shares risk with whom under different enforcement mechanisms?" *Economic Development and Cultural Change* 60, 677–706.

Barr, A. and G. Genicot (2008). "Risk sharing, commitment, and information: An experimental analysis." *Journal of the European Economic Association* 6, 1151–1185.

Beaman, L., A. BenYishay, J. Magruder, and A. M. Mobarak (2015). "Can network theory based targeting increase technology adoption?" Working paper.

Beaman, L., N. Keleher, and J. Magruder (2014). "Do job networks disadvantage women? Evidence from a recruitment experiment in Malawi," Working paper.

Beaman, L. and J. Magruder (2012). "Who gets the job referral? Evidence from a social networks experiment." *The American Economic Review* 102, 3574–3593.

Berg, J., J. Dickhaut, and K. McCabe (1995). "Trust, reciprocity, and social history." *Games and Economic Behavior* 10, 122–142.

Besley, T. and S. Coate (1995). "Group lending, repayment incentives and social collateral." *Journal of Development Economics* 46, 1–18.

Binswanger, H. P. (1980). "Attitudes toward risk: Experimental measurement in rural India." *American Journal of Agricultural Economics* 62, 395–407.

Binzel, C. and D. Fehr (2013). "Social distance and trust: Experimental evidence from a slum in Cairo." *Journal of Development Economics* 103, 99–106.

Binzel, C., E. Field, and R. Pande (2015). "Does the arrival of a formal financial institution alter informal sharing arrangements? Experimental evidence from India," Mimeo.

Blumenstock, J., N. Eagle, and M. Fafchamps (2012). "Risk and reciprocity over the mobile phone network: Evidence from Rwanda." Working paper.

Bohnet, I. and B. Frey (1999). "Social distance and other-regarding behavior in dictator games: Comment." *American Economic Review*, 335–339.

Bramoullé, Y., H. Djebbari, and B. Fortin (2009). "Identification of peer effects through social networks." *Journal of Econometrics* 150, 41–55.

Bramoullé, Y. and R. Kranton (2007). "Public goods in networks." *Journal of Economic Theory* 135, 478–494.

Branas-Garza, P., R. Cobo-Reyes, M. P. Espinosa, N. Jiménez, J. Kovářík, and G. Ponti (2010). "Altruism and social integration." *Games and Economic Behavior* 69, 249–257.

Breza, E. and A. Chandrasekhar (2015). "Social networks, reputation, and commitment: Evidence from a savings monitors experiment." NBER Working Paper.

Breza, E., A. G. Chandrasekhar, and H. Larreguy (2014). "Social structure and institutional design: Evidence from a lab experiment in the field," NBER Working Paper.

Bryan, G., S. Chowdhury, and A. M. Mobarak (2014). "Underinvestment in a profitable technology: The case of seasonal migration in Bangladesh." *Econometrica* 82, 1671–1748.

Bryan, G. T., D. Karlan, and J. Zinman (forthcoming). "Referrals: Peer screening and enforcement in a consumer credit field experiment." *American Economic Journal Microeconomics* 7(3), 174–204.

Cai, J., A. De Janvry, and E. Sadoulet (2015). "Social networks and the decision to insure." *American Economic Journal: Applied Economics* 7, 81–108.

Calvó-Armengol, A., E. Patacchini, and Y. Zenou (2009). "Peer effects and social networks in education." *The Review of Economic Studies* 76, 1239–1267.

Canay, I. A., J. P. Romano, and A. M. Shaikh (2014). "Randomization tests under an approximate symmetry assumption." *Technical Report No. 2014-13*, Stanford University.

Carpenter, J., S. Kariv, and A. Schotter (2012). "Network architecture and mutual monitoring in public goods experiments." *Review of Economic Design* 16, 175–191.

Chandrasekhar, A., H. Larreguy, and J. P. Xandri (2012). "Testing models of social learning on networks: Evidence from a framed field experiment." Working paper.

Chandrasekhar, A. G., C. Kinnan, and H. Larreguy (2014). "Social networks as contract enforcement: Evidence from a lab experiment in the field." NBER Working Paper.

Chandrasekhar, A. G. and R. Lewis (2011). "Econometrics of Sampled Networks," MIT Working Paper.

Choi, S., D. Gale, and S. Kariv (2005). "Behavioral aspects of learning in social networks: An experimental study." *Advances in Applied Microeconomics* 13, 25–61.

Choi, S., D. Gale, and S. Kariv (2012). "Social learning in networks: A quantal response equilibrium analysis of experimental data." *Review of Economic Design* 16, 135–157.

Choi, S., D. Gale, S. Kariv, and T. Palfrey (2011). "Network architecture, salience and coordination." *Games and Economic Behavior* 73, 76–90.

Coate, S. and M. Ravallion (1993). "Reciprocity without commitment: Characterization and performance of informal insurance arrangements." *Journal of Development Economics* 40, 1–24.

Cohen, J., P. Dupas, and S. Schaner (2015). "Price subsidies, diagnostic tests, and targeting of malaria treatment: Evidence from a randomized controlled trial." *The American Economic Review* 105, 609–645.

Comola, M. and S. Prina (2014). "Do interventions change the network? A dynamic peer effect model accounting for network changes." Working paper.

Conley, T. and C. Udry (2010). "Learning about a new technology: Pineapple in Ghana." *The American Economic Review* 100, 35–69.

Cunat, V. (2007). "Trade credit: Suppliers as debt collectors and insurance providers." *Review of Financial Studies* 20, 491–527.

Devoto, F., E. Duflo, P. Dupas, W. Pariente, and V. Pons (2011). "Happiness on tap: Piped water adoption in urban Morocco." NBER Working Paper, 16933.

Dieye, R., H. Djebbari, and F. Barrera-Osorio (2014). "Accounting for peer effects in treatment response," Working paper.

Duflo, E., R. Glennerster, and M. Kremer (2007). "Using randomization in development economics research: A toolkit." *Handbook of Development Economics* 4, 3895–3962.

Dupas, P. (2014). "Short-run subsidies and long-run adoption of new health products: Evidence from a field experiment." *Econometrica* 82, 197–228.

Dupas, P. and J. Robinson (2013a). "Savings constraints and microenterprise development: Evidence from a field experiment in Kenya." *American Economic Journal: Applied Economics* 5, 163–192.

Dupas, P. and J. Robinson (2013b). "Why don't the poor save more?" *American Economic Review* 103, 1138–1171.

Elliott, M. and B. Golub (2014). "A network approach to public goods." Working paper.

Fafchamps, M. and F. Gubert (2007). "Risk sharing and network formation." *The American Economic Review* 97, 75–79.

Fafchamps, M. and S. Quinn (2012). "Networks and manufacturing firms in Africa: Initial results from a randomised experiment." Working paper.

Fafchamps, M. and S. Quinn (2013). "Social networks and business practices: Evidence from a randomized experiment with microentrepreneurs." Working paper.

Feigenberg, B., E. Field, and R. Pande (2013). "The economic returns to social interaction: Experimental evidence from microfinance." *The Review of Economic Studies* 80(4), 1459–1483.

Fischer, G. (2013). "Contract structure, risk-sharing, and investment choice." *Econometrica* 81, 883–939.

Foster, A. D. and M. R. Rosenzweig (1995). "Learning by doing and learning from others: Human capital and technical change in agriculture." *Journal of Political Economy* 103, 1176–1209.

Foster, A. D. and M. R. Rosenzweig (2001). "Imperfect commitment, altruism, and the family: Evidence from transfer behavior in low-income rural areas." *Review of Economics and Statistics* 83, 389–407.

Gale, D. and S. Kariv (2003). "Bayesian learning in social networks." *Games and Economic Behavior* 45, 329–346.

Gale, D. M. and S. Kariv (2009). "Trading in networks: A normal form game experiment." *American Economic Journal: Microeconomics* 1(2), 114–132.

Garcia-Appendini, E. and J. Montoriol-Garriga (2013). "Firms as liquidity providers: Evidence from the 2007–2008 financial crisis." *Journal of Financial Economics* 109, 272–291.

Giné, X., P. Jakiela, D. Karlan, and J. Morduch (2010). "Microfinance games." *American Economic Journal: Applied Economics* 2, 60–95.

Glaeser, E., D. Laibson, J. Scheinkman, and C. Soutter (2000). "Measuring Trust." *Quarterly Journal of Economics* 115, 811–846.

Godlonton, S. and R. Thornton (2012). "Peer effects in learning HIV results." *Journal of Development Economics* 97, 118–129.

Goeree, J., M. McConnell, T. Mitchell, T. Tromp, and L. Yariv (2010). "The 1/d law of giving." *American Economic Journal: Microeconomics* 2, 183–203.

Grimm, V. and F. Mengel (2014). "An experiment on belief formation in networks," Working paper.

Harrison, G. W. and J. A. List (2004). "Field experiments." *Journal of Economic Literature* 42(4), 1009–1055.

Hoffman, E., K. McCabe, and V. Smith (1996). "Social distance and other-regarding behavior in dictator games." *The American Economic Review* 86, 653–660.

Hudgens, M. G. and M. E. Halloran (2008). "Toward causal inference with interference." *Journal of the American Statistical Association* 482, 832–842.

Jack, W. and T. Suri (2014). "Risk sharing and transactions costs: Evidence from Kenya's mobile money revolution." *The American Economic Review* 104, 183–223.

Jackson, M. O., T. Rodriguez-Barraquer, and X. Tan (2012). "Social capital and social quilts: Network patterns of favor exchange." *The American Economic Review* 102, 1857–1897.

Jakiela, P. and O. Ozier (2015). "Does Africa need a rotten kin theorem? Experimental evidence from village economies." *Review of Economic Studies*, forthcoming.

Karlan, D. (2005). "Using experimental economics to measure social capital and predict financial decisions." *The American Economic Review* 95, 1688–1699.

Kast, F., S. Meier, and D. Pomeranz (2012). "Under-savers anonymous: Evidence on self-help groups and peer pressure as a savings commitment device." NBER Working Paper.

Kimball, M. S. (1988). "Farmers' cooperatives as behavior toward risk." *American Economic Review* 78, 224–232.

Kocherlakota, N. R. (2004). "Figuring out the impact of hidden savings on optimal unemployment insurance." *Review of Economic Dynamics* 7, 541–554.

Kremer, M. and E. Miguel (2007). "The illusion of sustainability." *The Quarterly Journal of Economics* 122, 1007–1065.

Lee, L., X. Liu, and X. Lin (2010). "Specification and estimation of social interaction models with network structures." *The Econometrics Journal* 13, 145–176.

Leider, S., M. M. Möbius, T. Rosenblat, and Q.-A. Do (2009). "Directed altruism and enforced reciprocity in social networks*." *Quarterly Journal of Economics* 124, 1815–1851.

Ligon, E. and L. Schechter (2012). "Motives for sharing in social networks." *Journal of Development Economics* 99, 13–26.

Ligon, E., J. P. Thomas, and T. Worrall (2000). "Mutual insurance, individual savings, and limited commitment." *Review of Economic Dynamics* 3, 216–246.

Manski, C. (1993). "Identification of endogenous social effects: The reflection problem." *The Review of Economic Studies* 60, 531–542.

Manski, C. F. (2013). "Identification of treatment response with social interactions." *The Econometrics Journal* 16, S1–S23.

Möbius, M., T. Phan, and A. Szeidl (2015). "Treasure hunt: Social learning in the field." NBER Working Paper.

McMillan, J. and C. Woodruff (1999). "Interfirm relationships and informal credit in Vietnam." *The Quarterly Journal of Economics*, 114, 1285–1320.

Menzel, A. (2015). "Social networks and productivity spill-over within firms: evidence from Bangladeshi garment factories." Working paper.

Miguel, E. and M. Kremer (2007). "Worms: Identifying impacts on education and health in the presence of treatment externalities." *Econometrica* 72, 159–217.

Miller, G. and A. M. Mobarak (2015). "Learning about new technologies through social networks: Experimental evidence on non-traditional stoves in rural Bangladesh." *Marketing Science* 34(4), 480–499.

Mobarak, A. M. and M. R. Rosenzweig (2012). "Selling formal insurance to the informally insured," Working paper.

Mueller-Frank, M. and C. Neri (2013). "Social learning in networks: Theory and experiments." Working paper.

Munshi, K. (2003). "Networks in the modern economy: Mexican migrants in the US labor market." *The Quarterly Journal of Economics*, 549–599.

Ngatia, M. (2012). "Social interactions and individual reproductive decisions," Working paper.

Oster, E. and R. Thornton (2012). "Determinants of technology adoption: Peer effects in menstrual cup take-up." *Journal of the European Economic Association* 10, 1263–1293.

Ostrom, E. (1992). "Community and the endogenous solution of commons problems." *Journal of Theoretical Politics* 4, 343–351.

Pallais, A. and E. G. Sands (2014). "Why the referential treatment? Evidence from field experiments on referrals." Working paper.

Petersen, M. A. and R. G. Rajan (1994). "The benefits of lending relationships: Evidence from small business data." *The Journal of Finance* 49, 3–37.

Prina, S. (2015). "Banking the poor via savings accounts: Evidence from a field experiment." *Journal of Development Economics* 115, 16–31.

Rubin, D. B. (1978). "Bayesian inference for causal effects: The role of randomization." *The Annals of Statistics* 6(1), 34–58.

Schechter, L. (2007). "Traditional trust measurement and the risk confound: An experiment in rural Paraguay." *Journal of Economic Behavior & Organization* 62, 272–292.

Townsend, R. M. (1994). "Risk and insurance in village India." *Econometrica* 62, 539–591.

Udry, C. (1994). "Risk and insurance in a rural credit market: An empirical investigation in northern Nigeria." *The Review of Economic Studies* 61, 495–526.

CHAPTER 17

NETWORKS IN THE LABORATORY

SYNGJOO CHOI, EDOARDO GALLO, AND SHACHAR KARIV

17.1 Introduction

Social networks are an important determinant of individual behavior and aggregate outcomes in a variety of social and economic phenomena. Their study has been a central theme in the sociology literature, leading to a large body of theoretical and survey-based empirical findings that document their role in different domains.[1] In contrast, economics was largely oblivious to the role of social networks until two decades ago, when the early research on strategic network formation developed.[2] Theoretical work on social networks has witnessed an exponential growth ever since: economists have analyzed the role of networks in the context of different games, as well as their importance in markets and a variety of areas in economics including labor, development, international trade, and finance.[3]

Empirical research on social networks has lagged significantly behind the theory, and we still lack empirical validation for the bulk of the theoretical findings accumulated in the last two decades. The core reason for this lag is the challenges involved in the causal identification of the impact of network structure on behavior. A first hurdle is that observational data on social networks are usually unavailable or incomplete, and it remains challenging to identify the impact of network structure by proxying or imputing the network. Even when complete cross-sectional data is available, most

[1] See Wasserman (1994) for a slightly outdated but comprehensive review.

[2] Jackson and Wolinsky (1996) and Bala and Goyal (2000) are the seminal papers on network formation. As is often the case, there have been pioneers whose work preceded the systematic study of social networks in economics including Myerson (1977), Kirman (1983), and Montgomery (1991).

[3] See Goyal (2007) and Jackson (2008) for comprehensive reviews.

social networks are constantly evolving, leading to severe endogeneity issues. Finally, longitudinal data sets that include full information on the network are very rare, but even in these cases identification remains problematic.[4]

The experimental methodology allows the causal identification of the network structure by controlling for cofounding factors such as preferences and information. It enables the researcher to exogenously impose a network of interactions among a group of subjects, and then vary it in a different treatment to isolate the effect of the structure of the network on individual behavior and group-level outcomes. Recently there has been a rapid growth of experimental research on social networks, which is proving to be an invaluable tool to validate existing theoretical findings. Moreover, experiments are revealing how individuals actually use network information. They are generating behavioral data that relates network structure to choices, which can serve as an input for novel theoretical developments. The purpose of this chapter is to survey the literature on laboratory and online experiments on networks, and identify important directions for future research.[5]

The next section focuses on experimental investigations of how network structure influences behavior in a variety of games, which abstract away from prices and market interactions. One channel for incorporating network structure into a game is by making players' payoffs depend on the neighbors in the network. Sections 17.2.1–17.2.3 focus on this channel and review experimental work on different types of games played on networks including coordination, Prisoner's Dilemma, and games with strategic substitutes and complements. A second channel is for the network to determine the information players have to make choices. Sections 17.2.4 and 17.2.5 focus on this channel by reviewing experimental work on strategic communication and learning in a network context. In reality, full information about the social network is rarely available to the researchers as well as the individuals embedded in the network. While the former is mainly a challenge for researchers using observational data, the latter is a primary concern for theorists who need guidance on what are the realistic assumptions to be made about a player's information regarding the network, in order to construct models that generate behaviorally valid predictions. Section 17.2.6 reviews experimental studies of games on networks that vary the information subjects have about the network, and it discusses how they can generate valuable input for further theoretical work.

Section 17.3 reviews a more recent strand of the literature which explores the role of networks in markets. In many decentralized markets there are constraints on who can trade with whom, and networks are a natural tool to capture these buyer-seller relations and/or intermediation services. Section 17.3.1 reviews experimental work in which the network determines the trading opportunities available to market participants.

[4] See Manski (2000) for a slightly outdated treatment, and Chapter 12 in this handbook by Boucher and Fortin.

[5] We exclude experiments that test strategic network formation models, which are reviewed in Kosfeld (2004). Chapter 16 of this handbook by Breza discusses field experiments.

A second function of social networks is to provide information in markets where it is not available through prices or some external mechanism. Section 17.3.2 reviews experimental work on how social networks can be used to circulate information about traders' reputations. This information allows market participants to punish cheaters in environments with weak or nonexistent formal institutions to enforce contracts, and incomplete information about past transactions, thereby determining who has an informational advantage in the market.

We conclude this chapter in Section 17.4 by taking a holistic view of the current landscape of research on networks in economics. We identify directions for further experimental research that we deem important for several contexts in which social networks matter to determine individual behavior and aggregate outcomes.

17.2 Games on Networks

This section discusses experimental research on games on networks. We organize it according to the different contexts in which network structure has an impact on behavior: coordination, provision of local public goods, cooperation, communication and information exchange, and social learning. The last part discusses evidence on how the information individuals have about the network matters to determine how network structure affects behavior in these contexts.

17.2.1 Coordination and Strategic Complementarities

A primary purpose of social connections is to help us coordinate our choices to generate mutual benefits. For instance, we would like to pick the same phone messaging app to be able to communicate for free with our friends or acquaintances and avoid charges from the provider. In other contexts, individuals care not only about choosing the same action but also on the intensity of the actions because there are strategic complementarities between an individual and her neighbors' actions. For example, the value a user obtains from visiting and contributing to an online review blog (e.g., Yelp, Trip Advisor) is increasing the amount of information provided by other users. These situations are described by games with strategic complements, and the function of the network is to determine these strategic complementarities among specific individuals.[6]

A canonical example of games with strategic complements is coordination games with Pareto-ranked equilibria. To fix ideas, consider the two-player two-action coordination game in Table 17.1. If $a > c$, $d > b$ then (a,a) and (d,d) are the pure-strategy Nash equilibria, and if $d > a$ then the latter is efficient or payoff dominant. However, if

[6] Chapter 5 of this handbook by Bramoullé and Kranton discusses the theoretical work on network games.

Table 17.1 Two-player coordination game.

	X	Y
X	(a, a)	(b, c)
Y	(c, b)	(d, d)

Assume $a > c$, $d > b$, $d > a$ and $(a - c)^2 > (d - b)^2$ throughout.

$(a-c)^2 > (d-b)^2$ then the (a,a) equilibrium is risk dominant à la Harsanyi and Selten (1988). The experimental literature has extensively studied this type of coordination game even without the consideration of networks.[7] A major finding is that coordination failure is a common phenomenon in the laboratory.[8]

The theoretical literature has expanded the scope of coordination problems by considering a variety of local interaction structures and their impact on coordination. Each individual in the population plays the baseline coordination game playing a single action with all his neighbors in the network. The individual's utility depends on his own action as well as on the actions played by the neighbors. Ellison (1993) and Morris (2000) prove that, in an evolutionary learning framework, players converge to the risk-dominant equilibrium in some stylized structures of local interaction such as circles and lattice.

A handful of experimental studies has investigated the role of local interaction structure in coordination problems. Keser et al. (1998) reports an experimental study on the effects of local interaction on coordination. They consider a finite repetition of a three-player game in which an individual plays the game in Table 17.1 with payoffs $a = 80$, $b = 60$, $c = 10$, and $d = 90$. A player has to use a common strategy with each of the other two players and gets the minimum of the payoffs that his strategy gains from the two plays. There are two treatments that vary the network structure: a complete network of three players and a circle network of eight players. In the latter network, the player plays the coordination game with his two neighbors, while he is neither informed of the global structure of the network nor the size of the network. Subjects in the complete network converge quickly to coordinate on the payoff-dominant (d,d) equilibrium, while those in the circle network coordinate on the risk-dominant (a,a) equilibrium.

Berninghaus et al. (2002) investigate a game with the same payoffs as Keser et al. (1998) but they extend the experimental setup by enriching the size and structure of

[7] See Ochs (1995) for a comprehensive survey. Seminal experimental studies are Van Huyck et al. (1990, 1991) using order-statistic coordination games and Cooper et al. (1990) using variants of coordination games.

[8] Devetag and Ortmann (2007) provide a critical review of the literature and identify the major determinants affecting the success/failure of coordination in the laboratory. For instance, payoff structure matters in the sense of how attractive the secure strategy is and how risky the other actions are. Also, the group size can affect the incidence of coordination failure and success.

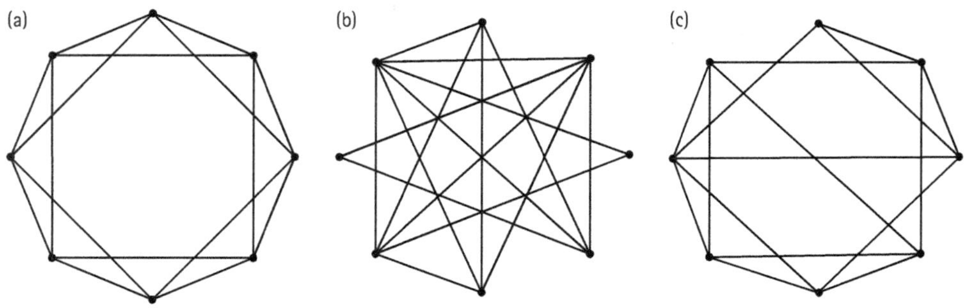

FIGURE 17.1 (a) Local, (b) random, and (c) small-world types of networks examined in Cassar (2007).

interaction and the payoff function. Specifically, they include local interaction in a lattice as well as in a circle with two neighbors and four neighbors. They also allow the payoff function to depend on the minimum as well as the average of the payoffs from local interactions. Berninghaus et al. (2002) find that while the payoff-dominant d action is an experimental regularity in the complete network, the risk-dominant action a is chosen more often in the circle network and the lattice. While the results of Keser et al. (1998) and Berninghaus et al. (2002) provide nice evidence on network effects on coordination, they keep constant the number of neighbors of each player but not the overall network size, making comparisons across network treatments problematic.

Cassar (2007) experimentally studies the coordination game with payoffs $a = 1$, $b = 4$, $c = -1$, and $d = 5$. She considers the three network structures in Figure 17.1 chosen to typify (a) local, (b) random, and (c) small-world types of networks. The networks keep constant relevant characteristics of the group such as the number of players and the total number of connections. She finds that 70% of subjects prefer the payoff-dominant d to the risk-dominant a action. Despite the high frequency of successful coordination, there are notable differences across networks. The small-world network achieves the highest level of coordination on the efficient outcome, while the random network has the lowest level of coordination. Also, the convergence rate of coordination differs across networks. Coordination on the payoff-dominant equilibrium occurs faster in the small-world network than in the random network.

The findings in Cassar (2007) on the relatively high incidence of the payoff-dominant d action seem at odds with those in Keser et al. (1998) and in Berninghaus et al. (2002). Frey et al. (2012) examine a coordination game with payoffs $a = 780$, $b = 660$, $c = 120$, and $d = 900$ on 10 different network structures of six nodes, including the complete, star, line, circle, and bipartite type of networks. In agreement with Cassar (2007) they find convergence to the payoff-dominant outcome in almost all cases. They also find no significant difference across network structures. The conflicting evidence on convergence to the payoff or the risk-dominant outcome across the studies may result from the difference in relative attractiveness/riskiness of the payoff-dominant action between the designs, and it can only be resolved with further experimentation.

A common drawback shared by the above studies is the focus on small networks of at most eight nodes, while in many contexts we are interested in coordination in large groups. Kearns et al. (2009) examine experimentally a simple coordination game in which groups of 36 subjects have one minute to coordinate either on blue or red, which are chosen to frame the experiment in the context of a presidential election in the United States.[9] The treatments were the assigned strength of preference for a color and the network structure. In all treatments subjects can only see their neighbors with their respective choices, but not the overall network. They find that coordination is more frequent in networks with a fat-tailed degree distribution compared to random networks. Moreover, a minority of well-connected subjects can make the whole network coordinate on their common preferred choice. In a follow-up study, Judd et al. (2010) examine a richer coordination problem: groups of 36 subjects have three minutes to coordinate on one color out of a set of nine choices. They run six network treatments which systematically vary the level of cliquishness (i.e., the extent of tight-knittedness of equally sized subgroups in the network). The intuitive result is that consensus is decreasing in the level of cliquishness.

Coordination games are a special case of the class of games with strategic complementarities in which a player's benefit from an action is increasing in the actions of her neighbors. In a seminal contribution, Ballester et al. (2006) analyze a game of strategic complements on any network. The model belongs to a class of games of strategic complements in which individuals have linear best replies. It offers a way of decomposing the payoff interdependence into a global and a local interaction component. The global interaction effect is uniform across all players and reflects a strategic substitutability in players' choices, whereas the local interaction effect varies across pairs of players and captures strategic complementarity in players' decisions. They show that there is a unique Nash equilibrium as long as the complementarity effects are lower than a threshold determined by the largest eigenvalue of the adjacency matrix of the network. The key result is the closed-form characterization of this equilibrium, which shows that a player's action is proportional to a (transformation of a) network centrality metric first defined by Bonacich (1987). In equilibrium a player's action increases in her connectedness, her neighbors' connectedness, her neighbors' neighbors connectedness, and so on. This result provides a tractable, rich, and intuitive relation between equilibrium play and network position.

Gallo and Yan (2015b) examine experimentally the Ballester et al. (2006) game of strategic complements on four different networks: circle and wheel networks with 9 nodes, a 15-node network with two completely connected clusters joined by a single node, and the 21-node network in Figure 17.2. A notable result of the experiment is that subjects depart from Nash equilibrium and coordinate on a more efficient outcome in the symmetric circle network, but on average they converge to the (inefficient) equilibrium predictions in the asymmetric networks. This is a demanding test of the

[9] Chapter 15 of this handbook by Aral also reviews this study as well as other large experiments with a particular focus on web-based experiments.

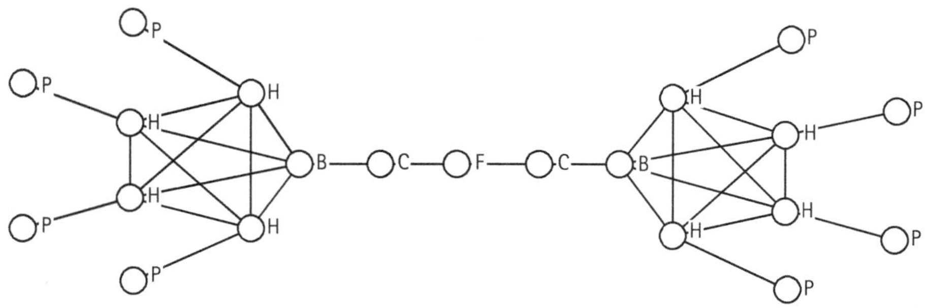

FIGURE 17.2 Twenty-one-node network in Gallo and Yan (2015b). Given the calibration in the experiment, the equilibrium predictions for subjects' efforts are $e(B) = 78$, $e(H) = 77$, $e(C) = 38$, $e(F) = 33$, and $e(P) = 33$.

equilibrium predictions in the Ballester et al. (2006) model, so the convergence on average is evidence of the behavioral validity of the theory. Looking more closely into subjects' behavior, Gallo and Yan (2015b) found that subjects appear not to incorporate the network interactions fully into their decisions and tend to base their choices on the local structure of the network as captured by the degree. For example, in the network in Figure 17.2, play according to Bonacich would predict that the ranking of subjects' efforts is $e(C) > e(F) = e(P)$, while play according to degree would predict $e(C) = e(F) > e(P)$. The results show that degree is a highly significant predictor of subjects' play on nodes F, P, and C while Bonacich centrality is not. The boundedly rational rule of only focusing on local network information may make sense in large, complex, and asymmetric networks. Whether human subjects adopt such a simple, boundedly rational rule of decision making in a complex network is an interesting question. A fuller investigation of this question will require one to use much larger networks than Gallo and Yan (2015b) used, with a careful selection of parameters of the games in order to drive a wedge between equilibrium choices fully reflecting the global structure of networks and choices made via such a bounded rational rule.

In all the studies reviewed in this section, the network structure is exogenously imposed by the experimenter. However, in reality coordination is facilitated by the fact that we actively choose our connections. Riedl et al. (2011) examine experimentally a weakest-link game played in two groups of 8 and 24 nodes, which we can think of as complete networks. They compare these treatments to an identical setup in which groups of 8 and 24 subjects can choose their connections endogenously. There is a clear difference in the dynamics of play irrespective of group size: subjects converge to the inefficient risk-dominant equilibrium in the complete network, but they overwhelmingly converge to the efficient payoff-dominant equilibrium when they can pick their connections. The resulting networks in the treatments with endogenous connections converge to the complete networks, but in the early rounds subjects can use the ability to exclude low contributors as a punishment mechanism to ensure convergence to the efficient equilibrium. This study shows that the inclusion of a network formation stage has a clear effect on behavior, and further experimental work in this vein is desirable both in coordination and other types of games as we will see in the next sections.

17.2.2 Public Goods and Strategic Substitutes

The problem of the provision of public goods has a central role in different areas of the social sciences (e.g., Ostrom 1990), including experimental research where there is a vast literature on experiments on public goods.[10] In the standard public goods game, subjects in a group have to decide how much of their initial endowment to contribute to a common pool. The sum of the contributions is scaled up by a common factor and distributed to all group members in equal shares independent of their contribution. A large number of studies have documented that subjects contribute significantly above Nash equilibrium in the one-shot version and the over-contribution pattern declines, but it does not converge to Nash when the public goods game is repeated.[11] The tendency to contribute above the Nash prediction is often attributed to social preferences and, in particular, a form of conditional cooperation in which the contribution to the public good is positively correlated with individuals' beliefs about other members' contributions. Experimental research has also investigated how subjects in the lab respond to a variety of environmental and institutional factors such as production technology, payoff structures, punishment, communication, and so forth.

Several types of public goods are local in the sense that an individual's contribution benefits others who are nearby either geographically or in a social network. Examples include neighbors who benefit from a well-kept garden, and non-excludable innovations which can be imitated by others who observe or learn about them. Bramoullé and Kranton (2007) analyze a public good game on a network in which links capture strategic substitutabilities between an agent's and her neighbors' actions. They show that there are a large number of equilibria, including specialized equilibria in which agents exert either no effort or provide the public good for all their neighborhood, distributed equilibria in which everyone exerts the same effort, and combinations of the two. Bramoullé and Kranton (2007) show that specialized equilibria are the only stable ones, and they relate the class of specialized equilibria to the graph theoretic concept of maximal independent sets, which allows one to check whether a specific action profile is a stable equilibrium just by checking if the set of specialists belongs to a maximal independent set of order 2.[12]

[10] See Ledyard (1995) for a comprehensive (although a bit outdated) review, and Chaudhuri (2011) for a recent survey.

[11] A series of early studies by Marwell and Ames (1981) show that contribution rates in the one-shot are in the 40%–60% range. See Fehr and Gächter (1999) for a study on a repeated public goods game.

[12] An *independent set* I of a network is a set of agents such that no two agents who belong to I are linked. An independent set is *maximal* when it is not a proper subset of any other independent set. A

Despite the nice correspondence between maximal independent sets and specialized equilibrium profiles, the size of the equilibrium set remains quite large. Experiments can therefore shed light on whether some of these equilibria are more salient and how saliency depends on the network structure. Rosenkranz and Weitzel (2012) examine experimentally the Bramoullé and Kranton (2007) local public good game on all six possible connected networks of four nodes. The first-order finding is that the frequency of equilibrium play is very low even on these simple and small networks, although it is higher than what one would expect with completely random play. Moreover, local coordination occurs 6–7 times more frequently than equilibrium play. Rosenkranz and Weitzel (2012) observe low frequency of equilibrium convergence in networks with stable specialized equilibria, but whenever convergence occurs it is almost always to stable specialized equilibria. An intriguing result is that equilibrium coordination varies with network structure in a nonmonotonic way, as it is highest in the complete and star networks, which are at opposite ends of the set of four node networks in terms of density and the spread of connectivity across nodes. Finally, subjects' actions are negatively correlated with their degree, which nicely mirrors the positive correlation found by Gallo and Yan (2015b) in the game of strategic complements on a network.

Galeotti and Goyal (2010) extend the Bramoullé and Kranton (2007) model by making the network endogenous. Agents play a network formation game à la Bala and Goyal (2000) as well as choosing a level of contribution to the local public good on the resulting network. The key result is that the introduction of a network formation stage drastically reduces the size of the equilibrium set: in every strict Nash equilibrium, the network has a core-periphery structure in which the agents at the core contribute while the agents at the periphery free ride. The introduction of a slight heterogeneity in the cost of providing the public good leads to a unique equilibrium, which is the star network with the central agent providing the public good and everyone else free riding. Moreover, the star network also maximizes welfare.

Goyal et al. (2014) test the predictions of the model for groups of four subjects. The baseline has homogeneous costs of providing the public good, and they compare this with a treatment in which one agent has a lower cost of providing the public good. The results provide only weak support for the theoretical prediction that the star network is the unique equilibrium. The low-cost individual is more likely to be a well-connected hub compared to the high-cost individuals, but the number of hubs is unchanged from the baseline, and the contribution to the public good of the low-cost individual is lower than predicted by the theory. Moreover, the resulting network has an average connectivity of about five links, which is significantly higher than the three links in the star network.

Further experimental work on public good games on networks is necessary to validate the theory and shed light on the features of network structure that determine equilibrium. A promising starting point is a paper by Bramoullé et al. (2014), which

maximal independent set of order r is a maximal independent set I such that any individual not in I is connected to at least r individuals in I.

provides a unified framework to analyze games of strategic complements and substitutes on networks, and nests the Bramoullé and Kranton (2007) and the symmetric case of the Ballester et al. (2006) models as special cases. The key result is the role of the lowest eigenvalue of the adjacency matrix of the network, which captures how much the network amplifies the direct effects of one individual's actions on his neighbors' actions. This result can inform the design of experiments to test behavior in network games with the presence of both strategic complements and substitutes, which would constitute a bridge between the work by Rosenkranz and Weitzel (2012) and Gallo and Yan (2015b). It can also inform the design of public good games on networks, which are larger than the four-node networks in Rosenkranz and Weitzel (2012) to explore the relation between structural features of the network and behavior in a richer setting.[13]

17.2.3 Cooperation

The emergence and sustenance of cooperative behavior is a defining characteristic of human societies. Social scientists and evolutionary biologists have extensively investigated the determinants of cooperation using the Prisoner's Dilemma game as an abstract representation of the trade-offs involved in an individual's decision to cooperate with or defect on others. In the simplest setting with two players interacting repeatedly, cooperation emerges if the players can condition their strategies on the other players' past behavior and the probability of another interaction is high enough.[14] An alternative mechanism for the emergence of cooperation is what is known as indirect reciprocity, which operates when there are repeated encounters within a group and there is a reputation mechanism that allows a player to know someone's past choices with a high probability.[15] Cooperation is costly but leads to the reputation of being a cooperative individual, and therefore may increase the chances of being the recipient of cooperative behavior.

An extensive number of experiments have tested these theoretical predictions. Murnighan and Roth (1983), among others, show that repetition in the Prisoner's Dilemma game between two individuals leads to cooperative behavior which increases the payoffs of the game as well as in the probability that the game will continue. Dal Bó and Fréchette (2011) validate these results and show that cooperation may prevail in infinitely repeated games, but the conditions under which this occurs are more stringent than the subgame perfect conditions usually considered, or even a condition based on risk dominance. In order to study indirect reciprocity, subjects play the game with randomly matched partners, and they are informed about the partner's choices in

[13] Suri and Watts (2011) report the results of a public good game played on networks of 24 nodes, but the choice of payoffs and the specific design they adopt makes it difficult to interpret their results in light of the Bramoullé et al. (2014) model.

[14] See, e.g., Fudenberg and Maskin (1986) and Binmore and Samuelson (1992).

[15] See Nowak and Sigmund (2005) for a review.

previous rounds. There is plenty of evidence that subjects condition their behavior on the partner's past choices, and thus individuals who have cooperated in the past tend to receive more cooperation.[16]

Different dimensions of social network structure play a role in the emergence of cooperative activity. As a first dimension, a crucial element for indirect reciprocity to emerge is the presence of a mechanism to share reputational information about third parties, and a natural channel to provide this information is communication through the social network. A second dimension is that in many instances cooperative activity is local, so the decision to cooperate or defect affects only an agent's neighbors in a geographical or social network rather than the whole society.

As the review in Chapter 6 of this handbook makes clear, there is no theoretical paper which provides a general characterization of a repeated Prisoner's Dilemma game on general network structures in the way that, for example, Ballester et al. (2006) and Bramoullé and Kranton (2007) do for one-shot games of strategic complements and substitutes, respectively. Haag and Lagunoff (2006) analyze the problem of a social planner who designs the optimal network for a group of individuals with heterogeneous discount factors. By restricting the attention to a specific type of trigger strategies, they show that greater social conflict may arise in more connected networks, and the optimal design exhibits a cooperative core and an uncooperative fringe when the individuals' discount factors are known to the planner. Several simulation-based studies[17] show that the structure of the social network has an effect on the level of cooperation, but they make specific behavioral assumptions on the agents' strategies, which would need validation in experimental data. Chapter 4 by Jackson identifies the study of the interplay between network structure and cooperation as one of the promising areas for future theoretical work, and we believe experiments can be an invaluable tool to provide guidance on the features of the network structure that are behaviorally relevant as well as the strategies that subjects use in this context.

The paper by Cassar (2007) we reviewed in Section 17.2.1 was one of the first experimental studies to examine the Prisoner's Dilemma game in Table 17.2 with payoffs $c = 5$, $a = 4$, $d = 1$, and $b = 0$ played on the three networks in Figure 17.1. Cooperation rates on all networks decrease to 20%–30% in the last rounds. There is some evidence that the cooperation rate is higher in the small world network compared to the local and random networks, and there is no difference between the local and random networks. Kirchkamp and Nagel (2007) report the results of a Prisoner's Dilemma game played on two different network structures of 18–20 subjects: regular networks of degree four and networks composed of two or three completely connected components. The first-order finding is that there is no effect of network structure on cooperation levels. Gracia-Lázaro et al. (2012) confirm that network structure has no impact on cooperation in a large-scale lab experiment with subjects playing a Prisoner's

[16] See, e.g., Wedekind and Milinski (2000), Milinski et al. (2002), and Seinen and Schram (2006).

[17] See, e.g., Ohtsuki et al. (2006) and Taylor et al. (2007).

Table 17.2 Two-player Prisoner's Dilemma game.

	Cooperate	Defect
Cooperate	(a, a)	(b, c)
Defect	(c, b)	(d, d)

Assume $c > a > d \geq b$ throughout.

Dilemma game with payoffs $c = 10$, $a = 7$, and $b = d = 0$ on two networks of more than 600 subjects, each with a regular and fat-tailed degree distribution, respectively.[18]

In reality, individuals choose their partners and homophily is pervasive in social networks, so we would expect cooperators to be more likely to be connected to other cooperators, which may generate a relation between cooperation and the structural properties of the network that is largely absent in experiments on fixed networks. Rand et al. (2011) examine experimentally a Prisoner's Dilemma game on an endogenous network with payoffs $c = 100$, $a = 50$, $d = 0$, and $b = -50$. In each round there is a first stage in which subjects can form or sever links followed by a Prisoner's Dilemma game played on the resulting network. The main treatment variable is the rate at which subjects can update the network: a baseline with a fixed network, a random mixing condition, and "viscous" and "fluid" conditions in which 10% and 30% of links can potentially be updated in each round, respectively. They find that cooperation level in the fluid condition stays at about 60%, which is significantly higher than in any of the other conditions indicating that the ability to choose connections has a positive impact on the level of cooperation.[19] Jordan et al. (2013) show that this is because the possibility to form new connections with cooperative individuals encourages defectors to switch to cooperative behavior even if many of their neighbors are defecting.

An interesting question is whether there are some structural properties of the emerging networks that are associated with high cooperation. Unfortunately, most of the studies choose asymmetric payoffs, which lead to the emergence of overconnected networks because in absolute terms the losses of being connected to a defector are lower than the gains of being connected to a cooperator. An exception is Gallo and Yan (2015a), who examine a prisoner's dilemma game with symmetric payoffs $c = 5$, $a = 3$, $d = -3$, and $b = -5$ in a setting where subjects can form or sever links in the first stage of each round at no cost. They examine how variations in the information about

[18] A recent contribution by Rand et al. (2014) finds an effect of network structure on cooperation by comparing several regular networks. The difference with previous studies may be driven by the particular payoffs chosen, the regularity of the networks, or the specification of the game. Grujić et al. (2014) carry out a comparative analysis of three previous studies and conclude that differences in payoffs and the network structure do not influence the level of cooperation. This is an area that deserves further investigation.

[19] Wang et al. (2012) and Cuesta et al. (2015) confirm the validity of these findings in similar studies.

the network and information about past actions of other subjects affect the emergence of cooperative activity.[20] They validate the findings in Rand et al. (2011) in a similar condition, and they show that the rate of cooperative activity is positively associated with the density and the level of clustering in the network.

17.2.4 Communication and Information Exchange

Communication and information exchange is common in many instances of social interaction. People exchange messages and information in order to avoid miscoordination and efficiency loss whenever a coordination problem is present. Game theorists model pure communication as "cheap talk": players' messages have neither direct payoff implications nor are binding for actions (Crawford and Sobel 1982). While theorists recognize the fact that cheap talk can play no role in strategic interaction because uninformative "babbling" equilibria always exist, they also provide conditions under which communication via cheap talk can signal players' intentions or private information to others and thus can improve upon how to play an underlying game. However, the problem of multiple equilibria is prevalent in cheap talk games, and standard refinement arguments cannot help much in resolving the issue.

Experimental research on cheap talk models has shown that communication can be effective in guiding subjects' behavior and information in equilibrium selection.[21] As a precursor of the experimental literature of communication networks, Cooper et al. (1989, 1992) study the effect of (one-way vs. two-way) pre-play communication structure on coordination in several two-player coordination games. An overall finding is that one-way communication increases coordination in the Battle of the Sexes, whereas two-way communication is more effective in coordination on an efficient outcome in games with Pareto-ranked equilibria. The reason is that each of the communication structures plays a distinct role and has a differential impact in resolving strategic uncertainty.

While Cooper et al. (1989, 1992) are seminal in considering the effect of communication structure on coordination, their settings are limited in terms of the scope of the network structure. Choi and Lee (2014) extend them by considering a richer set of communication networks in a multi-player game. Specifically, they consider a four-player version of the Battle of the Sexes game in which the four players have a common interest to coordinate but each player has his own preferred outcome. Prior to playing the underlying game, the players engage in finite periods of pre-play communication. In addition to varying communication length, Choi and Lee (2014) investigate four networks of communication—the complete, star, kite, and line networks. The complete network represents a horizonal structure of communication in which all players communicate with each other, whereas the star network describes a vertical,

[20] See Section 17.2.6 for further details.

[21] For a survey, see Crawford (1998).

centralized structure of communication in which one player takes the advantage of collecting information and influencing the group-level communication. The other two networks can be interpreted as representing communication structures lying in between, with less concentration of communication power on a single player. Because of the diversity in network positions in the setup, the experiment can address the effect of communication structure on equity of coordination outcomes as well as efficiency. Choi and Lee (2014) report substantial variations in both efficiency and equity across networks. Given the length of communication, the likelihood of efficient outcomes is highest in the complete network and lowest in the line network. Asymmetric networks tend to generate asymmetric coordination outcomes in favor of those who are better connected. However, the length of communication has an important influence on coordination outcomes. While increasing the length of communication improves the chance of coordination, it also makes the coordination outcome more equitable in the networks that produce asymmetric coordination outcomes.

The studies by Kearns et al. (2009) and Judd et al. (2010) we reviewed in Section 17.2.1 can also be seen as having a cheap talk element prior to the play in a coordination game. In Kearns et al. (2009), subjects have one minute to change their choice between blue and red, but these choices can be reversed at no cost until the end of the minute, so they can be interpreted as cheap talk about their final choice. Similarly, in Judd et al. (2010) they have three minutes and a choice among nine possible colors. Subjects differ in their preferences for the consensus color and are only informed about the current choices of their immediate neighbors. The results in Kearns et al. (2009) show that in networks generated using a preferential attachment process it is easier for subjects to reach global consensus than in random networks, and that the global consensus was frequently the preferred option of well-connected individuals.

Choi et al. (2011) explore the potential role of information networks in equilibrium selection in a dynamic game of public good provision. Networks are also used in describing various forms of observation structures of the history of play in dynamic games. The presence of asymmetric information about the play history can be an obstacle in achieving coordination. However, asymmetric information structure can make a certain outcome salient, as similar insights emerged from communication networks, and thus can make it easier for subjects to overcome coordination failure. Motivated by this idea, Choi et al. (2011) consider a simple dynamic game with three players in which players make voluntary contributions to the provision of a threshold-level public good over a finite number of periods. Players' contributions are irreversible and not refundable. The authors examine the empty network in which none of the players are informed of others' previous actions, and the complete network in which all players have full access to the history of play. In addition, they investigate a series of incomplete networks describing different asymmetric structures of information. While standard equilibrium analysis provides little guidance due to the multiplicity of equilibria, the experiments reveal that the degree to which subjects coordinate on efficient outcomes varies across different networks. Patterns emerging from the experimental data are overall consistent with two strategic incentives: those whose actions are observed may have an incentive to make contributions in early periods (strategic commitment), and those who can observe others' behavior delay their decisions (strategic delay). Asymmetries in the structure of information networks make these strategies salient.[22]

Despite still being relatively small, the experimental literature on communication and information networks has already accumulated insightful evidence on the role of network structure in equilibrium selection and coordination outcomes. A first direction to explore in future research is the role of communication network structure in multi-player coordination games with Pareto-ranked equilibria. In such underlying games, Peski (2010) proposes the concept of ordinal generalized-risk dominance to generalize Harsanyi and Selten's (1988) risk dominance notion. An open question is whether there is a relation between the structure of the communication network and the selection of the efficient over the generalized-risk dominant equilibrium. A second direction is an experimental test of the predictions in Hagenbach and Koessler (2010) and Galeotti et al. (2013), who extend the Crawford and Sobel (1982) cheap talk model to a network setting. Some key predictions of these models rely on the assumption that individuals' decisions to communicate depend on how many other individuals the recipient listens to (i.e., her in-degree). This requires individuals to take into account the network structure beyond their neighborhood and make inferences based on this information, which may not hold experimentally, as Section 17.2.6 will elaborate on. The relation between the in-degree distribution of the equilibrium communication network and the ranking of equilibria in terms of their efficiency is another example of a theoretical prediction that warrants an experimental investigation.

17.2.5 Social Learning

In many social and economic situations individuals learn from others by observing their decisions and/or learning about their beliefs on an underlying unknown state of the world. Economists use the umbrella term social learning to describe this phenomenon. A general message from the economics literature on social learning is the emergence of cascades that lead everyone in the society to converge to the same behavior. There may be inefficient information aggregation and convergence to a suboptimal outcome despite the fact that individuals maximize their own utility given beliefs formed in a Bayesian fashion.

The classical social learning model, introduced by Banerjee (1992) and Bikhchandani et al. (1992), and extended by Smith and Sørensen (2000), analyzes a sequence of agents making successive, once-in-a-lifetime decisions under incomplete and asymmetric information. That is, agents are uncertain about the underlying decision-relevant

[22] Kiss et al. (2014) experimentally examine a coordination game on networks of three players (two subjects and a computer) to investigate the role of the observability of actions in generating bank runs. They find similar results to Choi et al. (2011) in a setup with imperfect and incomplete information.

event, and the information about it is shared asymmetrically among them. The typical conclusion is that, despite the asymmetry of information, eventually every agent imitates her predecessor, even though she would have chosen a different action on the basis of her own information alone. In this sense, agents rationally "ignore" their own information and "follow the herd." Furthermore, since actions aggregate information poorly, herds often adopt an action that is suboptimal relative to the total information available to agents. This is an important result that helps us understand the basis for (possibly inefficient) uniformity of social behavior. Following Anderson and Holt (1997), a number of papers[23] investigate social learning experimentally and demonstrate that herd behavior can be replicated in the laboratory.

In practice, individuals are located in complex social networks and learn mainly from observing the decisions of their neighbors and/or learning their beliefs about the underlying state of the world. The classical model of social learning can be seen as the very special case of a directed line network, in which information flows and/or observations about others' decisions only happens once for each agent and in one direction from the beginning to the end of the line. The theoretical literature has explored the impact of social network structure on two different types of social learning: observational learning in which a link between two individuals represents their ability to observe each other's actions, and communication learning in which a link between two individuals indicates that they can (truthfully) share their beliefs about the underlying state of the world.[24]

Bala and Goyal (1998) and Gale and Kariv (2003) are the first theoretical models of observational social learning on networks. The key methodological difference in their approach is whether agents are fully Bayesian or there are exogenously imposed limitations in the agents' ability to make Bayesian inference on the network. Bala and Goyal (1998) assume a boundedly rational form of Bayesian updating in which agents only take into account actions and outcomes of neighbors' actions, and ignore any information that may be inferred by the sequence of neighbors' actions. Instead, Gale and Kariv (2003) investigate a fully Bayesian setup in which agents are able to make inferences about non-neighbors' actions from their observation of neighbors' actions and their knowledge of the overall social network.[25] A general result is the convergence to an equilibrium in which all agents play the same action. Moreover, this action is the optimal action as long as one imposes some restrictions on the network structure.[26]

Choi et al. (2005, 2012) and Choi (2012) have undertaken an experimental investigation of learning in three-person, directed networks and focus on using the theoretical framework of Gale and Kariv (2003) to interpret the data generated by the experiments.

[23] Selected contributions include Hung and Plott (2001), Kübler and Weizsäcker (2004), Çelen and Kariv (2004), Goeree et al. (2007), and Weizsäcker (2010).

[24] Chapter 19 by Golub and Sadler reviews learning in networks, and Chapter 16 by Breza surveys studies of social learning using field data.

[25] Other more recent contributions in this vein include Acemoglu et al. (2011) and Mueller-Frank (2013).

[26] Mueller-Frank (2014) shows that convergence may also fail if there is one fully Bayesian agent in a society of non-Bayesian agents due to the ability of the fully Bayesian agent to influence the consensus.

The experiment design utilizes three networks—the complete, the star, and the circle network—along with variations in the structure of private information about the unknown state of the world. In each period, players simultaneously choose which state is more likely to have occurred at the beginning. This guess is made on the basis of the individual's private signal and the history of play of their neighbors. As the game continues, the inference problem becomes more demanding because it requires a player to form higher order beliefs. Since noise in experimental data is inevitable, Choi et al. (2012) extend the Bayesian model to allow for the possibility of subjects making mistakes. This was done by adopting the model of Quantal Response Equilibrium of McKelvey and Palfrey (1995, 1998). While the Bayesian model overall performs well, there are instances of networks and information structures in which the Bayesian model has limitations in interpreting the data. Also, the heterogeneity of individual behavior in the data is hardly ignorable.

Choi (2012) develops a method for estimating a mixture model of heterogeneous rules of learning in networks. His approach is based on the observation that the sequence of tasks of learning constitutes a "cognitive hierarchy," which in turn suggests a natural hierarchy of cognitive types. Each cognitive type corresponds to the number of periods in which a player processes new information: starting from the lowest type who randomly guesses the state of nature, the next lowest type would only process his private signal but make no use of information obtained from the observations of his neighbors' decisions; the next lowest type would process his signal in period 1 and make an inference about his neighbors' signals from their decisions in the first period, but could not make any higher-order inferences from then on, and so on. The estimation results show that this structural approach does a very good job of interpreting subjects' behavior and accommodating the heterogeneity of individual behavior in the data.

In contrast to the observational learning literature, the prevalent approach in theoretical work on communication learning on networks has been the assumption of boundedly rational learning. The most widely used rule was first proposed by DeGroot (1974): each agent updates her beliefs by taking a weighted average of her neighbors' beliefs with the weight determined by the strength of the link in the communication network. DeMarzo et al. (2003) formulate a model in which agents receive signals at time 0, they truthfully communicate their belief to their neighbors at each time period, and they update their beliefs using DeGroot's (1974) rule. They show that in the long run all agents would converge to the same belief about the underlying state of the world, and the influence of each agent in determining the limit belief is tied to the agent's position in the communication network. This means that there will not be convergence to an unbiased aggregation of the initial signals except for the very special case in which the informativeness of each initial signal is exactly aligned with the influence of the recipient in the network.[27] Acemoglu et al. (2014) analyze a Bayesian communication

[27] Other more recent contributions using the DeGroot (1974) rule include Golub and Jackson (2010), Acemoglu et al. (2010), and Gallo (2014b).

learning model by assuming that agents can tag information, and they show that the presence of "information hubs" is a sufficient condition for asymptotic learning.

The predictive power of models based on the DeGroot (1974) setup hinges on the specific assumption of bounded rationality in the updating rule, which is ultimately an issue that can only be resolved empirically. The experimental method can be particularly helpful in shedding light on this question as it would be very challenging to identify the updating rule in observational data. Corazzini et al. (2012) examine experimentally how individuals learn in two networks of four nodes: a circle with directed links arranged in a clockwise pattern so that each individual has one incoming and one outgoing link, and a hub-type network obtained from the circle network by adding two links so that the choices of one subject are observed by all the others. In the first round each subject receives an integer signal drawn from a commonly known distribution, and in each one of 12 rounds she has to guess the mean of the four signals after learning her neighbors' guesses in the previous round. The predicted outcomes for Bayesian and DeGroot-type updating are the same in the circle, but they differ in the hub network: Bayesian updating gives each subject's signal the same weight, while DeGroot-type gives a clear ranking in the importance of signals depending on the network position of the subject who received the signal. The results clearly show that in the hub network subjects give different weights to signals in broad agreement with the predictions of the DeGroot dynamics. The authors also propose a generalized updating rule in which individuals give weight to individuals who are listened to, as well as listen to many others, which nests DeGroot as a special case, and they show that it gives a good fit to the data.

A drawback of Corazzini et al.'s setup (2012) is that they only investigate one network in which there is a difference between the outcomes of the Bayesian and DeGroot learning, making it difficult to generalize their findings. In a recent working paper, Grimm and Mengel (2014) report an experimental study testing the predictive power of Bayesian and DeGroot-type learning in five different networks with seven nodes. They find that subjects make decisions consistent with DeGroot-type updating in 80%–98% of the cases in which the predictions of the two models differ for specific positions in the network. However, the dynamics of convergence to a limit belief suggests that subjects may be using rules-of-thumb that are more sophisticated than simple DeGroot, and the authors propose an alternative non-Bayesian model of learning that extends the DeGroot model by allowing individuals to adjust the weight placed on their previous behavior according to their clustering coefficient, which captures the proportion of an individual's neighbors who are connected to each other. This adjusted model of non-Bayesian learning appears to perform better than the DeGroot model.[28]

[28] Using a similar setup, Mueller-Frank and Neri (2013) investigate networks of five and seven nodes. They find that individuals' decisions do not satisfy three properties which are required by a class of non-Bayesian updating rules in order to achieve consensus, which may explain the lack of convergence to consensus in their experiment.

A major theme in the literature of learning in social networks is understanding which model of updating best describes individuals' decisions, and, consequently, group outcomes. Bayesian updating is a natural benchmark case, but it has the drawback of not being very tractable, and it requires individuals to exercise increasingly demanding inferences from the observation of neighbors' behavior. DeGroot-type updating provides sharp predictions, but it makes ad hoc assumptions on the specific type of bounded rationality that individuals have when they process information. Further experimental research is required to identify the type of bounded rationality, which would be invaluable input for further theoretical work. A first step forward would be to identify which dimensions of information about the network the participants use in their updating, which is a topic we will discuss further in Sections 17.2.6 and 17.4. A second step would be to investigate how this updating varies with the size and complexity of the network, as the largest network explored so far has only seven individuals. Finally, there are econometric issues that require careful consideration: subjects in an experiment tend to make mistakes and display significant individual-level heterogeneity of learning behavior, which makes a clean identification strategy more challenging.

17.2.6 Incomplete Information about the Network

A common assumption of many theoretical and experimental studies that we have considered so far has been that individuals have complete information about the network structure. This is rarely the case when we consider applications, as individuals would usually have access to information about local features of the network (e.g., their degree) and aggregate statistics about the overall network structure, but no detailed information on the exact pattern of who is connected to whom. Even if the complete information about the network is available, a number of studies in social psychology show that the process of memorizing and recalling information about real social networks is affected by several biases,[29] some of which have been confirmed in an experimental setting.[30] These biases may influence how individuals make decisions in network games, especially in contexts in which equilibrium play requires knowledge of the network beyond the immediate neighborhood as in Bayesian learning in networks.

Galeotti et al. (2010) explore the role of incomplete information about the network in the context of games of strategic complements and substitutes, which we have reviewed in Sections 17.2.1 and 17.2.2. In their setup an agent knows her degree and the degree distribution of the whole network, but she does not have information on any other characteristic of the network including the identity of her neighbors. This is a rather severe form of incomplete information about the network, and an interpretation is that it applies to contexts in which an agent makes a decision before the specific identity of the

[29] Examples include Krackhardt (1987, 1990), and Kumbasar et al. (1994).

[30] The only two experimental studies we are aware of in network cognition are Janicik and Larrick (2005) and Dessi et al. (2016).

neighbors is realized. Their model defines a game of incomplete information in which a player's type is her degree, and it nests the incomplete information versions of the Ballester et al. (2006) and Bramoullé and Kranton (2007) setups. Recall from Sections 17.2.1 and 17.2.2 that in the complete information setup the game with strategic complements has a unique equilibrium in which an agent's play depends on her Bonacich centrality, while the game with strategic substitutes has a multiplicity of equilibria. Galeotti et al. (2010) show that the introduction of incomplete information allows us to prove the existence of monotone equilibria: actions are non-increasing (nondecreasing) in players' degrees under strategic substitutes (complements). Moreover, these are the unique symmetric equilibria if one puts some restrictions on the payoffs. This result is intuitive for the game of pure strategic complements, but in the case of strategic substitutes it reduces the equilibrium multiplicity present in the game with complete information, thereby significantly increasing the predictive power of the model.

Charness et al. (2014) test the predictions of the Galeotti et al. (2010) model in a series of experiments on a variety of networks with 5 nodes and a small set of networks with 20 nodes. Aside from the network structure, the two treatment variables are whether it is a game of strategic substitutes or complements, and the presence of complete or incomplete information about the network. They restrict their attention to active/inactive binary strategies, which implies that one of the binary actions leads to a secure outcome because a player receives a fixed payoff by choosing this action, regardless of her degree and the neighbors' decisions. In the incomplete information treatments with the small networks, subjects' play is in agreement with the predictions in Galeotti et al. (2010): subjects use threshold strategies and the frequency of active players is monotonically increasing (decreasing) with connectivity for the case of complements (substitutes). Whenever incomplete information induces a unique equilibrium, subjects almost always make choices that are consistent with the equilibrium. In the context of strategic complements, Charness et al. (2014) find that when there are multiple equilibria, network properties are predictive of subjects' behavior and thus serve as an equilibrium selection tool. Specifically, connectivity and clustering influence the likelihood of activity: high connectivity and more clustering tend to increase coordination on the efficient equilibrium rather than the secure but less efficient one. They also find evidence that the introduction of uncertainty drives play to the most secure equilibrium.

The experiment by Charness et al. (2014) is a very good illustration of how the comparison of treatments with complete and incomplete information about the network can help in understanding the role of uncertainty about the network as well as shed light on other experimental results in the complete information setup. In the context of strategic complements, the high frequency of equilibrium play when there is a unique equilibrium in the complete information setting is consistent with the results in Gallo and Yan (2015b), who find convergence on average to the equilibrium play on large networks when subjects have a large, non-binary set of actions at their disposal. The introduction of incomplete information about the network does not significantly alter the finding. In the context of strategic substitutes, the introduction of incomplete information helps to reduce the strategy space and acts as an equilibrium selection device both theoretically and experimentally: Charness et al. (2014) find high convergence to equilibria, in contrast to the results in Rosenkranz and Weitzel (2012), who find low frequency of equilibrium play. An important caveat is that subjects in Rosenkranz and Weitzel (2012) have a large set of actions to select from, so an open question is whether the results in Charness et al. (2014) hold in a setting with non-binary actions.

Gallo and Yan (2015a) examine the role of incomplete information about the network in the context of the Prisoner's Dilemma game on an endogenous network, which we reviewed in Section 17.2.3. In each round of a repeated game, subjects first form costless links with other subjects and then play a Prisoner's Dilemma game on the resulting network. The authors vary the information that subjects have about the network as well as the information about others' previous actions. In the baseline, subjects only know the identity and previous five actions of their neighbors. The network information treatment adds information on the full network to the baseline, the reputation treatment adds information on the previous five actions by everyone to the baseline, and the final treatment has full information on the network and others' previous five actions. Mouse-movement tracking data shows that subjects make active use of the network information, but the availability of full information about the network has no effect on the aggregate level of cooperation, which is solely driven by the availability of information on everyone's previous five actions. The availability of information about the network, in addition to information on everyone's actions, affects the distribution of cooperative activity: it allows cooperators to find each other and form their own separate community by excluding defectors to a separate community using the information about the network in the network formation process. Being part of the community of cooperators is highly beneficial, as it allows a subject in the cooperative community to earn a payoff per round that is 23% higher than if she were in a community of defectors of equal size. These results also show that the choice made in other experimental studies of the Prisoner's Dilemma game on an endogenous network to only give subjects information about neighbors, rather than the whole network, is not without consequence.

Experimental designs that vary the information about the network available to subjects can also be useful to differentiate between competing models. The experiment by Grimm and Mengel (2014), which we already described in Section 17.2.5, also varies information about network structure by allowing subjects to know only their own degrees in the network, or the degree distribution of the network as well as their own degree, or the complete structure of the network. The key insight is that fully Bayesian updating is responsive to the differential information about the network structure, but DeGroot-type updating is not because it disregards the knowledge about the network structure beyond the neighbors in the belief updating process. As we have seen in Section 17.2.5, subjects' decisions are very consistent with DeGroot updating. However, subjects make more correct guesses in some networks when they have more information about the network structure, which cannot be explained by

DeGroot updating. Grimm and Mengel (2014) show that if subjects have complete information about the network then the weight they place on their belief is increasing in their clustering coefficient, which captures the extent to which their neighbors are connected with each other. In other words, subjects take into account correlations in neighbors' beliefs in a rudimentary way rather than ignoring them as assumed by DeGroot updating.

The variation of the information about the network available to participants reveals novel insights about the network games reviewed in Sections 17.2.1, 17.2.2, 17.2.3, and 17.2.5. It sheds light on a range of issues, including how equilibrium selection depends on the network, how agents update their beliefs using network information, and how decisions are distributed in the population. A fertile avenue for further research would be a systematic examination of what information about the network individuals make use of and how it matters in their decisions: the studies we have reviewed vary the network information in an ad hoc fashion which is not grounded in evidence of how individuals memorize, recall, and use this type of information. We will explore further this theme in Section 17.4.

17.3 Markets and Networks

This section discusses existing experimental research on markets and networks. We organize it by two distinct strands of the literature. In the first strand, networks are used as a tool of representing the trading relation among market participants. The second strand reviews studies that investigate the impacts of communication and information networks on trading behavior and market outcomes.

17.3.1 Trading Frictions

The Walrasian theory of market equilibrium is a cornerstone of economics in understanding markets. It postulates that trade takes place on a centralized exchange mediated by a fictitious auctioneer. Competitive equilibrium in this frictionless economy has been a significant basis of understanding the workings of markets and economists' advice of public policy. Experimental research has also deepened our understanding on markets by investigating the properties of market institutions in a controlled environment. Starting from Chamberlin (1948) and Smith (1962, 1965), a large literature of market experiments has accumulated evidence that certain institutions in laboratory markets have remarkable properties of approximating an efficient allocation, predicted by the Walrasian theory, even with a small number of subjects.[31] One prominent such institution is the continuous double auction with a centralized process of trading.

[31] See Sunder (1992) for a slightly outdated but comprehensive survey.

In practice, there are many markets in which exchange is organized by decentralized trade and intermediation. In those environments, networks are a natural tool to represent the trading relationships among market participants. When the network is complete, every possible trading opportunity is present, and therefore there is no constraint on trading patterns. On the other hand, the incompleteness of the network signifies that some traders are unable to trade with each other. It implies either the pure loss of trading opportunities or the fact that an intermediation service is required for trading. When intermediation is costly, the incompleteness of the network becomes a source of trading frictions and a cause of inefficient allocation.

A number of theoretical studies use networks to understand the effects of network structure on market outcomes in a variety of situations, including two-sided networked markets with bargaining (e.g., Kranton and Minehart, 2003 and Corominas-Bosch, 2004), financial contagion (e.g., Allen and Gale, 2000), and intermediated trade (e.g., Condorelli and Galeotti, 2012).[32] A general takeaway from this body of work is that networks are a significant determinant of market efficiency and the division of trading surplus. Nevertheless, theory alone has limited predictive power, and it is not very informative for policy due to the complexities of networks and the multiplicity of equilibria. Experimental research can complement these theoretical advances by shedding some light on equilibrium selection and the behavioral rules individuals adopt when facing the complexities of networks.

A first branch of the experimental literature examines two-sided networked markets. Charness et al. (2007) is an experimental test of the model by Corominas-Bosch (2004). The market is described by a bipartite network of buyers and sellers representing the limited set of trading opportunities, and by a protocol of sequential alternating bargaining over a shrinking value of a homogeneous and indivisible good. Corominas-Bosch provides a theoretical method of decomposing any network of buyers and sellers into relatively simple subgraphs, plus some extra links. A nice feature of the decomposition result is that any network is decomposed into a union of smaller networks, each one either a complete network in which the short side of the market induced by that network receives all the surplus, or an even network in which traders split the surplus nearly evenly. Charness et al. (2007) employ two separate simple networks–a three-person network, which is competitive, and a four-person network, which is even—and combinations of these two resulting in a variety of seven-person networks. They observe such a high degree of bargaining efficiency that 75% of the possible agreements are reached in the first round and the total payoffs received are 96% of the maximum attainable. The decomposition result predicts stark difference in bargaining outcomes depending on how a link is added between two simple networks. The experimental data qualitatively validate the theoretical predictions.

[32] Applications of networked markets are presented in Chapter 20 by Cabrales, Gale, and Gottardi (for financial contagion), and Chapter 27 of the handbook by Condorelli and Galeotti discusses the theoretical literature on strategic intermediation in networks. Chapter 26 of the handbook by Manea discusses the theoretical literature on buyer and seller networks.

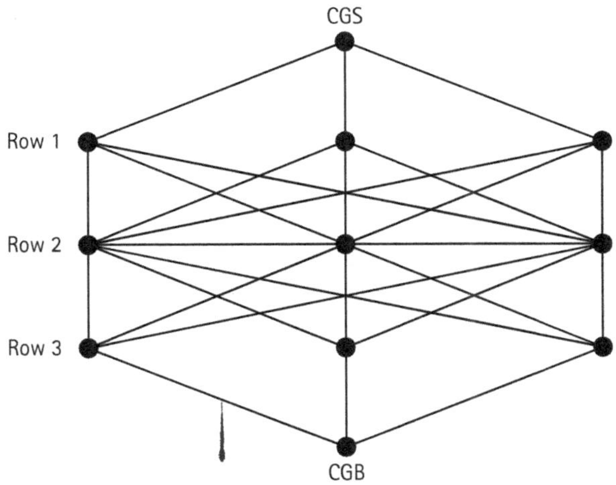

FIGURE 17.3 An example of a network in Gale and Kariv (2009).

Judd and Kearns (2008) also study experimentally bipartite exchange in large networked markets. The experiment examines a range of 36-person bipartite networks including regular and random networks as well as networks generated using a preferential attachment process, which means that the structure varies in terms of aggregate network properties, such as the degree distribution. The main focus of the experiment is testing the predictions on the mapping of structural asymmetries in network topology into pricing behavior and efficient outcomes. They find that the level of efficiency is quite high across all network treatments and those with more links (and with more trading opportunities) obtain higher benefits from trading. Nevertheless, there is evidence that subjects are inequity averse and try to equate the gains from trading, despite the fact that asymmetries introduced by the network ultimately lead to unequal distributions of gains from trading.

A second branch of the experimental literature explores the impact of networked intermediation on efficiency and surplus division. Gale and Kariv (2009) study a simultaneous bid-ask model of trading in networks. A buyer and a seller need to trade a commodity or asset through a set of intermediaries. Traders are located on a rectangular network consisting of rows and columns of intermediaries. Figure 17.3 shows an example with three columns and three rows of intermediaries connecting the seller (CGS) at the top with the buyer (CGB) at the bottom. Trades are restricted to adjacent rows and links represent potential trading opportunities. Each intermediary simultaneously chooses a bid (the price at which he is willing to buy the asset) and an ask (the price at which he is willing to sell the asset). Each member of traders in a given row can trade with every member of traders in an adjacent row with whom he has a link. The variations of trading networks in the design of Gale and Kariv (2009) feature essentially Bertrand competition among horizontally positioned traders. Thus, from a given network, adding rows increases the amount of intermediation required to capture

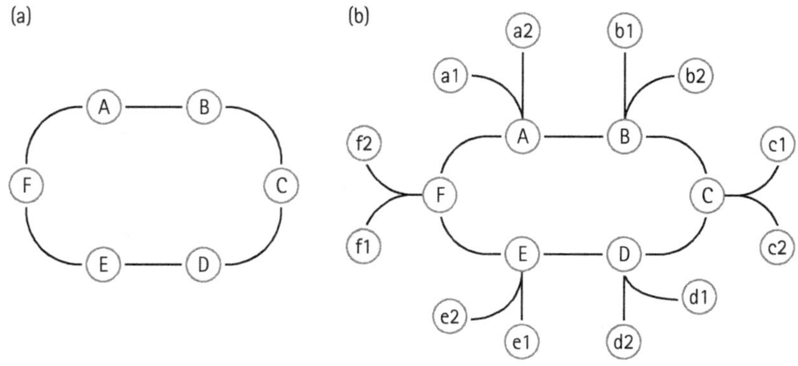

FIGURE 17.4 (a) Ring 6 and (b) ring with hubs and spokes in Choi et al. (2014).

the surplus available, whereas adding columns increases the amount of competition. Due to Bertrand competition, in an efficient equilibrium the asset's transaction price is equal to its value after traversing the first row. Gale and Kariv (2009) report that the level of efficiency is very high and that the pricing behavior observed in the experiment converges to competitive equilibrium behavior in a variety of treatments. However, the rate of convergence varies depending on networks and other parameters of the design.

Choi et al. (2014) propose a static model of posted prices in networks and test its empirical relevance in the laboratory. In their model, there are a set of intermediaries lying between a buyer and a seller. The passage of a commodity from the seller to the buyer generates value. Intermediaries simultaneously set a price to get a share of this value. The model deals with both a trading situation of complete information where intermediaries know the value of exchange, and a situation of incomplete information where intermediaries choose a price prior to knowing the value of exchange. Trading occurs through a least cost path and an intermediary earns payoffs only if he is located on it. Choi et al. (2014) offers a complete characterization of Nash equilibria under both information cases. Theory allows both efficient and inefficient equilibria and predicts that node criticality[33] is a necessary condition for the extraction of intermediation rents. Due to the multiplicity of equilibria, theory alone cannot make sharp predictions on efficiency and surplus division.

In the experimental part, Choi et al. (2014) examine several networks which vary in size and in the absence/presence of critical nodes. Figure 17.4 shows two networks with or without critical nodes: (a) ring 6 network and (b) ring with hubs and spokes. They also investigate variation of information on the value of exchange. The experimental data report a remarkably high level of efficiency across all networks in the benchmark model of complete information, in favor of an efficient equilibrium against an inefficient one. For instance, the efficient outcome occurs with probability 1 in the ring 6 network and with probability 0.95 in the ring with hubs and spokes. With regard to surplus

[33] A node is critical if it lies on all paths between the buyer and the seller.

division, the experimental results show that critical intermediaries set high prices and extract most of the surplus. As a result, intermediation costs are small in the ring 6 network (less than 15%) and are quite high in the ring with hubs and spokes (60% to over 95%). Thus, the model and the experiment taken together establish that the presence of critical intermediaries is both necessary and sufficient for large surplus extraction by intermediaries and that most of the intermediation rents accrue to critical intermediaries.

Experimental research on networked markets is an exciting research area. In experimental markets, one can control traders' preferences, technology, and private information, as well as network structure. It is practically impossible to achieve this level of control in observational market data. Because of such a methodological advantage, experiments on trading in networks can address issues that are hard to test using real market data.

17.3.2 Information Flows

Information plays a key role in the well-functioning of markets. As we have already seen in Sections 17.2.4 and 17.2.5, social networks are a channel for information to flow among individuals, and therefore the structural features of the communication network may be related to the outcomes that we observe in the market. Furthermore, the social network will create heterogeneities across individuals depending on their position in the network, which may result in some of them having an informational advantage. Here we focus on two functions of communication networks in markets. The first function is to monitor other market participants in a market environment in which contracts are not perfectly enforceable, and therefore information about other individuals' conduct is critical to ensure that cheaters are punished. The second function is to provide information about the value of goods in markets where this information is not common knowledge through publicly displayed prices, but is only shared privately by the participants in a market transaction.

The first function of communication networks in markets has received significant attention in the economic history and development literatures to explain the existence of active trading markets in contexts where there are no formal institutions to enforce contracts. For instance, Greif (1993) provides historical evidence that monitoring through communication networks allowed the Maghribis to become the main traders in the Mediterranean in the thirteenth century.

Cassar et al. (2010) report the results of an experiment to examine the role of information networks in trading behavior in a multi-market situation where contracts are not perfectly enforceable. The market institution is a continuous double auction in which buyers and sellers are randomly assigned and their values and costs are heterogeneous. There are two markets running simultaneously: a "local" market where contracts are strictly enforced, and a "distant" market where cheating is possible with a seller delivering a lower-quality good and a buyer paying less than promised. In addition to the structure of the two markets, traders are fully connected with a subset of traders in the distant market via a clique network, which enables them to observe and thus monitor the past play of their network members' trading including all bids, asks, and transactions made by them. Thus, traders know whether and which members of their network cheated, and can build up their reputation within their network. The clique network further varies with regard to the composition of values and costs to create networks with potentially high trading surplus and networks with low trading surplus. The baseline treatment has no network, so all trades in the distant market are anonymized. The results show that the presence of information networks significantly reduces cheating and increases efficiency, and that, due to the facilitation of monitoring within a network, networks lure high surplus traders out of the local market and into the distant market.

A second function of communication networks is to provide market information to traders in contexts where there is incomplete information because there are no publicly available prices and information about the value of goods is only circulated within social networks. For instance, Rauch and Trindade (2002) show that Chinese immigrant networks significantly increase international trade volumes, and this only happens for commodities whose prices are not publicly available, providing strong evidence that belonging to the network gives them an informational advantage. Gallo (2014a) extends the model in Young (1993) to capture this function of social networks in markets. In a decentralized market one buyer and one seller are randomly matched to play a Nash demand game in each time period, and before playing the game they receive information about past transactions through their social network. The process converges to a unique equilibrium where each buyer (seller) gets the same and the split between buyers and sellers depends on the degree of the least connected individual(s) in each network: the lower the degree of the least connected buyer (seller), the lower the share going to every buyer (seller). The testable predictions are that groups with high density and/or low variability in the number of connections across individuals allow their members to obtain a better deal.

Gallo (2014a) also reports the result of an experiment testing the predictions of the model. He examines four six-person networks of buyers which vary in density and distribution of connectivity: a regular network of degree 4, the circle, the star, and a 4-node circle network with two spokes. Subjects are assigned to a specific position in a network, which is unchanged for all the 50 rounds of the experiment, and they are told they are traders in a market and they will be trading with a seller played by a computer. At the beginning of a trading round, a subject receives a sample of information about the demands made by the seller in past transactions with the other subjects she is connected to. This information is randomly sampled by the computer from the history of play, and it is the only information a subject has prior to making a demand. The results of the experiment lend support to the theoretical predictions. Subjects in the regular network of degree 4, which has the highest density, converge to a significantly higher demand than subjects in other networks. Subjects in the star and circle with spokes networks, which are the only ones with a least connected node of degree 1, are undistinguishable and converge to a lower demand than the other two networks.

17.4 Future Directions

The previous sections have reviewed the main work in the literature on network experiments and identified open questions within specific topics that would benefit from further research. In this section we take a more holistic view of the current landscape of research on networks in economics, and identify directions for further experimental research that are important for several areas where networks matter.

The nature of theoretical modeling in the network literature varies significantly depending on the size of the network. At one end of the spectrum there are models describing phenomena in small networks of a few nodes: the standard game-theoretic approach applies well here as strategic considerations are paramount and the small set of players makes most problems tractable. At the other end of the spectrum, there are models describing phenomena on large networks: the prevailing approach is to use different types of stochastic processes with no strategic element or a boundedly rational approach based on heuristics. Theoretical models for the intermediate size case adopt a mix of the game-theoretic and stochastic approaches, and this is arguably the area where network structure has the most interesting effects and the literature is less developed. The social learning models we reviewed in Section 17.2.5 provide a good illustration of this spectrum with fully Bayesian and DeGroot-type models being particularly relevant for describing behavior on small and large networks, respectively, and a truly hybrid model between the two arguably still missing.

Up to now the experimental literature in economics has largely focused on small networks of at most a dozen nodes. This is a limitation to the general validity of the findings as some of the few experiments have compared intermediate and small networks show interesting evidence of the importance of network size.[34] A practical reason to focus on small networks is to keep session sizes manageable as well as the fact that if the network is the unit of analysis then the number of independent data points is divided by the network size, which means that large network experiments would require a large subject pool. However, these practical considerations have been overcome by several researchers outside of economics,[35] and a systematic study of how network structure affects behavior in intermediate and large sized networks is important to enrich our understanding of their impact on behavior.

[34] Examples include Choi et al. (2014) and Gallo and Yan (2015b).

[35] Among the studies covered in this review, Judd and Kearns (2008); Judd et al. (2010), and Kearns et al. (2009) conduct experiments on networks of 30–50 individuals, and Gracia-Lázaro et al. (2012) has networks of more than 100 nodes.

A related direction for future experimental research is improving our understanding of how individuals learn, memorize and recall information about the network, and what heuristics and potential consequent biases are involved in this process. An extensive literature in cognitive psychology has documented how individuals use heuristics to handle demanding cognitive tasks and how these heuristics may lead to systematic biases.[36] This is particularly relevant for networks of intermediate and large size where the complexity and sheer number of potential network architectures mean that individuals cannot possibly have complete information about the network they are embedded in. Dessi et al. (2016) provide some evidence that individuals tend to underestimate the mean degree and overestimate (underestimate) the number of rare (frequent) degrees in a 15-node network using a graphical methodology to generate the network in the lab, and show that these biases are also present in two real networks mapped through surveys. However, the cognitive processes we use to memorize and recall network information and the resulting biases are still largely unexplored. As we discussed in Section 17.2.6, the introduction of incomplete information about the network in theoretical models can provide novel insights, and experimental evidence on how to model incomplete information would be very valuable to avoid ad hoc assumptions and provide input to improve the behavioral validity and predictive power of the theory.

A prominent dimension of many social connections is their strength. The reduction of relations such as friendship, trust, the people we seek advice from, and communication to a binary variable is rather coarse and fails to capture the important role that the strength of links plays in relating network structure to behavior. The results in several theoretical models that we have reviewed in Section 17.2 apply to any weighted network (e.g., Ballester et al. 2006 and DeMarzo et al. 2003 among others). However, there is no paper we are aware of in the network experiments literature within and outside of economics that has investigated weighted networks. The creation of weighted networks in the lab presents its own challenges, but overcoming them would allow the exploration of a dimension of network structure, which plays an important role in many contexts where network structure affects behavior.

In recent years a growing number of experiments have shown that culture matters for play in different games.[37] Social relations are intertwined with culture, and we would expect the relation between social network structure and behavior to be dependent on culture in several contexts. For instance, Currarini et al. (2009) show that the tendency for individuals to form relations with others who are like them along some dimension, or homophily, shapes the network structure, and in turn this affects individual behavior (e.g., Golub and Jackson 2012). McPherson et al. (2001) review evidence that homophily varies along different dimensions, including ethnicity and culture. This suggests a relation between culture, the networks that form, and the way

[36] See Tversky and Kahneman (1974) for some examples of this body of work. Kahneman (2011) gives a comprehensive account accessible to the general public.

[37] Examples include Henrich et al. (2001) and Jackson and Xing (2014).

they impact behavior. Ideally, the investigation of the role of culture requires running an experiment with individuals in different geographical locations, which has become feasible only recently thanks to the development of web-based experiments.[38] The development and diffusion of web-based experiments opens up the opportunity of novel research on how culture and social networks jointly influence behavior.

Finally, in the introductory perspective on the literature in Chapter 3 of this handbook, Goyal argues that the economics of networks is transitioning to a "normal science" through the application of network models to competition, prices, and markets across different fields in economics. Some examples he gives of this transition include recent studies on the role of networks in product, financial, and labor markets which contribute to the gradual integration of networks into the standard economics framework, and policy-makers' growing awareness of their importance. A case in point is the prominence of networks in the discussions among academics, policy-makers and the general public in the aftermath of the 2008 financial crisis. A number of theory papers have been written on this topic since,[39] but we believe that the inclusion in theoretical models of realistic assumptions about the behavior of market agents is critical for the application of theoretical results to policy. In an ongoing project, Choi et al. (2016) examine experimentally how market freeze depends on network structure and the information agents have about the network in a standard trading market with a continuous double-auction. The findings in this experiment may shed light on the behavior of individuals in this environment, which can then be fed into theoretical models to generate predictions that can be tested experimentally. Our hope is that this type of two-way dialogue between theoretical and experimental work will continue to grow to increase our understanding of the relevance of networks in economics and policy-making.

References

Acemoglu, D., A. Ozdaglar, and A. ParandehGheibi (2010). "Spread of (mis) information in social networks." *Games and Economic Behavior* 70(2), 194–227.

Acemoglu, D., M. A. Dahleh, I. Lobel, and A. Ozdaglar (2011). "Bayesian learning in social networks." *The Review of Economic Studies* 78(4), 1201–1236.

Acemoglu, D., K. Bimpikis, and A. Ozdaglar (2014). "Dynamics of information exchange in endogenous social networks." *Theoretical Economics* 9(1), 41–97.

Allen, F. and D. Gale (2000). "Financial contagion." *Journal of Political Economy* 108(1), 1–33.

Anderson, L. R. and C. A. Holt (1997). "Information cascades in the laboratory." *The American Economic Review* 87(5), 847–862.

Bala, V. and S. Goyal (1998). "Learning from neighbours." *The Review of Economic Studies* 65(3), 595–621.

[38] Examples of web-based network experiments include Rand et al. (2011) and Gallo and Yan (2015a).

[39] See Chapters 20 and 21 of the handbook for a review.

Bala, V. and S. Goyal (2000). "A non-cooperative model of network formation." *Econometrica* 68(3), 1181–1229.

Ballester, C., A. Calvó-Armengol, and Y. Zenou (2006). "Who's who in networks. Wanted: The key player." *Econometrica* 74(5), 1403–1417.

Banerjee, A. V. (1992). "A simple model of herd behavior." *The Quarterly Journal of Economics* 107(3), 797–817.

Berninghaus, S. K., K.-M. Ehrhart, and C. Keser (2002). "Conventions and local interaction structures: Experimental evidence." *Games and Economic Behavior* 39(2), 177–205.

Bikhchandani, S., D. Hirshleifer, and I. Welch (1992). "A theory of fads, fashion, custom, and cultural change as informational cascades." *Journal of Political Economy* 100(5), 992–1026.

Binmore, K. G. and L. Samuelson (1992). "Evolutionary stability in repeated games played by finite automata." *Journal of Economic Theory* 57(2), 278–305.

Bonacich, P. (1987). "Power and centrality: A family of measures." *American Journal of Sociology* 92(5), 1170–1182.

Bramoullé, Y. and R. Kranton (2007). "Public goods in networks." *Journal of Economic Theory* 135(1), 478–494.

Bramoullé, Y., R. Kranton, and M. D'amours (2014). "Strategic interaction and networks." *The American Economic Review* 104(3), 898–930.

Cassar, A. (2007). "Coordination and cooperation in local, random and small world networks: Experimental evidence." *Games and Economic Behavior* 58(2), 209–230.

Cassar, A., D. Friedman, and P. H. Schneider (2010). "A laboratory investigation of networked markets*." *The Economic Journal* 120(547), 919–943.

Çelen, B. and S. Kariv (2004). "Distinguishing informational cascades from herd behavior in the laboratory." *American Economic Review* 94(3), 484–498.

Chamberlin, E. H. (1948). "An experimental imperfect market." *The Journal of Political Economy* 56(2), 95–108.

Charness, G., M. Corominas-Bosch, and G. R. Frechette (2007). "Bargaining and network structure: An experiment." *Journal of Economic Theory* 136(1), 28–65.

Charness, G., F. Feri, M. A. Meléndez-Jiménez, and M. Sutter (2014). "Experimental games on networks: Underpinnings of behavior and equilibrium selection." *Econometrica* 82(5), 1615–1670.

Chaudhuri, A. (2011). "Sustaining cooperation in laboratory public goods experiments: A selective survey of the literature." *Experimental Economics* 14(1), 47–83.

Choi, S. (2012). "A cognitive hierarchy model of learning in networks." *Review of Economic Design* 16(2–3), 215–250.

Choi, S. and J. Lee (2014). "Communication, coordination, and networks." *Journal of the European Economic Association* 12(1), 223–247.

Choi, S., D. Gale, and S. Kariv (2005). "Behavioral aspects of learning in social networks: An experimental study." *Advances in Applied Microeconomics* 13, 25–61.

Choi, S., D. Gale, S. Kariv, and T. Palfrey (2011). "Network architecture, salience and coordination." *Games and Economic Behavior* 73(1), 76–90.

Choi, S., D. Gale, and S. Kariv (2012). "Social learning in networks: A quantal response equilibrium analysis of experimental data." *Review of Economic Design* 16(2–3), 135–157.

Choi, S., A. Galeotti, and S. Goyal (2014). "Trading in networks: Theory and experiments." Cambridge-INET Working Paper 8.

Choi, S., E. Gallo, and B. Wallace (2016). "Systemic risk in financial networks: An experiment." Unpublished manuscript.

Condorelli, D. and A. Galeotti (2012). "Bilateral trading in networks." Unpublished manuscript, University of Essex, Department of Economics, Economics Discussion Papers 704.

Cooper, R., D. V. DeJong, R. Forsythe, and T. W. Ross (1989). "Communication in the battle of the sexes game: Some experimental results." *The RAND Journal of Economics* 20(4), 568–587.

Cooper, R., D. V. DeJong, R. Forsythe, and T. W. Ross (1992). "Communication in coordination games." *The Quarterly Journal of Economics* 107(2), 739–771.

Cooper, R. W., D. V. DeJong, R. Forsythe, and T. W. Ross (1990). "Selection criteria in coordination games: Some experimental results." *The American Economic Review* 80(1), 218–233.

Corazzini, L., F. Pavesi, B. Petrovich, and L. Stanca (2012). "Influential listeners: An experiment on persuasion bias in social networks." *European Economic Review* 56(6), 1276–1288.

Corominas-Bosch, M. (2004). "Bargaining in a network of buyers and sellers." *Journal of Economic Theory* 115(1), 35–77.

Crawford, V. (1998). "A survey of experiments on communication via cheap talk." *Journal of Economic Theory* 78(2), 286–298.

Crawford, V. P. and J. Sobel (1982). "Strategic information transmission." *Econometrica: Journal of the Econometric Society* 50(6), 1431–1451.

Cuesta, J. A., C. Gracia-Lázaro, A. Ferrer, Y. Moreno, and A. Sánchez (2015). "Reputation drives cooperative behaviour and network formation in human groups." *Scientific Reports* 5, article no. 7843.

Currarini, S., M. O. Jackson, and P. Pin (2009). "An economic model of friendship: Homophily, minorities, and segregation." *Econometrica* 77(4), 1003–1045.

Dal Bó, P. and G. R. Fréchette (2011). "The evolution of cooperation in infinitely repeated games: Experimental evidence." *The American Economic Review* 101(1), 411–429.

DeGroot, M. H. (1974). "Reaching a consensus." *Journal of the American Statistical Association* 69(345), 118–121.

DeMarzo, P. M., D. Vayanos, and J. Zwiebel (2003). "Persuasion bias, social influence, and unidimensional opinions." *Quarterly Journal of Economics* 118(3), 909–968.

Dessi, R., E. Gallo, and S. Goyal (2016). "Network cognition." *Journal of Economic Behavior & Organization*, forthcoming.

Devetag, G. and A. Ortmann (2007). "When and why? A critical survey on coordination failure in the laboratory." *Experimental Economics* 10(3), 331–344.

Ellison, G. (1993). "Learning, local interaction, and coordination." *Econometrica: Journal of the Econometric Society* 61(5), 1047–1071.

Fehr, E. and S. Gächter (1999). "Cooperation and punishment in public goods experiments." *Institute for Empirical Research in Economics*, Working paper no. 10 SSRN: 203194.

Frey, V., R. Corten, and V. Buskens (2012). "Equilibrium selection in network coordination games: An experimental study." *Review of Network Economics* 11(3).

Fudenberg, D. and E. Maskin (1986). "The folk theorem in repeated games with discounting or with incomplete information." *Econometrica: Journal of the Econometric Society* 54(3), 533–554.

Gale, D. and S. Kariv (2003). "Bayesian learning in social networks." *Games and Economic Behavior* 45(2), 329–346.

Gale, D. M. and S. Kariv (2009). "Trading in networks: A normal form game experiment." *American Economic Journal: Microeconomics* 1(2), 114–132.

Galeotti, A. and S. Goyal (2010). "The law of the few." *The American Economic Review* 100(4), 1468–1492.

Galeotti, A., S. Goyal, M. O. Jackson, F. Vega-Redondo, and L. Yariv (2010). "Network games." *The Review of Economic Studies* 77(1), 218–244.

Galeotti, A., C. Ghiglino, and F. Squintani (2013). "Strategic information transmission networks." *Journal of Economic Theory* 148(5), 1751–1769.

Gallo, E. (2014a). "Communication networks in markets." *Cambridge Working Papers in Economics* 1431.

Gallo, E. (2014b). "Social learning by chit-chat." *Journal of Economic Theory* 153, 313–343.

Gallo, E. and C. Yan (2015a). "The effects of reputational and social knowledge on cooperation." *Proceedings of the National Academy of Sciences* 112(12), 3647–3652.

Gallo, E. and C. Yan (2015b). "Efficiency and equilibrium in network games: An experiment." *Cambridge-INET Working Paper Series* 1503.

Goeree, J. K., T. R. Palfrey, B. W. Rogers, and R. D. McKelvey (2007). "Self-correcting information cascades." *The Review of Economic Studies* 74(3), 733–762.

Golub, B. and M. O. Jackson (2010). "Naive learning in social networks and the wisdom of crowds." *American Economic Journal: Microeconomics* 2(1), 112–149.

Golub, B. and M. O. Jackson (2012). "How homophily affects the speed of learning and best-response dynamics." *The Quarterly Journal of Economics* 127(3), 1287–1338.

Goyal, S. (2007). *Connections: An Introduction to the Economics of Networks*. Princeton University Press.

Goyal, S., S. Rosenkranz, U. Weitzel, and V. Buskens (2014). "Individual search and social networks." FEEM working paper no. 49.2014.

Gracia-Lázaro, C., A. Ferrer, G. Ruiz, A. Tarancón, J. A. Cuesta, A. Sánchez, and Y. Moreno (2012). "Heterogeneous networks do not promote cooperation when humans play a prisoners dilemma." *Proceedings of the National Academy of Sciences* 109(32), 12922–12926.

Greif, A. (1993). "Contract enforceability and economic institutions in early trade: The Maghribi traders' coalition." *The American Economic Review* 83(3), 525–548.

Grimm, V. and F. Mengel (2014). "An experiment on belief formation in networks." July 4. Available at SSRN: http://ssrn.com/abstract=2361007.

Grujić, J., C. Gracia-Lázaro, M. Milinski, D. Semmann, A. Traulsen, J. A. Cuesta, Y. Moreno, and A. Sánchez (2014). "A comparative analysis of spatial prisoner's dilemma experiments: Conditional cooperation and payoff irrelevance." *Scientific Reports* 4, article no. 4615.

Haag, M. and R. Lagunoff (2006). "Social norms, local interaction, and neighborhood planning*." *International Economic Review* 47(1), 265–296.

Hagenbach, J. and F. Koessler (2010). "Strategic communication networks." *The Review of Economic Studies* 77(3), 1072–1099.

Harsanyi, J. C. and R. Selten (1988). *A General Theory of Equilibrium Selection in Games. MIT Press Books*, June.

Henrich, J., R. Boyd, S. Bowles, C. Camerer, E. Fehr, H. Gintis, and R. McElreath (2001). "In search of homo economicus: Behavioral experiments in 15 small-scale societies." *American Economic Review* 91(2), 73–78.

Hung, A. A. and C. R. Plott (2001). "Information cascades: Replication and an extension to majority rule and conformity-rewarding institutions." *American Economic Review* 91(5), 1508–1520.

Jackson, M. O. (2008). *Social and Economic Networks*. Princeton University Press.

Jackson, M. O. and A. Wolinsky (1996). "A strategic model of social and economic networks." *Journal of Economic Theory* 71, 44–74.

Jackson, M. O. and Y. Xing (2014). "Culture-dependent strategies in coordination games." *Proceedings of the National Academy of Sciences* 111(Supplement 3), 10889–10896.

Janicik, G. A. and R. P. Larrick (2005). "Social network schemas and the learning of incomplete networks." *Journal of Personality and Social Psychology* 88(2), 348.

Jordan, J. J., D. G. Rand, S. Arbesman, J. H. Fowler, and N. A. Christakis (2013). "Contagion of cooperation in static and fluid social networks." *PloS ONE* 8(6), e66199.

Judd, J. S. and M. Kearns (2008). "Behavioral experiments in networked trade." In *Proceedings of the 9th ACM Conference on Electronic Commerce*, 150–159, ACM.

Judd, S., M. Kearns, and Y. Vorobeychik (2010). "Behavioral dynamics and influence in networked coloring and consensus." *Proceedings of the National Academy of Sciences* 107(34), 14978–14982.

Kahneman, D. (2011). *Thinking, Fast and slow*. Macmillan.

Kearns, M., S. Judd, J. Tan, and J. Wortman (2009). "Behavioral experiments on biased voting in networks." *Proceedings of the National Academy of Sciences* 106(5), 1347–1352.

Keser, C., K.-M. Ehrhart, and S. K. Berninghaus (1998). "Coordination and local interaction: Experimental evidence." *Economics Letters* 58(3), 269–275.

Kirchkamp, O. and R. Nagel (2007). "Naive learning and cooperation in network experiments." *Games and Economic Behavior* 58(2), 269–292.

Kirman, A. P. (1983). "Communication in markets: A suggested approach." *Economics Letters* 12(2), 101–108.

Kiss, H. J., I. Rodriguez-Lara, and A. Rosa-García (2014). "Do social networks prevent or promote bank runs? *Journal of Economic Behavior & Organization* 101, 87–99.

Kosfeld, M. (2004). "Economic networks in the laboratory: A survey." *Review of Network Economics* 3(1), 20–42.

Krackhardt, D. (1987). "Cognitive social structures." *Social Networks* 9(2), 109–134.

Krackhardt, D. (1990). "Assessing the political landscape: Structure, cognition, and power in organizations." *Administrative Science Quarterly* 35(2), 342–369.

Kranton, R. E. and D. F. Minehart (2003). "A theory of buyer-seller networks." In *Networks and Groups*, 347–378, Springer.

Kübler, D. and G. Weizsäcker (2004). Limited depth of reasoning and failure of cascade formation in the laboratory." *The Review of Economic Studies* 71(2), 425–441.

Kumbasar, E., A. K. Rommey, and W. H. Batchelder (1994). "Systematic biases in social perception." *American Journal of Sociology* 100(2), 477–505.

Ledyard, J. (1995). "Public goods: A survey of experimental research." In *Handbook of Experimental Economics*, Princeton University Press, Princeton, 111–194.

Manski, C. F. (2000). "Economic analysis of social interactions." *The Journal of Economic Perspectives* 14(3), 115–136.

Marwell, G. and R. E. Ames (1981). "Economists free ride, does anyone else? Experiments on the provision of public goods, IV." *Journal of Public Economics* 15(3), 295–310.

McKelvey, R. D. and T. R. Palfrey (1995). "Quantal response equilibria for normal form games." *Games and Economic Behavior* 10(1), 6–38.

McKelvey, R. D. and T. R. Palfrey (1998). "Quantal response equilibria for extensive form games." *Experimental Economics* 1(1), 9–41.

McPherson, M., L. Smith-Lovin, and J. M. Cook (2001). "Birds of a feather: Homophily in social networks." *Annual Review of Sociology* 27, 415–444.

Milinski, M., D. Semmann, and H.-J. Krambeck (2002). "Reputation helps solve the tragedy of the commons." *Nature* 415(6870), 424–426.

Montgomery, J. D. (1991). "Social networks and labor-market outcomes: Toward an economic analysis." *The American Economic Review* 81(5), 1408–1418.

Morris, S. (2000). "Contagion." *The Review of Economic Studies* 67(1), 57–78.

Mueller-Frank, M. (2013). "A general framework for rational learning in social networks." *Theoretical Economics* 8(1), 1–40.

Mueller-Frank, M. (2014). "Does one Bayesian make a difference?" *Journal of Economic Theory* 154, 423–452.

Mueller-Frank, M. and C. Neri (2013). "Social learning in networks: Theory and experiments." Available at SSRN 2328281, December.

Murnighan, J. K. and A. E. Roth (1983). "Expecting continued play in prisoner's dilemma games a test of several models." *Journal of Conflict Resolution* 27(2), 279–300.

Myerson, R. B. (1977). "Graphs and cooperation in games." *Mathematics of Operations Research* 2(3), 225–229.

Nowak, M. A. and K. Sigmund (2005). "Evolution of indirect reciprocity." *Nature* 437(7063), 1291–1298.

Ochs, J. (1995). "Coordination problems." *The Handbook of Experimental Economics* 195–251.

Ohtsuki, H., C. Hauert, E. Lieberman, and M. A. Nowak (2006). "A simple rule for the evolution of cooperation on graphs and social networks." *Nature* 441(7092), 502–505.

Ostrom, E. (1990). *Governing the Commons: The Evolution of Institutions for Collective Action.* Cambridge University Press.

Peski, M. (2010). "Generalized risk-dominance and asymmetric dynamics." *Journal of Economic Theory* 145(1), 216–248.

Rand, D. G., S. Arbesman, and N. A. Christakis (2011). "Dynamic social networks promote cooperation in experiments with humans." *Proceedings of the National Academy of Sciences* 108(48), 19193–19198.

Rand, D. G., M. A. Nowak, J. H. Fowler, and N. A. Christakis (2014). "Static network structure can stabilize human cooperation." *Proceedings of the National Academy of Sciences* 111(48), 17093–17098.

Rauch, J. E. and V. Trindade (2002). "Ethnic Chinese networks in international trade." *Review of Economics and Statistics* 84(1), 116–130.

Riedl, A., I. M. Rohde, and M. Strobel (2011). "Efficient coordination in weakest-link games." Unpublished manuscript.

Rosenkranz, S. and U. Weitzel (2012). "Network structure and strategic investments: An experimental analysis." *Games and Economic Behavior* 75(2), 898–920.

Seinen, I. and A. Schram (2006). "Social status and group norms: Indirect reciprocity in a repeated helping experiment." *European Economic Review* 50(3), 581–602.

Smith, L. and P. Sørensen (2000). "Pathological outcomes of observational learning." *Econometrica* 68(2), 371–398.

Smith, V. L. (1962). "An experimental study of competitive market behavior." *The Journal of Political Economy* 70(2), 111–137.

Smith, V. L. (1965). "Experimental auction markets and the walrasian hypothesis." *The Journal of Political Economy* 73(4), 387–393.

Sunder, S. (1992). *Experimental Asset Markets: A Survey.* Carnegie Mellon University.

Suri, S. and D. J. Watts (2011). "Cooperation and contagion in web-based, networked public goods experiments." *PLoS One* 6(3), e16836.

Taylor, P. D., T. Day, and G. Wild (2007). "Evolution of cooperation in a finite homogeneous graph." *Nature* 447(7143), 469–472.

Tversky, A. and D. Kahneman (1974). "Judgment under uncertainty: Heuristics and biases." *Science* 185(4157), 1124–1131.

Van Huyck, J. B., R. C. Battalio, and R. O. Beil (1990). "Tacit coordination games, strategic uncertainty, and coordination failure." *The American Economic Review* 80(1), 234–248.

Van Huyck, J. B., R. C. Battalio, and R. O. Beil (1991). "Strategic uncertainty, equilibrium selection, and coordination failure in average opinion games." *The Quarterly Journal of Economics* 106(3), 885–910.

Wang, J., S. Suri, and D. J. Watts (2012). "Cooperation and assortativity with dynamic partner updating." *Proceedings of the National Academy of Sciences* 109(36), 14363–14368.

Wasserman, S. (1994). *Social Network Analysis: Methods and Applications*, volume 8. Cambridge University Press.

Wedekind, C. and M. Milinski (2000). "Cooperation through image scoring in humans." *Science* 288(5467), 850–852.

Weizsäcker, G. (2010). "Do we follow others when we should? A simple test of rational expectations." *The American Economic Review* 100(5), 2340–2360.

Young, H. P. (1993). "An evolutionary model of bargaining." *Journal of Economic Theory* 59(1), 145–168.

PART V

DIFFUSION, LEARNING, AND CONTAGION

CHAPTER 18

DIFFUSION IN NETWORKS

P. J. LAMBERSON

This chapter examines how the topology of a network—the pattern of who is connected to whom—affects diffusion. Understanding network structures can be interesting in its own right, but we would like to know if these patterns impact behavior. Diffusion is one of the first outcomes for which we can definitively say network structure matters. The pattern of connections can make the difference between a hit product and a failure, a viral video and an obscure one, or a pandemic infection and a limited outbreak.

Many well-established economic models imply that agents influence one another's behavior. In models of social learning and information cascades, agents form beliefs about the value of taking an action by observing the behavior of other agents. In models of network externalities, the rewards to taking an action depend on how many other individuals take that or other complementary actions. And of course, in game-theoretic models, an agent's payoffs typically depend on the strategies of other players.

Before networks became commonplace in economics research, who influenced whom was often determined simply by the order in which agents took their decisions, or alternatively everyone influenced everyone else. But in reality we know that some people hold greater sway on our decisions than others, and who influences us depends on more than just the timing of our decisions. Networks provide a natural way to represent the structure of these interactions. Models of network diffusion consider the situation where a focal action spreads, or fails to spread, through the network.

The recipe for a network diffusion model involves two key ingredients. First, the model must specify the set of network structures under consideration. Often these networks are random with the exception of some fixed network features. For example, the widely employed configuration model networks discussed in Section 18.1.3 fix the degree distribution (more precisely, the degree sequence) of the network but are random otherwise. Second, there is a rule describing how agents in the network influence one another. The simplest rules mirror models of disease spread, and we explore these models first in Section 18.1. In Section 18.2 we turn to more sophisticated models of social influence. A key insight that emerges from the collection of models discussed in this chapter is that not only does network structure matter, but how the network matters depends on the way in which agents influence one another. Network features that facilitate contagion under one model of influence can inhibit diffusion in another. In particular, while the analogy between social contagions and disease spread provides many useful findings, we should be cautious about devising policies based on these findings because more sophisticated models of decision making often imply contradictory recommendations.

There are several topics relevant to network diffusion that we do not address in this chapter but can be found in other chapters in this volume. Identifying network contagion in data has proven to be a particularly difficult empirical challenge. The heart of the matter is that often multiple causal stories can explain the same pattern of outcomes, and separating diffusion from the other possible explanations often requires careful control of the network formation process or strong assumptions about the way in which individuals influence one another (Shalizi and Thomas 2011). The empirics of network diffusion are addressed in chapters 22 and 24. We also leave aside the question of identifying the most influential nodes in a network. Often, this question is framed in the context of a diffusion process: If you want a contagion to spread as widely as possible from a few initial "seed" nodes, which nodes should you infect first? The resulting "key player policies" are addressed in chapter 14. Finally, one way of representing how agents in a network influence one another is using a formal game. Games in networks are addressed in chapters 8, 9, and 10.

18.1 Simple Contagion Models

Malcolm Gladwell wrote, "... ideas and behavior and messages and products sometimes behave just like outbreaks of infectious disease" (2014). Indeed, many models of information and behavior diffusion in networks take inspiration from well-known "compartmental models" of disease spread such as the SI, SIS, and SIR models (Bailey 1975). This analogy is used so extensively that throughout the chapter we use the word *infected* to describe an individual who is exhibiting the behavior that we are interested in, regardless of whether that behavior is sharing a rumor, adopting a technology, watching a viral video, or turning out to vote. However, it turns out that some social contagions *do not* spread just like infectious diseases, and understanding what happens in these cases is one of the interesting contributions of economics to network science. We return to this point in Section 18.2, but before exploring more sophisticated models of social contagion it will be useful to understand how networks matter for simple contagions that do spread like diseases.

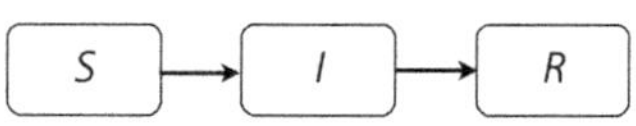

FIGURE 18.1 The SIR model of contagion.

18.1.1 Compartmental Models of Contagion: The SIR Model

The SI, SIS, SIR, and other compartmental models have been the standard models of disease spread for decades. Here we briefly describe the SIR model. The other models are defined similarly (see the books by Keeling and Rohani 2008 or Newman 2010 for details).

Suppose that every member of a large finite population is in one of three states—susceptible, infected, or recovered. Let S, I, and R denote the number of susceptible, infected, and recovered individuals, respectively, and let $N = S + I + R$ denote the total population. The model captures the idea that infected individuals can spread the infection to susceptible individuals when they come into contact (although not every contact need result in transmission of the infection), and after some time infected individuals recover, never to become infected again. Let k denote the average number of people any given person in the population contacts per unit of time. Then the expected number of contacts per unit of time that involve at least one susceptible individual is Sk. If individuals mix at random, the probability that any one of these contacts is with an infected individual is I/N, so the expected number of susceptible-infected contacts is SkI/N. Let β denote the probability that a given susceptible-infected contact results in transmission of the infection. In epidemiology this is known as the *infectivity* of the disease.[1] The resulting rate of new infections is $Sk\beta I/N$ people per unit of time. Let γ denote the probability that an infected individual recovers per unit of time, so on average $I\gamma$ individuals recover each time unit. Finally, let $\lambda = \beta/\gamma$ denote the ratio of the infection and recovery probabilities.

The dynamics of the contagion are determined by comparing the net infection rate to the net recovery rate. If the infection rate exceeds the recovery rate, the contagion will spread, otherwise it will die out. Thus, the contagion spreads if $Sk\beta I/N > I\gamma$, which after a little algebra becomes $(S/N)k\lambda > 1$. Assuming the infection is in its early stages and the population is large $S \simeq N$, so the fundamental inequality reduces to

$$k\lambda > 1. \tag{18.1}$$

In epidemiology the critical quantity $k\lambda$ is denoted R_0 and is called the *basic reproduction number* of the disease. It captures the average number of new infections created by each infected individual before they recover when the total number of infections is small relative to the size of the population.

As presented, the SIR model has no network structure. Individuals are assumed to mix randomly at each time step (we made this assumption when we set the probability of a given contact being infected at I/N). Compartmental models, like the SIR model, are sometimes called "mean field approximations" because the assumption is that each agent experiences the mean state of the system. Even this random mixing model gives

[1] Note that in some expositions of the SIR model the term β is used to represent the overall per person infection rate, $k\beta$. Here we keep the infectivity β and average number of contacts k separate because it simplifies the extension of the model to a networked setting.

some hint of the potential effects of network structure on diffusion in the term k, the average number of contacts. In particular, more contacts makes it more likely for the infection to spread.

The SI, SIS, and other compartmental models are defined similarly. In the SI model, once a susceptible agent becomes infected they remain infected forever. Assuming random mixing, any contagion with a positive infectivity will ultimately infect the entire population in the SI model. In the SIS model, agents that recover return to the susceptible state and can potentially become infected again rather than moving into a recovered state that is immune from reinfection. The contagion threshold in equation (18.1) applies to the SIS model as it does to the SIR model, but in the SIS model if (18.1) is satisfied then the infection becomes endemic in the population, while in the SIR model eventually all infected individuals recover. There are many additional variants including the SEIR (susceptible-exposed-infected-recovered), SIRS (SIS with a temporary immune period), MSIR (SIR with an initial maternally derived immune state, Hethcote 2000), and the SIR model with a carrier state (which has no convenient abbreviation, Keeling and Rohani 2008).[2]

Adapting the SIR model (and other compartmental models) described in this section to a networked population is straightforward; however, solving analytically for the behavior of the networked model is significantly more challenging than the random mixing model. In this context, we simply let β denote the probability that a susceptible individual *connected to* an infected individual becomes infected per unit of time, and suppose that each individual remains infected for a fixed length of time τ before recovering.

In general, the behavior of the model depends on the specific details of the network and which nodes are initially infected. The primary method that has emerged to deal with this difficulty is to examine expected outcomes of the contagion process where only some properties of the network (such as the degree distribution) are held fixed while others are assumed to be drawn from a given distribution. In some cases analytic results are possible, while in others results are obtained by simulating the contagion spreading through a network (or networks) many times and look at the average or distribution of outcomes.

As we will see in subsequent sections, many of the known results focus on the impact of the degree distribution of the network.[3] Roughly speaking, adding links to the network, in the sense of a first-order stochastic dominance shift in the degree distribution, or increasing the variability of degrees across nodes, in the sense of a mean preserving spread of the degree distribution, makes disease-like contagions more likely

[2] While epidemiological models inform much of the economic literature on diffusion, economics has made reciprocal contributions to the literature on disease spread by incorporating strategic behavior in the classical epidemiology models. Notable examples include the papers by Geoffard and Philipson (1996), Kremer (1996), Toxvaerd (2010ab), and Galeotti and Rogers (2013). The bulk of this literature focuses on equilibrium infection levels when agents have the option to vaccinate themselves against infection at a cost.

[3] The *degree* of a node is the number of links connected to that node and the *degree distribution* of a network is the probability distribution for these degrees over all nodes in the network.

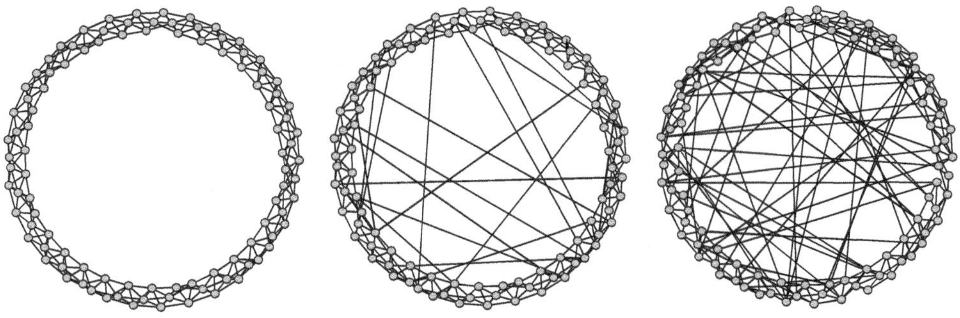

FIGURE 18.2 Three networks generated by the Watts and Strogatz small-world model. Moving from left to right, the rewiring probabilities are $p = 0$, $p = .05$, and $p = .2$.

to spread (Lamberson 2011). (See Section 18.1.4 for definitions and precise statements of the results.)

18.1.2 The Watts and Strogatz Small-World Model

Watts and Strogatz described one of the first findings on the systematic impact of network structure on diffusion in their seminal paper on small-world networks (Watts and Strogatz 1998, see also Chapter 14 in this volume). The Watts and Strogatz small-world (WS) network model has become a fundamental tool for understanding how network structure impacts diffusion, so we briefly recall it here.

For any fixed average degree k and number of nodes N, the WS model defines a family of networks parameterized by a single parameter $p \in [0, 1]$. To define the family, begin with a collection of N nodes arranged in a circle and connect each node to its k nearest neighbors. This defines a *ring lattice* of degree k, which is the WS network corresponding to $p = 0$. For $p > 0$, for each edge of the ring lattice with independent probability p, reconnect one end of the edge to a node chosen uniformly at random from the remaining nodes (not allowing for multiedges).[4] Each randomization of an edge is called a "rewiring," and thus p is referred to as the *rewiring probability* of the network. Figure 18.2 depicts three networks generated by the WS model with rewiring probabilities $p = 0$, $p = .05$, and $p = .2$.

In their 1998 paper, Watts and Strogatz examined the impact of the rewiring probability on a simulated SIR contagion spreading in the network. They found that the level of infectivity required to result in half of the network nodes becoming infected as well as the time for a highly infectious contagion to saturate the network both fall rapidly as the rewiring probability increases. In other words, infections spread more quickly and easily in the more random WS networks. It's not immediately clear, however,

[4] Notice that for each p the model actually defines a distribution of networks. A draw from this distribution is then called a Watts Strogatz small-world network.

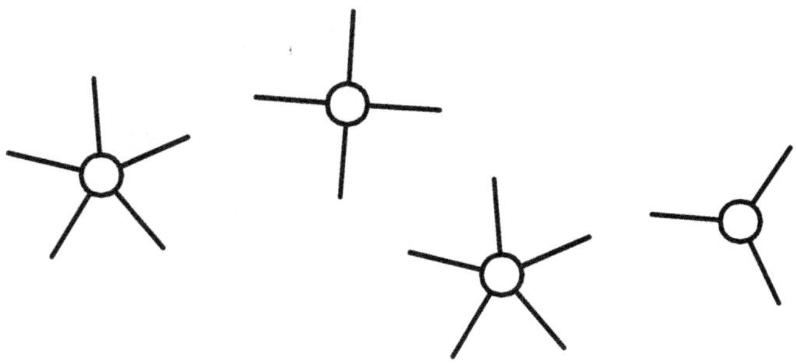

FIGURE 18.3 Nodes with stubs are connected at random to form a configuration model network.

exactly what accounts for the increase in diffusion that comes with increased rewiring because rewiring the WS networks affects several standard network metrics: it lowers the average clustering coefficient of the network, decreases the average path length, and increases the variance of the degree distribution.[5] In the next section, we describe analytic results that focus on the degree distribution alone.

18.1.3 Configuration Models

The configuration model generates a probability distribution over a collection of networks with a fixed *degree sequence* and has the property that the degrees of neighbors in the network are independent of one another. These features make the configuration model a prime example for several important papers on network diffusion in economics including Jackson and Rogers' and López-Pintado's models of network contagion (Jackson and Rogers 2007; López-Pintado 2008), the threshold models developed by Jackson and Yariv (2005, 2007), and Lamberson's model of social learning (2010).

To form a configuration model network begin with a collection of n nodes, and for each node i let k_i denote its degree. Then $k_1, k_2, \ldots$ is called the *degree sequence* of the network. Let $P(k)$ denote the degree distribution of the network and $\langle k \rangle$ the average degree. The degree sequence determines the degree distribution since $P(k) = \sum |\{k_i\}_{k_i=k}|/n$. To form a network with this degree sequence, imagine each node begins with k_i "stubs"—that is, half edges emanating from the node that are not connected to any other node, as shown in Figure 18.3. The total number of stubs equals twice the total number of links in the network, which we denote $2m$. Now, choose two of the stubs uniformly at random from the collection of $2m$ total stubs and connect them to form a complete edge. Then choose another two stubs from the remaining $2m - 2$ stubs

[5] The *clustering coefficient* of a network is the probability that two nodes with a common network neighbor are themselves connected. To compute the *average path length* of a network, for any pair of nodes count the number of links on the shortest path between the two nodes and then average this over all pairs of nodes in the network.

and connect them. Continue this process of connecting stubs until all pairs of stubs have been joined into edges. The resulting network will have the desired degree sequence. In general, results obtained for configuration model networks hold asymptotically as the number of nodes approaches infinity because this avoids certain difficulties that can arise in finite realizations of the model, such as nontrivial network clustering.[6]

In the economics literature the configuration model is often interpreted as a model of random matchings with incomplete information. Suppose that at some future time (or over the course of some future time period) each agent i is affected by k_i individuals. If each agent knows his or her own degree, but not the exact identities of their future neighbors, it is as if the agents know the configuration model but not the exact network that will be realized. This interpretation is perhaps best described in the paper on network games by Galeotti, Goyal, Jackson, Vega-Redondo, and Yariv (2010).

As we will see, one important feature of the configuration model is the *excess degree distribution*. This is the distribution of degrees of randomly chosen neighbors of randomly chosen nodes. The excess degree distribution plays a critical role in diffusion because the probability that a node becomes infected depends on the probability of having infected neighbors, and the probability that a neighbor is infected often depends on their degree: nodes with more neighbors typically have a higher probability of being infected.

Because neighboring nodes' degrees are independent in the configuration model, the probability that the neighbor of a given node has degree k is straightforward to calculate. If we start at a given node and follow a "stub" emanating from that node, there are $2m-1$ other possible stubs that could be connected to it. A node of degree k accounts for k of these stubs and there are $nP(k)$ nodes of degree k, so the probability of connecting to a degree k node is $knP(k)/(2m-1)$. For sufficiently large networks this is approximately the same as $knP(k)/2m$. Finally, since n is the total number of nodes and $2m$ is the sum of all the degrees of nodes in the network, $n/2m$ is one over the average degree. Thus, the excess degree distribution is given by

$$\tilde{P}(k) = \frac{P(k)k}{\langle k \rangle}. \tag{18.2}$$

Now, consider an SIR contagion spreading through a configuration model network. If β is the probability that a susceptible individual connected to an infected individual becomes infected per unit of time, and τ is the length of time they remain infected before recovering, then $e^{-\beta\tau}$ is the probability the infected individual recovers before

[6] There are three potential issues with this model worth noting. First, the sum of the degree sequence has to be even or else there will be a stub left over at the end of the process. Second, the process may generate self links. Third, the process may generate *multiedges*, where more than one edge connects the same two nodes. The even total degree requirement is usually unimportant, and for many applications self-edges and multiedges are of minimal concern because the probability that they occur approaches zero as the size of the network grows large. See Section 4.5.10 of Jackson (2008) and Section 13.2 of Newman (2010) for discussion of these issues and further details on the configuration model.

the infection is transmitted and $\phi = 1 - e^{-\beta\tau}$ is the probability the infection is passed on (see Newman 2010, Section 17.8 for details).

In some sense ϕ plays the role in the networked model that λ plays in the fully mixed model. For each network Γ there is a critical value $\phi_c(\Gamma)$ such that it is possible for a contagion to spread through the network if and only if ϕ exceeds $\phi_c(\Gamma)$.

One can use a bond percolation argument (originally due to Mollison 1977 and Grassberger 1983) to find the critical value, ϕ_c, that determines whether or not an epidemic is possible. This threshold is analogous to the threshold for R_0 given in equation (18.1). The idea of the bond percolation argument is as follows. Suppose that initially all edges of the network are white and we randomly color a fraction ϕ of the edges black. We can think of a black edge as an edge across which the infection spreads if it reaches that edge. The contagion has the potential to spread if the subnetwork consisting only of black edges contains a component large enough to span a positive fraction of the vertices from the whole network. Such a connected component of the network whose size grows in proportion to the total number of nodes is called a *giant component.* (For details on the definition of giant component, see Section 12.5 of Newman 2010.) A giant component comprised of the black edges exists, and thus an epidemic can occur, if and only if

$$\phi > \phi_c = \frac{\langle k \rangle}{\langle k^2 \rangle - \langle k \rangle}. \tag{18.3}$$

For intuition behind this equation, suppose that early in the epidemic an agent with k neighbors is infected by one of those neighbors. Because the epidemic is in an early stage all of her neighbors except the one that infected her are likely to be uninfected, so on average such an agent will have $k - 1$ uninfected neighbors.[7] Thus, this newly infected agent will herself create an average of $\phi(k - 1)$ additional infections. Because k is the degree of a randomly chosen neighbor of a node, its distribution is given by the excess degree distribution in equation (18.2). So, the expected number of new infections created by a node infected by one of her neighbors is given by

$$\sum_k \phi(k-1)\tilde{P}(k) = \phi \sum_k \frac{P(k)k(k-1)}{\langle k \rangle} = \phi \frac{\langle k^2 \rangle - \langle k \rangle}{\langle k \rangle}. \tag{18.4}$$

[7] Notice that this assumes that the infection status of the focal agent's neighbors are uncorrelated with one another, which is not necessarily true in general. In many real-world networks two agents with a friends in common are especially likely to be friends with each other (Watts and Strogatz 1998; Newman and Park 2003). In such a network, if an agent is infected by one of her neighbors, her other neighbors are more likely to be infected than an agent chosen randomly from the population because they are likely also connected to the infecting agent and thus may themselves been infected by him. However, in a configuration model network the probability that two agents with a friend in common are themselves connected goes to zero with the number of nodes, so we can safely assume that the infection status of the focal agent's neighbors are independent.

In order for the infection to grow, each newly infected agent has to create more than one additional new infection. Setting (18.4) to be greater than one and solving for ϕ, we obtain (18.3).[8]

One feature of equation (18.3) to note is the appearance of the second moment of the degree distribution $\langle k^2 \rangle$. This term appears repeatedly in analyses of network diffusion. The contribution that a given node makes to the spread of a contagion is roughly proportional to its degree squared since both the probability that a node becomes infected and the number of infections that node can create are proportional to the node's degree.

We can apply equation (18.3) to understand the epidemic tipping point for several well-known classes of networks. For example, for a Poisson random network $\phi_c = 1/k$, mirroring the random mixing equation (18.1).[9] For networks that have a power law degree distribution with exponent between two and three, ϕ_c approaches zero, and thus any contagion with positive ϕ spreads in such a network (Pastor-Satorras and Vespignani 2001).

18.1.4 Degree Distributions and Stochastic Dominance

Jackson and Rogers (2007) studied a variant of the SIS model using a technique known as a *degree-based mean field approximation* (Newman 2010) and showed how the concept of stochastic dominance applied to the degree distribution of the network could provide insight into which networks are more or less likely to permit an epidemic. Their method provided a template for analyzing the effects of stochastic dominance shifts in the degree distribution that several other authors apply to different models of influence including López-Pintado (2008), who also studies a variant of the SIS model; Jackson and Yariv (2007), who use a threshold model; and Lamberson (2010), who examines a model of social learning.

Before outlining the analysis of Jackson and Rogers' model, we briefly recall the definitions of first and second order stochastic dominance, which are key to the results.

Definition 1. *A distribution P first order stochastically dominates a distribution P′ if for every nondecreasing function* $u : \mathbb{R} \to \mathbb{R}$,

$$\sum_k u(k)P'(k) \leq \sum_k u(k)P(k). \tag{18.5}$$

[8] This intuition can be found in many places. The version here draws on Newman (2010) and Campbell (2013). For a rigorous derivation of equation (18.3), see Section 17.8.1 of Newman (2010).

[9] A *Poisson random network* is one in which any two nodes are connected with a fixed independent probability p. The name derives from the fact that the degree distribution is a Poisson distribution for large networks. See Barbour and Mollison (1990) for a full exploration of the relationship between the fully mixed and Poisson random network contagion models.

The concept of stochastic dominance is most familiar in the context of the valuation of risky assets. If P and P' are two lotteries then P first order stochastically dominates P' if the expected payoff from P is greater than the expected payoff from P' for any nondecreasing utility function u.

Definition 2. *A distribution P second order stochastically dominates a distribution P' if for every nondecreasing concave function $u : \mathbb{R} \to \mathbb{R}$,*

$$\sum_k u(k)P'(k) \leq \sum_k u(k)P(k). \tag{18.6}$$

If P and P' have the same mean then P second order stochastically dominates P' is equivalent to P' is a *mean preserving spread* of P. As with first-order stochastic dominance, second-order stochastic dominance has an interpretation in terms of risk: if a lottery P second order stochastically dominates a lottery P' then any risk averse individual prefers P to P'.

Jackson and Rogers apply stochastic dominance relations to network degree distributions (and excess degree distributions). In this setting, first-order stochastic dominance means the nodes are more likely to have higher degrees. A mean preserving spread of the degree distribution corresponds to an increase in the variability of degrees, so the network has more "hubs and spokes."

In Jackson and Rogers' model the probability that a node i becomes infected in a given period is given by

$$\beta(x_i k_i + \delta), \tag{18.7}$$

where k_i is the degree of node i, x_i is the fraction of i's neighbors that are infected, $\beta > 0$ is a constant controlling the diffusion rate, and $\delta \geq 0$ captures nonsocial infection. (Hill, Rand, Nowak, and Christakis 2010 later term an SIS model with the possibility of spontaneous infection the SISa model.) Agents recover with a constant probability γ per unit time.[10]

The starting point of Jackson and Rogers' analysis is the excess degree distribution $\tilde{P}(k)$ of equation (18.2). If $\rho(k)$ is the probability that a node of degree k is infected, then

$$x = \sum_k \tilde{P}(k)\rho(k) \tag{18.8}$$

is the probability that a randomly chosen neighbor is infected (assuming, as in the configuration model, that there is no correlation in neighboring agents' degrees).

One can then solve for steady states of the dynamic process under a continuous deterministic approximation where the assumption is that each agent i has a fraction x of infected neighbors (or, equivalently, each agent i acts as if their expected fraction of infected neighbors is x). Then, susceptible degree k agents become infected at a rate

[10] Note that throughout the chapter we deviate from the original notation of the models discussed when it is convenient for keeping notation consistent across models.

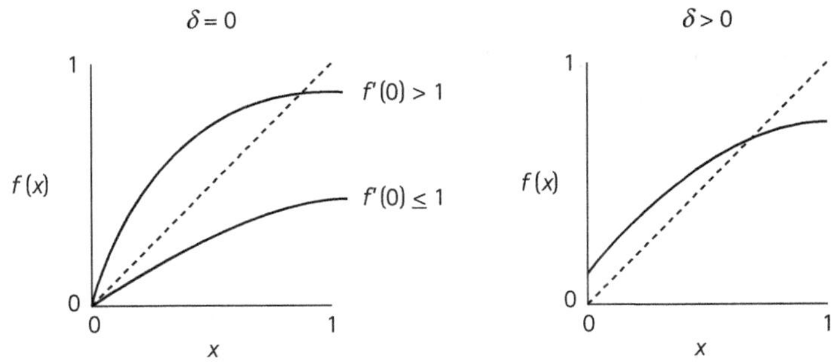

FIGURE 18.4 Possible dynamics in the SIS model of Jackson and Rogers (2007).

of $\beta(xk+\delta)$, and infected degree k agents recover at a rate of γ, so the overall rate of change of $\rho(k)$ is given by

$$\frac{\partial\rho(k)}{\partial t} = (1-\rho(k))\beta(xk+\delta) - \rho(k)\gamma. \tag{18.9}$$

Setting (18.9) equal to zero, solving for $\rho(k)$ and substituting into equation (18.8), we see that steady states of the process are given by fixed points of

$$f(x) = \sum_k \frac{\lambda(xk+\delta)\tilde{P}(k)}{1+\lambda(xk+x)} = \frac{1}{\langle k\rangle}\sum_k \frac{\lambda(xk^2+xk)P(k)}{1+\lambda(xk+x)}, \tag{18.10}$$

where as before $\lambda = \beta/\gamma$.

The function f is strictly increasing and strictly concave in x, so it must have either one or two fixed points. Figure 18.4 illustrates the possible dynamics. If there is no nonsocial infection (i.e., $\delta = 0$), then $f(0) = 0$, so no infection is always an equilibrium. Whether or not a second equilibria exists depends on

$$f'(0) = \frac{\lambda\langle k^2\rangle}{\langle k\rangle}. \tag{18.11}$$

If $f'(0) \leq 1$, no infection is the unique equilibrium, and if $f'(0) > 1$ there is a second non-zero equilibrium. If $\delta > 0$ so agents can become spontaneously infected, then $f > 0$ and has a unique non-zero fixed point, as illustrated in the right panel of Figure 18.4.

Equation (18.10) allows us to examine the impact of changes in the network, as encoded by the degree distribution P or the excess degree distribution $\tilde{P}$, on the steady-state infection levels. The main theorem from Jackson and Rogers (2007) shows that if both P and $\tilde{P}$ undergo a strict first-order stochastic dominance shift or if P undergoes a mean preserving spread, the highest steady-state equilibrium increases. Thus, networks that are more densely connected or in which there are more hubs and spokes are more likely to foster diffusion than sparser networks or networks in which all agents have similar degrees. The proof proceeds by applying the definitions of first-

and second-order stochastic dominance to shifts in the degree distribution or excess degree distribution in equation (18.10). The definitions imply that f increases under those transformations, and if f increases then all of the fixed points of f also increase. This method of analysis provides a template followed by several other diffusion models discussed below.

Jackson and Rogers' result on mean preserving spreads of the degree distribution echoes the computational findings of Watts and Strogatz discussed in Section 18.1.2. In a WS network with $p = 0$, all nodes have the same degree. Increasing p preserves the mean of the degree distribution, since there are the same total number of nodes and links in the network regardless of the value of p, but increases the variability of degrees. Suppose Γ_p and $\Gamma_{p'}$ are both WS networks with n nodes of average degree k, with rewiring probabilities p and p', respectively. If $p' > p$ the degree distribution of $\Gamma_{p'}$ is a mean preserving spread of the degree distribution of Γ_p. Thus, Jackson and Rogers' theorem implies that a contagion will spread more extensively in $\Gamma_{p'}$ than in Γ_p.

López-Pintado (2008) developed a networked SIS model similar to Jackson and Rogers'. The paper also uses a degree-based mean field approximation and examines the effect of stochastic dominance shifts in the degree distribution, but López-Pintado also considers the effect on the critical value of $\lambda = \beta/\gamma$ necessary to sustain a contagion as well as on equilibria of the process. In López-Pintado's model the probability that a susceptible agent of degree k with x infected neighbors becomes infected in a given time period is $\beta f(k,x)$, where f is non-negative and non-decreasing in x. Inclusion of the function f, which López-Pintado terms the *diffusion function*, generalizes the linear formulation given in equation (18.7) from Jackson and Rogers' model. López-Pintado shows that, as in the SIS model without network structure, there is a diffusion threshold λ_c such that if $\lambda > \lambda_c$ then the infection will spread from an initial infection to a non-zero steady state; if $\lambda \leq \lambda_c$ then the infection will die out. López-Pintado's main theorem states that the diffusion threshold λ_c for a network Γ will be lower than that for a network Γ' if any of the following hold:

- The degree distribution of Γ first order stochastically dominates the degree distribution of Γ', or
- the degree distribution of Γ is a mean preserving spread of the degree distribution of Γ' and $k^2 f(k,1)$ is convex for all $k \geq 1$, or
- the degree distribution of Γ' is a mean preserving spread of the degree distribution of Γ' and $k^2 f(k,1)$ is concave for all $k \geq 1$.

One consequence of generalizing the linear infection probability of Jackson and Rogers from equation (18.7) to the more flexible $\beta f(k,x)$ in López-Pintado's model is that the effect of a mean preserving spread of the degree distribution now depends on the functional form of f. However, in some sense a mean preserving spread of the degree distribution increases contagion in more cases than it decreases. For example, if the diffusion function f depends only on the number of infected neighbors x and not on an agent's degree k, then $k^2 f(k,1)$ is necessarily convex, and thus mean preserving spreads lower the contagion threshold. Alternatively, suppose that f is of the form

$f(k,x) = xk^{-\alpha}$. Then $k^2 f(k,1)$ is only concave for all $k \geq 1$ if α falls between one and two. For all other values of α, a mean preserving spread of the degree distribution lowers the diffusion threshold. (This example is examined in Corollary 4 of López-Pintado 2008.)

Taken together, the critical infectivity threshold for the configuration model, Watts and Strogatz (1998) small-world simulations, and the stochastic dominance results from Jackson and Rogers (2007) and López-Pintado (2008) paint a consistent picture of the impact of network structure on the spread of simple contagions: networks with more connections and more variable connections make diffusion easier, faster, and more extensive.[11]

18.2 More Sophisticated Agents

The disease spread analogy explored in Section 18.1 provides powerful insights into the relationship between network structure and diffusion, but it is also important to question the assumption that rational actors choosing to take behaviors can be well represented by models designed to capture the physical process of infection. As we will see in this section, more sophisticated models of interpersonal influence can change the model's predictions regarding which network features promote or inhibit contagion. Because the disease analogy no longer makes sense in these more sophisticated actor models, we use the term "adopter" rather than "infected" to describe agents that take the relevant action.

18.2.1 Threshold Models

One alternative to the epidemiological models discussed in the previous sections are threshold models, in which an individual adopts a behavior once some threshold number or fraction of her neighbors adopts the behavior (Granovetter 1978). This transmission rule arises naturally if one thinks of agents as having a cost to taking an action and a utility from the action that increases in the number (or fraction) of infected neighbors. Thresholds can also capture information cascades where agents infer the value of an action by observing the number or frequency with which their neighbors take that action.

Watts (2002) examines the conditions under which a large cascade can be triggered by a small initial seed in a threshold model where agents have a fractional threshold rule: each agent i has a threshold $\phi_i \in [0,1]$ such that the agent adopts when the fraction of i's adopting neighbors exceeds ϕ_i.

[11] With the one exception being a relatively narrow class of diffusion functions in López-Pintado's model for which a mean preserving spread of the degree distribution makes diffusion more difficult.

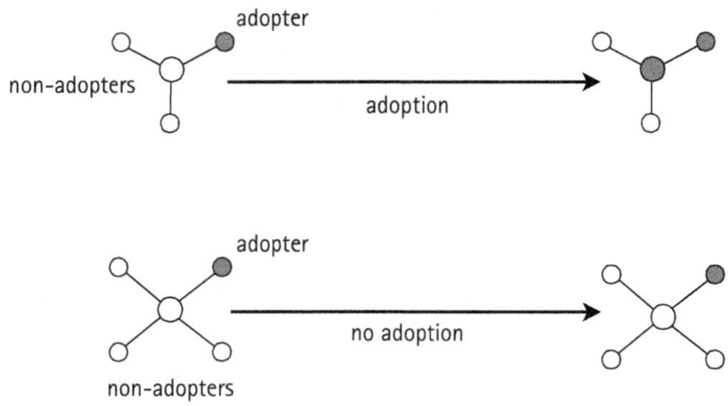

FIGURE 18.5 If the central node has an adoption threshold of .3, one adopting neighbor is sufficient to cause an adoption if the node has degree three but not if the node has degree four.

Using a fractional threshold as opposed to a threshold in terms of absolute number of adopting neighbors has the important consequence that additional connections can inhibit contagion. For example, suppose an agent has an adoption threshold of .3 and one adopting neighbor. As shown in Figure 18.5, if the agent has three total neighbors then the fraction of adopting neighbors, $1/3$, exceeds the agent's adoption threshold and she will adopt. But, if the agent has four total neighbors then the threshold is not met, so one adopting neighbor is insufficient to activate the agent. Watts calls nodes with thresholds sufficiently low so that one adopting neighbor results in adoption "vulnerable." The condition for a network to sustain a contagion is the existence of a large connected cluster of vulnerable nodes. By "large" we mean specifically a giant component as defined in Section 18.1.3.

Watts' analysis is similar to the bond percolation argument for the configuration model discussed in Section 18.1.3, but restricted to only the vulnerable vertices. Suppose the thresholds ϕ_i are independently distributed across the vertices with cumulative distribution F. Let $\rho_k = F(1/k)$ be the probability that a node with degree k is vulnerable. If $P(k)$ is the degree distribution for the network then $\rho_k P(k)$ is the probability that a randomly chosen node is vulnerable. Watts shows that a large connected cluster of vulnerable nodes exists if

$$\sum_k k(k-1)\rho_k P(k) > \langle k \rangle. \tag{18.12}$$

A key feature of (18.12) is that there are either no values of $\langle k \rangle$ for which the inequality is satisfied, or there is a continuous bounded interval of values of $\langle k \rangle$ that satisfy the equation. If $\langle k \rangle$ is too low, then the network is not connected enough to permit cascades. if $\langle k \rangle$ is too high, then there are not enough connected vulnerable nodes to get the contagion started. Watts refers to the interval of $\langle k \rangle$ values satisfying (18.12) as the *cascade window*. This stands in contrast to the disease like contagion models described

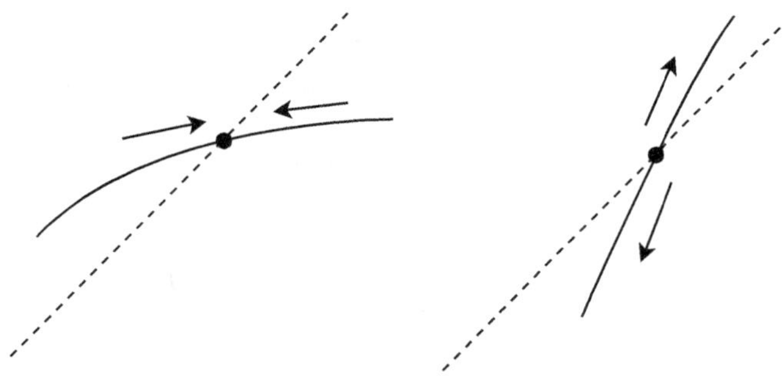

FIGURE 18.6 Local neighborhoods of stable (left) and unstable (right) equilibria. The dashed line corresponds to $f(x) = x$.

in Section 18.1.3, in which there is a single critical value of $\langle k \rangle$ above which contagion occurs.

Jackson and Yariv (2005, 2007) use a degree-based mean field approximation like those employed by Jackson and Rogers (2007) and López-Pintado (2008) to explore a threshold diffusion model. Each agent i has a cost to adoption c_i. Benefits to adoption are described by a function $\nu(k_i, x)$, where k_i is agent i's degree and x is the fraction of i's neighbors that i *expects* to adopt. Depending on the form of the benefits function ν, agents may have fractional or absolute adoption thresholds. If $\nu(k, x) = x$, Jackson and Yariv's model is equivalent to Watts'.

Jackson and Yariv explicitly interpret the approximation in terms of boundedly rational agents with incomplete information. Each agent is assumed to know their own costs and degree, the degree distribution in the population, and the fraction of adopting agents of each degree, but not the specific identities of their neighbors. Agents assume that costs and degrees are independent and that the distribution of their neighbors' actions will mirror the distribution of actions in the population.

Jackson and Yariv frame the model as a diffusion process where an initial fraction of agents adopt and then at each subsequent time step agents best respond to the distribution of actions in the population. If H is the cumulative distribution of costs and x_t is the probability that a random neighbor is an adopter at time t, then the fraction of degree k agents that would choose to adopt as a best response at time $t+1$ is given by $H(\nu(k, x_t))$. Thus, the probability that a random neighbor is an adopter at time $t+1$ is given by

$$f(x_t) = \sum_k \tilde{P}(k) H(\nu(k, x_t)). \qquad (18.13)$$

The function f defines a one-dimensional dynamical system that captures the diffusion process. Fixed points of f correspond to equilibria, which can be either *stable* or *unstable*. As illustrated in Figure 18.6, stable equilibria can be identified by the fact that f is above the forty-five degree line for values of x just below the equilibrium and below the forty-five degree line for values of x just above the equilibrium. At unstable

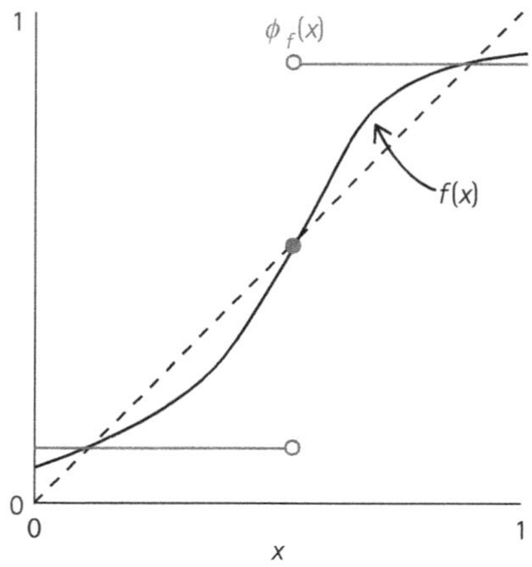

FIGURE 18.7 The function $\phi_f(x)$ describes the eventual resting point reached from an initial point x under the dynamical system f.

equilibria, popularly known as "tipping points" (Lamberson and Page 2012), f crosses the forty-five degree line from below to above.

We could imagine several scenarios that might create random fluctuations around the predicted deterministic dynamics of the model. For example, agents might not always best respond to their neighbor's actions, or they might not correctly observe the state of the population. As illustrated on the left side of Figure 18.6, if the system is at a stable equilibrium and a small perturbation moves it away from that point, the dynamics tend to push the system back to the equilibrium. As shown on the right side of Figure 18.6, if the equilibrium is unstable, any small deviation from the equilibrium will result in the system moving even further away from the equilibrium point.

To characterize which networks result in more or less diffusion, we introduce a function $\phi_f(x) : [0,1] \to [0,1]$ that describes where the system will come to rest as a function of the initial fraction of adopters. Specifically, for any $x \in [0,1]$ with $f(x) < x$, let $\phi_f(x)$ be the largest stable equilibrium of f less than x. For any $x \in [0,1]$ with $f(x) > x$ define $\phi_f(x)$ as the smallest stable equilibrium of f greater than or equal to x. If $x = f(x)$ then let $\phi_f(x) = x$. Figure 18.7 depicts an example. Let f_θ denote the dynamical system resulting from a set of model parameters θ (e.g., cost distribution and network). We say that a set of model parameters θ *generates greater diffusion* than another set θ' if the corresponding function $\phi_{f_\theta}(x) \geq \phi_{f_{\theta'}}(x)$ for all $x \in [0,1]$ and $\phi_{f_\theta}(x) > \phi_{f_{\theta'}}(x)$ for some $x \in [0,1]$. This formally captures the idea that for all initial conditions the resulting equilibrium adoption level is at least as high under θ as under θ', and some initial conditions result in greater adoption at equilibrium with parameters θ than with θ'. (This definition is equivalent to the those of Jackson and Yariv 2007 and Lamberson 2010.) Any change in model parameters that increases the values of the function f will result in greater diffusion.

Jackson and Yariv examine the consequence of first-order stochastic dominance shifts in the cost distribution, and first- and second-order stochastic dominance shifts in the degree distribution. Not surprisingly, when costs are lower more agents adopt.

Adding more connections to the network, in the sense of a first-order stochastic dominance shift in the excess degree distribution results in greater diffusion if $H(\nu(k,x))$ is increasing in k for all x and less diffusion if $H(\nu(k,x))$ is decreasing in k for all x. So, for example, if agents' payoffs depend on the absolute number of their adopting neighbors, kx, rather than the fraction x, more connected networks lead to greater diffusion. A mean preserving spread of the degree distribution increases diffusion if $H(\nu(k,x))$ is nondecreasing and convex in k.

Outside of economics, Centola and Macy (2007) examine a threshold model using simulations on the Watts-Strogatz family of networks. They coined the term *complex contagion* to describe contagions that require more than one contact to spread, as opposed to simple contagions like those described by the SIS or SIR models that can potentially spread with just one contact between an infected and susceptible individual. Recall that Watts and Strogatz (1998) found that a simple contagion spreads more easily and more quickly as the rewiring probability increased. In contrast, Centola and Macy demonstrated a nonmonotonic effect of the rewiring probability on the speed of diffusion for complex contagions: some rewiring increased the speed of diffusion, but higher levels of rewiring slowed the spread. The simulations also showed that while for low levels of rewiring the contagion nearly always saturated the network, too much rewiring made it nearly impossible for the contagion to spread. Using some toy examples, Centola and Macy argue that the inhibitory effects of high rewiring result from the narrowing of "bridges"—collections of links that span from one part of the ring lattice to another—which are necessary to allow a complex contagion to "hop" across the network.

18.2.2 Word of Mouth and Percolation

Campbell (2013) developed a model similar to the threshold models of Watts (2002) and Jackson and Yariv (2007), but Campbell introduces another strategic actor to the model: a seller that determines the price of the spreading product.

In Campbell's model the cost of adoption is a fixed price P set by a monopolist seller that is constant across agents, but the agents vary in how much they value the product. This contrasts with Jackson and Yariv's model where agents costs to adoption are heterogeneous but the benefits of adoption depend only on an agent's degree and number of adopting neighbors. Specifically, each agent i in Campbell's model has a valuation θ_i drawn independently from a uniform distribution on $[0,1]$. The utility to i from purchasing the product is $\theta_i - P$ and 0 from not purchasing, so agents with $\theta_i > P$ prefer to make the purchase. Thus, the probability that a randomly chosen agent prefers to adopt the product is $1 - P$.

In the basic version of the model agents are connected by a network drawn from the configuration model described in Section 18.1.3, and the results describe the behavior in expectation over all such networks. Given the social network, the model proceeds as follows. Initially only a small fraction ϵ of the agents are informed about the product. Agents that are informed and have valuations $\theta_i > P$ purchase the product and inform all of their neighbors. Agents with $\theta_i \leq P$ are a dead end for the diffusion process; they neither purchase the product nor inform their neighbors. Agents who become informed by their adopting neighbors make their own purchase decisions and, if they chose to purchase, go on to inform their neighbors and so on.[12]

The analysis of the model is similar to the percolation argument described for Watts' model in Section 18.2.1, but now the role of "vulnerable" nodes is played by agents who receive positive utility from adopting the product. We can think of removing all of the nodes from the network with $\theta_i \leq P$, so that only the agents that would be willing to purchase the product if they became informed remain. The question of whether the product will spread or not then reduces to determining whether or not this subnetwork consisting only of agents willing to adopt the product comprises a giant component of the network or not. The existence of such a giant component will depend both on the original friendship network structure as well as the price P. As the price increases, fewer nodes will be willing to adopt the product, and thus the likelihood that a giant component exists diminish. One of Campbell's first results, which follows from the percolation analysis described in Section 18.1.3, is that there is a critical price P^{crit} above which there is no such giant component of willing adopters.

Campbell goes on to examine the elasticity of demand and optimal pricing. In the baseline model described above, demand is more elastic and the optimal price is lower in this word-of-mouth model than if the population were fully informed (essentially because the price impacts both the probability of becoming informed as well as the probability of purchase). However, in an extension of the model where individuals' valuations for the product are positively correlated with their degree the opposite can occur, and the optimal price may be higher than in the fully informed case.

Campbell also examines shifts in the degree distribution similar to those studied in the papers discussed in Section 18.1.4. Adding links to the network moves the word-of-mouth model closer to the fully informed setting and thus increases the optimal price (Campbell's result is derived only for shifts in the mean degree of Poisson random networks rather than a general first-order stochastic dominance shift of the degree distribution). The effect of a mean preserving spread of the degree distribution on the optimal price depends on marginal costs: when marginal costs are low, the

[12] While they do not explore its theoretical properties in detail, it is worth mentioning that Banerjee, Chandrasekhar, Duflo, and Jackson (2013) develop an empirically motivated diffusion model with a structure similar to Campbell's. As in Campbell's model there are really two simultaneous diffusion processes: the spread of awareness of the product and the spread of adoption of the product (in this case the product is microfinance loans spreading in rural Indian villages). Agents in the model inform each of their neighbors about the product with independent probability q^P if they themselves have adopted the product, and probability q^N if they are not adopters. The adoption decision is modeled as a logistic function that depends on individual characteristics and the fraction of an agent's informed neighbors who choose to adopt. One can think of this logistic function combined with agent's individual characteristics as determining the valuations from Campbell's model.

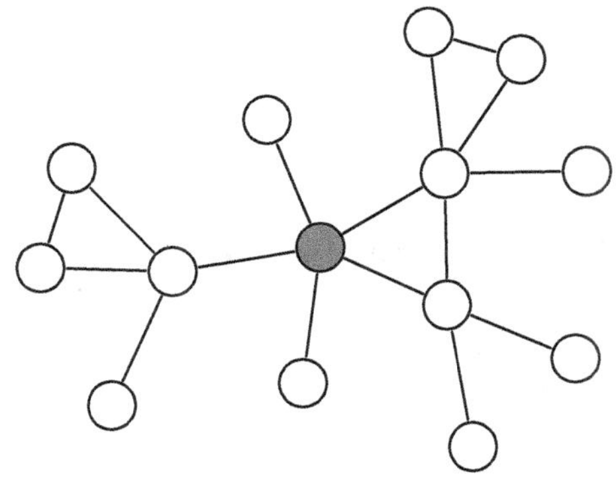

FIGURE 18.8 Random networks with clustering can be represented by the joint distribution of edges that are not part of triangles and pairs of edges that are part of closed triangles (Newman 2009). For example, the shaded node has three edges that are not part of closed triangles and one pair of edges that are part of a closed triangle.

optimal price decreases with a mean preserving spread of the degree distribution; when marginal costs are high, a mean preserving spread of the degree distribution increases the optimal price.

Campbell extends the model in several directions that go beyond most of the results discussed in this chapter. Namely, he introduces positive correlation between individuals' valuations and their degrees; homophily between individuals' valuations and their neighbors' valuations; and clustering in the network. Since our focus in this chapter is the relationship between network structure and diffusion, we will discuss the network clustering results and refer the reader to the original article for the other extensions.

Clustering in the network is captured using a technique developed by Newman (2009) in which the network is represented by a joint probability distribution describing the probability a node has a given number of edges that are not part of closed triangles and the number of pairs of edges that are part of closed triangles. Rather than simply having a degree given by the number of edges, you can think of each node as having an "edge degree" and a "triangle degree." For example, in the network depicted in Figure 18.8 the shaded node has edge degree three and triangle degree one because it has three edges that are not part of closed triangles and one pair of edges that are part of a closed triangle. Random networks with nontrivial clustering can then be constructed in a manner similar to the configuration model discussed in Section 18.1.3, but now nodes initially have edge stubs and triangle stubs that are randomly connected.

Introducing clustering does not qualitatively change Campbell's results. There continues to be a critical price for diffusion of the product, and the optimal price in the word-of-mouth model is still below the optimal price if agents are fully informed. However, for a given price increasing clustering (holding both degree and excess degree constant) decreases demand. The intuition is that increased clustering slows the spread of information because clustered ties are redundant. Every edge that closes a triangle is connecting two agents that already have a short path in the network between them (the path of length two through their mutual friend), while in an unclustered network this edge would have joined individuals with no common friends and thus allowed information to spread to nodes that might not have received it otherwise.

18.2.3 Social Learning

Lamberson (2010) used techniques similar to those employed by Jackson and Rogers (2007), López-Pintado (2008), and Jackson and Yariv (2005, 2007) in the papers discussed above to analyze a diffusion model of social learning in a network. Agents in this model choose whether or not to adopt an action, such as using a new technology, with an unknown payoff distribution. Each agent has prior beliefs about the unknown payoffs of the action, and initial adoption decisions are based on these prior beliefs alone. At each subsequent time step, adopting agents receive a payoff drawn from a fixed distribution and then communicate the payoffs they receive to their neighbors. All of the agents then update their beliefs based on the payoffs of their adopting neighbors from the previous time period(s) and revise their adoption decision. Then the process repeats with current adopters receiving new payoffs, communicating to their neighbors, and so on.

A key assumption is that agents' prior beliefs are treated differently than the information they receive from their neighbors. Rather than continuously updating their beliefs, agents only incorporate recently observed payoff information while their prior beliefs remain fixed and unchanged. Lamberson justifies the assumption by interpreting the prior beliefs as "representing a long-standing conviction or fundamental attitude" and supposes that agents store each payoff observation separately in their memory. Over time the agents forget older observations and replace them with more recent information.

While more complicated than the contagion and threshold models, the strategy of the analysis in Lamberson's model is the same as those discussed in Sections 18.1.4 and 18.2.1. First, using a degree-based mean field approximation, an equation for a one-dimensional dynamical system that depends on the model parameters, in this case the distribution of prior beliefs and the degree distribution, is derived. Then the definitions of first- and second-order stochastic dominance are applied to the equation governing the dynamical system to examine how the equilibria shift under changes to the network.

In Lamberson's model, as in the threshold models of Watts, and Jackson and Yariv, adding links to the network (again in the sense of a first-order stochastic dominance shift in the excess degree distribution) can *decrease* adoption in some cases. However, in the model Lamberson examines this can occur even when payoffs to adoption are positive on average (where payoffs to the status quo are fixed at zero), and thus adding links to the network reduces overall welfare. The result comes from the fact that if agents' prior beliefs are sufficiently high relative to the actual payoffs from adoption, more agents might adopt when they observe fewer payoffs than when they observe more. In this case more information results in worse decisions.

As the results in this section illustrate, when agents employ more sophisticated decision rules, the relationship between network structure and contagion may not follow the same pattern as predicted by models of simple disease like contagions. In particular, in both Watts' and Jackson and Yariv's threshold models and Lamberson's social learning model, increasing the connectedness of the network can inhibit diffusion.

18.3 Looking Forward

Nearly all of the models discussed in this chapter confine their representation of network structure to the degree distribution (Campbell's 2013 inclusion of network clustering is a notable exception). In some sense there is no network in these models, just agents with different degree types that correspond to how many other agents they randomly encounter in a given time period. But two networks with the same degree distribution can look quite different. For example, both of the networks shown in Figure 18.9 have one hundred nodes, and in each network all of the nodes have degree six. The network on the left is a lattice on a torus, while the network on the right is drawn from the random configuration model described in Section 18.1.3. By almost any measure of network structure that considers more than one-step neighborhoods, the networks differ significantly. The clustering coefficient of the left network is .4, while for the network on the right the clustering coefficient is just .0375. The average path length on the left is 3.95 and on the right is 2.74.

Because both networks in Figure 18.9 have the same degree distribution, many of the models discussed so far would treat them as effectively the same. One could ask then, are they effectively the same? A simple simulation exercise shows that they are not. Figure 18.10 shows the average percent of the population adopting from 1000 simulations of a threshold model (with thresholds uniformly distributed on $[1, 5]$) as a function of the initial percent adopting in each of the two networks. As we can see, in this example the "critical mass" required for diffusion is lower, and the extent of diffusion is greater, in the more clustered network.

To advance our understanding of how network structure beyond the degree distribution impacts diffusion, we will either have to develop new analytical methods or embrace one or both of two methods of analysis less common in the economics literature: simulations and numerical approximations. For example, while economists often interpret the degree-based mean field analysis discussed throughout this chapter as representing a random meeting process with incomplete information, epidemiologists, physicists, and biologists more often think of the model as simply an approximation to a contagion spreading through a fixed network. Taking the approximation interpretation opens the door to other approximations for which the corresponding meeting process or incomplete information interpretation is less clear. Nevertheless these approximations may be necessary to shed light on the role of deeper network characteristics in

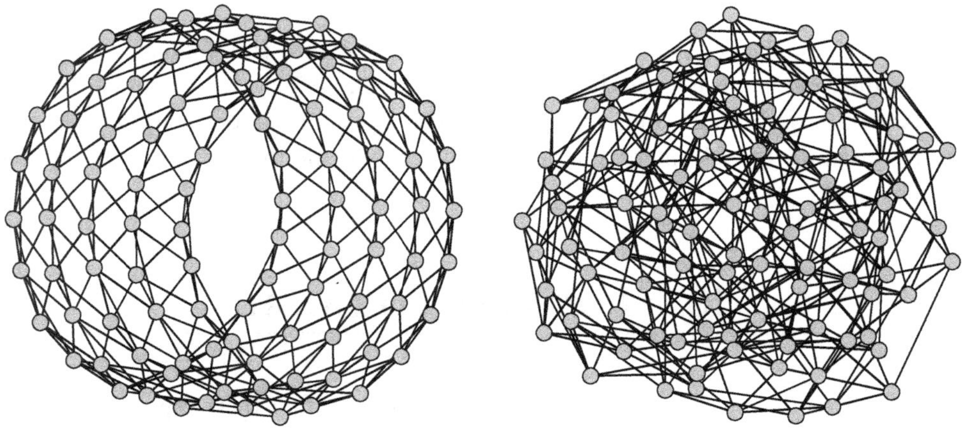

FIGURE 18.9 Two networks with the same degree distribution.

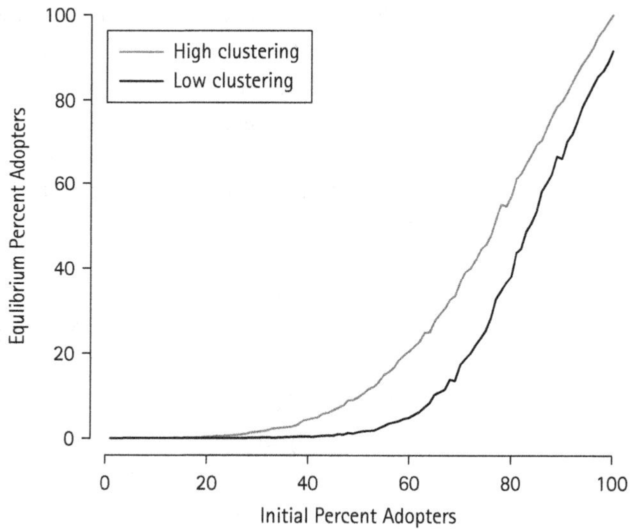

FIGURE 18.10 Average equilibrium adoption as a function of initial percent adopters from a threshold model in the two networks shown in Figure 18.9.

diffusion. For example, Lamberson (2015) employs a "pair approximation" (Morris 1997) to examine the effects of local correlation and network clustering in a diffusion model.

Simulations can be used both to analyze models that are too complex to be analytically tractable as well as to demonstrate that the simplifications necessary for analytic results provide good approximations. Such comparisons between analytic results and simulations are standard practice in other disciplines studying networks and diffusion such as mathematical biology and physics (for example, see the papers by Moore and Newman 2000; Ohtsuki, Hauert, Lieberman, and Nowak 2006; Hill et al. 2010). These simulations can also shed light on the conditions under which the analytic approximations are likely to diverge from the actual diffusion process. For example, Figure 17.5 of Newman (2010) illustrates that the configuration model approximation for the SI model is quite accurate on a network with a low clustering coefficient but differs substantially from simulation results on a more clustered network.

Another future direction for diffusion research is developing models that incorporate richer information about the individuals in the network and their patterns of interaction. For example, Galeotti and Rogers (2013) and Jackson and Lopez-Pintado (2013) both consider diffusion models with heterogeneous agents whose type may be correlated with their neighbors. In particular, these models allow for networks that exhibit homophily—the tendency for individuals to connect preferentially to others like themselves. Homophily is widely observed in empirical social networks (McPherson, Smith-Lovin, and Cook 2001), but before these models we had little understanding of its impact on contagion.

In the network diffusion models discussed in this chapter, agents' payoffs and choices depend on the behavior of their neighbors, but individuals do not strategically attempt to influence their contacts. However, we can imagine situations in which agents have a vested interest in their neighbors' behavior. For example, many viral marketing campaigns provide incentives so that if a social contact makes a purchase through a referral the sender of that referral receives some form of payment or discount (Leskovec, Adamic, and Huberman 2007; Aral, Muchnik, and Sundararajan 2013). In a political context voters may be influenced by their social contacts, and also attempt to influence the election outcome as a whole through their network. One paper that incorporates such strategic calculations is the model of rumor spreading developed by Bloch, Demange, and Kranton (2014). In their model some agents have an incentive to spread disinformation in order to influence a collective outcome.

Finally, while most contagion models treat the network of connections as fixed, real social networks are seldom static. Understanding how network dynamics interact with contagion dynamics is another open challenge for diffusion research.

References

Aral, S., L. Muchnik, and A. Sundararajan (2013). "Engineering social contagions: Optimal network seeding in the presence of homophily." *Network Science* 1(02), 125–153.

Bailey, N. (1975). *The Mathematical Theory of Infectious Diseases and its Applications*. London: Charles Griffin and Company Ltd.

Banerjee, A., A. G. Chandrasekhar, E. Duflo, and M. O. Jackson (2013). "The diffusion of microfinance." *Science* 341(6144).

Barbour, A. and D. Mollison (1990). *Epidemics and Random Graphs*, pp. 86–89. New York: Springer.

Bloch, F., G. Demange, and R. Kranton (2014). "Rumors and social networks." *Paris School of Economics, Working paper 2014* 15(1).

Campbell, A. (2013). "Word-of-mouth communication and percolation in social networks." *The American Economic Review* 103(6), 2466–2498.

Centola, D. and M. Macy (2007). "Complex contagions and the weakness of long ties." *American Journal of Sociology* 113(3), 702–734.

Galeotti, A., S. Goyal, M. O. Jackson, F. Vega-Redondo, and L. Yariv (2010). "Network games." *Review of Economic Studies* 77(1), 218–244.

Galeotti, A. and B. W. Rogers (2013). "Strategic immunization and group structure." *American Economic Journal: Microeconomics* 5(2), 1–32.

Geoffard, P.-Y. and T. Philipson (1996). "Rational epidemics and their public control." *International Economic Review* 37(3), 603–624.

Gladwell, M. (2014). "Q and A with Malcolm." http://gladwell.com/the-tipping-point/the-tipping-point-q-and-a/.

Granovetter, M. (1978). "Threshold models of collective behavior." *The American Journal of Sociology* 83(6), 1420–1443.

Grassberger, P. (1983). "On the critical behavior of the general epidemic process and dynamical percolation." *Mathematical Biosciences* 63(2), 157–172.

Hethcote, H. W. (2000). "The mathematics of infectious diseases." *SIAM Review* 42(4), 599–653.

Hill, A., D. Rand, M. Nowak, and N. Christakis (2010). "Infectious disease modeling of social contagion in networks." *PLoS Computational Biology* 6(11), e1000968.

Jackson, M. O. (2008). *Social and Economic Networks.* Princeton, NJ: Princeton University Press.

Jackson, M. O. and D. Lopez-Pintado (2013). "Diffusion and contagion in networks with heterogeneous agents and homophily." *Network Science* 1(1), 49–67.

Jackson, M. O. and B. W. Rogers (2007). "Relating network structure to diffusion properties through stochastic dominance." *The B.E. Journal of Theoretical Economics (Advances)* 7(1), 1–13.

Jackson, M. O. and L. Yariv (2005). "Diffusion on social networks." *Économie Publique* 16, 69–82.

Jackson, M. O. and L. Yariv (2007). "Diffusion of behavior and equilibrium properties in network games." *American Economic Review* 97(2), 92–98.

Keeling, M. and P. Rohani (2008). *Modeling Infectious Diseases in Humans and Animals.* Princeton University Press.

Kremer, M. (1996). "Integrating behavioral choice into epidemiological models of AIDS." *Quarterly Journal of Economics* 111(2), 549–73.

Lamberson, P. J. (2010). "Social learning in social networks." *The B.E. Journal of Theoretical Economics* 20(1), 36.

Lamberson, P. J. (2011). "Linking network structure and diffusion through stochastic dominance." *Connections* 31(1), 4–14.

Lamberson, P. J. (2015). Network games with local correlation and clustering. Available at SSRN: http://ssrn.com/abstract=2562567.

Lamberson, P. J. and S. E. Page (2012). "Tipping points." *Quarterly Journal of Political Science* 7(2), 175–208.

Leskovec, J., L. A. Adamic, and B. A. Huberman (2007). "The dynamics of viral marketing." *ACM Transactions on the Web* 1(1).

López-Pintado, D. (2008). "Diffusion in complex social networks." *Games and Economic Behavior* 62(2), 573–590.

McPherson, M., L. Smith-Lovin, and J. Cook (2001). "Birds of a feather: Homophily in social networks." *Annual Review of Sociology* 27(1), 415–444.

Mollison, D. (1977). "Spatial contact models for ecological and epidemic spread." *Journal of the Royal Statistical Society. Series B (Methodological)*, 283–326.

Moore, C. and M. E. Newman (2000). "Epidemics and percolation in small-world networks." *Physical Review E* 61(5), 5678.

Morris, A. J. (1997). *Representing Spatial Interactions in Simple Ecological Models*. Ph. D. thesis, University of Warwick.

Newman, M. and J. Park (2003). "Why social networks are different from other types of networks." *Physical Review E* 68(3), 36122.

Newman, M. E. (2009). "Random graphs with clustering." *Physical Review Letters* 103, 058701.

Newman, M. E. J. (2010). *Networks: An Introduction*. Oxford: Oxford University Press.

Ohtsuki, H., C. Hauert, E. Lieberman, and M. Nowak (2006). "A simple rule for the evolution of cooperation on graphs and social networks." *Nature* 441(7092), 502–505.

Pastor-Satorras, R. and A. Vespignani (2001). "Epidemic spreading in scale-free networks." *Physical Review Letters* 86(14), 3200–3203.

Shalizi, C. R. and A. C. Thomas (2011). "Homophily and contagion are generically confounded in observational social network studies." *Sociological Methods & Research* 40(2), 211–239.

Toxvaerd, F. (2010a). "Infection, acquired immunity and externalities in treatment." *CEPR Discussion Paper*, No. 8111.

Toxvaerd, F. (2010b). "Recurrent infection and externalities in prevention." *CEPR Discussion Paper*, No. DP8112.

Watts, D. (2002). "A simple model of global cascades on random networks." *Proceedings of the National Academy of Sciences* 99(9), 5766 –5771.

Watts, D. and S. Strogatz (1998). "Collective dynamics of 'small-world' networks." *Nature* 393, 440–442.

CHAPTER 19

LEARNING IN SOCIAL NETWORKS

BENJAMIN GOLUB AND EVAN SADLER

19.1 Introduction

Social ties convey information through observations of others' decisions as well as through conversations and the sharing of opinions. The resulting information flows play a role in a range of phenomena, including job search (Montgomery 1991), financial planning (Duflo and Saez 2003), product choice (Trusov et al. 2009), and voting (Beck et al. 2002). Understanding how individuals use information from their social environments, and the aggregate consequences of this learning, is therefore important in many contexts. Is dispersed information aggregated efficiently? Whose opinions or experiences are particularly influential? Can we understand when choices will be diverse, when people will choose to conform, and how the networks in which individuals communicate shape these outcomes?

This chapter surveys several approaches to these questions. We begin with a discussion of sequential social learning models in which each agent makes one decision; these models have admitted a rich analysis within a canonical Bayesian paradigm. We next discuss the DeGroot model of repeated linear updating. This theory employs a simple heuristic learning rule, delivering a fairly complete characterization of learning dynamics as a function of network structure. Finally, we review work that studies repeated Bayesian (or quasi-Bayesian) updating in general networks.

19.2 The Sequential Social Learning Model

An important early branch of the social learning literature arose to explain widespread conformity within groups, referred to as herd behavior. Banerjee (1992) and

Bikhchandani et al. (1992) independently proposed models in which players each take a single action in sequence. Before making a choice, each player observes all previous actions and a private signal. Though individual payoffs are independent of other players' actions, others' choices provide information about their signals, and therefore about what action is best. This information constitutes an externality. Oftentimes, individuals acting later will optimally ignore their private signals and copy the crowd instead. If this happens, the choices of these players cease to reveal new information, and the population can herd on a suboptimal action. Variations on this insight have been used to explain phenomena ranging from fads and fashions to stock bubbles, and there is a rich literature on extensions and applications.[1]

In this section we cover more recent efforts, adapting the classical herding models to a network structure that encodes partial observation of the history. The distinguishing feature in this line of work is the sequential structure, and we collectively refer to these models as the *sequential social learning model* (SSLM). While Bayesian learning is often difficult to analyze in a network, sequential models have proved particularly tractable. Since each player makes only one choice, there is no scope for strategic interactions, and researchers can study the information externality in isolation from other concerns.

Two key principles provide intuition for long-run learning outcomes in the SSLM: the *improvement principle* and the *large-sample principle*. The improvement principle notes that a player always has the option to copy one of her neighbors, so the payoff from imitating provides a lower bound on her expected utility. Since the player additionally has a private signal, she might be able to improve upon imitation. One family of learning results characterizes how well the population learns through this mechanism. The intuition of the improvement principle is generally robust to features of the network, but the extent of learning depends heavily on the private signals because continued improvement on imitation requires the possibility of a strong signal. If players observe many actions, we can additionally employ the large-sample principle: If observations are at least partly independent, then a large sample collectively conveys much information, even if each signal is individually weak. In these networks, the exact distribution of private signals matters less, and learning is far more robust.

19.2.1 The SSLM with Homogeneous Preferences

Our first pass at the SSLM assumes that players differ only in their initial information, not in their preferences. Our formulation includes as special cases the classical models of Banerjee (1992), Bikhchandani et al. (1992), and Smith and Sørensen (2000), which assume that all players observe the entire history. However, a decision maker's information set is typically far less complete. Authors have taken different approaches to represent limited observation of other players' choices. One approach, adopted by

[1] See Chamley (2004) for an overview. Bose et al. (2008) and Ifrach et al. (2013) offer recent contributions.

Smith and Sørensen (2008), is for players to take anonymous random samples from the past. We focus more explicitly on the network of connections between players; players know the identities of those they observe, and we consider arbitrary sampling processes. Our framework encompasses recent work by Çelen and Kariv (2004), Acemoglu et al. (2011), Arieli and Mueller-Frank (2014), and Lobel and Sadler (2015). We can also include the model of Eyster and Rabin (2011) if we relax the assumption that players' signals are identically distributed.

19.2.1.1 Information and Strategies

Each player $n \in \mathbb{N} = \{1,2,3,\ldots\}$ makes a binary choice of action $x_n \in \{0,1\}$ in sequence. The state of the world is $\theta \in \{0,1\}$, and players share a common prior $q_0 = \mathbb{P}(\theta = 1)$. Players prefer to choose the action that matches the state: We assume a common utility function $u(x,\theta)$ with

$$u(1,1) > u(0,1) \text{ and } u(0,0) > u(1,0).$$

Player n observes a signal s_n taking values in an arbitrary metric space $\mathcal{S}$, and the signals $\{s_n\}_{n\in\mathbb{N}}$ are independent and identically distributed conditional on the underlying state θ. We use $\mathbb{F}_\theta$ to denote the signal distribution conditional on the state. We assume $\mathbb{F}_0$ is not almost everywhere equal to $\mathbb{F}_1$, so some signal realizations provide information.

For concreteness, consider some examples of signal structures. We could have binary signals, with each s_n taking values in $\mathcal{S} = \{0,1\}$ and where

$$\mathbb{P}(s_n = 0 \mid \theta = 0) = \mathbb{P}(s_n = 1 \mid \theta = 1) = g > \frac{1}{2}.$$

The realization $s_n = 0$ provides evidence in favor of $\theta = 0$, while $s_n = 1$ provides evidence in favor of $\theta = 1$. This is the signal structure studied by Banerjee (1992) and Bikhchandani et al. (1992). We could also consider real-valued signals with $\mathbb{F}_0$ and $\mathbb{F}_1$ being probability measures on $\mathbb{R}$. For instance, suppose $\mathbb{F}_0$ and $\mathbb{F}_1$ are both supported on $\mathcal{S} = [0,1]$, where $\mathbb{F}_0$ has density $2-2s$ and $\mathbb{F}_1$ has density $2s$. In this case, lower signal realizations provide stronger evidence in favor of $\theta = 0$: the signal $s \in [0,1]$ induces the likelihood ratio $\frac{s}{1-s}$.

In addition to the signal s_n, player n observes the choices of those in her neighborhood $B(n) \subseteq \{1,2,\ldots,n-1\}$. That is, player n observes the value x_m for each $m \in B(n)$. This neighborhood is a subset of players who have already acted before player n, and the sequence of neighborhoods $\{B(n)\}_{n\in\mathbb{N}}$ constitutes the observational network. Each neighborhood is randomly drawn according to a distribution, and we use $\mathbb{Q}$ to denote the joint probability distribution of the sequence of neighborhoods. We assume the distribution $\mathbb{Q}$, which we call the network, is common knowledge among the players.

The following are examples of networks:

- For each $n \in \mathbb{N}$, $B(n) = \{1,2,\ldots,n-1\}$ with probability 1. This is the complete network of Banerjee (1992), Bikhchandani et al. (1992), and Smith and Sørensen (2000), in which each player observes all of her predecessors.

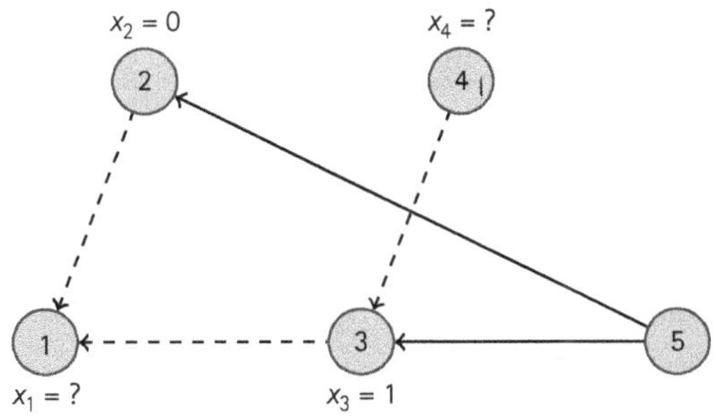

FIGURE 19.1 A realized network. Player 5 observes players 2 and 3, but is uncertain about other decisions and links.

- For each $n \in \mathbb{N}$, $B(n) = \{n-1\}$ with probability 1. This is the network of Çelen and Kariv (2004) in which each player observes only her immediate predecessor.
- For each $n \in \mathbb{N}$, $B(n)$ contains one element drawn uniformly at random from $\{1,2,\ldots,n-1\}$, and all these draws are independent. This example of observing a random predecessor demonstrates how neighborhood realizations can be stochastic.
- With equal probability, $B(2) = \{1\}$ or $B(2) = \emptyset$. If $B(2) = \emptyset$, then $B(n) = \emptyset$ for all n; otherwise $B(n) = \{n-1\}$ for all n. Here, every agent observes only her immediate predecessor or every agent observes nobody. So neighborhood realizations are again random and, in contrast to the previous example, correlated.

The information player n observes is then $I_n = \{s_n, x_m, m \in B(n)\}$: her private signal and the actions of all the predecessors in her realized neighborhood. This is a natural generalization of the early sequential learning literature, which implicitly assumed a complete network.

We often reference player n's *private belief* $p_n = \mathbb{P}(\theta = 1 \mid s_n)$ separately from player n's *social belief* $q_n = \mathbb{P}(\theta = 1 \mid x_m, m \in B(n))$, and we use $\mathbb{G}_\theta$ to denote the distribution function of p_n conditional on the state. A strategy σ_n for player n maps each possible realization of her information I_n to an action $x_n \in \{0,1\}$. A strategy profile $\sigma \equiv \{\sigma_n\}_{n\in\mathbb{N}}$ induces a probability distribution $\mathbb{P}_\sigma$ over the sequence of actions. The profile σ is a *perfect Bayesian equilibrium* if each player maximizes her expected utility, given the strategies of the other players:

$$\mathbb{E}_\sigma\left[u(\sigma_n,\theta) \mid I_n\right] \geq \mathbb{E}_\sigma\left[u(\sigma_n',\theta) \mid I_n\right]$$

for any strategy σ_n'. Since each player acts once in sequence, an inductive argument establishes the existence of an equilibrium, though in general this is non-unique since some players may be indifferent between the two actions.

19.2.1.2 Long-Run Learning Metrics

Our study of equilibrium behavior centers on asymptotic outcomes. In particular, we consider two metrics—diffusion and aggregation—based on players' expected utility as the index n approaches infinity. *Aggregation* occurs if players' utility approaches what they would obtain with perfect information:

$$\lim_{n\to\infty} \mathbb{E}_\sigma \left[u\left(x_n,\theta\right)\right] = q_0 u(1,1) + (1-q_0)u(0,0).$$

This represents the best asymptotic outcome we can hope to achieve: For later players, it is as though the private information of those that came before them is aggregated into a single, arbitrarily precise signal.

The definition of *diffusion* depends on the signal distribution, and the support of private beliefs in particular. The support of the private beliefs is the region $[\underline{\beta},\overline{\beta}]$, where

$$\underline{\beta} = \inf\{r \in [0,1] \,|\, \mathbb{P}(p_1 \le r) > 0\} \text{ and } \overline{\beta} = \sup\{r \in [0,1] \,|\, \mathbb{P}(p_1 \le r) < 1\}.$$

It can be shown that there is a unique binary signal $\tilde{s} \in \{0,1\}$, a random variable such that

$$\mathbb{P}(\theta = 1 \,|\, \tilde{s} = 0) = \underline{\beta} \text{ and } \mathbb{P}(\theta = 1 \,|\, \tilde{s} = 1) = \overline{\beta}.$$

We shall call $\tilde{s}$ the *expert signal*. Diffusion occurs if we have

$$\liminf_{n\to\infty} \mathbb{E}_\sigma \left[u\left(x_n,\theta\right)\right] \ge \mathbb{E}\left[u(\tilde{s},\theta)\right] \equiv u^*.$$

Intuitively, we have diffusion if players perform as though they were guaranteed to receive one of the strongest possible signals.

In any particular network $\mathbb{Q}$, diffusion or aggregation might occur or not, depending on the signal structure or the equilibrium. To focus our attention on the network's role, we say that a network diffuses or aggregates information only if this occurs for *any* signal structure and *any* equilibrium.

Definition 1. We say that a network $\mathbb{Q}$ **aggregates (diffuses) information** if aggregation (diffusion) occurs for every signal distribution and every equilibrium strategy profile.

Note that aggregation is generally a stronger criterion than diffusion; if $1-\underline{\beta}=\overline{\beta}=1$, the two metrics coincide. In this case, we say that *private beliefs are unbounded*, whereas private beliefs are bounded if $\underline{\beta} > 0$ and $\overline{\beta} < 1$. Much work on the SSLM studies conditions under which aggregation occurs. In their seminal paper, Smith and Sørensen (2000) demonstrate that aggregation occurs in a complete network—a network in which all players observe the entire history—if and only if private beliefs are unbounded. This means there are informative signal structures for which the complete network fails to aggregate information, so the complete network does not aggregate according to Definition 1. Acemoglu et al. (2011) find this characterization holds much more generally when the neighborhoods $\{B(n)\}_{n\in\mathbb{N}}$ are mutually independent. Our metrics provide an alternative perspective that emphasizes the *network*'s role in social learning. Often, when aggregation turns on whether beliefs are bounded or unbounded, the network diffuses information according to Definition 1, but it does not aggregate information.

19.2.1.3 *Necessary Conditions for Learning*

Connectedness is the most basic requirement for information to spread in a network. In the SSLM, this corresponds to each player having at least indirect access to a large number of signals. If there exists a chain of players $m_1, m_2, \ldots, m_k$ such that $m_i \in B(m_{i-1})$ for each $i \geq 2$ and $m_1 \in B(n)$, we say that m_k is in player n's *personal subnetwork* $\hat{B}(n)$. A necessary condition for a network to aggregate or diffuse information is that the size of $\hat{B}(n)$ should grow without bound as n becomes large.

Definition 2. The network $\mathbb{Q}$ features **expanding subnetworks** if for any integer K, we have

$$\lim_{n \to \infty} \mathbb{P}\left(|\hat{B}(n)| < K\right) = 0.$$

Proposition 1. If $\mathbb{Q}$ diffuses information, then $\mathbb{Q}$ features expanding subnetworks.

To see why this condition is necessary, suppose $|\hat{B}(n)| < K$ for some player n. We can bound this player's expected utility by what she could attain with access to K independent signals. Since this need not reach the level u^*, we cannot guarantee diffusion with infinitely many such players.[2]

With a basic necessary condition for diffusion in hand, we organize further analysis of the SSLM according to our two key principles. It will become clear that each learning metric corresponds to a particular learning principle: diffusion to the improvement principle and aggregation to the large-sample principle. Though we focus on utility-based metrics, we comment first on behavioral outcomes and belief evolution.

19.2.2 Herding and Cascades

Historically, the SSLM literature focuses on long-run patterns of behavior and belief dynamics: The central phenomena are herding and informational cascades. Herding means that all players conform in their behavior after some time, while an informational cascade occurs if all players ignore their private signals after some time. Formally, herding occurs if there is a random variable x supported on $\{0,1\}$ such that the sequence of actions $\{x_n\}_{n \in \mathbb{N}}$ converges almost surely to x. Defining a cascade requires additional notation. For some thresholds $\underline{\alpha}$ and $\overline{\alpha}$, player n optimally ignores her signal s_n whenever $q_n \in C \equiv [0, \underline{\alpha}] \cup [\overline{\alpha}, 1]$. We call C the *cascade set* of beliefs, and if $q_n \in C$ we say that player n cascades. An informational cascade occurs if, with probability 1, all players with sufficiently high indices cascade. In a complete network, a cascade implies herding. If the social beliefs $\{q_n\}_{n \in \mathbb{N}}$ ever reach the cascade set, no new information is revealed, and q_n remains constant thereafter.

[2] See Lobel and Sadler (2015, Theorem 1) for more detail.

Though herding and informational cascades are clearly related, Smith and Sørensen (2000) make clear these are distinct notions.[3] Players can herd even while it remains possible for some signal realization to change their behavior, because, with some positive probability, such signals may never occur. Indeed, if all players observe the entire history, herd behavior is guaranteed. This is because the sequence of social beliefs q_n is a martingale, and the martingale convergence theorem implies that it converges almost surely. At the same time, the "overturning principle" of Smith and Sørensen (2000) says that whenever a sufficiently late-moving player takes a different action from her predecessors, the social belief must change substantially. These facts combined show that herding must occur, whether or not there is a cascade, and whether or not aggregation occurs. Now, when *is* there a cascade? If private beliefs are unbounded, then clearly there is never a cascade, because signals can be arbitrarily strong. However, Smith and Sørensen note that the martingale $\{q_n\}$ must converge to a random variable supported on the cascade set, and then it is said that a *limit cascade* occurs. In this limiting sense, both herding and cascades always emerge in the complete network, regardless of the signal structure.[4]

In a general network, we need not find such regular patterns of long-run behavior or belief evolution. In a version of the SSLM with $B(n) = \{n - 1\}$ for all n, Çelen and Kariv (2004) demonstrate that, even though herd-like behavior appears for long stretches, true herds or cascades may be absent. Lobel and Sadler (2015) show that social beliefs along a subsequence of players could converge almost surely to a random variable that puts positive probability strictly outside the cascade set. If these beliefs converge to a point outside the cascade set, actions along the subsequence will follow an i.i.d. sequence of random variables. Nevertheless, networks that diffuse information display outcomes that are closely related to informational cascades. In a network that diffuses information, players' ex ante utility is at least as high as if they were in a cascade, even if social beliefs do not converge to the cascade set.

19.2.3 The Improvement Principle

We now return to welfare in the SSLM, focusing first on the improvement principle and information diffusion.

19.2.3.1 Two Lemmas on the Improvement Principle

There are two basic steps in any application of the improvement principle: choosing whom to imitate and determining if improvement is possible. We express the selection

[3] Herrera and Hörner (2012) give a precise characterization of when cascades occur.

[4] Herding also occurs under alternative sampling rules. For instance, Banerjee and Fudenberg (2005) study a sequential decision model in a continuum of players, in which each decision-maker samples several predecessors uniformly at random; in their model, herding is a robust outcome.

component through *neighbor choice functions*.[5] A neighbor choice function $\gamma_n : 2^{\{1,\ldots,n\}} \rightarrow \mathbb{N}$ for player n selects a neighbor from any realization of $B(n)$. We require that $\gamma_n(S) \in S$ if S is nonempty; otherwise, we take $\gamma_n(S) = 0$. In what follows, we use γ_n to refer to both this function and its (random) value. Given our original network $\mathbb{Q}$ and a sequence of neighbor choice functions $\{\gamma_n\}_{n\in\mathbb{N}}$, we implicitly define a new random network $\mathbb{Q}_\gamma$. In the network $\mathbb{Q}_\gamma$, we include only those links in $\mathbb{Q}$ that are selected by the neighbor choice functions. In any realization of this network, each player has at most one neighbor.

We say that an improvement principle holds if, for some sequence of neighbor choice functions, the following heuristic procedure leads to diffusion. Each player n discards all observations of neighbors' decisions except the observation of player γ_n's decision. Player n then chooses an action to maximize expected utility given this single observation and her private signal. Proving that an improvement principle holds entails showing that player n can earn strictly higher utility than her chosen neighbor m whenever $\mathbb{E}_\sigma[u(x_m,\theta)] < u^*$.

Lemma 1 (Improvement Principle). Suppose there exists a sequence of neighbor choice functions $\{\gamma_n\}_{n\in\mathbb{N}}$ and a continuous, increasing function $\mathcal{Z}$ such that:

(a) The network $\mathbb{Q}_\gamma$ features expanding subnetworks.
(b) For all $u < u^*$, we have $\mathcal{Z}(u) > u$.
(c) For any $\epsilon > 0$, there exists N_ϵ such that for any $n \geq N_\epsilon$, with probability at least $1-\epsilon$,

$$\mathbb{E}_\sigma[u(x_n,\theta) \mid \gamma_n] > \mathcal{Z}(\mathbb{E}_\sigma[u(x_{\gamma_n},\theta)]) - \epsilon \tag{19.1}$$

Then the network $\mathbb{Q}$ diffuses information.

This result extends Lemma 4 of Lobel and Sadler (2014), and its proof is essentially identical. Condition (c) expresses the key intuition: For all neighbors except some that γ_n selects with negligible probability, player n can make an improvement. To apply Lemma 1, we must construct a suitable improvement function $\mathcal{Z}$.

Lemma 2. There exists a continuous, increasing function $\mathcal{Z}$, with $\mathcal{Z}(u) > u$ for all $u < u^*$, such that

$$\mathbb{E}_\sigma\left[u(x_n,\theta)\,|\,\gamma_n = m\right] > \mathcal{Z}\left(\mathbb{E}_\sigma\left[u(x_m,\theta)\,|\,\gamma_n = m\right]\right). \tag{19.2}$$

Proof Sketch: We describe a heuristic procedure that a player could follow to obtain the desired improvement; since the players are Bayesian, the actual improvement is at least as high. Suppose player n copies the action of m unless she receives a signal inducing a private belief p_n that is very close to an extreme point $\underline{\beta}$ or $\overline{\beta}$. In the latter

[5] The concept of a neighbor choice function is implicit in the work of Acemoglu et al. (2011), which builds improvements on the neighbor with the highest index. The formalization used here was introduced by Lobel and Sadler (2015).

case, player n chooses the action the signal suggests. Conditional on following her signal, it is as though player n receives very nearly the expert signal $\tilde{s}$, so her expected utility is an average of her neighbor m's utility and something arbitrarily close to u^*. Thus, improvements can accumulate up to the level u^* in the long run.

This argument highlights the significance of the expert signal and why we cannot count on further improvements. Improving on imitation requires following at least the extreme values of the signal, and the expert signal represents an upper bound on what a player can obtain when following an extreme signal.

19.2.3.2 Sufficient Conditions for Diffusion

There is one more step to connect Lemmas 1 and 2 into a general result, namely bounding the difference between $\mathbb{E}_\sigma[u(x_m,\theta)]$ and $\mathbb{E}_\sigma[u(x_m,\theta) \mid \gamma_n = m]$. Player n can imitate player m only if player m is contained in $B(n)$. Therefore, player n's expected utility conditional on imitating player m is not the same as player m's expected utility: Imitation earns m's expected utility *conditional on n choosing to imitate player m*—that is, conditional on $\gamma_n = m$. If $\mathbb{E}_\sigma[u(x_m,\theta)]$ and $\mathbb{E}_\sigma[u(x_m,\theta) \mid \gamma_n = m]$ are approximately equal for large n, then Lemmas 1 and 2 immediately imply information diffusion.

Proposition 2 (Diffusion). Suppose there exists a sequence of neighbor choice functions $\{\gamma_n\}_{n\in\mathbb{N}}$ such that $\mathbb{Q}_\gamma$ features expanding subnetworks, and for any $\epsilon > 0$, there exists N_ϵ such that for any $n \geq N_\epsilon$, with probability at least $1-\epsilon$,

$$\mathbb{E}_\sigma[u(x_{\gamma_n},\theta) \mid \gamma_n] > \mathbb{E}_\sigma[u(x_{\gamma_n},\theta)] - \epsilon.$$

Then diffusion occurs.

A number of conditions on the network can ensure that

$$\mathbb{E}_\sigma\left[u(x_m,\theta) \mid \gamma_n = m\right] = \mathbb{E}_\sigma\left[u(x_m,\theta)\right],$$

and any of these immediately implies information diffusion.

Corollary 1. The network $\mathbb{Q}$ diffuses information if any of the following conditions holds:

(a) The neighborhoods $\{B(n)\}_{n\in\mathbb{N}}$ are mutually independent, and $\mathbb{Q}$ features expanding subnetworks.
(b) There exists a sequence of neighbor choice functions $\{\gamma_n\}_{n\in\mathbb{N}}$ such that $\mathbb{Q}_\gamma$ is deterministic and features expanding subnetworks.
(c) There exists a sequence of neighbor choice functions $\{\gamma_n\}_{n\in\mathbb{N}}$ such that $\mathbb{Q}_\gamma$ features expanding subnetworks and the random vector $\{B(i)\}_{i=1}^m$ is independent of the event $\gamma_n = m$ for all $n > m$.

Proposition 2 unifies and extends earlier results from Acemoglu et al. (2011) and Lobel and Sadler (2015), applying broadly whenever we can bound the difference

$$\mathbb{E}_\sigma\left[u(x_m,\theta)\mid \gamma_n=m\right]-\mathbb{E}_\sigma\left[u(x_m,\theta)\right]$$

using properties of the network.[6] Though the improvement principle is generally robust to features of the network $\mathbb{Q}$, we can construct examples, using highly correlated neighborhoods, in which information fails to diffuse. Lobel and Sadler (2015) highlight through several examples that asymmetric information about the overall network can disrupt the improvement principle even if connectivity is not an issue.

19.2.3.3 *Failure to Aggregate*

Although the improvement principle can ensure only information diffusion, it is worth asking whether we can do better and aggregate information in many of these networks, particularly if players have multiple neighbors. This question is incompletely answered at present, but results in the literature suggest that aggregation generally fails unless some players have large neighborhoods.

Proposition 3 (Failure to Aggregate). The random network $\mathbb{Q}$ fails to aggregate information if $\mathbb{Q}$ satisfies any of the following conditions:

(a) $B(n)=\{1,2,\ldots,n-1\}$ for all n.
(b) $|B(n)|\leq 1$ for all n.
(c) $|B(n)|\leq M$ for all n and some constant M, the neighborhoods $\{B(n)\}_{n\in\mathbb{N}}$ are mutually independent, and

$$\lim_{n\to\infty}\max_{m\in B(n)} m=\infty \quad \text{almost surely.}$$

This result is Theorem 3 of Acemoglu et al. (2011). The complete network—and any network in which players have at most one neighbor—will fail to aggregate information. Recall that this means there is *some* signal structure for which, with positive probability, late movers' actions do not approach optimality given all of society's information. In this case, aggregation fails for any signal structure leading to bounded private beliefs. Part (c) says that, in general, aggregation fails if there is a bound on neighborhood size and players' observations are independent. We cannot dispense with the condition of independence, because we can construct example networks, with correlated neighborhoods and $|B(n)|\leq 2$ for all n, that aggregate information. The degree to which we can relax independence is an open question, and a more detailed

[6] Lobel and Sadler (2015) define a measure of network distortion to bound this difference. Assuming that $\{B(n)\}_{n\in\mathbb{N}}$ are conditionally independent given the state of an underlying Markov chain with finitely many states, they further generalize part (a) of the corollary by applying an improvement principle to the minimum utility across all states of the Markov chain.

understanding of the boundary between diffusion and aggregation would constitute a valuable contribution to this literature.

19.2.4 The Large-Sample Principle

Part (a) of Proposition 3 already demonstrates that large samples alone are insufficient to ensure aggregation, and indeed classical papers (Banerjee 1992; Bikhchandani et al. 1992; Smith and Sørensen 2000) that study sequential learning in a complete network focus on this failure and associated behavioral patterns. To aggregate information from a large sample, infinitely many observations must contain at least some new information, which means that infinitely many players must respond to their private signals. However, if all players observe a large sample, social information will overwhelm the private signals. In some sense the complete network is a knife-edge case in which the large-sample principle fails because no one is forced to rely on a private signal. If we disrupt some of the connections in the network, creating a subsequence of "sacrifical lambs" with no social information, this group can provide enough information for the rest of the network to learn the true state.

Proposition 4 (Aggregation). Suppose there exists a sequence of players $\{m_i\}_{i\in\mathbb{N}}$ such that $\{B(m_i)\}_{i\in\mathbb{N}}$ are mutually independent,

$$\sum_{i\in\mathbb{N}} \mathbb{P}\left(B(m_i) = \emptyset\right) = \infty,$$

and

$$\lim_{n\to\infty} \mathbb{P}\left(m_i \in B(n)\right) = 1$$

for each i. Then the network $\mathbb{Q}$ aggregates information.

Proposition 4 follows from standard martingale convergence arguments, examples of which are found in several papers on social learning.[7] The sequence of players $\{m_i\}_{i\in\mathbb{N}}$ provides enough information to fully reveal the state asymptotically, and all players observe those in this sequence with probability approaching 1. The sequence may consist of an arbitrarily small portion of the network, highlighting an important discontinuity in learning outcomes when we compare a complete network with an almost-complete network.

The basic insight of this proposition generalizes in several directions. Depending on the signal distribution, the sacrificial lambs could have nonempty neighborhoods as long as some signal realizations still dominate the available social information. We also need not have everyone in the network observe the sacrificial lambs: As long as some players observe the large sample and aggregate information, others can learn

[7] See for instance, Goeree et al. (2006), Acemoglu et al. (2011), and Lobel and Sadler (2014).

through imitation. Theorem 4 in Acemoglu et al. (2011) explores both possibilities, giving a more general result. However, the bounds of applicability for the large-sample principle are imprecisely known. Given an infinite set of players following their private signals, there are potentially many ways a network can collect, aggregate, and disperse this information. The literature is still missing a general characterization of networks that aggregate information.

19.2.5 The SSLM with Heterogeneous Preferences

The improvement principle and the large-sample principle respond differently when we introduce preference heterogeneity. On an intuitive level, the improvement principle should suffer because imitation no longer guarantees the same payoff that a neighbor obtains. If a neighbor's preferences are sufficiently different, copying could result in a relatively lower payoff; long-run learning requires not only improvement but also compensation for this gap. However, heterogeneity also raises the prospect that more players will choose to follow private information, which suggests that the large-sample principle has more room to operate as long as neighborhoods are sufficiently large.

We extend the model of the previous section to allow preferences with common and private components.[8] Each player n privately observes a type $t_n \in (0,1)$, and a player of type t will earn utility

$$u(x,\theta,t) = \begin{cases} 1-\theta+t & \text{if } x=0 \\ \theta+1-t & \text{if } x=1. \end{cases}$$

A player's type t neatly parameterizes her trade-off between error in state 0 and error in state 1. Action 1 is chosen only if the player believes that $\theta = 1$ with probability at least t. Hence, to choose action 1, players with high types require more convincing information.

19.2.5.1 No Improvement Principle

Using an example, we illustrate how heterogeneous preferences disrupt the improvement principle. Suppose the signal structure is such that $\mathbb{G}_0(r) = 2r - r^2$ and $\mathbb{G}_1(r) = r^2$, and consider the network topology $\mathbb{Q}$ in which each agent observes her immediate predecessor with probability 1. Suppose player 1 has type $t_1 = \frac{1}{5}$, and all other players n have type $t_n = 1 - t_{n-1}$ with probability 1. Even though the network satisfies our connectivity condition, information diffusion fails.

An inductive argument will show that all players with odd indices err in state 0 with probability at least $\frac{1}{4}$, and likewise players with even indices err in state 1 with probability at least $\frac{1}{4}$. For the first player, observe that $\mathbb{G}_0\left(\frac{1}{5}\right) = \frac{9}{25} < \frac{3}{4}$, so the base

[8] Smith and Sørensen (2000) consider a case in which players may have completely opposed preferences, leading to an outcome they call "confounded learning."

case holds. Now suppose the claim holds for all players of index less than n, and n is odd. The social belief q_n is minimized if $x_{n-1} = 0$, taking the value

$$\frac{\mathbb{P}_\sigma(x_{n-1} = 0 \mid \theta = 1)}{\mathbb{P}_\sigma(x_{n-1} = 0 \mid \theta = 1) + \mathbb{P}_\sigma(x_{n-1} = 0 \mid \theta = 0)} \geq \frac{\frac{1}{4}}{\frac{1}{4} + 1} = \frac{1}{5}.$$

It follows that player n will choose action 1 whenever $p_n > \frac{1}{2}$. We obtain the bound

$$\mathbb{P}_\sigma(x_n = 1 \mid \theta = 0) \geq 1 - \mathbb{G}_0\left(\frac{1}{2}\right) = \frac{1}{4}.$$

An analogous calculation proves the inductive step for players with even indices. Hence, all players err with probability bounded away from 0, and since private beliefs are unbounded, diffusion fails.

In the example, each player has a single neighbor whose preferences are substantially different from her own. The difference in preferences means that the choice of this neighbor provides less useful information. Suppose a player and her neighbor are diners choosing between an Italian and a Japanese restaurant. If the neighbor prefers Japanese food, then observing this neighbor choose the Japanese restaurant is a weak signal of quality, but if the neighbor chooses the Italian restaurant, the choice provides strong information. If our player prefers Italian food, she would benefit more if the signal qualities were reversed.

Lobel and Sadler (2014) offer more general results on this model. If types are i.i.d. random variables and there is a uniform bound on neighborhood size, we can always find a type distribution with support $(0, 1)$ such that diffusion fails. Whether an improvement principle holds depends on the relative frequency of strong signals versus strong preferences. For an improvement principle to hold, the signal distribution must have a thicker tail than the type distribution, meaning that strong signals should be far more common than strong preferences.

19.2.5.2 Robust Aggregation

Preference heterogeneity has a very different impact on the large-sample principle. If players have large neighborhoods, then we can eliminate the need for sacrificial lambs because, with rich enough support in the type distribution, there is always a chance that preferences roughly balance against available social information, making the private signal relevant to a player's decision. Goeree et al. (2006) first noted this effect in a complete network, but the argument can apply much more broadly. Analogous to Proposition 4 of the previous section, we have the following result.[9]

[9] Note that players need not know precisely who is responding to a private signal. Knowing that some players have a positive probability of following a private signal is enough to statistically identify the state.

Proposition 5. Suppose preference types are i.i.d. with full support on $(0, 1)$, and there exists a sequence of players $\{m_i\}_{i\in\mathbb{N}}$ such that

$$\lim_{n\to\infty} \mathbb{P}(m_i \in B(n)) = 1$$

for each i. Then information aggregates.

19.2.6 Remarks

Asymptotic outcomes of sequential observational learning are now well understood, but important challenges remain. A significant gap in our knowledge concerns short-run dynamics and rates of learning in these models. Lobel et al. (2009) study learning rates in special cases with $|B(n)| = 1$ for each player n, linking the rate of learning to the tail of the private belief distribution. However, a more general characterization of learning rates and short-run outcomes is absent outside of particular examples. The complexity of Bayesian updating in a network makes this difficult, but even limited results would offer a valuable contribution to the literature.

There is also little consideration in this literature of where information comes from. Mueller-Frank and Pai (2014) and Ali (2014) are exceptions. Each studies a version of the SSLM with a complete network in which players, rather than being endowed with a signal, must pay a cost to learn about the payoffs of the available actions. If these costs are arbitrarily low for some players, then the players learn the true state asymptotically; essentially, we can trade an assumption of strong exogenous signals for an assumption of sufficiently low costs to acquire strong signals.

Another challenge arises from the strong assumptions in the SSLM. Bayesian rationality demands much from players in these models, raising questions about how realistic such a representation of behavior is. Even in the sequential framework, general networks necessitate extremely complex reasoning. Though this is certainly cause for concern, the proofs of our results suggest these models can still provide a useful benchmark. Analysis centers on relatively simple heuristics, and we show that Bayesian players must perform at least as well. Selecting a neighbor to imitate and following the most popular choice among a large group are intuitive procedures that do not require perfect rationality to succeed. Thus, implicit in these results about Bayesian agents is a broader class of results on a whole range of heuristics. The absence of repeated decisions or strategic interactions in the SSLM is a more fundamental limitation, which forces a departure from this framework to address certain questions.

19.3 Repeated Linear Updating (DeGroot) Models

The sequential social learning models that we discussed in the previous section are so tractable because each agent makes only one decision, even though different agents make their decisions at different times. Models of this sort provide many insights, but they also constrain influence to flow in only one direction. In contrast, many of the most fundamental substantive questions about social learning are inherently about the *dynamics* of individual opinions and choices in a world where individuals make many decisions and can repeatedly influence one another. For example: Can network structure explain lasting disagreements in a segregated society, in which members of two groups repeatedly observe each other's views but persist in holding different opinions? When everyone's initial opinion can influence everyone else, whose opinions end up being particularly influential?

One fruitful approach to these questions, which we focus on in this section, models the dynamics of repeated updating without giving up tractability by having equations of motion that are linear in agents' estimates and stationary over time.

19.3.1 Framework

19.3.1.1 The Basic DeGroot Model

The basic idea behind *DeGroot repeated linear updating models* is that agents start out with some initial estimates, and then all agents update those estimates simultaneously at discrete times $t = 1, 2, 3, \ldots$ Someone's estimate in a given period is obtained by taking a weighted average of some others' estimates. More formally, let X be a convex subset of a vector space,[10] and let $N = \{1, 2, \ldots, n\}$ be a set of agents. The *estimate* or *opinion* of agent i at time t is written $x_i(t)$, and the updating rule for estimates is

$$x_i(t) = \sum_j W_{ij} x_j(t-1) \tag{19.3}$$

for all positive integers t; the initial values $x_i(0)$ are exogenous. Here $\mathbf{W}$ is an n-by-n matrix with nonnegative entries, with the property that every row sums to 1: for each i, we have $\sum_j W_{ij} = 1$. A typical entry, W_{ij}, is called the *weight* agent i places on agent j. One interpretation is that $\mathbf{W}$ represents a social network: each agent has immediate access, not to everyone's estimates, but only to those of a subset of agents. Those agents are the only j's for which W_{ij} is positive, and the weights represent how an agent averages the opinions she can observe. The simplest case is that she places equal weight on the previous opinion of everyone she observes, so that $W_{ij} = 1/d_i$ whenever W_{ij} is nonzero, where d_i is the number of agents whose estimates i can observe.

We can also go in the opposite direction, from an updating matrix to a network. Given any $\mathbf{W}$, we can view it as a graph with N as the set of nodes. There is a directed link, or arrow, from i to j if i places nonzero weight on j according to $\mathbf{W}$; we label that arrow by the corresponding weight W_{ij}. (See Figure 19.2 for an example.) Thus, abusing

[10] The simplest example is $X = \mathbb{R}$, and for simplicity, we will sketch many proofs only for this special case. But one of the virtues of DeGroot's formulation of this process is that we can also think of X as consisting, for example, of a set of probability distributions.

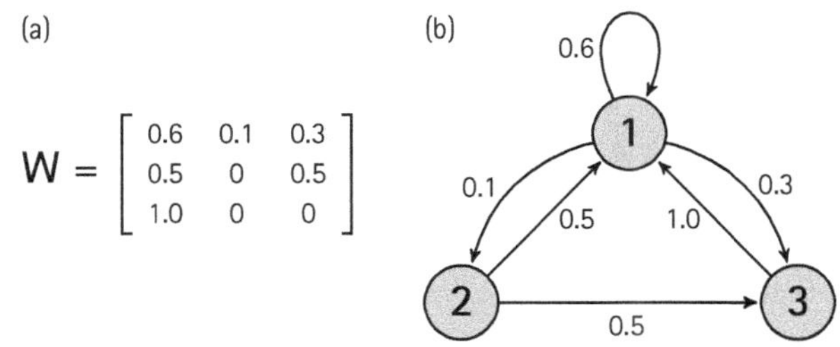

FIGURE 19.2 A weight matrix (a) and the corresponding weighted directed graph (b).

terminology slightly, we sometimes identify a matrix with the corresponding network (weighted directed graph) and speak of a link in $\mathbf{W}$ from i to j.

19.3.1.2 Variations

Some variations on this basic model immediately come to mind. In (19.3), we can allow time-dependent weights, resulting in an updating rule $\mathbf{x}(t) = \mathbf{W}(t)\mathbf{x}(t-1)$ (see, for example, Chatterjee and Seneta 1977). The weights can also be stochastic, with a distribution that is either fixed or changing over time—a case we will discuss in detail later. Another variation permits each agent to hold some persistent "original" or "private" estimate $y_i \in X$, on which she always puts some weight, resulting in the updating rule of Friedkin and Johnsen (1999):

$$x_i(t) = \alpha_i \sum_j W_{ij} x_j(t-1) + (1-\alpha_i) y_i.$$

Finally, there are related models with a discrete set of possible opinions, sometimes called "voter" models (see Mossel and Tamuz 2014 for details).

19.3.1.3 Matrix Powers and the Connection with Markov Chains

We can write the updating rule (19.3) in matrix notation as $\mathbf{x}(t) = \mathbf{W}\mathbf{x}(t-1)$, where $\mathbf{x}(t) = (x_i(t))_{i \in N} \in X^n$ is a vector stacking everyone's time-t estimates. Iterating this shows that

$$\mathbf{x}(t) = \mathbf{W}^t \mathbf{x}(0). \tag{19.4}$$

Thus, the evolution of the estimate vector $\mathbf{x}(t)$ essentially reduces to the dynamics of the matrix powers $\mathbf{W}^t$. Recall that each row of $\mathbf{W}$ sums to 1, and that the entries of this matrix are nonnegative. In other words, $\mathbf{W}$ is a row-stochastic (or Markov) matrix. The iterates $\mathbf{W}^t$ of Markov matrices have been extensively studied, because they capture the t-step transition probabilities of a Markov chain in which the probability of a transition from i to j is W_{ij}. (For introductions, see Seneta 2006 and Meyer 2000, Chapter 8.) Indeed, readers who know Markov chains can quickly absorb the main facts we present about the DeGroot model by reducing them to familiar facts about Markov chains.

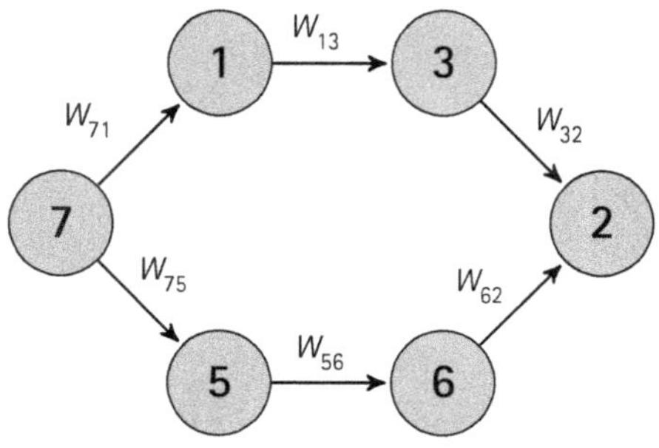

FIGURE 19.3 The sequences $s = (7,1,3,2)$ and $s' = (7,5,6,2)$ are both paths of length 3 from agent $i = 7$ to agent $j = 2$, and are therefore both members of the set N^4_{72}. If these are the only such paths, $(\mathbf{W}^3)_{ij} = w(\mathbf{W},s) + w(\mathbf{W},s') = W_{71}W_{13}W_{32} + W_{75}W_{56}W_{62}$.

There are two useful ways to think of the entry $(\mathbf{W}^t)_{ij}$ in addition to its definition as an entry of a matrix power. First, equation (19.4) shows that $(\mathbf{W}^t)_{ij} = \frac{\partial x_i(t)}{\partial x_j(0)}$; in words, $(\mathbf{W}^t)_{ij}$ is the derivative of *i*'s time-*t* estimate with respect to *j*'s initial estimate. In this sense, $(\mathbf{W}^t)_{ij}$ measures how much *j* influences *i* in *t* steps. The second way of thinking of $\mathbf{W}^t$ is as a sum over various paths of indirect influence. Let N^{t+1}_{ij} be the set of all sequences of $t+1$ agents (i.e., elements of N) starting at i and ending at j. Any such sequence s is called a *walk* of t steps from i to j. We can associate this sequence with a product of t elements of $\mathbf{W}$: the weights in the network $\mathbf{W}$ we meet as we walk from i to j along the sequence s. For example, the sequence $s = (7,1,3,2)$ in the set N^4_{72} corresponds to the product $w(\mathbf{W},s) := W_{71}W_{13}W_{32}$, which we call the *weight of s in* $\mathbf{W}$. We can prove inductively that

$$(\mathbf{W}^t)_{ij} = \sum_{s \in N^{t+1}_{ij}} w(\mathbf{W},s). \tag{19.5}$$

In other words, $(\mathbf{W}^t)_{ij}$ adds up all the weights of t-step ($(t+1)$-agent) sequences that take us from i to j. These sequences are the conduits of *j*'s influence on i in t steps of updating: i pays attention to somebody, who pays attention to somebody, ..., who pays attention to j. Each such sequence contributes to *j*'s influence on i according to its weight in $\mathbf{W}$. A simple example of how all this looks in a concrete case is depicted in Figure 19.3.

A theme of the rest of this section is to study the powers of $\mathbf{W}$—when they converge to a limit, how fast, what this limit looks like—to derive substantive conclusions about social learning.

19.3.1.4 Foundations

In contrast to the sequential learning literature, the mechanics of DeGroot models came first and efforts at microeconomic foundations came later. Here we present two economic rationales for DeGroot-type rules. We defer a discussion of the history—and of some other rationales for the DeGroot model—to Section 19.3.5.1.

Persuasion Bias

To our knowledge, DeMarzo et al. (2003) were the first to discuss in detail how the DeGroot updating rule might arise in a quasi-Bayesian way from an imperfect optimization within a standard microeconomic model. (However, see Lehrer and Wagner 1981, who give some axiomatic foundations for iterated weighted averaging schemes.) Suppose agent i starts out with a private signal $\mu + \epsilon_i$, where μ is a normally distributed state that the agents are trying to estimate, and the ϵ_i are mean-zero, normally distributed errors that are independent of the state and of each other. Consider a case in which the prior distribution the agents hold about μ is diffuse, so that they have very imprecise prior information—indeed, consider an (improper) prior over μ that is uniform over the real line. Then the posterior expectations of μ conditional on the private signals alone are $x_i(0) = \mu + \epsilon_i$. Now suppose that before forming a time-t estimate, each agent observes the previous estimates, $x_j(t-1)$, of some subset of the others. Then Bayesian updating at $t = 1$ corresponds to (19.3) with suitable weights (Gelman et al. 2013, Sections 2.5 and 3.5; DeGroot 2005, Section 9.9): An agent optimally pools others' estimates with her own past estimate by taking a linear combination of all those estimates, with coefficients accounting for the different precisions of different agents' information.

In future periods, there is still something left to learn: If an agent is not connected to everyone, then her contacts' revised estimates convey information about what is known elsewhere in the network. It turns out that linear averaging is also the Bayesian updating rule in future periods, but the optimal weights change over time.[11] DeMarzo et al. (2003) motivate the DeGroot process with unchanging weights as a behavioral heuristic: It is complicated to revise the weights optimally, so agents stick with the weights of the first period. DeMarzo et al. (2003) also suggest that this captures a *persuasion bias* or *echo chamber* effect, much studied by social psychologists, in which people tend to be unduly swayed by things they hear repeatedly.

Myopic Best-Reply Dynamics

Another microfoundation imagines agents learning to play a game. Suppose agents take actions $x_i \in X$ and have payoffs making reaction functions linear in the actions of the others. For instance, suppose $\mathbf{W}$ is a matrix each of whose rows sums to 1; $W_{ii} = 0$; X is a normed space, and agents have the payoffs

$$u_i(x_1, x_2, \ldots, x_n) = -\sum_{j \neq i} W_{ij} \|x_i - x_j\|^2.$$

Then agent i's best response to a profile $\mathbf{x}_{-i} = (x_j)_{j \neq i}$ is $\sum_j W_{ij} x_j$. This is a simple coordination game: It is costly to make a choice (e.g., of a technology or a language) that

[11] For instance, if i heard j's initial information and j is not connected to anyone except i, then there is no reason for i to put any weight on j in the future. More generally, Bayesian agents adjust the weights they place on others in order to optimally incorporate new information, while trying not to overweight the old. See Mossel and Tamuz (2010) for details.

differs from that of one's neighbors. Any action profile in which all agents take the same action is a Nash equilibrium of such a game. What equilibrium will be reached, and how long will it take? One approach to answering these questions is to consider myopic best-reply dynamics: In each period, players select best responses to last-period actions. This is a simple way for them to incorporate information from their surroundings and attempt to coordinate with their neighbors (i.e., to learn how to play the game). Such a rule gives rise to exactly the dynamics of the basic DeGroot model, described by (19.3); see Golub and Jackson (2012).[12]

Other Foundations and a Critique

The microfoundations discussed for DeGroot updating are not fully satisfying. There is a lot of myopia and limited cognition—not to mention quite particular functional form assumptions—inherent in both the persuasion bias and the best-response foundations. Whether actual people behave as if these assumptions hold is an important empirical question about which the evidence is only beginning to come in (Corazzini et al. 2012; Chandrasekhar et al. 2015).

The DeGroot model is a continuing focus of study, despite these important caveats, because it allows us to understand the evolution of beliefs rather completely—both via intuitions and via formal results—as a function of the underlying network. A reason for this tractability is the connection with matrix powers and Markov chains described in Section 19.3.1.3, and the rest of this section explores some of what that connection opens up. Looking beyond this analysis, the hope is that it provides a useful benchmark for the study of other processes of repeated learning in networks.

19.3.2 The Long-Run Limit: Consensus Estimates and Network Centrality

Let us start with the basic DeGroot model of Section 19.3.1.1 with a fixed matrix **W** of updating weights. There are three basic questions concerning the long-run behavior of individuals' beliefs that we answer in this section:

(i) Does each individual's estimate settle down to a long-run limit? That is, does $\lim_{t\to\infty} x_i(t)$ exist for every i?
(ii) When there is convergence, are the limiting estimates the same across agents? In other words, *is there consensus in the long run* for all possible vectors of initial estimates?[13]

[12] In some ways it is more appealing to think of each node i as a continuum of players, who all make the same observations and move simultaneously. What is observable about a node is the average action of its members. In this case, there is no need for the restriction $W_{ii} = 0$, because individuals can care about coordinating with those in their own node. Also, the assumption of myopic actions is more reasonable when each agent is negligible; see Section 19.4.1.

[13] Berger (1981) discusses a necessary and sufficient condition for a weighting matrix to generate consensus *for a single* $\mathbf{x}(0)$; this condition is in terms of both **W** and $\mathbf{x}(0)$. Our discussion here

(iii) When there is a consensus in the long run, what is this consensus? How does it depend on the matrix **W** and the initial estimates?

It is question (iii) that offers the richest connection between outcomes and network structure; the answers to the first two questions set the stage for characterizing that connection.

19.3.2.1 The Strongly Connected Case: Convergence to a Consensus

We first answer the prior questions in an important special case—that of *strongly connected* networks; the general case reduces to this one. The network **W** is *strongly connected* (or the matrix is *irreducible*) if any agent i has a directed path in the network **W** to any agent j.[14] Equivalently, it is impossible to partition the agents into two nonempty sets so that at least one of the sets has no directed link to the other in **W**.

In strongly connected networks, the answer to (i) and (ii) is affirmative once we rule out a "small" set of network structures. To give an idea of the sort of sufficient condition we need, suppose some agent k has a positive self-weight, so that $W_{kk} > 0$. We can prove via (19.5) in Section 19.3.1.3 that strong connectedness of the network and this positive self-weight assumption imply that, *for some* q, *we have* $(\mathbf{W}^q)_{ij} > 0$ *for every* i *and* j. Any strongly connected matrix **W** having this property is called *primitive* (Seneta 2006, Definition 1.1).[15] It is a standard fact that **W** is primitive if and only if $\lim_{t\to\infty} x_i(t)$ exists for each i and the value of the limit is independent of i.[16]

19.3.2.2 The Strongly Connected Case: Influence on the Consensus

Answering (iii) highlights the most interesting connection between the DeGroot updating process and the structure of the network in which agents communicate (i.e., **W**): that limiting consensus estimates are a linear combination of various agents' initial estimates, weighted by those agents' *network centralities.*

Heuristically, from (19.4) we can write the following equation for the (constant) vector of limiting estimates: $\mathbf{x}(\infty) = \mathbf{W}^{\infty}\mathbf{x}(0)$. We can make this rigorous: There is a limit $\mathbf{W}^{\infty}$ of the sequence $(\mathbf{W}^t)_t$ that satisfies this equation for any $\mathbf{x}(0)$. Moreover, since we have already seen that, for any starting estimates, all agents converge to some

characterizes the **W** under which consensus is reached for all $\mathbf{x}(0)$, and thus this condition depends only on **W**.

[14] Recall the interpretation of **W** as a network from Section 19.3.1.1. For basic definitions of graph-theoretic notions, see Jackson (2010) or other textbooks that discuss directed graphs.

[15] The general necessary and sufficient condition for primitivity of a strongly connected **W** is aperiodicity: The greatest common divisor of all cycles (walks that return to their starting point) is 1.

[16] To show that primitivity implies convergence, first note that since agents form their estimates by taking convex combinations, the sequences $\max_i x_i(t)$ and $\min_i x_i(t)$ are each monotone in t. Thus the maximum and minimum must converge to the same point as long as the distance between them gets arbitrarily small. Let $W_{\min} > 0$ be the minimum of the entries in $\mathbf{W}^q$. As the two most extreme agents put weight at least $W_{\min}$ on each other in q steps of updating, the difference between the maximum and minimum entries in **x** decreases by at least a factor of $1 - W_{\min} < 1$ every q steps. For more discussion, as well as a proof of the converse, see statements 8.3.10 and 8.3.16 in Meyer (2000).

consensus, it follows that all rows of $\mathbf{W}^{\infty}$ must be equal to the *same* row vector, which we will call $\boldsymbol{\pi}^{\mathrm{T}}$. Thus, a typical entry of $\mathbf{x}(\infty)$ is equal to

$$x(\infty) = \sum_i \pi_i x_i(0). \tag{19.6}$$

In words, the consensus estimate is a *linear* combination of initial estimates, and the coefficients π_i do not depend on the initial estimates $\mathbf{x}(0)$, but only on the network. The coefficient π_i measures how much i's initial estimate affects the consensus.

We explore in some detail how these *influence coefficients* π_i relate to the underlying network $\mathbf{W}$. The vector $\boldsymbol{\pi}$ satisfies the following equation[17] with $\lambda = 1$:

$$\boldsymbol{\pi}^{\mathrm{T}}\mathbf{W} = \lambda\boldsymbol{\pi}^{\mathrm{T}}. \tag{19.7}$$

Indeed, $\boldsymbol{\pi}^{\mathrm{T}}$ is the *unique* nonnegative, nonzero vector that sums to 1 and satisfies this equation (for any value of λ). Such a vector is called a left-hand *eigenvector centrality* of $\mathbf{W}$, and its entries are called agents' (left-hand) eigenvector centralities. A typical row of the system of equations (19.7), with $\lambda = 1$, reads $\pi_i = \sum_j W_{ji}\pi_j$. In words, i's influence is a weighted sum of the influences of those who put weight on i, with π_j weighted by how much weight j puts on i.

In short, one of the nicest features of the DeGroot model is that we can express the influence of each agent's initial estimate on the final consensus in terms of agents' centralities according to a natural and much-studied measure.[18] Recalling the connection with Markov chains, we also observe that $\boldsymbol{\pi}$ is the unique stationary distribution of the Markov chain described by $\mathbf{W}$.

We now summarize our findings on both convergence and the form of the limit.

Proposition 6. The following hold for all values of $\mathbf{x}(0)$ if $\mathbf{W}$ is strongly connected and primitive:

(i) Each $x_i(t)$ converges to a limit as $t \to \infty$.
(ii) They all converge to the *same* limit.
(iii) This limit is equal to $\sum_i \pi_i x_i(0)$, where π_i is i's left-hand eigenvector centrality in $\mathbf{W}$.

See DeGroot (1974) or Statement 8.3.10 in Meyer (2000) for a formal proof. There is a case in which influence can be calculated very explicitly—that of *reversible weights*:

17 That it should satisfy this equation is intuitive if we believe (at least heuristically) that $\mathbf{W}^{\infty}\mathbf{W} = \mathbf{W}^{\infty}$, and then recall that a typical row of $\mathbf{W}^{\infty}$ is $\boldsymbol{\pi}^{\mathrm{T}}$.

18 Eigenvector centrality is defined rather abstractly, as a vector $\boldsymbol{\pi}$ that satisfies (19.7), possibly with a proportionality constant. See Bonacich (1987) and Jackson (2010, Chapter 2) for some classic motivations for this kind of definition. A lot is known about the structure of such $\boldsymbol{\pi}$. For some insight on how agents' centralities relate to simple properties of the network (and also an application of network centrality ideas to macroeconomics), see Acemoglu et al. (2012). For the general comparative statics of $\boldsymbol{\pi}$ in $\mathbf{W}$, see Schweitzer (1969) and Conlisk (1985).

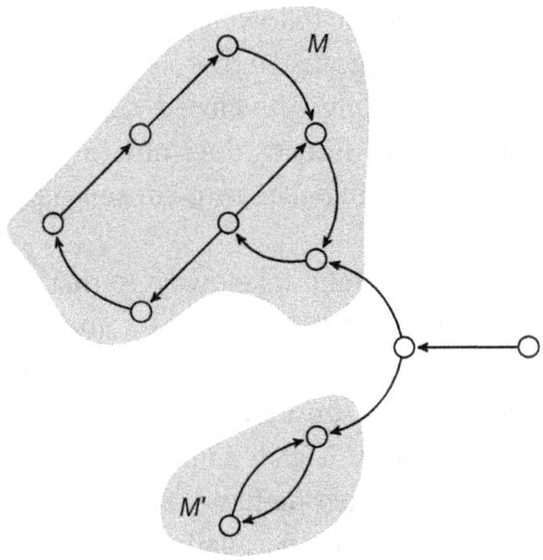

FIGURE 19.4 An illustration of a network that is not strongly connected. Here M and M' are closed communicating classes (inside the class, one can follow a directed path from any node to any other), and the remaining nodes are in no closed communicating class.

Suppose we are given some connected weighted graph $\mathbf{G}$ in the form of a symmetric adjacency matrix—which may describe, say, how much time various pairs spend interacting bilaterally. Assume $W_{ij} = G_{ij}/\sum_i G_{ij}$, so that the influence of j on i is equal to the share of i's time that is spent with j. In that case, one can check that $\pi_i = \sum_j W_{ij}/\sum_{j,k} W_{jk}$.[19] In other words, an agent's centrality is proportional to the amount of her interaction as a fraction of the total interaction in the network.[20]

19.3.2.3 Beyond Strong Connectedness

If a network is not strongly connected, we can reduce the study of its steady state to the study of convergence to consensus in suitable strongly connected subgraphs. In particular, we can view any directed graph as a disjoint union of strongly connected subgraphs that have no links exiting them (called *closed communicating classes*) and some remaining nodes. (See Figure 19.4 for an example.) A closed communicating class, by definition, cannot be influenced by anything that goes on outside of it, so the analysis of how its agents' estimates converge reduces to what we have studied above, restricted to that class.

Moreover, we can see that, for any i and large enough q, the weight $(\mathbf{W}^q)_{ij}$ is positive only if j is in a closed communicating class.[21] Thus, the estimates of agents outside all the closed communicating classes are eventually convex combinations of the consensus beliefs of the agents in various closed communicating classes. The details of how this works are given in Meyer (2000, p. 698) and Golub and Jackson (2010, Theorem 2).

[19] One can calculate directly that $\boldsymbol{\pi}^{\mathrm{T}}\mathbf{W} = \boldsymbol{\pi}^{\mathrm{T}}$ holds, and then use the fact that a strongly connected $\mathbf{W}$ can have only one eigenvector (up to scale) corresponding to the eigenvalue 1.

[20] The reason for the name *reversible weights* is that, in Markov chain language, this corresponds to the case of $\mathbf{W}$ being a reversible chain. See Levin et al. (2009, Section 9.1).

[21] In view of the discussion of Markov chains in Section 19.3.1.3, this corresponds to the statement that if a Markov chain starts outside the closed communicating classes and proceeds with transition probabilities given by $\mathbf{W}$, eventually it will be found, with arbitrarily high probability, inside some closed communicating class. Proving this is a good exercise.

From this it follows that if there are two or more closed communicating classes, long-run consensus (the issue contemplated by question (ii)) no longer obtains. If there is only one closed communicating class and **W** restricted to that class is primitive, consensus does obtain. Convergence of estimates within individual closed communicating classes to consensus requires only primitivity when **W** is restricted to those classes.

An important take-away is that ignoring anyone outside itself—being a closed communicating class—gives a group a lot of power under the DeGroot model. Such a group can certainly sustain its own views in the sense that its long-run consensus depends only on its own initial opinions. Moreover, if that group receives attention from outside, it also gains a decisive influence on the more malleable agents in its society. On the one hand, this feature highlights a quirk of the DeGroot model: Very small differences in updating weights (e.g., whether an inward-looking group gets a tiny amount of attention or no attention from the outside world) can make a huge—indeed, a discontinuous—difference in the model's asymptotic predictions.[22] On the other hand, this feature cleanly captures the fact that stubborn, inward-looking groups have a particularly durable internal inertia and, as long as they are not totally ignored by those outside them, substantial external influence. This observation seems consistent with some observations of political and academic persuasion (DeMarzo et al. 2003).

19.3.2.4 The Large-Population Limit: When is Consensus Correct?

Let us assume networks are strongly connected and primitive (so that there is consensus), and consider again the setting introduced in the discussion of persuasion bias (Section 19.3.1.4). There is a true state μ, and we assume agents start out with noisy estimates of it and are interested in learning its true value. (Normality is not important for this application.) Are large societies of DeGroot updaters able to aggregate information so well that the consensus becomes concentrated tightly around the truth? More formally, take an infinite sequence of networks $(\mathbf{W}^{(n)})_{n=1}^{\infty}$, with $\mathbf{W}^{(n)}$ having n agents making up a set $N^{(n)}$. Suppose initial estimates $x_i^{(n)}(0)$ in network n are noisy estimates of the true state of the world, according to the stochastic specification given in the discussion of persuasion bias in Section 19.3.1.4, and let $x^{(n)}(\infty)$—now a random variable—be the consensus reached in network n. Can we say that the $x^{(n)}(\infty)$ converge in probability to μ as n grows? If we can, then in a certain asymptotic sense the agents are as good at learning as someone who had access to everyone's initial information and aggregated it optimally.

Let us assume that the variances of the noise terms ϵ_i are bounded both above and below. Then we have the following result.

[22] This is perhaps a reason to focus more on predictions of the DeGroot model for large but fixed values of t—which are nicely continuous in **W**—than on the literal $t = \infty$ predictions, whose behavior is analytically cleaner but in many ways more peculiar.

Proposition 7 (Golub and Jackson 2010). Under the model just described with random initial beliefs, the $x^{(n)}(\infty)$ converge in probability to μ if and only if $\lim_{n\to\infty} \max_i \pi_i^{(n)} = 0$.

To prove the "if" direction, we use the expression for the limit consensus estimate given by (19.6) to say that $\mathrm{Var}[x^{(n)}(\infty) - \mu] = \sum_i (\pi_i^{(n)})^2 \mathrm{Var}[\epsilon_i]$. This converges to 0 if and only if $\lim_{n\to\infty} \max_i \pi_i^{(n)} = 0$ (here we are using our assumption about the variances of the ϵ_i); then, using Chebyshev's inequality, we conclude that $x^{(n)}(\infty)$ converges in probability to its mean, μ. The converse uses the same variance calculation, and is left as an easy exercise.

To summarize the key point, large societies achieve asymptotically exact estimates of the truth if and only if the influence of the most influential agent decays to 0 as society grows large. Without this, the idiosyncratic noise in someone's initial belief plays a nontrivial role in everyone's asymptotic belief, and asymptotic correctness is not achieved.

To make the condition for failure of good aggregation more concrete, we give a corollary.

Corollary 2. Suppose we can find an m and $\epsilon > 0$ so that, for each n, there is a group $L^{(n)} \subseteq N^{(n)}$ of m or fewer "opinion leaders," with each individual $i \in N^{(n)} \setminus L^{(n)}$ giving some leader $\ell \in L^{(n)}$ at least ϵ weight ($W_{i\ell}^{(n)} \geq \epsilon$). Then individuals' estimates do not converge in probability to μ.

The proof works by manipulating (19.7) to show that $\lim_{n\to\infty} \max_i \pi_i^{(n)} = 0$ does not hold, and then using Proposition 7. Thus, societies with a small group that influences everyone cannot achieve asymptotic correctness of beliefs. These issues are explored in more detail in Golub and Jackson (2010).

19.3.3 Speed of Convergence to the Long-Run Limit: Segregation and Polarization

When estimates converge to a consensus (or to some other steady state we can characterize), it is important to know how fast this happens. For practical purposes, consensus is often irrelevant unless it is reached reasonably quickly. If it takes thousands of "rounds" of updating to reach a consensus, the model's limit predictions are unlikely to be useful. To say it another way, even if a network satisfies conditions (e.g., strong connectedness) that ensure convergence in the long run, we may still empirically observe disagreement. In this case, it is the medium-run (as opposed to $t = \infty$) behavior of the system that is practically relevant, and we would like to theoretically understand how network structure relates to medium-run disagreement. Some obvious questions arise:

(i) How long does it take for differences in estimates to become "small"?
(ii) What do agents' estimates look like as they are converging to consensus?
(iii) What network properties correspond to fast or slow convergence?

As a preliminary, note that how fast consensus is reached depends on both the network and the starting estimates $\mathbf{x}(0)$. In the trivial case where the initial estimates are all identical, consensus is reached instantly, regardless of the network. In the general case, a full analysis of the time it takes estimates to converge would deal with the dependence of this outcome on $\mathbf{W}$ and on $\mathbf{x}(0)$ jointly. We focus on understanding how properties of the *network* affect convergence time, and for that reason we will often think about worst-case convergence time: roughly, how many rounds of updating it takes for differences of opinion to be guaranteed to be "small" for *any* $\mathbf{x}(0)$ we might start with.

19.3.3.1 A Spectral Decomposition of the Updating Matrix

We saw earlier, in Section 19.3.1.3, that powers of $\mathbf{W}$ are important. We now build on that with a convenient decomposition. For simplicity, we restrict attention to strongly connected, primitive updating matrices $\mathbf{W}$, as in Section 19.3.2.1.

Lemma 3. For generic[23] $\mathbf{W}$, we may write

$$\mathbf{W}^t = \sum_{\ell=1}^{n} \lambda_\ell^t \mathbf{P}_\ell, \tag{19.8}$$

where the following properties are satisfied:

(a) The eigenvalues $\lambda_1 = 1, \lambda_2, \lambda_3, \ldots, \lambda_n$ are the n distinct eigenvalues of $\mathbf{W}$, ordered from greatest to least according to modulus.
(b) The matrix $\mathbf{P}_\ell$ is a projection operator corresponding to the nontrivial, one-dimensional eigenspace of λ_ℓ.
(c) $\mathbf{P}_1 = \mathbf{W}^\infty$ and $\mathbf{P}_1\mathbf{x}(0) = \mathbf{x}(\infty)$;
(d) $\mathbf{P}_\ell \mathbf{1} = \mathbf{0}$ for all $\ell > 1$, where $\mathbf{1}$ is the vector of all 1's.

The import of (c) is that we have encountered $\mathbf{P}_1$ before, in Section 19.3.2.1. It is equal to $\mathbf{W}^\infty$ and corresponds to the eigenvalue $\lambda_1 = 1$ (which, as (a) states, is always an eigenvalue of any row-stochastic matrix $\mathbf{W}$). All other eigenvalues are strictly smaller. In other words, the leading term of (19.8) corresponds to the steady state we studied in

[23] In this lemma, drawing a $\mathbf{W}$ from some measure absolutely continuous with respect to Lebesgue measure buys us a lot: for instance, the luxury of not having to deal with repeated eigenvalues, which—as will become apparent—would require substantial extra bookkeeping. This is mostly a convenience for exposition; Proposition 8, for example, holds with only minor adjustments without assuming any genericity (see Debreu and Herstein 1953, Theorem V and Meyer 2000, Section 7.9 for a flavor of the arguments). However, some of the results in DeMarzo et al. (2003) that we will discuss do break down under highly symmetric network structures. It is a good exercise to replicate the ensuing discussion, replacing uses of this lemma with the corresponding facts about the Jordan canonical form, to see what survives and what requires major adjustment.

Section 19.3.2.1, and the other terms are deviations from that asymptotic steady-state weighting matrix.

19.3.3.2 Speed of Convergence to Consensus

How fast the steady-state summand $\mathbf{P}_1$ comes to dominate depends mainly on $|\lambda_2|$, the magnitude of the second-largest eigenvalue of $\mathbf{W}$. When this number is not too large (say, $|\lambda_2| = 0.6$), all the terms after the first term in (19.8) become negligible as soon as t is at all large (e.g., $t \geq 10$). Thus, the quantity $1 - |\lambda_2|$, called the *absolute spectral gap*, is an important measure of this system's tendency to equilibrate (Levin et al. 2009, Section 12.2). Systems with a small spectral gap (large second eigenvalue) exhibit very slow decay of the nonstationary part in the worst case. The following is a simple formal version of this statement:

Proposition 8. Consider the DeGroot updating process given by (19.3). For generic $\mathbf{W}$,

$$\frac{1}{2}|\lambda_2|^t - (n-2)|\lambda_3|^t \leq \sup_{\mathbf{x}(0)\in[0,1]^n} \|\mathbf{x}(t) - \mathbf{x}(\infty)\|_\infty \leq (n-1)|\lambda_2|^t.$$

Here $\|\cdot\|_\infty$ is the supremum norm, so that $\|\mathbf{x}(t) - \mathbf{x}(\infty)\|_\infty$ is the largest deviation from consensus experienced by any agent. The proof of Proposition 8 is a good exercise in matrix analysis.[24] The result provides a succinct answer to question (i): $|\lambda_2|^t$ is a precise estimate of how much deviation from consensus this network permits at time t. This basic insight has been applied in a variety of ways; for example, in an elaboration of the DeGroot model, Acemoglu et al. (2010) use it to characterize the long-run influence of "forceful" agents (who sway others much more than they themselves are swayed).

We will now explore the λ_2 term of (19.8) in more detail, to better understand both its magnitude and its structure in terms of the underlying network.

19.3.3.3 One-Dimensionality of Deviations from Consensus: A Left–Right Spectrum

Just as the $\ell = 1$ term in (19.8) describes the steady-state component to which estimates are converging, the $\ell = 2$ term of (19.8) describes the dominant term in the *deviation* from consensus (i.e., in what is left over after we subtract the steady-state component). This $\ell = 2$ term can be seen as corresponding to a *metastable* or *medium-run* state in which most disagreement is gone but a critical persistent part remains.[25]

[24] The key equation is $\mathbf{x}(t) - \mathbf{x}(\infty) = \sum_{\ell=2}^{n} \lambda_\ell^t \mathbf{P}_\ell \mathbf{x}(0)$, which follows from (19.8) and Lemma 3. The upper bound follows by applying the triangle inequality to this equation. The lower bound is obtained by using the same equation for a suitable choice of $\mathbf{x}(0) \in [0,1]^n$: Take a nonzero vector $\mathbf{z} \in \text{span}(\mathbf{P}_2)$, and assume (by scaling it) that its largest entry is equal to exactly $1/2$; let $\mathbf{x}(0) = \frac{1}{2}\mathbf{1} + \mathbf{z}$, and use Lemma 3 along with standard norm inequalities.

[25] We are in this regime after enough time has passed for the $\ell \geq 3$ terms in (19.8) to die away, but not the $\ell = 2$ term.

In view of this, let us now dig deeper into the structure of $\mathbf{P}_2$. Since it is an operator that projects onto a one-dimensional space, we may write $\mathbf{P}_2 = \boldsymbol{\sigma}\boldsymbol{\rho}^{\mathrm{T}}$, where $\boldsymbol{\rho}^{\mathrm{T}}$ is a left-hand eigenvector of $\mathbf{W}$ corresponding to eigenvalue λ_2 and $\boldsymbol{\sigma}$ is a right-hand eigenvector of $\mathbf{W}$ corresponding to eigenvalue λ_2.[26]

As a consequence, once t is large enough that $|\lambda_3|^t$ is small relative to $|\lambda_2|^t$, the difference $\mathbf{x}(t) - \mathbf{x}(\infty)$ is essentially $\lambda_2^t \boldsymbol{\sigma}(\boldsymbol{\rho}^{\mathrm{T}}\mathbf{x}(0))$. Thus, if λ_2 is a positive real number, *individual i's deviation from consensus is proportional to* σ_i*, irrespective of* $\mathbf{x}(0)$.[27] DeMarzo et al. (2003) note a striking interpretation of this: Across many different issues, the ordering of agents' medium-run views is determined by a single, network-based number—a position σ_i on a left-right spectrum. For example, if estimates are real numbers (so $X = \mathbb{R}$), then agents' deviations from consensus on a given issue are either ordered the same as σ_i or in the opposite order (depending on whether $\boldsymbol{\rho}^{\mathrm{T}}\mathbf{x}(0)$ is positive or negative). More generally, $\boldsymbol{\rho}^{\mathrm{T}}\mathbf{x}(0) \in X$ determines the "axis" of disagreement if initial estimates are given by $\mathbf{x}(0)$. If $X = \mathbb{R}$, this is a scalar, but in general $\boldsymbol{\rho}^{\mathrm{T}}\mathbf{x}(0)$ is a vector; in the medium run, all deviations from consensus are proportional to it.

19.3.3.4 How Does the Deviation from Consensus Depend on Network Structure?

The decomposition presented in Section 19.3.3.1 opens the door to many neat mathematical results. It yields a rich set of statistics we can use to think about polarization and deviation from consensus, and we can compute these statistics efficiently. On the other hand, what we have seen so far leaves something to be desired. We would like more concrete, hands-on insights about how the "visible," geometric structure of the social graph relates to the persistence of disagreement. Fortunately, there is a large field of applied mathematics, particularly probability theory, devoted to studying this.[28] The basic take-away is that what makes networks slow to converge is being *segregated*. There are many different ways to capture this. We mention two of them informally, and refer the reader to relevant studies for the technical details.

The *bottleneck ratio*, *Cheeger constant*, or *conductance* is defined as

$$\Phi_\star(\mathbf{W}) = \min_{\substack{M \subseteq N \\ \pi(M) \leq \frac{1}{2}}} \frac{\sum_{i \in M, j \notin M} \pi_i W_{ij}}{\sum_{i \in M} \pi_i}.$$

The bottleneck ratio is small if there is some group (having at most half the influence in society) that pays a small amount of attention outside itself, relative to its influence. The

[26] See Meyer (2000, statements 7.2.9 and 7.2.12) for details. Since $\mathbf{P}_2\mathbf{1} = \mathbf{0}$ by Lemma 3(d), we know that $\boldsymbol{\rho}^{\mathrm{T}}\mathbf{1} = \mathbf{0}$ (i.e., that the entries in $\boldsymbol{\rho}$ add up to 0).

[27] A correction of DeMarzo et al. (2003) that is due to Taubinsky (2011).

[28] In probability theory and statistics, this question comes up in seeking to characterize the *mixing time*, a measure of how quickly Markov chains equilibrate and reach their stationary distributions—which is important in Markov Chain Monte Carlo statistical methods. See Levin et al. (2009) for details.

attention or weight summed in the numerator is weighted by influence. The situation where the bottleneck ratio is small corresponds to the existence of a bottleneck. In the case of reversible communication (recall the end of Section 19.3.2.2), the second-largest eigenvalue of **W** can be bounded on both sides in terms of this bottleneck ratio (Levin et al. 2009, Theorem 13.14). That, in turn, yields bounds on the decay of disagreement (recall Proposition 8). For a sophisticated use of these sorts of bounds in a paper on the DeGroot model, see Acemoglu et al. (2009).

Another approach is to think of segregation probabilistically. Imagine, for example, that there are two groups of equal size (say, boys and girls). Friendships happen within a group with probability p_s and between groups with probability p_d. Given these probabilities, friendships are independent across pairs. The matrix **W** is then formed based on the friendship graph as described at the end of Section 19.3.2.2. In that case, it turns out that we can characterize the rate of convergence quite precisely. In particular, λ_2 converges in probability to $\frac{p_s}{p_d} - 1$ as the random network grows large. By Proposition 8, we can convert this into a bound on the worst-case disagreement at any particular time. This gives a clean way of saying that in a simple model of social segregation, it is segregation—and not network density, or anything else—that makes all the difference for the speed of convergence (Golub and Jackson 2012; Chung and Radcliffe 2011).

One more fact worth noting: The metastable structure of disagreement discussed in Section 19.3.3.3 is related to network structure. For example, in the two-group random graph just discussed, agents within each group converge to something resembling an internal consensus (with essentially all agents in one group being on the same side of the eventual consensus). Then we can approximate the situation with just two nodes (corresponding to the two groups) communicating bilaterally until the whole network reaches a consensus. For more on this, see Golub and Jackson (2012).

The connection between the iteration of Markov matrices and the structure of the corresponding network has spawned a huge body of literature, surveyed in book-length treatments. For more, see Levin et al. (2009), Montenegro and Tetali (2006), and Aldous and Fill (2002).

19.3.4 Time-Varying Updating Matrices

The most restrictive and unrealistic feature of the DeGroot model as we have presented it is that updating weights remain fixed over time and are deterministic. Relaxing this assumption is a major concern of the literature on this model, especially in statistics and engineering, going back to Chatterjee and Seneta (1977).

It turns out that it is fairly easy to extend the DeGroot model to a richer one, the *stochastic DeGroot model*, in which the weights agents use are stochastic. For this section, we will assume that X is a compact, convex subset of a normed vector space. Suppose that $\mathbf{x}(t) = \mathbf{W}(t)\mathbf{x}(t-1)$ and the matrices $\mathbf{W}(t)$ are independent and identically distributed random variables. Now the $\mathbf{x}(t)$ are also random variables, but they can be analyzed using what we already know. Indeed, define $\overline{\mathbf{x}}(t) = \mathbb{E}[\mathbf{x}(t)]$ and $\overline{\mathbf{W}} = \mathbb{E}[\mathbf{W}(t)]$. Because expectation commutes with matrix multiplication for independent matrix-valued random variables, we have

$$\overline{\mathbf{x}}(t) = \mathbb{E}[\mathbf{x}(t)] = \mathbb{E}[\mathbf{W}(t)\mathbf{W}(t-1)\cdots\mathbf{W}(1)\mathbf{x}(0)] = \overline{\mathbf{W}}^t\mathbf{x}(0). \tag{19.9}$$

In words, the *expectation* of the DeGroot process follows the law of motion of a nonrandom DeGroot process with updating matrix $\overline{\mathbf{W}}$. Thus, we know immediately that if the process $\overline{\mathbf{x}}(t)$ does not converge to a consensus vector $\overline{\mathbf{x}}(\infty)$ for all profiles of initial estimates, then neither can $\mathbf{x}(t)$ converge to any random vector of consensus beliefs.[29] Remarkably, the converse also holds: If the $\overline{\mathbf{x}}(t)$ converge to a consensus $\overline{\mathbf{x}}(\infty)$ for all profiles of initial estimates, then all agents' estimates in the random updating process converge almost surely to some (random) consensus.[30] There is even a neat condition characterizing when $\overline{\mathbf{x}}(t)$ converges to a consensus vector for all starting beliefs: that all eigenvalues of $\overline{\mathbf{W}}$ except λ_1 (which is equal to 1) are less than 1 in modulus. To summarize:

Proposition 9. The following are equivalent:

(i) All eigenvalues of $\overline{\mathbf{W}}$ except λ_1 (which is equal to 1) are less than 1 in modulus.
(ii) The process described by (19.9) converges to a consensus limit for all values of $\overline{\mathbf{x}}(0)$.
(iii) For any value of $\mathbf{x}(0)$, the random variables $\mathbf{x}(t)$ in the stochastic DeGroot model converge almost surely to a (random) consensus $\mathbf{x}(\infty)$.

For detailed discussions of all this, see Tahbaz-Salehi and Jadbabaie (2008). Using (19.9) and parallelling Section 19.3.2.2, we can say that $\mathbb{E}[x_i(\infty)] = \sum_i \overline{\pi}_i x_i(0)$, where $\overline{\boldsymbol{\pi}}^{\mathrm{T}}$ is the left-hand eigenvector of $\overline{\mathbf{W}}$ corresponding to the eigenvalue 1 (i.e., an influence vector). This gives us some information about the consensus.

To analyze the medium-run behavior of the random process, one can build on the analysis of Section 19.3.3, moving between expectations and realized random estimates in a way analogous to the above treatment of the long run.

Independent and identically distributed updating matrices are the simplest way to relax the assumption of constant weights, but there are many other directions that have

[29] That is, the random beliefs cannot converge to a consensus even in the weakest sense of convergence for random variables. This fact relies on the equivalence between L^1 convergence and convergence in probability for random variables taking values in a compact set.

[30] We will give the flavor of a direct argument, assuming $\overline{\mathbf{W}}$ is strongly connected. The essential idea is to focus on the expectation of the difference between the maximum and minimum estimates in $\mathbf{x}(t)$. Since $\overline{\mathbf{W}}$ is primitive (recall from Section 19.3.2.1 that this is equivalent to $\overline{\mathbf{x}}(t)$ converging), a stochastic version of the argument we gave for convergence in Section 19.3.2.1 shows that this expected difference decreases over time to 0. Since the difference is a bounded random variable, it must also converge to 0 almost surely (see, for example, Acemoglu et al. 2010, Theorem 1). This is easy to extend to the case in which $\overline{\mathbf{W}}$ is not strongly connected because the assumed condition on $\overline{\mathbf{W}}$ means it can have only one closed communicating class (recall Section 19.3.2.3).

been explored. DeMarzo et al. (2003) study a version of the DeGroot model in which the weights agents place on themselves change over time, while the relative weights placed on others remain the same. Chatterjee and Seneta (1977) explore some basic issues of convergence in a model where weights are changing over time. As we note in the next section, there are elaborations of the model in which the weights actually depend on others' opinions.

19.3.5 Remarks

19.3.5.1 Some History and Related Literatures

To our knowledge, the social psychologist John French (1956) was the first to discuss a special case of the DeGroot model—one with each agent placing equal weights on her contacts—as a way to think about the evolution of opinions over time. He motivated his early formulation using ideas from physics, particularly the balance of "forces" of influence. French's ideas were developed by Harary (1959), who recognized that French's process was related to the mathematics of Markov chains. Harary generalized French's results on convergence by using the theory of directed graphs, but continued working with a model in which all agents place equal weights on their contacts. DeGroot (1974) appears to have been the first to write a fully general version of the process, with arbitrary weights, as a model of opinion updating, and to point out the connection between consensus opinions and the stationary distribution of a corresponding Markov chain (recall Section 19.3.2.2).

There are a variety of other names for the DeGroot model, corresponding to expositions in other literatures. In philosophy, it is often called the Lehrer–Wagner model, for Lehrer and Wagner (1981) worked on a related model around the same time as DeGroot. They focused on justifying the influence coefficients (recall Section 19.3.2.2) as a normatively reasonable scheme for aggregating views in a network of peers. As we have mentioned, Friedkin and Johnsen (1999), motivated by sociological theories of disagreement, studied versions of this model in which each agent persistently weights an "initial" or otherwise fixed opinion.

A recent literature studies versions of the DeGroot model in which opinions that are too far from one's own are weighted little or not at all (Hegselmann and Krause 2002; Lorenz 2007). In engineering and control theory, there is a large literature on "gossip" algorithms, which use DeGroot-type rules as a means of pooling information or synchronizing behavior across devices (see Shah 2009 for a survey). This literature extends and generalizes many aspects of the classic models presented above. For example, Moreau (2005) considers nonlinear updating rules, and gives generalizations of the conditions for consensus discussed above.

19.3.5.2 A Review and a Look Ahead

In some ways, the DeGroot model is an intuitive and reasonable heuristic—as we've seen formally in Sections 19.3.1.4 and 19.3.2.4—but we do not have a tight characterization of when it is, in some sense, the "right" or "best" feasible heuristic.[31] Our empirical knowledge is also incomplete: The literature has only begun to explore how well the DeGroot rule fits the behavior of real people. Both of these issues present obvious avenues for further study, theoretical and empirical. In this section we have tried to demonstrate the DeGroot model's great tractability and its capacity to produce social learning dynamics with a rich but manageable structure. Because of these features, the DeGroot model has become an important benchmark with which to compare other learning dynamics, and a starting point for more sophisticated imperfectly rational models. For an example of the latter, see Section 19.4.2.

19.4 Repeated Bayesian Updating

Like the DeGroot model studied in Section 19.3, the models we are about to present have agents revising actions and beliefs repeatedly. But they are different from the DeGroot model in seeking, like the sequential models of Section 19.2, to accommodate potential coarseness of communication (e.g., observations of a binary choice) and more rational learning rules. While the dynamics become more complicated, some of the main substantive conclusions—such as convergence to a common opinion or action—survive.

We adopt a standard notation across each model in this section. Time is discrete, the possible states of the world are $\theta \in \Theta$, and each player n chooses an action $x_n(t)$ in period t. We represent the network as a directed graph $G = (V, E)$, where the set of vertices V is either finite or countably infinite, and we write $B(n) = \{m : nm \in E\}$ for player n's set of neighbors.

19.4.1 Myopic Updating

One popular approach to modeling repeated updating of beliefs assumes that players ignore future effects of their decisions. Players myopically choose the best action today based on current beliefs without regard for any effects on other players or future information availability. Beliefs are still rational given the information players observe, so we can think of these models as an approximation to fully rational behavior with heavy discounting. Alternatively, we can think of each node in a network as corresponding to a continuum of identically informed agents, with only the average action of the continuum being observed by others. In this case, individuals cannot affect anyone's observation or decision, and again agents simply take the best action given their beliefs.

[31] Carroll (2015) is an inspiring example of how nonstandard modeling of agents' optimization problems can rationalize linear rules in quite a different (incentive theory) setting.

19.4.1.1 *Continuous Actions: Revealing Beliefs*

Geanakoplos and Polemarchakis (1982) and Parikh and Krasucki (1990) introduced the first models of this type. In every period, each player takes an action $x_n(t) \in [0,1]$ that perfectly reveals her belief at the beginning of that period about the probability of some event E. Players have a common prior and are endowed with different information about the state. A central insight (subsequently developed further in work by Mueller-Frank (2013, 2014)) is that each player's beliefs allow her neighbors to infer what that player could have seen last period, and therefore narrow down the set of states. When this process stops, it is common knowledge between any pair of neighbors what their beliefs are. By an extension of Aumann's (1976) reasoning on the impossibility of agreeing to disagree, it cannot be common knowledge that these beliefs are different, so consensus is reached in any connected network. Moreover, when there are only finitely many states in Θ, a player's belief can change at most finitely many times, so the opinions converge in finite time. In the Gaussian environment of Section 19.3.1.4, in which agents receive normal signals about a normally distributed state, similar reasoning works. There, finiteness of the state space is replaced by finite dimensionality of the unknowns (see Theorem 3 of DeMarzo et al. 2003 and Mossel and Tamuz 2010).

We can even say something about the correctness of the consensus in each of the settings just discussed. When there are finitely many states, it takes nongeneric priors for different knowledge to lead to the same posterior probability of any event. Therefore, it holds generically that agents' beliefs perfectly reveal their knowledge, and all information is perfectly aggregated as soon as consensus is reached (Mueller-Frank 2014). For somewhat different reasons, perfect aggregation also holds in the Gaussian environment just discussed. There, agents end up holding the beliefs that would be held by someone who observed all the private information initially dispersed in society. (This is another aspect of a result mentioned above, Theorem 3 of DeMarzo et al. 2003).

19.4.1.2 *Discrete Actions*

The conclusions of the previous section depend on a very strong assumption about belief revelation, but we can obtain conformity with far coarser communication protocols. For the rest of Section 19.4, we assume that $x_n(t) \in \{0,1\}$.[32] Two observations allow us to characterize long-run behavior in this family of models. First, an imitation principle holds: Any player asymptotically earns at least as high a payoff as any neighbor because imitation is an available strategy. Second, beliefs evolve according to a martingale, so each individual's belief must converge almost surely to a (random) limit. Together, these observations imply that long-run behavioral conformity is a robust outcome in any connected network. Barring indifference, all players converge on the same action in finite time. An important question is then whether the limiting action is optimal.

[32] Many of the models we discuss can accommodate more general action spaces, but the binary case captures the key insights in each.

Bala and Goyal (1998) provide a seminal contribution to this literature, studying a model of social experimentation in arbitrary networks.[33] Let Y denote an arbitrary space of outcomes. Conditional on the state θ, each action $x \in \{0,1\}$ is associated with a distribution $F_{x,\theta}$ over outcomes. Players share a common utility function $u : \{0,1\} \times Y \to \mathbb{R}$; if player n chooses action $x_n(t)$ in period t, she earns expected utility

$$\int u(x_n(t), y) \mathrm{d}F_{x_n(t),\theta}(y).$$

In every period, a player has beliefs about the underlying state and chooses an action to maximize current-period expected utility.

Each player n observes the outcome of her action in each period as well as the actions and outcomes of all players $m \in B(n)$. Given these observations, player n updates her beliefs about the underlying state and carries these into the next period. This updating is imperfect in that players use information only from neighbors' realized outcomes in Y; they do not infer additional information from the actions neighbors choose. A player learns the true payoff distribution for any action that a neighbor takes infinitely often, which immediately implies an imitation principle, and players will eventually converge on the same action. In general, players can converge on a suboptimal action, but in large networks, relatively mild conditions will ensure they learn the optimal one.

One sufficient condition is having enough diversity in initial beliefs. Any player with a prior close to the truth will choose the optimal action for several periods, generating information that can persuade others to adopt this action. If some players have priors arbitrarily close to the truth, this guarantees that the optimal action is played for arbitrarily many periods by a single player, and this implies convergence to the optimal action throughout the network.

A condition relating more closely to the network structure is that infinitely many players have locally independent neighborhoods. Two neighborhoods $B(m)$ and $B(n)$ are locally independent if they are disjoint; this means that players m and n have independent information. As long as the distribution of initial beliefs is such that each player has some positive probability of selecting the optimal action in the first period, then infinitely many players with locally independent neighborhoods will sample the optimal action. Some of these players will obtain positive results and continue using the optimal action, gathering more information and ensuring that all players eventually learn the best action.

The key intuition behind both conditions is to make sure that some player samples the optimal action infinitely often. This guarantees that someone will obtain positive results, and this knowledge will spread to the rest of the network via the imitation principle. The second condition highlights the importance of independent information sources.

[33] Within the literature on learning in games, there are antecedents considering agents who can observe their neighbors in particular networks; see Ellison and Fudenberg (1993) for an example with agents arranged on a line using a simple rule of thumb based on popularity weighting.

If a small group is observed by everyone,[34] then all players receive highly correlated information, and this can cause convergence to a suboptimal action, even in an infinite network.

Gale and Kariv (2003) respond more directly to the limitations of the SSLM we discussed earlier. The authors study a similar model of observational learning in which players each receive a single informative signal at the beginning of the game, but they eliminate the sequential structure: Each player in the game makes a separate decision in each period. Strategic interactions still pose technical challenges, so they assume that in each period, players choose the myopically optimal action given their current beliefs. Given this behavior, belief updating based on observed neighbors' choices is fully rational. In this context, the imitation principle is directly analogous to the improvement principle in the SSLM: Players can guarantee the same expected utility as a neighbor through imitation, and they may improve based on their other information. Similar results have been established by studying the improvement principle in more general settings. Rosenberg, Solan, and Vielle (2009) relax the assumption that agents observe all their neighbors every period, allowing intermittent observation. Mueller-Frank (2013) generalizes beyond the case of decision rules that maximize expected utility, considering arbitrary choice correspondences; he also permits the decision rules not to be common knowledge.

Results on the optimality of eventual outcomes in the observational learning model are less complete but follow a similar intuition to those of Bala and Goyal (1998). Examples suggest that behavior converges quickly in a dense network, limiting the available information and making suboptimal long-run behavior more likely. In a sparse network, learning takes longer but leads to better asymptotic outcomes.

19.4.2 Local Heuristics

An alternative way to simplify repeated updating supposes that players heuristically incorporate information from their neighbors. Jadbabaie et al. (2012) offer a canonical example of this approach that is closely related to the DeGroot model studied in the previous section. Players have priors over a finite set of possible states, and they attempt to learn the true state over time. At the beginning of each discrete period, a player observes an exogenous private signal and the current beliefs of her neighbors. Signals are i.i.d. across time, but they may correlate across players within a single period. A player incorporates her signal into her beliefs via Bayesian updating, and subsequently the player takes a weighted average of this posterior with her neighbors' beliefs to obtain the prior for the following period. The weights are given in a stochastic matrix $\mathbf{W}$ that is fixed over time.

[34] Bala and Goyal (1998) refer to this group as the "Royal Family."

Suppose the state space is $\Theta = \{0,1\}$, so we can represent the belief of player n at time t as a number $p_n(t) = \mathbb{P}(\theta = 1)$. On receiving the signal $s_n(t)$, player n updates her belief to

$$p'_n(t) \equiv \mathbb{P}(\theta = 1 \mid s_n(t)) = \frac{\mathbb{P}(s_n(t) \mid \theta = 1) p_n(t)}{\mathbb{P}(s_n(t) \mid \theta = 1) p_n(t) + \mathbb{P}(s_n(t) \mid \theta = 0) \left(1 - p_n(t)\right)}.$$

Players assign weights to each other according to a stochastic matrix $\mathbf{W}$ that is fixed over time. We say m is a neighbor of n if $W_{nm} > 0$, and let G denote the corresponding graph. The player combines her updated belief with the reported beliefs of her neighbors to arrive at

$$p_n(t+1) = W_{nn} p'_n(t) + \sum_m W_{nm} p_m(t).$$

This leads to belief dynamics similar to the DeGroot model, and in fact we may view the DeGroot model as a special case in which signals are always uninformative.

If some signals are informative, then this model leads to the most robust learning outcomes of any we have considered. If $G_{nn} = 1$ for each n, and G is strongly connected, then beliefs converge almost surely to the truth. This occurs regardless of correlations between players' signals and regardless of how influential any player is in the network. Crucial to this finding is the continual flow of new information. As long as someone in the network receives new information in each period, this will spread throughout the network. Jadbabaie et al. (2013) provide an extended study of how the distribution of information across individuals interacts with network structure to determine the speed of learning.

19.4.3 Rational Expectations

Recently, Mossel et al. (2015) offer a significant contribution to this literature, studying a model of repeated observational learning in a network with fully rational expectations. The state space is $\Theta = \{0,1\}$, and each player n receives a single informative signal s_n before the first period. Player n observes the choices of each $m \in B(n)$ in every period; let $h_n(t)$ denote the history player n observes by the beginning of period t. Player n earns utility in each period equal to the probability that her action matches the state in that period:

$$u(x_n(t), h_n(t), s_n) = \mathbb{P}(\theta = x_n(t) \mid h_n(t), s_n).$$

The information structure is essentially identical to the model of Gale and Kariv (2003), but a crucial difference is that players discount future payoffs at a rate $\lambda \in (0,1)$, and they play a perfect Bayesian equilibrium.

The authors characterize a general class of networks in which players learn the true state almost surely. We say the graph G is *L-locally-connected* if an edge from n to m implies the existence of a path from m to n with length at most L. If an infinite graph is L-locally-connected and there is a bound d on the number of neighbors any player observes, then all players converge on the optimal action almost surely. The proof builds on familiar imitation and martingale convergence arguments, but it also requires several technical innovations. We can interpret the bounded-degree and L-local-connectedness conditions as a way of ensuring that no one player is too influential. In this sense, the intuition here is similar to earlier models with myopic players.

19.5 Final Remarks

Research on learning in social networks enjoys a diversity of approaches, providing a rich set of answers to our motivating questions. Long-run consensus is a central finding throughout this literature, occurring for a wide range of information structures and decision rules in large classes of networks. Typically, the main assumption needed is an appropriately defined notion of connectedness.

The consistency of this finding may cause some discomfort because we often observe disagreement empirically, even about matters of fact. Explaining such disagreement is an important task for this literature going forward. Of course, we can get disagreement in these models by assuming that networks are disconnected or that agents' preferences are opposed, but such assumptions are not always appropriate. The DeGroot model, with the detailed predictions it makes about a metastable state with one-dimensional deviations from the consensus, comes closest to providing an account of disagreement in terms of network properties. A theory explaining long-run disagreement, especially one with rational foundations and appropriate sensitivity to network structure, would constitute a valuable contribution.

A central theme throughout this literature is that influential individuals have a negative impact on long-run learning, and the different models offer complementary insights as they elaborate on this point. The tractability of sequential models allows us to separate the improvement principle and the large-sample principle as distinct learning mechanisms. This distinction provides intuition for the extent of learning in different networks and a nuanced understanding of how preference heterogeneity impacts learning. DeGroot models deliver a precise grasp of individuals' influence, as well as global learning rates, allowing more detailed comparisons between networks. The significance of strategic interactions for the correctness of eventual consensus is still poorly understood, presenting an important direction for future work.

References

Acemoglu, D., V. M. Carvalho, A. Ozdaglar, and A. Tahbaz-Salehi (2012). "The network origins of aggregate fluctuations." *Econometrica* 80, 1977–2016.

Acemoglu, D., M. Dahleh, I. Lobel, and A. Ozdaglar (2011). "Bayesian learning in social networks." *The Review of Economic Studies* 78, 1201–1236.

Acemoglu, D., A. Ozdaglar, and A. ParandehGheibi (2009). "Spread of (mis)information in social networks." Available at arXiv http://arxiv.org/abs/0906.5007.

Acemoglu, D., A. Ozdaglar, and A. ParandehGheibi (2010). "Spread of (mis)information in social networks." *Games and Economic Behavior* 70, 194–227.

Aldous, D. and J. A. Fill (2002). "Reversible Markov chains and random walks on graphs." Unfinished monograph, recompiled 2014, available at http://www.stat.berkeley.edu/~aldous/RWG/book.html.

Ali, N. S. (2014). "Social learning with endogenous information." Working paper.

Arieli, I. and M. Mueller-Frank (2014). "Non-sequential learning and the wisdom of crowds." Working Paper.

Aumann, R. J. (1976). "Agreeing to disagree." *Annals of Statistics* 4, 1236–1239.

Bala, V. and S. Goyal (1998). "Learning from neighbours." *The Review of Economic Studies* 65, 595–621.

Banerjee, A. (1992). "A simple model of herd behavior." *The Quarterly Journal of Economics* 107, 797–817.

Banerjee, A. and D. Fudenberg (2005). "Word-of-mouth learning." *Games and Economic Behavior* 46, 1–22.

Beck, P. A., R. J. Dalton, S. Greene, and R. Huckfeldt (2002). "The social calculus of voting: Interpersonal, media, and organizational influences on presidential choices." *American Political Science Review* 96, 57–73.

Berger, R. L. (1981). "A necessary and sufficient condition for reaching a consensus using DeGroot's method." *Journal of the American Statistical Association* 76, 415–418.

Bikhchandani, S., D. Hirshleifer, and I. Welch (1992). "A theory of fads, fashion, custom, and cultural change as information cascades." *The Journal of Political Economy* 100, 992–1026.

Bonacich, P. (1987). "Power and centrality: A family of measures." *American Journal of Sociology* 92, 1170–1182.

Bose, S., G. Orosel, M. Ottaviani, and L. Vesterlund (2008). "Monopoly pricing in the binary herding model." *Economic Theory* 37, 203–241.

Carroll, G. D. (2015). "Robustness and linear contracts." *American Economic Review* 105, 536–563.

Çelen, B. and S. Kariv (2004). "Observational learning under imperfect information." *Games and Economic Behavior* 47, 72–86.

Chamley, C. (2004). *Rational Herds: Economic Models of Social Learning*. Cambridge, UK: Cambridge University Press.

Chandrasekhar, A. G., H. Larreguy, and J. P. Xandri (2015). "Testing models of social learning on networks: Evidence from a lab experiment in the field." *National Bureau of Economic Research working paper* 21468, available at: http://www.nber.org/papers/w21468.

Chatterjee, S. and E. Seneta (1977). "Towards consensus: Some convergence theorems on repeated averaging." *Journal of Applied Probability* 14, 89–97.

Chung, F. and M. Radcliffe (2011). "On the spectra of general random graphs." *The Electronic Journal of Combinatorics* 18, P215.

Conlisk, J. (1985). "Comparative statics for Markov chains." *Journal of Economic Dynamics and Control* 9, 139–151.

Corazzini, L., F. Pavesi, B. Petrovich, and L. Stanca (2012). "Influential listeners: An experiment on persuasion bias in social networks." *European Economic Review* 56, 1276–1288.

Debreu, G. and I. N. Herstein (1953). "Nonnegative square matrices." *Econometrica* 21, 597–607.

DeGroot, M. H. (1974). "Reaching a consensus." *Journal of the American Statistical Association* 69, 118–121.

DeGroot, M. H. (2005): *Optimal Statistical Decisions*. Hoboken, New Jersey: Wiley Classics Library, Wiley.

DeMarzo, P. M., D. Vayanos, and J. Zwiebel (2003). "Persuasion bias, social influence, and unidimensional opinions." *The Quarterly Journal of Economics* 118, 909–968.

Duflo, E. and E. Saez (2003). "The role of information and social interactions in retirement plan decisions: Evidence from a randomized experiment." *The Quarterly Journal of Economics* 118, 815–842.

Ellison, G. and D. Fudenberg (1993). "Rules of thumb for social learning." *Journal of Political Economy* 101, 612–643.

Eyster, E. and M. Rabin (2011). "Rational observational learning." Working Paper.

French Jr, J. R. (1956). "A formal theory of social power." *Psychological Review* 63, 181.

Friedkin, N. E. and E. C. Johnsen (1999). "Social influence networks and opinion change." *Advances in Group Processes* 16, 1–29.

Gale, D. and S. Kariv (2003). "Bayesian learning in social networks." *Games and Economic Behavior* 45, 329–346.

Geanakoplos, J. and H. Polemarchakis (1982). "We can't disagree forever." *Journal of Economic Theory* 28, 192–200.

Gelman, A., J. B. Carlin, H. S. Stern, D. B. Dunson, A. Vehtari, and D. B. Rubin (2013). *Bayesian Data Analysis*, 3rd ed. Chapman & Hall/CRC Texts in Statistical Science. Hoboken, New Jersey: Taylor & Francis.

Goeree, J., T. Palfrey, and B. Rogers (2006). "Social learning with private and common values." *Economic Theory* 28, 245–264.

Golub, B. and M. O. Jackson (2010). "Naïve learning in social networks and the wisdom of crowds." *American Economic Journal: Microeconomics* 2, 112–149.

Golub, B. and M. O. Jackson (2012). "How homophily affects the speed of learning and best-response dynamics." *The Quarterly Journal of Economics* 127, 1287–1338.

Harary, F. (1959). "A criterion for unanimity in French's theory of social power." In *Studies in Social Power*, ed. by D. Cartwright, Ann Arbor, MI: Institute for Social Research, 168–182.

Hegselmann, R. and U. Krause (2002). "Opinion dynamics and bounded confidence models, analysis, and simulation." *Journal of Artificial Societies and Social Simulation* 5, 1–24.

Herrera, H. and J. Hörner (2012). "A necessary and sufficient condition for information cascades." Working paper.

Ifrach, B., C. Maglaras, and M. Scarsini (2013). "Bayesian social learning from consumer reviews." Working paper.

Jackson, M. O. (2010). *Social and Economic Networks*. Princeton, NJ: Princeton University Press.

Jadbabaie, A., P. Molavi, A. Sandroni, and A. Tahbaz-Salehi (2012). "Non-Bayesian social learning." *Games and Economic Behavior* 76, 210–225.

Jadbabaie, A., P. Molavi, and A. Tahbaz-Salehi (2013). "Information heterogeneity and the speed of learning in social networks." Working paper.

Lehrer, K. and C. Wagner (1981). *Rational Consensus in Science and Society*. Dordrecht-Boston: Reidel Publishing Company.

Levin, D. A., Y. Peres, and E. L. Wilmer (2009). *Markov Chains and Mixing Times*, Providence, RI: American Mathematical Society; with a chapter on coupling from the past by James G. Propp and David B. Wilson.

Lobel, I., D. Acemoglu, M. Dahleh, and A. Ozdaglar (2009). "Rate of convergence of learning in social networks." *Proceedings of the American Control Conference*.

Lobel, I. and E. Sadler (2014). "Preferences, homophily, and social learning." Forthcoming in *Operations Research*.

Lobel, I. and E. Sadler (2015). "Information diffusion in networks through social learning." *Theoretical Economics* 10, 807–851.

Lorenz, J. (2007). "Continuous opinion dynamics under bounded confidence: A survey." *International Journal of Modern Physics C* 18, 1819–1838.

Meyer, C. D. (2000). *Matrix Analysis and Applied Linear Algebra*. Philadelphia: SIAM.

Montenegro, R. R. and P. Tetali (2006). *Mathematical Aspects of Mixing Times in Markov Chains*. Foundations and Trends in Theoretical Computer Science Series. Boston, MA: Now Publishers.

Montgomery, J. D. (1991). "Social networks and labor-market outcomes: Toward an economic analysis." *The American Economic Review* 81, 1408–1418.

Moreau, L. (2005). "Stability of multiagent systems with time-dependent communication links." *IEEE Transactions on Automatic Control* 50, 169–182.

Mossel, E., A. Sly, and O. Tamuz (2015). "Strategic learning and the topology of social networks." *Econometrica* 83, 1755–1794.

Mossel, E. and O. Tamuz (2010). "Efficient Bayesian learning in social networks with Gaussian estimators." Available at arXiv http://arxiv.org/abs/1002.0747.

Mossel, E. and O. Tamuz (2014). "Opinion exchange dynamics." Available at http://arxiv.org/abs/1401.4770.

Mueller-Frank, M. (2013). "A general framework for rational learning in social networks." *Theoretical Economics* 8, 1–40.

Mueller-Frank, M. (2014). "Does one Bayesian make a difference?" *Journal of Economic Theory* 154, 423–452.

Mueller-Frank, M. and M. Pai (2014). "Social learning with costly search." *American Economic Journal*: Microeconomics forthcoming.

Parikh, R. and P. Krasucki (1990). "Communication, consensus, and knowledge." *Journal of Economic Theory* 52, 178–189.

Rosenberg, D., E. Solan, and N. Vieille (2009). "Informational externalities and emergence of consensus." *Games and Economic Behavior* 66, 979–994.

Schweitzer, P. J. (1968). "Perturbation Theory and Finite Markov Chains." *Journal of Applied Probability* 5, 401–413.

Seneta, E. (2006). *Non-negative Matrices and Markov Chains*. Springer Series in Statistics. New York: Springer.

Shah, D. (2009). "Network gossip algorithms." In *IEEE International Conference on Acoustics, Speech and Signal Processing, 2009. ICASSP 2009.* 3673–3676.

Smith, L. and P. Sørensen (2000). "Pathological outcomes of observational learning." *Econometrica* 68, 371–398.

Smith, L. and P. Sørensen (2008). "Rational social learning with random sampling." Working paper.

Tahbaz-Salehi, A. and A. Jadbabaie (2008). "A necessary and sufficient condition for consensus over random networks." *IEEE Transactions on Automatic Control* 53, 791–795.

Taubinsky, D. (2011). "Network architecture and the left-right spectrum." *The B.E. Journal of Theoretical Economics* 11, 1–25.

Trusov, M., R. E. Bucklin, and K. Pauwels (2009). "Effects of word-of-mouth versus traditional marketing: Findings from an Internet social networking site." *Journal of Marketing* 73, 90–102.

CHAPTER 20

FINANCIAL CONTAGION IN NETWORKS

ANTONIO CABRALES, DOUGLAS GALE, AND PIERO GOTTARDI

20.1 Introduction

The aim of this chapter is to provide an introduction to the literature on financial contagion in networks. We aim to do this by focusing on a limited number of papers in some formal detail, trying to illustrate their analogies and differences as much as possible within a common framework. We divide the discussion in two parts. In the first one, we consider contagion via transmission of shocks (i.e., an abrupt drop in the flow of revenue to one firm), which affects other firms connected to it through financial linkages. We then study informational contagion, by which we mean the process whereby a shock to one market is transmitted to other markets by means of information revealed in the first market.

In the first part we will consider mostly symmetric firms, connected via financial linkages, and symmetric networks of different kinds. The symmetry makes the analysis simpler and easier to present, and it also shows that systemic problems can arise even with symmetry. This is important to note because much of the popular discussion about financial firms has been directed at firms with "special" features, such as "too-big" or "too-interconnected" to fail. We also mention the extension of the results to more general kinds of networks, and to heterogeneous firms. And we end by describing in less detail the results in a wider class of papers.

The second part analyzes informational contagion. It is divided in two subsections: a discussion of contagion between markets, then of contagion between financial firms.

Informational contagion is in some cases related to contagion occurring through tranmission of shocks but it often amplifies the effect of the former. For example, imagine that a shock can travel in a network through a path of at most length k. Then, if the financial network and the origin of a shock are common knowledge among all market participants, and if k is small with respect to the average network distance, the effects of a shock will in general be quite limited. But with imperfect information about the origin of a shock and the network topology, a majority of firms could optimally take protective measures to avoid contagion once a shock arrives, something they would not do with complete information.

Informational contagion can also be independent from shock transmission. This happens, for example, when a price movement in one country affects prices in a different country for informational reasons. This is because traders in the latter country can infer some information related to common shocks from the price movements in the former.

20.2 Contagion through Shock Transmission

20.2.1 The Model

Let there be N financial firms (say, banks). Each firm has liabilities equal to l towards external investors, and assets given by claims to the returns on "projects." There are N projects, and the return on each project i is subject to shocks: it is equal to R if there is no shock, and to $R - s_i \in [0, R)$ if a shock hits. The analysis investigates the effects of the presence of financial linkages among firms on their financial situation, and in particular on their solvency. We can portray these linkages in a general, abstract way by saying that the value v_i of the assets of a firm i may be related to the value v_j of the assets of any other firm $j \neq i$ as they both depend on the vector $\mathbf{r}$ which describes the realizations of the returns of the N projects (for each i, $r_i = R - s_i$ if a shock hits the return of project i and to R otherwise). This relationship is modeled as follows:

$$v = f(A; \mathbf{r}),$$

where $f : \mathbb{R}_+^{N \times N} \times \mathbb{R}_+^N \to \mathbb{R}_+^N$ and A is a $N \times N$, non-negative matrix with generic entry a_{ij}. The matrix A describes the pattern of the linkages among the N firms, and the function $f(.)$ the effect of these linkages on the value of the firms' assets.

If the value v_i of the assets of firm i is lower than the value l of its liabilities, the firm defaults. Because of the presence of linkages among firms, default events are correlated. Since default of a firm is costly (because of the destruction in value of the firms' assets due to the termination of its activities or to liquidation costs[1]), one of

[1] We can think of r and l as revenues and payments due the subsequent period. The firm's assets also generate revenue in later periods, but when liquidated the value of the firm's assets is zero, for simplicity (in any case lower than the present value of future returns, hence the cost of default).

the main objectives of the analysis is indeed to analyze the extent of default in the system and whether generalized default, or contagion, may occur in the presence of such linkages.

In the literature we find different microfoundations for the map $f(.)$ and the matrix A, leading to different interpretations of the elements of this matrix and to different properties of $f(.)$. In particular, we present in what follows some alternative microfoundations that lead $f(.)$ to be a linear function of $\mathbf{r}$, or nonlinear but still continuous, or even discontinuous, and we will examine the consequences of these different properties.

In these microfoundations, the linkages among firms may have a different nature. In particular, in Cabrales, Gottardi, and Vega-Redondo [CGV] (2013) and Elliott, Golub, and Jackson [EGJ] (2014) they arise from the mutual ownership of the claims to the returns of the underlying projects: that is, the returns on the assets of a generic firm i are given by a certain linear combination of the returns of the N projects, with weights given by the i-th row of the matrix A: $\sum_j a_{ij} r_j$. In CGV, the term a_{ij} describes the ownership by firm i of claims entitling the owner to a fraction of the returns of project j. These claims are obtained via a sequence of rounds of exchanges of assets by each firm i, initially endowed with full ownership of project i, with a subset of other firms (constituting its immediate neighbors). The pattern of exchanges at each round is described by the matrix B, where the nonzero elements of row i describe i's trades with its immediate neighbors. Hence we have (when the number of rounds of these exchanges is given by K)

$$A = B^K \text{ and } f(A;\mathbf{r}) = A\mathbf{r}. \tag{20.1}$$

EGJ, on the other hand, consider a situation where firms engage in exchanges of equity among them (again starting from a situation where each firm i fully owns project i). Letting c_{ji} denote the fraction of the outstanding equity of firm i sold to firm j, and $\hat{c}_{ii}$ the fraction that remains owned by external investors, we have $\hat{c}_{ii} = 1 - \sum_{j \neq i} c_{ji}$. As shown by EGJ (in line with Brioschi, Buzzacchi, and Colombo 1989, and Fedenia, Hodder, and Triantis 1994), this ownership structure again entitles the owners of equity of the N firms to a linear combination of the returns of the underlying projects with weights given by the matrix

$$A = \widehat{C}(I - C)^{-1}, \tag{20.2}$$

where $\hat{C}$ is the diagonal matrix with generic entry $\hat{c}_{ii}$ and C the matrix with generic entry c_{ij} (and zero diagonal terms). The fact that the mutual ownership now takes the form of equity and that, as we said above, the default of a firm entails a cost implies that any other firm that is owning equity of the bankrupt firm must bear part of this cost, in proportion to the level of its equity ownership. Hence, letting β denote the (immediate) cost of default[2] for a firm, we have:

$$v = f(A;\mathbf{r}) = A(\mathbf{r} - \beta \mathbf{1}_{\{v_i<l\}}), \tag{20.3}$$

where $\mathbf{1}_{\{v_i<l\}}$ denotes (with a slight abuse of notation) the vector of indicator functions taking value 1 if $v_i < l$ and 0 otherwise, for $i = 1,\ldots,N$. We see from (20.3) that, as long as $\beta > 0$, the value v of the firms' assets is now a discontinuous function of $\mathbf{r}$ that jumps whenever some firm defaults. Moreover, v is determined as a fixed point of the map defined by (20.3), since the level of v also determines whether a firm is solvent or not; if insolvent a firm has to pay the additional default cost, which in turn affects the value of v, and so on. To see this, take $N = 2$,

$$A = \begin{bmatrix} 0.7 & 0.3 \\ 0.3 & 0.7 \end{bmatrix},$$

$R = 1$ if no shock hits and $r_i = 0.5$ if a shock hits, $l = 0.8$ and $\beta = 0.5$. Then, if project 2 is hit by a shock ($r_2 = 0.5$), not only firm 2 defaults, but also firm 1, because of the fraction of default costs it must bear. In contrast, if $r_i = R$ for each i, there are two fixed point solutions of (20.3)—one where both firms are solvent, and the other where both firms default. That is, simply the belief that a firm may default suffices to generate losses in the firms' value that trigger a default even when the returns on the underlying projects have not been hit by any shock.

As we have seen, the approach in CGV and EGJ looks at the presence of linkages among the assets of the different firms given by cross-holdings of claims to the returns on projects.[3] In contrast, Glasserman and Young [GY] (2015) and Acemoglu, Ozdaglar, and Tahbaz-Salehi [AOT] (2014) consider linkages among both assets and liabilities of firms, arising from mutual lending and borrowing relationships among them, via standard debt contracts. In this case a_{ij} denotes the payments due from firm i to firm j. The value of firm i's liabilities is then augmented now to include the value of the payments due to other firms, $l_i + \sum_{j\neq i} a_{ij}$. Firm i is again endowed with a 100% share of the returns generated by project i. The book-value value of its assets is similarly augmented by the book-value of the loans it granted, $r_i + \sum_{j\neq i} a_{ji}$. To establish the solvency of firm i, however, what matters is not the book-value but the actual value of the firm's assets, which reflects the actual payments made by its debtors and may be less than the due payments when they are in default. Hence, the actual payments to firm i depend on the value of the assets of this firm's debtors; that is, on the payments they in turn receive from their own debtors. This creates an interdependence among the actual value of the assets of all firms in the system.[4]

[2] Strictly speaking, in line with footnote 1, β should be viewed as the part of the cost that must be born immediately (more precisely, in the subsequent period). Also, a limited liability constraint could be added in (20.3) with no change in the substance of the results.

[3] These claims could also be given by derivatives whose return is a more general, nonlinear function of the yield of these projects.

[4] Such an environment is also considered by Shin (2009) to study the effects of changes in credit volumes.

AOT and GY resort to the analysis of Eisenberg and Noé (2011) to find a consistent set of payments, which can be made by firms linked by mutual borrowing and lending relationships when some of them default, while respecting the possible differences in the seniority of the firms' liabilities. Suppose that all the liabilities among firms have equal seniority[5] so that the total payments made by any given firm to other firms are distributed to them pro rata, on the basis of the value of their claims towards this firm. For any vector $\mathbf{r}$ of realized returns, a final (equilibrium) repayment vector is then obtained as a solution of the following system:

$$p_i(\mathbf{r})\left(\sum_{j\neq i} a_{ij}+l\right)=\left(\sum_{j\neq i} a_{ij}+l\right)\wedge\left(\sum_{j\neq i} p_j(\mathbf{r})\,a_{ji}+r_i\right) \tag{20.4}$$

for all i. The actual payments made by firm j to firm i are then

$$p_{ij}(\mathbf{r})=p_i(\mathbf{r})\,a_{ij},\ \text{for all } j.$$

Note that $p_i(\mathbf{r})\in[0,1]$ and $p_i(\mathbf{r})=1$ if firm i does not default. Firm i defaults if the actual value of its assets $\sum_{j\neq i} p_j(\mathbf{r})\,a_{ji}+r_i$ falls below the value of its liabilities $\sum_{j\neq i} a_{ij}+l$. As shown by Eisenberg and Noé (see also AOT and GY), in the environment considered there is (generically) a unique solution of system (20.4) for a vector $p(\mathbf{r})=(\ldots,p_i(\mathbf{r}),\ldots)$. This solution is a continuous function of $\mathbf{r}$; it is however a nonlinear function of $\mathbf{r}$, more precisely it is piecewise linear. Consider for instance the case where only firm i defaults: we have $p_i(\mathbf{r})=\left(\sum_{j\neq i} a_{ji}+r_i\right)/\left(\sum_{j\neq i} a_{ij}+l\right)$ and $p_j(\mathbf{r})=1$ for all $j\neq i$.

In the situation considered by AOT and GY the value of the firms' assets net of their internal liabilities is then given by

$$v=f(A;\mathbf{r})=\mathbf{r}+(A^T-A)p(\mathbf{r}) \tag{20.5}$$

20.2.2 Contagion

The environment presented in the previous section can be used to study how the presence of linkages among firms may generate contagion; that is, shocks hitting one firm may propagate through the whole system and possibly generate widespread default. More specifically, what we mean by a shock hitting firm i is that $r_i<R$; the size of the shock is then given by the magnitude of s_i. The mechanism of contagion may then vary with the nature of the linkages among firms. In CGV, it occurs because the presence of linkages implies that firms are exposed to the same shocks[6] through a

[5] In AOT, external liabilities are assumed to be senior to internal liabilities. This has only minor qualitative implications, which we discuss later.

[6] E. Lazear, in http://online.wsj.com/news/articles/SB10001424052970203554104577003924075089102, refers to this as the 'popcorn mechanism'.

correlation in their portfolios. The same is true in EGJ, but in this case there is also an element of a domino effect, as the default of one firm constitutes an additional negative shock for the firms with whom the firm is linked. A similar domino effect constitutes the only contagion mechanism in AOT and GY.[7] The analysis in what follows will show, however, that the effect of the pattern of linkages on contagion, and more generally on the way in which shocks propagate through the system, does not depend in a fundamental way on the mechanism that generates contagion.

Following the papers mentioned above, we address this issue by considering first the case where all the N firms are connected, either directly or indirectly. The presence of such connections implies that all firms in the system will be affected, to various degrees, by a shock hitting any one firm, and for sufficiently large shocks, one or more firms will default. We also restrict our attention to *regular* networks where each firm is equally exposed to other firms in the system—that is, $\sum_{j\neq i} a_{ji} = \sum_{j\neq i} a_{ij}$ for all i, and the pattern of exposure is also the same (the matrix A is symmetric). In such a situation, the identity of the firm directly hit by the shock does not matter, and we can take it, w.l.o.g., to be firm 1.

More precisely, following EGJ we can compute the minimal size of the shock $s(j)$ hitting firm 1 such that $j \in \{1, \dots, N\}$ firms default.[8] Let $\mathbf{r}_j$ be the vector with elements $(R - s(j), R, \dots, R)^T$ such that $f(A; \mathbf{r}_j)$ has j components less or equal to l (at least one of them equal) and the other $N - j$ strictly greater than l. This vector is uniquely defined, except in the discontinuous framework of EGJ. Hence a shock of size $s(j) + \varepsilon$, for ε small, will lead to the default of j firms. Studying how the values $s(1), \dots, s(N)$ vary with A—that is, with the pattern of linkages among the firms—allows us to see how well different network structures do in limiting the extent of contagion. Again following the papers mentioned above, we focus most of our attention on the "extreme" cases—as far as the density of connections is concerned—of a *complete network* structure, where each firm is directly linked to any other firm in the system, and of the (one-directional) *ring network*, where each firm is directly linked to only one other firm in the system.

We show in what follows that, with a complete network structure, either only one firm defaults (the one directly hit by the shock), or all firms default. In contrast, with the ring network structure we may have different thresholds for the size of the shocks leading to $1, 2, \dots, N$ firms defaulting. We will then compare the ring and the complete structure in terms of their ability to limit defaults in the system and see which role the form of $f(.)$, that is, whether it is linear, piecewise linear or discontinuous, as discussed in the previous section, plays in this comparison.

[7] Another mechanism of contagion sometimes mentioned in the literature (see, e.g., Cifuentes et al. 2005) is the one occurring via the fire sales of the assets of defaulting firms, which depress the value of other firms.

[8] We assume that $R > l$ so that in normal circumstances no firm defaults.

Let us now be more precise. In the CGV and EGJ framework, the complete and the ring network are described,[9] respectively, by the following specifications of the matrix A:

$$A^C = \begin{bmatrix} \alpha & \frac{1-\alpha}{N-1} & \cdots & \frac{1-\alpha}{N-1} \\ \frac{1-\alpha}{N-1} & \alpha & \cdots & \frac{1-\alpha}{N-1} \\ \vdots & \vdots & \ddots & \vdots \\ \frac{1-\alpha}{N-1} & \frac{1-\alpha}{N-1} & \cdots & \alpha \end{bmatrix},$$

$$A^R = \begin{bmatrix} \alpha & \frac{(1-\alpha)^2}{(1-\alpha^{N-1})} & \cdots & \frac{(1-\alpha)^2}{(1-\alpha^{N-1})}\alpha^{N-2} \\ \frac{(1-\alpha)^2}{(1-\alpha^{N-1})}\alpha^{N-2} & \alpha & \cdots & \frac{(1-\alpha)^2}{(1-\alpha^{N-1})}\alpha^{N-3} \\ \vdots & \vdots & \ddots & \vdots \\ \frac{(1-\alpha)^2}{(1-\alpha^{N-1})} & \frac{(1-\alpha)^2}{(1-\alpha^{N-1})}\alpha & \cdots & \alpha \end{bmatrix}. \tag{20.6}$$

It is natural to assume that $\alpha > (1-\alpha)^2 / (1-\alpha^{N-1})$, which ensures that in both cases[10] each firm is more exposed to the returns of its own project than to the projects of the other firms.

Let $s^C(1)$ and $s^R(1)$ denote the minimal size of the shock leading to one firm defaulting (under the above assumption, the one directly hit by the shock), respectively for the matrix A^C and A^R. It is immediate to see that the value of both $s^C(1)$ and $s^R(1)$ is obtained as a solution to

$$\alpha\,(R - s(1)) + (1-\alpha)R = l, \tag{20.7}$$

hence $s^C(1) = s^R(1) > R - l$. This property shows that in this framework the presence of linkages to other firms allows a firm to withstand larger shocks to the returns on its project without defaulting (when $\alpha = 1$, $s(1) = R - l$), that is, offers some insurance against these shocks.

With a complete structure, since all the off-diagonal terms of A^C are the same, we have $s^C(2) = \ldots = s^C(N)$. There is then only one other threshold, $s^C(N)$, defining the minimal shock leading to all the N firms defaulting. This is obtained from[11]

$$\frac{1-\alpha}{N-1}\left(R - s^C(N) - \beta\right) + \left(1 - \frac{1-\alpha}{N-1}\right) R = l. \tag{20.8}$$

In contrast, for the ring the off-diagonal terms of A^R have different values (and decrease with distance from the diagonal), and so we have a different threshold $s^R(j)$ for each

[9] See the Appendix for details.

[10] We have in fact $(1-\alpha)^2 / (1-\alpha^{N-1}) > (1-\alpha)/(N-1)$, since this is equivalent to $1/(N-1) < 1/\left(1 + ... + \alpha^{N-2}\right)$ and hence to $\alpha + \ldots + \alpha^{N-2} < N-2$.

[11] Because of limited liability there is a natural upper bound at R for the admissible values of the size of the shock $s(j)$. If a solution satisfying the constraint $s(j) \leq R$ does not exist, it is not possible that j firms default.

$j = 2, \ldots N$, obtained as a solution of

$$\left(\left(\sum_{i=0}^{j-3} \alpha^i\right) \frac{(1-\alpha)^2}{\left(1-\alpha^{N-1}\right)} (R-\beta) + \frac{(1-\alpha)^2}{\left(1-\alpha^{N-1}\right)} \alpha^{j-2} \left(R - s^R(j) - \beta\right) + \left(\sum_{i=j-1}^{N-2} \alpha^i\right) \frac{(1-\alpha)^2}{\left(1-\alpha^{N-1}\right)} R + \alpha R\right) = l \quad (20.9)$$

From the above expressions we readily obtain some properties of the pattern of contagion in the different network structures. The following result follows from the property that the off diagonal coefficients of both matrices A^C and A^R are decreasing in N.

Result 1. *The minimal size of the shock leading to all firms defaulting both in the complete $s^C(N)$ and the ring structure $s^R(N)$ increases*[12] *as N increases. The same property clearly holds in the ring for all other values $s^R(j)$, for $N > j > 1$.*

Hence the larger is the system (N), the more diffuse is the exposure of each firm to the returns on the projects of other firms, and the more difficult it is that a shock to the return on the project of one firm leads to contagion, with other firms defaulting. That is, larger systems are better at buffering large shocks, but when shocks are so large that they cannot be buffered, the extent of default is clearly also larger.

The key difference between the extent of contagion in the complete and ring structure is that generalized default, by all firms in the system, is harder with the ring than with the complete structure, but with the ring intermediate levels of default are also possible. Formally, recalling that β is the cost of default for a firm, we have:

Result 2. *If β equals zero or is sufficiently small, then $s^R(2) < s^C(N) < s^R(N)$.*

Since, as we saw $s^C(N) = s^C(j)$ for all $j > 1$, this result shows that a ring structure is better able to withstand large shocks, but does worse (i.e., has a larger number of firms defaulting) for shocks of intermediate size.[13] Thus when we consider the extent of contagion in ring and complete structures we face a trade-off.

On the other hand, when the cost β is large (that is, when the "jumps" in the value of $f(.)$ are sufficiently significant), the situation is rather different:

Result 3. *If β is sufficiently large, as soon as one firm defaults all firms default as well, both for the ring and the complete structure: $s^k(N) \leq s^k(1)$ for $k = R, C$.*

[12] This is true as long as N is not so large that the system is able to withstand a shock of maximal size ($s = R$) without having generalized default, in which case $r(N)$ is not defined.

[13] The result follows immediately from the fact that, as noticed in footnote 10, the largest off-diagonal term in A^R is larger than the (common) off diagonal term of A^C, while the smallest off-diagonal term in A^R is smaller.

That is, in this case a default of any firm means an immediate default of all the firms in the system. To understand this result, and the difference with respect to Result 2, note that, when $\beta = 0$, the size of the shock that needs to be absorbed by the system is given by s, equal to the size of the shock hitting firm 1. In contrast, when $\beta > 0$ the size of the shock the system must bear is bigger, since it also includes the cost of firms who default, and is larger the greater the number of defaults in the system. Thus we see clearly the amplification effect generated by the presence of default costs that must be borne (immediately) when the claims to projects' returns mutually owned by firms are given by equity. Furthermore, we can show that the property established in Result 2 is no longer valid when β is large: generalized default is now *easier* in the ring than in the complete structure. This is due to the fact that contagion in the ring is triggered by the default of the direct neighbor of a firm. This entails the obligation to face not only a share of the shock, which declines with distance from the firm directly hit, but also a share of the default costs. These do not decrease but actually increase with distance (as we see from [20.9]).[14]

In the AOT and GY framework, regular structures can be similarly described: for the complete structure we have $a_{ij} = a_{ji} = a/(N-1)$ for some $a > 0$ all i, j, while for the ring $a_{ii+1} = a$ for all i and $a_{ij} = 0$ otherwise. Focusing the attention, as in these papers, on the case where $\beta = 0$, we can show that similar results to the ones obtained above still hold; in particular, Results 1 and 2 are both still valid.[15]

Result 4. *In the AOT/GY framework, both in the complete and the ring structure the minimal size of the shock $s^C(N)$ and $s^R(N)$ that induces default of all firms is larger the larger the number of firms N, and the larger the "capital buffer" $R - l$. We also have (when all claims have the same seniority), $s^R(N) > s^C(N) > s^R(2)$.*

Hence the properties of contagion in these network structures are analogous to the case where linkages are given by cross-holdings of claims. There are two main differences and we shall focus here on them. The first one is an immediate consequence of (20.4):

Result 5. *In the AOT/GY framework we have $s^C(1) = s^R(1) = R - l$, while in CGV/EGJ $s^C(1) = s^R(1) > R - l$.*

As already noticed, this property in CGV/EGJ readily follows from (20.7): it tells us that, in that framework, a firm can withstand a larger shock than the one that would trigger its default in the absence of any linkage to other firms ($R - l$), hence the presence of linkages provides some insurance against the shock that may hit a firm. This is not true in AOT/GY as the entire value of the return r_i generated by project i shows in

[14] The details are in the Appendix.

[15] See the Appendix. In light of the previous discussion, it is useful to point out that the size of the shock that needs to be absorbed by the system is still given by s (actually, now strictly smaller than s in the presence of external claims with equal priority).

the asset side of firm i, not only a fraction α as in (20.7); we can also view this as due to the use of debt, instead of equity, in forming linkages. Hence in AOT/GY the presence of linkages among firms does not provide any insurance to a firm against the shocks hitting the returns on the firm's own project. These linkages cannot be motivated by risk-sharing considerations, but from other (e.g., technological or trade related) requirements.[16]

The second difference is a consequence of the assumption in AOT that external claims are senior to internal claims (that is, to other firms in the network). Under that assumption we have:[17]

Result 6. *In the AOT/GY framework, when external claims are senior the minimal size of the shock inducing all firms to default is the same in the complete and the ring structure:* $s^C(N) = s^R(N)$.

Since for the ring we also have, as shown in Result 4, $s^R(1) < s^R(j) < s^R(N)$ for $1 < j < N$, for shocks of intermediate size, between $s^R(1) = R - l$ and $s^R(N)$, there are multiple failures in the ring and only a single failure in the complete structure. Thus in the environment considered by AOT there is no trade-off, and the extent of default and contagion is always (weakly) larger in the ring than the complete structure. In this respect, the ring structure is always dominated by the complete one. This is one place where the seniority of external claims in AOT makes a difference. Note the similarity of this result to Result 3, obtained in the EGJ framework when β is large.

The extension of the above results to other network structures, exhibiting "intermediate" densities of connections between the case of the ring and the complete structure, is discussed by AOT and EGJ. In particular, AOT consider specifications of the matrix A given by convex combinations of the matrix for the ring and the complete structure, obtaining results somewhere in between those found for the ring and the complete (see, e.g., their Proposition 3 c).

20.2.3 Optimality

Once we have established the properties of networks with different densities of connections, in terms of their ability to withstand shocks of different sizes, the obvious next step is to determine the optimal network structure in a given environment. This has clearly important implications in guiding policy interventions aiming to affect the pattern of financial linkages among firms.

Regarding the notion of optimality to be considered, the papers reviewed have settled mostly for considering the number of defaulting firms as the key criterion to assess

[16] These requirements must also preclude netting, which would otherwise be beneficial in limiting the spread of default.

[17] See Proposition 3 (a) and (b) in AOT. We also show this for completeness in the Appendix.

optimality. This can be justified in light of the previous observation that the default of financial firms entails a deadweight cost, given by the destruction in value of the firms' assets. Under the assumption that the cost of default is the same for each firm, the number of defaults is thus proportional to the total welfare loss in the economy.

The results of the previous section yield a number of immediate corollaries when the distribution of shocks has a single element in its support, denoted s^*. From Results 2 and 4 we have both in the CGV/EGJ and the AOT/GY frameworks:

Result 7.

(a) *If* $s^* \leq s^R(2)$ *or* $s^* > s^R(N)$, *then the ring and the complete networks are equivalent.*
(b) *If* $s^R(2) < s^* \leq s^C(N)$, *then the complete network dominates the ring.*
(c) *If* $s^C(N) < s^* \leq s^R(N)$, *then the ring network dominates the complete.*

The intuition for the result is quite simple. If the shock is small enough that all firms (possibly except the one directly hit by the shock) survive, or big enough that they all fail, independent of the network structure, there is no reason to prefer one structure over the other. If the shock is of intermediate size, there are two possibilities: it can be sufficiently large that some firms fail in the ring network, but not large enough to make more than one firm fail in the complete one, which then proves superior. Alternatively, the shock can be larger so all firms in the complete network fail, while some firms survive in the ring, so the latter is better in this case.

In the previous result we compared network structures that only differed in terms of the density of connections among firms. There are, however, other important aspects of the network that play an important role in determining the number of firms defaulting for different shock sizes. The first one is the amount of integration of any firm with (equivalently, its degree of exposure to) the rest of the firms in the economy, as captured by $1 - \alpha$ in the EGJ/CGV framework, and by a in AOT/GY. In this respect it is immediate to verify, given (20.7)–(20.9) and (20.10), (20.12), (20.13), that[18]

Result 8. *The minimal size of the shock* $s(1)$ *inducing one failure increases with* $1 - \alpha$ *in EGJ/CGV while it is invariant w.r.t.* a *in AOT/GY. In contrast, both in EGJ/CGV and AOT/GY* $s(j)$ *decreases with* $1 - \alpha$ *(resp.* a*), for all* $j > 1$.

In other words, in the EGJ/CGV framework the presence of linkages allows firms to lower the probability of default when a shock hits them directly, thereby providing insurance against this event, but these linkages also enhance the probability of contagion, that is the probability that one firm default when other firms are hit. This trade-off is similar to the one observed when we discussed Result 2. The result is related to Proposition 2 in EGJ, which states that higher integration makes contagion more likely. On the other hand, as already noted after Result 5, in AOT/GY the degree of

[18] The claim holds both for the complete and the ring structure, hence we omit the superscript R or C.

integration a unambiguously increases the level of defaults in the system. Hence a cannot be viewed as a meaningful policy parameter which could be varied, while α can.

In the analysis so far, we ignored another important element that may allow to reduce the extent of contagion, that is the possibility of segmenting the system of N firms into disjoint components. The possibility of segmentation may prove a more effective instrument than the weakening of connections in limiting contagion. It is in fact easy to verify that in situations like the one of Result 7c, the segmentation of the system of N firms into disjointed, complete components is typically superior to a single, ring structure. This poses the more general question of whether the choice of weaker connections but in integrated systems, versus that of denser connections in segmented systems proves more effective in limiting the possibility of default by many firms in the presence of intermediate but large shocks. As the previous discussion makes clear, to properly analyze the pros and cons of segmentation versus density of connection, it is important to go beyond the simple shock distribution considered in Result 7 and to consider richer probability structures.

To fix attention, we follow CGV and focus on the case where shocks have a Pareto distribution. More precisely, $r = R - \rho$, where ρ has support[19] $[1, \infty)$ and density $\gamma / \rho^{\gamma+1}$.Under these conditions, Propositions 1 and 2 in CGV show that:

Result 9. *When $\gamma < 1$ the optimal network structure in CGV exhibits maximal segmentation (i.e., components of minimal size, equal to 2, in which case ring and complete networks are equivalent). When $\gamma > 1$ the optimal network structure has minimal segmentation, with a single, completely connected component.*

This is again easy to understand in the context of Result 7. A Pareto distribution with $\gamma < 1$ exhibits fat tails, hence a sufficiently large mass of the distribution is concentrated on large shocks s^* such that $s^* > s^R(N)$, and the only way to limit contagion and defaults in these cases is to divide the system into components of minimal size. Similarly, when $\gamma > 1$, the mass of the distribution is concentrated on smaller shocks, and hence it is likely to be the case that $s^* < s^C(N)$ so that a single, fully connected component allows to minimize the extent of default in the system (as in this case the opportunities for risk-sharing are high). For these shock distributions, the breakup of the system into small components turns out again to be more effective than a lower density of the connection among firms in limiting contagion. Similar results hold in the AOT/GY framework, as Proposition 5 in AOT shows, for example, that when shocks are large, more segmented networks always dominate the less segmented ones.

More nuanced results can be obtained with other distributions. For example, Proposition 4 in CGV states:

Result 10. *When the shock distribution is given by a mixture of a Pareto distribution with parameter $\gamma > 1$ and another Pareto distribution with parameter $\gamma' < 1$, the optimal*

[19] This distribution could be suitably truncated so as to satisfy the non-negativity constraint on r.

pattern of segmentation for the completely connected structure in CGV is intermediate and symmetric, with identical components of intermediate size.[20]

Also, in this case, interestingly, the complete structure still dominates the ring, but Proposition 5 in CGV finds yet other, less "regular" distributions for which a single ring component is optimal—that is, for which less dense connections are more effective than segmentation.

EGJ in turn use random networks, which in principle encompass networks with varying degrees, and allow both for segmentation and different density of connections. The main result where this is studied is their Proposition 1, where it is shown that a necessary condition for widespread contagion is that the average degree is bigger than 1, a well-known condition for a random network to have a giant component,[21] but not too high.[22] If in fact the average degree is sufficiently high, one basically reverts to a complete network, which can be very robust and impedes large cascades if shocks are not too large.

Decentralized network formation may lead to departures from optimality because of the usual disconnect between efficiency and stability of a network due to non-internalized externalities (see, e.g., Goyal 2007, Jackson 2009). In the present context CGV show (Proposition 6) that these externalities may be induced by the feasibility constraint that operates on admissible configurations: in the absence of side payments, firms may then deviate to form components of individually optimal size without internalizing the fact that the firms which are left out will be forced to be in inefficiently small components.

20.2.4 Heterogeneity, Asymmetries and other Extensions

Even though in the previous discussion we focused our attention on the case where firms are, essentially, homogenous and on symmetric network structures, it is clearly important to extend the analysis to allow for heterogeneity among firms and asymmetries in the network. Battiston et al. (2012) show in fact that real financial networks are quite asymmetric, with, for example, a core-periphery structure.

As already mentioned in the previous section, EGJ explore contagion in random networks, where asymmetries may also arise. EGJ also deal with core-periphery, homophily grouped industries, and correlated shocks in their section IV. AOT also characterize the pattern of contagion in general networks (in Proposition 7) and briefly consider the impact on contagion of the presence of firms of different sizes, studying (in Proposition 10) the threshold for contagion in that case.

[20] More precisely, the result holds for an open set of values of the weights of the mixture.

[21] See, e.g., Jackson (2009), Ch. 4.

[22] To properly relate this result to the previous ones in this section, it is useful to point out that in a random network segmentation and density of the connections cannot be separately controlled via the degree.

CGV study the properties of the optimal network structure when firms differ for the probability distribution of the shocks they face, or for their sizes (with probability of shocks that are proportional to size). In both cases they find (propositions 7 and 8) that default rates are minimized if firms form homogeneous components; that is, there should be positive assortative matching.

Loepfe, Cabrales, and Sánchez (2013) extend CGV through numerical methods to environments with more realistic features. They find that the implications of Results 2 and 9 extend to situations with heterogeneity in the firms' size and in the distribution of linkages among firms,[23] in the sense that less (more) dense networks are more robust when shocks come from unbounded distributions that put more weight on large (small) shocks even with asymmetric configurations. They also find that in the real-life network of corporate control studied by Battiston et al. (2012) the consequences of link removal (moving so to a less dense network) on the vulnerability of the system depend on the nature of the shock distribution as suggested by the results in CGV.

Why real networks do exhibit asymmetries, and what are the consequences for contagion and optimality is an important question to address in the future. For example, risk-sharing may not be the only motive why firms are connected. They also share information, which can lead to a very centralized structure (see, e.g., Guimerà et al. 2002); they provide services to one another; and joint stakes are useful in the presence of incomplete contracts (Grossman and Hart 1986, Hart and Moore 1990). This suggests that an important avenue for future research would be to study optimal network formation in environments where risk-sharing and contagion are not the only trade-offs facing the firms.

Also, a network is a complicated object and understanding the effects of shocks could thus be cognitively as well as informationally demanding. Another avenue for further research is then to properly recognize these features in studying the formation of linkages and the behavior of firms in response to shocks. This might possibly lead to effects that are similar to the market freezes discussed in Section 20.3.2, but it needs to be explored thoroughly.

Finally, in the papers considered the arrival of a shock is independent of the network structure. But this need not be true in general; the network could influence the likelihood of a shock. To illustrate this problem, EGJ have an example (in the Online Appendix, Section 3) in which the firms manipulate the failure threshold to take advantage of neighbors. This issue needs to be explored in more depth.

20.2.5 Additional Literature

Allen and Gale (2000) is a seminal contribution to the study of contagion in interconnected financial systems. They analyze a model in the Diamond and Dybvig (1983)

[23] The results prove also robust to the possibility that shocks hit simultaneously more than one firm, and to other kinds of shock distributions.

tradition. In their model a single, completely connected component is the network structure that minimizes the extent of default, but they study the trade-offs that occur when that structure is not possible. With respect to the papers discussed in Sections 20.2.2 and 20.2.3, the model considered by Allen and Gale (2000) has more detail in terms of the microfoundation of the linkages among banks and of the analysis of their decisions, but this is done at the price of a rather simplified network structure, as only four banks are present. Freixas, Parigi, and Rochet (2000) consider an environment where a lower density of connections, even though it limits risk-sharing, has the positive consequence of reducing the incentives for deposit withdrawal. They study an aspect of the problem we have not discussed so far, since they show that a ring induces more bank discipline than a complete structure, as the higher exposure to a single debtor increases the incentives to control moral hazard.

Cifuentes, Ferrucci, and Shin (2005) explore the issue of contagion due to liquidity shocks in an Eisenberg and Noé (2001) framework where banks are required to maintain a certain liquidity ratio, and sales by distressed banks can generate a cascade of failures through their effect on prices when the market's demand for illiquid assets is less than perfectly elastic. They do not explore however the impact of different structures of exposures among firms on the likelihood of contagion. Greenwood, Landier, and Thesmar (2015) also explore contagion through fire sales. In addition to exploring this problem theoretically, they estimate its impact using balance sheet data for European banks, and show how these estimates can be useful to formulate policies. Another dynamic equilibrium model with cascading defaults (through fire sales of assets) is Bluhm, Faia, and Krahmen (2013), which also provides a measure of systemic importance of institutions based on Shapley values.

Cohen-Cole, Patacchini, and Zenou (2012) study contagion without defaults in a dynamic model with strategic lending and borrowing. They show that in equilibrium small changes in uncertainty, or behavior, can propagate through the network even without defaults, and provide a measure of total systemic risk related to Katz-Bonacich centrality[24] (Chapter 11 in this handbook, by Yves Zenou, discusses other uses of Katz-Bonacich centrality in "key player" policies). They also calculate the contribution of individual banks to this measure, and verify its empirical usefulness.

Another line of literature studies the issue of contagion in the context of large, usually random, networks. In many of these papers, the approach is numerical, based on large-scale simulations. Gai, Haldane, and Kapadia (2011), for example, study a model of interbank lending and show that complexity and concentration of the network of lending relationships can amplify the fragility of the system.[25] Blume et al. (2011) add to these models a strategic analysis of network formation. They assume that agents benefit from the number of direct contacts they have. But they also have a cost, arising from the fact that a shock to an agent in the network is transmitted to all its contacts, direct but also indirect. They find that social optimality is attained around the threshold where

[24] Katz (1953), Bonacich (1987).

[25] See also Nier et al. (2007), Kapadia et al. (2012), and Anand et al. (2012).

a large component emerges, but individuals will generally want to connect beyond this point. This is due to the fact that agents do not internalize the effect on others of adding new channels (i.e., linkages) that facilitate the spread of shocks.[26] Gofman (2013) calibrates a network-based model of over-the-counter bilateral trades in the federal funds market. He compares the calibrated architecture to nine counterfactuals and shows the presence of nonmonotonicities in the risk of contagion with respect to the maximum number of trading partners of an institution.

Finally, we should briefly mention a large empirical literature whose main objective is policy related. These papers explore which summary statistics of the network of firms' relationships are better able to predict contagion of shocks. For example, Battiston et al. (2012) propose a measure of centrality (*Debtrank*) that is inspired by the measure of *Pagerank* centrality that Google uses to rank webpages. Denbee et al. (2011) propose a measure of centrality á la Katz-Bonacich and apply it to English data. Bonaldi, Hortaçsu, and Kastl (2014) propose a measure of systemicity based on the estimation of spillovers between funding costs of individual banks. Of particular interest in this respect is Elsinger, Lehar, and Summer (2011) who, using Austrian data, show that correlation in banks' asset portfolios is the main source of systemic risk. A related emphasis on cross-ownership and investment has been pursued as well by the literature on "balance sheet effects" to understand the Asian financial turmoil in the late 1990s (Krugman 1999) as well as the current crisis (Ahrend and Goujard 2011).

20.3 Informational Contagion

20.3.1 Contagion between Markets

Informational contagion refers to the process whereby information about one market has an impact on another market. The study of informational contagion did not originally require the use of an explicit network, but the concept of a network is a natural framework for the study of informational contagion. Two of the earliest studies, Kodres and Pritzker [KP] (2002) and Pavlova and Rigobon [PR] (2008), illustrate the network structure that is implicit in the study of informational contagion.

[KP] make use of the familiar rational expectations model of financial markets, in which private information is aggregated in asset prices. They consider a model consisting of two asset markets, A and B, and assume that traders in one market can observe the asset prices in both markets. The presence of noise traders prevents prices from revealing information perfectly. In fact, prices will reflect a combination of noise and fundamentals. The fundamental values of the assets in the two markets are affected by a combination of common and market-specific shocks. Traders in market

[26] This form of inefficiency is different from the one discussed in Section 20.2.3.

A, observing changes in the prices of assets in market *B*, will try to infer information that is relevant to the valuation of assets in market *A*. In this way, an increase in prices in market *B* may cause prices to rise in market *A* if traders infer that the price increases in market *B* are due to a common shock that has raised fundamental values in both markets. Since prices do not reveal information perfectly, traders in market *A* may be fooled by price increases in market *B* that result either from noise trading or from fundamental shocks that are irrelevant to market *A*. Thus, the spillover from market *B* to market *A* may be unjustified by the information originally received by traders in market *B*.

Now suppose that there are three markets, *A*, *B*, and *C*, and that *A* and *B* have a common factor, *B* and *C* have a common factor, but that *A* and *C* have no factors in common. Contagion is possible between *A* and *B* and between *B* and *C* for the usual reasons explained above. What is more surprising is that portfolio rebalancing allows financial contagion between market *A* and market *C*. Even if there is no information asymmetry in market *B* and no market specific shock, a shock to market *A* will spill over to market *B* and affect prices there. These price changes will be interpreted as the result of a shock common to *B* and *C* and will thereby influence prices in market *C*.

In the simplest cases, the idea of a network structure is almost superfluous, but if one thinks of a larger number of markets, for example, capital markets in different countries, the network concept arises quite naturally. Suppose the markets are indexed by $i = 1,\ldots,m$ and the factors affecting asset values are indexed by $j = 1,\ldots,n$. Each market i is affected by a set of factors $J(i)$. A pair of markets i_A and i_B will share some common factors $J_{i_A} \cap J_{i_B}$ but each will also be affected by specific factors $J_{i_A} \backslash J_{i_B}$ and $J_{i_B} \backslash J_{i_A}$ that do not directly affect each other. Two markets, i_A and i_B, are directly connected if $J_{i_A} \cap J_{i_B}$ is non-empty. Of course, the "connection" between two markets is more complicated than the existence of an edge between two nodes in a graph, but the analogy is clear.

[PR], like [KP], use the rational expectations framework, but seek to show how informational contagion can affect economies that have no common shocks. They illustrate this idea with the Brazilian financial crisis that followed the Russian default of 1998. The link between the apparently unrelated crises in Russia and Brazil was the New York market, which was linked to both. The Russian default caused New York banks to adjust their portfolios because of institutional constraints designed to control risk exposures. This reallocation of wealth in turn caused changes in prices in the Brazilian market that signaled a bad shock. Here the network is salient: New York constitutes the **center** of the network and Brazil and Russia represent two **peripheral** nodes.

These two papers also illustrate the difference between **pure information externalities** and **payoff externalities.** In [KP] there is a pure informational externality: the actions of traders in market *B* do not affect the payoffs of traders in market *A*. The only reason that traders in market *A* pay attention to the actions of traders (or prices) in market *B* is because of the information those actions (or prices) reveal. In [PR], on the other hand, the actions of the traders in New York have a direct affect on the traders in Brazil, through their effect on prices, in addition to the information revealed by the change in prices. Pure information externalities are somewhat unusual in economics, but they are characteristic of models of herd behavior (Banerjee 1992; Bikhchandani, Hirshleifer, and Welsh 1992). In those models, a sequence of agents chooses a discrete action after observing a private signal. One can think of the sequence of agents as forming a kind of network in which agent $N + 1$ observes the actions of agents $n = 1,\ldots,N$, but not their private information. The discrete action chosen by the preceding agents is a coarse signal of their private information. This makes informational cascades possible in which agents ignore their private information and follow their predecessors, and herd behavior in which agents choose the same action indefinitely. These ideas have also been applied in market settings. Avery and Zemsky (1998) have shown that the informational role of prices can prevent the occurrence of informational cascades and incorrect herds. Cipriani and Guarino (2008a,b) have found conditions under which cascades can exist in financial markets (this requires transaction costs or gains from trade) and showed that contagion may occur between asset markets (information revealed by prices in one market may cause a cascade in another market).

In [KP] and [PR], traders are assumed to observe prices in all markets, but prices are not perfectly revealing because of the presence of noise. Other writers assume that agents in one location can only observe a subset of other locations. Examples include Gale and Kariv (2003), Acemoglu, Dahleh, Lobel, and Ozdaglar (2011), and Mueller-Frank (2013). This literature is discussed in Chapter 19 by Golub and Sandler in this volume.

20.3.2 Contagion between Banks

Informational contagion between banks has been explored in settings that do not call for an explicit network structure. One example is the paper by Ahnert and Georg (2012), in which the release of bad information about one bank provides information about other banks because of common asset holdings and common counterparty exposures. The paper by Ahnert and Bertsch (2014) makes use of a different type of contagion, the "wake-up call." Bad information about one bank causes depositors to obtain costly information about another bank because the depositors suspect a common shock. Even if it transpires that the second bank was not subject to a common shock, the information revealed about the second bank may, coincidentally, provoke a run. Thus, even in the absence of a common shock, there is informational contagion.

These models, like those of [KP] and [PR], are interesting because they provide an account of informational contagion between banks but they do not make use of an explicit network. It is easy to see how these informational contagion channels could be embedded in a network framework to give a richer account of financial contagion. Two examples of financial contagion in networks are described next.

Caballero and Simsek [CS] (2013) have exploited an explicit network formulation to show how informational contagion can amplify the mechanical contagion that results from counterparty exposures. Imagine a sequence of banks, indexed by $i = 1,\ldots,m$, arranged in a circular network. Bank i owes one unit to bank $(i-1) \bmod m$. Banks are funded by deposits and capital (equity), and the latter allows banks to suffer some losses without failing. If bank i fails completely (its assets become worthless), it cannot pay what it owes bank $i-1$. Then bank $i-1$ can pay bank $i-2$ the amount $\min\{1,e\}$, where e represents the bank's equity, bank $i-2$ will be able to pay bank $i-3$ the amount $\min\{\min\{1,e\}+e,0\}$ and so on. If bank k is far enough away from bank i so that $(i-k+1)e \geq 1$,[27] it will receive the full amount and bank k will not fail. Thus, a complete failure of bank i only causes limited contagion within the network.

Now suppose that all banks know that some bank has failed completely but the identity of the failing bank is unknown. Then every bank will have a positive probability of being hit by the contagion and failing. In fact, if the probability distribution of any bank failing is unknown and depositors have an extreme form of uncertainty aversion, they may assume the worst case (i.e., that bank i is close to the failing bank and will also fail). In effect, uncertainty aversion collapses the network, making the failed bank every bank's neighbor. As a result, the banks take precautionary actions in the situation with uncertainty about the identity of the failing bank, such as having more liquidity available, and they do it to a considerably larger extent compared to what they do in the situation where they know the structure of the network and which bank got hit by a shock.

Notice that the CS model is related to the ones in Section 20.2.1 (most closely, to AOT and GY), with the main substantial difference being the incomplete information about the network structure. Thus, as mentioned earlier, informational contagion in CS is an amplification mechanism, to be added to the one caused by exposure to other firms in Section 20.2.1.

Alvarez and Barlevy [AB] (2014) have undertaken a more complex and nuanced analysis of uncertainty about the location of bad banks. In their analysis, banks are randomly assigned to locations on a network. Some of the banks are bad banks, which have received a shock that makes it impossible for them to repay their debts in full. Other banks have not received such a shock but may have counterparty exposure, directly or indirectly, to one of the bad banks. As in [CS], bank i's exposure depends on the distance between bank i and the nearest bad bank. The complexity of the analysis derives partly from the generality of the networks considered and the difficulty of characterizing the distribution of good and bad banks. Each of the good banks has an investment option that can produce enough revenue to keep the bank solvent but it requires investment that has to be externally funded. The problem is that, because of the uncertainty of the location of good and bad banks, the debt overhang discourages investors from investing in the good banks. [AB] show that, under certain conditions, mandatory disclosure of banks' financial condition can permit some banks to obtain external funding and return to solvency, whereas nondisclosure ensures that none will receive the new investment they need to survive.

[27] The contagion ends at bank k if bank $k-1$ can repay its debt in full and banks $k-2,\ldots,i$ cannot. In that case, bank $k-1$ can pay at most $(i-k+1)e$ and k will not fail if and only if $(i-k+1)e \geq 1$.

These two examples illustrate how financial contagion depends not only on the network structure but also on the banks' knowledge and beliefs about that structure. This suggests that providing information about or ensuring the transparency of the network structure, may be a useful regulatory tool. Of course information and transparency may also have implications for competition, as it may make tacit collusion easier. The interaction between systemic regulation, whose main objective is financial stability and competition policy, is another potentially important avenue for further research.

References

Acemoglu, Daron, Munther A. Dahleh, Ilan Lobel, and Asuman Ozdaglar (2011). "Bayesian learning in social networks." *Review of Economic Studies* 78, 1201–1236.

Acemoglu, Daron, Asuman Ozdaglar, and Alireza Tahbaz-Salehi (2014). "Systemic risk and stability in financial networks." *American Economic Review*, in press.

Ahnert, Toni and Co-Pierre Georg (2012). "Information contagion and systemic risk." LSE Working Paper.

Ahnert, Toni and Christoph Bertsch (2013). "A wake-up call: information contagion and strategic uncertainty." unpublished.

Ahrend, Rudiger and Antoine Goujard (2011). "Drivers of systemic banking crises: The role of bank-balance-sheet contagion and financial account structure." OECD Working Paper, 71(902).

Alvarez, Fernando and Gadi Barlevy (2014). "Mandatory disclosure and financial contagion." Federal Reserve Bank of Chicago Working Paper 2014-04.

Anand, Kartik, Prasanna Gai, Sujit Kapadia, Simon Brennan, and Matthew Willison (2012). "A network model of financial system resilience." Bank of England Working Paper No. 458.

Allen, Franklin, and Douglas Gale (2000). "Financial contagion." *Journal of Political Economy* 42, 1–33.

Banerjee, Abhijit V. (1992). "A simple model of herd behavior." *The Quarterly Journal of Economics* 107, 797–817.

Bikhchandani, Sushil, D Hirshleifer, and Ivo Welch (1992). "A theory of fads, fashion, custom, and cultural change as informational cascades." *Journal of Political Economy* 100, 992–1026.

Battiston, Stefano, Michelangelo Puliga, Rahul Kaushik, Paolo Tasca, and Guido Caldarelli (2012). "DebtRank: Too central to fail? Financial networks, the FED and systemic risk." *Scientific Reports* 2, 541.

Bluhm, Marcel, Ester Faia, and Jan Pieter Krahnen (2013). "Endogenous banks' networks, cascades and systemic risk." mimeo.

Bonacich, Philipp (1987). "Power and centrality: A family of measures."*American Journal of Sociology* 92, 1170–1182.

Blume, Larry, David Easley, Jon Kleinberg, Robert Kleinberg, and Éva Tardos (2011). "Network formation in the presence of contagious risk." *Proceedings of the 12th ACM Conference on Electronic Commerce.*

Bonaldi, Pietro, Ali Hortaçsu, and Jakub Kastl (2014). "An empirical analysis of systemic risk in the EURO-zone." mimeo.

Brioschi, Francesco, Luigi Buzzacchi, and Massimo G. Colombo (1989). "Risk capital financing and the separation of ownership and control in business groups." *Journal of Banking & Finance* 13, 747–772.

Caballero, Ricardo and Alp Simsek (2013). "Fire sales in a model of complexity." *The Journal of Finance* 68, 2549–2587.

Cabrales, Antonio, Piero Gottardi, and Fernando Vega-Redondo (2013). "Risk-sharing and contagion in networks." UC3M working paper, Economics 13.

Cifuentes, Rodrigo, Gianluigi Ferrucci and Hyun Song Shin (2005). "Liquidity risk and contagion." *Journal of the European Economic Association* 3(2–3), 556–566.

Cipriani, Marco and Antonio Guarino (2008a). "Transaction costs and informational cascades in financial markets." *Journal of Economic Behavior & Organization* 68, 581–592.

Cipriani, Marco and Antonio Guarino (2008b). "Herd behavior and contagion in financial markets." *The B.E. Journal of Theoretical Economics* 8, 1935–1704.

Cohen-Cole, Ethan, Eleonora Patacchini and Yves Zenou (2014). "Systemic risk and network formation in the interbank market." mimeo.

Dasgupta, Amil (2004). "Financial contagion through capital connections: A model of the origin and spread of bank panics." *Journal of the European Economic Association* 2, 1049–1084.

Denbee, Edward, Christian Julliard, Ye Li, and Kathy Yuan (2011). "Network risk and key players: A structural analysis of interbank liquidity." mimeo.

Diamond, Douglas W. and Philip H. Dybvig (1983). "Bank runs, deposit insurance, and liquidity." *Journal of Political Economy* 91(5), 401–419.

Eisenberg, Laurence and Thomas Noé (2001). "Systemic risk in financial systems." *Management Science* 47(2), 236–249.

Elliott, Matthew L., Matthew O. Jackson, and Benjamin Golub (2014). "Financial networks and contagion." *American Economic Review* 104(10), 3115–3153.

Elsinger, Helmut, Alfred Lehar and Martin Summer (2006). "Risk assessment for banking systems." *Management Science* 52, 1301–1314.

Fedenia, Mark, James E. Hodder, and Alexander J. Triantis (1994). "Cross-holdings: estimation issues, biases, and distortions." *Review of Financial Studies* 7, 61–96.

Freixas, Xavier, Bruno M. Parigi, and Jean-Charles Rochet (2000). "Systemic risk, interbank relations, and liquidity provision." *Journal of Money, Credit, and Banking* 32, 611–638.

Gai, Prasanna, Andrew Haldane, and Sujit Kapadia (2011). "Complexity, concentration and contagion." *Journal of Monetary Economics* 58, 453–470.

Gale, Douglas and Shachar Kariv (2003). "Bayesian learning in social networks." *Games and Economic Behavior* 45, 329–346.

Glasserman, Paul and H. Peyton Young (2015). "How likely is contagion in financial networks?" *Journal of Banking & Finance* 50, 383–399.

Gofman, Michael (2013). "Efficiency and stability of a financial architecture with too-interconnected-to-fail institutions." mimeo.

Golub, Ben and Evan Sadlery (2015). "Learning in social networks." Chapter 19 in this volume.

Goyal, Sanjeev (2007). *Connections: An Introduction to the Economics of Networks.* Princeton: Princeton University Press.

Greenwood, Robin, Augustin Landier, and David Thesmar (2015). "Vulnerable banks." *Journal of Financial Economics*, in press.

Grossman, Sanford J. and Oliver D. Hart (1986). "The costs and benefits of ownership: A theory of vertical and lateral integration." *Journal of Political Economy* 94(4), 691–719.

Guimerà, Roger, Albert Díaz-Guilera, Fernando Vega-Redondo, Antonio Cabrales and Àlex Arenas (2002). "Optimal network topologies for local search with congestion." *Physical Review Letters* 89(24), 248701.

Hart, Oliver and John Moore (1990). "Property rights and the nature of the firm." *Journal of Political Economy* 98(6), 1119–1158.

Jackson, Matthew O. (2009). *Social and Economic Networks*. Princeton: Princeton University Press.

Kapadia, Sujit, Mathias Drehmann, John Elliott, and Gabriel Sterne (2012). "Liquidity risk, cash-flow constraints and systemic feedbacks." Bank of England Working Paper No. 456.

Katz, Leo (1953). "A new status index derived from sociometric analysis." *Psychometrika* 18, 39–43.

Kodres, Laura E. and Matthew Pritsker (2002). "Rational expectations model of financial contagion." *Journal of Finance* 57, 769–799.

Krugman, Paul (1999). "Balance sheets, the transfer problem, and financial crises." *International Tax and Public Finance* 6, 459–472.

Loepfe, Lasse, Antonio Cabrales, and Angel Sánchez (2013)."Towards a proper assignment of systemic risk: The combined roles of network topology and shock characteristics." *PLoS ONE* 8, e77526. doi:10.1371/journal.pone.0077526.

Mueller-Frank, Manuel (2013). "A general framework for rational learning in social networks." *Theoretical Economics* 8, 1–40.

Nier, Erlend, Jing Yang, Tanju Yorulmazer, and Amadeo Alentorn (2007). "Network models and financial stability. " *Journal of Economic Dynamics and Control Volume* 31, 2033–2060.

Pavlova, Anna, and Roberto Rigobon (2008). "The role of portfolio constraints in the international propagation of shocks." *Review of Economic Studies* 75, 1215–1256.

Song Shin, Hyun (2009). "Securitization and financial stability." *Economic Journal* 119, 309–332.

Vega-Redondo, F. (2007). *Complex Social Networks*, Econometric Society Monograph Series. Cambridge: Cambridge University Press.

Zenou, Yves (2015). "Key players." Chapter 11 in this volume.

APPENDIX

We present here some further details of the argument behind some of the claims in the main text.

The expressions in (20.6)

It is immediate to verify that the expression of A^C is obtained from (20.2) with the following specification of the matrix of the cross-ownership of shares among firms[28]:

$$C = \begin{bmatrix} 0 & c/(N-1) & \cdots & c/(N-1) \\ c/(N-1) & 0 & \cdots & c/(N-1) \\ \vdots & \vdots & \ddots & \vdots \\ c/(N-1) & c/(N-1) & \cdots & 0 \end{bmatrix}$$

[28] The same is true for the specification in (20.1), when the matrix B has equal off-diagonal terms.

Similarly, A^R is obtained from[29]

$$C = \begin{bmatrix} 0 & c & \cdots & 0 \\ 0 & 0 & \cdots & 0 \\ \vdots & \vdots & \ddots & \vdots \\ c & 0 & \cdots & 0 \end{bmatrix}.$$

Contagion in ring and complete structures when β is large

The relationship between $s^C(1)$ and $s^C(N)$ depends on the values of β and N. By subtracting (20.8) from (20.7) we get

$$-\alpha s^C(1) + \frac{1-\alpha}{N-1} s^C(N) = -\frac{1-\alpha}{N-1}\beta.$$

From this expression it is clear that there exists some $\bar{\beta}^C > 0$ such that $s^C(N) < s^C(1)$ for $\beta > \bar{\beta}^C$, so that once the first default occurs, all firms default. More explicitly, $\bar{\beta}^C$ is defined by the value of β for which $s^C(1) = s^C(N)$, and hence satisfies

$$\frac{1-\alpha}{N-1}\left(s^C(N) - s^C(1)\right) = \left(\alpha - \frac{1-\alpha}{N-1}\right) s^C(1) - \frac{1-\alpha}{N-1}\bar{\beta}^C = 0.$$

Similarly for the ring, subtracting the expression of (20.9) for j from that for $j-1$ we get

$$\alpha s^R(j) - s^R(j-1) = -\alpha\beta.$$

Proceeding by induction it is then easy to see that if $s^R(2) < s^R(1)$, we have $s^R(j) < s^R(j-1)$ for all $j > 2$, so the first firm defaulting brings all other firms down. Building on this observation, the threshold value $\bar{\beta}^R$ such that $s^R(1) > s^R(N)$ for $\beta > \bar{\beta}^R$ is obtained as the solution of

$$s^R(1) - \alpha s^R(1) = \alpha\bar{\beta}^R.$$

Recalling that $s^R(1) = s^C(1)$, we get

$$\frac{\bar{\beta}^C}{\bar{\beta}^R} = \frac{\alpha\,(N\alpha - 1)}{(1-\alpha)^2}.$$

Hence $\bar{\beta}^C > \bar{\beta}^R$ iff

$$\alpha > \frac{\sqrt{4N-3} - 1}{2N-2}$$

[29] Similarly for (20.1) when

$$B = \begin{bmatrix} \theta & 1-\theta & \cdots & 0 \\ 0 & \theta & \cdots & 0 \\ \vdots & \vdots & \ddots & \vdots \\ 1-\theta & 0 & \cdots & \theta \end{bmatrix}$$

and $K = N - 1$.

and this inequality always holds, under our assumption that $\alpha > (1-\alpha)^2 / \left(1-\alpha^{N-1}\right)$ when $N \geq 6$. Hence we can say that, provided N is not too small, $\bar{\beta}^C > \bar{\beta}^R$, and so generalized default is easier in the ring than in the complete structure—that is, it obtains for smaller values of β in the ring structure.

Results 1 and 2 in the AOT/GY framework (Result 4)

The minimal shock $s^C(N)$ leading to all firms defaulting in a complete network structure in the AOT/GY environment is again obtained by considering the case where only firm 1 defaults, so that

$$p_1(a+l) = a + R - s^C(N)$$

while all the other firms have just enough resources to ensure they can repay ($p_i = 1$ for all $i \neq 1$):

$$R + \left(\frac{N-2}{N-1}\right)a + a\frac{a+R-s^C(N)}{(N-1)(a+l)} = a + l \tag{20.10}$$

Solving[30] (20.10) yields the value of $s^C(N)$.[31] As one would expect the threshold $s^C(N)$ increases (i.e., the shock has to be bigger) the larger the number of firms N, and the larger the "capital buffer" $R-l$ (i.e., the difference between the outside liabilities and the outside assets of a firm).

In a ring structure, denoting by $\mathbf{r}_1 = (R-s, R, .., R)$ the vector of realized returns where firm 1 is hit by a shock which leads it to default, we have

$$p_i(\mathbf{r}_1) = \min\left(\frac{ap_{i-1}(\mathbf{r}_1) + R}{a+l}, 1\right) \tag{20.11}$$

for $i > 1$, with $p_1(\mathbf{r}_1) = (a+R-s)/(a+l)$. It is possible to verify that $p_i(\mathbf{r}_1) \geq p_{i-1}(\mathbf{r}_1)$ (with the inequality being strict as long as $i-1$ defaults). When the shock is at the threshold level $s^R(N)$ so that all the first $N-1$ firms default and the N-th firm has just enough resources to be able to repay, we also have $ap_{N-1}(\mathbf{r}_1) + R = a + l$ and, for all $i = 2, \ldots, N-1$:

$$p_i(\mathbf{r}_1) = \frac{a}{a+l}p_{i-1}(\mathbf{r}_1) + \frac{R}{a+l}.$$

[30] Note that an admissible ($s \leq R$) solution of (20.10) exists iff

$$\frac{a}{a+l} \geq \frac{R-l}{l}(N-1).$$

[31] The expression is stated for the case $\beta = 0$. It would not be difficult to include extra costs of reorganization with default (so that $\beta > 0$) in this framework. The effects would be similar in this case—higher threshold for default (smaller shocks suffice to generate default) and multiple payment equilibria.

Hence

$$p_{N-1}(\mathbf{r}_1) = \left(\frac{a}{a+l}\right)^{N-2}\left(\frac{a+r}{a+l}\right) + \frac{R}{a+l}\left(1 + \ldots + \left(\frac{a}{a+l}\right)^{N-3}\right)$$
$$= \left(\frac{a}{a+l}\right)^{N-2}\left(\frac{a+r}{a+l}\right) + \frac{R}{l}\left(1 - \left(\frac{a}{a+l}\right)^{N-2}\right)$$

so that $s^R(N)$ is obtained as solution of the following equation

$$R + a\left(\left(\frac{a}{a+l}\right)^{N-2}\left(\frac{a+R-s^R(N)}{a+l}\right) + \frac{R}{l}\left(1 - \left(\frac{a}{a+l}\right)^{N-2}\right)\right) = a + l. \tag{20.12}$$

This clearly implies that $s^R(N)$ is also increasing in N, as well as in $R - l$.

From the above we also see that $s^R(2)$ is obtained as a solution of

$$R + a\frac{a+R-s^R(2)}{a+l} = a + l. \tag{20.13}$$

Subtracting equation (20.13) from (20.10) yields

$$\frac{(N-1)(a+l) + R - s^C(N) - l}{(N-1)(a+l)} - \left(\frac{a+R-s^R(2)}{a+l}\right) = 0,$$

or

$$(N-2)\,l - s^C(N) - (N-2)\,(R - s_2^R) + s^R(2) = 0,$$

and hence

$$s^R(2) - s^C(N) = (N-2)\left(R - s^R(2) - l\right) < 0,$$

thus verifying the property $s^C(N) > s^R(2)$.

Finally, we verify the property $s^C(N) < s^R(N)$. From equations (20.10) and (20.12) we get

$$\left(\frac{a}{a+l}\right)^{N-2}\left(\frac{a+R-s^R(N)}{a+l}\right) + \frac{R}{l}\left(1 - \left(\frac{a}{a+l}\right)^{N-2}\right)$$
$$= \left(\frac{N-2}{N-1}\right) + \frac{1}{N-1}\,\frac{a+R-s^C(N)}{(a+l)}$$

and hence

$$\left(\frac{a}{a+l}\right)^{N-2}\left(\frac{s^R(N)-s^C(N)}{a+l}\right) = \frac{R}{l}\left(1 - \left(\frac{a}{a+l}\right)^{N-2}\right) - \left(\frac{N-2}{N-1}\right)$$
$$+ \frac{a+R-s^C(N)}{a+l}\left[\left(\frac{a}{a+l}\right)^{N-2} - \frac{1}{N-1}\right].$$

Thus $s^C(N) < s^R(N)$ iff

$$\left(\frac{R}{l}-1\right)\left(1-\left(\frac{a}{a+l}\right)^{N-2}\right)+\left(1-\frac{a+R-s^C(N)}{a+l}\right)\left(\frac{1}{N-1}-\left(\frac{a}{a+l}\right)^{N-2}\right)>0,$$

or, substituting for $s^C(N)$ and simplifying

$$\left(\tfrac{R-l}{l}\right)\left(1-\left(\tfrac{a}{a+l}\right)^{N-2}\right)+\left(\tfrac{(N-1)(R-l)}{a}\right)\left(\tfrac{1}{N-1}-\left(\tfrac{a}{a+l}\right)^{N-2}\right)>0$$
$$\Leftrightarrow \left(\tfrac{a}{l}\right)\left(1-\left(\tfrac{a}{a+l}\right)^{N-2}\right)+\left(1-(N-1)\left(\tfrac{a}{a+l}\right)^{N-2}\right)>0$$

The above inequality can be equivalently rewritten as

$$a+l>\left(\frac{a}{a+l}\right)^{N-2}(a+(N-1)\,l)\Leftrightarrow(a+l)^{N-1}>a^{N-1}+(N-1)\,la^{N-2}. \qquad (20.14)$$

Note that when $l=0$ (20.14) holds as equality, and its derivative w.r.t. l is strictly positive for all $N>1$, $a>0$, $l\geq 0$, which establishes the validity of (20.14) and hence of the claimed property.

Ring and complete in AOT/GY when external claims have priority (Result 6)

In this case the expression determining $r^C(N)$ has to be modified as follows:

$$R+\left(\frac{N-2}{N-1}\right)a+\frac{a+r^C(N)-l}{a}\frac{a}{N-1}=a+l, \qquad (20.15)$$

yielding

$$r^C(N)=l-(N-1)\,(R-l)\,.$$

Similarly, in a ring structure (20.11) has to be modified as follows

$$p_i(\mathbf{r})=\min\left(\frac{ap_{i-1}(\mathbf{r})+R-l}{a},1\right)$$

for $i>1$, with $p_1(r)=(a+r-l)/a$. At $\mathbf{r}_N$ we have $p_i(\mathbf{r}_N)=p_{i-1}(\mathbf{r}_N)+(R-l)/a$ for all $i=2,\ldots,N-1$ so that $p_{N-1}(\mathbf{r}_N)=(a+r-l)/a+(N-2)(R-l)/a$ and hence $r^R(N)$ is now determined as a solution of the following equation

$$R+(N-2)\,(R-l)+a+r^R(N)-l=a+l \qquad (20.16)$$

or

$$r^R(N)=l-(N-1)\,(R-l)=r^C(N).$$

CHAPTER 21

NETWORKS, SHOCKS, AND SYSTEMIC RISK

DARON ACEMOGLU, ASUMAN OZDAGLAR, AND ALIREZA TAHBAZ-SALEHI

21.1 Introduction

The recent financial crisis, often attributed in part to contagion emanating from pervasive entanglements among financial institutions, has rekindled interest in the role of complex economic, financial, or social interlinkages as channels for propagation and amplification of shocks. In the words of Charles Plosser, the president of the Federal Reserve Bank of Philadelphia:

> due to the complexity and interconnectivity of today's financial markets, the failure of a major counterparty has the potential to severely disrupt many other financial institutions, their customers, and other markets.
>
> (Plosser 2009)

Similar ideas on the role of interconnections and the possibility of cascades have also surfaced in a variety of other contexts. For instance, Acemoglu et al. (2012 2014b) and Jones (2013) have argued that idiosyncratic shocks at the firm or sectoral level can propagate over input-output linkages within the economy, with potentially significant implications for macroeconomic volatility and economic growth, while Caplin and Leahy (1993) and Chamley and Gale (1994) have emphasized the spread of economic shocks across firms due to learning and imitation.

Though the domains studied by these and other related papers are often different, their underlying approaches share important economic and mathematical parallels. Most importantly, in each case, the problem is one of a set of interacting agents who influence each other, thus opening the way for shocks to one agent to propagate to the rest of the economy. Furthermore, on the methodological side, almost all these papers rely on a network model to capture the pattern and extent of interactions between agents. Despite these parallels, there is a bewildering array of different (and sometimes even contradictory) results, often presented and developed with little linkage to other findings in the literature.

The disparity in the predictions and results of different studies in the literature can be best illustrated by focusing on a concrete setting, namely that of financial interactions. The models of financial interactions studied in a variety of papers, such as Allen and Gale (2000), Giesecke and Weber (2006), Blume et al. (2011), Battiston et al. (2012), Elliott, Golub, and Jackson (2014), Cabrales, Gottardi, and Vega-Redondo (2014), and Acemoglu, Ozdaglar, and Tahbaz-Salehi (2015c) are, at least on the surface, very similar. In each case, a financial institution's "state" which, for example, captures its health or ability to meet its obligations, depends on the state of other financial institutions to which it is connected.[1] Consequently, shocks to a given institution can propagate to other institutions within the economy, potentially snowballing into a systemic crisis. Despite such commonalities, the predictions of many of the papers in this literature are quite different or sometimes even contradictory. For example, in the models of Allen and Gale (2000) and Freixas, Parigi, and Rochet (2000), denser interconnections mitigate systemic risk, whereas several other papers, such as Vivier-Lirimont (2006) and Blume et al. (2011), have suggested that such dense interconnections can act as a destabilizing force.

Our aim in this chapter is to unify and improve the understanding of the key economic and mathematical mechanisms in much of the literature on the effects of network interactions on the economy's aggregate performance. We start with a general reduced-form model in which n agents interact with one another. Each agent is assigned a real-valued variable known as its *state* which, depending on the context, may capture her choice of actions (e.g., output or investment) or some other economic variable of interest. Our reduced-form model consists of three key ingredients: (i) a fairly general *interaction function* that links each agent's state to a summary measure of the states of other agents; (ii) an (*interaction*) *network* that specifies how these summary measures are determined as a function of other agents' states; and (iii) an *aggregation function* that describes how agent-level states collectively shape the macroeconomic variable of interest.

We first show that our general framework nests a wide variety of problems studied in the literature, including those mentioned above. We also show that under fairly general conditions on the interaction function, an equilibrium—defined as a mutually consistent set of states for all agents in the network—always exists and is generically unique. We then use our framework to study how the nature of inter-agent interactions shape various measures of aggregate performance. Our analysis not only nests the main

[1] For instance, in the context of counterparty relationships considered by Acemoglu et al. (2015c), the connections capture the extent of prior interbank lending and borrowing and each bank's state captures its ability to meet those obligations. As highlighted in Cabrales, Gale, and Gottardi (2015), other forms of interlinkages operate in a similar fashion.

results obtained in several papers in the literature, but also clarifies where the sources of differences lie.

In order to obtain sharp and analytical predictions for the role of network interactions in shaping economic outcomes, we focus on an economy in which agent-level shocks are small. This assumption enables us to approximate the equilibrium state of each agent and the economy's macroeconomic state by the first few terms of their Taylor expansions. Our results show that the impact of network structure depends on the properties of the economy's *Leontief matrix* corresponding to the underlying interaction network. This matrix, which is defined in a manner analogous to the same concept used in the literature on input-output economies, accounts for all possible direct and indirect effects of interactions between any pair of agents. Using this characterization, we show that the curvatures of the interaction and aggregation functions play a central role in how the economy's underlying network translates microeconomic shocks into macroeconomic outcomes.

As our first characterization result, we show that as long as the interaction and aggregation functions are linear, the economy exhibits a "*certainty equivalence*" property from an ex ante perspective, in the sense that the expected value of the economy's macro state is equal to its unperturbed value when no shocks are present. This observation means that, in a linear world, the economy's aggregate performance, in expectation, does not depend on the intricate details of its underlying interaction network.

Our next set of results illustrates that this certainty equivalence property may no longer hold if either the aggregation or interaction function is nonlinear. Rather, in the presence of a nonlinear interaction or aggregation function, the exact nature of these nonlinearities is central to determining how the economy's underlying interaction network affects its ex ante performance.

We show that with a nonlinear aggregation function, the economy's ex ante performance depends on the heterogeneity in the extent to which agents interact with one another. In particular, if the aggregation function is concave—for example, to capture the idea that volatility is detrimental to the economy's aggregate performance—a more uniform distribution of inter-agent interactions increases macroeconomic performance in expectation. An important corollary to this result establishes that with a concave aggregation function, *regular* economies (in which the overall influence of each agent on the rest of the agents is identical across the network) outperform all other economies. These results are consistent with, and in some ways generalize, those of Acemoglu et al. (2012), who, in the context of input-output economies, show that the volatility of the economy's aggregate output increases in the extent of heterogeneity in the role of different firms as input-suppliers. Our results thus clarify that it is the concavity of economy's aggregation function—resulting from the focus on volatility—that lies at the heart of the results in Acemoglu et al. (2012).

We then focus on understanding how nonlinearities in the interaction function shape the economy's ex ante performance. Our results illustrate that when the interaction function is concave, economies with denser interconnections outperform those whose interaction networks are more sparse. In particular, the complete network, in which interlinkages are maximally dense, outperforms all other (symmetric) economies. Furthermore, we show that with a convex interaction function, this performance ordering flips entirely, making the complete network the worst performing economy. This flip in the comparative statics of aggregate performance with respect to the network structure parallels the findings in Acemoglu, Ozdaglar, and Tahbaz-Salehi (2015c), who show that, in the context of financial interactions, whether the complete network fosters stability or instability depends on the size and number of shocks: with a few small shocks, the complete network is the most stable of all economies, whereas when shocks are numerous or large, there is a phase transition, making the complete network the least stable financial arrangement. Our results here clarify that the findings of Acemoglu et al. (2015c) are essentially due to the fact that increasing the size or the number of shocks corresponds to a shift from a concave to a convex region of the interaction function, thus reversing the role of interbank connections in curtailing or causing systemic risk. They also highlight that similar phase transitions transforming the role of network interconnections in shaping aggregate performance can emerge in other settings with nonlinear interactions.

Overall, our results highlight that the relationship between the economy's aggregate performance and its underlying network structure depends on two important economic variables: (i) the nature of economic interactions, as captured by our interaction function; and (ii) the properties of the aggregate performance metric, as captured by the notion of aggregation function in our model.

We also use our framework to provide a characterization of how the nature of interactions determine the agents' relative importance in shaping aggregate outcomes. As long as agent-level interactions are linear, the well-known notion of *Bonacich centrality* serves as a sufficient statistic for agents' "systemic importance": negative shocks to an agent with a higher Bonacich centrality lead to larger drops in the economy's macro state. We also demonstrate that, in the presence of small enough shocks, this result generalizes to economies with nonlinear interactions, but with one important caveat: even though a strictly larger Bonacich centrality means that the agent has a more pronounced impact on the economy's macro state, two agents with identical Bonacich centralities are not necessarily equally important. This is due to the fact that Bonacich centrality only provides a first-order approximation to the agents' impact on aggregate variables. Therefore, a meaningful comparison of systemic importance of two agents with identical Bonacich centralities (as in a regular network) requires that we also take their higher-order effects into account. As our final result, we provide such a characterization of agents' systemic importance in regular economies. We show that the second-order impact of an agent on the economy's macro state is summarized via a novel notion of centrality, called *concentration centrality*, which captures the concentration of an agent's influence on the rest of the agents (as opposed to its overall influence captured via Bonacich centrality).

These characterization results thus highlight that relying on standard and off-the-shelf notions of network centrality (such as Bonacich, eigenvector, or betweenness centralities) for the purpose of identifying systemically important agents may be misleading. Rather, the proper network statistic has to be informed by the nature of microeconomic interactions between different agents.

21.1.1 Related Literature

As already indicated, this chapter relates to several strands of literature on social and economic networks, such as the literature on network games, various models of systemic risk, and the literature that studies microeconomic foundations of macroeconomic fluctuations. Many of the papers related to our setup are discussed in the next section, when we describe how different models are nested within our general framework. Here, we provide a brief overview of the literature and some of the key references.

The critical building block of our general framework is an interaction network whereby each player's "state" is a function of the state of its neighbors in a directed, weighted network. These interlinked states could be thought of as best responses of each player to the actions of her neighbors. As such, our setup builds on various different contributions on the network games literature, such as Calvó-Armengol and Zenou (2004), Ballester, Calvó-Armengol, and Zenou (2006), Candogan, Bimpikis, and Ozdaglar (2012), Allouch (2012), Badev (2013), Bramoullé, Kranton, and D'Amours (2014), and Elliott and Golub (2015), several of which can be cast as special cases of our general framework.[2] Several papers consider applications of network games to various specific domains. For example, Calvó-Armengol, Patacchini, and Zenou (2009) study peer effects and education decisions in social networks; Calvó-Armengol and Jackson (2004) study the role of referral networks in the labor market; and Galeotti and Rogers (2013), Acemoglu, Malekian, and Ozdaglar (2014a), and Dziubiński and Goyal (2014) consider a network of interlinked players making endogenous security investments against an infection or an attack. Jackson and Zenou (2015) and Bramoullé and Kranton (2015) provide thorough surveys of the network games literature.

Even though the literature on network games does not generally consider the propagation of idiosyncratic shocks, our results highlight that, depending on the specific economic question at hand, the interaction models at the heart of this literature could be used for the study of such propagation.

A related literature has directly originated from the study of cascades. Various models have been developed in the computer science and network science literatures, including the widely used threshold models (Granovetter 1978) and percolation models (Watts 2002). A few works have applied these ideas to various economic settings, including

[2] Network games of incomplete information are studied in Galeotti et al. (2010).

Durlauf (1993) and Bak et al. (1993) in the context of economic fluctuations; Morris (2000) in the context of contagion of different types of strategies in coordination games; and more recently, Gai and Kapadia (2010) and Blume et al. (2011) in the context of spread of an epidemic-like financial contagion.

The framework developed in this chapter is also closely linked to a small literature in macroeconomics that studies the propagation of microeconomic shocks over input-output linkages. This literature, which builds on the seminal paper by Long and Plosser (1983), has witnessed a recent theoretical and empirical revival. On the theoretical side, Acemoglu et al. (2012 2014b) and Jones (2013) argue that the propagation of idiosyncratic shocks and distortions over input-output linkages can have potentially significant implications for macroeconomic volatility and economic growth.[3] On the empirical side, Foerster, Sarte, and Watson (2011), Carvalho (2014), di Giovanni, Levchenko, and Méjean (2014), Acemoglu, Autor, Dorn, Hanson, and Price (2015a), and Carvalho, Nirei, Saito, and Tahbaz-Salehi (2015) provide evidence for the relevance of such propagation mechanisms in different countries.

As mentioned earlier, this chapter is also closely related to the growing literature on the spread of financial shocks over a network of interconnected financial institutions. The seminal papers of Allen and Gale (2000) and Freixas, Parigi, and Rochet (2000) developed some of the first formal models of contagion over financial networks. The recent financial crisis resulted in further attention to this line of work. Some of the more recent examples include Gai, Haldane, and Kapadia (2011), Battiston et al. (2012), Alvarez and Barlevy (2014), and Glasserman and Young (2015).

Within this literature, four recent papers deserve further discussion. The first, which is our own work (Acemoglu, Ozdaglar, and Tahbaz-Salehi 2015c), considers a network of banks linked through unsecured debt obligations and studies the emergence of financial cascades resulting from counterparty risk. This paper, which in turn builds on and extends Eisenberg and Noe's (2001) seminal framework of financial interlinkages, is explicitly treated as a special case of our general framework here. The second is the related paper by Elliott, Golub, and Jackson (2014), which also considers financial contagion in a network, though based on microfoundations linked to cross-shareholdings across institutions as opposed to counterparty risk. The third is Cabrales, Gottardi, and Vega-Redondo (2014), which is closely connected to Elliott et al. (2014) and in addition considers the endogenous formation of the financial network.[4] Finally, Cabrales, Gale, and Gottardi (2015) provide a unified treatment of the previous three papers, highlighting various commonalities as well as some important differences between them. The key distinction between their unified treatment and ours is that they start with the fixed point equation resulting from the interactions in the various

[3] Relatedly, Gabaix (2011) argues that microeconomic shocks can lead to aggregate fluctuations if the firm-size distribution within the economy exhibits a heavy enough tail, even in the absence of input-output linkages.

[4] Other papers that study network formation in related contexts include Bala and Goyal (2000), Babus (2014), Zawadowski (2013), Acemoglu, Ozdaglar, and Tahbaz-Salehi (2014c), Farboodi (2014), and Erol and Vohra (2014).

financial network models, whereas we develop a more general framework starting from the best response equations or the equations linking each agent's state to her neighbors'. This formulation enables us to nest not only existing models of financial networks but a wider array of network interactions, use first- and second-order approximations to provide a sharper characterization of the structure of equilibrium, and clarify the role of interaction and aggregation functions in transforming small, agent-level shocks into differences in aggregate performance or volatility.

21.1.2 Outline

The rest of this chapter is organized as follows. In Section 21.2, we provide our general framework for the study of network interactions and present a few examples of how our setup maps to different applications. In Section 21.3, we provide a second-order approximation to the macro state of the economy in terms of the economy's underlying interaction network. Section 21.4 uses these results to characterize how the nature of interactions between different agents impacts the macro state of the economy from an ex ante perspective, whereas Section 21.5 provides a characterization of the systemic importance of different agents. Section 21.6 concludes.

21.2 General Framework

Consider an economy consisting of n agents indexed by $N = \{1, \ldots, n\}$. Of key interest to our analysis is each agent i's *state*, $x_i \in \mathbb{R}$, which captures the agent's choice of action (e.g., output or investment) or some other economic variable of interest (such as the solvency of a financial institution). In the next three subsections we will provide concrete examples clarifying the interpretation of these states. For the time being, however, we find it convenient to work with a general, reduced-form setup without taking a specific position on how to interpret the agents or their states.

The key feature of the environment is that the states of different agents are interlinked. Such interdependencies may arise due to strategic considerations, contractual agreements, or some exogenous (e.g., technological) constraints on the agents. Formally, the state of any given agent i depends on the states of other agents via the relationship

$$x_i = f\left(\sum_{j=1}^{n} w_{ij}x_j + \epsilon_i\right), \tag{21.1}$$

where f is a continuous and increasing function, which we refer to as the economy's *interaction function*. As the name suggests, this function represents the nature of interactions between the agents in the economy. The variable ϵ_i is an "agent-level" shock, which captures stochastic disturbances to i's state. We assume that these shocks are independently and identically distributed (so that they correspond to "idiosyncratic" shocks) and have mean zero and variance σ^2.

The constant $w_{ij} \geq 0$ in (21.1) captures the extent of interaction between agents i and j. In particular, a higher w_{ij} means that the state of agent i is more sensitive to the state of agent j, whereas $w_{ij} = 0$ implies that agent j does not have a direct impact on i's state. Without much loss of generality, we assume that $\sum_{j=1}^{n} w_{ij} = 1$, which guarantees that the extent to which the state of each agent depends on the rest of the agents is constant. We say the economy is *symmetric* if $w_{ij} = w_{ji}$ for all pairs of agents i and j.

For a given f, the interactions between agents can be also represented by a weighted, directed graph on n vertices, which we refer to as the economy's *interaction network*. Each vertex in this network corresponds to an agent and a directed edge from vertex j to vertex i is present if $w_{ij} > 0$, that is, if the state of agent i is directly affected by the state of agent j.

Finally, we define the *macro state* of the economy as

$$y = g(h(x_1) + \cdots + h(x_n)), \tag{21.2}$$

where $g, h : \mathbb{R} \to \mathbb{R}$. As we will clarify in what follows, y represents some macroeconomic outcome of interest that is obtained by aggregating the individual states of all agents. Throughout the paper, we refer to g as the economy's *aggregation function*.

An equilibrium in this economy is defined in the usual fashion by requiring each agent's state to be consistent with those of others. Formally:

Definition 1. Given the realization of the shocks $(\epsilon_1, \ldots, \epsilon_n)$, an *equilibrium* of the economy is a collection of states $(x_1, \ldots, x_n)$ such that equation (21.1) holds for all agents i simultaneously.

As the above definition clarifies, our solution concept is an ex post equilibrium notion, in the sense that agents' states are determined after the shocks are realized. This notion enables us to study how the equilibrium varies as a function of the shock realizations.

Throughout the paper, we assume that $f(0) = g(0) = h(0) = 0$. This normalization guarantees that, in the absence of shocks, the equilibrium state of all agents and the economy's macro state are equal to zero.

We next show how a wide variety of different applications can be cast as special cases of the general framework developed above.

21.2.1 Example: Network Games

Our framework nests a general class of network games as a special case. Consider, for example, an n-player, complete information game, in which the utility function of agent

i is given by

$$u_i(x_1,\ldots,x_n) = -\frac{1}{2}x_i^2 + x_i f\left(\sum_{j=1}^{n} w_{ij}x_j + \epsilon_i\right),$$

where x_i denotes the action of player i and ϵ_i is realization of some shock to her payoffs. That is, the payoff of player i depends not only on her own action, but also on those of her neighbors via the interaction function f. In this context, the underlying network, encoded in terms of coefficients w_{ij}, captures the pattern and strength of strategic interactions between various players in the game.

It is immediate to verify that as long as the interaction function f satisfies certain regularity conditions—essentially to ensure that one can use the first-order conditions—and that $w_{ii} = 0$ for all i, the best-response of player i as a function of the actions of other players is given by equation (21.1). Consequently, the collection $(x_1,\ldots,x_n)$ that solves the system of equations (21.1) corresponds to the Nash equilibrium of the game.

The game described above nests a wide variety of models studied in the literature. Note that since f is increasing, the players face a game of strategic complements over the network: the benefit of taking a higher action to player i increases the higher the actions of her neighbors are. Examples of such network games include research collaboration among firms (Goyal and Moraga-González 2001), crime networks (Ballester, Calvó-Armengol, and Zenou 2006), peer effects (Calvó-Armengol, Patacchini, and Zenou 2009), and local consumption externalities (Candogan, Bimpikis, and Ozdaglar 2012). On the other hand, had we assumed that the interaction function f is decreasing, the players would have faced a network game of strategic substitutes, as in Bramoullé and Kranton (2007) who study information sharing and the provision of local public goods.[5]

An important subclass of network games is the case in which players' payoff functions are quadratic,

$$u_i(x_1,\ldots,x_n) = -\frac{1}{2}x_i^2 + \alpha\sum_{j=1}^{n} w_{ij}x_i x_j + \alpha x_i\epsilon_i, \tag{21.3}$$

where $\alpha \in (0,1)$ is some constant.[6] Under such a specification, the corresponding interaction function is given by $f(z) = \alpha z$, hence, implying that the equilibrium of the game can be characterized as a solution to a system of linear equations.

We end our discussion by pointing out two natural candidates for the economy's macro state in this context. The first is the sum (or the average) of the agents' equilibrium

[5] Allowing both for strategic complementarities and substitutabilities, Acemoglu, Garcia-Jimeno, and Robinson (2015b) develop an application of these models in the context of local municipalities' state capacity choices, and estimate the model's parameters using Colombian data.

[6] See Zenou (2015) for a discussion and a variety of extensions of the baseline network game with quadratic payoffs.

actions,

$$y_{\text{agg}} = x_1 + \cdots + x_n,$$

representing the aggregate level of activity in the economy. In our general framework, this corresponds to the assumption that $g(z) = h(z) = z$. The second is the total or average utility (or equivalently total social surplus) in the equilibrium, given by $y_{\text{sw}} = \sum_{i=1}^{n} u_i$. Although summing both sides of equation (21.3) over all players i shows that social surplus depends not only on the agents' states, but also on weights w_{ij} and the realizations of the shocks ϵ_i, using the fact that equilibrium actions satisfy (21.1) enables us to write y_{sw} in the form of equation (21.2) as

$$y_{\text{sw}} = \frac{1}{2}\sum_{i=1}^{n} x_i^2,$$

which corresponds to $g(z) = z$ and $h(z) = z^2/2$ in our general framework.

21.2.2 Example: Production Networks

Our general setup also nests a class of models that focus on the propagation of shocks in the real economy. In this subsection, we provide an example of one such model along the lines of Long and Plosser (1983) and Acemoglu, Carvalho, Ozdaglar, and Tahbaz-Salehi (2012), and show that it can be cast as a special case of our general framework.

Consider an economy consisting of n competitive firms (or sectors) denoted by $\{1, 2, \ldots, n\}$, each of which produces a distinct product.[7] Each product can be either consumed by a mass of consumers or used as an input for production of other goods. All firms employ Cobb-Douglas production technologies with constant returns to scale that transform labor and intermediate goods to final products. Production is subject to some idiosyncratic technology shock. More specifically, the output of firm i, which we denote by X_i, is equal to

$$X_i = b_i A_i^{\alpha} l_i^{1-\alpha} \left(\prod_{j=1}^{n} X_{ij}^{w_{ij}} \right)^{\alpha}, \tag{21.4}$$

where A_i is the corresponding productivity shock; l_i is the amount of labor hired by firm i; X_{ij} is the amount of good j used for production of good i; b_i is a constant; and $\alpha \in (0, 1)$ is the share of intermediate goods in production. The exponent $w_{ij} \geq 0$ in (21.4) captures the share of good j in the production technology of good i: a higher w_{ij} means that good j is more important in producing i, whereas $w_{ij} = 0$ implies that good j is not a required input for i's production technology. The assumption that firms employ constant returns to scale technologies implies that $\sum_{j=1}^{n} w_{ij} = 1$ for all i.

[7] Since each one of these firms is supposed to act competitively, they can also be interpreted as "representative firms" standing in for a set of competitive firms within each of the n sectors.

The economy also contains a unit mass of identical consumers. Each consumer is endowed with one unit of labor which can be hired by the firms for the purpose of production. We assume that the representative consumer has symmetric Cobb-Douglas preferences over the n goods produced in the economy. In particular,

$$u(c_1,\ldots,c_n) = \tilde{b}\prod_{i=1}^{n} c_i^{1/n},$$

where c_i is the amount of good i consumed and $\tilde{b}$ is some positive constant.

One can naturally recast the interactions between different firms in such an economy in terms of a network, with each vertex corresponding to a firm and the factor shares w_{ij} capturing the intensity of interactions between them. Furthermore, given the log-linear nature of Cobb-Douglas production technologies, the equilibrium (log) output of each firm can be written in the form of equation (21.1), linking it to the outputs of its input suppliers and the productivity shocks in the economy.

To see this, consider the first-order conditions corresponding to firm i's problem:

$$X_{ij} = \alpha w_{ij} p_i X_i / p_j \tag{21.5}$$

$$l_i = (1-\alpha) p_i X_i / \omega, \tag{21.6}$$

where ω denotes the market wage and p_i is the price of good i. The market clearing condition for good i, given by $X_i = c_i + \sum_{j=1}^{n} X_{ji}$, implies that

$$s_i = \omega/n + \alpha \sum_{j=1}^{n} w_{ji} s_j,$$

where $s_i = p_i X_i$ is the equilibrium sales of firm i. Note that in deriving the above expression, we are using the fact that the first-order condition of the consumer's problem requires that $c_i = \omega/np_i$. Given that the above equality defines a linear system of equations in terms of the equilibrium sales of different firms, it is straightforward to show that $s_i = p_i X_i = \zeta_i \omega$ for some constant ζ_i.[8] Therefore, replacing for equilibrium price p_i in equations (21.5) and (21.6) in terms of the output of firm i yields $X_{ij} = \alpha w_{ij} \zeta_i X_j / \zeta_j$ and $l_i = (1-\alpha)\zeta_i$. Plugging these quantities back into the production function of firm i leads to

$$X_i = b_i \zeta_i (1-\alpha)^{1-\alpha} A_i^{\alpha} \prod_{j=1}^{n} \left(\alpha w_{ij} X_i / \zeta_j\right)^{\alpha w_{ij}}.$$

[8] To be more precise, $\zeta_i = v_i/n$, where v_i is the i-th column sum of matrix $(I - \alpha W)^{-1}$. In Section 21.3, we show that this quantity coincides with the notion of Bonacich centrality of firm i in the economy.

Now it is immediate that with the proper choice of constants b_i, the log output of firm i, denoted by $x_i = \log(X_i)$, satisfies

$$x_i = \alpha \sum_{j=1}^{n} w_{ij} x_j + \alpha \epsilon_i, \tag{21.7}$$

where $\epsilon_i = \log(A_i)$ is the log productivity shock to firm i. In other words, the interactions between different firms can be cast as a special case of our general framework in equation (21.1) with linear interaction function $f(z) = \alpha z$.

We end our discussion by remarking that the logarithm of real value added in the economy, which is the natural candidate for the economy's macro state y, can also be expressed in terms of our general formulation in (21.2). Because of the constant returns to scale assumption, firms make zero profits in equilibrium, all the surplus in the economy goes to the consumers, and as a consequence, value added is simply equal to the market wage ω. Choosing the ideal price index as the numeraire, that is, $\frac{n}{\tilde{b}}(p_1 \dots p_n)^{1/n} = 1$, and using the fact that $p_i = \zeta_i \omega / X_i$, we obtain that the log real value added in the economy is equal to

$$\log(\omega) = \frac{1}{n} \sum_{i=1}^{n} \log(X_i) - \frac{1}{n} \sum_{i=1}^{n} \log(\zeta_i) + \log(\tilde{b}/n).$$

Therefore, with the appropriate choice of $\tilde{b}$, we can rewrite log(GDP) as

$$y = \log(\text{GDP}) = \frac{1}{n} \sum_{i=1}^{n} x_i, \tag{21.8}$$

as in (21.2) in our general framework with $g(z) = z/n$.

21.2.3 Example: Financial Contagion

As a final example, we show that our general framework also nests models of financial contagion over networks. As a concrete example, we focus on a variant of a model along the lines of Eisenberg and Noe (2001) and Acemoglu, Ozdaglar, and Tahbaz-Salehi (2015c), who study how the patterns of interbank liabilities determine the extent of financial contagion.

Consider an economy consisting of n financial institutions (or banks), which are linked to one another via unsecured debt contracts of equal seniority. Each bank i has a claim of size $\xi_{ij} = w_{ij}\xi$ on bank j, where we assume that $\sum_{j=1}^{n} w_{ij} = \sum_{j=1}^{n} w_{ji} = 1$, thus guaranteeing that all banks have identical total claims (of size ξ) on the rest of the banking system. In addition to its interbank claims and liabilities, bank i has an outside asset of net value a and is subject to some liquidity shock ϵ_i.

Following the realizations of these liquidity shocks, banks need to repay their creditors. If a bank cannot meet its liabilities in full, it defaults and repays its creditors on a pro rata basis. Let z_{im} denote the repayment of bank m on its debt to bank i. The cash flow of bank i is thus equal to $c_i = a + \epsilon_i + \sum_{m=1}^{n} z_{im}$. Therefore, as long as $c_i \geq \xi$, bank i can meet its liabilities in full, guaranteeing that $z_{ji} = w_{ji}\xi$ for all banks j. If, on the other hand, $c_i \in (0, \xi)$, the bank defaults and its creditors are repaid in proportion to the face value of their contracts (i.e., $z_{ji} = w_{ji}c_i$). Finally, if $c_i \leq 0$, bank i's creditors receive nothing, that is, $z_{ji} = 0$. Putting the above together implies that the repayment of bank i on its debt to a given bank j is equal to

$$z_{ji} = \max\left\{\min\left\{w_{ji}\left(a + \epsilon_i + \sum_{m=1}^{n} z_{im}\right), w_{ji}\xi\right\}, 0\right\}.$$

Summing both sides of the above equation over the set of banks j and letting $x_i = \sum_{j=1}^{n} z_{ji}$ denote the total out-payment of bank i to its creditors implies

$$x_i = \max\left\{\min\left\{\sum_{m=1}^{n} w_{im}x_m + a + \epsilon_i, \xi\right\}, 0\right\}, \tag{21.9}$$

where we are using the fact that $z_{im} = w_{im}x_m$.

It is then straightforward to see that the interactions between different banks can be represented as a network, with each vertex corresponding to a bank and the size of bank i's obligation to bank j representing the intensity of interactions between the two. Furthermore, the specific nature of interbank repayments can be cast as a special case of our general model (21.1) with interaction function $f(z) = \max\{\min\{z + a, \xi\}, 0\}$.

Note that unlike the examples presented in Sections 21.2.1 and 21.2.2, this interaction function does not satisfy the normalization assumption $f(0) = 0$ if $a > 0$. Nevertheless, this is not of major consequence, as a simple change of variables would restore the original normalization: redefining the state of agent i as $\hat{x}_i = x_i - \xi$ leads to the modified interaction function $\hat{f}(z) = \max\{\min\{z + a, 0\}, -\xi\}$, which satisfies $\hat{f}(0) = 0$ whenever $a > 0$. Given that all our results and their corresponding economic insights are robust to the choice of normalization, we find it easier to work with the original model.

Finally, assuming that each default results in a deadweight loss of size A (for example, because of the cost of early liquidation of long-term projects), the social surplus in the economy is equal to

$$y = A\sum_{i=1}^{n} \mathbf{1}\{x_i \geq \xi\},$$

corresponding to $g(z) = z$ and $h(z) = A\mathbf{1}\{z \geq \xi\}$ in our general framework.

21.2.4 Existence and Uniqueness of Equilibrium

We now return to the general framework introduced above and establish the existence and (generic) uniqueness of equilibrium. In general, the set of equilibria not only depends on the economy's interaction network, but also on the properties of the interaction function. We impose the following regularity assumption on f:

Assumption 1. There exists $\beta \leq 1$ such that $|f(z) - f(\tilde{z})| \leq \beta|z - \tilde{z}|$ for all $z, \tilde{z} \in \mathbb{R}$. Furthermore, if $\beta = 1$, then there exists $\delta > 0$ such that $|f(z)| < \delta$ for all $z \in \mathbb{R}$.

This assumption, which is satisfied in each of the economies discussed in Sections 21.2.1–21.2.3 as well as in most other natural applications of this framework, guarantees that the economy's interaction function is either (i) a contraction with Lipschitz constant $\beta < 1$; or alternatively, (ii) a bounded non-expansive mapping. Either way, it is easy to establish that an equilibrium always exists. In particular, when $\beta < 1$, the contraction mapping theorem implies that (21.1) always has a fixed point, whereas if f is bounded, the existence of equilibrium is guaranteed by the Brouwer fixed point theorem.

Our first formal result shows that beyond existence, Assumption 1 is also sufficient to guarantee that the equilibrium is uniquely determined over a generic set of shock realizations.

Theorem 1. *Suppose that Assumption 1 is satisfied. Then, an equilibrium always exists and is generically unique.*

A formal proof of the above result is provided in the Appendix. Intuitively, when $\beta < 1$, the contraction mapping theorem ensures that the economy has a unique equilibrium. The economy may have multiple equilibria, however, when $\beta = 1$ (for example, as in the financial contagion example in Section 21.2.3). Nevertheless, Theorem 1 guarantees that the equilibrium is generically unique, in the sense that the economy has multiple equilibria only for a measure zero set of realizations of agents-level shocks.

21.3 Smooth Economies

In the remainder of this chapter, we study how the economy's underlying network structure, as well as different properties of the aggregation and interaction functions, shape economic outcomes. In particular, we are interested in characterizing how these features determine the extent of propagation and amplification of shocks within the economy.

To achieve this objective, we impose two further assumptions on our model. First, we assume that the underlying economy is *smooth*, in the sense that functions f, g, and h are continuous and at least twice differentiable. The class of smooth economies nests many of the standard models studied in the literature, such as variants of the network games and the production economy presented in Sections 21.2.1 and 21.2.2. On the other hand, the model of financial interactions presented in Section 21.2.3 is not nested within this class, as the corresponding interaction function is not differentiable everywhere. Nevertheless, this non-smoothness is not of major consequence, as the interaction function f can be arbitrarily closely approximated by a smooth function $\tilde{f}$ in such a way that economic implications of the model under this smooth approximation are identical to those of the original model.[9]

As our second assumption, we focus on the case where agent-level shocks are small. This assumption enables us to approximate the equilibrium state of each agent and the economy's macro state by the first few terms of their Taylor expansions. Even though it may appear restrictive, our following results highlight that such a "small-shock analysis" can lead to fairly general and robust insights on how different network interactions shape economic outcomes.

21.3.1 First-Order Approximation

We start our analysis by providing a first-order (that is, linear) approximation to the agents' equilibrium states around the point where $\epsilon_i = 0$ for all i. If agent-level shocks are small, such an approximation captures the dominant effects of how shocks shape the economy's macro state.

Let us first use the implicit function theorem to differentiate both sides of the interaction equation (21.1) with respect to the shock to agent r:

$$\frac{\partial x_i}{\partial \epsilon_r} = f'\left(\sum_{m=1}^{n} w_{im} x_m + \epsilon_i\right)\left(\sum_{m=1}^{n} w_{im}\frac{\partial x_m}{\partial \epsilon_r} + \mathbf{1}\{r=i\}\right). \tag{21.10}$$

Evaluating the above equation at the point $\epsilon = (\epsilon_1, \dots, \epsilon_n) = 0$ yields

$$\left.\frac{\partial x_i}{\partial \epsilon_r}\right|_{\epsilon=0} = f'(0)\left(\sum_{m=1}^{n} w_{im}\left.\frac{\partial x_r}{\partial \epsilon_i}\right|_{\epsilon=0} + \mathbf{1}\{r=i\}\right),$$

where we are using the fact that in the absence of shocks $x_m = 0$ for all m. This equation can be rewritten in matrix form as $\partial x/\partial \epsilon_r|_{\epsilon=0} = f'(0) W \partial x/\partial \epsilon_r|_{\epsilon=0} + f'(0) e_r$, where $x = (x_1, \dots, x_n)'$ is the vector of agents' states and e_r represents the r-th unit vector. It is therefore immediate that the derivative of the agents' states with respect to the shock

[9] More specifically, it is sufficient for $\tilde{f}$ to satisfy Assumption 1 and, as f, be initially concave and then convex.

to agent r is given by

$$\left.\frac{\partial x}{\partial \epsilon_r}\right|_{\epsilon=0} = f'(0)\left[I - f'(0)W\right]^{-1} e_r. \tag{21.11}$$

Note that, as long as $f'(0) < 1$, the matrix $I - f'(0)W$ is invertible, implying that the right-hand side of (21.11) is well-defined. We find it useful to define the following concept:

Definition 2. The *Leontief matrix* of the economy with parameter $\alpha \in [0,1)$ is $L = (I - \alpha W)^{-1}$, where $W = [w_{ij}]$ is the economy's interaction matrix.

In view of the above definition, we can rewrite equation (21.11) as

$$\left.\frac{\partial x_i}{\partial \epsilon_r}\right|_{\epsilon=0} = \alpha \ell_{ir}, \tag{21.12}$$

where $\alpha = f'(0)$ and ℓ_{ir} is the (i,r) element of the economy's Leontief matrix with parameter α. The equilibrium state of agent i around the point $\epsilon = 0$ can then be linearly approximated as

$$x_i = \alpha \sum_{r=1}^{n} \ell_{ir} \epsilon_r. \tag{21.13}$$

In other words, when the agent-level shocks are small (so that we can rely on a linear approximation), the economy's Leontief matrix serves as a sufficient statistic for the network's role in determining the state of agent i. More specifically, the impact of a shock to agent r on the equilibrium state of agent i is simply captured by ℓ_{ir}.

Before continuing with our derivations, a few remarks are in order. First, note that Definition 2 generalizes the well-known concept of the Leontief input-output matrix to an economy with a general form of interaction among agents. In particular, the (i,r) element of the matrix not only captures the direct interaction between agents i and r but also accounts for all possible indirect interactions between the two. To see this, note that ℓ_{ir} can be rewritten as

$$\ell_{ir} = 1 + \alpha w_{ir} + \alpha^2 \sum_{k=1}^{n} w_{ik} w_{kr} + \ldots, \tag{21.14}$$

where the higher-order terms account for the possibility of indirect interactions between i and r. Thus, essentially, equation (21.14) shows that a shock to agent r impacts agent i not only through their direct interaction term w_{ir}, but also via indirect interactions with the rest of the agents: such a shock may impact the state of some agent k and then indirectly propagate to agent i. However, note that the impact of a shock to agent r on i's state is deflated by a factor $\alpha < 1$ whenever the length of the indirect interaction chain between the two agents is increased by one.

FIGURE 21.1 The star interaction network.

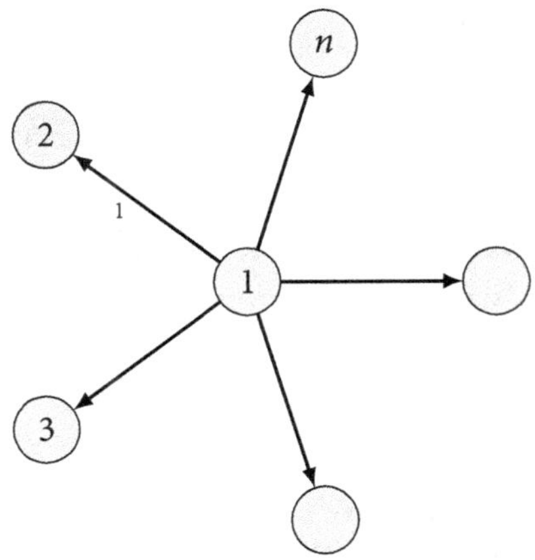

In view of the interpretation that ℓ_{mi} captures the equilibrium impact of agent i on the state of agent m, it is natural to interpret $\sum_{m=1}^{n} \ell_{mi}$ as the extent of agent i's overall influence on the rest of the agents in the economy. We define the following concept, which is well-known in the study of social and economic networks:

Definition 3. For a given parameter $\alpha \in [0,1)$, the *Bonacich centrality* of agent i is $v_i = \sum_{m=1}^{n} \ell_{mi}$, where $L = [\ell_{ij}]$ is the corresponding Leontief matrix of the economy.

To see how the above concept captures an intuitive notion of network centrality as well as the overall extent of agents' influence on one another, consider the *star interaction network* depicted in Figure 21.3.1. As the figure suggests, a shock to agent 1, which takes a more central position in the network, should have a larger impact on other agents' states compared to a shock to agent $i \neq 1$. Indeed, it is easy to verify that the Bonacich centrality of agent 1 is equal to $v_1 = 1 + \alpha n/(1-\alpha)$, whereas $v_i = 1$ for $i \neq 1$.

More generally, in any given interaction network, agent i's Bonacich centrality can be written recursively in terms of the centralities of the rest of the agents in the economy:

$$v_i = 1 + \alpha \sum_{j=1}^{n} v_j w_{ji}. \tag{21.15}$$

This expression shows that i has a higher centrality (and hence a more pronounced impact on the rest of the agents) if it interacts strongly with agents that are themselves central.

Returning to our derivations, we next provide a linear approximation to the economy's macro state y in the presence of small shocks. Differentiating (21.2) with respect to ϵ_r yields

$$\frac{\partial y}{\partial \epsilon_r} = g'(h(x_1) + \cdots + h(x_n)) \sum_{m=1}^{n} h'(x_m) \frac{\partial x_m}{\partial \epsilon_r}. \tag{21.16}$$

Evaluating this expression at $\epsilon = 0$ and replacing for the derivative of agent i's state from (21.12), we obtain

$$\left.\frac{\partial y}{\partial \epsilon_r}\right|_{\epsilon=0} = \alpha g'(0) h'(0) \sum_{m=1}^{n} \ell_{mr}, \tag{21.17}$$

where we have again used that fact that, in the absence of shocks, $x_m = 0$ for all m and that $h(0) = 0$. Putting Definition 3 and equation (21.17) together leads to the following linear approximation to the economy's macro state as a function of its underlying interaction network, the interaction and aggregation functions, and the agent-level shocks:

Theorem 2. *Suppose that $f'(0) < 1$. Then, the first-order approximation to the macro state of the economy is*

$$y^{1\text{st}} = f'(0) g'(0) h'(0) \sum_{i=1}^{n} v_i \epsilon_i, \tag{21.18}$$

where v_i is the Bonacich centrality of agent i with parameter $f'(0)$.

The above result highlights that, as long as one is concerned with the first-order effects, the agents' Bonacich centralities serve as sufficient statistics for how shocks impact the economy's macro state. In particular, shocks to agents who take more central roles in the economy's interaction network have a more pronounced influence on economy's macro state. The intuition underlying this result can be understood in terms of the recursive definition of agents' centralities in (21.15): a shock to an agent with a higher Bonacich centrality impacts the states of other relatively central agents, which in turn propagate the shock further to other agents, and so on, eventually leading to a larger aggregate impact.

We end our discussion by remarking that when the interaction and aggregation functions are linear, Theorem 2 provides an exact characterization of—as opposed to a linear approximation to—the macro state of the economy in terms of the agent-level shocks. For example, recall the special case of network games with quadratic utilities studied in Section 21.2.1. By Theorem 2, the aggregate level of activity in such an economy is proportional to a convex combination of agent-level shocks, with weights given by each agent's Bonacich centrality in the network:

$$y_{\text{agg}} = \alpha \sum_{j=1}^{n} v_j \epsilon_j.$$

This result coincides with those of Ballester et al. (2006) and Calvó-Armengol et al. (2009), to cite two examples. Similarly, in the context of production economies with Cobb-Douglas production functions studied in Section 21.2.2, recall from (21.7) that the log output of any firm i is a linear function of the log-output of its suppliers. Using the log-value added, defined in (21.8), as the macro state of the economy, Theorem 2 implies that

$$\log(\text{GDP}) = \frac{\alpha}{n}\sum_{j=1}^{n} v_j \epsilon_j,$$

where ϵ_j is the log productivity shock to firm j, confirming a representation used in Acemoglu et al. (2012).

21.3.2 Second-Order Approximation

The linear approximation provided in the previous section characterizes how, in the presence of small shocks, the nature and strength of interactions between agents shape the economy's macro state. An important limitation of such an approximation is that the solution exhibits a *certainty equivalence* property, in the sense that the expected value of the economy's macro state is equal to its unperturbed value when no shocks are present. More specifically, as Corollary 1 below will show, $\mathbb{E}[y^{1\text{st}}] = 0$ regardless of the economy's interaction network or the shape of the interaction and aggregation functions.[10] Consequently, even though potentially useful from an ex post perspective, the first-order approximation provided in Theorem 2 is not particularly informative about how the economy's interaction network shapes aggregate outcomes from an ex ante point of view. In order to go beyond this certainty equivalence property, we next provide a second-order approximation to the economy's macro state. As our results in the following sections will show, taking the second-order effects into account provides a more refined characterization of how agent-level shocks shape economic outcomes.

We start by differentiating both sides of equation (21.10) with respect to the shock to agent j:

$$\frac{\partial^2 x_i}{\partial \epsilon_r \epsilon_j} = f''\left(\sum_{m=1}^{n} w_{im} x_m + \epsilon_i\right)\left(\sum_{m=1}^{n} w_{im}\frac{\partial x_m}{\partial \epsilon_r} + \mathbb{1}\{i=r\}\right)\left(\sum_{m=1}^{n} w_{im}\frac{\partial x_m}{\partial \epsilon_j} + \mathbb{1}\{i=j\}\right)$$
$$+ f'\left(\sum_{m=1}^{n} w_{im} x_m + \epsilon_i\right)\left(\sum_{m=1}^{n} w_{im}\frac{\partial^2 x_m}{\partial \epsilon_r \partial \epsilon_j}\right).$$

[10] See Schmitt-Grohé and Uribe (2004) for a similar argument in the context of a general class of discrete-time rational expectations models.

Evaluating this expression at $\epsilon = 0$ implies

$$\left.\frac{\partial^2 x_i}{\partial \epsilon_r \epsilon_j}\right|_{\epsilon=0} = f''(0)\left(\alpha \sum_{m=1}^{n} w_{im}\ell_{mr} + \mathbf{1}\{i=r\}\right)\left(\alpha \sum_{m=1}^{n} w_{im}\ell_{mj} + \mathbf{1}\{r=j\}\right) + \alpha \sum_{m=1}^{n} w_{im} \left.\frac{\partial^2 x_m}{\partial \epsilon_r \partial \epsilon_j}\right|_{\epsilon=0},$$

where, once again, we are using the fact that $x_m = 0$ for all m and that the first derivative of the agents' states with respect to the shocks can be written in terms of the economy's Leontief matrix, as given by (21.12). On the other hand, one can show that $\alpha \sum_{m=1}^{n} w_{im}\ell_{mr} = \ell_{ir} - \mathbf{1}\{i=r\}$.[11] Therefore, the previous expression can be simplified to

$$\left.\frac{\partial^2 x_i}{\partial \epsilon_r \epsilon_j}\right|_{\epsilon=0} = f''(0)\ell_{ir}\ell_{ij} + \alpha \sum_{m=1}^{n} w_{im} \left.\frac{\partial^2 x_m}{\partial \epsilon_r \partial \epsilon_j}\right|_{\epsilon=0},$$

leading to

$$\left.\frac{\partial^2 x_i}{\partial \epsilon_r \epsilon_j}\right|_{\epsilon=0} = f''(0) \sum_{m=1}^{n} \ell_{im}\ell_{mr}\ell_{mj}, \tag{21.19}$$

where we are using the definition of the Leontief matrix. The above equation thus provides the second-order derivates of agents' equilibrium states as a function of the interaction function and the Leontief matrix of the economy.

To obtain a second-order approximation to the macro state of the economy, we need to also differentiate (21.16) with respect to ϵ_j:

$$\frac{\partial^2 y}{\partial \epsilon_r \partial \epsilon_j} = g''(h(x_1) + \cdots + h(x_n)) \sum_{m=1}^{n}\sum_{i=1}^{n} h'(x_m)h'(x_i)\left(\frac{\partial x_m}{\partial \epsilon_r}\right)\left(\frac{\partial x_i}{\partial \epsilon_j}\right) + g'(h(x_1) + \cdots + h(x_n)) \sum_{m=1}^{n}\left[h'(x_m)\frac{\partial^2 x_m}{\partial \epsilon_r \epsilon_j} + h''(x_m)\frac{\partial x_m}{\partial \epsilon_r}\frac{\partial x_m}{\partial \epsilon_j}\right].$$

Replacing for the first-order and second-order derivates of agents' equilibrium states from (21.13) and (21.19), respectively, leads to

$$\left.\frac{\partial^2 y}{\partial \epsilon_r \partial \epsilon_j}\right|_{\epsilon=0} = \alpha^2 g''(0)\left[h'(0)\right]^2 \sum_{m=1}^{n}\sum_{i=1}^{n} \ell_{mr}\ell_{ij} + g'(0)\sum_{m=1}^{n}\left[h'(0)f''(0)\sum_{k=1}^{n}\ell_{mk}\ell_{kr}\ell_{kj} + \alpha^2 h''(0)\ell_{mr}\ell_{mj}\right],$$

[11] To see this, recall that the Leontief matrix can be rewritten as $L = \sum_{k=0}^{\infty} \alpha^k W^k$, which implies that $\alpha WL = L - I$.

which can be further simplified to

$$\left.\frac{\partial^2 y}{\partial\epsilon_r\partial\epsilon_j}\right|_{\epsilon=0} = g'(0)h'(0)f''(0)\sum_{m=1}^{n} v_m\ell_{mr}\ell_{mj} + \alpha^2 g''(0)\left[h'(0)\right]^2 v_r v_j + \alpha^2 g'(0)h''(0)\sum_{m=1}^{n}\ell_{mi}\ell_{mj},$$

where v_m is the Bonacich centrality of agent m with parameter α. Combining the above with (21.17) leads to the following result:

Theorem 3. *Suppose that $f'(0) < 1$. Then, the second-order approximation to the macro state of the economy is given by*

$$\begin{aligned} y^{2\text{nd}} &= f'(0)g'(0)h'(0)\sum_{i=1}^{n} v_i\epsilon_i \\ &+ \frac{1}{2}g''(0)\left[f'(0)h'(0)\right]^2\sum_{i=1}^{n}\sum_{j=1}^{n} v_i v_j\epsilon_i\epsilon_j \\ &+ \frac{1}{2}g'(0)\sum_{i=1}^{n}\sum_{j=1}^{n}\left(h'(0)f''(0)\sum_{m=1}^{n} v_m\ell_{mi}\ell_{mj} + \left[f'(0)\right]^2 h''(0)\sum_{m=1}^{n}\ell_{mi}\ell_{mj}\right)\epsilon_i\epsilon_j, \end{aligned} \tag{21.20}$$

where $L = [\ell_{ij}]$ is the economy's Leontief matrix with parameter $\alpha = f'(0)$ and v_i is the corresponding Bonacich centrality of agent i.

This result thus refines Theorem 2 by providing a second-order approximation to the role of agent-level shocks in shaping the economy's macro state. Note that the first line of (21.20) is simply the first-order approximation, $y^{1\text{st}}$, characterized in (21.18). The rest of the terms, which depend on the curvatures of the interaction and aggregation functions, capture the second-order aggregate effects. The second line, in particular, corresponds to additional terms resulting from the nonlinearity of the aggregation function, g. Note that these terms depend simply on Bonacich centralities, the v_i terms. This is due to the fact that as long as the interaction function f is linear, the total influence of agent i on the rest of the agents in the economy is given by the Bonacich centrality of agent i, $v_i = \sum_{m=1}^{n}\ell_{mi}$. The third line, on the other hand, shows that if either the interaction function f or the h function is nonlinear, the centrality measures are no longer sufficient statistics for the shocks' second-order effects. Rather, other network statistics—in particular, $\sum_{m=1}^{n}\ell_{mi}\ell_{mj}$ and $\sum_{m=1}^{n} v_m\ell_{mi}\ell_{mj}$—also play a key role in how shocks propagate throughout the economy.

It is also worth noting that as long as shocks are small enough and the linear approximation is nontrivial, the second-order terms in (21.20) are dominated by the effect of the first-order terms. However, as our following results will show, in many applications the linear terms are equal to zero (reflecting the above-mentioned certainty equivalence property), and hence are uninformative about the nature of the economy's macro state, making the second-order approximation essential for a meaningful characterization of the aggregate impact of microeconomic shocks.

21.4 Ex Ante Aggregate Performance

In the remainder of this chapter, we use Theorems 2 and 3 to characterize how network interactions translate small, agent-level shocks into aggregate effects measured by the economy's macro state.

This section provides a comparative study of the role of the economy's underlying network structure—as well as its interaction and aggregation functions—in shaping y from an ex ante perspective, by interpreting the expectation of the macro state y as the economy's "performance metric." Formally:

Definition 4. An economy *outperforms* another if $\mathbb{E}[y]$ is larger in the former than the latter.

A natural first step to obtain a comparison between the performance of different economies in the presence of small shocks is to compare their first-order approximations. Recall from Theorem 2 that the first-order approximation of an economy's macro state is equal to a linear combination of agent-level shocks with the corresponding weights given by the agents' Bonacich centralities (i.e., $y^{1\text{st}} = f'(0)g'(0)h'(0)\sum_{i=1}^{n} v_i\epsilon_i$), leading to the following immediate corollary:

Corollary 1. $\mathbb{E}[y^{1\text{st}}] = 0$.

This simple corollary shows that the economy exhibits a *certainty equivalence property* from an ex ante perspective up to a first-order approximation: the expected value of the economy's macro state is equal to its unperturbed value when no shocks are present, regardless of the nature of pairwise interactions or the shape of the interaction and aggregation functions. The more important implication, however, is that the linear approximation provided in Theorem 2 is not informative about the comparative performance of different economies, even in the presence of small shocks. Rather, a meaningful comparison between the ex ante performance of two economies requires that we also take the higher-order terms into account.

Thus, a natural next step is to use the second-order approximation provided in Theorem 3. Equation (21.20) shows that once second-order terms are taken into account, the ex ante performance of the economy, $\mathbb{E}[y^{2\text{nd}}]$, depends on the curvatures of the interaction and aggregation functions. In order to tease out these effects in a transparent manner, in the remainder of this section, we focus on how nonlinearities in each of these functions shape the economy's macro state, while assuming that the rest of the functions are linear.

21.4.1 Nonlinear Aggregation: Volatility

We first consider an economy with a general, potentially nonlinear aggregation function g, while assuming that f and h are increasing, linear functions. In this case, the ex ante performance of the economy is given by

$$\mathbb{E}[y] = \mathbb{E}g(x_1 + \cdots + x_n).$$

This observation highlights that the curvature of g essentially captures the extent to which society cares about volatility, for instance, because of risk-aversion at the aggregate level. To see this, suppose that g is concave. In this case, the economy's performance is reduced the more correlated agents' states are with one another. In fact, if $g(z) = -z^2$, the economy's ex ante performance simply captures the volatility of $x_1 + \cdots + x_n$. On the other hand, a convex g corresponds to the scenario in which performance increases with volatility.

In either case, Theorem 3 implies that the expected value of the economy's macro state, up to a second-order approximation, is given by

$$\mathbb{E}[y^{2\text{nd}}] = \frac{1}{2}\sigma^2 g''(0)\left[f'(0)h'(0)\right]^2 \sum_{i=1}^{n} v_i^2, \tag{21.21}$$

where we are using the assumption that all shocks are independent with mean zero and variance σ^2 and the assumption that functions f and h are linear. Equation (21.21) shows that, in contrast to Corollary 1, not all economies have identical performances once second-order terms are taken into account. Rather, the economy's ex ante performance depends on $\sum_{i=1}^{n} v_i^2$, which in turn, can be rewritten as

$$\sum_{i=1}^{n} v_i^2 = n \cdot \text{var}(v_1, \ldots, v_n) + \frac{n}{(1-\alpha)^2},$$

where $\alpha = f'(0)$, thus leading to the following result:

Proposition 4. *Suppose that the aggregation function g is concave (convex). An economy's ex ante performance decreases (increases) in* $\text{var}(v_1, \ldots, v_n)$.

This proposition implies that, if g is concave, networks in which agents exhibit a less heterogenous distribution of Bonacich centralities outperform those with a more unequal distribution. This is due to the fact that a more equal distribution of Bonacich centralities means that shocks to different agents have a more homogenous impact on the economy's macro state, and thus wash each other out more effectively at the aggregate level. On the other hand, a more unequal distribution of centralities implies that shocks to some agents play a disproportionally larger role in shaping y, and as a result are not canceled out by the rest of the agent-level shocks, increasing the overall volatility and reducing the value of $\mathbb{E}[y]$ whenever g is concave.

To see the implications of Proposition 4, consider an economy with the underlying star interaction network depicted in Figure 21.1. As already mentioned, the Bonacich centralities of agents in such an economy are highly unequal as agent 1 has a disproportionally large impact on the states of the rest of the agents. In fact, it is easy to show that $\sum_{i=1}^{n} v_i^2$ is maximized for the star interaction network. This implies that when g is concave, the star network has the least ex ante performance (and hence, the highest level of volatility) among all economies.[12]

At the other end of the spectrum are regular economies in which the extent of interaction of each agent with the rest of the agents is constant. More formally,

Definition 5. An economy is *regular* if $\sum_{j=1}^{n} w_{ji} = 1$ for all agents i.

Figures 21.2(a) and 21.2(b) depict two regular networks, known as the *ring* and *complete* interaction networks, respectively. Because they are symmetric, all agents in both economies should have identical Bonacich centralities. In fact, summing both sides of (21.14) over i in an arbitrary regular economy implies that

$$v_r = 1 + \alpha + \alpha^2 + \cdots = 1/(1-\alpha)$$

for all r, where recall that $\alpha = f'(0)$. This implies the following result:

Lemma 1. *In any regular economy, all agents have identical Bonacich centralities.*

Therefore, $\text{var}(v_1, \ldots, v_n)$ is minimized for all regular economies, implying that with a concave g, they outperform all other economies from an ex ante perspective: all agent-level shocks in such an economy take symmetric roles in determining the macro state, and minimize the overall volatility of $x_1 + \cdots + x_n$ and thus increase $\mathbb{E}[y]$. This implies the following corollary to Proposition 4:

Corollary 2. *Suppose that aggregation function g is concave (convex). Any regular economy outperforms (underperforms) all other economies, whereas the economy with the star interaction network underperforms (outperforms) all others.*

This corollary and Proposition 4 are closely connected to the results in Acemoglu et al. (2012), who show that in the context of the production economies presented in Section 21.2.2, aggregate output volatility is increasing in the extent of heterogeneity in the firms' centralities and is maximized (minimized) for the star (regular) network. This parallel can be better appreciated by noting that the logarithm of output of a given firm i satisfies linear equation (21.7) and that $\log(\text{GDP}) = (1/n)\sum_{i=1}^{n} x_i$. Therefore,

[12] Note that by Hölder's inequality, $\sum_i v_i^2 \leq (\max_i v_i)(\sum_i v_i) = n \max_i v_i/(1-\alpha) \leq n(1-\alpha+\alpha n)/(1-\alpha)^2$, regardless of the economy's interaction network, where recall that $\alpha = f'(0)$. This inequality is tight for the star network, implying that $\sum_i v_i^2$ obtains its maximal value.

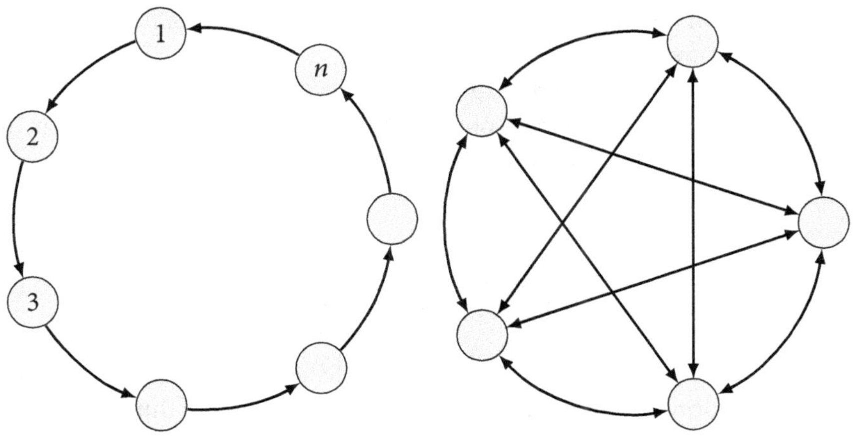

(a) The ring interaction network (b) The complete interaction network

FIGURE 21.2 Two regular economies

the volatility of log value added is simply

$$\mathrm{var}\left(\log(\mathrm{GDP})\right) = \frac{1}{n^2}\mathbb{E}(x_1 + \cdots + x_n)^2.$$

Setting $g(z) = -(z/n)^2$ implies that economies that have a higher ex ante performance in the sense of Definition 4 are less volatile at the aggregate level. Hence, Proposition 4 and Corollary 2 guarantee that any economy in which firms exhibit more heterogeneity in terms of their roles as input-suppliers exhibits higher levels of aggregate (log) output volatility due to idiosyncratic firm-level shocks. Our results thus show that it is the concavity of economy's aggregation function that lies at the heart of the findings of Acemoglu et al. (2012).

21.4.2 Nonlinear Interactions

We now focus on the role of nonlinear interactions in shaping the economy's ex ante performance. To illustrate this role in a transparent manner, we consider an economy with a general, nonlinear interaction function f, while assuming that g and h are increasing, linear functions.[13] The ex ante performance of such an economy is given by

$$\mathbb{E}[y] = \sum_{i=1}^{n}\mathbb{E}[x_i] = \sum_{i=1}^{n}\mathbb{E}f\left(\sum_{j=1}^{n} w_{ij}x_j + \epsilon_i\right).$$

[13] The results, and in fact the expressions, are essentially identical when h is also nonlinear.

The above equation highlights that the curvature of the interaction function f captures the extent of "risk-aversion" at the micro-level.

To understand the role of interlinkages in affecting economic performance, we focus on the set of symmetric, regular economies.[14] Recall from Theorem 3 that, in the presence of small shocks, the expected value of the economy's macro state can be approximated by

$$\mathbb{E}[y^{2\mathrm{nd}}] = \frac{1}{2}\sigma^2 g'(0)h'(0)f''(0)\sum_{i=1}^{n}\sum_{m=1}^{n} v_m \ell_{mi}^2. \tag{21.22}$$

Note that as before, we need to rely on a second-order approximation, as the first-order terms are not informative about the comparative performance of different economies; that is $\mathbb{E}[y^{1\mathrm{st}}] = 0$ regardless of the shape of f or the economy's interaction network. Given that all agents in a regular network have identical Bonacich centralities, equation (21.22) shows that the economy's performance depends on the value of $\sum_{i,m} \ell_{mi}^2$. On the other hand, it is easy to verify that

$$\sum_{m=1}^{n} \ell_{mi}^2 = v^2/n + \mathrm{var}(\ell_{1i}, \ldots, \ell_{ni}), \tag{21.23}$$

where $v = \sum_{m=1}^{n} \ell_{mi} = 1/(1-\alpha)$ is the agents' (common) Bonacich centrality, thus suggesting that the term $\sum_{i=1}^{n}\sum_{m=1}^{n} \ell_{mi}^2$ decreases if inter-agent influences ℓ_{mi} are more evenly distributed. The following result, which is proved in the Appendix, captures this idea formally:

Corollary 3. *Suppose that there are no self-interaction terms, that is,* $w_{ii} = 0$. *If the interaction function f is concave (convex), then the complete network outperforms (underperforms) all other symmetric economies.*

This corollary is related to the findings of Acemoglu, Ozdaglar, and Tahbaz-Salehi (2015c), who, in the context of the model of financial interactions presented in Section 21.2.3, show that the complete financial network exhibits a "phase transition": when the total net asset value of the financial system is large enough, the complete network is the financial network with the least number of defaults. However, as the net asset value of the financial system is reduced, beyond a certain point, the complete network flips to be the economy with the maximal number of bank failures.[15]

To see the connection between their results and Corollary 3, recall that the corresponding interaction function in such an economy is given by $f(z) = \max\{\min\{z+$

[14] Recall that an economy is said to be symmetric if $w_{ij} = w_{ji}$ for all $i \neq j$.

[15] To be more precise, Acemoglu et al. (2015c) state their results in terms of whether exogenous shocks that hit financial institutions are small or large. Nevertheless, given that such shocks simply impact the net asset value of the banks, their results can be equivalently stated in terms of the size of the net asset value of the banks, a.

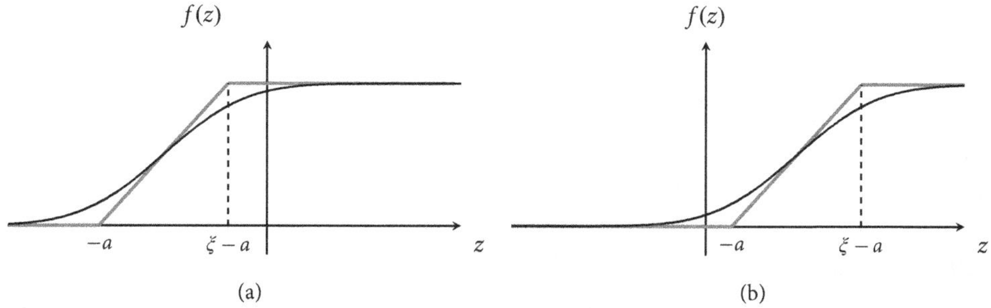

FIGURE 21.3 The interaction function $f(z) = \max\{\min\{z + a, \xi\}, 0\}$ corresponding to the model of financial interactions in Section 21.2.3. Panels (a) and (b) plot the function for the case that $a > \xi$ and $a < 0$, respectively. The thin red line in each panel depicts a smooth approximation to the interaction function.

$a, \xi\}, 0\}$. As depicted in Figure 21.3(a), for large enough values of a (in particular, when $a > \xi$), this interaction function is concave in the neighborhood of 0. Therefore, Corollary 3 implies that the complete network outperforms all other economies.[16] In contrast, once the banks' net asset value a becomes small enough, the interaction function is locally convex around 0, as depicted in Figure 21.3(b). In stark contrast to the former case, Corollary 3 now implies that all other economies would outperform the complete network. Thus, our characterization results clarify that the findings of Acemoglu et al. (2015c) are due to the fact that reducing the banks' net asset values (for example, due to some exogenous shocks) essentially corresponds to a shift from the concave to the convex region of the interaction function, thus reversing the role of interbank connections.

In addition to providing a different perspective on the results of Acemoglu et al. (2015c), Corollary 3 presents a partial answer to the question posed in the Introduction, related to the sometimes contradictory claims on the role of dense network interconnections in creating systemic risk and instability. It shows that when economic (financial) interactions correspond to a concave f, denser interconnections are stabilizing (as in Allen and Gale 2000), whereas they play the role of generating systemic risk when these interactions correspond to a convex f function. Finally, our result that more densely interconnected networks are more unstable in the presence of convex interactions is akin to similar results in the epidemic-like cascade models (such as Blume et al. 2011), in which a bank fails once the number of its defaulting counterparties passes a certain threshold.

[16] As already noted, even though the corresponding interaction function is not smooth, it can be arbitrarily closely approximated by a smooth function in such a way that the economic implications of the model under this smooth approximation are identical to those of the original model. Figure 21.3 depicts one such smooth approximation.

21.5 Systemically Important Agents

A central concern in many analyses of economic and social networks is the identification of "key players" or "systemically important agents" (e.g., Ballester et al. 2006 and Zenou 2015). Loosely speaking, these are entities that have a disproportionally high impact on some aggregate statistic of interest. For example, Banerjee, Chandrasekhar, Duflo, and Jackson (2013 2014) study how the social network position of the first individual to receive information about a new product within a village can increase the extent of information diffusion within that community. Similarly, in the context of multi-agent contracting in the presence of externalities, Bernstein and Winter (2012) are interested in obtaining an ordering of agents who when subsidized induce the maximal level of participation by other agents. Relatedly, in the context of the example presented in Section 21.2.3, Acemoglu et al. (2015c) characterize the set of systemically important institutions in a financial network, a shock to whom would lead to a large cascade of defaults.

In this section, we utilize Theorems 2 and 3 to study how different features of the environment determine the impact of each agent on the macro state of the economy and provide a characterization of the set of agents that are more important from a systemic perspective. We start by defining this concept formally:

Definition 6. Agent i is said to be *systemically more important* than agent j if $y_{(i)} < y_{(j)}$, where $y_{(i)}$ denotes the macro state of the economy when agent i is hit with a negative shock.

In other words, agent i is systemically more important than agent j if a shock to i leads to a larger drop in the economy's macro state. Note that in general, the relative systemic importance of an agent may depend on the size of the negative shock. Nevertheless, we can use our results in Section 21.3 to provide a characterization of the systemic importance of different agents for small enough shocks.

We should also remark that our notion of systemically important agents is related to, but distinct from, the notion of "key players" studied by Ballester et al. (2006) and Zenou (2015). Whereas our focus is on how a shock to a given agent impacts some macroeconomic variable of interest, these papers study the impact of the removal of an agent from the network.

21.5.1 Linear Interactions

We start by focusing on economies where the interaction and aggregation functions are linear. This enables us to highlight, in a transparent manner, how the presence of nonlinearities can shape equilibrium outcomes.

Recall that when the interaction and aggregation functions are linear, Theorem 2 provides an exact characterization of the economy's macro state in equilibrium. More specifically, it shows that y is a linear combination of the idiosyncratic, agent-level shocks, with the weights proportional to the Bonacich centralities of the corresponding agents, leading to the following result:

Proposition 5. *Suppose that the economy's interaction function is linear. Then agent i is more systemically important than agent j if* $v_i > v_j$*, where* v_i *is the Bonacich centrality of agent i.*

In other words, in an economy with linear interactions, a negative shock to the agent with the highest Bonacich centrality leads to the largest drop in the economy's macro state. The intuition underlying this result is simple and well-known in the literature: shocks to more central agents propagate more extensively over the network and as a result have larger impacts on the economy's macro state.

To see the implications of the above result, consider the economies depicted in Figures 21.1 and 21.2. Given that the ring and complete networks depicted in Figure 21.2 are regular, Proposition 5 suggests that in the presence of linear interactions, all agents in such economies are equally systemically important. In contrast, in the economy depicted in Figure 21.1, agent 1 takes a more central position with respect to the rest of the agents, leading to the intuitive result that it is the most systemically important agent within the economy.

Proposition 5 also has sharp predictions for the set of systemically important agents in the class of network games with quadratic payoffs discussed in Section 21.2.1. Recall that the first-order conditions in such games can be represented in the form of a linear interaction function. Thus, by Proposition 5, the player with the highest Bonacich centrality would be the most influential player in the game. This is indeed in line with the observations of Candogan, Bimpikis, and Ozdaglar (2012), who argue that subsidizing players with the highest centrality would induce the largest increase in the level of aggregate activity in the economy.

Similarly, in the context of production economies with Cobb-Douglas (and hence, log-linear) production technologies discussed in Section 21.2.2, Acemoglu et al. (2012) show that productivity shocks to firms with higher centralities have a larger impact on the economy's aggregate output, an observation consistent with the predictions of Proposition 5. More specifically, in line with the examples we discussed above, they also argue that, compared to a shock of equal size to one of the more peripheral firms, a shock to firm 1 in the star network depicted in Figure 21.1 would have a much larger impact on the log value added of the economy.

Finally, Proposition 5 also echoes some of the results in the literature on social learning that studies the long-run implications of different learning rules. In particular, Golub and Jackson (2010) show that if agents update their beliefs as a linear combination of their neighbors' opinions (what is commonly known as DeGroot-style learning), the information available to those with higher centralities plays a more prominent role in the eventual beliefs in the society. Relatedly, Jadbabaie et al. (2012 2013) show that the rate of information aggregation in a social network is more sensitive to the quality of the signals observed by the more central agents.[17]

21.5.2 Nonlinear Interactions

Our previous results show that if the economy's interaction function is linear, Bonacich centrality provides a comprehensive measure for agents' systemic importance. This observation also means that, as long as agent-level shocks are small enough, more central agents would play a more prominent role in shaping the economy's macro state, even if the interactions are nonlinear. This is due to the fact that by Theorem 2, the economy's macro state can be linearly approximated by

$$y_{(i)}^{1\text{st}} = f'(0)g'(0)h'(0)v_i\epsilon_i,$$

leading to the following result:

Corollary 4. *If $v_i > v_j$, then agent i is systemically more important than agent j for all interaction functions f.*

This conclusion is subject to an important caveat: even though $v_i > v_j$ implies that i is more systemically important than j in the presence of small shocks, $v_i = v_j$ does not guarantee that the two agents are equally systemically important. Rather, in such a scenario, a meaningful comparison of the agents' systemic importance requires that we also take their higher-order effects into account. Thus, Corollary 4 is simply not applicable to regular economies, in which all agents have identical Bonacich centralities.

In order to obtain a meaningful measure for agents' systemic importance in a regular economy, a natural step would be to utilize Theorem 3 to compare the second-order effects of agent-level shocks on the economy's macro state. From (21.20), we have that, in any regular economy,

$$y_{(i)}^{2\text{nd}} = f'(0)g'(0)h'(0)v\epsilon + \frac{g''(0)}{2}\left(f'(0)h'(0)\right)^2 v^2\epsilon^2 + \frac{1}{2}g'(0)\left(vh'(0)f''(0) + [f'(0)]^2 h''(0)\right)\left(\sum_{m=1}^{n} \ell_{mi}^2\right)\epsilon^2,$$

where $v = 1/(1-\alpha)$ is the agents' (common) Bonacich centrality, thus implying that agent i's systemic importance is determined by the value of $\sum_{m=1}^{n} \ell_{mi}^2$. On the other

[17] The main results in this literature are in terms of agents' eigenvector centralities, defined as a limiting case of Bonacich centrality. In particular, the eigenvector centrality of agent i satisfies $\hat{v}_i = \lim_{\alpha\to 1}(1-\alpha)v_i$. See Jackson (2008) for a discussion on other notions of centrality and their relationships to one another.

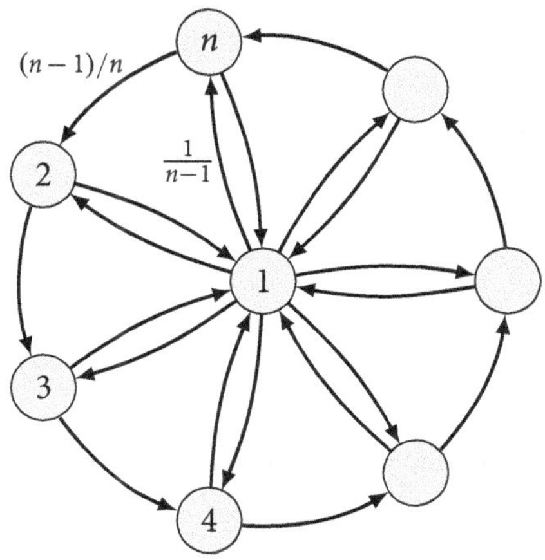

FIGURE 21.4 A regular economy where agents have identical centralities, but differ in their concentration centralities.

hand, recall from (21.23) that $\sum_{m=1}^{n} \ell_{mi}^2$ essentially measures the variation in the extent to which agent i influences other agents in the economy. We define the following concept:

Definition 7. The *concentration centrality* of agent i is $d_i = \text{stdev}(\ell_{1i}, \ldots, \ell_{ni})$, where $L = [\ell_{ij}]$ is the economy's Leontief matrix.

Thus, a smaller d_i means that agent i's influence is more evenly distributed throughout the economy. In other words, whereas an agent's Bonacich centrality captures its overall influence, concentration centrality measures how evenly the agent's influence is distributed across the rest of the agents. As an example, consider the economy depicted in Figure 21.4. It is easy to verify that the depicted network corresponds to a regular economy, implying that all agents have identical Bonacich centralities. However, the extent of dispersion is not identical across agents. Rather, for large enough values of n, $d_1 < d_i$ for all $i \neq 1$: compared to all other agents, agent 1's interactions are more evenly distributed throughout the economy.

This discussion is summarized in the next proposition.

Proposition 6. *Suppose that the economy's interaction network is regular.*

(a) If f is concave, then i is systemically more important than j if and only if $d_i > d_j$.
(b) If f is convex, then i is systemically more important than j if and only if $d_i < d_j$.

Taken together, Proposition 6 and Corollary 4 suggest that while Bonacich centralities summarize the first-order effects of agent-level shocks on aggregate outcomes, the second-order effects are captured by the agents' concentration centralities. These second-order effects become critical in a regular network, where first-order terms are simply uninformative about agents' systemic importance.

Proposition 6 also reenforces an observation made by Acemoglu et al. (2015c) that relying on standard and off-the-shelf notions of network centrality (such as eigenvector or betweenness centralities) for the purpose of identifying systemically important agents may be misleading. As Proposition 6 suggests, the proper notion of network centrality has to be informed by the nature of microeconomic interactions between different agents.

21.6 Conclusion

This chapter presented a unified framework nesting a wide variety of network interaction models, such as various classes of network games, models of macroeconomic risk built up from microeconomic shocks, and models of financial interactions. Under the assumption that shocks are small (and the relevant interactions are smooth), our main results provide a fairly complete characterization of the equilibrium, highlighting the role of different types of network interactions in affecting the macroeconomic performance of the economy. Our characterization delineates how microeconomic interactions function as a channel for the propagation of shocks. It also provides a comparative study of the role of the economy's underlying network structure—as well as its interaction and aggregation functions—in shaping macroeconomic outcomes. In addition to clarifying the relationship between disparate models (for example, those focusing on input-output linkages, financial contagion, and general cascades), our framework highlights some of the reasons behind the apparently contradictory conclusions in the literature on the role of network interactions in the emergence of systemic risk.

Our hope is that the framework provided here will be useful in future work on understanding network interactions in general and the study of network games, macroeconomic risk and financial contagion in particular. We believe that several important issues remain open to future research. First, our framework focuses on an environment in which shock realizations are common knowledge. Generalizing this setup to environments with incomplete and private information would enable us to understand the interplay between network interactions and information asymmetries. A second direction for future research would be to apply similar analyses to economies that exhibit richer strategic interactions (e.g., general imperfect competition rather than competitive or monopolistically competitive economies). Finally, a systematic investigation of endogenous network formation in the presence of rich propagation and cascade dynamics remains an important area for future research.

A. Technical Appendix

Lemma 2. *Suppose that* $|f(\tilde{z}) - f(z)| = |\tilde{z} - z|$ *for a pair of points* $\tilde{z} > z$*. Then, the interaction function* f *is linear in the interval* $[z, \tilde{z}]$ *with a unit slope.*

Proof. Pick an arbitrary point $\hat{z} \in [z, \tilde{z}]$. Given Assumption 1 and the monotonicity of the interaction function, it must be the case that

$$f(\tilde{z}) - f(\hat{z}) \le \tilde{z} - \hat{z}$$
$$f(\hat{z}) - f(z) \le \hat{z} - z.$$

Summing the above inequalities immediately implies that both inequalities have to be tight simultaneously. Therefore, for any $\hat{z}$ in the interval $[z, \tilde{z}]$, it must be the case that $f(\hat{z}) = \hat{z} + f(z) - z$. □

Lemma 3. *The interaction function f has at most countably many discontinuity points.*

Proof. Let D denote the set of points where f is discontinuous. For any $z \in D$, define

$$f(z^-) = \lim_{t \uparrow z} f(t)$$
$$f(z^+) = \lim_{t \downarrow z} f(t).$$

Given the fact that f is nondecreasing, it must be the case that $f(z^-) < f(z^+)$. Therefore, there exists a rational number $F(z) \in \mathbb{Q}$ such that

$$f(z^-) < F(z) < f(z^+).$$

Furthermore, for any pair of points $z, \tilde{z} \in D$ satisfying $z < \tilde{z}$, it is immediate that $F(z) < F(\tilde{z})$. Consequently, $F : D \to \mathbb{Q}$ has to be an injection, proving that D is at most countable. □

Proof of Theorem 1

We prove this result for two separate cases depending on whether (i) $\beta < 1$ or (ii) $\beta = 1$. Throughout, we assume that the economy's interaction network is *strongly connected* in the sense that there exists a directed path from each agent to any other agent in the economy. In case of a disconnected interaction network, the proof would apply to any connected component separately.

Case (i)

First, suppose that $\beta < 1$. Define the mapping $\Phi : \mathbb{R}^n \to \mathbb{R}^n$ as

$$\Phi_i(x_1, \ldots, x_n) = f\left(\sum_{j=1}^{n} w_{ij} x_j + \epsilon_i\right). \tag{21.24}$$

For any $x, \tilde{x} \in \mathbb{R}^n$, we have

$$\begin{aligned} |\Phi_i(x_1, \ldots, x_n) - \Phi_i(x_1, \ldots, x_n)| &\le \beta \left| \sum_{j=1}^{n} w_{ij}(x_j - \tilde{x}_j) \right| \\ &\le \beta \sum_{j=1}^{n} w_{ij} \left| x_j - \tilde{x}_j \right|, \end{aligned}$$

where the first inequality is a consequence of Assumption 1 and the second inequality follows from a simple application of the triangle inequality. The fact that $\sum_{j=1}^{n} w_{ij} = 1$ implies that

$$|\Phi_i(x_1,\ldots,x_n) - \Phi_i(x_1,\ldots,x_n)| \leq \beta \max_j |x_j - \tilde{x}_j|,$$

and as a consequence,

$$\max_i |\Phi_i(x_1,\ldots,x_n) - \Phi_i(x_1,\ldots,x_n)| \leq \beta \max_j |x_j - \tilde{x}_j|.$$

In other words,

$$\left\| \Phi(x) - \Phi(\tilde{x}) \right\|_\infty \leq \beta \left\| x - \tilde{x} \right\|_\infty.$$

Therefore, the mapping Φ is a contraction with respect to the infinity norm with a Lipschitz constant $\beta < 1$. The contraction mapping theorem then immediately implies that the mapping has a unique fixed point $x^* = \Phi(x^*)$, for all shock realizations $(\epsilon_1,\ldots,\epsilon_n) \in \mathbb{R}^n$. □

Case (ii)

Next, suppose that $\beta = 1$. In this case, Assumption 1 guarantees that there exists $\delta > 0$ such that $|f(z)| < \delta$ for all z.

Recall mapping Φ from (21.24). By assumption, it is continuous and maps the compact and convex set $[-\delta,\delta]^n$ to itself. Therefore, by the Brouwer fixed-point theorem, there exists $x^* \in [-\delta,\delta]^n$ such that $\Phi(x^*) = x^*$, thus proving the existence of an equilibrium.

Next, we prove that this equilibrium is generically unique. Suppose that the economy has two distinct equilibria, denoted by x and $\tilde{x}$. Let $e = |x - \tilde{x}| \in \mathbb{R}^n$ be the element-wise difference between the two equilibria, which by assumption is a non-zero vector. By definition, for any given agent i, we have

$$\begin{aligned} e_i &= \left| f\Big(\sum_{j=1}^{n} w_{ij}x_j + \epsilon_i\Big) - f\Big(\sum_{j=1}^{n} w_{ij}\tilde{x}_j + \epsilon_i\Big) \right| \\ &\leq \Big|\sum_{j=1}^{n} w_{ij}(x_j - \tilde{x}_j)\Big| \qquad (21.25) \\ &\leq \sum_{j=1}^{n} w_{ij}e_j, \qquad (21.26) \end{aligned}$$

where the first inequality is a consequence of Assumption 1. We now show that both inequalities above are tight for all agents i.

Suppose that either inequality holds strictly for some agent i, implying that $e_i < \sum_{j=1}^{n} w_{ij}e_j$. Let $q \in \mathbb{R}^n$ denote the left eigenvector corresponding to the top eigenvalue of matrix W. By the Perron-Frobenius theorem, vector q is element-wise strictly positive.[18] Multiplying both

[18] For more on the Perron-Frobenius theorem, see Chapter 2 of Berman and Plemmons (1979).

sides by q_i and summing over all agents i implies that

$$\sum_{i=1}^{n} q_i e_i < \sum_{j=1}^{n}\sum_{i=1}^{n} q_i w_{ij} e_j = \sum_{j=1}^{n} q_j e_j,$$

leading to a contradiction. Therefore, it is immediate that (21.25) and (21.26) hold as equalities, thus implying that $e_i = \sum_{j=1}^{n} w_{ij} e_j$ for all agents i, or in matrix notation, $We = e$.

Consequently, by the Perron-Frobenius theorem, e has to be proportional to the Perron vector of matrix W, which is the vector of all ones. In other words, $e_i = c$ for all i and some strictly positive constant c. Furthermore, the fact that (21.26) holds as an equality implies that $x_i - x_i'$ has the same sign for all i. Assuming that $x_i < \tilde{x}_i$, it must be the case that $\tilde{x}_i = x_i + c$ for all agents i.

Summarizing the above implies that for all agent i,

$$x_i = f\left(\sum_{j=1}^{n} w_{ij} x_j + \epsilon_i\right)$$

and

$$x_i + c = f\left(\sum_{j=1}^{n} w_{ij} x_j + \epsilon_i + c\right).$$

Letting $z_i = \sum_{j=1}^{n} w_{ij} x_j + \epsilon_i$ and subtracting both sides of the above equalities lead to

$$|f(z_i + c) - f(z_i)| = c.$$

Thus, by Lemma 2, the interaction function f has to be linear with a unit slope within the interval $[z_i, z_i + c]$. Consequently, there exists some constant b_i such that $f(z) = z + b_i$ for all $z \in [z_i, z_i + c]$. Therefore,

$$x_i = \sum_{j=1}^{n} w_{ij} x_j + \epsilon_i + b_i$$

for all i. Multiplying both sides of the above equality by q_i and summing over all agents i lead to

$$\sum_{i=1}^{n} q_i \epsilon_i = -\sum_{i=1}^{n} q_i b_i, \tag{21.27}$$

where once again we are using the fact that $\sum_{i=1}^{n} q_i w_{ij} = q_j$. Therefore, the economy has two distinct equilibria if and only if the agent-level shocks satisfy (21.27). Now, Lemma 3 guarantees that there are at most countably many of such values b_i, as otherwise the interaction function f would have uncountably many points of discontinuity. In other words, for the economy to have multiple equilibria, the term $\sum_{i=1}^{n} q_i \epsilon_i$ has to belong to a countable set B. This coupled with the observation that $q_i > 0$ guarantees that the economy has a unique equilibrium for a generic set of shock realizations. □

Proof of Corollary 3

Suppose that f is concave. The proof for the case in which f is convex is identical. Recall from Equation (21.22) that the ex ante performance the economy is decreasing in $\sum_{i=1}^{n}\sum_{m=1}^{n}\ell_{mi}^2$, which can be rewritten as

$$\sum_{i=1}^{n}\sum_{m=1}^{n}\ell_{mi}^2 = \text{trace}(L^2).$$

Denoting the k-th largest eigenvalue of a generic matrix X with $\lambda_k(X)$, we have:

$$\text{trace}(L^2) = \sum_{k=1}^{n}\lambda_k^2(L) = \sum_{k=1}^{n}(1-\alpha\lambda_k(W))^{-2},$$

where the second inequality is a consequence of the fact that $L=(I-\alpha W)^{-1}$.

On the other hand, the assumption that $w_{ii}=0$ implies that $\text{trace}(W)=\sum_{k=1}^{n}\lambda_k(W)=0$, whereas $\sum_{j=1}^{n}w_{ij}=1$ guarantees that $\lambda_1(W)=1$. Putting these two observation together implies that $\sum_{k=2}^{n}\lambda_k(W)=-1$. Therefore,

$$\text{trace}(L^2) = \frac{1}{(1-\alpha)^2} + \sum_{k=2}^{n}(1-\alpha\lambda_k(W))^{-2} \tag{21.28}$$

$$\geq \frac{1}{(1-\alpha)^2} + (n-1)\left(1+\frac{\alpha}{n-1}\right)^{-2}, \tag{21.29}$$

where the second equality is due to the fact that function $Q(z)=(1-\alpha z)^{-2}$ is convex. On the other hand, it easy to show that for the complete network, $\lambda_k(W)=-1/(n-1)$ for all $k\neq 1$. Therefore, the complete network obtains the lower bound in (21.29), and hence, has maximal ex ante performance when the interaction function is concave. □

References

Acemoglu, Daron, David Autor, David Dorn, Gordon H. Hanson, and Brendan Price (2015a). "Import competition and the Great U.S. Employment Sag of the 2000s." Forthcoming in *Journal of Labor Economics*.

Acemoglu, Daron, Vasco M. Carvalho, Asuman Ozdaglar, and Alireza Tahbaz-Salehi (2012). "The network origins of aggregate fluctuations." *Econometrica* 80, 1977–2016.

Acemoglu, Daron, Camilo Garcia-Jimeno, and James A. Robinson (2015b). "State capacity and economic development: A network approach." *American Economic Review* 105(8), 2364–2409.

Acemoglu, Daron, Azarakhsh Malekian, and Asuman Ozdaglar (2014a). "Network security and contagion." NBER Working Paper No. 19174.

Acemoglu, Daron, Asuman Ozdaglar, and Alireza Tahbaz-Salehi (2014b). "Microeconomic origins of macroeconomic tail risks." NBER Working Paper No. 20865.

Acemoglu, Daron, Asuman Ozdaglar, and Alireza Tahbaz-Salehi (2014c). "Systemic risk in endogenous financial networks." Columbia Business School Working Paper.

Acemoglu, Daron, Asuman Ozdaglar, and Alireza Tahbaz-Salehi (2015c). "Systemic risk and stability in financial networks." *American Economic Review* 105, 564–608.

Allen, Franklin and Douglas Gale (2000). "Financial contagion." *Journal of Political Economy* 108, 1–33.

Allouch, Nizar (2012). "On the private provision of public goods on networks." Fondazione Eni Enrico Mattei Working Paper 40.2012.

Alvarez, Fernando and Gadi Barlevy (2014). "Mandatory disclosure and financial contagion." Federal Reserve Bank of Chicago Working Paper No. 2014-04.

Babus, Ana (2014). "The formation of financial networks." Discussion Paper 06-093, Tinbergen Institute.

Badev, Anton (2013). "Discrete games in endogenous networks: Theory and policy." Population Studies Center Working Paper 13-05, University of Pennsylvania.

Bak, Per, Kan Chen, José Scheinkman, and Michael Woodford (1993). "Aggregate fluctuations from independent sectoral shocks: self-organized criticality in a model of production and inventory dynamics." *Ricerche Economiche* 47, 3–30.

Bala, Venkatesh and Sanjeev Goyal (2000). "A noncooperative model of network formation." *Econometrica* 68, 1181–1229.

Ballester, Coralio, Antoni Calvó-Armengol, and Yves Zenou (2006). "Who's who in networks. Wanted: The key player." *Econometrica* 74, 1403–1417.

Banerjee, Abhijit, Arun G. Chandrasekhar, Esther Duflo, and Matthew O. Jackson (2013). "The diffusion of microfinance." *Science* 341, 1236498.

Banerjee, Abhijit, Arun G. Chandrasekhar, Esther Duflo, and Matthew O. Jackson (2014). "Gossip: Identifying central individuals in a social network." Working paper.

Battiston, Stefano, Domenico Delli Gatti, Mauro Gallegati, Bruce Greenwald, and Joseph E. Stiglitz (2012). "Liaisons dangereuses: Increasing connectivity, risk sharing, and systemic risk." *Journal of Economic Dynamics and Control* 36, 1121–1141.

Berman, Abraham and Robert J. Plemmons (1979). *Nonnegative Matrices in the Mathematical Sciences*. New York: Academic Press.

Bernstein, Shai and Eyal Winter (2012). "Contracting with heterogeneous externalities." *American Economic Journal: Microeconomics* 4, 50–76.

Blume, Lawrence, David Easley, Jon Kleinberg, Robert Kleinberg, and Éva Tardos (2011). "Which networks are least susceptible to cascading failures?" In *52nd IEEE Annual Symposium on Foundations of Computer Science (FOCS)* 393–402.

Bramoullé, Yann and Rachel Kranton (2007). "Public goods in networks." *Journal of Economic Theory* 135, 478–494.

Bramoullé, Yann and Rachel Kranton (2015). "Network games." In *Oxford Handbook on the Economics of Networks*, Yann Bramoullé, Brian W. Rogers, and Andrea Galeotti, eds., Oxford, UK: Oxford University Press.

Bramoullé, Yann, Rachel Kranton, and Martin D'Amours (2014). "Strategic interaction and networks." *American Economic Review* 104, 898–930.

Cabrales, Antonio, Douglas Gale, and Piero Gottardi (2015). "Financial contagion in networks." In *The Oxford Handbook on the Economics of Networks*, Yann Bramoullé, Andrea Galeotti, and Brian Rogers, eds., Oxford, UK: Oxford University Press.

Cabrales, Antonio, Piero Gottardi, and Fernando Vega-Redondo (2014). "Risk-sharing and contagion in networks." CESifo Working Paper No. 4715.

Calvó-Armengol, Antoni and Matthew O Jackson (2004). "The effects of social networks on employment and inequality." *American Economic Review* 426–454.

Calvó-Armengol, Antoni, Eleonora Patacchini, and Yves Zenou (2009). "Peer effects and social networks in education." *Review of Economic Studies* 76, 1239–1267.

Calvó-Armengol, Antoni and Yves Zenou (2004). "Social networks and crime decisions: The role of social structure in facilitating delinquent behavior." *International Economic Review* 45, 939–958.

Candogan, Ozan, Kostas Bimpikis, and Asuman Ozdaglar (2012). "Optimal pricing in networks with externalities." *Operations Research* 60, 883–905.

Caplin, Andrew and John Leahy (1993). "Sectoral shocks, learning, and aggregate fluctuations." *Review of Economic Studies* 60, 777–794.

Carvalho, Vasco M. (2014). "From micro to macro via production networks." *Journal of Economic Perspectives* 28, 23–48.

Carvalho, Vasco M., Makoto Nirei, Yukiko Saito, and Alireza Tahbaz-Salehi (2015). "Supply chain disruptions: Evidence from the Great East Japan Earthquake." Working paper.

Chamley, Christophe and Douglas Gale (1994). "Information revelation and strategic delay in a model of investment." *Econometrica* 62, 1065–1085.

di Giovanni, Julian, Andrei A. Levchenko, and Isabelle Méjean (2014). "Firms, destinations, and aggregate fluctuations." *Econometrica* 82, 1303–1340.

Durlauf, Steven N. (1993). "Nonergodic economic growth." *Review of Economic Studies* 60, 349–366.

Dziubiński, Marcin and Sanjeev Goyal (2014). "How to defend a network?" Cambridge-INET Institute Working Paper No: 2014/01.

Eisenberg, Larry and Thomas H. Noe (2001). "Systemic risk in financial systems." *Management Science* 47, 236–249.

Elliott, Matthew and Benjamin Golub (2015). "A network approach to public goods." Working paper.

Elliott, Matthew, Benjamin Golub, and Matthew O. Jackson (2014). "Financial networks and contagion." *American Economic Review* 104, 3115–3153.

Erol, Selman and Rakesh Vohra (2014). "Network formation and systemic risk." PIER Working Paper No. 14-029.

Farboodi, Maryam (2014). "Intermediation and voluntary exposure to counterparty risk." Working paper.

Foerster, Andrew T., Pierre-Daniel G. Sarte, and Mark W. Watson (2011). "Sectoral versus aggregate shocks: A structural factor analysis of industrial production." *Journal of Political Economy* 119, 1–38.

Freixas, Xavier, Bruno M. Parigi, and Jean-Charles Rochet (2000). "Systemic risk, interbank relations, and liquidity provision by the central bank." *Journal of Money, Credit and Banking* 32, 611–638.

Gabaix, Xavier (2011). "The granular origins of aggregate fluctuations." *Econometrica* 79, 733–772.

Gai, Prasanna, Andrew Haldane, and Sujit Kapadia (2011). "Complexity, concentration and contagion." *Journal of Monetary Economics* 58, 453–470.

Gai, Prasanna and Sujit Kapadia (2010). "Contagion in financial networks." *Proceedings of the Royal Society A: Mathematical, Physical and Engineering Science* 466, 2401–2423.

Galeotti, Andrea, Sanjeev Goyal, Matthew O. Jackson, Fernando Vega-Redondo, and Leeat Yariv (2010). "Network games." *Review of Economic Studies* 77, 218–244.

Galeotti, Andrea and Brian W. Rogers (2013). "Strategic immunization and group structure." *American Economic Journal: Microeconomics* 5, 1–32.

Giesecke, Kay and Stefan Weber (2006). "Credit contagion and aggregate losses." *Journal of Economic Dynamics and Control* 30, 741–767.

Glasserman, Paul and H. Peyton Young (2015). "How likely is contagion in financial networks?" *Journal of Banking & Finance* 50, 383–399.

Golub, Benjamin and Matthew O. Jackson (2010). "Naïve learning in social networks and the wisdom of crowds." *American Economic Journal: Microeconomics* 2, 112–149.

Goyal, Sanjeev and José Luis Moraga-González (2001). "R&D networks." *The RAND Journal of Economics* 32, 686–707.

Granovetter, Mark (1978). "Threshold models of collective behavior." *American Journal of Sociology* 83, 1420–1443.

Jackson, Matthew O. (2008), *Social and Economic Networks*. Princeton University Press, Princeton, NJ.

Jackson, Matthew O. and Yves Zenou (2015). "Games on networks." In *Handbook of Game Theory with Economic Applications*, H. Peyton Young and Shmuel Zamir, eds., volume 4, 91–157. Amsterdam: Elsevier.

Jadbabaie, Ali, Pooya Molavi, Alvaro Sandroni, and Alireza Tahbaz-Salehi (2012). "Non-Bayesian social learning." *Games and Economic Behavior* 76, 210–225.

Jadbabaie, Ali, Pooya Molavi, and Alireza Tahbaz-Salehi (2013). "Information heterogeneity and the speed of learning in social networks." Columbia Business School Working Paper No. 13-28.

Jones, Charles I. (2013). "Misallocation, economic growth, and input-output economics." In *Proceedings of Econometric Society World Congress*, Daron Acemoglu, Manuel Arellano, and Eddie Dekel, eds., Cambridge University Press.

Long, John B. and Charles I. Plosser (1983). "Real business cycles." *Journal of Political Economy* 91, 39–69.

Morris, Stephen (2000). "Contagion." *Review of Economic Studies* 67, 57–78.

Plosser, Charles I. (2009). "Redesigning financial system regulation." Speech at the New York University Conference, "Restoring Financial Stability: How to Repair a Failed System." http://www.phil.frb.org/publications/speeches/plosser/2009/03-06-09_nyu-restoring-financial-stability.pdf.

Schmitt-Grohé, Stephanie and Martín Uribe (2004). "Solving dynamic general equilibrium models using a second-order approximation to the policy function." *Journal of Economic Dynamics and Control* 28, 755–775.

Vivier-Lirimont, Sébastian (2006). "Contagion in interbank debt networks." Working Paper.

Watts, Duncan J. (2002). "A simple model of global cascades on random networks." *Proceedings of the National Academy of Sciences of the United States of America* 99, 5766–5771.

Zawadowski, Adam (2013). "Entangled financial systems." *Review of Financial Studies* 26, 1291–1323.

Zenou, Yves (2015). "Key players." In *The Oxford Handbook on the Economics of Networks*, Yann Bramoullé, Andrea Galeotti, and Brian Rogers, eds., Oxford University Press, Oxford.

PART VI

COMMUNITIES

CHAPTER 22

INFORMAL TRANSFERS IN SOCIAL NETWORKS

MARKUS MOBIUS AND TANYA ROSENBLAT

22.1 Introduction

Many fundamental economic institutions facilitate transfers across time. Banks take deposits and repackage them as loans. Insurance companies mitigate idiosynchratic risk across a large pool of agents. These institutions can only function if they are able to control moral hazard: debtors have to repay their loans, and victims of adverse shocks have to be able to rely on their insurance policy to make transfers. Developed economies usually rely on the legal system to protect lenders and policy holders.

However, developing countries often lack a reliable legal system, which raises the transaction costs of providing loans or insurance. Even developed economies often require lenders to own physical collateral that can be used to secure loans and thus lower transaction costs. Certain groups of borrowers, such as entrepreneurs starting a business, frequently lack physical collateral. In these situations, *informal lending* within close-knit communities can substitute for formal transfers even in the absence of physical collateral. But what mechanisms make such informal arrangements work and control for moral hazard? How can a lender trust that an informal loan will be repaid in the future? How much assistance can a farmer with a bad harvest realistically expect from his extended family and friends? Social networks provide a natural structure to study these phenomena: intuitively we expect that households receive greater assistance from socially close neighbors with whom they share stronger ties. Moreover, agents in social networks play a repeated supergame with their friends and neighbors that can help support informal lending and risk-sharing arrangements.

In this chapter we describe the *social collateral* approach introduced by Karlan, Mobius, Rosenblat, and Szeidl (2009) and Ambrus, Mobius, and Szeidl (2014b) as a simple way to model informal transfers within social networks. This approach views social links as "collateral" that can be used to control moral hazard: if a borrower does not repay a loan (in case of borrowing) or if a an agent refuses to help a neighbor in need (in case of risk-sharing), they risk losing the social link and its associated benefits. In this class of models, the function of social capital (which is exactly the social network) is analogous to the role function of physical collateral in formal transfer arrangements.

The social collateral approach makes two key simplifying assumptions that keep the model analytically tractable and empirically relevant. First of all, the entire network supergame is collapsed into a two-period game: borrowing or risk-sharing arrangements are implemented in period 1 and out-of-equilibrium punishments occur in period 2. Second, there is no explicit group punishment (such as ostracism)—not repaying a loan, for example, only jeopardizes the direct link between the borrower and the lender (or intermediary). However, it can be shown that group punishments that are robust to *coalitional deviations* can reduce the set of implementable outcomes (under certain conditions) to the same set that is implementable through direct punishment schemes.

22.2 Empirical Facts

Before discussing the social collateral approach in detail, it is useful to summarize some of the main findings of the empirical literature on informal transfers.

First of all, informal transfers are remarkably successful in mitigating income risk on the village level. In particular, Townsend (1994) finds that the full insurance model where all agents pool their income provides a good benchmark for understanding consumption in Indian villages.[1]

Second, transfers are highly *local* and tend to occur between socially close households. Fafchamps and Lund (2003) find that transfers take place primarily through networks of friends and relatives rather than at the village level. Similarly, De Weerdt and Dercon (2006) collect detailed data on insurance networks within a single village in Tanzania. They find that networks are not clustered but largely overlapping. They also confirm the effectiveness of these insurance networks in mitigating income risk: they cannot reject full insurance at village level for food insurance, and find partial insurance of non-food consumption in the networks. Angelucci, De Giorgi, and Rasul (2012) exploit a natural experiment where the Mexican government made cash transfers to a subset of households in Mexican villages. They find that receiving households share these transfers with relatives in the village (but not with non-relatives). This paper contributes by using exogenous income variation to examine the risk-sharing network

[1] However, some papers have also documented substantial deviations from the full risk-sharing benchmark. Kazianga and Udry (2006) find little evidence of consumption smoothing during a period of severe drought in Burkina Faso. Gertler and Gruber (2002) find that informal insurance against several illness is very imperfect. Dercon and Krishnan (2000) show that risk-sharing does not always occur even *within* households as women in rural Ethiopia bear the brunt of adverse shocks. Hayashi, Altonji and Kotlikoff (1996) also reject full risk-sharing within families using PSID data.

and extent of full insurance. Attanasio, Barr, Cardenas, Genicot and Meghir (2012) use a field experiment with 70 Colombian villages show that pairs of participants who are friends or relatives are more likely to form risk-sharing groups which suggests that social obligations to neighbors in the social network are an important determinant of informal transfers.

While social proximity is typically highly correlated with geographic distance, remittances between relatives who live far apart are an important exception because they can help mitigate local shocks. Gubert (2002) finds evidence that insurance is an important motive in remittances. Yang and Choi (2007) show that remittances respond to income shocks which is consistent with an insurance motivation. They cannot reject the hypothesis that households with a migrant worker are fully insured against income shocks, and can strongly reject for households without a migrant worker. Technological advances have reduced the transaction costs of remittances. Jack and Suri (2014) show that M-pesa cash transfers through mobile phones insulate households from shocks thanks to an increase in remittances from a diverse pool of senders.

Third, informal transfers often take the form of loans. Udry (1994) was one of the first papers to demonstrate that informal credit contracts play a direct role in pooling risk, as repayments owed by borrowers depend on the random shocks faced by both the borrowers and lenders. Fafchamps and Lund (2003) use detailed data on gifts, loans, and asset sales to examine which methods are used to cope with income shocks. They find that gifts and loans are the most common mechanisms.

22.3 Informal Lending and Trust

We consider a situation where a borrower needs the asset of a lender to produce social surplus.[2] In the absence of legal contract enforcement, borrowing must be secured by an informal arrangement supported by the social network: connections in the network have associated consumption value, which serve as *social collateral* to enable borrowing.

22.3.1 Motivating Example

To understand the basic logic of the model, consider the examples in Figure 22.1, where agent s would like to borrow an asset, like a car, from agent t, in an economy with no formal contract enforcement. In Figure 22.1a, the network consists only of s and t; the value of their relationship, which represents either the social benefits of friendship or the present value of future transactions, is assumed to be 2. As in standard models of

[2] This asset might represent a factor of production, such as a farming tool, vehicle, or animal; it could also be an apartment, a household durable good, or simply a cash payment.

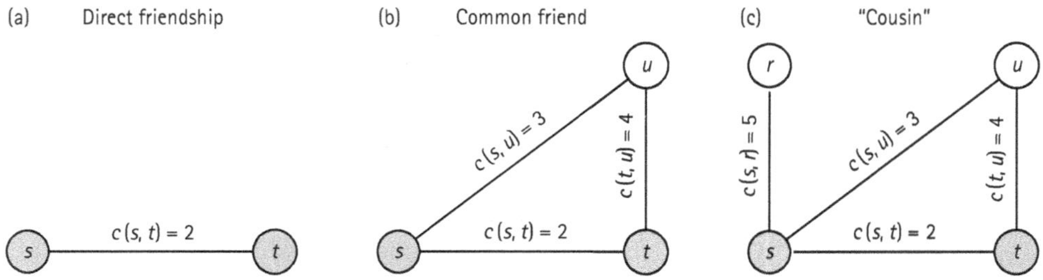

FIGURE 22.1 Informal borrowing in some sample social networks.

informal contracting, t will only lend the car if its value does not exceed the relationship value of 2.

More interesting is Figure 22.1b, where s and t have a common friend u, the value of the friendship between s and u is 3, and that between u and t is 4. Here, the common friend increases the borrowing limit by $\min[3,4] = 3$, the weakest link on the path connecting borrower and lender through u, to a total of 5. The logic is that the intermediate agent u vouches for the borrower, acting as a guarantor of the loan transaction. If the borrower chooses not to return the car, he is breaking his promise of repayment to u, and therefore loses u's friendship. Since the value of this friendship is 3, it can be used as collateral for a payment of up to 3. For the lender t to receive this amount, u must prefer transmitting the payment to losing the friendship with him, explaining the role of the weakest link.

Finally, Figure 22.1c illustrates the limits of ostracism under a coalitional refinement. Assume that the lender also has a "cousin" r with whom he has a relationship valued at 5. If the cousin could also act as a guarantor for s then the borrowing limit might increase by an additional 5 to a total of 10. However, the cousin's threat to break off her relationship with the borrower is not credible: for any loan amount exceeding 5, the borrower could propose a "side-deal" to intermediary u and her cousin such that u can reimburse the lender for her guaranteed amount (which is at most 3) while transferring 0 to the cousin in case of default. The cousin and intermediary are not worse off as a result of this side-deal. The borrower will use her friendship and therefore incur a combined loss of at most 5—but since she borrowed an amount exceeding 5 she is strictly better off under such a side-deal. Hence, group-level punishment of the borrower that involves agents that are unconnected to the lender (such as the cousin in panel C) is not credible under coalitional refinement.

22.3.2 Model

Formally, a social network $G = (W,E)$ consists of a set W of agents (vertices or nodes) and a set E of edges (links), where an edge is an unordered pair of distinct vertices.

Each link in the network represents a friendship or business relationship between the two parties involved. We formalize the strength of relationships using an exogenously given capacity $c(u,v)$.

Definition 1. *A capacity is a function* $c: W \times W \to \mathbb{R}$ *such that* $c(u,v) > 0$ *if* $(u,v) \in E$ *and* $c(u,v) = 0$ *otherwise.*

The capacity measures the utility benefits that agents derive from their relationships. For ease of presentation, we assume that the strength of relationships is symmetric, so that $c(u,v) = c(v,u)$ for all u and v.

The model consists of five stages.

Stage 1: Realization of needs. Two agents s and t are randomly selected from the social network. Agent t, the lender, has an asset that agent s, the borrower, desires. The lender values the asset at V, and it is assumed that V is drawn from some prior distribution F over $[0,\infty)$. The identity of the borrower and the lender as well as the value of V are publicly observed by all players.

Stage 2: Borrowing arrangement. At this stage, the borrower publicly proposes a *transfer arrangement* to all agents in the social network. The role of this arrangement is to punish the borrower and compensate the lender in the event of default. A transfer arrangement consists of a set of transfer payments $h(u,v)$ for all u and v agents involved in the arrangement. Here $h(u,v)$ is the amount u promises to pay v if the borrower fails to return the asset to the lender. Once the borrower has announced the arrangement, all agents involved have the opportunity to accept or decline. If all involved agents accept, then the asset is borrowed and the borrower earns an income $\omega(V)$, where ω is a nondecreasing function with $\omega(0) = 0$. If some agents decline, then the asset is not lent, and the game moves on directly to stage 5.

Stage 3: Repayment. Once the borrower has made use of the asset, he can either return it to the lender or steal it and sell it for a price of V. If the borrower returns the asset then the game moves to the final stage 5.

Stage 4: Transfer payments. All agents observe whether the asset was returned in the previous stage. If the borrower did not return the asset, then the transfer arrangement is activated. Each agent has a binary choice: either he makes the promised payment $h(u,v)$ in full or he pays nothing. If some agent u fails to make a prescribed transfer $h(u,v)$ to v, then he loses his friendship with agent v (i.e., the (u,v) link "goes bad"). If (u,v) link is lost, then the associated capacity is set to zero for the remainder of the game. We let $\tilde{c}(u,v)$ denote the new link capacities after these changes.

Stage 5: Friendship utility. At this stage, agents derive utility from their remaining friends. The total utility enjoyed by an agent u from his remaining friends is simply the sum of the values of all remaining relationships (i.e., $\sum_v \tilde{c}(u,v)$).

22.3.3 Analysis

We are interested in characterizing the *maximum* amount $T^{st}(c)$ that agent s can borrow from lender t for a given social network that is characterized by the capacity function c. We will refer to $T^{st}(c)$ as the *borrowing limit*.

The model is a multi-stage game with observed actions. We focus on the set of pure strategy subgame perfect equilibria. In order to rule out non-credible equilibria (as shown in Figure 22.1c) we require that all equilibria are "side-deal proof".

Consider the subgame starting in stage 2, after the identities of the borrower and the lender and the value of the asset are realized, and for any pure strategy σ, let $U_u(\sigma)$ denote the total utility of agent u in this subgame. We formalize the idea of a side-deal as an alternative transfer arrangement $\widetilde{h}(u,v)$ that s proposes to a subset of agents $S \subset W$ after the original arrangement is accepted. If this side-deal is accepted, agents in S are expected to make transfer payments according to $\widetilde{h}$, while agents outside S continue to make payments described by h. In order for the side-deal to be credible to all participating agents, it must be accompanied by a proposed path of play that these agents find optimal to follow. This motivates the following definition.

Definition 2. *A side-deal with respect to a pure strategy profile σ is a set of agents S, a transfer arrangement $\widetilde{h}(u,v)$ for all $u,v \in S$, and a set of continuation strategies $\{\widetilde{\sigma}_u | u \in S\}$ proposed by s to agents in S at the end of stage 2, such that*

(i) $U_u\left(\widetilde{\sigma}_u, \widetilde{\sigma}_{S\setminus u}, \sigma_{-S}\right) \geq U_u\left(\sigma'_u, \widetilde{\sigma}_{S\setminus u}, \sigma_{-S}\right)$ *for all* σ'_u *and all* $u \in S$,
(ii) $U_u(\widetilde{\sigma}_S, \sigma_{-S}) \geq U_u(\sigma_S, \sigma_{-S})$ *for all* $u \in S$,
(iii) $U_s(\widetilde{\sigma}_S, \sigma_{-S}) > U_s(\sigma_S, \sigma_{-S})$.

Condition (i) says that all agents u involved in the side-deal are best-responding on the new path of play (i.e., that the proposed path of play is an equilibrium for all agents in S conditional on others playing their original strategies σ_{-S}). Condition (ii) says that if any agent $u \in S$ refuses to participate in the side-deal, then play reverts to the original path of play given by σ. Finally, (iii) ensures that the borrower s strictly benefits from the side-deal.

Definition 3. *A pure strategy profile σ is a side-deal proof equilibrium if it is a subgame perfect equilibrium that admits no side deals.*

We are now almost ready to state the main result in Karlan et al. (2009). The final pre-requisite is a definition of *maximum flow*, which is a well-known concept in optimization theory and computer science (Ford, Jr. and Fulkerson 1956).

Definition 4. *An $s \to t$ flow with respect to capacity c is a function $f : G \times G \to \mathbb{R}$ that satisfies*

(i) *Skew symmetry:* $f(u,v) = -f(v,u)$.
(ii) *Capacity constraints:* $f(u,v) \leq c(u,v)$.
(iii) *Flow conservation:* $\sum_w f(u,w) = 0$ *unless* $u = s$ *or* $u = t$.

The value of a flow is the amount that "leaves" the borrower s, given by $|f| = \sum_w f(s,w)$. Let $T^{st}(c)$ denote the maximum value among all $s \to t$ flows.

The maximum flow captures the intuitive notion of "sum of weakest links" for all distinct paths that connect a borrower and a lender. For example, the maximum flow between borrower and lenders in Figure 22.1 is equal to 2, 5, and 5 in panels a to c. It turns out that the maximum flow exactly characterizes the borrowing limit.

Theorem 1. *There exists a side-deal proof equilibrium that implements borrowing between s and t if and only if the asset value V satisfies*

$$V \leq T^{st}(c). \tag{22.1}$$

22.3.4 Empirical Application

The social collateral model can be easily applied to empirical applications where the social network is known. In particular, the maximum flow can be efficiently calculated using the Ford-Fulkerson algorithm (Ford, Jr. and Fulkerson 1956). Karlan et al. (2009) report on one empirical application that uses data from two Peruvian shantytowns. In 2005, the authors collected social network data on 299 households. In particular, the authors recorded, for each link, how much time the subject spends on average with the friend or acquaintance per week and whether the subject ever borrowed money from each social link.

The amount of time provides a convenient proxy for the strength of a relationship. In the data, the distribution of time spent together is skewed: the average responder spends fewer than 6 minutes with the bottom 10% of his/her friends and more than 3 hours with the top 10%. To obtain a more homogenous measure, the authors define *normalized time* for two connected agents u and v as the value, for the amount of time they spend together, of the empirical cumulative distribution function of time spent together in their community. With this definition, the empirical distribution of normalized time $\tau(u,v)$ across all connected pairs is a discretized uniform distribution on the unit interval in each community.

The authors also assume that link capacities are created by an increasing production function g such that $c(u,v) = c \cdot \tau(u,v)$ (i.e., spending more time together results in stronger links). They also restrict attention to the subgraph that includes all direct links of s and t (hence borrowing arrangement can only involve common friends). This allows for a simple decomposition of the trust flow between s and t as

$$T^{st}(c) = c \cdot \tau(s,t) + c \cdot \sum_{v \in N_s \cap N_t} \min(\tau(s,v), \tau(v,t)), \tag{22.2}$$

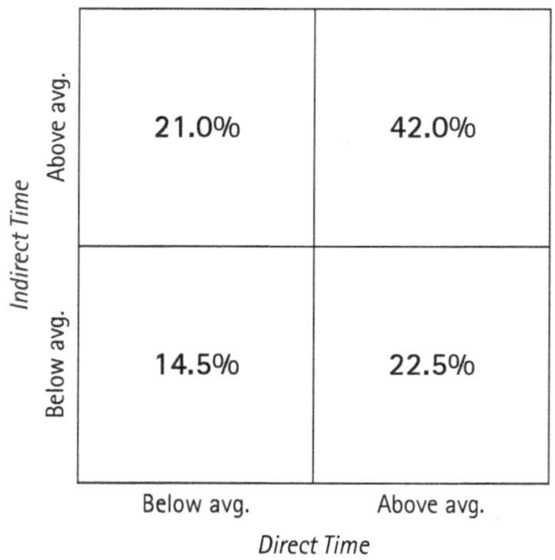

FIGURE 22.2 Propensity to borrow as a function of direct and indirect flow.

where the first term represents the *direct flow* and the second term is the *indirect flow*. Here N_s is the set of direct friends of agent s.

Figure 22.2 groups all social links of each borrower into four categories along two dimensions: whether the direct flow between borrower and friend is below or above the average direct flow, and whether the indirect flow between borrower and friend is below or above the average indirect flow. The authors then calculate the share of loans that fall into each of the resulting four categories. About 14.5% of loans involve borrower/lender pairs with both below-average direct flow and below-average indirect flow. Almost double as many loans involve borrower/lender pairs with either above-average direct or above-average indirect flow. About three times as many loans involve borrower and lenders with both above-average direct and above-average indirect flow. Indirect paths appear to play an important role in creating social collateral for borrowing.[3]

22.4 Consumption Risk-Sharing

We now turn our attention to risk-sharing (Ambrus et al. 2014b). The application of the social collateral model in this context is very similar to borrowing.

22.4.1 Motivating Example

To gain some intuition, consider the three networks in Figure 22.1. We assume that with probability $\frac{1}{2}$, agent s (previously the "borrower") experiences a negative endowment

[3] Jackson, Rodriguez-Barraquer, and Tan (2012) find a similar result in their analysis of data from Indian villages, however, they rely on a different model.

shock $-x$, while agent t (previously the "lender") experience a positive shock $+x$. With probability $\frac{1}{2}$ the shocks are reversed. All other agents in the economy experience no shocks. If agents have standard concave utility over consumption, the egalitarian social planner would optimally ensure that everyone in the economy consumes 0.

However, the planner's ability to redistribute endowments might be limited by the social network. For example, consider panel A of Figure 22.1 and the state of the world where t has the positive endowment shock. Intuitively, the planner should not expect agent t to agree to any transfer that exceeds 2 since the worst punishment that could be inflicted on her would be to lose her link which is worth 2. More generally, we should not expect that t will ever transfer more than the maximum flow between s and t: by the Ford-Fulkerson theorem any set of agents that includes t and excludes s will "cut" a set of links whose sum is greater or equal to the maximum flow $T^{st}(c)$. Moreover, there is at least one such set where the sum of cut links is exactly equal to $T^{st}(c)$. Therefore, whenever the social planner requires agent t to transfer more than $T^{st}(c)$, then agent t could assemble a coalition of agents such that the cost of potentially lost links (a proxy for the worst punishment that can be imposed by the planner on the group of deviators) is lower than t's transfer. In other words, agent t could reimburse members of her coalition for lost links instead of making a requested payment. This limits the extent of transfers between agents s and t in panels b and c of Figure 22.1 to a maximum of 5.

22.4.2 Risk-Sharing Arrangements

We now turn to the formal model, which allows us to analyze risk-sharing when more than two agents receive shocks. In our model, agents face income uncertainty due to factors such as weather shocks and crop diseases. We denote the vector of endowment realizations by $e = (e_i)_{i \in W}$, which is drawn from a commonly known joint distribution. The vector of endowments is observed by all agents.

A *risk-sharing arrangement* specifies a collection of bilateral transfer payments $t^e = \left(t^e_{ij}\right)$, where t^e_{ij} is the net dollar amount transferred from agent i to agent j in state of the world e, so that $t^e_{ij} = -t^e_{ji}$ by definition. The risk-sharing arrangement t^e implements a consumption allocation x^e where $x^e_i = e_i - \sum_j t^e_{ij}$. For simplicity, we suppress in notation the dependence of the transfers t^e_{ij} and consumption allocation x^e on e.

An agent who consumes x_i enjoys utility $U_i(x_i, c_i)$, where $c_i = \sum_j c(i,j)$ denotes the total value that agent i derives from all his relationships in the network, and U is strictly increasing and concave. The case where consumption and friendship are perfect substitutes is analytically convenient but the qualitative results can be extended to the case of imperfect substitutes. The agent's ex-ante expected payoff is $EU_i(x_i + c_i)$, where the expectation is taken over the realization of endowment shocks.

We say that a risk-sharing arrangement is *incentive compatible* if every agent i prefers to make each of his promised transfers t_{ij} rather than lose the (i,j) link and its associated value. Because consumption and friendships are perfect substitutes, incentive compatibility implies $t_{ij} \leq c(i,j)$.

By construction, risk-sharing arrangements are robust to coalitional deviations. To see this, we need some definitions. For any group of agents F, we define the *perimeter* $c[F]$ of F to be sum of the values of all links between the group and the rest of the community:

$$c[F] = \sum_{i \in F,\, j \notin F} c(i,j) \tag{22.3}$$

Intuitively, the perimeter is the maximum extent to which the rest of the community could punish group F using ostracism. Similarly, we define the total endowment of the group as e_F and their total consumption under a risk-sharing arrangement as x_F.

Definition 5. *A consumption allocation x is coalition-proof if* $e_F - x_F \leq c[F]$ *holds for all groups of agents F.*

It is easy to see that a risk-sharing arrangement implements a consumption allocation which is coalition-proof. Hence, no group of agents has an incentive to deviate: the net transfer between any group of agents and the rest of the community, defined as the difference between the group's total endowment and total consumption, does not exceed the sum of the values of all links connecting the group and the rest of the community.

22.4.3 Equivalence Result

The definition of an informal risk-sharing arrangement looks at first quite restrictive because the social network not just constrains the feasible consumption allocations but also serves as a conduit for transfers. For example, could a village elder (as a stand-in for the constrained social planner) achieve superior consumption allocations by simply taxing households with positive endowment shocks and redistribute the proceeds to households with negative shocks? This elder would not have to worry about finding a set of bilateral transfers to implement her preferred allocation and instead would only have to ensure that the final allocation is coalition-proof.

Surprisingly, the answer to this question is negative.

Theorem 2. *A consumption allocation x that is feasible* ($\sum x_i = \sum e_i$) *and coalition-proof can be implemented by an incentive-compatible informal risk-sharing arrangement.*

The theorem states that the elder implements *exactly the same* insurance arrangements as is possible with link-level punishment. The proof builds again on the mathematical theory of network flows.[4]

[4] In particular, Ambrus et al. (2014b) show that finding a transfer representation for a coalition-proof allocation is equivalent to finding a flow in an auxiliary network with two additional nodes s and t

Theorem 2 has two main implications. First of all, the results shows that it is sufficient to study risk-sharing arrangements. Links matter not because they act as conduits for transfer, but because they define the costs of deviations, and hence the pattern of obligations in the community. A second implication of the theorem is that it relates the *geometry* of the network to its effectiveness for risk-sharing.

22.4.4 Limits of Risk-Sharing

How effective are typical social networks in sharing risk? Can local obligations to help close neighbors, relatives, and friends aggregate to effective risk-sharing on the village level? The research of Townsend (1994) suggests the answer should be affirmative since the full insurance model provides a surprisingly good benchmark even though it is typically rejected in the data.

It turns that *full* risk-sharing under any endowment realization is generally impossible unless the social network is extremely *expansive*. To measure expansiveness, we define the *perimeter-area ratio* for a set of agents F as $a[F] = c[F]/|F|$, where area stands for the number of agents in F. Intuitively, $a[F]$ represents the group's maximum obligation to the community relative to the group's size. Figure 22.3 shows typical sets F for three distinct geometries. Panel A shows a line which has very low expansion properties: a large connect set will have a perimeter-area ratio equal to $\frac{2}{|F|}$. In contrast, the plane in panel B has significantly better expansion properties as its typical perimeter is of order $\sqrt{|F|}$. Finally, the binary tree in panel C is an expander graph whose perimeter-area ratio is always bounded away from 0 for arbitrary sets F.

Intuitively, we expect that networks with better expansion properties allow for more risk-sharing because it is more difficult to find a blocking coalition as alluded to in theorem 2. The next proposition makes this precise. To simplify the exposition, we focus from now on the special case of i.i.d. uniform endowment shocks over the interval $[-1, 1]$.[5]

Proposition 1. *[Limits to full risk-sharing] Under the aboive assumptions, equal sharing is supported by an incentive-compatible risk-sharing arrangement* ***if and only if*** *for every subset of agents F the perimeter-area ratio satisfies* $a[F] \geq 2\left(1 - \frac{|F|}{|W|}\right)$.

The condition implies that $a[F]$ must be greater than the constant $S/2$ for any set of a size at most half of the community. In particular, an implication for large

added. According to the theorem of Ford and Fulkerson (1956), the maximum flow equals to the value of the minimum cut (i.e., the smallest capacity that must be deleted so that s and t end up in different components). They prove that each cut in the flow problem corresponds to a coalition, and then the coalition-proofness condition ensures that the cut values are high enough so that the desired flow can be implemented.

[5] The results of Ambrus et al. (2014b) apply to general endowment distributions as long as the tails of the distributions are not too thick and shocks are not too correlated.

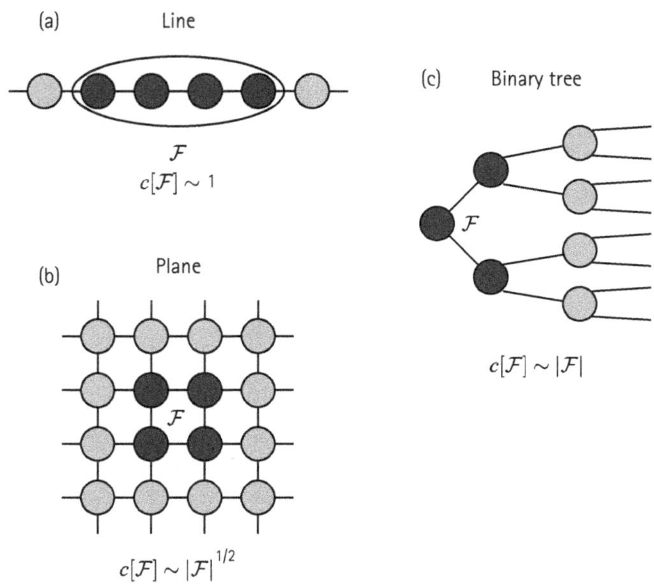

FIGURE 22.3 Expansion properties of three sample networks.

networks is that $a[F]$ must be bounded away from zero since the members of F must be willing to provide resources to the rest of the community even when they all get the highest possible realization while everyone outside gets the minimum. This implies that essentially the only type of graph that allows full risk-sharing for *any* endowment realization are expander graphs such as the binary tree.

22.4.5 Line and Plane

We now show that risk-sharing on the plane and similar networks is very good, and substantially better than on the line. It is helpful to first develop an intuition for this result.

Plane networks turn out to be just sufficiently well-connected to generate very good risk-sharing in most states of the world. The key insight is that with a two-dimensional structure, outcomes in which the coalitional constraint binds under equal sharing become rare. To see the logic, consider the regular plane with the i.i.d. $[-1, 1]$ shocks. As we have seen, equal sharing fails because households in a large n by n square F would need to give up n^2 resources if all of them get a positive $+1$ shock, which is an order of magnitude larger than the perimeter $c[F] \sim n$.

The key is that for large n, such extreme realizations are unlikely, and in typical realizations the required transfers do not exceed the perimeter. With i.i.d. shocks, the standard deviation of the group's endowment is only n, which is only of order n even though it is the sum of n^2 random variables—intuitively, a lot of the idiosyncratic shocks cancel out within the group. Thus the *typical shock* in F has the same order of magnitude as the maximum pledgeable amount, and hence potentially deviating coalitions are rare. By way of contrast, the argument breaks down for the line, since the perimeter of even large interval sets is only 2, a constant.

To formulize these ideas, we assume that agents have quadratic utility function such that we can express the average utility loss relative to the benchmark of equal sharing as

$$SDISP\,(x) = \left[\mathrm{E}\frac{1}{|W|}\sum_{i\in W}(x_i - \overline{e})^2 \right]^{1/2}, \tag{22.4}$$

which is the square-root of the expected cross-sectional variance of x. For non-quadratic utilities, $SDISP\,(x)$ can be interpreted as a second-order approximation of the utility based measure.

Proposition 2. *There exist positive constants K, K' and K'' such that*

(i) On the infinite line with capacities c and i.i.d. shocks, we have $SDISP\,(x) \geq K/c$ for all incentive-compatible risk-sharing arrangements.

(ii) On the infinite plane with capacities c, we have $SDISP\,(x) \leq K' \exp\left[-K''c^{2/3}\right]$ for some incentive-compatible risk-sharing arrangement.

This proposition characterizes the rate of convergence to full risk-sharing as capacities increase. The contrast between the line and plane is remarkable. Risk-sharing is relatively poor on the line: *SDISP* goes to zero at a slow polynomial rate of $1/c$ as c goes to infinity. In contrast, the rate of convergence for the plane is exponentially fast, confirming our intuition that agents are able to share typical shocks due to the more expansive structure.

The difference in the rates of convergence become quickly apparent in simulations. Figure 22.4 compares risk-sharing on equally sized line and plane (100 agents each) while fixing initial endowments and the total capacity per agent across both networks.[6] SDISP declines rapidly on the plane and full risk-sharing is already achieved at a capacity of 1.4 per agent. In contrast, SDISP declines more slowly on the line and full risk-sharing requires capacity per agent far exceeding 2.

Ambrus et al. (2014b) extend the result for the plane to *geographic networks* that exhibit a two-dimensional sub-structure but are less regular than the plane. Using data from Peruvian social networks they show that real-world networks are geographic networks but not expander graphs. Hence, the social collateral model can explain very good risk-sharing in real-world social networks where agents only have local obligations to a small subset of the population (such as close neighbors, friends, and relatives).

[6] If we think of capacities as social collateral then this exercise compared a linear and planar network with the same amount of social capital per agent.

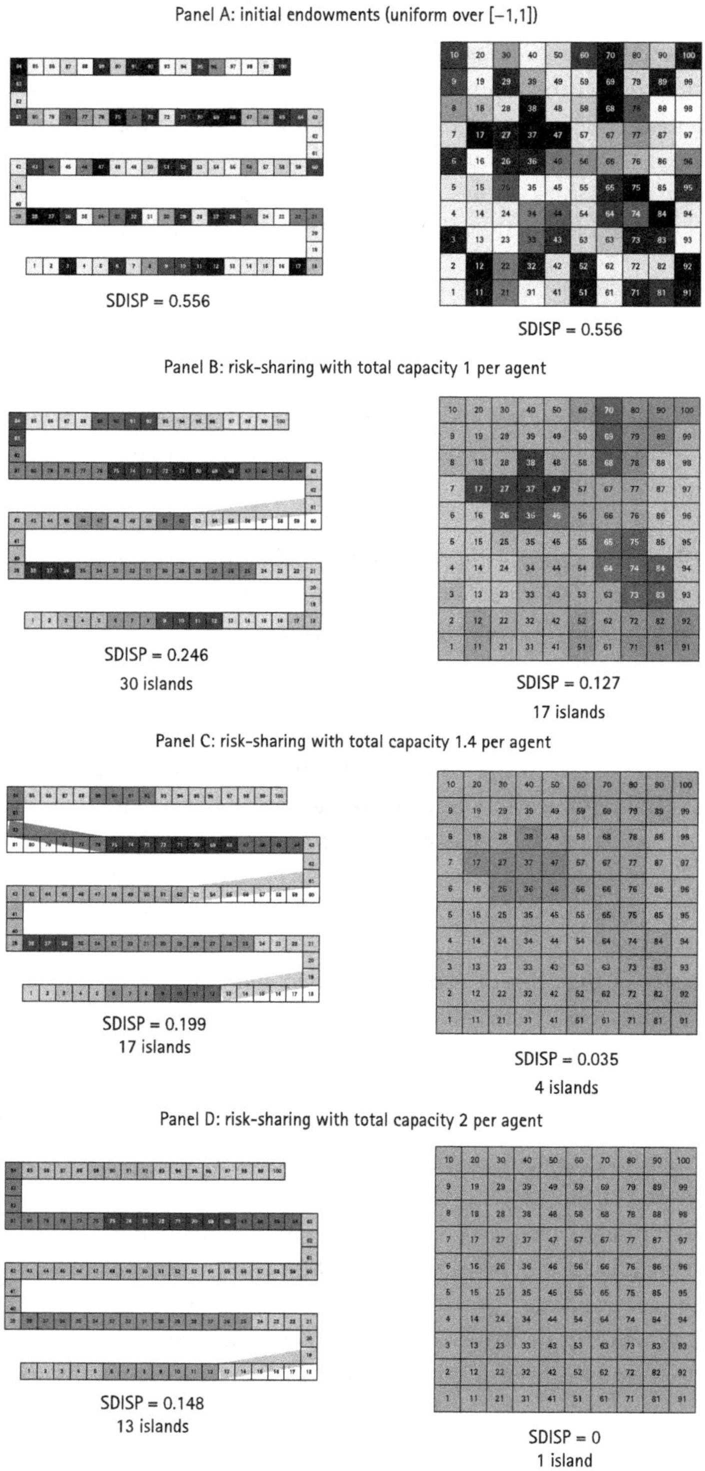

FIGURE 22.4 Risk-sharing simulations on line and plane for different capacities.

22.4.6 Risk-Sharing Islands

Ambrus et al. (2014b) also characterize the micro-structure of specific risk-sharing arrangements. Intuitively, the network is partioned into a set of contiguous "risk-sharing islands," as shown in Figure 22.4, such that agents within the same island, consume the same amount while agents in neighboring islands consume either more or less. Moreover, the IC constraints for transfer across islands bing while they are slack within islands.

This phenomenon has two important implications.

Local sharing. When an agent in the interior of an island receives an endowment shock, she will share the risk first *locally* with other neighbors in her own risk-sharing island. If consumption within the island increases (or decreases) sufficiently so that IC constraints to neighboring islands no longer bind, then the shock will also be shared with neighboring islands. In that sense, risk-sharing in the social collateral model is local.

Endogenous socialization. Compare two geometries, such as line and plane, where risk-sharing islands have the same average size (which translates into comparable risk-sharing in both networks). Then the more expansive network will tend to have a higher share of agents at the boundary of a risk-sharing island. Therefore, the incentive of agents to invest in socialization is at the margin greater in the more expansive network. Hence, the very features which create good risk-sharing on expansive networks such as the plane *also* make these networks more stable because they increase agents' incentives to invest in socialization.

22.5 Other Mechanisms

At this point, it is useful to contrast the social collateral approach to other theory frameworks.

22.5.1 Altruism

The social collateral approach assumes that agents are selfish and attributes any lending or risk-sharing to repeated game effects. However, we might expect that people feel particularly altruistic towards family, friends, and neighbors and might for this reason alone provide help. Leider, Mobius, Rosenblat, and Do (2007) analyze this question by matching student subjects to direct and indirect friends at various social distances and have these pairs of subjects play a series of dictator games. They use a within-subject design where the recipient either finds out or does not find out the dictator's identity. This allows them to distinguish *directed altruism* (being nice to one's friends due to "warm glow") from repeated game effects. There is substantial directed altruism towards direct friends, but indirect friends are treated similarly to randomly picked nameless individuals within the network. Hence, directed altruism might not be able to explain transfers within the network that do not involve direct friends. In a closely related paper, Ligon and Schechter (2012) have villagers in rural Paraguay play variants of the dictator game to examine motives for sharing. They correlate behavior measured in the experiment with the real-world sharing outside the experiment to find that repeated-game motives seem to better explain sharing in the real world.

Earlier work on lifetime (inter-vivos) transfers also rejects the altruism hypothesis. For example, Altonji, Hayashi, and Kotlikoff (1997) find that parents increase transfers to children by only 13 cents for every dollar that is redistributed from child to parent. Cox, Eser, and Jimenez (1998) and Cox and Rank (1992) analyze the patterns of intergenerational transfers and also find that they are consistent with exchange motives rather than altruism.

22.5.2 Sharing Rules and Bargaining

Bramoullé and Kranton (2007b) abstract away from the enforcement problem within connected networks and instead assume that agents within a connected component share resources equally (for example, social neighbors might repeatedly pool and share their income, which eventually results in equal sharing across the component). Ambrus, Chandrasekhar, and Elliott (2014a) extend this basic idea and analyze an environment where agents within a connected component bargain over the joint surplus. They show that the surplus is allocated according to the Myerson value where more central agents receive higher shares. In these papers, the network defines the bargaining position of agents rather than enforcement, as in the social collateral approach.[7]

22.5.3 Other Repeated Game Models

There are a number of papers in the literature that also study enforcement in a repeated game context.

Ligon, Thomas, and Worrall (2002) formulate a theory of limited commitment to explain deviations from full risk-sharing in village economies. They test their model with data from three Indian villages and find that the model can fully explain the dynamic response of consumption to income, but cannot explain the distribution of consumption across households (also see Barr and Genicot (2008) for empirical evidence). The social collateral model can be thought as a special case of the limited

[7] In related empirical work, Altonji, Hayashi, and Kotlikoff (1992) use PSID data to show that risk-sharing within the extended family is not independent of the distribution of resources.

commitment model where the social network defines the set of constraints on bilateral transfers.

Ali and Miller (2013) analyze a model where agents' needs and transfers are not perfectly observed (unlike in the social collateral model). This raises a new set of questions such as whether agents have an incentive to communicate deviations and how quickly information spreads within the network. They show that equilibria with severe group punishments (permanent ostracism) are difficult to sustain because cheated-upon agents might not have an incentive to communicate truthfully.

Finally, Jackson et al. (2012) propose a "social quilt" model where agents play continuous-time Prisoner's Dilemma games with their social neighbors under perfect information. The authors focus on equilibria that are renegotiation-proof and characterize the set of stable networks. They show that these networks only include "supported links" such that any two friends who exchange favors have a common friend. The empirical predictions of their model are in fact very similar to the ones reported in Section 22.3.4. The biggest difference is that the social collateral approach takes the social network as exogenously given and then finds the set of feasible borrowing or risk-sharing arrangements within that context.

22.5.4 Endogenous Network

Another strand of the literature analyzes network formation when agents form links to mitigate risk. Endogenous network formation can be inefficient because agents do not fully internalize the benefits and costs of forming links (Bramoullé and Kranton 2007a; Bramoullé and Kranton 2007b). Rosenzweig and Stark (1989) provide evidence from rural India that families arranges marriages of daughters to distant locations to mitigate income shocks.[8] Fafchamps and Gubert (2007b) find that geographic proximity is a strong correlate of risk-sharing networks, likely because it facilitates monitoring and enforcement (also see Fafchamps and Gubert 2007a).

22.6 Conclusion

The social collateral approach provides an analytically tractable and empirically relevant way of modeling informal transfers in social networks. This approach has been applied to analyze (1) borrowing and trust in networks and (2) consumption risk-sharing.

[8] However, Munshi and Rosenzweig (2009) argue the existence of sub-caste networks that provide mutual insurance to their members restrict marriage mobility.

References

Ali, S. Nageeb and David A. Miller (2013). "Ostracism." Unpublished manuscript, University of California at San Diego.

Altonji, Joseph G., Fumio Hayashi, and Laurence J. Kotlikoff (1992). "Is the extended family altruistically linked? Direct tests using micro data." *The American Economic Review* 82(5).

Altonji, Joseph G., Fumio Hayashi, and Laurence J. Kotlikoff (1997). "Parental altruism and inter vivos transfers: Theory and evidence." *Journal of Political Economy* 105(6).

Ambrus, Attila, Arun G. Chandrasekhar, and Matt Elliott (2014). "Social investments, informal risk sharing, and inequality." Technical Report, National Bureau of Economic Research.

Ambrus, Attila, Markus Mobius, and Adam Szeidl (2014). "Consumption risk-sharing in social networks." *American Economic Review* 104(1), 149–182.

Angelucci, Manuela, Giacomo De Giorgi, and Imran Rasul (2012). "Resource Pooling Within Family Networks: Insurance and Investment."

Attanasio, Orazio, Abigail Barr, Juan Camilo Cardenas, Garance Genicot, and Costas Meghir (2012). "Risk pooling, risk preferences, and social network." *American Economic Journal: Applied Economics* 4(2), 134–167.

Barr, Abigail and Garance Genicot (2008). "Risk sharing, commitment, and information: An experimental analysis." *Journal of the European Economic Association* 6(6), 1151–1185.

Bramoullé, Yann and Rachel Kranton (2007a). "Risk sharing across communities." *The American Economic Review*, 70–74.

Bramoullé, Yann and Rachel Kranton (2007b). "Risk-sharing networks." *Journal of Economic Behavior & Organization* 64(3), 275–294.

Cox, Donald and Mark R. Rank (1992). "Inter-vivos transfers and intergenerational exchange." *The Review of Economics and Statistics*, 305–314.

Cox, Donald, Zekeriya Eser, and Emmanuel Jimenez (1998). "Motives for private transfers over the life cycle: An analytical framework and evidence for Peru." *Journal of Development Economics* 55(1), 57–80.

De Weerdt, Joachim and Stefan Dercon (2006). "Risk-sharing networks and insurance against illness." *Journal of Development Economics*, 81(2), 337–356.

Dercon, Stefan and Pramila Krishnan (2000). "In sickness and in health: Risk sharing within households in rural Ethiopia." *Journal of Political Economy* 108(4), 688–727.

Fafchamps, Marcel and Flore Gubert (2007). "The formation of risk sharing networks." *Journal of Development Economics* 83(2), 326–350.

Fafchamps, Marcel and Flore Gubert (2007). "Risk sharing and network formation." *American Economic Review Papers and Proceedings* 97(2).

Fafchamps, Marcel and Susan Lund (2003). "Risk-sharing networks in rural Philippines." *Journal of Development Economics* 71, 261–287.

Ford, Jr., Lester Randolph and Delbert Ray Fulkerson (1956). "Maximal flow through a network." *Canadian Journal of Mathematics* 8, 399–404.

Gertler, Paul and Jonathan Gruber (2002). "Insuring consumption against illness." *The American Economic Review* 51–70.

Gubert, Flore (2002). "Do migrants insure those who stay behind? Evidence from the Kayes area (Western Mali)." *Oxford Development Studies* 30(3), 267–287.

Hayashi, Fumio, Joseph Altonji, and Laurence Kotlikoff (1996). "Risk-sharing between and within families." *Econometrica: Journal of the Econometric Society* 261–294.

Jack, William and Tavneet Suri (2014). "Risk sharing and transactions costs: Evidence from Kenya's mobile money revolution." *American Economic Review* 104(1), 183–223.

Jackson, Matthew O., Tomas Rodriguez-Barraquer, and Xu Tan (2012). "Social capital and social quilts: Network patterns of favor exchange." *The American Economic Review* 102(5), 1857–1897.

Karlan, Dean, Markus Mobius, Tanya Rosenblat, and Adam Szeidl (2009). "Trust and Social Collateral." *Quarterly Journal of Economics*.

Kazianga, Harounan and Christopher Udry (2006). "Consumption smoothing? Livestock, insurance and drought in rural Burkina Faso." *Journal of Development Economics* April 79(2), 413–446.

Leider, Stephen, Markus M. Mobius, Tanya S. Rosenblat, and Quoc-Anh Do (2007). "Directed altruism and enforced reciprocity in social networks." working paper 13135, NBER May.

Ligon, Ethan and Laura Schechter (2012). "Motives for sharing in social networks." *Journal of Development Economics* 99(1), 13–26.

Ligon, Ethan, Jonathan P. Thomas, and Tim Worrall (2002). "Informal insurance arrangements with limited commitment: Theory and evidence from village economies." *Review of Economic Studies* 69(1), 209–244.

Munshi, Kaivan and Mark Rosenzweig (2009). "Why is mobility in India so low? Social insurance, inequality, and growth." Technical Report, National Bureau of Economic Research.

Rosenzweig, Mark R. and Oded Stark (1989). "Consumption smoothing, migration, and marriage: Evidence from rural India." *Journal of Political Economy*, August 97(4), 905–926.

Townsend, Robert (1994). "Risk and insurance in village India." *Econometrica* 62, 539–591.

Udry, Chris (1994). "Risk and insurance in a rural credit market: An empirical investigation in Northern Nigeria." *Review of Economic Studies* 61(3), 495–526.

Yang, Dean and HwaJung Choi (2007). "Are remittances insurance? Evidence from rainfall shocks in the Philippines." *The World Bank Economic Review* 21(2), 219–248.

CHAPTER 23

COMMUNITY NETWORKS AND MIGRATION

KAIVAN MUNSHI

23.1 Introduction

The process of migration, within and between countries, has historically been characterized and continues to be characterized by the movement of groups of individuals. Depending on the context, these groups are drawn from the same caste, clan, parish, or village in the origin. Preexisting social connections within these groups help migrants in different ways when they move. These connections provide psychological and social support. More importantly, from an economist's perspective, they provide various forms of economic support at the destination.

Migrants, being newcomers to the destination economy, are especially vulnerable to information and commitment problems that prevent them from participating fully in the market. For example, prospective employers will not hire migrants because they do not know their worth. Banks will not lend to migrants because they are unable to provide security and have yet to establish a personal reputation. Under these conditions, communities can harness preexisting social ties to form migrant networks that overcome these constraints to economic activity. Migrants who have been in the destination labor market for a while and have established a reputation within their firms can provide referrals to capable new arrivals who belong to their network. The community network can also pool its wealth and provide credit to members with promising investment opportunities, with the knowledge that these individuals will repay their loans in the future (to avoid the social sanctions they would face if they reneged on their obligations).

As described in Section 23.2 of this chapter, there is a voluminous literature in social history, sociology, and economics that describes the role played by migrant community networks in supporting their members, both historically and in the contemporary economy. While a rapidly emerging theoretical literature in economics examines how network structure determines the extent to which information and commitment problems can be resolved, in this chapter I will restrict attention to simpler network properties such as size and connectedness. Nevertheless, we will see that it is difficult to provide credible statistical evidence that networks support migration and improve the outcomes of their members at the destination. This is because the size and composition of the migrant network will respond to (unobserved) changes in the destination economy that directly determine the outcomes of its members. Any observed correlation between individual outcomes and network characteristics could in that case be entirely spurious.

Much progress has been made over the past decade in tackling this problem and I will present two approaches in Section 23.3 that have been used in previous research. The first approach exploits exogenous variation in economic conditions at the *origin* to construct a statistical instrument for network characteristics at the *destination*. This approach is only valid under very special conditions. The second approach exploits exogenous variation in the *population* characteristics of the community, together with restrictions from the theory, to identify a role for community networks. This approach can be implemented more generally, and is feasible even when the migrant community network is not observed directly.

While economists traditionally modeled migration as a choice based on differential wages at the origin and the destination, credible statistical evidence that networks matter for migration decisions and outcomes using the methods described above has transformed the literature. It is now common practice for researchers to add a "networks" variable to reconcile the pattern of migration across destinations. The networks variable is typically measured by the stock of migrants from the individual's origin community at each destination. The pitfalls of such an approach have been noted above, and while networks may well play a crucial role in migration, it is important to incorporate networks appropriately in models of migration and to adequately address the identification problems that are associated with their empirical analysis.

Migrant networks will increase the income of those individuals who move by providing them with credit for business investment and with jobs. This will naturally generate inequality within the population from which the network is drawn as well as across communities. Destination networks will form and grow over time in certain communities, favoring their members, while equally capable migrants in other communities will remain at a persistent disadvantage. Within communities, inequality will increase or decrease depending on who benefits from the migrant network. Because these networks grow over time, inequality within the community will also change over time, not necessarily monotonically, as described in Section 23.4.

The discussion up to this point has focused on migrant networks at the destination. If the community can support a network far away, we would imagine that it could support networks serving different purposes at the origin as well. Once we add origin networks to the mix, the analysis of migration becomes more complex. As described in Section 23.5, while the community may support the movement of *groups* of its members to a new location, it will discourage the movement of *individuals*, either because they lose the services such as insurance that are provided by the origin network when they move independently, or because there are explicit restrictions on mobility. This tension could persist even at the destination, once networks span multiple generations, with established members discouraging the next generation from pursuing opportunities elsewhere. A complete characterization of the inter-generational evolution of migrants community networks, across space and occupations, is noticeably lacking in the economics literature. The concluding section will briefly discussion possible directions that this research could take.

23.2 Migrant Networks in Historical Perspective

The development of the United States is associated with the first large-scale movement of workers across national boundaries. During the Age of Mass Migration (1850–1913), the United States received 30 million European immigrants. Abramitzky, Boustan, and Eriksson (2014) calculate that this resulted in 38% of workers in northern cities being foreign-born in 1910. Labor markets in the nineteenth century could be divided into three segments: a stable segment with permanent employment, an unstable segment with periodic short-term unemployment, and a marginal but highly flexible segment characterized by spells of long-term and short-term unemployment (Gordon, Edwards, and Reich 1982). Migrants being newcomers to the U.S. market typically ended up in the unstable and marginal segments, where the uncertain labor demand and the lack of information about their ability and diligence naturally provided an impetus for the formation of ethnic job networks (Conzen 1976; Hoerder 1991).

Accounts by contemporary observers and an extensive social history literature indicate that friends and kin from the origin community in Europe played an important role in securing jobs for migrants in the U.S. labor market in the nineteenth century and the first quarter of the twentieth century. As an immigrant put it, "The only way you got a job [was] through somebody at work who got you in" (Bodnar, Simon, and Weber 1982: 56). Early historical studies used census data, which provide occupations and country of birth, to identify ethnic clusters in particular locations and occupations (Hutchinson 1956; Gordon, Edwards, and Reich 1982). More recently, social historians have linked parish registers and county data in specific European sending communities to census and church records in the United States to construct the entire chain of migration from those communities as it unfolded over time. This research has documented the formation of new settlements by pioneering migrants, the subsequent channeling of migrants from the origin community in Europe to these settlements, as well as the movement of groups from the original settlement to new satellite colonies elsewhere in the United States (Gjerde 1985; Kamphoefner 1987; Bodnar 1985).

Migration from Europe ceased in 1913, but it was soon replaced by the movement of African-Americans from the rural South to northern cities. The first major movement of blacks out of the South commenced in 1916. Over 400,000 blacks moved to the North between 1916 and 1918, exceeding the total number who moved in the preceding 40 years. During the first phase of the Great Migration, running from 1916 to 1930, over one million blacks (one-tenth the black population of the United States) moved to northern cities (Marks 1983). This movement was driven by both pull and push factors. The increased demand for labor in the wartime economy coupled with the closing of European immigration gave blacks new labor market opportunities (Mandle 1978; Gottlieb 1987). Around the same time, the boll weevil beetle infestation reduced the demand for labor in southern cotton-growing counties (Marks 1989). Adverse economic conditions in the South, together with segregation and racial violence, encouraged many blacks to leave (Tolnay and Beck 1990). Their movement was facilitated by the penetration of the railroad into the deep South (Wright 1986). A confluence of favorable and unfavorable circumstances thus set the stage for one of the largest internal migrations in history.

Although external sources of information such as newspapers and recruiting agents played an important role in jump-starting the migration process, and agencies such as the Urban League provided migrants with housing and job assistance at the destination, networks linking southern communities to specific northern cities, and to neighborhoods within those cities, soon emerged (Gottlieb 1987; Marks 1991; Carrington, Detragiache, and Vishwanath 1996). "[These] networks stimulated, facilitated, and helped shape the migration process at all stages from the dissemination of information through the black South to the settlement of black southerners in northern cities" (Grossman 1991: 67).

The large-scale movement of labor in the United States, supported by migrant networks, was being replicated in other parts of the world as economies industrialized and cities grew in the nineteenth century. For example, Mumbai's industrial economy in the late nineteenth century and through the first half of the twentieth century was characterized by wide fluctuations in the demand for labor (Chandavarkar 1994). Frequent job turnover will naturally give rise to labor market networks, particularly when the quality of a freshly hired worker is difficult to assess and performance-contingent wage contracts cannot be implemented. The presence of such recruitment networks has indeed been documented by numerous historians studying Mumbai's economy prior to independence in 1947 (Chandavarkar 1994; Morris 1965; Burnett-Hurst 1925). These networks appear to have been organized around the jobber, a foreman who was in charge of a work gang in the mill, factory, dockyard, or construction site, and more importantly also in charge of labor recruitment.

Given the information and enforcement problems that are associated with the recruitment of short-term labor, it is not surprising that the "jobber had to lean on social connections outside his workplace such as his kinship and neighborhood connections" (Chandavarkar 1994: 107). Here the endogamous caste served as a natural social unit from which to recruit labor. The primary marriage rule in Hindu society, which recent genetic evidence indicates is over 2,000 years old (Moorjani et al. 2013), is that individuals must marry within their caste or *jati*. Muslims follow the same pattern of endogamous marriage within their *biradaris*, while converts to Christianity continue to marry within their original *jatis*. Marriage ties within these kinship groups, formed over many generations, strengthen information flows and improve enforcement. The resulting formation of networks drawn from these kinship communities led to a fragmentation of Mumbai's labor market along social lines.

The presence of caste clusters in the textile mills, for example, has been well documented. Gokhale's (1957) survey of textile workers in a single mill in the 1950s showed that Maratha and Kunbi (both middle caste) men were evenly distributed throughout the mill, while Harijan (low caste) men were employed for the most part in the spinning section. Consistent with the presence of local (jobber-specific) networks, the caste clusters that were observed in particular mills often differed from the general pattern for the industry as a whole. The same sort of caste-based clustering has been documented among Mumbai's dock workers (Cholia 1941), construction workers, and in the railway workshops (Burnett-Hurst 1925), the leather and dyeing industries, and in the Bombay Municipal Corporation and the Bombay Electric Supply and Transportation Company (Chandavarkar 1994).

Although most historical accounts of caste-based networking in Indian cities are situated prior to independence in 1947, a few studies conducted over the subsequent decades in India indicate that these patterns persisted over many generations. For example, Patel (1963) surveyed 500 mill workers in Mumbai in 1961–62 and found that 81% had relatives or members of their caste in the textile industry. 50% of the workers got jobs in the mills through the influence of their relatives, and 16% through their friends, many of whom would have belonged to the same caste. Forty years later, Munshi and Rosenzweig (2006) surveyed the parents of schoolchildren residing in the same area of the city. Sixty-eight percent of the fathers employed in working-class occupations reported that they received help from a relative or member of their caste in finding their first job, while 44% of fathers in white-collar occupations reported such help.

Labor market networks continue to be active in cities throughout the world, most often among migrant populations. For example, Rees (1966) reports that informal sources accounted for 80% of all hires in blue-collar occupations and 50% of all hires in white-collar occupations in an early study set in Chicago. We would expect social ties to play an even stronger role for migrants in the United States Indeed, over 70% of the undocumented Mexicans, and a slightly higher proportion of the Central Americans, that Chavez (1992) interviewed in 1986 found work through referrals from friends and relatives. Similar patterns have been found in contemporary studies of Salvadoran immigrants (Menjivar 2000), Guatemalan immigrants (Hagan 1994), and Chinese immigrants (Nee and Nee 1972; Zhou 1992). Individual respondents in the Mexican Migration Project (MMP), discussed in greater detail below, were asked how they obtained employment on their last visit to the United States; relatives (35%) and friends or *paisanos* from the origin village in Mexico (35%) account for the bulk of job referrals.

23.3 Identifying Migrant Networks

While direct evidence on the support provided by migrant networks is sometimes available, it will often be the case that a causal link between observed network characteristics and individual outcomes will need to be established. The statistical problem that arises when attempting to make this connection (which is also discussed by Boucher and Fortin in this handbook) is parsimoniously summarized by the following equation,

$$y_{ict} = \beta X_{ct} + \omega_i + \epsilon_{ct}, \tag{23.1}$$

where y_{ict} is an outcome for migrant i, belonging to community c, in period t, such as employment or wages, that is potentially determined by the network and X_{ct} measures the size of the community network in that period. Individuals benefit from larger networks because more referrals from fellow members are available. At the same time, there will be more competition for available referrals in a larger network. If the first effect dominates, then the hypothesis is that $\beta > 0$. ω_i measures the individual's ability, which directly determines his outcome in the destination labor market, and ϵ_{ct} is an exogenous labor demand shock. Both ω_i and ϵ_{ct} are unobserved by the econometrician. Notice that the demand shock has a community, c, subscript. This reflects the idea that individual migrants from a given origin location could be endowed with specific skills that channel them into particular segments of the labor market even when networks are absent.

The estimated β coefficient in equation (23.1) will be biased. This is because the size of the network, X_{ct}, will respond to demand shocks at the destination, ϵ_{ct}, violating the orthogonality condition. Selective migration will also bias the estimated network effect. If there is positive selection on ability (i.e., higher ability individuals are more likely to migrate), then an increase in X_{ct} implies that the marginal migrant will have lower ability: $E(\omega_i)$ is decreasing in X_{ct}. If there is negative selection, $E(\omega_i)$ will be increasing in X_{ct}. Least squares estimation of the preceding equation will thus be associated with both omitted variable and selection bias.

The difficulty in interpreting the estimated β coefficient as a network effect, discussed above, plagues much of the recent cross-country literature on migration; for example, Beine et al. (2011), Bertoli and Fernandez-Huertas Moraga (2012), Docquier et al. (2014). This literature includes the stock of migrants from the origin country in each destination country to measure the location-specific network size. The same approach has been used to examine networks within the United States (Patel and Vella 2013). The limitation of this strategy is that the stock could instead reflect the unobserved match between skills acquired at the origin and skills needed at the destination. Where this match is better, the flow of migrants over time and, hence, their stock, will be larger. With panel data, it is possible to include fixed effects for each origin-destination pair. However, changes in the stock of migrants at a given destination will still reflect unobserved demand shocks, as discussed above, complicating the network inference problem.

One solution to this problem is to find a statistical instrument that determines X_{ct} but is uncorrelated with ϵ_{ct}. A major advantage of working with migration data is that the size of the network will be determined by pull factors from the destination as well as push factors from the origin. Origin characteristics or shocks that generate exogenous variation in the size of the migrant network, but are uncorrelated with demand shocks at the destination, will thus be valid instruments. Note, however, that we would still need to include individual fixed effects when estimating equation (23.1) because $E(\omega_i)$ will respond to changes in X_{ct}, whether or not they are exogenously determined.

Munshi (2003) shows how network effects can be consistently estimated in the context of immigrant Mexican labor in the United States. Migration from Mexico tends to be recurrent, with individuals working in the United States for spells of three to four years and then returning. Panel data from the MMP can be used to study the labor market outcomes in the United States of a sample of individuals drawn from different Mexican origin communities (villages) over multiple migration spells. The idea is to assess whether the same individual does better in spells where he has access to a larger network in the United States. The MMP collected information from a large number of communities (see Massey et al. 1987 for a description of these data). Each community was surveyed once only and retrospective information over many years was collected from approximately 200 individuals. This information included the location of the individual in each year (United States or Mexico) and his labor market outcome (employment, job-type). Munshi measures the size of the community network in the United States in a given year by the fraction of sampled individuals in the community who were located in the United States in that year. To test for network effects, the sample is restricted to person-years in the United States. In the most basic specification, corresponding to equation (23.1), we would regress each individual's labor market outcome on the contemporaneous size of his community network, including fixed effects in the regression.

Once fixed effects are included, we are effectively assessing the effect of changes in network size on changes in labor market outcomes (i.e., is the individual more likely to be employed [and holding a better job] in years in which his network in the United States is relatively large). However, we know from the discussion above that even if a positive correlation is obtained, this correlation could be entirely spurious if individual labor market outcomes and the size of the community network are jointly determined by (unobserved) economic conditions in the United States. To estimate the causal effect of networks on individual outcomes, we need to find a statistical instrument for network size. A valid instrument in this context will generate changes in network size but will be uncorrelated with direct determinants of individual labor market outcomes in the United States. Munshi's innovation is to use rainfall in Mexican origin communities, or more correctly rainfall shocks once fixed effects are included, as instruments for network size in the United States.

In practice, network effects will depend on their size and their vintage, since migrants who have been in the United States longer are more established and better positioned to provide referrals. Instead of simply including the size of the network as the key regressor, a more sophisticated specification would thus include the fraction of sampled individuals who recently arrived in the United States and the corresponding fraction for established migrants, separately as regressors.

Table 23.1, Column 1 reports the estimated network effects, based on the instrumental variable regression described above. The number of established migrants in the destination network has a strong effect on the individual's labor market outcome, which is measured by employment. In contrast, the number of recent migrants does not significantly affect his employment. Column 2 reports the corresponding reduced-form estimates. The established migrants would have moved in response to negative rainfall shocks many years ago and, as expected, we see that distant-past rainfall has a negative and significant effect on employment. In contrast, recent-past rainfall, which determines the number of recent migrants, does not significantly affect employment.

The implicit assumption in the preceding argument is that if there is a drought in the origin village, the demand for labor will decline, with an accompanying increase in migration from Mexico to the United States. Providing support for this assumption, we see in Column 3 that recent-past rainfall has a *positive* and significant effect on employment in the Mexican origin village, whereas the effect of distant-past rainfall is smaller in magnitude. Local rainfall in Mexican communities far from the border has no impact on the U.S. labor market. However, it has a strong effect on the number of migrants, and these migrants, in turn, improve outcomes for their network members years later when they are established. The estimated network effect is large in magnitude; if the networks were shut down but migration flows remained unchanged, unemployment would increase from 4% to 10%. Complementing this finding, the prevalence of preferred (more remunerative) non-agricultural jobs would decline from 51% to 32%.

One alternative interpretation of the results in Table 23.1 is that they reflect an individual experience effect; the individuals who moved in response to the negative rainfall shock years ago are now doing better themselves. However, when Munshi restricts the sample to individuals who arrived recently in the United States, he finds that the estimated network effects are even larger. This is exactly what the theory would predict, since newcomers to the foreign labor market benefit the most from referrals.

The preceding example provides a general framework for identifying network effects. Panel data (and fixed effects) allow the econometrician to control for selection into the network. Rainfall shocks in the origin location generate exogenous variation in the size and the vintage of the network in the destination labor market. The theory is used to place additional restrictions on the data that rule out the alternative explanation based on an individual experience effect. As predicted, recent arrivals benefit more from the network, while established migrants contribute disproportionately to the network. Munshi's application is exceptionally well suited to testing for network effects because both panel data and a clean source of variation in network size (by vintage) is available.

It is, however, possible to identify network effects even when this is not the case, as long as there is exogenous variation in the population characteristics of the community, by deriving and testing additional predictions from the theory. The example that follows shows how this can be done.

The setting for this example is the American South in the decades after Emancipation in 1865. Chay and Munshi's (2014) objective is to assess whether and where African Americans were able to overcome centuries of social dislocation and form new networks once they were free. Their analysis starts with the observation that cropping patterns varied substantially across southern counties, during and after slavery. Where labor-intensive crops such as tobacco, cotton, rice, and sugarcane were grown, black slaves would have lived and worked in large plantations. Their numbers on these plantations would have been large enough to support cooperative arrangements even during slavery. In contrast, where crops such as wheat and corn were grown, blacks were dispersed more widely, living and working in small family farms. Restricted social interaction across these farms would have prevented cooperative arrangements from forming. Blacks could interact without restriction after Emancipation, but the strength and frequency of these interactions would have been limited by spatial proximity, which was determined, once again, by cropping patterns. Social connectedness would thus have been greater in southern counties where labor intensive crops were grown, both during and after slavery. Greater connectedness would have supported higher levels of cooperation, resulting in larger networks drawn from the population. These larger networks would, in turn, have allowed blacks to work more effectively as a group to achieve common objectives.

Southern blacks had two significant opportunities to work together in the decades after Emancipation. First, blacks were able to vote and to elect their own leaders during and just after Reconstruction, 1870–1890. Second, blacks were able to leave the South and find jobs in northern cities during the Great Migration, 1916–1930. Based on the theory, more connected populations would have supported the formation of larger networks of black activists during Reconstruction and larger networks of black workers moving together to northern cities during the Great Migration. This, in turn, would have given rise to greater overall political participation and migration.

Population connectedness → network size → political participation (migration)

While a positive relationship between population connectedness and particular outcomes during Reconstruction and the Great Migration is consistent with the presence of underlying (unobserved) black networks, other explanations are available. For example, racial conflict could have been greater in counties where labor-intensive plantation crops were grown, encouraging individual black voters to turn out during Reconstruction and to move independently to northern cities during the Great Migration. Alternatively, adverse economic conditions in these counties could have encouraged greater migration without requiring a role for black cooperation. Chay and Munshi's strategy to identify the presence of underlying networks takes advantage of the additional prediction of their theory, which is that networks will only form above a threshold level of population connectedness. There should thus be no association between the outcomes of interest—political participation and migration—and population connectedness up to a threshold and a positive association thereafter.

Figure 23.1 reports the relationship between population connectedness and (separately) black political participation and migration. Population connectedness is measured by the fraction of cultivated land in the county that was allocated to labor-intensive plantation crops in 1890, midway between Reconstruction and the Great Migration, adjusting for differences in labor intensity across those crops. Plantation size in 1860 (before slavery) is smoothly increasing in this measure of population connectedness. That measure is by construction equal to the population density of black farm workers in 1890. It thus represents both the social capital that was carried forward from the period of slavery as well as the strength and frequency of social interactions in the period after slavery. Black political participation is measured by the number of Republican votes in the 1872 presidential election, since blacks would have voted almost exclusively for the Republican Party (the party of the Union) at that time (Morrison 1987). Some whites would also have voted for the Republican Party at this time. This would confound the test of the theory if the number of white votes varied systematically with our measure of black population connectedness. Robustness tests reported below will rule out this potential source of confounding variation. The black migration measure is derived from inter-censal changes in the black population between 1910 and 1930 (recall that the Great Migration commenced in 1916), adjusting for natural changes due to births and deaths. It appears from Figure 23.1 that the specific nonlinearity implied by the theory, characterized by a slope discontinuity at a threshold, is obtained for both political participation and migration.

Chay and Munshi construct a statistical estimator that allows them to formally test whether the data-generating process underlying a particular outcome is consistent with the theory. Based on this test they verify that both relationships reported in Figure 23.1 are consistent with the theory. In addition, they show formally that the specific nonlinearity implied by their theory of network formation is also obtained for the following outcomes: (i) the election of black leaders during Reconstruction, which complements the pattern of voting and which would not be obtained if the results were driven by white Republican votes, (ii) church congregation size in black denominations, which is the most direct available measure of network size, and (iii) the clustering of black migrants in northern destination cities. In contrast, this nonlinearity is not obtained for (i) Republican votes after Reconstruction when blacks were effectively disfranchised, (ii) black migration prior to 1916, (iii) white migration, and (iv) church congregation size in non-black denominations.

No single alternative can explain the specific nonlinear relationship between population connectedness and outcomes associated with underlying networks, obtained for blacks alone at particular points in time. The nonlinear relationship that is obtained for black church congregation size and the clustering of black migrants in northern destinations, in particular, provide direct support for the hypothesis that blacks were able to work together to achieve common objectives in counties where population connectedness exceeded a threshold. If black migration decisions were based on factors that did not include a coordination externality, then the probability of moving to the same destination would not track migration levels so closely. Once again, the (implied) magnitude of the estimated network effect is large; for example, over half of the migrants to the North came from the third of Southern blacks who lived in the most connected counties, while less than 10% came from the third in the least-connected counties.

Empirical analyses of migrant networks have utilized both strategies described above. A series of papers on Mexican migration to the United States use historical migration patterns, which in turn were determined by initial railroad placement, to predict current migration (Woodruff and Zenteno 2007; McKenzie and Rapoport 2007, 2012). Beaman (2012) uses variation in the placement of refugees across locations in the United States to estimate network effects. It is possible that railroads were not placed randomly and that historical migration patterns are associated with unobserved community characteristics that directly determine labor market outcomes in the United States today. Similarly, it is possible that refugees were assigned to cities where they had a comparative advantage, resulting in spatial clustering by national origin and, hence, a potentially spurious network effect. Although these instruments plausibly satisfy the exclusion restriction, additional support is required to credibly identify network effects. Chay and Munshi (2014) face the same problem in their analysis, since pre-determined cropping patterns could be associated with individual or county characteristics that directly determine migration decisions. Their strategy to identify a role for networks is to exploit additional predictions from the theory; specifically, the non linear relationship between black population connectedness and migration when networks are active. In general, what is needed are additional theoretical predictions on variation in the network effect over cohorts of migrants, as in Munshi (2003), or across subpopulations, as in Chay and Munshi (2014). The studies listed above do derive and test such restrictions, providing credible evidence that networks are active in different contexts.

23.4 Migrant Networks and Inequality

The discussion up to this point has focused on the role of migrant networks in improving the economic outcomes of their members. If these networks are effective, they will naturally have consequences for inequality; across communities with networks of varying strength, as well as within communities depending on who migrates.

Migrant selection is the subject of a large and active literature in international migration. The starting point for this literature is Borjas (1987), who extends the Roy (1951) model to derive predictions for selection by education (or ability) that depends on wage inequality in the sending and receiving country. If wages are increasing more (less) steeply in education in the sending country, then there will be negative (positive) selection on education. As summarized in Abramitzky, Boustan, and Eriksson (2012), empirical tests from across the world find mixed support for this prediction. Moreover, the nature of migrant selection from Mexico to the United States, a topic of great policy interest that has received much research attention, remains unresolved (Chiquar and Hanson 2005; Cuecuecha 2005; Orrenius and Zavodny 2005; Mishra 2007; Ibarran and Lubotsky 2007; Fernandez-Huertas Moraga 2011, 2013).

One way to resolve this apparent ambiguity is to allow migration costs to vary with education. For example, Chiquar and Hanson document positive selection on education for migrants from Mexico to the United States despite the fact that wages are increasing more steeply with education in Mexico. Their response is to augment the Borjas model by allowing moving costs to be decreasing in education. This is evidently unsatisfactory; once moving costs are allowed to vary with education, any result can be explained. In addition, the augmented Borjas model cannot explain the dynamics of selection from Mexican origin communities. As documented in McKenzie and Rapoport (2007), while there is positive selection initially in their sample of communities, this is replaced by negative selection over time. As I show below, all the results in the literature can be easily reconciled by adding networks to the Roy model.

For ease of exposition, suppose that there are two education levels: low (L) and high (H). Less educated workers are channeled into low-skill occupations, while more educated workers are channeled into high-skill occupations. Without networks, the wages at the origin (O) and the destination (D) for the two types of workers are denoted by W_e^O, W_e^D, $e \in \{L,H\}$, respectively. A worker will choose to migrate if $W_e^D - c \geq W_e^O$, where the distribution of the moving cost, c, is independent of education and is characterized by the function $F(c)$. Because we have assumed that the distribution of moving costs is the same for both types of workers, there will be positive selection if $W_H^D - W_H^O > W_L^D - W_L^O$; i.e. if $W_H^D - W_L^D > W_H^O - W_L^O$, and negative selection if the sign is reversed, as in the Borjas model.

A natural way to introduce migrant networks into this simple model is to allow them to increase wages at the destination. As noted earlier, labor market networks tend to be concentrated in blue-collar occupations. This is because educational credentials are a good indicator of competence in white-collar occupations, but not necessarily so in blue-collar occupations. Moreover, production tends to take place in teams in these occupations, making it difficult for the firm to attribute effort or competence to individuals on the job. Networks of socially connected workers can overcome both the information and the enforcement problems that arise with team production. This is incorporated in the theoretical framework by allowing the low-skill wage at the destination, W_L^D, to be increasing in the size of the migrant network, which, in turn, is increasing over time.

Suppose that $W_H^D - W_H^O > W_L^D - W_L^O$ to begin with, before migrant networks have had a chance to form. This implies that there will be positive migrant selection out of all communities. However, if a migrant network does subsequently form in a given community, the right-hand side of the preceding inequality will increase over time, possibly resulting in a switch in the sign of the inequality. As documented by McKenzie and Rapoport, there will be positive migrant selection and an increase in inequality within the community in the early stages, followed by negative selection and a decline in inequality once the destination network exceeds a threshold size. Note, however, that this dynamic pattern will not be obtained in all communities. Recall from Chay and Munshi (2014) that migrant networks will only form when population connectedness exceeds a threshold. Migrant networks will thus increase inequality across communities, and this inequality will increase over time as migrant networks strengthen in some communities but not others. Moreover, we could obtain positive or negative selection on average in a sample of communities at a given point in time, depending on their population characteristics and the stage of development of their migrant networks. Our Roy model with migrant networks is thus easily able to generate the dynamic and the cross-sectional patterns of migrant selection that have been documented in the literature.

McKenzie and Rapoport (2007, 2012) also add networks to the Roy model to generate the observed patterns of selection. However, they choose to embed the network effect in the moving cost, following past work by Carrington, Detragiache, and Vishwanath (1996). One limitation of this approach is that it is at odds with the functions that migrant networks perform. These networks primarily solve labor and credit market imperfections at the *destination*, once the migrant has arrived. While the community may support the movement of migrants, by providing credit or insurance, these services will be provided by networks situated at the *origin*. If the empirical analysis exploits exogenous variation in community characteristics, as in Chay and Munshi, this distinction is less relevant; both origin and destination networks will form when population connectedness exceeds a threshold level. However, if the destination network is measured directly, then there will be a disconnect between a theory based on moving costs and the empirical analysis. A second limitation of the moving-cost approach is that its predictions—for example, as in McKenzie and Rapoport (2012)—are sensitive to a number of auxiliary assumptions, such as the distribution of schooling in the population. In contrast, the observed patterns of migrant selection are obtained with minimal and defensible assumptions when the migrant network serves more naturally to improve individual outcomes at the destination.

23.5 The Intergenerational Dynamics of Migration

The discussion up to this point has focused on migrant networks at the destination. However, community networks at the origin will also shape migration. On the one hand, networks at the origin can support migration by financing the move. On the other hand, mutual insurance networks at the origin can restrict the movement of individuals. If an individual moves on his own from the village to the city, he always has an incentive to under-report his income realizations as a way of getting transfers to flow in his direction. Even if this information problem is resolved, he will still have an incentive to renege on his obligation to make transfers to less fortunate community members far away in the village. Munshi and Rosenzweig (2015) show theoretically and empirically that these information and commitment problems can restrict individual migration when formal insurance at the destination is unavailable. These restrictions will be paradoxically amplified when the rural insurance network is well functioning because individual movers have more to lose.

The preceding discussion highlights the important role that origin-based networks can play in supporting or restricting migration. Losing access to the origin network is yet another reason why migrants tend to move in groups, since the presence of a group at the destination will solve the information and enforcement problems discussed above. In some contexts, as with mutual insurance, the origin network will discourage individuals from moving, but may encourage migration by groups of individuals because this diversifies income risk. In other contexts, the origin network will discourage any migration. This will be the case, as in farming communities, when the performance of the origin network depends on the number of members who are *based* at the origin. When we extend our analysis of destination-based community networks past one generation, an analogous argument applies. A network that has been established at a particular destination will discourage the next generation from moving elsewhere or into new occupations if its performance depends on its size.

Munshi and Rosenzweig (2006) show theoretically that the restrictions on mobility described above can be welfare enhancing when they are established, but can result in a dynamic inefficiency in the long run. To see why this is the case, return to our setup with two types of occupations, skilled and unskilled, and two levels of associated education, high (H) and low (L). Consider a migrant network that has already been established at a particular destination, so moving costs can be ignored. Individual heterogeneity is now in ability, which determines the cost of education. It costs $\overline{C}_e$ for low-ability individuals to attain high education, whereas the corresponding cost for high-ability individuals is $\underline{C}_e < \overline{C}_e$. We normalize so that the cost of attaining low education is zero.

As above, the unskilled wage, W_L, is increasing in the size of the network. We assume that members of the network in a given generation provide referrals to the next generation. Depending on the size of the network in the previous generation, the following conditions are satisfied:

$$\text{C1.}\quad W_H - \overline{C}_e > W_L(0)$$

$$\text{C2.}\quad W_H - \underline{C}_e < W_L(N).$$

The first condition says that if a network was not active in the previous generation, then individuals of both types would invest in high education and end up in skilled jobs. The second condition says that if everyone in the community selected into the network in the previous generation (N is the size of the community) then individuals of both types would select low education and end up in the unskilled occupation.

Suppose that the ability distribution is the same across all communities, but communities are exogenously assigned to one equilibrium or the other in the initial period. If conditions C1 and C2 are satisfied, it follows that communities will stay in the initial equilibrium from one generation to the next, with everyone either investing or not investing in education. Now suppose that the high-skill wage starts to increase. When condition C2 just binds, high-ability individuals will be indifferent between investing and not investing in education in those communities where everyone has traditionally selected into the low-skill occupation. If the high-skill wage increases marginally above that level, high-ability individuals will deviate from the traditional equilibrium and invest in education. Overall welfare in the community will decline in this case because the low-ability individuals who remain in the (smaller) network now have a substantially lower wage. This is one reason why blue-collar communities and farming communities, where traditional occupations require a high degree of networking, are often characterized by cultures that discourage occupational and spatial mobility (Elder and Conger 2000; Gans 1962; Kornblum 1974).

While a culture that restricts mobility may have been welfare-enhancing when it was put in place, its persistence can result in a dynamic inefficiency if the skilled wage increases sufficiently. Culture and social norms are very persistent, explaining the common perception that farming and blue-collar communities often stubbornly resist change. This provides a new intergenerational perspective on mobility, since networks in these communities would have historically supported the occupational and spatial mobility of their members. Today, they hold them back. A complete characterization of the relationship between community networks and migration requires attention to networks at the origin and the destination, as well as the dynamic process through which migrant networks form, become established, and then serve as the point of departure for further migration in subsequent generations.

23.6 Conclusion

This chapter describes the role that community networks have played, and continue to play, in facilitating migration. Establishing that these networks improve the outcomes of their members is a challenging statistical problem. Exogenous variation in either network or community characteristics, together with restrictions from the theory, can be used to credibly identify network effects.

Networks are characterized by their size and connectedness (broadly defined) in this chapter. This contrasts with the focus on more detailed measures, often revolving around the centrality of members, in the rapidly growing literature on networks in economics. For example, Nava (2015) surveys an emerging theoretical literature on repeated games and networks that is concerned with the relationship between network structure and equilibrium payoffs in environments where commitment problems are relevant. Similarly, Chaney (2015) surveys the theoretical literature on information diffusion in trade networks in which the structure of these networks determines the extent to which informational frictions can be reduced. One possible direction for future research on migration networks would be to incorporate a more detailed network structure. However, it is important to exercise caution when venturing down this path. There is no powerful empirical fact that lends credence to the idea that these more detailed network structures will add substantially to our understanding of the process of migration. Moreover, credible inference with these detailed network structures is likely to be even more challenging than it is with the simpler structures, based on size and connectedness.

A possibly more fruitful direction for future research would be to more completely characterize the process of network formation and development, as it unfolds over multiple generations within a community, with new networks sometimes breaking away and existing networks dissolving. The theoretical challenges and data requirements for such analyses are substantial. However, in this case, the payoffs are more visible; it is easy to see how this analysis would complement existing models of growth and shed new light on the dynamics of inequality.

References

Abramitzky, Ran, Leah Platt Boustan, and Katherine Eriksson (2012). "Europe's tired, poor, huddled masses: Self-selection and economic outcomes in the age of mass migration." *American Economic Review* 102(5), 1832–1856.

Abramitzky, Ran, Leah Platt Boustan, and Katherine Eriksson (2014). "A nation of immigrants: Assimilation and economic outcomes in the age of mass migration." *Journal of Political Economy*, forthcoming.

Beaman, Lori A. (2008). "Social networks and the dynamics of labor market outcomes: Evidence from refugees resettled in the U.S." Northwestern University, Department of Economics, typescript.

Beine, Michel, Frederic Docquier, and Caglar Ozden (2011). "Diasporas." *Journal of Development Economics* 95, 30–41.

Bertoli, Simone and Jesus Fernandez-Huertas Moraga (2012). "Visa policies, networks and the cliff at the border," IZA Discussion Papers 7094, Institute for the Study of Labor (IZA).

Bodnar, John (1985). *The Transplanted: A History of Immigrants in Urban America*. Bloomington: Indiana University Press.

Bodnar, John, Roger Simon, and Michael P. Weber (1982). *Lives of Their Own: Blacks, Italians, and Poles in Pittsburgh, 1900–1960*. Urbana: University of Illinois Press.

Borjas, George (1987). "Self-selection and the earnings of immigrants." *American Economic Review* 77(4), 531–553.

Boucher, Vincent and Bernard Fortin (2015). "Some challenges in the empirics of the effects of networks," in Jan Bramoulle, Andrea Galleotti, and Brian W. Rogers, eds. *Oxford Handbook in the Economics of Networks*. Oxford: Oxford University Press.

Burnett-Hurst, A.R. (1925). *Labour and Housing in Bombay: A Study in the Economic Conditions of the Wage-Earning Classes of Bombay*. London: P.S. King and Son.

Carrington, William J., Enrica Detriagiache, and Tara Vishwanath (1996). "Migration with endogenous moving costs." *American Economic Review* 86(4), 909–930.

Chandavarkar, Rajnarayan (1994). *The Origins of Industrial Capitalism in India: Business Strategies and the Working Classes in Bombay, 1900–1940*. Cambridge: Cambridge University Press.

Chaney, Thomas (2015). "Networks in International Trade." in Jan Bramoulle, Andrea Galleotti, and Brian W. Rogers, eds. *Oxford Handbook in the Economics of Networks*. Oxford: Oxford University Press.

Chavez, Leo (1992). *Shadowed Lives: Undocumented Immigrants in American Society*. Harcourt Brace Jovanovich College Publishers.

Chay, Kenneth and Kaivan Munshi (2014). "Black networks after emancipation: Evidence from Reconstruction and the Great Migration," University of Cambridge mimeo.

Cholia, R. P. (1941). Dock Labourers in Bombay. Bombay.

Chiquiar, Daniel and Gordon H. Hanson (2005). "International migration, self-selection, and the distribution of wages: Evidence from mexico and the united states." *Journal of Political Economy* 113(2), 239–281.

Conzen, Kathleen Neils (1976). *Immigrant Milwaukee 1836–1860: Accommodation and Community in a Frontier City*. Cambridge, MA: Harvard University Press.

Cuecuecha, Alfredo (2005). "The immigration of educated Mexicans: The role of informal social insurance and migration costs," ITAM mimeograph.

Docquier, Frederic, Giovanni Peri, and Ilse Ruyssen (2014). "The cross-country determinants of potential and actual migration." *International Migration Review*, forthcoming.

Elder Jr., Glen H. and Rand D. Conger (2000). *Children of the Land: Adversity and Success in Rural America*. Chicago and London: University of Chicago Press.

Fernandez-Huertas Moraga, Jesus (2011). "New evidence on emigrant selection." *Review of Economics and Statistics* 93(1), 72–96.

Fernandez-Huertas Moraga, Jesus (2013). "Understanding different migrant selection patterns in rural and urban Mexico." *Journal of Development Economics* 103, 182–201.

Gans, Herbert (1962). *The Urban Villagers: Group and Class in the Life of Italian-Americans*. New York: The Free Press of Glencoe.

Gjerde, Jon (1985). *From Peasants to Farmers: The Migration from Balestrand, Norway to the Upper Middle West*. Cambridge, England: Cambridge University Press.

Gokhale, R. G. (1957). *The Bombay Cotton Mill Worker*. Bombay.

Gordon, David M., Richard Edwards, and Michael Reich (1982). *Segmented Work, Divided Workers: The Historical Transformation of Labor in the United States*. Cambridge England: Cambridge University Press.

Gottlieb, Peter (1991). "Rethinking the Great Migration," in Joe William Trotter Jr. (Ed.), *The Great Migration in Historical Perspective: New Dimensions of Race, Class, and Gender*. Bloomington: Indiana University Press.

Grossman, James (1989). *Land of Hope: Chicago, Black Southerners, and the Great Migration*. Chicago: The University of Chicago Press.

Hagan, Jacqueline Maria. *Deciding to be Legal: A Maya Community in Houston*. Philadelphia: Temple University Press, 1994.

Hoerder, Dirk (1991). "International labor markets and community building by migrant workers in the Atlantic economies," in *A Century of European Migrants*, 1830–1930, eds. Rudolph J. Vecoli and Suzanne M. Sinke, Urbana: University of Illinois Press.

Hutchinson, E. P. (1956). *Immigrants and Their Children, 1850–1950*. New York: John Wiley and Sons, Inc.

Kamphoefner, Walter D. (1987). *The Westfalians: From Germany to Missouri.* Princeton, NJ: Princeton University Press.

Ibarran, Pablo and Darren Lubotsky (2007). "Mexican Immigration and self-selection: New evidence from the 2000 Mexican Census." in George Borjas (Ed.), *Mexican Immigration in the United States*, Chicago: University of Chicago Press.

Kornblum, William (1974). *Blue Collar Community.* Chicago and London: University of Chicago Press.

Mandle, Jay R. (1978). *The Roots of Black Poverty: The Southern Plantation Economy After the Civil War.* Durham, NC: Duke University Press.

Marks, Carole (1983). "Lines of Communication, Recruitment Mechanisms, and the Great Migration of 1916–1918." *Social Problems* 31(1), 73–83.

Massy, Douglas, Rafael Alarcon, Jorge Durand, and Humberto Gonzalez (1987). *Return to Aztlan: The Social Process of International Migration from Western Mexico.* Berkeley: University of California Press.

McKenzie, David and Hillel Rapoport (2007). "Network effects and the dynamics of migration and inequality: Theory and evidence from Mexico." *Journal of Development Economics* 84, 1–24.

McKenzie, David and Hillel Rapoport (2012). "Self-Selection patterns in Mexico-U.S. migration: The role of migration networks." *Review of Economics and Statistics* 92(4), 811–821.

Menjivar, Cecilia (2000). *Fragmented Ties: Salvadoran Immigrant Networks in America.* Berkeley: University of California Press.

Mishra, Prachi (2007). "Emigration and wages in source countries: Evidence from Mexico." *Journal of Development Economics* 82(1), 180–199.

Morris, Morris David (1965). *The Emergence of an Industrial Labor Force in India: A Study of the Bombay Cotton Mills, 1854–1947.* Berkeley: University of California Press.

Morrison, Minion K.C. (1987). *Black Political Mobilization: Leadership, Power, and Mass Behavior.* Albany: State University of New York Press.

Moorjani, Priya, Kumarasamy Thangaraj, Nick Patterson, Mark Lipson, Po-Ru Loh, Periyasamy Govindaraj, Bonnie Berger, David Reich, and Laiji Singh (2013). "Genetic evidence for recent population mixture in India." *American Journal of Human Genetics* 93(3), 422–438.

Munshi, Kaivan (2003). "Networks in the modern economy: Mexican migrants in the U.S. labor market." *Quarterly Journal of Economics* 118(2), 549–597.

Munshi, Kaivan and Mark Rosenzweig (2006). Traditional institutions meet the modern world: Caste, gender and schooling choice in a globalizing economy." *American Economic Review* 96(4), 1225–1252.

Munshi, Kaivan and Mark Rosenzweig (2015). "Networks and misallocation: Insurance, migration, and the rural-urban wage gap." *American Economic Review*, forthcoming.

Nava, Francesco (2015). "Repeated games and networks." In Jan Bramoulle, Andrea Galleotti, and Brian W. Rogers, eds. *Oxford Handbook in the Economics of Networks.* Oxford: Oxford University Press.

Nee, Victor G. and Brett de Bary Nee (1972). *Longtime Californ': A Documentary Study of an American Chinatown.* New York: Pantheon Books.

Orrenius, Pia M. and Madeline Zavodny (2005). "Self-Selection among undocumented immigrants from mexico." *Journal of Development Economics* 78(1), 215–240.

Patel, Krishna and Francis Vella (2013). "Immigrant networks and their implications for occupational choice and wages." *Review of Economics and Statistics* 95(4), 1249–1277.

Patel, Kunj (1963). *Rural Labor in Industrial Bombay*. Bombay: Popular Prakashan.
Rees, Albert (1966). "Information networks in labor markets." *American Economic Review (Papers and Proceedings)* 56, 559–66.
Roy, A. D. (1951). "Some thoughts on the distribution of earnings." *Oxford Economic Papers* 3(2), 135–146.
Tolnay, Stewart E. and E. M. Beck (1990). "Lethal violence and the Great Migration, 1900–1930." *Social Science History* 14(3), 347–370.
Woodruff, Christopher and Renee Zenteno (2007). "Migration networks and microenterprises in Mexico." *Journal of Development Economics* 82(2), 509–528.
Wright, Gavin (1986). *Old South, New South: Revolutions in the Southern Economy Since the Civil War*, New York: Basic Books, Inc.
Zhou, Min (1992). *Chinatown: The Socioeconomic Potential of an Urban Enclave*. Philadelphia: Temple University Press, 1992.

CHAPTER 24

SOCIAL NETWORKS AND THE LABOR MARKET

LORI BEAMAN

24.1 Introduction

The importance of social networks in the labor market has long been emphasized by sociologists, going back to early work by Rees (1966). A large fraction of jobs—up to 50%—are attained through informal channels, including employee referrals (Ioannides and Loury 2004). This figure is strikingly consistent across different contexts, from the United States to Europe to even developing countries. In a small sample in Kolkata, 40% of employees reported that they helped a friend or relative get a job with their current employer (Beaman and Magruder 2012). Among Bangladeshi garment factory workers, 32% reported receiving a referral for their current job, half of which come from people living in the same extended family compound (Heath 2014).

This chapter discusses recent theoretical and empirical work seeking to understand why social networks seem to play such an important role in the labor market, and the consequences of the use of social networks in the labor market. Section 24.2 starts by elaborating three main reasons networks may be important: search frictions, imperfect information about worker productivity or the match quality of the worker with the firm, and peer effects in the workplace. A number of robust predictions come out of this theoretical work, but there remain a number of open questions about modelling assumptions, and some distinct models generate similar empirical predictions.

Recent advances in the empirical literature are discussed in Section 24.3, starting with a brief discussion on evidence demonstrating a causal relationship between the size or quality of a worker's social network and his labor market outcomes. The majority of the section focuses on empirical work which attempts to provide insights into the underlying theoretical models. The literature uses a broad range of empirical techniques, data sources, and empirical contexts, including data from developed and developing countries. Nevertheless, the literature still faces key limitations, including limited data on firm practices and the ubiquitous problem of unobserved characteristics. Although there is robust evidence that many workers find their jobs through informal channels, the data is often silent on the details of how social connections are used. The models described in Section 24.2 will highlight how these details may be key to uncovering the underlying model of why workers and firms are using informal search, and isolating the underlying model is important for understanding the broader economic implications of network-based search.

The primary models in this literature focus on the potential for social networks to address underlying labor market frictions. Therefore, the labor market becomes more efficient thanks to the role of social networks, at least in most models. The chapter also discusses a few potential costs to using social networks to address these market inefficiencies in Section 24.4. Individuals in groups with initially disadvantaged labor market outcomes, such as blacks or other disadvantaged groups, may be worse off when firms rely on referrals (Calvo-Armengol and Jackson 2004). Social networks may also prevent their network members from taking advantage of new opportunities as the economy changes and evolves (Munshi and Rosenzweig 2006). The availability of private information within social networks can also generate negative externalities to other firms, leading to an inefficient unraveling of the labor market (Fainmesser 2013). Section 24.5 highlights some key challenges and potential ways forward in the literature.

24.2 Role of Networks in the Labor Market

In the literature there are three broad categories of reasons why workers and firms may use informal social networks as part of labor market search and recruitment. The first comes out of the large literature on labor market search. The existence of search frictions, for example in Mortensen and Pissarides (1994), may mean that social networks serve as a low-cost way of disseminating information about a job. Workers' social networks may be useful for passing on information about job vacancies (Calvo-Armengol 2004), or through direct referrals to employers. The second arises from the idea that existing employees may have useful information about either the match quality or unobserved characteristics of applicants, giving rise to a role for referrals in particular. The seminal work in this area is by Montgomery (1991), with related models about firms learning about match quality by Dustmann et al. (2014); Simon and Warner (1992) and Galenianos (2013). Finally, firms may also benefit from social interactions in the workplace through some form of peer effects. This may arise from on-the-job monitoring (Kugler 2003) or through joint team production (Bandiera et al. 2013). I describe each broad class of model in turn below.

24.2.1 Search Frictions

The models in Calvo-Armengol (2004) and Calvo-Armengol and Jackson (2004) allow job information to arrive to workers through two primary channels: (i) there is a random probability of hearing about a new job, which can be a function of that worker's current employment status, depending on the variant of the model; and (ii) information can be passed along from someone in the worker's social network.[1] In both papers, information passing within the social network generates a positive correlation in employment patterns and wages across time and people. It can also predict that workers who have been unemployed for a long time have a lower probability of becoming employed, since in this situation it is likely that unemployed agents' networks have a low average employment rate.

The setup of Calvo-Armengol and Jackson (2004) does not explicitly model firms. There is a growing literature that incorporates firms and looks at equilibrium outcomes in a search framework, but in contrast to Mortensen and Pissarides (1994), job information originates from two courses as in Calvo-Armengol and Jackson (2004). In Mortensen and Vishwanath (1994), workers are equally productive but there is an uneven distribution of wages based on how a particular worker receives his job information. In the equilibrium wage distribution, workers with more employed contacts earn more. Both Calvo-Armengol and Zenou (2005) and Ioannides and Soetevent (2006) also use a search framework to look at equilibrium outcomes and model more explicitly social network structure. Calvo-Armengol and Zenou (2005) show that an increase in network size increases the job match rate. However, the relationship is nonmonotonic: after a critical value, job matches decrease with network size. The intuition for this result is that in larger networks, unemployed network members have to compete more with other unemployment members—since overall a higher fraction of jobs are found through the network channel. While a larger network has the advantage of pulling in more job information, the congestion effect eventually dominates and suppresses the job finding rate.[2] The model also allows coordination failures within the network, so that the social network-based job dissemination process need not be socially efficient. Building on Calvo-Armengol and Zenou (2005), Ioannides and Soetevent (2006) represent social connections as a random graph where there is a distribution in the number of contacts (degree) that individuals have, instead of everyone having the same network size. The network is exogenously determined in both cases, an issue we will return to in Section 24.2.3. This model shows that workers with a larger number of contacts will have lower unemployment rates and

[1] Boorman (1975) is early theoretical work allowing for workers to receive information through family and friends.

[2] This congestion effect is also highlighted in Calvo-Armengol (2004). When network formation is endogenous, networks can become inefficiently large since individual network members do not internalize the externality they create when adding links. The net advantage of direct and indirect links will also depend on the specific network topology that arises from the network formation game.

receive higher wages. In contrast to Calvo-Armengol and Zenou (2005), however, they find no evidence of the nonmonotonicity in network density when using a Poisson degree distribution in particular. Both Wabha and Zenou (2005) and Beaman (2012) provide empirical support for a nonmonotonic relationship between network size and employment outcomes, using data from urban Egypt and refugees resettled in the United States, respectively. Therefore the sensitivity of the theoretical prediction to different assumptions about the underlying network formation process is an area which needs to be further explored.

Galenianos (2014) also uses an equilibrium search model, where workers are homogenous in terms of productivity and their network. Firms and workers meet in the frictional labor market or through a referral. There is a worker-firm match and wages are determined through Nash bargaining. The model predicts that the use of referrals generates higher aggregate matching efficiency. When the formal market is more efficient, however, referrals will be used less. The model helps explain differences across industries in aggregate matching efficiency, since there are significant differences in the use of referrals across industries. Fontaine (2008) also models wages which are bargained in a labor market with search frictions, and the paper helps to explain why workers with very similar observable characteristics receive different wages. The use of social networks in the labor markets generates unequal arrival rates among otherwise similar workers, as workers' outcomes depend on the employment rate within their network in any given period. Heterogeneity in the job arrival rate combined with endogenous wage setting leads to wage dispersion, as individuals with more offers have stronger outside options and can bargain higher wages. This mechanism is thus an additional explanation for equilibrium wage dispersion, even without heterogeneity in worker productivity.[3]

24.2.1.1 Sensitivity of Prediction on Wages

While these models pretty consistently predict that a larger network will increase a worker's employment rate and a better aggregate matching rate, at least up to a certain threshold in network size, the predictions on wages are very sensitive to particular modelling assumptions. In Mortensen and Vishwanath (1994), jobs acquired through the formal channel are draws from the full wage offer distribution. Job information from the network, however, is drawn from the distribution of wages held by network members. Since the accepted wage distribution will always stochastically dominate the offer distribution, jobs acquired through the network will have higher wages than jobs attained through formal search. This modeling choice may be intuitive if the source of job information is primarily network members' own firms, and there is heterogeneity in the productivity of firms for example. By contrast, Beaman (2012) uses the same basic set up as in Calvo-Armengol and Jackson (2004) to look at how cohorts entering the labor market are affected by the size of their network. In this model, wage offers coming

[3] Note that in order to generate persistent differences across groups, one needs to endogenize search intensity as in Calvo-Armengol and Jackson (2004). In Fontaine (2008), all networks are converging to the aggregate unemployment rate over time.

from network members are offers that were passed along—and therefore rejected—by those who originally received the offer. The distribution of wages through the network channel is therefore dominated by the offer wage distribution. On net, the prediction is ambiguous on the effect of a larger network on wages. A larger network means more offers, where low offers can be declined, but network jobs are associated with lower wages than offers received through formal search. Both Calvo-Armengol and Zenou (2005) and Ioannides and Soetevent (2006) preclude the possibility that the wage distribution of jobs within the network is worse than the original offer distribution by assuming another alternative wage structure where an employed worker would never gain from directly using job information received about another position. The assumption is that after a one-period probationary period, workers who keep their job receive a higher wage, reflecting firm-specific human capital. All of these assumptions are reasonable and can be defended, but they have important implications for what we will observe in the data—even within this same class of model. It will be hard to use cross-sectional differences in wages to disentangle different models, when even within one type of model you can generate any empirical pattern.

24.2.2 Information

The seminal paper by Montgomery (1991) dramatically increased attention on the role of social networks in the labor market among economists, though there had long been interest in this topic among sociologists. Montgomery (1991) puts forth a model where there is asymmetric information about job applicants, and firms have a hard time screening candidates. There are two periods in the model, and workers are each in the labor market for only one period. Their type is revealed to the firm once they start work, but there is no scope for learning and adjusting wages since workers are employed for only one period. Firms pay a wage based on the expected productivity of workers in the market. The model assumes homophily within social networks: high-ability workers are more likely to have high-ability friends. Firms know the ability of their incumbent workers, and in equilibrium will only ask high-ability workers to make referrals. If they have no high-ability workers, they will recruit from the market directly. The model predicts that unemployed workers who exogenously have more high-ability friends will get more offers (through their contacts plus possibly the market) and end up with higher wages, since they can select the highest wage out of all offers they receive. The equilibrium wage distribution will therefore exhibit a positive correlation between wages and ability, but the correlation is driven by the homophily in workers' social networks. The stronger the homophily—i.e., if networks are more homogenous in ability—the worse off unconnected workers will be. This is because firms will assume the pool of individuals who do not get hired through referrals are low quality, offering low wages to reflect their expected productivity.

The homophily assumption in the original work by Montgomery (1991) is strong. The assumption can be weakened, and the model predictions hold as long as high-ability employees have information about who is also high ability in their network and they have proper incentives to reveal that information to their employers. As highlighted in Beaman and Magruder (2012), workers also face incentives generated within their social network. They may want to help out friends or family in need, or assist individuals who would do them favors in the future—especially if one's labor market network overlies with the social network used for credit. A worker's reputation with the firm may be sufficient to induce the worker to refer only good-quality workers. In Section 24.3, I describe recent papers using firm data, which show that many firms also provide explicit bonus payments if referred workers stay on the job for at least a certain amount of time.[4] These bonuses, if properly designed, can also provide sufficient incentive for employees to refer only good-quality candidates to their employers.

The Montgomery model is about screening candidates on absolute ability; firms also may be looking to maximize match quality. There are a number of papers in the literature which look at match quality in a general equilibrium search framework, and are therefore very related to the search friction models discussed in Section 24.2.1. This includes Simon and Warner (1992), Dustmann et al. (2014), and Galenianos (2013). In these models, referred workers need not be better on average than nonreferred workers, as in the case of Montgomery (1991). I view these models as distinct from the search friction models in Section 24.2.1 because these models assume that social network members provide an information advantage to fellow members and/or to their employers.

In Galenianos (2013), when a firm is looking to hire a new worker, the firm and the worker receive a signal of the match quality. Expected output of a match depends on the firm's own productivity and also the match quality. There is a more accurate signal of the match quality of referred workers than workers met through the formal labor market. The model then predicts that referred workers are more likely to be hired, and their wages will be higher due to the better expected match quality (conditional on being hired). Referred workers also have higher productivity and lower separation rates. These differentials decline over time, though, as nonreferred workers with low match quality leave the firm. The predictions of the model in Dustmann et al. (2014) are similar: referred workers will have higher wages and are less likely to quit than nonreferred workers. The wage premium also falls over time. They use this model to make specific

[4] Fafchamps and Moradi (2015) provide empirical evidence consistent with the notion that incumbent workers may bring in low-quality candidates in order to get the bonus payment, which was about one week's worth of wages in their context. Related work by Zinovyeva and Bagues (2015) also highlights the tradeoff between efficiency and bias that comes with the use of social connections in the labor market. Their context is promotions of professors in Italian academia, where evaluators are randomly assigned to candidates. They find that candidates who get an evaluator who is a strong connection are more likely to be promoted. These candidates have worse observable characteristics but perform worse in the years following promotion than promoted candidates who were evaluated by noncontacts. This suggests that evaluators' bias is responsible for the promotion. Candidates evaluated by weak connections are also more likely to be promoted despite worse observable characteristics, but they are similarly productive going forward as other promoted professors. This work highlights that social contacts may hold better information about candidates, but relying on information from social contacts also risks bringing in bias into the labor market.

predictions on hiring patterns for ethnic minorities and show that a firm is more likely to hire an applicant from a given minority when there are a higher fraction of existing workers from that minority employed at the firm.

24.2.3 Importance of Heterogeneity

24.2.3.1 Endogenous Network Formation

Most models in the literature assume that a worker's network, particularly its size, is determined exogenously. Calvo-Armengol (2004) is an exception, as it models strategic network formation when personal contacts can generate job information. This results in different possible network structures, and aggregate unemployment depends on what the entire network structure looks like—not just the total number of links. This arises even though individuals in the Calvo-Armengol (2004) model are all ex ante identical. We should further consider that there are multiple dimensions of worker heterogeneity which may be important. Some groups of individuals, say unskilled workers, may use family and friends for job search more intensively than others (Ioannides and Loury 2004). Galeotti and Merlino (2014) uses an equilibrium search model and allows workers to choose investments in their network, thereby endogenizing the use of informal search. The returns to investing contacts vary with the business cycle: when the job separation rate is low, workers do not need to invest since they are unlikely to need contacts to find another job. When the separation rate is high, there is a lot of competition within social networks and an additional contact is not very likely to generate a new job offer. The model then predicts an inverted U-shape relationship between bad economic times and network investments. This generates accordingly an inverted U-shape relationship between the job separation rate and the likelihood a worker gets a job through his network. An important implication of the model is that if unskilled workers and minorities use their networks more for search, the group that relies more on network-based search will be disproportionately affected by negative economic shocks.

Another example is Bramoullé and Saint-Paul (2010), in which agents form connections in their network over time and those connections display significant homophily—in this case, in terms of employment status. The probability that a link is created is higher among two employed individuals than among one employed and one unemployed worker. This type of homophily generates a correlation in employment outcomes over time: those who are unemployed will be less likely to find employment since they will also be gaining very few useful network links over time. This is consistent with the negative duration dependence of exit rates from unemployement observed in many datasets, and the endogenity of the network in the model generates a stronger duration dependence than in Calvo-Armengol and Jackson (2004), who assume a fixed network (though they allow for dropping out of the labor market). The model nevertheless suggests that even for the unemployed, an increase in the number of contacts is beneficial for labor market outcomes.

24.2.3.2 Heterogeneity of Workers and Firms

Heterogeneity in worker and firm productivity can also affect the model predictions even without allowing for endogenous network formation. The predictions on wages are already sensitive to the assumptions of how job information is generated and then passed. Bentolila et al. (2010) further highlights the difficulty in predicting the empirical patterns in wages. Workers are heterogeneous in an attribute which makes them well suited for some occupations but not others. They have a network of social contacts, some of which are employed in their own occupation but many of which are in different occupations. This means that search through contacts may generate a worse quality match, because a job offer received through a friend may require changing to a less suitable occupation. Here, social contacts are more useful when the labor market is tight.[5] This model generates the prediction that workers who found their job through informal search would have shorter unemployment spells but would have lower wages than those who used formal search, since the latter are more likely to have a good occupational match.

A second example of how the introduction of heterogeneity can alter our theoretical predictions is the model by Galenianos (2013) already described above. When heterogeneity in firm productivity is added into the model, Galenianos then finds that higher productivity firms will find it cost effective to invest in screening technologies to get a more accurate signal of match quality, and thus use referrals less often. This can weaken or even reverse the empirical prediction that referred workers would earn higher wages than nonreferred workers. Data on firm productivity is therefore needed to test the prediction of the model that the higher match quality associated with referred workers leads to higher wages. With data on firm productivity, the econometrician could make sure that only referred and nonreferred workers from firms with similar productivity are compared. Unfortunately, good measures of firm productivity may be very challenging to find.

To resume, most of the models discussed thus far predict that referred workers should be more likely to get hired by a firm and have shorter unemployment spells. Finding this empirical pattern for employment rates in the data will therefore not help differentiate the models. Using data on wages will also not get us much further: both the search friction class of models and the learning model can have ambiguous predictions on wages, particularly once heterogeneity is allowed for.

24.2.4 Peer Effects

The models up to now have focused on firms wanting to reduce costs or improve the inherent quality of their workers. Firms may also use existing employees to recruit

[5] This also creates a feedback mechanism since the economy may get stuck in a state where workers are not fully productive since they are in the wrong occupations.

new workers because they want to exploit positive social interactions on the job. Kugler (2003) develops a model where efficiency wages are used to induce effort, but referrals can provide on-the-job monitoring and thus lower monitoring costs. Firms can choose to hire through referrals or through formal methods, operating in an otherwise standard Pissarides (1990) setting. Firms that use employee referrals can pay lower efficiency wages than would otherwise be necessary and seek to induce high effort. Alternatively, firms that hire through formal channels are more likely to offer lower wages and accept a higher probability of shirking. Firms with larger networks have a larger pool of referrals to choose from and are more likely to use referrals. This generates an equilibrium prediction that high-paying firms will use referrals while low-paying firms will use formal search methods. This is in contrast to the prediction in Galenianos (2013), who assumed more productive firms invested in technologies to get a more accurate signal and would use referrals less often.

Heath (2014) puts forth a different moral hazard model, along with supportive empirical evidence. Her empirical context, which motivates the theory, is Bangladeshi garment workers. In this industry, there is a binding minimum wage. This limits the ability of firms to use wage increases over time in order to induce effort from workers: they cannot lower the pay enough in the first period, given the minimum wage law and workers' productivity, to provide sufficient incentives for effort. Instead, they can use the social relationships between existing employees and potential new hires to induce effort from the new worker. The firm can offer a contract where the incumbent worker—the referral provider—is punished if the new worker has low output. With this contract, the firm can hire new workers who would otherwise be unemployed, because the first-best initial wage is below the minimum wage. The most direct testable implication of the model is a positive correlated wage structure between the referred and incumbent worker pairs. Second, incumbent workers who make referrals should be more productive than workers who do not make referrals. However, both of these predictions could arise from a homophily model. There are, however, predictions from this model which diverge from many screening models. First, referred workers would have *worse* skills than nonreferred workers, since nonreferred workers had expected productivity high enough to be hired given the minimum wage. Second, the relative wages of referred over nonreferred workers should increase over time with tenure in the firm, in contrast to the prediction of learning models where wages should converge over time.

In summary, the moral hazard models of Kugler (2003) and Heath (2014) have severable testable predictions, at least two of which differs in substantive ways from learning and selection models. We will return to this in Section 24.3.

There is an additional class of models which fall under "peer effects," which is different than moral hazard. There may be complementarities in the production process when individuals who are socially connected work together. Bandiera et al. (2013) put forth a model where teams comprised of friends are less likely to free-ride. There may also be more scope for cooperation among social contacts, as in Bandiera et al. (2005).

Social connections within teams may therefore increase productivity for the firm.[6] Therefore, some firms may choose to use referrals not because referred workers are on average higher quality, but because they can take advantage of on-the-job synergies. This mechanism is likely to depend on the exact nature of the production function within the firm, or perhaps even more narrowly for some occupations within a given firm.

24.3 Empirical Evidence

24.3.1 Causal Impacts

The focus of this survey is not on *whether* social networks influence the labor market but *why* and what are the consequences. There are now a number of papers in the literature which provide solid evidence that the quantity and quality of social contacts affect workers' employment status.[7] The fundamental identification problem is that individuals who are friends usually share common unobserved (or unobservable) characteristics (Manski 1993). It is therefore difficult to disentangle whether individuals with different quality networks have different outcomes because of the network or because of unobserved characteristics that directly affect labor market outcomes and are correlated with the quality of their network. Munshi (2003) uses lagged rainfall shocks to generate exogenous variation in the size of a Mexican migrant's network within the United States. The key is that those rainfall shocks have no effect on a migrant's labor market outcome within the United States in subsequent periods. The data used, which are from the Mexican Migration Project, ask respondents retrospective questions about stays in the United States over a long time period, and this facilitates a fixed effect estimation strategy to hold constant the unobserved ability of the migrant himself. The paper establishes a causal link between network size and the probability of employment: a larger number of established migrants in the network increases the probability of employment of newly arrived migrants. There is no direct information on wages in

[6] Empirically, however, when the authors implemented a tournament team incentive scheme, workers chose to join teams without their friends. Theoretically, high-powered incentives may generate (i) an overall increase in effort but it may trade off with (ii) inducing workers to sort based on ability and thereby lose the productivity gains of friends working together. The net effect of those two forces will depend on the details of the incentives provided at the firm and the gains of increased effort are larger or small than losses due to an increase in free-riding. In Bandiera et al. (2009), the same authors find that social connections lead to managers inefficiently targeting effort towards their friends and is detrimental to overall firm performance. It is therefore far from obvious that social connections have a positive effect on on-the-job performance.

[7] The evidence on the impact on wages, however, is less robust, possibly because there may be heterogeneity across industries, occupations, countries, etc., in how and why networks are used—and the predicted effect on wages, as seen in Section 24.2, is sensitive to a variety of factors.

the dataset; instead, Munshi shows that workers with a larger network of established migrants are more likely to work in manufacturing—as opposed to agriculture, which is generally associated with higher wages.

Two additional papers use natural experiments with immigrants to provide causal evidence on the impact of social networks. Edin et al. (2003) uses refugee resettlement in Sweden as a source of exogenous variation in the size of an individual's social network. Initial placement of refugees was determined centrally by the government, limiting the ability of refugees to sort themselves. The authors then use the initial placement as an instrument for the share of an immigrant's ethnic group in their locality. They find that a larger network size increases earnings of low-education immigrants, and that an increase in the quality of the network—captured by average income—increases earnings overall. Beaman (2012) uses the placement of refugees into U.S. cities to show that a larger social network influences the labor market outcomes of newly arrived refugees. While a larger number of tenured refugees increases the probability of employment for a newly arrived refugee, having more network members who arrived around the same time decreases the probability of employment due to congestion effects.

Laschever (2009) looks at the labor market outcomes of World War I veterans and uses their assignment to a specific military company—which is essentially random—as a natural experiment. The data on outcomes come primarily from the 1930 Census and contains information on employment status, industry and occupation. Laschever uses the veterans' names in the Census to match to military records on which companies soldiers were assigned to. The analysis shows that individuals with more employed peers are more likely to be employed themselves. The author further argues that this effect is driven by endogenous, not contextual (in the language of Manski) effects, which would be consistent with a model based on referrals or informal job information passing. These papers only provide a brief snapshot of the literature, highlighting the variety of creative empirical strategies used to uncover causal effects, and Topa (2011) provides a more comprehensive review.[8]

24.3.2 Mechanisms

In order to move the literature forward, we need better evidence on the underlying mechanism(s). Much of the literature demonstrating a causal relationship between a worker's social network and his labor market outcomes uses what can be called a "potential network" approach. The advantage of this approach is gaining exogenous

[8] There is also a literature looking at neighborhood effects and the role of residential sorting. This literature has shown that local social interactions are important for explaining employment patterns across neighborhoods in urban areas (Topa 2001; Conley and Topa 2002) and that neighbors play an important role in labor market search (Bayer et al. 2008; Hellerstein et al. 2008). See Topa (2011) for an excellent review.

variation in individuals' networks, but the drawback is that there may be limited information on whether and how workers are interacting with the "potential" network. This drawback makes it particularly challenging to uncover the underlying mechanisms. In Munshi (2003), there is survey evidence that points to established migrants providing a wide variety of assistance to new migrants, ranging from housing to information about jobs. However, we do not know which activities were critical for the improved labor market outcomes. Does housing assistance free up time for more labor market search? Do established migrants pass along job information? Both of these stories would be broadly consistent with search friction models. Or, perhaps established migrants provide new migrants with referrals at their jobs, and it is only with the improved information about the worker provided through the network that migrants can get good jobs. Beaman (2012) has the same limitation: the data does not provide any direct evidence of how refugees help one another. The theoretical model in the paper is based on information passing, but the information could originate in some cases from an employed worker's firm or from indirectly hearing about a position. In order to look at underlying mechanisms and test specific models, innovation in both data and methodology are needed. The "potential network" empirical strategy has successfully established a causal link. It has therefore been very useful as a first step in this literature. However, a combination of better data which reveals exactly how a network member has provided assistance and more nuanced empirical strategies will be needed going forward. In this section, I discuss in more detail the available empirical evidence related to mechanisms and its relationship to existing theoretical models. The section is organized around the type of data used, though this fundamentally does not restrict how the data can be used.

24.3.2.1 Proprietary Firm Data

Most likely, reliable information about candidates will come from a firm's existing employees through referrals. In the empirical literature, there is a severe dearth of data directly on referrals. Burk et al. (2015) and Brown et al. (2015) are two examples of exciting new data generated recently on referrals specifically, and in both cases the data comes from firm administrative records. Burk et al. (2015) has data on nine firms from three industries: high-tech, trucking, and call centers. Brown et al. (2015) has data from one mid-sized U.S. corporation in the financial services industry. Using this data we learn a tremendous amount about how firms use referrals, and a number of regularities come out of the data on these 10 firms. In both Burk et al. (2015) and Brown et al. (2015), candidates who are referred are more likely to be hired and have lower turnover.[9] Burk et al. (2015) attempts to calculate the overall profit benefits of referrals versus nonreferrals and find that referrals are less costly, primarily due to lower turnover.

[9] In the financial services firm in Brown et al. (2015), the referred applicants in positions with low education requirements experience a larger benefit in the probability of getting an offer than applicants to positions with higher educational requirements, consistent with the idea that social networks play a more important role in low-skill industries and occupations.

Overall differences in productivity are small, though referrals do perform better in a few aspects, particularly low-frequency events like accidents among truckers and patents in the high-tech firm. What is also striking from these studies is that there appears to be no uniform finding—even among this relatively small sample of 10 firms—on whether referred workers earned more than nonreferred workers. In call centers and trucking, wages are similar between these two types of workers. In the high-tech firm, referred workers earned 1.7% more, and this wage premium is present even conditional on observable characteristics, and referred workers in the financial services firm in Brown et al. (2015) experienced a 2.1% salary premium.[10] As described above, the models are quite mixed on the predictions on wages, and the broader empirical literature is also very mixed. The majority of the literature relies on across-firm comparisons, but Bentolila et al. (2010) found that workers who found their jobs through social contacts (which may or may not be referrals directly) experienced a 2.5% reduction in wages. Bridges and Villemez (1986), Holzer (1987), and Marsden and Gorman (2001) find no relationship.

While these data are very helpful in generating important descriptive facts about how firms use referrals and we can compare the stylized facts to equilibrium predictions of the above models, it is not easy to uncover causal relationships with this data. Nevertheless, these two papers are valuable steps forward and point out some empirical facts that are hard to accord with existing models. While some of the stylized facts are consistent with a screening model, such as lower turnover, it is striking that we don't observe large productivity differences between referred and nonreferred workers, as one would predict from a screening (either based on learning or homophily) model.[11] Brown et al. (2015) also find no difference in promotion rates between referred and nonreferred workers, suggesting little to no productivity differences. Burk et al. (2015) find that referred applicants do not have better hard-to-observe characteristics than nonreferred applicants. Yet, in the high-tech firms in Burk et al. (2015), the wage premium of referred workers is present even conditional on observable characteristics, again consistent with learning models.

Recall that peer benefit models, such as Heath (2014), predict that the level and variance of wages should increase with tenure, while screening models predict a decline with tenure. In Brown et al. (2015), the initial wage premium of referred workers diminishes over time, as would be predicted by learning and screening models, but this is not the case in the high-tech firm in Burk et al. (2015). In the context of Bangladeshi garment workers, the wage premium of referrals increases over time.

[10] Brown et al. (2015) speculate that the lack of referral worker wage premium in call centers and trucking is driven by a lack of flexibility in wage setting in those institutional environments.

[11] As highlighted in Beaman and Magruder (2012), the incentive of workers to refer high-quality candidates may be undermined by worker's own preferences to help out individuals in their network who may not be the most qualified. If this were the case, however, you would anticipate that firms would cease to use referrals over time.

24.3.2.2 *Experimental Data*

Another fruitful direction of research is the use of experiments, which generates data precisely designed for the research question at hand, and the experiment can help with issues surrounding identification. Pallais and Sands (2014) use a series of experiments on the online labor market, oDesk, where the authors themselves can essentially function as a firm. They find evidence consistent with screening models but also peer benefits, in the form of team production. Referred workers performed better than nonreferred workers and—as in the traditional firms—had lower turnover. They also performed better with the workers who referred them when asked to conduct a task which required teamwork (generating a slogan for a public service announcement). Beaman and Magruder (2012) test the validity of one specific component of the screening model using a hybrid laboratory-field experiment in peri-urban Kolkata. Network members must have useful information about their network members in order for employees to be helpful in screening candidates either on absolute ability or on match quality. Participants complete a cognitive task in the laboratory. They are then asked to refer a family or friend for a future session at the laboratory and are offered a bonus for making a referral. The terms of the bonus are randomized across individuals, and the main treatments offer either a fixed bonus—paid as long as a referred subject completes the task—or a bonus with a performance component, where a higher amount is paid if the referred participant does well. The results suggest that the performance incentive leads to better referred participants, which is consistent with individuals possessing information about their network members' abilities but also needing incentives to reveal this information.

Both Beaman and Magruder (2012) and Pallais and Sands (2014) highlight that experiments, both field, laboratory and hybrid methods of combining lab and field experiments, may be quite helpful. Both papers can most cleanly identify the precise question of interest. However, there remains a gap between the reasons that could work for a firm (i.e., can existing employees refer workers who are higher quality?) and what firms actual motivations are. We need data on real-world firms in order to accomplish this. However, limited data on representative firms and unobservable factors about workers and firms create challenges for achieving this goal.

24.3.2.3 *Large Administrative Data*

Administrative data, particularly those which match firms with employees for a large fraction of a given population, provide a number of opportunities to learn about social networks. However, other data sources or creative empirical strategies are needed to identify who is likely a referral or network-based hire. Hensvik and Skans (2015) look at the predictions of the Montgomery model using Swedish administrative data. They are able to identify likely referrals through the work histories of current coworkers: those who previously worked together at another firm are likely to have known one another. These (likely) referred workers have higher cognitive ability, as measured by an AFQT-equivalent test not readily available to employers, and noncognitive skills.

Dustmann et al. (2014) use German matched employee-employer administrative data linked to survey data to test a learning model. The key empirical strategy is to look at minority workers, and assume that two individuals of the same ethnicity are likely to have a preexisting social connection. The ethnicity of the worker therefore serves as a proxy for a referral in the data. The administrative data is also linked to two surveys to validate the assumption with a sub-sample of the data. Combining the survey data with the comprehensive administrative records is a crucial innovation in this paper. The results show that a firm is more likely to hire an individual from a particular minority group when there is a large number of workers from that group already employed in the firm. Moreover, workers who are at firms with a higher share of co-ethnics experience higher wages initially but experience slower wage growth than minority workers.

Using unique archival records, Fafchamps and Moradi (2015) highlight an environment where referred workers, in this case soldiers of the Ghana army during World War I, were actually lower quality than nonreferred workers and were even more likely to abandon their position. This is particularly true among high-ranked officers, who would be harder to penalize for poor recruits. This suggests that in this setting, the army was not using soldiers to recruit high-quality soldiers, and it is likely that soldiers were enticed by cash bonuses to make referrals.

24.3.3 Weak versus Strong Ties

A substantial literature has followed the seminal work of Granovetter (1973) on the "strength of weak ties." Granovetter's insight was that individuals who are weakly connected are most likely to bring in novel information about the labor market. Within a network of close friends, who are all themselves connected to one another, there will rarely be new information about job opportunities. However, friends who connect two almost-independent networks are likely to bring new information from one group to another. These friends are described as bridges in Granovetter's original work. A bridge is formally "a line in a network which provides the only path between two points" (Granovetter 1973). Bridges are more likely to be weak ties, which generated Granovetter's hypothesis that weak ties are more important in the labor market than strong ties. Weak versus strong ties can be measured in two distinct ways: the first way is by measuring how close two individuals are, in terms of time spent together or some other proxy. The second approach uses network structure: a node can be directly characterized as a bridge. However, this is hard to do empirically (Nepusz et al. 2008; Saha et al. 2011), and there are rarely true bridges since most networks in the world exhibit small world properties. More often, strong ties are measured based on the fraction of one's friends who are also connected to each other. A higher level of connectedness would indicate a strong tie. Recent theoretical work by Zenou (2015) formalizes the idea of weak and strong ties in a social interaction model and looks at labor market dynamics. Ties between dyads in the labor market that do not change over time are considered strong, and weak ties occur periodically through random encounters. In a given period, an individual will either meet a strong or weak tie and have an opportunity to gain job information from that connection. This notion of a weak tie is analogous to defining it based on time spent together. Similar to Galeotti and Merlino (2014), workers can endogenously choose to invest in their network, but in this case the worker is choosing who to spend time with and thereby choosing between weak and strong ties. Zenou finds that workers will use weak ties more in bad economic ties (when either the job arrival rate is low or the job destruction rate is high).

Empirical work on whether weak or strong ties are more useful for labor market search is notably mixed. Early work in sociology found no correlation between tie strength and wages (Bridges and Villemez 1986; Marsden and Hurlbert 1988). Montgomery (1992), however, argues that it is flawed to empirically test Granovetter's hypothesis using only data on the wages and the origin of the jobs that are accepted. Ideally the full set of job offers should be evaluated. For example, suppose everyone is likely to receive an offer through a weak tie. Only some individuals then also receive an offer through a strong tie. If the job that came from the strong tie was accepted, it must have been because it was a high wage draw. Therefore, comparing the wages of accepted jobs acquired through weak versus strong ties can be misleading. There is certainly empirical work in support of the weak ties hypothesis (Lin et al. 1981; Grabowicz et al. 2012; Bakshy et al. 2012). Other work has, though, questioned the value of weak ties. Heath (2014), Magruder (2010), and Kramarz and Skans (2007) all show the importance of family members, an obvious example of strong ties, in facilitating jobs. Gee et al. (2012) use innovative data from Facebook and look at whether strong or weak ties are more likely to generate a new job. They select pairs of members who are social connected, who share the same employer, and whose timing of employment makes a referral plausible. Among these matches, the majority arose between individuals who would be considered weak ties: these are dyads who do very little posting on each other's walls and do not share many friends in common. This is because individuals on Facebook have many more weak ties than strong ties. Using a fixed effects specification, however, an individual is more likely to be working with his strong ties than his weak ties. van der Leij and Goyal (2011) uses data from co-authorship patterns among economists. Overall, they reject the hypothesis that weak ties are more important than strong ties because strong ties actually lie on more shortest paths than weak ties, inconsistent with the idea that only weak ties serve as bridges.

In order to understand the heterogeneity in these results, we must again return to the underlying reason firms or workers are using social networks in the first place. Granovetter's (1973) original work assumed, if implicitly, that search frictions were the fundamental reason for the use of networks. Bridges are essential for information to widely disseminate. However, if firms are seeking information through their employees, strong ties may be essential. This is recognized in other work by Granovetter (1995), and this possibility is shown formally in the work by Karlan et al. (2009). The model is set up such that there is a value of the relationship between each node within a network.

Favors can only be done within the network such that no one in the network has the incentive to deviate and renege on their promise, since they value their relationships more than the value gained by reneging. This generates network-based trust that can be useful for things like informal borrowing and lending. The complete network structure is important for facilitating credit since indirect links can be used to leverage more trust through multi-party promises, thereby increasing the value of assets that can be lent in the network. In the context of labor market search, employers are connected to their workers, who have connections with potential applicants. Though the employer wants the worker to reveal an applicant's type, the applicant can bribe his connection to reveal false information. Credible information on applicant type can only be communicated when the employer's trust of their incumbent workers is larger than the highest bribe an applicant is willing to pay (based on the Nash bargaining process determining wages). The model also suggests that referrals will be used more often when worker ability is more important to firm productivity. This is likely to occur in high-skill jobs, though noncognitive skills may also be quite important in low-skill jobs and difficult to observe by firms. This model suggests that the underlying reason firms in a given empirical sample use networks may alter our expectation on whether weak or strong ties will be more important in the labor market. This may help to explain why the empirical literature is mixed.

24.4 Unintended Consequences

The models in Section 24.2 highlight multiple ways in which social networks can smooth away labor market frictions, increasing efficiency potentially in the labor market. This can be beneficial to both firms and workers. Reducing search frictions should reduce the length of unemployment spells for workers and reduce search costs for firms. Heath (2014) highlights that the employee-worker tie can help reduce the inefficiency created by a minimum wage law in Bangladesh. This enables workers who would otherwise not be employed to enter the formal labor market. However, there may be costs to relying on informal institutions—which are largely outside the influence of policy-makers—to improve the functioning of the labor market. Calvo-Armengol and Jackson (2004) highlighted this potential in their theoretical work related to the black-white wage gap. In the model, initial differences in the employment rate across networks can lead to long-term inequality across groups as individuals in low-employment networks have an incentive to drop out of the labor market. The model is consistent with empirical data showing that blacks in the United States are less likely to be employed and less likely to be in the labor market than whites in the United States. The model therefore highlights that inequality may be exacerbated through the use of social networks in the labor market.

Networks may also play a role in the gender employment and wage gap that is observed worldwide. A number of papers in sociology, and to a lesser extent economics, have signaled that the use of employee referrals may drive segregation of jobs across genders (Doeringer and Piore 1971; Mouw 2006; Rubineau and Fernandez 2010; Tassier and Menczer 2008). Empiricaly, most social networks show properties of homophily, particularly gender homophily. Coupled with significant differences in the distribution of occupations between men and women, this could make it hard for women to break into high-wage industries, where women are traditionally under-represented. Consistent with this, Lalanne and Seabright (2011) find that women executives in the United States and Europe don't leverage their contacts into higher salaries as well as their male counterparts. Loury (2006), using NLSY data, found that male workers referred by women get lower on average wages than those who applied through formal channels. Mortensen and Vishwanath (1994) also show theoretically how network-based job information dissemination can disadvantage women, even if men and women are are equally productive but men have more contacts.

Beaman et al. (2013) use a field experiment with a survey firm in Malawi to demonstrate the possibility that the use of employee referrals can further disadvantage an initially disadvantaged group. Referrals generated about 30% fewer women than traditional recruitment methods. This is largely driven by the fact that men tend to refer other men, consistent with the gendered nature of social networks worldwide but particularly in developing countries. However, men do know women they could refer: in a treatment arm of the experiment, men were asked to refer women specifically. The men were just as likely to make a referral as when they could choose a man, and the women they referred were similar quality. A similar finding is in Fernandez and Sosa (2005), who use observational data from a call center. In their environment, men are the initially disadvantaged group in the sense that a majority of the firm's workers are women. The data show that employee referrals are overwhelmingly female candidates.

Informal institutions, such as labor market referral networks, which are initially advantageous to network members may also create inefficiencies when local economies change and evolve. The work by Munshi and Rosenzweig (2006) provides a counterpoint to the argument that the use of social networks in the labor market disadvantage women. In this setting—Bombay, India—men traditionally used social networks to secure good blue collar jobs. However, liberalization in the 1990s generated a dramatic increase in white collar jobs, increasing the returns to education in English (compared to the local vernacular language). The paper puts forth a model in which communities had an incentive to prevent boys from exiting the blue collar sector to maximize aggregate utility within the social network. When the returns to English education rose, families allowed their girls to switch from vernacular to English language schooling at a faster rate than boys—since girls were traditionally excluded from the blue collar networks and not subject to the restrictions within the network. The authors collected household survey data in one area of Bombay and demonstrate empirical patterns consistent with the model, though the paper does not provide any direct evidence on the types of restrictions which may have been in place to prevent boys from exiting the blue collar sector.

Fainmesser (2013) highlights that information signals which are only within a network can lead to unraveling of labor markets, which is inefficient for the market as a whole. A canonical example is the market for some specialities within medicine, where offers for speciality training happen too early in a student's educational process. Private but noisy signals are observed by the student's current teachers, and this can lead to early offers to students with a positive signal. Since the remaining students in the market are assumed to be "lemons," this leads to unraveling. This is inefficient since the signals of quality become much more precise later in a student's education. The paper further describes how changes in network structure will accentuate or reduce the unraveling, and which market rules can prevent the unraveling.

24.5 Conclusion

This chapter summarizes research, primarily in economics, looking to explain the role of social networks in the labor market. While there are a number of high-quality empirical papers establishing a causal relationship between the size and quality of workers' social networks on their labor market outcomes, far less is known on why firms and workers rely on informal channels for recruitment. There are a number of possible explanations analyzed in the theoretical literature. One possibility is that there are significant search frictions, and social network-based job information dissemination lowers search costs for both workers and firms. In most of these models, workers with better networks will therefore enjoy better labor market outcomes. A second explanation stems from missing information in the labor market. Either firms are not able to observe whether a worker is a high-ability type (Montgomery 1991), or there is uncertainty about the match quality between a worker and the firm (Simon and Warner 1992; Dustmann et al. 2014; Galenianos 2013). Both mechanisms will predict that referred applicants are more likely to be hired than formal applicants and will earn higher wages once hired, though the introduction of heterogeneity in workers or firms can undermine this prediction. We would also expect referred workers to have better hard-to-observe characteristics than nonreferred workers. A third class of models argues that firms want to exploit positive peer effects on the job, in the form of either improved monitoring (Kugler 2003; Heath 2014) or in team production (Bandiera et al. 2013). When addressing moral hazard to induce effort from workers is a primary concern of firms, employee referrals may allow workers to be hired who would otherwise be considered too low quality or too costly to monitor. In this case, referred workers may have worse observable characteristics than those hired through formal channels, in contrast to models stemming from missing information. While a number of mechanisms have been clearly described in this theoretical literature, we still need more comprehensive work on what assumptions are really critical in driving the main predictions in each class of models, and how heterogeneity in workers and firms may alter these predictions. Given the literature's current state, it is challenging to empirically test the role of each mechanism since no one has of yet fully characterized the models' predictions for different types of firms and workers. And of course, these mechanisms are not mutually exclusive, and the same firm may have multiple motivations for using informal methods of recruitment. This poses an even greater challenge to the empirical literature.

The empirical literature faces significant challenges in identifying which models are most relevant. In many data sets, particularly the large administrative data, we do not observe how an individual gets a job (Dustmann et al. 2014; Glitz 2013). Even if an informal search is reported, we do not know the nature of the help received from family and friends. Did they hear about the job or did the network member actually provide a recommendation to the employer? Models based on screening suggest the importance of direct employee referrals, not just information dissemination within a social network. New data has been used which contains detailed information on whether a referral took place and contains information on applicants and hired workers (Burk et al. 2015; Brown et al. 2015); however, this covers a very small number of firms (a maximum of nine), which may or may not be representative of the labor market at large. Field experiments have provided key insights into whether workers can provide useful information on applicants (Beaman and Magruder 2012; Pallais and Sands 2014), as required by screening models. However, field experiments will not likely be able to tell us what firms do themselves outside of the context of the study.

This is an area of research which will benefit most from an iterative process between theoretical and empirical work, with theory guiding the empirical tests, which can then help further refine our models. There also seems to be great promise in working with firms to generate new data and potentially incorporate experiments or natural experiments in the analysis. Linking survey data with large administrative data also appears a fruitful direction for new research.

References

Bakshy, E., I. Rosenn, C. Marlow, and L. Adamic (2012). "The role of social networks in information diffusion." In *Proceedings of the 21st International Conference on World Wide Web*, 519–528.

Bandiera, O., I. Barankay, and I. Rasul (2005). "Social preferences and the response to incentives: Evidence from personnel data." *Quarterly Journal of Economics* 120, 917–962.

Bandiera, O., I. Barankay, and I. Rasul (2009). "Social connections and incentives in the workplace: Evidence from personnel data." *Econometrica* 77, 1047–1094.

Bandiera, O., I. Barankay, and I. Rasul (2013). "Team incentives: Evidence from a firm level experiment." *Journal of the European Economic Association* 11, 1079–1114.

Bayer, P., S. Ross, and G. Topa (2008). "Place of work and place of residence: Informal hiring networks and labor market outcomes." *Journal of Political Economy* 116, 1150–1196.

Beaman, L. (2012). "Social networks and the dynamics of labor market outcomes: Evidence from refugees resettled in the U.S." *Review of Economics Studies* 79, 128–161.

Beaman, L., N. Keleher, and J. Magruder (2013). "Do job networks disadvantage women? Evidence from a recruitment experiment in malawi." Mimeo, Northwestern University.

Beaman, L. and J. Magruder (2012). "Who gets the job referral? Evidence from a social networks experiment." *American Economic Review* 102, 3574–3593.

Bentolila, S., C. Michelacci, and J. Suarez (2010). "Social contacts and occupational choice." *Economica* 77, 20–45.

Boorman, S. (1975). "A combinatorial optimization model for transmission of job information through contact networks." *Bell Journal of Economics* 6, 216–249.

Bramoullé, Y. and G. Saint-Paul (2010). "Social networks and labor market transitions." *Labour Economics* 17(1), 188–195.

Bridges, W. P. and W. J. Villemez (1986). "Informal hiring and income in the labor market." *American Sociological Review* 51(4), 574–582.

Brown, M., E. Setren, and G. Topa (2015). "Do informal referrals lead to better matches? Evidence from a firm's employee referral system." Forthcoming in *Journal of Labor Economics* 34(1), January 2016.

Burk, S., B. Cowgill, M. Hoffman, and M. Housman (2015). "The value of hiring through employee referrals." *Quarterly Journal of Economics* 130(2), 805–839.

Calvo-Armengol, A. (2004). "Job contact networks." *Journal of Economic Theory* 115, 191–206.

Calvo-Armengol, A. and M. Jackson (2004). "The effects of social networks on employment and inequality." *American Economic Review* 94, 426–454.

Calvo-Armengol, A. and Y. Zenou (2005). "Job matching, social network and word-of-mouth communication." *Journal of Urban Economics* 57, 500–522.

Conley, T. and G. Topa (2002). "Socio-economic distance and spatial patterns in unemployment." *Journal of Applied Econometrics* 17, 303–327.

Doeringer, P. B. and M. J. Piore (1971). *Internal Labor Markets and Manpower Analysis.* New York: M.E. Sharpe.

Dustmann, C., A. Glitz, and U. Schoenberg (2014). "Referral-based job search networks." Working paper, University College London.

Edin, P.-A., P. Fredriksson, and O. Aslund (2003). "Ethnic enclaves and the economic success of immigrants: Evidence from a natural experiment." *Quarterly Journal of Economics* 118, 329–357.

Fafchamps, M. and A. Moradi (2015). "Referral and job performance: Evidence from the Ghana colonial army." *Economic Development and Cultural Change* 63(4), 715–751.

Fainmesser, I. (2013). "Social networks and unraveling in labor markets." *Journal of Economic Theory* 148, 64–103.

Fernandez, R. and M. Sosa (2005). "Gendering the job: Networks and recruitment at a call center." *American Journal of Sociology* 111, 859–904.

Fontaine, F. (2008). "Why are similar workers paid differently? The role of social networks." *Journal of Economic Dynamics and Control* 32(12), 3960–3977.

Galenianos, M. (2013). "Learning about match quality and the use of referrals." *Review of Economic Dynamics* 16, 668–690.

Galenianos, M. (2014). "Hiring through referrals." *Journal of Economic Theory* 152, 304–323.

Galeotti, A. and L. Merlino (2014). "Endogenous job contact networks." *International Economic Review* 55, 1201–1226.

Gee, L., J. Jones, and M. Burke (2012). "Social networks and labor markets: How strong ties relate to job transmissions using facebook's social network." Working paper, Tufts University.

Glitz, A. (2013). "Coworker networks in the labour market." CESifo Working Paper Series.

Grabowicz, P. A., J. J. Ramasco, E. Moro, J. M. Pujol, and V. M. Eguiluz (2012). "Social features of online networks: The strength of intermediary ties in online social media." *PloS One* 7(1), e29358.

Granovetter, M. (1973). "The strength of weak ties." *American Journal of Sociology* 78, 1360–1380.

Granovetter, M. (1995). *Getting a Job: A Study of Contacts and Careers*. University of Chicago Press.

Heath, R. (2014). "Why do firms hire using referrals? Evidence from Bangladeshi garment factories." Mimeo, University of Washington.

Hellerstein, J., M. McInerney, and D. Neumark (2008). "Neighbors and co-workers: the importance of residential labor market networks." Working Paper 14201, National Bureau of Economic Research.

Hensvik, L. and O. Skans (2015). "Social networks, employee selection and labor market outcomes." Forthcoming in *Journal of Labor Economics* 34(4), October 2016.

Holzer, H. J. (1987). "Job search by employed and unemployed youth." *Industrial & Labor Relations Review* 40(4), 601–611.

Ioannides, Y. and L. Loury (2004). "Job information networks, neighborhood effects, and inequality." *Journal of Economic Literature* 42, 1056–1093.

Ioannides, Y. M. and A. R. Soetevent (2006). "Wages and employment in a random social network with arbitrary degree distribution." *The American Economic Review* 96(2), 270–274.

Karlan, D., M. Mobius, T. Rosenblat, and A. Szeidl (2009). "Trust and social collateral." *Quarterly Journal of Economics* 124, 1307–1361.

Kramarz, F. and O. Skans (2007). "With a little help from my parents? Family networks and youth labor market entry." Working paper, Center for Research in Economics and Statistics.

Kugler, A. (2003). "Employee referrals and efficiency wages." *Labour Economics* 10, 531–556.

Lalanne, M. and P. Seabright (2011). "The old boy network: Gender differences in the impact of social networks on remuneration in top executive jobs." IDEI Working Paper No. 689.

Laschever, R. (2009). "The doughboys network: Social interactions and labor market outcomes of World War I veterans." Mimeo, Purdue University.

Lin, N., W. M. Ensel, and J. C. Vaughn (1981). "Social resources and strength of ties: Structural factors in occupational status attainment." *American Sociological Review* 46(4), 393–405.

Loury, L. (2006). "Some contacts are more equal than others: Informal networks, job tenure, and wages." *Journal of Labor Economics* 24, 299–318.

Magruder, J. (2010). "Intergenerational networks, unemployment, and persistent inequality in South Africa." *American Economic Journal: Applied Economics* 2, 62–85.

Manski, C. (1993). "Identification of endogenous social effects: The reflection problem." *Review of Economics Studies* 60, 531–542.

Marsden, P. V. and E. H. Gorman (2001). "Social networks, job changes, and recruitment." In *Sourcebook of Labor Markets: Evolving Structures and Processes*, Ivar Berg and Arne L. Kalleberg, eds., 467–502. New York: Kluwer Academic/Plenum.

Marsden, P. V. and J. S. Hurlbert (1988). "Social resources and mobility outcomes: A replication and extension." *Social Forces* 66(4), 1038–1059.

Montgomery, J. (1991). "Social networks and labor market outcomes: Toward an economic analysis." *American Economic Review* 81, 1408–1418.

Montgomery, J. (1992). "Job search and network composition: Implications of the strength-of-weak-ties hypothesis." *American Sociological Review* 57, 586–596.

Mortensen, D. and T. Vishwanath (1994). "Personal contacts and earnings: It is who you know!" *Labour Economics* 1, 187–201.

Mortensen, D. T. and C. A. Pissarides (1994). "Job creation and job destruction in the theory of unemployment." *The Review of Economic Studies* 61(3), 397–415.

Mouw, T. (2006). "Estimating the causal effect of social capital: A review of recent research." *Annual Review of Sociology* 32, 79–102.

Munshi, K. (2003). "Networks in the modern economy: Mexican migrants in the US labor market." *Quarterly Journal of Economics* 118, 549–599.

Munshi, K. and M. Rosenzweig (2006). "Traditional institutions meet the modern world: Caste, gender, and schooling choice in a globalizing economy." *American Economic Review* 96, 1225–1252.

Nepusz, T., A. Petróczi, L. Négyessy, and F. Bazsó (2008). "Fuzzy communities and the concept of bridgeness in complex networks." *Physical Review E* 77(1), 016107.

Pallais, A. and E. Sands (2014). "Why the referential treatment? Evidence from field experiments on referrals." Working paper, Harvard University.

Pissarides, C. A. (1990). *Equilibrium Unemployment Theory*. MIT Press.

Rees, A. (1966). "Information networks in labor markets." *American Economic Review* 56, 559–566.

Rubineau, B. and R. Fernandez (2010). "Missing links: Referrer behavior and job segregation." Working Paper 4784-10, MIT Sloan School.

Saha, T., C. Domeniconi, and H. Rangwala (2011). "Detection of communities and bridges in weighted networks." *Machine Learning and Data Mining in Pattern Recognition*, 584–598.

Simon, C. and J. Warner (1992). "Matchmaker, matchmaker: The effect of old boy networks on job match quality, earnings, and tenure." *Journal of Labor Economics* 10, 306–330.

Tassier, T. and F. Menczer (2008). "Social network structure, equality, and segregation in a labor market with referral hiring." *Journal of Economic Behavior and Organization* 66, 514–528.

Topa, G. (2001). "Social interactions, local spillovers and unemployment." *Review of Economics Studies* 68, 261–295.

Topa, G. (2011). "Labor markets and referrals." *Handbook of Social Economics* 1, 1193–1221.

van der Leij, M. and S. Goyal (2011). "Strong ties in a small world." *Review of Network Economics* 10(2), 1–20.

Wabha, J. and Y. Zenou (2005). "Density, social networks and job search methods: Theory and application to Egypt." *Journal of Development Economics* 88, 443–473.

Zenou, Y. (2015). "A dynamic model of weak and strong ties in the labor market." *Journal of Labor Economics* 33(4), 891–932.

Zinovyeva, N. and M. Bagues (2015). "The role of connections in academic promotions." *American Economic Journals: Applied Economics* 7(2), 264–292.

PART VII

ORGANIZATIONS AND MARKETS

CHAPTER 25

ATTENTION IN ORGANIZATIONS

WOUTER DESSEIN AND ANDREA PRAT

25.1 Introduction

Organizations—private firms, government agencies, and nonprofit organizations—can be modeled as networks of agents who are working together toward a common set of goals. Arrow (1974) views organizations as ways to overcome the limits of individual agents. By bringing together multiple workers, organizations can perform tasks that are outside the reach of any individual. While this creates production opportunities, it also poses a challenge. In order to be productive, workers must coordinate their actions. Often this requires communicating information that is dispersed throughout the organization.

However, we humans face cognitive limits: transmitting and absorbing information requires time and energy. Managers spend a considerable part of their work time communicating with other workers. Bandiera et al. (2011) report that over 80% of the work time of executive managers is spend in communication activities, such as meetings, phone conversations, events, conferences, and so on. Mankins et al. (2014) find that senior executives devote more than two days every week to meetings involving three or more coworkers, and 15% of an organization's collective time is spent in meetings.

As Arrow (1974) noted, given the importance of communication both as an opportunity and as a cost, organizations will strive to optimize information flows between workers. This leads to two important predictions in organizational economics. First, communication patterns within an organization will not be random but they will, at least in part, be shaped by the goals of the organization. Second, the cost of communication will be an important factor in designing the organization. Different organizational charts imply different information flows, and hence different costs. A number of scholars have developed and extended Arrow's (1974) insights into formal models. This literature forms a bridge between theories of rational inattention (Sims 2003), whereby agents pay a price to transmit or receive information from other agents, and network economics, where typically links between agents are not explicitly stated in terms of information transmission and agents' payoffs are not expressed in terms of actions to be taken with incomplete information. The term *attention network* appears appropriate for this type of model.

Attention networks are most closely related to the field of organizational economics (surveyed by Gibbons and Roberts 2013). Limits to attention play a crucial role in other theories in organizational economics, such as Bolton and Dewatripont (1994) and Garicano (2000). We focus here on an explicit network-theoretic approach, leaving a more general discussion to the survey of organizational economics with cognitive costs by Garicano and Prat (2013). Attention networks can be used to discuss a central theme of organizational economics: coordination. Organizations exist to coordinate specialized workers. As emphasized by Adam Smith, breaking up a production process into specialized tasks allows to dramatically increase productivity. But while the division of labor has resulted in huge productivity gains in modern economies, it also creates a need for coordination of specialized activities. The main role of organizational networks, therefore, is to achieve coordination in the presence of the division of labor. The key feature of attention networks in organizations—the one highlighted by Arrow (1974)—is further that they are endogenous. Attention networks in organizations are designed, shaped, and optimized for the goal of coordination. Communication is costly, and the decision to invest in communication is made consciously by agents, either individually or as a group.

The typical attention network model contains the following elements:

- A set of agents, each of which observes some information from the environment, may choose to transmit (at a cost) information to other agents, and may choose to receive (at a cost) information from other agents.
- A set of tasks, which must be allocated to agents, who must make decisions on the basis of information that is available to them.
- A payoff function for individuals and for the organization, which depends on how well the decisions that are taken fit with the state of the world and with each other. Some models like Dessein and Santos (2006) and Dessein, Galeotti, and Santos (2014) take a team-theoretic approach and assume that agents have a common objective. Other models, like Calvó, De Martí, and Prat (2015), instead follow the game-theoretic tradition of assuming different objectives for the different agents.[1]
- An attention cost function that models the payoff implications of information transmission. Attention costs can relate to active communication when they are sustained by the sender (speaking, writing, providing product samples, etc.) or passive communication when they are incurred by the receiver (listening, reading,

[1] However, most of the results surveyed in this chapter hold whether one uses a team-theoretic or a game-theoretic approach. See Section 25.4 for a discussion.

examining product samples, etc.), or they can be formulated at the group or organizational level (for example, the time agents spend in meetings).

The equilibrium of an attention network describes how communication flows and how decisions are made within the organization. The models also predict who influences whom within an organization. When an agent receives new information, this affects not just his decision but also what signals he transmits to other agents, and hence what actions they choose. Thus, agent *i*'s influence has a natural meaning in this setting, as how other agents' actions are affected by a change in *i*'s information.[2]

While this literature is recent, there are already a number of results of relevance to organizational economics:

1. A decentralized attention network or a set of standardized operating procedures are two alternative ways of achieving coordination among members of an organization. The former is more likely to be optimal when local information is more important and communication costs are lower (Dessein and Santos 2006).[3]
2. Influence and communication patterns can be highly asymmetric even starting from a perfectly symmetric interaction function. When attention is scarce, it is optimal for an organization to direct their members' attention to a small set of key agents (Dessein, Galeotti, and Santos 2014).
3. Influence and communication patterns within an organization are highly interrelated. If we observe communication patterns—for instance, through electronic records—we can use eigenvector centrality to rank the influence of the members of the organization (Calvó et al. 2015).

Models of attention networks are distinct from the rest of network economics in that they must include two elements: (i) Nodes represent Bayesian decision-makers; (ii) Links represent endogenous costly communication between nodes. Element (i) is shared with a number of economic network theories, including models of learning in networks (Chapter 19 by Golub and Sadler). Regarding element (ii), a number of network models contain endogenous link formation (Chapter 8 by Vannetelbosch and Mauleon), which often may be interpretable as a reduced form of communication (e.g., two nodes connected by the link obtain higher payoffs by exchanging information). However, what characterizes attention networks models is that information transmission is modeled explicitly with a Bayesian setup, as a costly, endogenous activity, whose benefit can be computed within the model in terms of better decision-making.

[2] This brief survey focuses on the intensity of communication, assuming that the mode of communication—namely language—is given. As Arrow (1974) noted, we should expect organizations to affect the mode of communication as well, by developing a technical language, a *code*, that is suited to the type of problems they face. Cremer, Garicano, and Prat (2007) propose a model of codes and analyze its implications for the theory of the firm.

[3] In our model, decision-making is always decentralized. See Alonso, Dessein, and Matouschek (2008, 2015) for two related models who study when centralized decision-making (with vertical communication) is preferred over decentralized decision-making (with horizontal communication).

Attention networks are closely related to models of cheap talk in networks, like Koessler and Hagenbach (2010) and Galeotti, Ghiglino, and Squintani (2013). Those models include element (i). However, communication between nodes is endogenous but not costly. The focus is therefore on whether agents find it in their interest to reveal private information to other agents. Instead, in the models discussed in this chapter, agents would disclose everything if communication were free. The assumption that attention is costly is therefore crucial to all the results that we will soon discuss. Dewan and Myatt (2008) explore costly endogenous communication in a political economy setting.

Finally, networks have been central to sociology and have found a number of important applications in the sociology of organizations, as discussed in Burt (2005).

The chapter is organized as follows. Section 25.2 first focuses on the building block of attention networks, endogenous communication between agents, and discusses the role of attention networks in achieving coordination among specialized agents (Dessein and Santos 2006). This section also compares the merits of attention networks relative to a more bureaucratic way of coordinating economic activity: coordination through centrally imposed standard operating procedures. Whereas Section 25.2 imposes attention networks to be symmetric in nature, Sections 25.3 and 25.4 explore optimal attention networks and asymmetric communication patterns.

The shape of the information cost function is crucial in determining the properties of the endogenous communication network. If communication costs are convex, but not excessively so, even an ex ante symmetric set of agents will choose a corner solution that results in an asymmetric communication network (Section 25.3, based mainly on Dessein, Galeotti, Santos, 2014) If instead communication costs are sufficiently convex, the communication network will correspond to an interior solution. It will then be possible to characterize equilibrium networks in a general setting (which includes active and passive communication) and to establish a connection with eigenvector centrality (Section 25.4, based mainly on Calvó, De Martí, and Prat, 2015).

Section 25.5 concludes by providing a short discussion of the empirical literature on attention networks, a fast-growing field thanks to the increasing availability of data on behavior within organizations.

25.2 The Role of Attention Networks within Organizations

We discuss attention networks in the context of the Dessein and Santos (2006; hereafter DS) model of an organization, in which multiple specialized agents work together and must coordinate their individual tasks. Coordination is made difficult by the need to adapt those tasks to a changing environment. We will use the DS model to study *endogenous* attention networks in organizations. Attention networks facilitate coordinated adaptation to a changing environment. Since organizational attention is scarce, attention networks are optimized to make optimal use of this scarce resource. Since attention is naturally modeled as a communication process, we will use the terms attention and communication interchangeably.

25.2.1 The Dessein-Santos Model

Production in DS requires the combination of n tasks, each performed by one agent $i \in \mathcal{N} = \{1,2,\ldots,n\}$. The profits of the organization depend on (i) how well each task is *adapted* to the organizational environment, and (ii) how well each task is *coordinated* with the other tasks. For this purpose, agent i must take a primary action, $a_{ii} \in \mathbb{R}$, and a coordinating action, $a_{ij} \in \mathbb{R}$, for each task $j \in \mathcal{N} \setminus \{i\}$.

Payoffs. Ideally, agent $i \in \mathcal{N}$ should set his primary action a_{ii} as close as possible to local information θ_i, a random variable with variance σ_θ^2 and mean $\hat{\theta}_i$. One can interpret $\hat{\theta}_i$ as the status quo or the standard operation procedure (SOP) for task i, where we assume that $\hat{\theta}_i$ is known to all agents.[4] In contrast, only agent i observes θ_i. We refer to θ_i as the local information pertaining to task i and assume its realization is independent across tasks. Agent $j \neq i$, in turn, should set the coordinating action a_{ji} as close as possible to action a_{ii}. The expected misadapation and miscoordination losses to the organization then amount to $\Gamma = \sum_i \Gamma_i$ where

$$\Gamma_i = E\left[\phi(a_{ii} - \theta_i)^2\right] + \beta \sum_{j \neq i} E\left[(a_{ii} - a_{ji})^2\right] \tag{25.1}$$

where ϕ is the weight given to misadaptation and β the weight given to miscoordination. The parameter $\beta > 0$ can be interpreted as measuring task-interdependence. We take a team-theoretic perspective so that all agents choose their primary and coordinating actions in order to minimize expected coordination and adaptation losses, as captured by Γ.

Communication and timing. Agents send a message to each other about their primary action. Communication is assumed to be imperfect: Agent i's message is received and understood by agent j with probability p_i, while agent j learns nothing with probability $1 - p_i$. The timing of the game is assumed as follows. In stage 1, agent $i \in \mathcal{N}$ observes θ_i, chooses a_{ii}, and communicates it to all agents $j \neq i$. In stage 2, agent j receives agent i's message with probability p_i and sets $a_{ji} = a_{ii}$ when he learns the value of a_{ii} and sets $a_{ji} = E(a_{ii}) = \hat{\theta}_i$ otherwise. It follows that agent i chooses a_{ii} to minimize $E[\phi(a_{ii} -$

[4] We endogenize the quality and precision of standard operating procedures in section 25.2.2.

$\theta_i)^2 + \beta(1-p_i)(n-1)(\hat{\theta}_i - a_{ii})^2]$:

$$a_{ii} = \hat{\theta}_i + \frac{\phi}{\phi + \beta(n-1)(1-p_i)}(\theta_i - \hat{\theta}_i) \tag{25.2}$$

where we verify that, indeed, $E(a_{ii}) = \hat{\theta}_i$.

Communication frictions and coordination-adaptation trade-offs. Note that if agent i can perfectly communicate his primary action to agent j, there is no trade-off between adaptation and coordination: agent i then optimally sets $a_{ii} = \theta_i$ and agent $j \neq i$ ensures coordination by setting $a_{ji} = a_{ii}$. In the presence of communication frictions, however, adaptation-coordination trade-offs arise. Indeed, assume $p_i = 0$ so that agent j receives no information about a_{ii} and therefore sets $a_{ji} = \hat{\theta}_i$. Then the more agent i adapts a_{ii} to θ_i the larger are the coordination costs with tasks $j \neq i$. By ignoring his local information θ_i and sticking to the standard operating procedure or the status quo $a_{ii} = \hat{\theta}_i$, however, agent i can ensure perfect coordination with tasks $j \neq i$. The ratio

$$\alpha_i \equiv \phi/[\phi + \beta(n-1)(1-p_i)]$$

can be interpreted as the optimal degree of adaptiveness or discretion for agent i given the need for coordination β and the quality of communication p_i.

Substituting the equilibrium choices for primary and coordinating actions given communication frictions p_i for tasks $i \in \mathcal{N}$ yields expected losses

$$\Gamma = \sum_{i=1}^{n} \frac{\phi\beta(n-1)(1-p_i)}{\phi + \beta(n-1)(1-p_i)} \sigma_\theta^2. \tag{25.3}$$

Note that equilibrium costs are increasing in the variance σ_θ^2 (unexpected contingencies, change in the environment), the importance of adaptation as measured by ϕ, the importance of coordination β, as well as the division of labor n. In contrast, equilibrium losses are decreasing in the quality of communication p_i between agents.

As pointed out by DS, extensive specialization results in organizations that are increasingly inflexible and ignore local knowledge: from (25.2), a_{ii} is less correlated with θ_i as n increases. As DS show, the division of labor within organizations is therefore limited by the need for adaptation. We refer to DS for a study and analysis of the optimal degree of specialization in organizations. In particular, DS allows each agent to undertake several tasks in the production process, where a broader task-allocation improves coordination but reduces the gains from specialization. In the remainder of this chapter, we will take both the number of tasks n and the task-specialization of agents as given.

25.2.2 Attention Networks versus Standard Operating Procedures

We now extend the DS model to highlight two very distinctive ways of coordinating economic activity in organizations:

- Coordination through (horizontal) communication networks.
- Coordination through (centrally imposed) Standard Operating Procedures (SOPs).

On the one hand, the firm can improve coordination by fostering bilateral communication between agents. For example, the organization can carve out plenty of time for meetings and information exchange between agents. The firm can further invest in communication networks and intranets, and improve horizontal communication by training its employees in interpersonal communication skills, instituting a collaborative culture, recruiting employees with similar technical backgrounds, and so on. Formally, we will assume that at a cost $F_N > 0$, the organization can create a high-functioning communication network where $p_i = p_H > 0$ for all $i \in \mathcal{N}$. If the organization does not invest in a communication network, then $p_i = p_L < p_H$ for all $i \in \mathcal{N}$. Wlog, we will set $p_L = 0$. We denote $d_C = 1$ when the firm invests in a horizontal communication network with $p = p_H$, and $d_C = 0$ otherwise.

Alternatively, the firm can improve coordination by investing in commonly understood SOPs. This creates a role for a center or headquarter manager to clearly communicate task guidelines and procedures to agents, and to update those procedures as the environment changes. If agents largely stick to such commonly understood task instructions, coordination can be achieved without any need for communication. How adaptive the organization is then depends on how well operating procedures capture task-specific information and how quickly the organizational environment changes, which may result in outdated SOPs. To capture the role of standard operating procedures, we extend the DS model by adding two ingredients:

1. The organization lives for two periods. The local information in the first period, denoted by $\theta_{i,1}$, is normally distributed with mean $\hat{\theta}_i$ and variance σ_θ^2. The local information in the second period, $\theta_{i,2}$, has a mean $\hat{\theta}_i + \varepsilon$ and variance σ_θ^2, where ε is normally distributed with mean 0 and variance σ_ε^2. The variance σ_ε^2 then reflects the amount of environmental change or turbulence there is in the environment.
2. Agent j does not observe the mean $\hat{\theta}_i$ directly, but knows that the mean $\hat{\theta}_i$ itself is a random variable with mean 0 and variance $\hat{\sigma}_\theta^2 > \sigma_\varepsilon^2$. At at a cost F_S per task, a headquarter manager can learn $\hat{\theta}_i$ in the first period and (perfectly) communicate it to the organization. The role of headquarters is thus to establish

or improve standard operating procedures for each task and communicate those to the organization.[5]

We denote $d_S = 1$ when the firm establishes standard operating procedures for each task so that all agents observe $\hat{\theta}_i$ for $i \in \mathcal{N}$, and $d_S = 0$ otherwise. Abusing notation, we denote expected losses over two periods given investment choices (d_C, d_S) by $\Gamma(d_C, d_S)$.

We distinguish between four cases.

Case 1: $(d_C, d_S) = (1,1)$. The organization invests both in establishing a high-functioning communication network and establishing standard operating procedures. This case is almost identical to the benchmark DS model, except that the variance of the local information in the second period is given by $\sigma_\theta^2 + \sigma_\varepsilon^2$ rather than σ_θ^2. Expected losses to the organization over the two periods are given by

$$\Gamma(1,1) = n\frac{\phi\beta(n-1)(1-p_H)}{\phi+\beta(n-1)(1-p_H)}\sigma_\theta^2 + n\frac{\phi\beta(n-1)(1-p_H)}{\phi+\beta(n-1)(1-p_H)}(\sigma_\theta^2+\sigma_\varepsilon^2) + F_S + F_C.$$

Case 2: $(d_C, d_S) = (0,0)$. There is no communication between agents and no standard operating procedures are established. From the perspective of agent j, the random variable $\theta_{i,1}$ has a mean 0 and variance $\hat{\sigma}_\theta^2 + \sigma_\theta^2$; the random variable $\theta_{i,1}$ has a mean 0 and variance $\hat{\sigma}_\theta^2 + \sigma_\theta^2 + \sigma_\varepsilon^2$. Expected adaptation and coordination losses are given by

$$\Gamma(0,0) = n\frac{\phi\beta(n-1)}{\phi+\beta(n-1)}(\sigma_\theta^2+\hat{\sigma}_\theta^2) + n\frac{\phi\beta(n-1)}{\phi+\beta(n-1)}(\sigma_\theta^2+\sigma_\varepsilon^2+\hat{\sigma}_\theta^2).$$

Case 3: $(d_C, d_S) = (0,1)$. There is no communication between agents, but there are standard operating procedures. From the perspective of agent j, the random variable $\theta_{i,1}$ has a mean $\hat{\theta}_i$ and variance σ_θ^2; the random variable $\theta_{i,2}$ has a mean $\hat{\theta}_i$ and variance $\sigma_\theta^2 + \sigma_\varepsilon^2$. Expected losses to the organization are given by

$$\Gamma(0,1) = n\frac{\phi\beta(n-1)}{\phi+\beta(n-1)}\sigma_\theta^2 + n\frac{\phi\beta(n-1)}{\phi+\beta(n-1)}(\sigma_\theta^2+\sigma_\varepsilon^2) + F_S.$$

Case 4: $(d_C, d_S) = (1,0)$. The firm invests in a communication network but not in standard operating procedures. From the perspective of agent j, the random variable $\theta_{i,1}$ has a mean 0 and variance σ_θ^2. If communication was not successful in the first period, the random variable $\theta_{i,2}$ has a mean 0 and variance $\sigma_\theta^2 + \sigma_\varepsilon^2$ in the second period. If communication was successful in the first period, we assume wlog that the mean $\hat{\theta}_i$ is also known to agent j, so that information about $\theta_{i,1}$ is also informative about

[5] To simplify the analysis, we assume that headquarters can only establish operating procedures in the first period. If σ_ε^2 is large and F_S is small, however, it may be optimal to update operating procedures in the second period.

$\theta_{i,2}$. Expected losses to the organization are given by

$$\Gamma(1,0) = n\frac{\phi\beta(n-1)(1-p_H)}{\phi+\beta(n-1)(1-p_H)}(\sigma_\theta^2+\hat{\sigma}_\theta^2)$$
$$+n\frac{\phi(1-p_H)(n-1)\beta}{\phi+(1-p_H)(n-1)\beta}(\sigma_\theta^2+(1-p_H)\hat{\sigma}_\theta^2+\sigma_\varepsilon^2)+F_C.$$

Fixing $d_S \in \{0,1\}$, the benefits of *investing in a communication network* ($d_C = 1$) equal

$$\Gamma(0,d_S) - \Gamma(1,d_S) \equiv \Delta_C(d_S).$$

Fixing $d_C \in \{0,1\}$, the benefits of *investing in standard operating procedures* ($d_S = 1$) equal

$$\Gamma(0,d_C) - \Gamma(1,d_C) \equiv \Delta_S(d_C).$$

It now easy to show that

(i) Communication networks and standard operating procedures are substitutes: $\Delta_C(0) > \Delta_C(1)$ and, similarly, $\Delta_S(0) > \Delta_S(1)$. Thus, investing in communication networks is less attractive if one also invests in standard operating procedures, and vice versa. Intuitively, better operating procedures reduce the value of horizontal communication as agent *j* can better predict the local information of agent *i* which, in turn, allows agent *i* to be more adaptive to his local information even in the absence of communication.

(ii) $\Delta_C(d_S)$ is increasing in σ_θ^2, σ_ε^2, $-F_C$, and p_H whereas $\Delta_S(d_C)$ is not affected by changes in σ_θ^2, σ_ε^2 and F_C and is decreasing in p_H. Hence, an increase in σ_θ^2, σ_ε^2, $-F_C$ or p_H makes investing in communication networks more attractive, whereas it does not affect or decreases the benefits of investing in standard operating procedures. Intuitively, σ_θ^2 reflects local information held by agent *i*, which is not captured in high-quality operating procedures, whereas σ_ε^2 reflects how quickly operating procedures become obsolete. Both therefore make communication networks more valuable.[6] Standard operating procedures are further less valuable as the communication quality p_H improves, as SOPs are only useful in case communication fails.

(iii) $\Delta_S(d_C)$ is increasing in $-F_S$ whereas $\Delta_C(d_S)$ is not affected by F_S.

The following proposition follows directly from the above observations:[7]

[6] Note, however, that an increase in $\hat{\sigma}_\theta^2$, that is the variance of the optimal standard operating procedures both makes standard operating procedures and communication networks more attractive. Since attention networks and SOPs are substitutes, the comparative statics with respect to $\hat{\sigma}_\theta^2$ are ambiguous.

[7] Our results are similar to those obtained in Aoki (1986) in a different team-theoretic setup. Building on Cremer (1980), Aoki compares the efficiency of vertical and horizontal information structures in coordinating operational decisions among interrelated units (shops) whose cost conditions are uncertain. Aoki then uses his model to compare stylized differences in the internal organization of large Japanese

Proposition 1. *$\Gamma(d_C, d_S)$ is supermodular in $d_C, -d_S, \sigma_\theta^2, \sigma_\varepsilon^2, p_H, -F_C$ and F_S. Hence:*

1. *Coordination through communication networks (standard operating procedures) is more (less) likely when*

   (i) *local information is more important—that is, σ_θ^2 is larger, and/or there is more environmental change—that is, σ_ε^2 is larger.*
   (ii) *communication quality p_H is higher or the cost of communication networks, F_C, is lower.*
   (iii) *the cost of implementing high-quality standard operating procedures, F_S, is higher.*

2. *Communication networks ($d_C = 1$) and standard operating procedures ($d_S = 1$) are substitutes:*

   (i) *A decrease in the cost of communication networks F_C can result in a change from $d_S^* = 1$ to $d_S^* = 0$, but never the other way around.*
   (ii) *A decrease in the cost of establishing standard operating procedures F_S can result in a change from $d_C^* = 1$ to $d_C^* = 0$, but never the other way around.*

25.3 Organizational Focus: Convexities in Attention Networks

In the previous section, it was assumed that communication networks are symmetric—all agents observe or learn each other's actions with the same probability p. Drawing upon Dessein, Galeotti, and Santos (2014, DGS hereafter), we now relax the assumption that communication networks are symmetric and allow for $p_i \neq p_j$ and let an organization designer optimize over $[p_i, \ldots, p_n]$. Our starting point is that organizational attention is scarce, and communication networks optimally distribute this attention among the agents of the organization. We show that even in a symmetric environment where all agents are ex ante identical, optimal attention networks and information flows are often asymmetric ex post because of the complementarity between attention and decision-making. In particular, when organizational attention is scarce, a hybrid approach to coordinating economic activity is optimal where attention networks coordinate the tasks of a select number of agents and the remaining tasks are coordinated using standard operating procedures. Scarce attention thus creates convexities in the optimal allocation of attention, where all attention is (optimally) monopolized by a few

and U.S. manufacturing firms. A horizontal information structure, similar to "coordinating through communication networks" in our model, is said to be more represenative of how Japanese firms coordinate production in the 1970s and early 1980s. In contast, it is observered how U.S. manufacturing firm tend to rely more on the use of a vertical information structure or what we refer to as "standard operating procedures".

agents. We first discuss this result in our baseline model and then discuss the robustness of our results to alternative communication technologies.

25.3.1 Baseline Model

Our starting point is that p_i—the quality of the communication about agent i's action—is increasing in the organizational attention t_i devoted to agent i. Organizational attention is scarce, however, in that there is a fixed attention budget: $\sum_i t_i \leq T$. We can think of t_i as the "air-time" or "attention" agent i receives and T could be the length of time agents spend in meetings as opposed to production. For our analysis, we revert to the original DS model where there is only one period and where the mean $\hat{\theta}_i$ of the random variable θ_i is common knowledge. Relative to DS, however, we add an additional stage 0 where the organizational designer optimally chooses the qualities $[p_1,\ldots,p_n]$ of the communication links. We assume that p_i follows a Poisson process—that is, $p_i = 1 - e^{-\lambda t_i}$ where λ is the constant hazard rate that any agent $j \in \mathcal{N} \setminus \{i\}$ correctly learns the primary action taken by agent i. We can interpret $1/\lambda$ as a measure of the complexity of tasks. Note that the communication cost or attention t_i required to achieve a given communication quality p_i is increasing and convex in p_i. This reflects decreasing marginal returns to attention. Denoting by $P \equiv 1 - e^{-\lambda T}$ the maximum communication quality that can be achieved by focusing all organizational attention on one agent, the organizational attention constraint $\sum_i t_i \leq T$ can be rewritten as

$$\sum_{i \in \mathcal{N}} \log(1 - p_i) \geq \log(1 - P). \tag{25.4}$$

At stage 0, an organization designer then optimally chooses $[p_1,\ldots,p_n]$ subject to constraint (25.4), which will be binding at the optimum.

Two tasks, two agents. Assume now first that $n = 2$, so that the organization consists of two agents and (25.4) is equivalent to $(1 - p_1)(1 - p_2) \geq 1 - P$. In order to increase the communication quality p_1, the organization then needs to reduce the communication quality p_2. The larger is the maximal communication probability P, the less there is a trade-off between good communication on task 1 versus task 2. Note that P will be large when attention is not scarce and/or tasks are not very complex.

Given that all agents are ex ante symmetric, all tasks are equally important, and interdependencies are symmetric, one may conjecture that the optimal communication network will be symmetric as well. Moreover, it is easy to verify that given constraint (25.4), $p_1 + p_2$ is uniquely maximized when $p_1 = p_2$. DGS show, however, that when organizational attention is scarce—that is, whenever $P < \bar{P}$—it is optimal to focus all attention on one of the two agents, $(p_1^*, p_2^*) \in \{(P,0),(0,P\}$. Intuitively, having a high-quality communication link to agent i and letting agent i be very adaptive to his local information are complementary choices. The more agent i is adaptive to his local information, the less valuable are standard operating procedures in coordinating tasks, and the more important is communication to achieve coordination. More attention should therefore be focused on agent i. By the same token, if agent i is not responsive to his local information, then it is a waste of time to devote attention to agent i, as coordination is achieved appropriately by the common knowledge and adherence to standard operating procedures.

Assume that equilibrium primary actions are linear in θ_i and $\hat{\theta}_i$, that is

$$a_{ii} = \hat{\theta}_i + \alpha_i(\theta_i - \hat{\theta}_i),$$

where α_i can be interpreted as the adaptiveness (or discretion) of agent i. Substituting a_{ii} in (25.1) and taking expectations, expected losses equal

$$\Gamma = \phi(1-\alpha_1)^2\sigma_\theta^2 + \phi(1-\alpha_2)^2\sigma_\theta^2 + \beta(1-p_1)\alpha_1^2\sigma_\theta^2 + \beta(1-p_2)\alpha_2^2\sigma_\theta^2. \qquad (25.5)$$

Inspecting (25.5), it is immediate that α_1 and p_1 are complementary choices. The more adaptive is agent 1, the larger are the benefits of improving communication about agent 1 in order to minimize Γ. If we had chosen an attention constraint with a constant rate of substitution between p_1 and p_2, for example, $p_1 + p_2 \leq P$, then it would always be optimal to focus all attention on one agent so that either $p_1 = 0$ or $p_2 = 0$. More naturally, however, there are decreasing marginal returns to attention, as captured by the constraint (25.4). Indeed, from (25.4), the higher is p_1, the more one needs to reduce p_2 for any additional increase in p_1. Such decreasing marginal returns create a countervailing force against focusing all attention on the same task. The following proposition, taken directly from DSG, shows that an asymmetric attention networks is optimal if and only if attention is scarce:

Proposition 2. *Suppose $\beta > 1$. There exists a $\bar{P}(\beta)$ such that:*

(i) An asymmetric (focused) attention network is optimal, $(p_1^*, p_2^*) \in \{(P,0),(0,P)\}$ if and only if $P \leq \bar{P}(\beta)$
(ii) A symmetric (balanced) attention network is optimal, $(p_1^*, p_2^*) = (\tilde{p}, \tilde{p})$ if and only if $P > \bar{P}(\beta)$, where $2\log(1-\tilde{p}) = \log(1-P)$
(iii) $\bar{P}(\beta)$ is increasing in the importance of coordination, β.

To summarize the above proposition, if organizational attention is scarce (T is small) or the environment are very complex (λ is small), then $P = 1 - e^{-\lambda T}$ is small as well, and it optimal to focus all attention on one agent, say agent 1. Agent 1 is then allowed to be very adaptive to his task, and coordination with agent 1 will be achieved through the attention network. In contrast, agent 2 will be forced to largely ignore his local information, and coordination with this agent's tasks will be achieved through adherence to the commonly known standard operating procedure, $\hat{\theta}_2$.

If, on the other hand, organizational attention is abundant (T is large) or the environment is not very complex (λ is large), then P will be large, and it will be optimal to have a symmetric attention network where both agents divide attention equally. Intuitively, it is then feasible for each agent to communicate his primary action almost perfectly. Both agents can then be responsive to their local information, and coordination will be achieved through the attention network for both agents. Standard operating procedures play a limited role in coordinating activity.

Large organizations. DGS extend their setup to incorporate $n > 2$ tasks, in which case they show that the optimal communication network consists of ℓ leaders and $n - \ell$ followers. All attention is equally split among the ℓ leaders, whereas no attention is devoted to the $n - \ell$ followers. The ℓ leaders are very responsive to their local information, and coordination with their task is achieved through the attention network. In contrast, coordination with the tasks of the $n - \ell$ followers is achieved by letting those agents stick closely the commonly known standard operating procedures. In other words, when attention is scarce, a hybrid approach to coordinating economic activity is optimal. Attention networks coordinate the tasks of a select number of agents and the remaining tasks are coordinated using standard operating procedures. The better the communication technology (the larger is λ), the larger is ℓ and the more the organization relies on attention networks rather than standard operating procedures. In contrast, the more interdependent are tasks, and the more important is the avoidance of coordination losses, the smaller is ℓ and the less the organization relies on networks.[8]

25.3.2 Alternative Communication Technologies

We now discuss the robustness of our results to alternative communication technologies. Let us denote by m_i the information received by agent j regarding θ_i, and define the *residual variance* about θ_i as $\text{Var}(\theta_i|m_i) \equiv E(\theta_i - E(\theta_i|m_i))^2$. In our *baseline model,* we have assumed that agent $j \neq i$ observes θ_i with a probability $p_i = 1 - e^{-\lambda t_i}$ where t_i is the attention devoted to task i. Given this communication technology

$$E[\text{Var}(\theta_i|m_i)] = \sigma_\theta^2(1 - p_i) = \sigma_\theta^2 e^{-\lambda t_i}. \tag{25.6}$$

Substituting (25.6) into (25.3), expected organizational losses can be written as:

$$\Gamma(t) = \sum_{i=1}^{i=n} \frac{\phi\beta(n-1)E(\text{Var}(\theta_i|m_i))}{\phi\sigma_\theta^2 + \beta(n-1)E(\text{Var}(\theta_i|m_i))}\sigma_\theta^2. \tag{25.7}$$

One can verify that $-\Gamma(t)$ is convex in attention t_i when t_i is small. As discussed above and shown by DGS, whenever organizational attention is scarce ($T < \bar{T}$), it follows that $-\Gamma(t)$ is maximized by setting $t_i = T/\ell$ for ℓ agents $i \in \mathcal{L} \subset \mathcal{N}$ and set $t_j = 0$ for

[8] In DGS, the number of leaders and who is a leader is determined in equilibrium. A number of other papers, such as Bolton, Brunnermeier, and Veldkamp (2012), Dessein and Santos (2014), and Van den Steen (2014), also build on DS in order to study how a leader can achieve coordination among members of an organization. Communication networks are not endogenized, however, as communication is always between the exogenously appointed leader and the remainder of the organization.

$n - \ell$ agents $j \in \mathcal{N} \backslash \mathcal{L}$. Put differently, there are convexities in the optimal allocation of attention: tasks should either receive a lot of attention or no attention at all.

Instead of the above binary communication technology, assume now that θ_i is independently normally distributed for $i \in \mathcal{N}$ and that agents $j \neq i$ observe a noisy message $m_i = \theta_i + \epsilon_i$ about θ_i, where ϵ_i is (independently) normally distributed. Given linear decision rules $a_{ii} = \hat{\theta}_i + \alpha_i(\theta_i - \hat{\theta}_j)$ and $a_{ji} = E(\theta_i|m_i)$, one can show (see DGS) that expected organizational losses given an attention network t are again given by (25.7). We now consider *two alternative communication technologies* who only differ in how fast $E(\mathrm{Var}(\theta_i|m_i))$ decreases as a function of the attention t_i devoted to θ_i.

25.3.2.1 Rational Inattention and Entropy Information Costs

Information Theory and the literature on Rational Inattention (Sims 2003) posit that the communication costs or "communication capacity" $C(m)$ required to send a message $m = (m_1,\ldots,m_n)$ about $\theta = (\theta_1,\ldots,\theta_n)$ is equal to the reduction in entropy of θ following the observation of m. Following this literature, we posit that communication is optimized under the constraint

$$C(m) = H(\theta) - H(\theta|m) \leq T, \tag{25.8}$$

where $H(\theta)$ is the (differential) entropy of θ and $H(\theta|m)$ the entropy of θ conditional upon observing m. In other words, the attention capacity T of the organization puts a constraint on the total reduction in entropy following communication.[9] Given that m_i and the conditional distributions $F(\theta_i|m_i)$ are independently normally distributed for $i \in \mathcal{N}$, attention constraint (25.8) can be rewritten as

$$\sum_{i \in \mathcal{N}} \left(2 \ln \sigma_\theta^2 - 2 \ln Var(\theta_i|m_i)\right) \leq T$$

or still: $\sum_{i \in \mathcal{N}} t_i \leq T$, with

$$\mathrm{Var}(\theta_i|m_i) = \sigma_\theta^2 e^{-2t_i}. \tag{25.9}$$

Since (25.9) and (25.6) are equivalent up to a rescaling of the attention capacity, we obtain identical results as in our baseline model. Hence, whenever organizational attention T is scarce, normally distributed information and entropy information costs imply that the optimal attention network is asymmetric where a few agents monopolize all attention.

[9] Formally, this is equivalent with assuming that T is the (Shannon) capacity of the Guassian communication channel. The capacity of a channel is a measure of the maximum data rate that can be reliably transmitted over the channel. Shannon capacity has proven to be an appropriate concept for studying information flows in a variety of disciplines: probability theory, communication theory, computer science, mathematics, and statistics, as well as in both portfolio theory and macroeconomics.

25.3.2.2 *Sampling from a Normal Distribution*

An alternative way of modeling noisy communication is to assume that the number of i.i.d. signals agent j receives about θ_i is linear in the attention t_i devoted to task i. Let m_i be the average realization of t_i signals $s_{ik} = \theta_i + \varepsilon_{ik}$, with $k \in \{1,..,t_i\}$, where ϵ_{ik} is i.i.d. normally distributed with variance σ_ε^2 : $m_i = \frac{1}{t_i}\sum s_{ik}$. Then $m_i = \theta_i + \epsilon_i$ where $\sigma_\epsilon^2 = \sigma_\varepsilon^2/t_i$ so that

$$\mathrm{Var}(\theta_i|m_i) = \frac{\sigma_\varepsilon^2}{\sigma_\varepsilon^2 + t_i\sigma_\theta^2}\sigma_\theta^2 \tag{25.10}$$

Whereas for t_i small, we have that $-\Gamma(t)$ is convex in t_i for communication technologies (25.9) and (25.6), it is now easy to verify that given communication technology (25.10), $-\Gamma(t)$ is always concave in t_i. Hence, given (25.10), the optimal allocation of attention is symmetric, that is, $t_i^* = T/n$ for all $i \in \mathcal{N}$. Intuitively, regardless of the communication technology, the complementarity between attention and decision-making results in a convexity in the value of information. Indeed, from (25.7), $-\Gamma(t)$ is convex in $-E(\mathrm{Var}(\theta_i|m_i))$. Convexities in the cost of communication, however, provide a countervailing force to focus all attention on a few tasks. Indeed, technologies (25.9), (25.6), and (25.10) all exhibit convex communication costs to reduce $\mathrm{Var}(\theta_i|m_i)$—that is, $\partial^2\mathrm{Var}(\theta_i|m_i)/\partial t_i^2 > 0$. It is only for communication technology (25.10), however, that this convexity in the cost of communication dominates the convexity in the value of information for any value of t_i. In contrast, for technologies (25.9) and (25.6), the convexity in the cost of communication only dominates when t_i is sufficiently large.

In the next section, we study optimal attention networks in more complex environments: (i) tasks and interdependencies between tasks are asymmetric, (ii) communication has both an active and a passive component, and (iii) agents do not necessarily maximize a common objective function. To simplify the analysis, we will assume that communication costs are sufficiently convex, as in technology (25.10), allowing us to focus on interior solutions.

25.4 Communication and Influence in Attention Networks: Interior Solutions

DGS showed that an ex-ante symmetric environment can give rise to a highly skewed communication patterns. This section considers asymmetric environments and studies how the ex-ante asymmetry determines an ex-post asymmetry in terms of communication and influence. The analysis will rely on Calvó, de Martí, and Prat (2015), henceforth CDP. The main outcome will be a set of predictions on communication and influence flows, which can be seen as a formalization of Arrow's (1974) theory of organizational communication discussed in the Introduction.

Consider a set of N agents. Each agent i faces a local state of the world:

$$\theta_i \sim \mathcal{N}(0, 1/s_i),$$

where s_i denotes the precision of θ_i, i.e. $s_i = 1/Var(\theta_i)$. If the local states were correlated, agents' actions may be correlated in equilibrium even if agents do not communicate. We prefer to abstract from this form of direct correlation in order to focus on the role of communication. Therefore, we assume that θ_i is independent across agents.

Each agent i observes only θ_i and can then engage in communication with all other agents. Information transmission requires effort from both the sender and the receiver. The signal is more precise if the sender invests more in *active communication* (e.g., speaking or writing) and the receiver invests more in *passive communication* (e.g., listening or reading). Namely, agent i receives message y_{ij} from agent j, such that

$$y_{ij} = \theta_j + \varepsilon_{ij} + \eta_{ij},$$

where ε_{ij} and η_{ij} are two normally distributed noise terms

$$\varepsilon_{ij} \sim \mathcal{N}\left(0, 1/r_{ij}\right), \qquad (25.11)$$

$$\eta_{ij} \sim \mathcal{N}\left(0, 1/p_{ij}\right), \qquad (25.12)$$

and r_{ij} (resp. p_{ij}) is the precision of ε_{ij} (resp. η_{ij}). We assume that all stochastic terms are mutually independent (and independent of the θ's).

In the first stage of the game every agent chooses how much to invest in communication. Namely, agent i selects the values of two vectors: (i) The precision of the active communication part of all the signals he sends: $\left(r_{ji}\right)_{j\neq i}$, for which he incurs cost $k_r^2 \sum_{j\neq i} r_{ji}$, where $k_r \geq 0$ is a parameter; (ii) The precision of the passive communication part of all the signals he receives, $\left(p_{ij}\right)_{j\neq i}$, for which he incurs cost $k_p^2 \sum_{j\neq i} p_{ij}$, where $k_p \geq 0$ is a parameter (p is mnemonic for passive).

In the second stage of the game, every agent observes the signals he has received from the other agents and chooses the value of action $a_i \in (-\infty, \infty)$.

CDP can be formulated in three ways: as a noncooperative game where each agent maximizes his own expected payoff, as a team-theoretical problem where each agent maximizes the sum of expected payoffs of all agents, or as a hybrid problem where communication investments are made cooperatively while actions are chosen selfishly. An earlier version of CDP (Calvó, de Martí, and Prat 2009) considered all three versions and show they produce qualitatively similar solutions. The present chapter focuses exclusively on the first version.

The payoff of agent i is a classic quadratic objective function

$$u_i = -\left(d_{ii}\left(a_i - \theta_i\right)^2 + \sum_{j\neq i} d_{ij}\left(a_i - a_j\right)^2 + k_r^2 \sum_{j\neq i} r_{ji} + k_p^2 \sum_{j\neq i} p_{ij}\right), \qquad (25.13)$$

where the term d_{ii} measures the adaptation motive (i.e., the importance of tailoring *i*'s action to the local state), and the term d_{ij} represents the coordination motive, namely the interaction between the action taken by agent *i* and the action taken by agent *j*. For the rest of the paper we assume that the interaction terms are positive ($d_{ij} \geq 0$ for all *i* and all *j*).

The game has two versions according to whether investment in communication occurs before or after the agent observes his local state θ_i. The "before" version captures the idea that investment has a long-term component (e.g., two firms appoint liaison officers). The "after" version represents a shorter-term investment, like the direct cost of writing and reading a report. As Calvó. et al. (2015) discuss, both versions have the same linear pure-strategy equilibrium. For concreteness, in this chapter, we focus on the "before" version. We refer to this game as $\Gamma(\boldsymbol{D},\boldsymbol{k},\boldsymbol{s})$, where $\boldsymbol{D} = \left(d_{ij}\right)_{i,j}$, $\boldsymbol{k} = \left(k_r, k_p\right)$ and $\boldsymbol{s} = (s_i)_i$.

This can be seen as a game of communication and influence. In equilibrium, agents communicate with each other, and they influence each other's decisions through the signals they communicate. The analysis of this game is divided in two parts. First, we provide a closed-form characterization of equilibrium. Second, we show that influence in equilibrium is approximated with an appropriately defined notion of eigenvector centrality, which can be computed directly on the interaction matrix $\boldsymbol{D}$.

Let us begin by characterizing equilibrium play. To do this, consider first a game with just two players. First, normalize the interaction matrix of agent *i* by dividing it by the sum of all interaction terms

$$\omega_{ij} = \frac{d_{ij}}{d_{i1} + d_{i2}}.$$

The payoff—net of communication costs—of, say, agent 1 can now be written as

$$-\omega_{11}\overbrace{(a_1 - \theta_1)^2}^{\text{adaptation}} - \omega_{12}\overbrace{(a_1 - a_2)^2}^{\text{coordination}}.$$

Focus on the second stage of the game. Given investments in communication, how do agents select their actions as functions of signals they receive? One can check this stage has a linear equilibrium of the following form:

$$a_1^* = b_{11}\theta_1 + b_{12}y_2$$
$$a_2^* = b_{21}y_1 + b_{22}\theta_2$$

where the *b*-coefficients solve

$$\begin{array}{ll} b_{11} = \omega_{11} + \omega_{12}b_{21} & b_{22} = \omega_{22} + \omega_{21}b_{12} \\ b_{12} = \omega_{12}b_{22}\frac{r_{12}p_{12}}{s_2 r_{12} + s_2 p_{12} + r_{12}p_{12}} & b_{21} = \omega_{21}b_{11}\frac{r_{21}p_{21}}{s_1 r_{21} + s_1 p_{21} + r_{21}p_{21}}. \end{array}$$

Note that b_{12} and b_{21} represent the influence of agents on each other. The influence of *i*'s signal on *j* decision depends on how informative *i*'s signal is as well as how much *j* cares about coordinating with *i*.

Once we know what happens in the second stage, we can use backward induction to solve for equilibrium communication investments in the first stage. We find that active communication and passive communication are, respectively:

$$r_{21} = b_{21}\frac{\sqrt{d_{12}}}{k_r}, \tag{25.14}$$

$$p_{12} = b_{12}\frac{\sqrt{d_{11}+d_{12}}}{k_p}. \tag{25.15}$$

Investment in active and passive communication is not equal. The numerator of the right-hand side of equation (25.14) contains only d_{12} while its counterpart in (25.15) contains both d_{11} and d_{12}. Passive communication offers a more direct return: the listener can make use of the signal he receives. Active communication is instead more indirect: the speaker makes an investment in order for the listener to use the resulting signal. Therefore, in this type of model passive communication has an intrinsic advantage. Of course this advantage can be undone by other considerations, like a lower cost of investing in active communication or the presence of economies of scope. However, everything else equal, agents will invest relatively more in listening than in speaking.

Once the two-agent case is solved, one can more easily understand the general n-agent case. The logic is similar but we must add one important element: the possibility of indirect effects among agents. For example, if there are three agents, i may want to learn about j's state because he cares about m's action and he knows that m cares about j's state.

Two additional pieces of notation are needed. Let $\mathbf{\Omega}$ be the matrix of normalized interactions with typical element ω_{ij}. Let the matrix of normalized benefits be given by

$$h_{ij} = \begin{cases} \omega_{jj} & \text{if } i = j \\ -s_j\left(\frac{k_p}{\sqrt{D_i}} + \frac{k_r}{\sqrt{d_{ji}}}\right), & \text{otherwise.} \end{cases}$$

Provided that the cost of communication parameters k_r and k_p are sufficiently low (to avoid corner solutions), we have:

Proposition 3. *The game* $\Gamma(\mathbf{D}, \mathbf{k}, \mathbf{s})$ *has a linear equilibrium where:*

(i) Decisions are given by

$$\mathbf{b}_{\cdot j} = (\mathbf{I} - \mathbf{\Omega})^{-1} \cdot \mathbf{h}_{\cdot j} \quad \textit{for all } j;$$

(ii) Active communication is

$$r_{ij} = \frac{\sqrt{d_{ji}}b_{ij}}{k_r} \quad \textit{for all } i \neq j;$$

(iii) Passive communication is

$$p_{ij} = \frac{\sqrt{D_i} b_{ij}}{k_p} \qquad \text{for all } i \neq j.$$

Proposition 3 is the generalization of the two-agent case. The inverse matrix $(I - \Omega)^{-1}$, which captures the direct and indirect interactions of agents' actions on one another. It can be understood as an infinite series of higher-order normalized effects:

$$(I - \boldsymbol{\Omega})^{-1} = I + \boldsymbol{\Omega} + \boldsymbol{\Omega}^2 + \boldsymbol{\Omega}^3 + \cdots = \sum_{l \geq 0} \boldsymbol{\Omega}^l.$$

Proposition 3 can be seen as one way of formalizing Arrow's (1974) idea that communication and decisions pattern are shaped by the objectives of the members of the organization. Given underlying parameters that describe complementarities, information cost, and uncertainty, we can predict how much each agent will communicate in equilibrium and who much he will be influenced by other agents.

The second part of the analysis focuses on influence. Proposition 3 characterizes influence as a game-theoretic phenomenon. It turns out that this strategic approach is approximately equal to a much simpler network centrality concept.

To see, we need to additional definitions. First, define a sequence of games as follows. Fix D, s, k_r, and k_p, and define the payoff function:

$$u_i = -\left(d_{ii} (a_i - \theta_i)^2 + \frac{1}{t} \sum_{j \neq i} d_{ij} \left(a_i - a_j\right)^2 + t^{\lambda} k_r^2 \sum_{j \neq i} r_{ji} + t^{\lambda} k_p^2 \sum_{j \neq i} p_{ij} \right),$$

where $t \in (0, \infty)$ and $\lambda > 1$. For every value of t we define a different game, which we can call $\Gamma\left(D, s, k_r, k_p, t\right)$. As t goes to zero, coordination becomes relatively more important than adaptation (and communication costs go down in order to guarantee that the solutions do not run into non-negativity constraints on communication intensities).

For every value of the parameter t, we have a natural definition of an agent's influence as the effect on all agent's actions (including his own) of an increase in his own state. Namely, the global influence of agent i, that we denote by $\mathcal{I}_i$, is

$$\mathcal{I}_i(t) = \sum_{j=1}^{n} b_{ji} \qquad k = 1, \ldots, n.$$

Second, let us introduce an axiomatic network centrality concept, which seeks to assign an "importance" index to every node of a network purely on the basis of the strength of links between agents. The concept, referred to as eigenvector centrality, has been known since the 1950s and has found a larger number of applications in a number of fields. Palacios-Huerta and Volij (2004) provided an axiomatization of the index, and Golub and Jackson (2010) used it in network economics.

Let $\tilde{G}$ be the matrix with entries $\gamma_{ii} = 0$ for all i, and $\gamma_{ij} = \frac{d_{ij}}{\sum_{k \neq i} d_{ik}}$. The eigenvector index of agent i is ι_i, defined as the i-th component of the vector that solves:

$$\iota = \tilde{G}' \iota$$

and that satisfies $\sum_j \iota_j = 1$.

We can show that the game-theoretic influence index tends to the axiomatic influence index when t goes to zero:

Proposition 4. *As $t \to 0$, the relative global influence of agents converges to the ratio of eigenvector centrality indices weighted by an adaptation vs coordination ratio. Namely, for any i and j,*

$$\lim_{t \to 0} \frac{\mathcal{I}_i(t)}{\mathcal{I}_j(t)} = \frac{\iota_i}{\iota_j} \frac{\frac{d_{ii}}{D_{-i}}}{\frac{d_{jj}}{D_{-j}}}$$

In particular, if $d_{ii} = d_{jj}$ and $D_{-i} = D_{-j}$ for all $i, j \in N$, then we obtain that

$$\lim_{t \to 0} \frac{\mathcal{I}_i(t)}{\mathcal{I}_j(t)} = \frac{\iota_i}{\iota_j}$$

This result implies that, when t is sufficiently small, namely when coordination is more important than adaptation, eigenvector centrality is a good approximation of game-theoretic influence. This result is useful in practice because eigenvector centrality is easier to compute. It also creates a conceptual link between equilibrium influence in organizations and influence as defined in other contexts where eigenvector centrality is often used, such as search engines and bibliometrics.

25.5 Conclusions

The three previous sections discussed key findings of the theory of attention networks. First, a decentralized attention network or a set of standardized operating procedures are two alternative ways of achieving coordination among members of an organization. The former is more likely to be optimal when local information is more important and communication costs are lower. Second, influence and communication patterns can be highly asymmetric even starting from a perfectly symmetric interaction function. When attention is scarce, it is optimal for an organization to direct their members' attention to a small set of key agents. Third, influence and communication patterns within an organization are highly interrelated. If we observe communication patterns—for instance, through electronic records—we can use eigenvector centrality to rank the influence of the members of the organization.

The rest of this section discusses two promising applications of the endogenous communication network framework. We first discuss empirical analyses, and then models that combine endogenous communication and behavioral biases.

The theories discussed in this chapter yield testable predictions on communication patterns within organizations. Both Dessein et al. (2014) and Calvó et al. (2015) characterize information flows as a function of the underlying primitives. These predictions can be tested, or used as a basis for estimation, provided one has data on communication patterns.

Until the 1990s, communication within organizations could only be measured through ethnographic studies. Mintzberg (1973) used personal observation to study how five top executives allocated their time to various activities. However, the IT revolution has created a wealth of data on communication patterns within organizations, such as email records and calendar information.

Palacios-Huerta and Prat (2011) analyze two data sets containing email communication traffic between all the executives of the same company (a European retailer). As suggested by Theorem 4 of Calvó et al. (2015), they compute the predicted influence of agents on the basis of their eigenvector centrality in the email traffic network. The email-based influence index of an agent turns out to be strongly correlated to the agent's influence as proxied by standard organizational variables, such as income and rank. Moreover, the discrepancies between the current email-based index of an agent and his or her current income and rank predict future promotions and dismissals.

Bandiera et al. (2011, 2014) analyze communication patterns within organizations from a different angle. They collect information on how chief executive officers of hundreds of companies around the world utilize their work time. In particular, they observe who the CEO spends time with. This includes internal constituencies such as the finance division or the marketing division, or external constituencies such as customers or investors. The theories reviewed here predict that the allocation of this very scarce resource, CEO attention, should reflect the priorities of the company and the CEO.[10]

Let us finally turn our attention to behavioral theories that incorporate elements of endogenous communication within organizations. The models reviewed in this chapter build on Bayesian agents. However, the idea of organizations with endogenous information transmission can be extended to settings where agents have cognitive biases. In fact, an important question is how organizations will structure themselves in order to minimize the potentially detrimental effects of biased information processing. Bénabou (2013) considers a network of agents with anticipatory bias, which affects how they process and recall the information they observe. In such a setting, information avoidance may be beneficial or detrimental to welfare. If it makes bad news even worse for other agents (as in the case of risk spillovers), then it is detrimental.

[10] Halac and Prat (2014) study a principal-agent model, where the principal may invest in a non-directly observable attention technology that provides recognition and reward to a hard-working agent. The model is analyzed in a dynamic setting.

If it dampens the effect of bad news (as in the case of group morale), then it is beneficial.

Sethi and Yildiz (2013) consider endogenous communication networks where individual agents have subjective prior beliefs. In each period an agent receives a private signal and chooses to observe the opinion of another agent in the network. Observing an opinion provides information both about the state of the world and the prior of the agent whose opinion is observed. Long-run behavior may be history-dependent and inefficient, with some agents emerging as opinion leaders.

Makarov (2011), finally, studies an organization where employees display present-bias preferences and communication (e.g., email) can be high-priority or low-priority. In equilibrium, the organization suffers from social procrastination as agents spend excessive time on low-priority communication. In this setting, the organization may benefit from policies that restrict communication.

References

Alonso, Ricardo, Wouter Dessein, and Niko Matouschek (2008). "When does coordination require centralization?" *American Economic Review* 98(1), 145–179.

Alonso, Ricardo, Wouter Dessein, and Niko Matouschek (2015). "Organizing to adapt and compete." *American Economic Journal: Microeconomics* 7(2), 158–87.

Aoki, Masahiko (1986). "Horizontal vs. vertical information structure of the firm." *American Economic Review* 76, 971–983.

Arrow, Kenneth J. (1974). *The Limits of Organization*. New York: Norton.

Bandiera, Oriana, Luigi Guiso, Andrea Prat, and Raffaella Sadun (2011). "What do CEOs do?" CEPR Discussion Paper 8235, February 2011.

Bandiera, Oriana, Andrea Prat, and Raffaella Sadun (2014). "Managerial capital at the top: Evidence from the time use of CEOs." Working paper, Columbia University, December 2014.

Bénabou, Roland (2013). "Groupthink: Collective delusions in organizations and markets." *Review of Economic Studies* 80, 429–462.

Bolton, Patrick, Markus Brunnermeier, and Laura Veldkamp (2013). "Leadership, coordination and corporate culture." *Review of Economic Studies* 80(2), 512–537.

Bolton, Patrick and Matthias Dewatripont (1994). "The firm as a communication network." *Quarterly Journal of Economics* 109(4), 809–839.

Burt, Ronald S. (2005). *Brokerage & Closure*. New York: Oxford University Press.

Calvó-Armengol, Antoni, Joan de Martí, and Andrea Prat (2009). "Endogenous communication in complex organizations." Working paper, London School of Economics.

Calvó-Armengol, Antoni, Joan de Martí, and Andrea Prat (2015). "Communication and influence." *Theoretical Economics* 10(2), 649–690.

Cremer, Jacques (1980). "A partial theory of the optimal organization of a bureaucracy." *The Bell Journal of Economics* 11(2), 683–693.

Cremer, Jacques, Luis Garicano, and Andrea Prat (2007). "Language and the theory of the firm." *Quarterly Journal of Economics* 122(1), 373–407.

Dessein, Wouter, Andrea Galeotti, and Tano Santos (2014). "Rational inattention and organizational focus." American Economic Review. Forthcoming.

Dessein, Wouter and Tano Santos (2006). "Adaptive organizations." *Journal of Political Economy* 114(50), 956–995.

Dessein, Wouter and Tano Santos (2014). "Managerial style and attention." Columbia Business School, Mimeo.

Dewan, Torun and David Myatt (2008). "The qualities of leadership: Direction, communication, and obfuscation." *American Political Science Review*, 102(3), 351–368.

Galeotti, Andrea, Christian Ghiglino, and Francesco Squintani (2013). "Strategic information transmission in networks." *Journal of Economic Theory* 148(5), 1751–1769.

Garicano, Luis (2000). "Hierarchy and the organization of knowledge in production." *Journal of Political Economy* 108(5), 874–904.

Garicano, Luis and Andrea Prat (2013). "Organizational economics with cognitive costs." In *Advances in Economics and Econometrics: Theory and Applications, Proceedings of the Tenth World Congress of the Econometric Society*. Cambridge University Press.

Gibbons, Robert and John Roberts (eds.) (2013). *The Handbook of Organizational Economics*. Princeton University Press.

Halac, Marina and Andrea Prat (2014). "Managerial attention and worker engagement." CEPR Discussion Paper 10035.

Hagenbach, Jeanne and Frédéric Koessler (2010). "Strategic communication networks." *Review of Economic Studies* 77(3), 1072–1099.

Golub, Benjamin and Matthew O. Jackson (2010). "Naive learning in social networks and the wisdom of crowds." *American Economic Journal: Microeconomics* 2(1), 112–149.

Mankins, Michael C., Chris Brahm, and Gregory Caimi (2014). "Your scarcest resource." *Harvard Business Review* 92(5), 74–80.

Makarov, Uliana (2011). "Networking or not working: A model of social procrastination from communication." *Journal of Economic Behavior and Organization* 80(3), 574–585.

Palacios-Huerta, Ignacio and Andrea Prat (2010). "Measuring the impact factor of agents within an organization using communication patterns." CEPR Discussion Paper 8040.

Palacios-Huerta, Ignacio and Oscar Volij (2004). "The measurement of intellectual influence." *Econometrica* 72(3), 963–977.

Sims A. Christopher (2003). "Implications of rational inattention." *Journal of Monetary Economics* 50(3), 665–690.

Sethi, Rajiv and Muhammet Yildiz (2013). "Perspectives, opinions, and information flows." MIT Department of Economics Working Paper 13–23.

Van den Steen, Eric (2014). "A formal theory of strategy." Harvard Business School Working Paper.

CHAPTER 26

MODELS OF BILATERAL TRADE IN NETWORKS

MIHAI MANEA

26.1 Introduction

The study of prices and allocations in markets constitutes a central part of economics. The classical paradigm of general equilibrium presumes large economies in which no individual trader has market power. Under this approach, goods are homogeneous and infinitely divisible, and the law of one price holds—that is, identical goods have the same price. Moreover, traders take prices as given. The market clearing conditions implicitly assume that all buyers can trade freely with all sellers without delays or transaction costs. The actual trading procedure and the price formation mechanism are, however, left unmodeled. The theory predicts that efficient allocations emerge in such frictionless competitive economies.

General equilibrium theory, as such, is silent on what the underlying trading process is and how exactly prices emerge. Stahl (1972) and Rubinstein (1982) applied the tools of noncooperative game theory to model explicitly the bargaining process that determines prices. Their pioneering noncooperative bargaining models consider bilateral monopoly situations in which both parties have market power. Building on the noncooperative bargaining approach, Rubinstein and Wolinsky (1985, 1990), Gale (1986a, b, 1987), and McLennan and Sonnenschein (1991) later developed the analysis of trade via bilateral exchanges in dynamic markets with a large number of traders. The primary motivation of this early work on bargaining in markets was to provide noncooperative foundations for general equilibrium theory. The research identifies conditions on the composition of the economy and the matching process that ensure convergence to a competitive equilibrium as bargaining frictions vanish.

However, the assumptions that deliver noncooperative foundations for general equilibrium theory do not reflect the realities of many economic activities. Price-taking behavior is unrealistic in markets that involve only a small number of traders. Social and business relationships, geography, and technological compatibility determine which pairs of traders can engage in exchange. Some trades may entail prohibitive transaction costs. Certain goods and services are tailored for specific segments of the market. Trade is dynamic and market conditions change over time. These departures from the foundations of general equilibrium theory give rise to complex market interactions. Local demand and supply influence the market power of every trader, yet trading activity in remote areas of the economy may have significant spillover effects. Hence, the bargaining power of every trader depends on both the local topology of trading opportunities and the global market architecture. The varying nature of competition in different parts of the economy leads to systematic deviations from the law of one price. Furthermore, decentralized bargaining may generate inefficient outcomes because local incentives for trade are not necessarily aligned with global efficiency.

The asymmetries among both traders and goods described above naturally call for a network formulation. A link between a pair of traders indicates that they can trade (or form a partnership) with each other. Many questions emerge: how do local prices depend on the global network structure? When does the law of one price hold? How do competitive forces in different segments of the market determine the bargaining power of every position in the network? Are trading outcomes efficient? How does the network evolve over time as traders reach agreements and exit the market? What networks are expected to form if traders have to invest in their links? How does the underlying market mechanism affect payoffs and allocations? These questions constitute the focus of an active area of research in network economics. In this chapter we survey some important contributions to the literature, focusing mainly on noncooperative models of bilateral bargaining in networks.

Section 26.2 defines some key network concepts and reviews two classic results from graph theory, Hall's marriage theorem and the Gallai-Edmonds decomposition theorem. In Section 26.3 we discuss bargaining models in which every pair of traders that reaches an agreement is removed from the network without replacement. Thus the pool of bargaining partners for each trader active in the market shrinks over time. In this setting, traders need to anticipate how the network of trading opportunities will evolve and how their future bargaining position will improve or deteriorate when agreements are forged in different parts of the network. The initial work of Corominas-Bosch (2004) and Polanski (2007) in this area assumes that linked traders are matched in pairs according to a centralized mechanism that maximizes the total surplus that can be achieved from exchange. As a consequence, these models predict that all agreements occur simultaneously, at the beginning of the game, and that equilibrium outcomes are efficient. Both Corominas-Bosch and Polanski find a close relationship between equilibrium payoffs and the Gallai-Edmonds decomposition. In contrast to the research of Corominas-Bosch and Polanski, Abreu and Manea (2012a, b) study markets with decentralized matching in which a single pair of linked traders bargains at a time. Decentralized bargaining leads to richer market dynamics: not all matches result in agreement, a pair of linked players may refuse to trade at some stage yet agree to trade at a later one, and multiple equilibria may coexist. Moreover, decentralization creates incentives for inefficient trade when we restrict attention to Markov perfect equilibria. A construction of a subgame perfect equilibrium developed by Abreu and Manea (2012b) demonstrates that efficiency may nonetheless be attained in every network using a complex design of punishments and rewards. We compare the balance of bargaining power across the models covered in Section 26.3 and reflect on sources of divergence.

Section 26.4 presents the stationary bargaining model of Manea (2011a). In that model, players who reach agreements exit the market and new players fill the positions left vacant, so that the network of trading opportunities does not change over time. A partition of the network into oligopolies emerges endogenously in equilibrium. The oligopolies with the lowest seller-to-buyer ratio have the greatest market power and drive market outcomes. The tractability of the model permits a detailed analysis of both price dispersion in the network and strategic network formation. We conclude the section with Nguyen's (2014) alternative characterization of equilibrium payoffs based on convex programing.

Section 26.5 reviews the celebrated assignment game of Shapley and Shubik (1971) along with its network applications. After summarizing the key results of Shapley and Shubik, we describe two price formation mechanisms—one proposed by Crawford and Knoer (1981), the other by Demange, Gale, and Sotomayor (1986)—both of which converge to the buyer-optimal core allocation in the assignment game. We then turn to a seminal contribution to the literature on buyer-seller networks due to Kranton and Minehart (2001). In that setting, buyers have private information about their valuations and sellers run simultaneous price-ascending auctions for their linked buyers à la Demange, Gale, and Sotomayor. For every network, the buyer-optimal core outcome of the associated assignment game emerges in equilibrium. Furthermore, a network that maximizes total expected welfare forms if buyers make linking investments before their values are realized. We briefly discuss work by Elliott (2015) showing that the latter conclusion relies critically on the assumption that one side of the market incurs all linking costs. Lastly, we analyze the noncooperative model of Elliott and Nava (2014), which introduces bargaining frictions in the assignment game. In contrast to the other work explored in Section 26.5, Elliott and Nava find that trade is typically inefficient. Section 26.6 provides concluding remarks and suggests directions for future research.

26.2 Framework

A finite set N of *players* interacts in a market. In examples, we assume that $N = \{1,2,\ldots,n\}$ for some positive integer n. Each player can forge at most one agreement with some other player. Agreements may entail forming a partnership or trading a single unit of an indivisible good. The set of feasible bilateral agreements is described by a network. Formally, a *network* formed by the set of players (or *nodes*) N is a collection of *links* $G \subseteq \{(i,j) \mid i \neq j \in N\}$ with $(i,j) \in G$ if and only if $(j,i) \in G$. To avoid double counting in summations and set cardinalities, we identify the pairs (i,j) and (j,i) and

FIGURE 26.1 A network example.

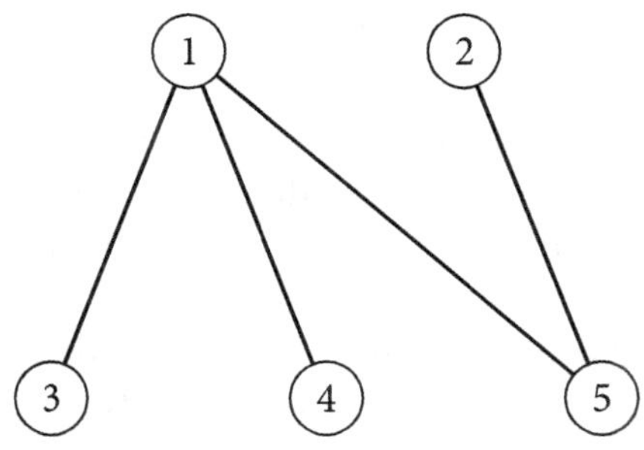

use the shorthand ij for the corresponding undirected link between i and j. When $ij \in G$, we say that i is *linked* to (or is a *neighbor* of) j in G. A node $i \in N$ is *isolated* in G if it is not linked to any node in G. For any network G and subset of nodes $M \subseteq N$, let $\mathcal{L}^G(M)$ denote the set of all neighbors in G of nodes from M, that is, $\mathcal{L}^G(M) = \{i \mid \exists j \in M, ij \in G\}$. A set $M \subseteq N$ is called *G-independent* if no pair of nodes in M is linked in G, i.e., $\mathcal{L}^G(M) \cap M = \emptyset$.

In the models of Sections 26.3 and 26.4, the existence of a link ij in G indicates that players i and j can generate a *unit* of surplus by reaching an agreement with each other. The assumption of symmetric link values allows us to focus exclusively on how the structure of the network G affects the bargaining power of every node in G. For some of the models we discuss, players are partitioned into *buyers* and *sellers*, with the corresponding sets denoted by B and S ($B \cup S = N$ and $B \cap S = \emptyset$). A network G is *bipartite* with the partition (B, S) if every link in G connects a buyer to a seller or, using the notation above, $\mathcal{L}^G(B) \subseteq S$ and $\mathcal{L}^G(S) \subseteq B$. The underlying assumption is that buyers have unit demand and sellers have unit supply.

A network G' is a *subnetwork* of G if it consists of a subset of the links in G, i.e., $G' \subseteq G$. The subnetwork G' of G *covers* a set of nodes $M \subseteq N$ if every node in M has at least one link in G'. The *subnetwork* of G *induced* by a subset of nodes M is the network $\{ij \in G \mid i, j \in M\}$. Players i and j are (path) *connected* in a network G if there exists a sequence $i_0 = i, i_1, \ldots, i_{\bar{k}} = j$ such that $i_k i_{k+1} \in G$ for $k = \overline{0, \bar{k} - 1}$. Connectedness is an equivalence relation; its equivalence classes are called *connected components* of G. Thus all players in a given connected component are connected by a path in the network, and there are no links between distinct components.

A *matching* in a network G is a subnetwork H of G in which every node has at most one link (i.e., $H \subseteq G$ and there do not exist three distinct nodes i, j, k in G such that both links ij and ik belong to H). If $ij \in H$, then we say that i and j are *matched* (with each other) under H. A matching of G is *perfect* if it covers all the nodes in G. Matchings of G that contain the largest number of links are called *maximum matchings*. The cardinality of the maximum matchings of G defines the *maximum total surplus* in G. For example, in the five-player network from Figure 26.1, there are two maximum matchings, $\{(1,3),(2,5)\}$ and $\{(1,4),(2,5)\}$, each generating a maximum total surplus of 2.

A simple condition due to Hall (1935) characterizes the bipartite networks which admit perfect matchings. More generally, Hall's theorem provides a necessary and sufficient condition for such networks to contain matchings that cover all the buyers (or sellers). The condition requires that every group of buyers be collectively linked to a set of sellers that has at least the same cardinality.

Theorem 1 (Hall's Marriage Theorem). *Suppose that G is a bipartite network with the partition (B,S). Then there exists a matching of G that covers B if and only if*

$$|\mathcal{L}^G(M)| \geq |M|, \forall M \subseteq B. \tag{26.1}$$

The network G has a perfect matching if and only if it has an equal number of buyers and sellers ($|B| = |S|$) and satisfies condition (26.1).

Maximum matchings determine the total surplus that can be generated when bilateral trades are organized efficiently. The following characterization of maximum matchings, independently discovered by Gallai (1964) and Edmonds (1965), proves useful in the welfare analysis of the models we consider. The result relies on the following partition of the set of nodes N in a network G. The set of *under-demanded* nodes in G, denoted by U, consists of all nodes i with the property that there exists at least one maximum matching of G in which i is not matched. The set of *over-demanded* nodes in G, denoted O, consists of nodes that do not belong to U and are linked to at least one node in U. The set of *perfectly matched* nodes in G is formed by the remaining nodes, $P = N \setminus (O \cup U)$. For an illustration, in the network from Figure 26.1, nodes 3 and 4 are under-demanded, 1 is over-demanded, and 2 and 5 are perfectly matched.

Theorem 2 (Gallai-Edmonds Decomposition). *Fix a network G with the sets of perfectly matched, over-demanded, and under-demanded nodes denoted by P, O, and U, respectively.*

1. *The following statements hold for every maximum matching H of G:*
    - *every node in O is matched under H to a node in U;*
    - *every element of P is matched under H to another element of P;*
    - *every connected component of the subnetwork induced by U in G has an odd number of nodes, and in each such component all nodes except at most one are covered by H.*
2. *The network obtained by removing any single node from every connected component of the subnetwork induced by U in G admits a perfect matching.*
3. *If G is bipartite, then U is G-independent.*[1]

[1] Part (3) of the result is not regularly stated with the decomposition theorem but follows immediately from part (2). For a proof, assume without loss of generality that one connected component C of the subnetwork induced by U in G includes more buyers than sellers. If C consists of more than one buyer,

The comprehensive monograph on matching theory of Lovász and Plummer (2009) provides a modern reference for Theorems 1 and 2.

A few special networks come up in our analysis and examples. In a *line* network, the set of nodes is ordered linearly and every node is linked only to its immediate predecessor and successor in the order (if any); we refer to the first and last nodes in the order as the *endpoints* of the line. A *cycle* is a network obtained by adding a link between the endpoints of a line network. A *star* is a network in which one node, called the *center*, is linked to all the other nodes, the *spokes*, and there are no links between spokes.

26.3 Non-Stationary Bargaining Models

In this section we discuss several noncooperative bargaining models in which pairs of players linked in a network G forge agreements and exit the market. Every model assumes that players bargain with their neighbors in G at discrete dates $t = 0, 1, \ldots$ over an infinite time horizon and have a common discount factor $\delta \in (0, 1)$. Pairs of players that reach agreements trade and leave the network. The remaining players continue to bargain in the resulting subnetwork. Unless otherwise stated, at every stage, players have perfect information of all past actions (including moves by nature).

Let v_i^δ denote the (expected equilibrium) *payoff* of player $i \in N$ for a given discount factor δ. We say that a payoff profile $(v_i^\delta)_{i \in N}$ (and the equilibrium behavior that generates it) is *efficient* for the underlying network G if $\sum_{i \in N} v_i^\delta$ is equal to the maximum total surplus in G. Efficiency requires that all agreements take place without any delay. In settings where matching is decentralized and agreements cannot occur simultaneously, the following welfare criterion is more suitable. The family of payoffs $((v_i^\delta)_{i \in N})_{\delta \in (0,1)}$ (and the corresponding equilibrium outcomes) is *asymptotically efficient* if $\sum_{i \in N} v_i^\delta$ converges to the maximum total surplus in G as $\delta \to 1$.

26.3.1 Bargaining with Public Offers

Corominas-Bosch (2004) studies a game on a bipartite network in which buyers and sellers alternate in making public offers. In periods $t = 0, 2, \ldots$ each (remaining) seller simultaneously posts a price in the interval $[0, 1]$ at which he is willing to trade the good, and then every buyer observes the posted prices and simultaneously announces a price level he is willing to accept (without naming a specific trading partner). For every $p \in [0, 1]$, consider the subnetwork induced by the buyers and sellers who expressed willingness to trade at (the exact) price p. A maximum matching H of this subnetwork is selected according to a deterministic procedure, and the buyer-seller pairs matched

then its connectedness implies that it includes at least one seller i. Then removing seller i from C leads to a set of nodes with a different number of buyers and sellers, which clearly cannot have a perfect matching. This contradicts part (2).

under H trade at price p and exit the market. Players who have not completed a transaction remain in the game for period $t+1$. In periods $t = 1, 3, \ldots$ roles are reversed, with buyers posting prices and sellers responding. Players have a common discount factor $\delta \in (0, 1)$. The main finding of Corominas-Bosch (2004) is that this game has an efficient subgame perfect equilibrium for every discount factor, which yields payoffs closely tied to the Gallai-Edmonds decomposition.

Theorem 3 (Corominas-Bosch 2004). *In any bipartite network, there exists an efficient subgame perfect equilibrium for every discount factor δ in which over- and under-demanded players receive payoffs of 1 and 0, respectively, while perfectly matched sellers and buyers obtain payoffs of $1/(\delta + 1)$ and $\delta/(\delta + 1)$, respectively.*

To prove this result, Corominas-Bosch constructs an equilibrium in which all agreements take place in the first period and the trades carried out form a maximum matching. In the first period of the constructed equilibrium, over- and under-demanded sellers post prices of 1 and 0, respectively. Perfectly matched sellers ask for a price of $1/(\delta + 1)$, which is derived from Rubinstein's (1982) two-player bargaining game with alternating offers. Given these equilibrium offers, buyers who are perfectly matched, over-demanded, and under-demanded accept prices of $1/(\delta + 1)$, 0, and 1, respectively. The specification of strategies off the equilibrium path is more elaborate.

We now provide some intuition for the constructed equilibrium. Consider a bipartite network G with buyer set B and seller set S. Let P, O, and U denote the sets of perfectly matched, over-demanded, and under-demanded nodes in G, respectively. Under the strategies (partially) described above, every under-demanded player receives a zero payoff (under-demanded players who do not reach a first-period agreement become isolated). To see why buyer $i \in U$ has no profitable deviations in the first period, suppose that i does not accept a price of 1. First, note that by part (3) of Theorem 2, the under-demanded buyer i is linked only to over-demanded sellers, so i does not have any neighbor who is willing to accept a price different from 1 in the first period under the prescribed strategies. Hence buyer i does not trade in the first period if he deviates from his prescribed strategy. We next argue that i becomes isolated following the deviation. Given the prescribed strategies and i's deviation, the set of sellers who post a price of 1 and buyers who accept price 1 in the first period is $(O \cap S) \cup (U \cap B) \setminus \{i\}$. Let G' denote the subnetwork of G induced by this set of nodes. Since $i \in U$, there exists a maximum matching of G that does not cover $\{i\}$. By the Gallai-Edmonds decomposition theorem, such a matching links every node in $O \cap S$ to one in $(U \cap B) \setminus \{i\}$. It follows that maximum matchings of G' consist of at least $|O \cap S|$ links. Since every link in the bipartite network G includes one seller, no matching of G' may contain more than $|O \cap S|$ links, and every matching with $|O \cap S|$ links must cover $O \cap S$. It follows that every maximum matching of G' consists of exactly $|O \cap S|$ links and covers $O \cap S$. Hence, under the assumed market clearing mechanism, the entire set of over-demanded sellers $O \cap S$ trades and exits the market following i's deviation. Since i can be linked in G only to over-demanded sellers, the deviation leaves i isolated, which makes i indifferent between following the prescribed strategy and deviating from it.

To understand the relationship between the constructed equilibrium and competitive outcomes in a corresponding economy where buyers and sellers can trade freely, fix a bipartite network G in which there are more buyers than sellers. Then demand exceeds supply and general equilibrium theory predicts that all trades take place at a price of 1. The subgame perfect equilibrium identified by Corominas-Bosch converges to a competitive outcome as $\delta \to 1$ if and only if all sellers are over-demanded and all buyers are under-demanded. Combining theorems 1 and 2, we can prove that this condition is equivalent to $|\mathcal{L}^G(M)| > |M|$ for every subset of sellers M. Therefore, Corominas-Bosch's equilibrium implements a competitive outcome if and only if every group of traders on the short side of the overall economy has collective market power in the sense that they have access to a group of trading partners of larger size.

Charness, Corominas-Bosch, and Frechette (2007) test the predictions of a simplified version of Corominas-Bosch's model in an experimental setting. Although the payoff asymmetries observed in the laboratory are consistent with the direction of relative bargaining power suggested by the theory, the division of surplus in the experiment is not as extreme. We next discuss bargaining models that predict a more moderate balance of bargaining power.

26.3.2 Bargaining with Efficient Centralized Matching

Polanski (2007) considers a model of bargaining in a network with an efficient centralized matching procedure. In every period a maximum matching of the remaining network is drawn with equal probability, and for each matched pair either party is selected as the proposer with probability $1/2$. In each match, the proposer makes an offer to his partner specifying a split of the unit surplus generated by their link. If the partner accepts the offer, then the two players receive the agreed shares and are removed from the network. The game proceeds to the next round in the subnetwork induced by the players who have not previously reached an agreement. Players have a common discount factor $\delta \in (0, 1)$. Polanski assumes that at the bargaining stage each player observes only his own partner (if matched) and the offer made by his partner (when the partner acts as the proposer). Players learn all past actions at the beginning of every new period.

Polanski states that this bargaining game has a unique subgame perfect equilibrium, but the proof he provides is incomplete. However, this does not substantially change the message of his paper. For the sake of accuracy, we prove in the Appendix that the game admits a unique Markov perfect equilibrium (MPE).[2, 3] Polanski shows that in bipartite

[2] In this setting, the natural notion of a Markov state for a player i active in the market at some stage is given by the set of players who have not reached an agreement by that stage, along with the identity of i's match and the offer received by i in the current period. An *MPE* is a sequential equilibrium in which each player i's behavior after any history depends only on i's state induced by that history. This nonstandard adaptation of the definition of MPEs based on private states reflects the nature of imperfect information in Polanski's game.

[3] The uniqueness of subgame perfect and sequential equilibria is an open problem.

networks the unique MPE payoffs are related to the Gallai-Edmonds decomposition in a similar fashion to the equilibrium identified by Corominas-Bosch (2004). However, the payoffs of over- and under-demanded players are usually less extreme.

Theorem 4 (Polanski 2007). *For every network and discount factor, the bargaining game has a unique MPE, in which all matches result in agreement at every stage. In bipartite networks, the MPE payoffs belong to the interval* $(0, 1/2)$ *for under-demanded players and to* $(1/2, 1)$ *for over-demanded players, while those of perfectly matched players are equal to* $1/2$.

The first part of the result is proven in the Appendix. The proof relies critically on the assumption that a maximum matching is selected for bargaining in the subnetwork remaining at every stage. We will see that the main conclusions of the argument—the uniqueness and the efficiency of the MPE, along with the fact that every match results in an agreement in the MPE—fail to extend to the setting with decentralized matching analyzed by Abreu and Manea (2012a, b), where it is assumed that a single match forms at a time.

For a sketch of the proof for the second part of Theorem 4, fix a bipartite network G and let P, O, and U denote the sets of perfectly matched, over-demanded, and under-demanded nodes in G, respectively. Consider first a player $i \in P$. By Theorem 2, every maximum matching links player i to another player $j \in P$. Suppose that i and j do not reach an agreement when matched to bargain with each other in the first period. All players in $(P \setminus \{i,j\}) \cup O$ are covered by every maximum matching and, by the first part of the result, must trade in the first period of the MPE. By the construction of the Gallai-Edmonds decomposition, the perfectly matched players i and j do not have any links to under-demanded players. Thus, i and j are left in a subnetwork that contains the single link ij following their disagreement. Then the continuation payoffs of both i and j conditional on failing to reach an agreement with each other in the first period are equal to $\delta/2$. In the unique MPE, the proposer in the match (i,j) offers a payoff of $\delta/2$ to his partner and the partner accepts the offer with probability 1. Hence player i's expected payoff conditional on being matched to bargain with j in the first period equals $1/2(1 - \delta/2 + \delta/2) = 1/2$. Since i is always matched to a player in P in the first period and expects a payoff of $1/2$ conditional on being matched with any such player, player i's MPE payoff must be $1/2$.

Consider next a player $i \in O$. By Theorem 2, every maximum matching pairs player i with some $j \in U$. Assume as before that i and j do not reach an agreement when matched with each other in the first period. Theorem 2 implies that player j's connected component C in the subnetwork induced by U in G consists of an odd number of vertices and that every maximum matching that contains the link ij covers all the nodes in C. Hence all players different from j in C—including all of j's neighbors within U—trade in the first period of the MPE. As argued above, all players in $O \setminus \{i\}$ also reach agreements in the first period of the MPE. Therefore, ij constitutes j's only link left in the aftermath of his disagreement with i. Since any maximum matching of the ensuing subnetwork has a single link, this subnetwork consists of the link ij and possibly other links ik for $k \in N$. Then the second-period expected MPE payoffs conditional on i being matched with j and not trading in the first period are at least $1/2$ for i and at most $1/2$ for j. This immediately implies that player i receives an expected payoff of at least $1/2$ by forging an agreement with j in the first period of the MPE. The bound carries over to i's MPE payoffs in the original network because i is always matched to an under-demanded player in the first period.[4]

Finally, consider a player $i \in U$. Since G is bipartite, part (3) of Theorem 2 implies that i is not linked to any player in U. By Theorem 2, in every maximum matching, player i is either matched with a player $j \in O$ or not matched with any player. In the former event, an argument that mirrors the steps establishing the payoff bounds for over-demanded players proves that i's conditional expected payoff cannot exceed $1/2$. In the latter event, all first-period matches result in agreement, leading to a subnetwork in which i is isolated and receives a zero payoff.

The proof that perfectly matched players in G receive payoffs of exactly $1/2$ and over-demanded players in G obtain payoffs greater than $1/2$ in the MPE does not invoke the hypothesis that G is bipartite, so these conclusions hold for general networks. However, Polanski provides an example of a non-bipartite network in which an under-demanded player receives a payoff above $1/2$. In networks that are not bipartite, maximum matchings may contain links between pairs of under-demanded players. Since under-demanded players can have asymmetric payoffs, the balance of bargaining power in matches between such players relative to the even split in which the involved parties receive payoffs of $1/2$ is ambiguous.

26.3.3 Bargaining with Decentralized Matching

In the models of both Corominas-Bosch and Polanski, all agreements take place in the first period and equilibrium outcomes are efficient. The efficiency result is driven by the assumption of maximum matchings embedded in the two models. By contrast, Abreu and Manea (2012a) study a network setting with decentralized matching and sequential bargaining. In every period a single pair of linked players is matched according to some probability distribution (which depends on the set of remaining players), and each of the two parties is chosen with probability $1/2$ to make an offer to his partner specifying a division of the unit surplus. If the partner accepts the offer, then the two parties exit the game with the proposed shares. If the offer is rejected, then the match dissolves and the two players remain in the game for the next period. As in the earlier models, players have a common discount factor $\delta \in (0, 1)$.

Abreu and Manea (2012a) prove that the bargaining game admits an MPE for every discount factor (the payoff-relevant variables are the set of remaining players, along

[4] In certain events that occur with positive probability, we obtain a strict lower bound of $1/2$ for player i.

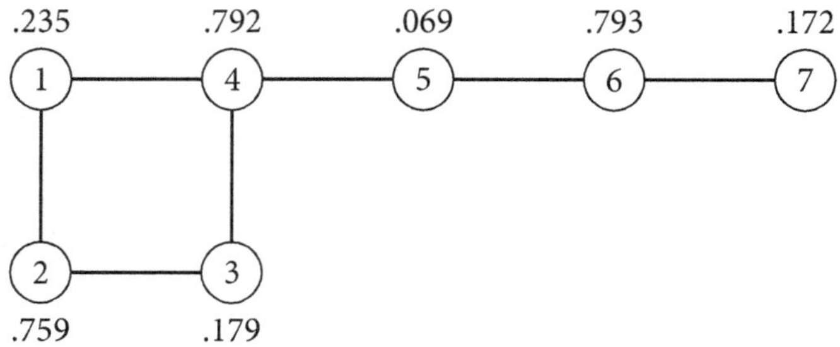

FIGURE 26.2 One of three profiles of MPE limit payoffs.

with nature's selection of a link and a proposer, and the proposer's offer). However, in contrast to the centralized bargaining model of Polanski, multiple MPEs, which yield distinct payoffs, may exist. Moreover, not all matches lead to agreements. It is also possible that a matched pair of players declines to trade at some stage but forges an agreement thereafter (following a sequence of equilibrium trades). The following example demonstrates these possibilities.

Consider the bargaining game on the network from Figure 26.2, in which each link is equally likely to be activated for bargaining in every subnetwork. Abreu and Manea (2012a) show that for high δ this game has an MPE in which players 1 and 4 do not trade when matched with each other in the original network but an agreement emerges when the first link different from $(1,4)$ is activated for bargaining. The limit MPE payoffs as $\delta \to 1$ are represented next to corresponding nodes in the figure.

In the constructed equilibrium, the first agreement may induce the following subgames. If players 1 and 2 (2 and 3) reach the initial agreement, then players $3,4,5,6,7$ $(1,4,5,6,7)$ are left in a bargaining game on the five-player line network. If instead players 3 and 4 (4 and 5) reach the first agreement, the remaining subnetwork has two connected components, $\{1,2\}$ and $\{5,6,7\}$ ($\{1,2,3\}$ and $\{6,7\}$). Players 1 and 2 are then involved in a two-player game in which they are matched to bargain with probability $1/3$ in any period before another agreement is reached (in the complementary event the link $(5,6)$ or $(6,7)$ is selected for bargaining) and with probability 1 following an agreement across the link $(5,6)$ or $(6,7)$. Similarly, players $5,6$, and 7 bargain in a version of the game on a three-player line network (or, alternatively, a star with player 6 at the center) in which matching frequencies at every stage depend on whether players 1 and 2 traded up to that stage. The subnetworks ensuing after an agreement takes place across the links $(5,6)$ and $(6,7)$ in the original network are depicted at the bottom half of Figure 26.3. The limit MPE payoffs as $\delta \to 1$ in these subgames as well as other subnetworks that may arise following an agreement are represented next to each node. We compute the limit MPE payoffs in the five-player line network in Section 26.3.4. The reader can consult Abreu and Manea (2012a) for more details on MPE outcomes in the other subgames. One robust property emerging from the inspection of these payoffs is that even indexed players are substantially stronger than odd indexed ones.

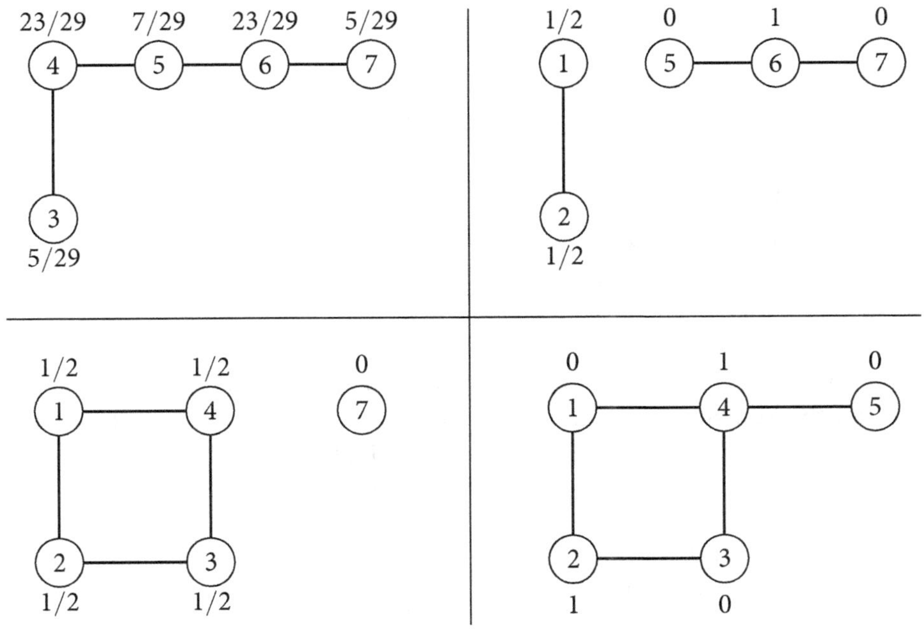

FIGURE 26.3 MPE payoffs in four subgames.

The intuition for this equilibrium specification is that even though players 1 and 3 occupy symmetric positions in the original network, their asymmetric behavior in matches with player 4 enhances the bargaining power of player 1. Indeed, when 3 and 4 forge the first agreement, players 1 and 2 are left in a bilateral bargaining game (see the top-right panel in Figure 26.3), in which player 1 obtains a limit continuation payoff of 1/2, the highest payoff available to odd indexed players in any subgame. This boosts the payoff of player 1 relative to other odd indexed players. In particular, player 3 does not benefit from a similarly favorable scenario because 1 does not trade with 4 in the original network. The fact that players 3 and 4 trade when matched in the original network further lowers 3's bargaining power since player 4 has a high payoff relative to player 2. The substantial difference between the payoff of player 2 and those of 4 and 6 can be attributed to the initial agreement between players 3 and 4, which undermines player 2's strong position by leaving him stranded in the bilateral monopoly with player 1.

The difference in the bargaining power of players 1 and 3 induced by their asymmetric treatment of player 4 turns out to be sufficiently large to make the conjectured agreements and disagreements incentive compatible. Players 1 and 4 do not have incentives to trade with each other in the original network because either of them can benefit from waiting to be matched with a weaker neighbor. All other initial matches result in agreements since no link different from (1,4) connects a pair of similarly

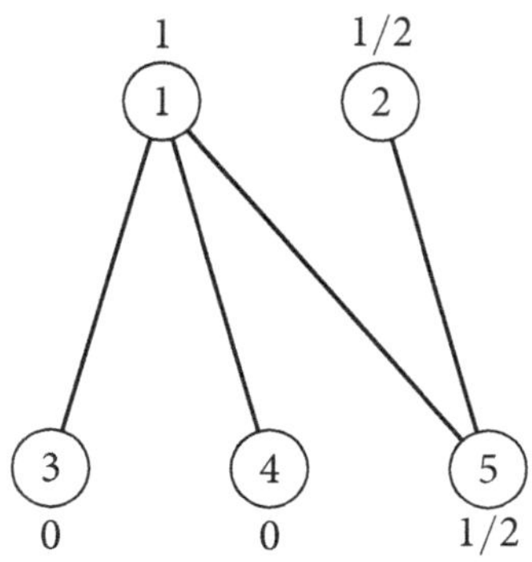

FIGURE 26.4 Common limit equilibrium payoffs for the three models.

strong players (in the constructed MPE) with odd and even indices. An interesting feature of this equilibrium is that players 1 and 4 do not trade when matched in the original network, but reach an agreement with each other in subgames that arise with positive probability (for instance, following an agreement between either 2 and 3 or 5 and 6).

We can obtain another MPE for high δ by simply interchanging the roles of players 1 and 3 in the equilibrium described above. For sufficiently high δ, Abreu and Manea argue that the game has a third MPE, in which the matches $(1,4)$ and $(3,4)$ lead to trade with a common probability close to 0.529, while all other matches result in agreement with probability 1, in the original network. Therefore, for high δ the bargaining game has at least three MPEs with distinct payoffs.

26.3.4 Model Comparisons

It is instructive to compare the equilibrium outcomes in the three models discussed in this section thus far. For the network depicted in Figure 26.4, all three models predict that players 1 and 5 do not trade with each other.[5] The intuition is that player 1 has monopoly power over players 3 and 4, while player 5 acts a monopolist for player 2. Then players 1 and 5 are better off trading with their weaker neighbors than forging an agreement with each other. The three models yield identical limit payoffs as players become patient. The common limit payoffs are given by 1 for player 1, 0 for players 3 and 4, and $1/2$ for players 2 and 5.

However, each model generates substantially different payoffs in the five-player line network shown in Figure 26.5. This can be viewed as a bipartite network if we label players $\{1,3,5\}$ as buyers and $\{2,4\}$ as sellers. In Corominas-Bosch's equilibrium, the under-demanded players $1,3$, and 5 receive payoffs of 0, while the over-demanded

[5] Note that in Polanski's model, players 1 and 5 never get the opportunity to bargain with each other.

Corominas-Bosch	0	1	0	1	0
Polanski	1/6	5/6	0	5/6	1/6
Abreu and Manea	5/29	23/29	2/29	23/29	5/29

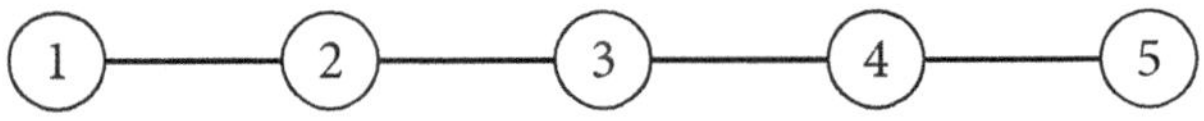

FIGURE 26.5 Distinct limit equilibrium payoffs in the three models.

players 2 and 4 obtain a payoff of 1. These extreme payoffs are explained by the strong competitive forces induced by public offers.

To compute the limit MPE payoffs in Polanski's model, note that there are three maximum matchings in this network: $\{(1,2),(3,4)\},\{(1,2),(4,5)\},\{(2,3),(4,5)\}$. Recall that all matches lead to trade in the MPE. When players 1 and 2 are matched with each other, they believe that the other match is equally likely to be $(3,4)$ or $(4,5)$ and anticipate that either of these matches results in an agreement. Thus if players 1 and 2 fail to reach an agreement, then they are left in a bilateral bargaining game or in a game on the line network formed, in order, by players $1,2,$ and 3. Either subgame arises with probability $1/2$ conditional on the match $(1,2)$ being formed. The limit continuation payoffs of player 1 are $1/2$ in the former subgame and 0 in the latter. Then player 2 offers 1 an amount equal to 1's continuation payoff conditional on the match $(1,2)$ ending in disagreement, which converges to $1/2 \times 1/2 = 1/4$ as players become patient; player 1 accepts the offer for every discount factor in the MPE. Similarly, player 1 offers 2 an amount that converges to $3/4$, which 2 accepts. Hence the limit expected payoff of player 1 conditional on being matched (with his sole neighbor, player 2) in the first round is $1/4$. Since player 1 becomes isolated when the matching $\{(2,3),(4,5)\}$ is drawn in the first period, an event which occurs with probability $1/3$, his limit equilibrium payoff is $2/3 \times 1/4 = 1/6$. By symmetry, player 5 obtains the same payoff. Similar calculations show that players 2 and 4 receive limit MPE payoffs of $5/6$, while player 3 gets 0 in the limit.

Abreu and Manea (2012a) provide a computational method for identifying MPEs for high discount factors in their model. This method proves that if each remaining link is activated for bargaining with equal probability at every stage, then the decentralized bargaining game on the five-player line network admits an MPE in which all matches result in agreement for high δ.[6] The limit MPE payoffs (v_i) as $\delta \to 1$ solve the following system of equations:

$$v_1 = \frac{1}{4}\frac{1}{2}(1 - v_2 + v_1) + \frac{1}{4}0 + \frac{1}{4}\frac{1}{2} + \frac{1}{4}0$$

[6] In this example, one can show that for any discount factor there exists a unique MPE and that every match results in agreement in the MPE.

$$v_2 = \frac{1}{4}\frac{1}{2}(1 - v_1 + v_2) + \frac{1}{4}\frac{1}{2}(1 - v_3 + v_2) + \frac{1}{4}\frac{1}{2} + \frac{1}{4}1$$
$$v_3 = \frac{1}{4}0 + \frac{1}{4}\frac{1}{2}(1 - v_2 + v_3) + \frac{1}{4}\frac{1}{2}(1 - v_4 + v_3) + \frac{1}{4}0$$
$$v_4 = v_2$$
$$v_5 = v_1.$$

For example, the first term in the payoff equation for player 1 captures the fact that the link $(1,2)$ is activated for bargaining in the first period with probability $1/4$, in which case player 1 obtains a limit payoff of $1 - v_2$ or v_1, depending on nature's selection of a proposer; the second term corresponds to the probability $1/4$ event that players 2 and 3 are matched to bargain with each other in the original network and forge an agreement that isolates player 1. The last two equations reflect network (and equilibrium) symmetries. The unique solution to the system of linear equations is given by $v_1 = v_5 = 5/29, v_2 = v_4 = 23/29, v_3 = 2/29$.

The discrepancy between the payoffs in Polanski's model and Abreu and Manea's in the example above can be assigned to the following strategic differences. In the model of Polanski, two pairs of players are matched simultaneously and an agreement in one of the matches influences the outside options of the players involved in the other. By contrast, in the model of Abreu and Manea, only one match forms at a time and the matched players continue to bargain in the same network if they fail to reach an agreement. In other words, the assumption of centralized efficient matching speeds up the process of isolating under-demanded players and reduces their bargaining power.

26.3.5 Efficiency with Decentralized Bargaining

In both the equilibrium analyzed by Corominas-Bosch and the unique MPE of Polanski, the centralized organization of matches for trade ensures efficient outcomes. Under the decentralized bargaining protocol of Abreu and Manea, links that are not part of any maximum matching may be activated for bargaining and result in inefficient agreements. Indeed, incentives for bilateral agreements are not always aligned with global welfare maximization in a setting with bargaining frictions created by decentralized random matching. The following example from Abreu and Manea (2012a) illustrates this problem. Consider the game induced by the four-player network from Figure 26.6 in which each link is equally likely to be selected for bargaining as long as no agreement has been reached. In this network there exists a single maximum matching, formed by the links $(1,2)$ and $(3,4)$, which generates the maximum total surplus of 2. Trade between player 2 and either 3 or 4 leaves the other two players isolated, so agreements in the matches $(2,3)$ and $(2,4)$ are inefficient. However, Abreu and Manea prove that there exists a unique MPE for every discount factor, in which

FIGURE 26.6 Asymptotically inefficient MPE.

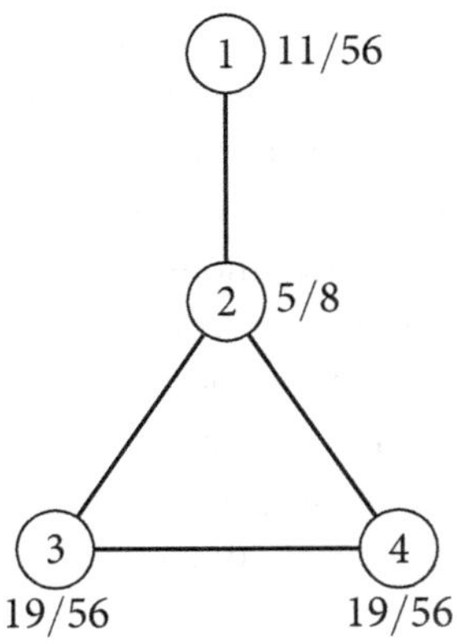

every match materializes in an agreement.[7] The unique MPE payoffs are shown to converge as $\delta \to 1$ to $11/56$ for player 1, $5/8$ for player 2, and $19/56$ for players 3 and 4. The limit payoffs sum to $3/2$, which reflects the fact that with probability $1/2$ one of the inefficient matches (2,3) and (2,4) forms and in this event only one unit of surplus is created along the equilibrium path. Hence the MPE is asymptotically inefficient, generating a total welfare of $3/2$, which is smaller than the available total surplus of 2.[8]

This example raises the following question: if we allow for non-Markovian behavior, is it possible to structure incentives in order to construct an asymptotically efficient subgame perfect equilibrium? The tension between the global derivation of maximum matchings and the local nature of bilateral interactions makes this a challenging question. Pairs of players who reach inefficient agreements are removed from the network permanently, so cooperative behavior that generates the maximum total surplus cannot be enforced via standard repeated game threats. An additional complication is that, when multiple maximum matchings exist, links that are part of a maximum matching in the network prevailing at some stage may cease to have this property after a series of efficient agreements takes place. Thus the notion of efficient agreements is history dependent. Despite these obstacles, Abreu and Manea (2012b) establish that an asymptotically efficient equilibrium always exists.

Theorem 5 (Abreu and Manea 2012b). *The bargaining game admits a family of asymptotically efficient subgame perfect equilibria.*

The proof constructs asymptotically efficient equilibria in which players who resist the temptation to reach inefficient agreements are rewarded by certain neighbors and players who do not conform to the rewarding procedure are punished via the threat

[7] We can immediately rule out any MPE in which only efficient agreements take place. In such an equilibrium, all players would expect payoffs below $1/2$ (the MPE could be decomposed into two separate bilateral bargaining games in which each of the pairs (1,2) and (3,4) is matched with a probability that evolves stochastically), creating strict incentives for the pairs (2,3) and (2,4) to trade when matched. Ruling out MPEs in mixed strategies, which could generate asymptotically efficient outcomes if the probability of inefficient agreements vanishes as $\delta \to 1$, requires more meticulous arguments.

[8] In this example, Polanski's model assumes that only the maximum matching can form and predicts an efficient equilibrium outcome in which every player expects a payoff of $1/2$.

of a sequence of trades that isolates them from the network. The evolving nature of the network structure as prescribed agreements take place complicates the design of punishment and reward schemes. Unlike in Polanski's model, it is a priori unclear which matches should lead to an agreement (the example from Section 26.3.3 speaks to this issue) and how the resulting stochastic evolution of the network affects the outside options in every match. Since it is difficult to specify any subgame perfect equilibrium, the construction of asymptotically efficient equilibria starts from an implicitly defined Markov strategy profile. The idea is to modify the payoffs in the original game by imposing prohibitive fines for players who reach inefficient agreements. Fines are also used to incentivize players to forge efficient agreements in certain matches. Such "forced" agreements pave the way for the isolation of deviators. An MPE of the modified game is then employed as the reference point for punishments and rewards in the original game. For matches unaffected by the payoff modifications, the strategies prescribed by this MPE are sequentially rational. The incentives to deviate resulting from the modifications of the original game need to be adjusted via explicit constructions of punishments and rewards.

While the precise calibration of reward and punishment paths for any given network structure is intricate, it is worth pointing out a key step in the proof that relates to the other models discussed in this section. Abreu and Manea prove that the limit MPE payoffs of perfectly matched and over-demanded players in the modified bargaining game for every network are greater than or equal to $1/2$.[9] To understand the role that this finding plays in the equilibrium construction, consider a link ij that is not part of a maximum matching in the subnetwork remaining at some stage, so an agreement between i and j would be inefficient. Then neither i nor j is under-demanded in the Gallai-Edmonds decomposition for the remaining subnetwork. By the finding mentioned above, the limit MPE payoffs of both i and j at the stage under consideration in the modified game are at least $1/2$. It follows that the gains that i and j can obtain by reaching an inefficient agreement with each other, relative to the reference payoffs derived from the modified game, converge to 0 as $\delta \to 1$. Thus a small reward bounded away from 0 is sufficient to deter i from accepting a tempting offer from j (and vice versa). The proof shows how such a reward can be delivered by one of i's neighbors, a suitably chosen player k, following a series of agreements that occur with probability bounded away from 0 if i turns down a tempting offer (relative to the reference payoffs) from j. Player k is incentivized to offer the reward to i via the threat of a sequence of agreements that leaves k vulnerable to isolation. Players i and j deliver the ultimate punishment by forging an agreement (that is incentive compatible with respect to the reference payoffs) following such a history, which isolates k in the remaining network.

[9] This observation is reminiscent of Polanski's result but is technically more involved because, as already argued, under the decentralized matching process it is difficult to predict which matches lead to an agreement and how the network evolves.

We lastly record that Abreu and Manea (2012b) test the robustness of their conclusion with respect to the bargaining protocol. They extend the construction of asymptotically efficient equilibria to an alternative model, which assumes that in every period a player, rather than a link, is selected stochastically and that the selected player can activate any of his links for bargaining. Once a link is activated, either party is chosen with equal probability to propose a division of the surplus as in the benchmark model.

26.4 Bargaining in Stationary Networks

In each of the models discussed in the previous section, there is a finite number of players and a finite set of feasible trades. Players who reach agreements are removed from the network and the pool of potential trading partners for each remaining player shrinks over time. These assumptions are realistic in small specialized markets, in which the entry of new traders is impossible (for the relevant time horizon). In this section, we consider the distinct setting of an economy that is continually replenished with new traders, so that the market composition is constant over time. Rubinstein and Wolinsky (1985) studied the first such bargaining model in the case of a stationary market with symmetric buyers and sellers. Gale (1987) expanded the analysis to markets with heterogeneous buyers and sellers. The focus of this research was to provide noncooperative foundations for general equilibrium theory. In particular, Gale found that the law of one price holds in the sense that all prices at which trade takes place in equilibrium converge to the same level as bargaining frictions disappear.

As discussed in the introduction, we should not expect the law of one price to apply to economies with network asymmetries (even as players become patient). It is nevertheless interesting to understand how the network architecture shapes local prices and bargaining power in a stationary market. Manea (2011a) pursues this matter in the context of the following bargaining model. Consider a network G connecting the finite set of nodes N, and let $(p_{ij} > 0)_{ij \in G}$ be a probability distribution over the links in G. In every period $t = 0, 1, \ldots$, a link ij in G is selected with probability p_{ij}, and each of the players i and j is chosen with equal probability to make an offer to the other player specifying a division of the unit surplus available to the pair. If the other player accepts the offer, then the two players exit the game with their agreed shares. The market is maintained in a steady state as follows: if i and j reach an agreement, then in period $t+1$ two new players assume the same positions in the network as i and j. If the offer is rejected, then the match is dissolved and the two players remain in the market for the next matching and bargaining round. Players share a discount factor $\delta \in (0, 1)$.

The stationarity of the economy simplifies the equilibrium analysis considerably as we do not need to keep track of payoffs in subnetworks that arise endogenously following agreements in the original network. A result of Manea (2014a) implies that every subgame perfect equilibrium of the stationary bargaining game generates the same expected payoffs, which we denote by v_i, for each player active at position i in the network at the beginning of any period.[10] The equilibrium payoffs constitute the unique solution to the following system of equations:

$$v_i = \sum_{\{j|ij \in G\}} \frac{p_{ij}}{2} \max(1 - \delta v_j, \delta v_i) + \left(1 - \sum_{\{j|ij \in G\}} \frac{p_{ij}}{2}\right) \delta v_i, \forall i \in N.$$

The payoff equations capture the following equilibrium conditions. In any subgame perfect equilibrium, when i is selected to propose to j, if $1 - \delta v_j > \delta v_i$ then i offers δv_j to j, an offer which j accepts with probability 1. If $1 - \delta v_j < \delta v_i$ then i makes an offer that j rejects with probability 1.[11] Given the stationarity of the environment and the fact that all subgame perfect equilibria are payoff equivalent, player i expects a continuation payoff of δv_i if he does not forge an agreement in the current period. Therefore, in any subgame following the selection of a link and a proposer, player i obtains a payoff different from δv_i only in the event that he is selected to make an offer to a player j for which $1 - \delta v_j > \delta v_i$. This event occurs with probability $p_{ij}/2$ and yields a payoff of $1 - \delta v_j$ for i.

Manea (2011a) shows that there exist a discount factor threshold $\underline{\delta} \in (0, 1)$ and a subnetwork G^* of G such that in any equilibrium of the bargaining game for any discount factor $\delta \in (\underline{\delta}, 1)$, trade takes place with probability 1 across all links in G^* and with probability 0 for the remaining links.[12] This finding is used to demonstrate that equilibrium payoffs converge to a limit $(v_i^*)_{i \in N}$ as $\delta \to 1$. The following concepts, which rely on the endogenous network of agreements G^*, prove helpful in characterizing the limit equilibrium payoffs. A nonempty set of players is *mutually estranged* if it is G^*-independent. The set of *partners* for a mutually estranged set M is defined as $\mathcal{L}^{G^*}(M)$.

Fix a mutually estranged set M with partner set L. For high δ, the set L spans the relevant bargaining opportunities of the players in M. Then the group M is collectively weak if L has a relatively small cardinality. The relevant measure of M's strength turns out to be the *shortage ratio* of M, defined as the ratio of the number of partners to estranged players, $|L|/|M|$. Manipulating the payoff equations, we find that for every mutually estranged set M with partner set L, the equilibrium payoffs $(v_i)_{i \in N}$ for sufficiently high δ satisfy

$$\sum_{j \in L} v_j \geq \sum_{i \in M} v_i.$$

[10] More generally, Manea (2014a) establishes the payoff equivalence of equilibria in markets with multiple player types in which the matching frequencies for every pair of types are exogenous and time-dependent.

[11] When $1 - \delta v_j = \delta v_j$, player i is indifferent between offering j the minimum amount v_j necessary for an agreement and making an unacceptable offer.

[12] Formally, for every $\delta \in (\underline{\delta}, 1)$ and $ij \in G$, equilibrium payoffs satisfy $\delta(v_i + v_j) < 1$ if $ij \in G^*$ and $\delta(v_i + v_j) > 1$ otherwise (it is impossible that $\delta(v_i + v_j) = 1$).

Loosely speaking, M cannot have more "collective bargaining power" than its partner set L. The inequality above implies that the ratio of the limit equilibrium payoffs of the worst-off player in M and the best-off player in L does not exceed the shortage ratio of M. Based on this observation, Manea (2011a) establishes the following bounds on limit equilibrium payoffs:

$$\min_{i \in M} v_i^* \leq \frac{|L|}{|M| + |L|}$$
$$\max_{j \in L} v_j^* \geq \frac{|M|}{|M| + |L|}.$$

A key step in the analysis shows that the bounds on limit equilibrium payoffs corresponding to a mutually estranged set M and its partner set L need to bind unless the worst-off player in M belongs to a mutually estranged set with a lower shortage ratio. The intuition is that in every connected component of G^* (where not all players have limit payoffs of $1/2$), some mutually estranged players and their partners share the unit surplus according to the corresponding shortage ratio as $\delta \to 1$. Therefore, the bounds above must bind for any mutually estranged set that achieves the lowest shortage ratio.

Define $r_1 = \min_{M \text{ is } G\text{-independent}} |\mathcal{L}^G(M)|/|M|$ and let M_1 be the union of all G-independent sets M that minimize the expression $|\mathcal{L}^G(M)|/|M|$. Set $L_1 = \mathcal{L}^G(M_1)$. Manea (2011a) shows that if $r_1 < 1$, then r_1 constitutes the minimum shortage ratio over all mutually estranged sets, while M_1 represents the largest mutually estranged set minimizing the shortage ratio, and L_1 serves as its partner set. This means that when the lowest shortage ratio is smaller than 1, it can be computed by restricting attention to sets that are independent in the original network G instead of the a priori unknown agreement network G^*, with corresponding sets of partners formed by neighbors in G rather than G^*. The analysis reveals that all members of M_1 obtain minimum limit equilibrium payoffs, given by $r_1/(r_1+1)$, and all members of L_1 achieve the maximum equilibrium limit payoffs of $1/(r_1+1)$. By definition, the players in M_1 are only linked to players in L_1 in the original network G. Moreover, since M_1 consists of all players with minimum limit payoffs and every member of L_1 has at least one link to M_1, players in L_1 do not have incentives to trade with players outside M_1 for high δ. Hence G^* does not contain any links from $L_1 \cup M_1$ to the remaining set of nodes. Then the nodes in $L_1 \cup M_1$ can be removed from G and the network induced by the remaining nodes can be analyzed as a separate submarket.

The arguments above lead to the following network decomposition algorithm. Define the sequence $(r_s, M_s, L_s, N_s, G_s)_s$ recursively as follows. Let $N_1 = N$ and $G_1 = G$. For $s \geq 1$, the algorithm terminates if $N_s = \emptyset$. Otherwise, let

$$r_s = \min_{M \subseteq N_s,\ M \text{ is } G\text{-independent}} \frac{|\mathcal{L}^{G_s}(M)|}{|M|}. \tag{26.2}$$

If $r_s \geq 1$, then the algorithm stops. Else, let M_s be the union of all minimizers M in (26.2) and define $L_s = \mathcal{L}^{G_s}(M_s)$. Set $N_{s+1} = N_s \setminus (M_s \cup L_s)$ and let G_{s+1} be the subnetwork of G induced by the nodes in N_{s+1}. Denote by $\bar{s}$ the (finite) step at which the algorithm ends.

A key lemma proves that if $r_s < 1$, then M_s is G-independent and minimizes the expression in (26.2). Hence, at every step, the algorithm determines the largest mutually estranged set minimizing the shortage ratio in the subnetwork induced by the remaining players and removes the corresponding estranged players and partners. This definition ensures that all players with extremal limit payoffs in the remaining network are identified simultaneously, so that the removed players trade only with one another in equilibrium for high δ. Then we can treat the residual subnetwork as a separate market.

The algorithm ends when all players have been removed or $|\mathcal{L}^{G_s}(M)| \geq |M|$ for every G-independent set $M \subseteq N_s$. The sequence $(r_s)_s$ is strictly increasing and the sets $M_1, L_1, \ldots, M_{\bar{s}-1}, L_{\bar{s}-1}, N_{\bar{s}}$ partition the set of nodes N. The main result of Manea (2011a) establishes that the limit equilibrium payoff of each player is determined by his cell in this partition.

Theorem 6 (Manea 2011a). *The limit equilibrium payoffs as $\delta \to 1$ in the stationary bargaining game are given by*

$$v_i^* = \frac{r_s}{r_s+1}, \qquad \forall i \in M_s, \forall s < \bar{s}$$

$$v_j^* = \frac{1}{r_s+1}, \qquad \forall j \in L_s, \forall s < \bar{s}$$

$$v_k^* = \frac{1}{2}, \qquad \forall k \in N_{\bar{s}}.$$

As an illustration, for the network from Figure 26.7, the algorithm ends in $\bar{s} = 2$ steps. The relevant outcomes are $r_1 = 1/2, M_1 = \{3,4\}, L_1 = \{1\}$ and $r_2 = 1, N_2 = \{2,5\}$. Thus the limit equilibrium payoffs are given by 2/3 for player 1, 1/3 for players 3 and 4, and 1/2 for players 2 and 5. Recall that all the models discussed in Section 26.3 predict that in this network player 1 receives a payoff of 1, while players 3 and 4 obtain a payoff of 0 in the limit $\delta \to 1$. To examine this discrepancy, consider a scenario in which players 1 and 3 are matched and reach an agreement with each other in the first period. Then, in the setting of Section 26.3, players 1 and 3 are removed from the network without replacement, and player 4 is left isolated, in which case his continuation payoff is 0. By contrast, in the stationary economy, a new trader enters the market at node 1 in the second period, and player 4 may get a chance to bargain with this trader. In equilibrium, player 4 capitalizes on the perpetual rebirth of bargaining partners at node 1 to secure a limit payoff of 1/3. Relatedly, limit payoffs in the stationary case sum to 7/3, which is greater than the maximum total surplus of 2 in this network. The reason is that the entrance of a new player at node 1 following the first-period agreement across the link (1,3) creates the potential for realizing a unit of surplus in a match with the player present at node 4 from the first period. In general, since newly born players may share surplus with older ones, the total sum of limit payoffs generated by the positions in

FIGURE 26.7 Limit equilibrium payoffs in the stationary model.

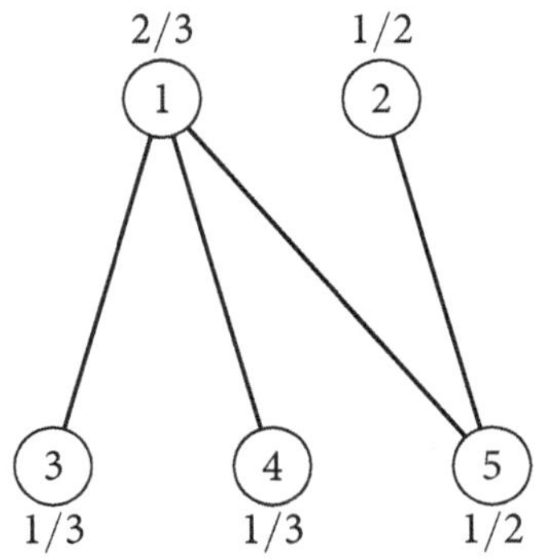

the network does neither reflect the size of any matching in the static network, nor constitute a suitable welfare measure in the stationary setting.

An important implication of the characterization of limit equilibrium payoffs is that submarkets emerge endogenously in equilibrium. The group L_s acts as an oligopoly for the players in M_s. The *oligopoly subnetwork* induced by the set of nodes $L_s \cup M_s$ is a union of connected components of the agreement network G^*. For high δ, every equilibrium transaction takes place between members of the same oligopoly subnetwork. The limit prices as $\delta \to 1$ are uniform within every such subnetwork. In equilibrium, each trader self-selects into the most favorable submarket to which he has access.

Note that limit equilibrium payoffs do not depend on the relative probabilities $(p_{ij} > 0)_{ij \in G}$ with which links are activated for bargaining. Furthermore, the conclusions of the model extend to a setting in which multiple disjoint links are activated for bargaining in every period. Relatedly, the model and the results admit an alternative interpretation whereby there is a continuum of players at each node and bargaining proceeds simultaneously for a positive mass of heterogeneous matches. Manea (2014b) provides foundations for the steady state assumption in a setting with a continuum of players where new traders optimally decide whether to enter the market for a (small) cost.

26.4.1 Network Formation

To approach the question of network formation in the stationary bargaining model, it is useful to first define and characterize equitable networks. A network G is called *equitable* if all players receive limit equilibrium payoffs of 1/2 in G.[13] Theorem 6 implies that a network G is equitable if and only if $|\mathcal{L}^G(M)| \geq |M|$ for every G-independent set M. Manea (2011a) shows that the latter condition is equivalent to G containing a

[13] Recall that limit equilibrium payoffs depend on the collection of links activated for matching with positive probability but not on the relative activation probabilities.

subnetwork that covers all of G's nodes and consists of a disjoint union of a matching and cycles with odd numbers of nodes. Since bipartite networks cannot contain odd cycles, the necessary and sufficient condition for a bipartite network G to be equitable boils down to G admitting a perfect matching (cf. Theorem 1).

Manea (2011b)—the online appendix to Manea (2011a)—shows that if the players active in the first-period market are responsible for forming the network before they engage in the stationary bargaining game, and there are no linking costs, then (1) adding a new link to a network cannot decrease the limit payoffs of either of the players it connects; and (2) a network is pairwise stable[14] if and only if it is equitable. Gauer (2014) studies network formation in the stationary bargaining model under the assumption that every link entails positive costs for both parties involved. Clearly, in this setting equitable networks in which some links are "redundant" for attaining equitability cannot be pairwise stable. In the class of equitable networks, only "skeleton" networks—formed by disjoint unions of a matching and cycles of odd length—can survive pairwise stability. Gauer confirms that such skeleton networks are indeed pairwise stable for small linking costs. However, he finds that some non-equitable networks (in particular, networks formed by odd cycles and a single isolated player) are also pairwise stable for small costs. Gauer then proceeds to partially characterize pairwise stable networks for arbitrary linking costs and identifies a richer collection of stable structures for intermediate costs.

In Manea (2011b), the definition of pairwise stability is adjusted to bipartite networks to account for the fact that only buyer-seller pairs can contemplate forming new links. The modified solution concept is coined *two-sided pairwise stability.* We say that a bipartite network is *non-discriminatory* if all transactions take place at the same limit price (i.e., all buyers obtain identical limit equilibrium payoffs). When linking costs are zero, Manea (2011b) proves that a bipartite network is two-sided pairwise stable if and only if it is non-discriminatory. Manea (2011a) develops a simplified version of the network decomposition algorithm for the case of bipartite networks. The payoff characterization derived from that alternative decomposition reveals that a bipartite network is non-discriminatory if and only if for every buyer subset M, the ratio $|\mathcal{L}^G(M)|/|M|$ is greater than or equal to the ratio of seller to buyer nodes in the network. Polanski and Vega-Redondo (2014) extend the result that two-sided pairwise stable networks are non-discriminatory to a setting in which buyers and sellers have heterogeneous valuations. They also generalize the necessary and sufficient condition for non-discriminatory pricing.

26.4.2 Coalitional Bargaining

Nguyen (2014) extends the model of Manea (2011a) to a setting in which coalitions of various sizes can create different amounts of surplus. A coalition along with an ordering

[14] See Jackson and Wolinsky (1996) for the definition of pairwise stability.

of its members is randomly drawn, and then the first player in the order proposes a split of the surplus available to the coalition and the other members decide, in order, whether to accept or reject the offer. An agreement is reached only if the proposal is unanimously accepted. The members of the agreeing coalition exit the market and are replaced by clones. Players have a common discount factor δ.

Nguyen shows that all stationary equilibria are payoff equivalent and characterizes the unique equilibrium payoffs as the solution to a convex optimization problem. For the special case of Manea (2011a), the convex program reduces to

$$\begin{aligned} \min_{(v_i)_{i\in N},(z_{ij})_{ij\in G}} \quad & 2\delta(1-\delta)\sum_{i\in N} v_i^2 + \sum_{ij\in G} p_{ij} z_{ij}^2 \\ \text{s.t.} \quad & \delta(v_i+v_j)+z_{ij} \geq 1, \forall ij \in G. \end{aligned}$$

The main idea for this characterization is that the dual optimization problem boils down to solving the equilibrium payoff equations. Nguyen proves that the limit equilibrium payoffs as $\delta \to 1$ constitute the optimal solution for the following convex program:

$$\begin{aligned} \min_{(v_i)_{i\in N}} \; & \sum_{i\in N} v_i^2 \\ \text{s.t.} \; & v_i + v_j \geq 1, \forall ij \in G. \end{aligned}$$

This finding leads to an alternative proof for Theorem 6. Nguyen also applies his result to the study of multilateral bargaining in the context of intermediation in networks and cooperation within overlapping communities.

26.5 The Assignment Game and Related Noncooperative Models

Shapley and Shubik (1971) provide an elegant and powerful analysis of two-sided markets based on cooperative game theory. In their assignment game, as in the other models discussed in this chapter, buyers have unit demand and sellers have unit supply for a heterogeneous good. The sets of buyers and sellers, which we denote by B and S, respectively, are finite. The good supplied by seller j, which we call *good j* for brevity, is worth nothing to j and $a_{ij} \geq 0$ to buyer i. An *assignment* is a mapping $\mu : B \cup S \to B \cup S$ with $\mu(i) \in S \cup \{i\}$ for $i \in B$ and $\mu(j) \in B \cup \{j\}$ for $j \in S$, which satisfies $\mu(\mu(k)) = k$ for all $k \in B \cup S$. For a buyer-seller pair (i,j), the condition $\mu(i) = j$ is equivalent to $\mu(j) = i$ and indicates that buyer i is *assigned* good j under μ. If $\mu(i) = i$ ($\mu(j) = j$) for buyer i (good j), then i (j) is left *unassigned* by μ. An assignment μ is *efficient* if it maximizes the *total surplus* $\sum_{i\in B,\mu(i)\neq i} a_{i\mu(i)}$. We refer to the achieved maximum as the *maximum total surplus*. Assuming that every coalition can organize trade efficiently, this setting generates a *cooperative game* with transferable utility in which the value of each coalition is determined by the maximum total surplus available in the economy consisting only of its members. Shapley and Shubik establish several facts about the structure of the core of this cooperative game, which is coined the *assignment game*.[15]

Theorem 7 (Shapley and Shubik 1971).

1. *The core of the assignment game is nonempty.*
2. *Any profile of core payoffs corresponds to a competitive equilibrium outcome in an economy where each good j is assigned an individual price p_j.*
3. *The set of core payoffs forms a lattice with the partial order under which a payoff profile dominates another if all buyers (weakly) prefer the former profile, and all sellers prefer the latter.*
4. *The core contains a payoff profile that all buyers prefer to any other point in the core—the buyer-optimal core payoff—which corresponds to the minimum competitive equilibrium price vector. Similarly, there exists a seller-optimal core payoff corresponding to the maximum competitive equilibrium prices.*

This result lends an interesting interpretation to the equilibrium constructed by Corominas-Bosch (2004), which we discussed in Section 26.3.1. A bipartite network G naturally defines an assignment game in which $a_{ij} = 1$ if $ij \in G$ and $a_{ij} = 0$ otherwise. In this assignment game, every coalition M creates a value equal to the maximum total surplus in the subnetwork induced by M in G. Corominas-Bosch (1999) points out that at every core allocation of the associated assignment game, under-demanded players in G must receive a zero payoff, which then implies that over-demanded players in G obtain a payoff of 1. The buyer-optimal core allocation yields payoffs of 1 and 0 for perfectly matched buyers and sellers, respectively. The payoffs of perfectly matched buyers and sellers are reversed in the seller-optimal core outcome. Consequently, Corominas-Bosch (1999) concludes that the limit payoffs in the equilibrium described by Theorem 3 represent the average of the buyer- and the seller-optimal core allocations.

26.5.1 A Decentralized Price Adjustment Mechanism

The existence of extremal competitive prices for the assignment game facilitates the design of tatonnement processes that describe some intuitive price adjustment mechanisms. Crawford and Knoer (1981) develop a *decentralized* price formation process that converges to the lowest competitive prices. Their approach resembles the deferred acceptance algorithm of Gale and Shapley (1962). Let ε be a small positive number, which serves as the minimum unit of price increments. At every stage in the

[15] For an exhaustive survey of the research on the assignment game, refer to Section 8 of Roth and Sotomayor (1990). The reader is also encouraged to peruse the original paper of Shapley and Shubik, which is superbly written. Koopmans and Beckman (1957) provide an early analysis of the connection between efficient allocations and competitive outcomes in the assignment game.

price adjustment mechanism, buyer i faces a personalized price p_{ij} for good j; all prices are initialized at 0. Each buyer i demands a good j that maximizes his payoff $a_{ij} - p_{ij}$ given the current price levels. A buyer does not submit any demand if no good yields positive payoffs at the prevailing prices. Each seller j tentatively assigns good j to one of the buyers i who demand it at the highest price p_{ij}. Then the personal price p_{ij} for every buyer i who demanded, but was not assigned, good j is increased by ε. The other prices do not change. The price adjustment process ends when no new demand is submitted, at which point the existing assignments become definitive and each buyer i assigned a good j pays his personal price p_{ij} to seller j. Crawford and Knoer prove that the final prices converge to the minimum competitive prices as $\varepsilon \to 0$.[16]

Note that the "decisions" at every step in the mechanism of Crawford and Knoer rely only on "local" information available to the relevant traders. At each step, buyer i's demand depends solely on his personal valuations $(a_{ij})_{j \in S}$ and prices $(p_{ij})_{j \in S}$, and similarly seller j's tentative assignment is determined by the set of buyers B_j who demand good j and the prevailing price levels $(p_{ij})_{i \in B_j}$. The steady escalation in personal prices can be interpreted as voluntary bidding by buyers whose price offers are rejected. However, it is not clear whether buyers have incentives to participate in this type of gradual Bertrand competition, particularly in situations in which their valuations are private information. A working paper by Gautier and Holzner (2015) explores this issue in the context of a labor market in which firms bid against one another in order to attract workers and there is asymmetric information regarding the network formed by job applications. Each worker observes and interacts directly only with the set of firms to which he applied and vice versa. In a departure from the mechanism of Crawford and Knoer, the wage tentatively accepted by a worker is revealed to all of the firms to which the worker submitted applications. It is assumed that all worker-firm pairs are equally productive, which is captured by an assignment game with $a_{ij} = 1$ if worker j applied to firm i and $a_{ij} = 0$ otherwise. Gautier and Holzner construct a noncooperative game of offers and counteroffers that (approximately) implements the firm-optimal core outcome for the associated assignment game.

26.5.2 Centralized Simultaneous Auctions

Demange, Gale, and Sotomayor (1986) propose a *centralized* mechanism based on *simultaneous ascending-price auctions* that converges to the buyer-optimal core payoff. Again, $\varepsilon > 0$ denotes a small unit for price increments. At every stage $t = 0, 1, \ldots$, the auctioneer announces a price p_{jt} for each good j. Initial prices are set at $p_{j0} = 0$ for all j. At stage t, every buyer i submits his optimal demand set, which consists of the goods j that maximize i's payoff $a_{ij} - p_{jt}$ at stage t prices; i submits an empty demand

[16] Kelso and Crawford (1982) generalize the result to a setting in which buyers have multi-unit demand and substitutable preferences over goods.

set if $a_{ij} - p_{jt} < 0$ for all j. Submitted demands induce a network G_t in which i is linked to j if buyer i demands good j. If there exist matchings of G_t that cover all buyers with nonempty demand sets, then the auction ends and the market is cleared according to one of these matchings at stage t prices. In the absence of such matchings, Hall's Theorem implies the existence of a subset of buyers M with the property that $|M| > |\mathcal{L}^{G_t}(M)|$. We say that any set of goods $\mathcal{L}^{G_t}(M)$ demanded by such a group of buyers M is *over-demanded*. The auctioneer selects an over-demanded set that is minimal with respect to inclusion and increases the price of every good j in the set by ε for stage $t+1$, i.e. $p_{j(t+1)} = p_{jt} + \varepsilon$. The prices of all other goods j' are left unchanged, $p_{j'(t+1)} = p_{j't}$.

Theorem 8 (Demange, Gale, and Sotomayor 1986). *The simultaneous auction ends in a finite number of steps and the final prices it generates converge as $\varepsilon \to 0$ to the minimum competitive prices in the associated economy.*

In an interpretation of the assignment game in the context of a bipartite network of buyers and sellers, Kranton and Minehart (2001) consider a noncooperative game in which sellers supply identical goods and each buyer has the same value for all the goods offered by his neighbors. Buyers first decide simultaneously with which sellers to form links; the cost $c > 0$ of every linked is paid by the buyer who forms it. The ensuing pattern of links defines a network G, which is observed by all buyers. Then nature draws the private value v_i of each buyer i independently from an identical distribution F.[17] Only buyer i learns the value v_i. The realized network and values generate an assignment game a in which $a_{ij} = v_i$ if $ij \in G$ and $a_{ij} = 0$ otherwise. Buyers participate in a simultaneous ascending-price auction for this assignment game. The auction proposed by Kranton and Minehart is similar to the one developed by Demange, Gale, and Sotomayor (1986) except for a few design differences.[18] Kranton and Minehart prove a strong efficiency result.

Theorem 9 (Kranton and Minehart 2001). *There exists a perfect Bayesian equilibrium with the following properties:*

1. *in any subgame following the formation of a network, each buyer i competes in the auction of every seller he is linked to until the price reaches his valuation v_i, at which point he drops out of the auction, and the resulting allocation for every realization of buyer values constitutes a minimum price competitive equilibrium outcome in the associated economy;*

[17] The conclusions of this model extend immediately to heterogeneous linking costs and value distributions.

[18] Prices increase continuously and each buyer decides at every point whether to continue competing in or drop out of the auction of any linked seller. Maximal submarkets induced by the network of buyer-seller pairs (i,j) with the property that buyer i has not dropped out of seller j's auction are cleared as prices increase.

2. *at the ex-ante network formation stage, buyers form a network that maximizes total expected welfare net of linking costs.*

This efficiency result is related to a key finding of Demange (1982) and Leonard (1983), which reveals that at the buyer-optimal core allocation of any assignment game $(a_{ij})_{i\in B, j\in S}$, each buyer receives the marginal value he contributes to the grand coalition.[19] Demange and Leonard independently considered the situation in which the vector of valuations $(a_{ij})_{j\in S}$ constitute buyer i's private information and analyzed incentives for truthful revelation of preferences in a mechanism in which every buyer reports a vector of valuations and then the mechanism designer implements a minimum price competitive equilibrium of the economy with reported valuations. Using the characterization of the buyer-optimal core allocation in terms of welfare externalities, they proved that this direct revelation mechanism is *strategy-proof*: it is a weakly dominant strategy for every player to report his true valuations. The first part of the result of Kranton and Minehart refines this conclusion by addressing the issue of interim incentives in the *dynamic* environment of the simultaneous auction mechanism.[20] The intuition for the result on efficient network formation is that the positive welfare externalities created by adding a link to any network are fully captured in the auction by the buyer who forms and bears the cost of that link.

To test the limits of Kranton and Minehart's efficiency result, Elliott (2015) considers a network formation game in which (1) payoffs are determined by a fixed convex combination of the buyer- and the seller-optimal core outcomes and (2) establishing a link requires investments from both buyer and seller. When linking costs are shared according to an exogenous rule, hold-up problems may lead to underinvestment in link formation, which in some cases eliminates all the potential gains from trade; overinvestment in links that do not affect the maximum total surplus (but change its division) is also possible and can dissipate half of the gains from trade. The alternative assumption that buyers and sellers negotiate the split of linking costs endogenously eliminates underinvestment but may exacerbate overinvestment.

[19] Elliott (2015) develops a characterization of the buyer-optimal core payoffs which provides further economic insight. A buyer i and a seller j are said to be an *efficient match* for each other if i receives good j at some efficient assignment. Buyer i constitutes an *outside option* for seller j if i is an efficient match for j in the assignment game obtained by eliminating a buyer who is an efficient match for j in the original game. An *opportunity path* originating at buyer i is a sequence $i_0, j_0, i_1, j_1, \ldots$ with $i = i_0$ such that for $k \geq 0$ seller j_k is an efficient match for buyer i_k and buyer i_{k+1} is an outside option for seller j_k (the last player in the sequence is either a buyer with no efficient match or a seller with no outside option). Elliott proves that the buyer-optimal core payoff of each buyer i can be computed by alternatively adding and subtracting the values along any opportunity path $i_0, j_0, i_1, j_1, \ldots$ (i.e., $a_{i_0 j_0} - a_{i_1 j_0} + a_{i_1 j_1} - \ldots$) originating at i. This finding generalizes a result of Kranton and Minehart (2000).

[20] While Kranton and Minehart restrict attention to assignment games derived from networks with homogeneous goods, Gul and Stacchetti (2000) obtain a general version of this result for settings with multi-unit demand for heterogeneous goods and utility functions that satisfy the gross substitutes condition.

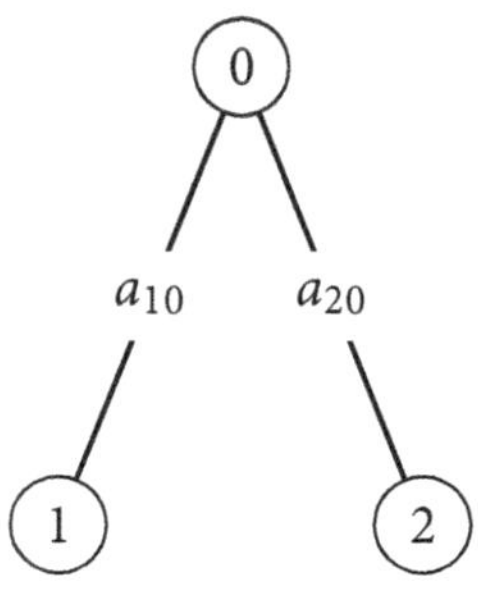

FIGURE 26.8 An example with one seller and two buyers.

26.5.3 A Noncooperative Bargaining Game

All models discussed thus far in this section predict that an efficient allocation will arise. Elliott and Nava (2014) challenge this conclusion by analyzing a noncooperative bargaining model derived from an assignment game $(a_{ij})_{i\in B, j\in S}$. Let $(p_k > 0)$ denote a probability distribution over the set of players. At every date $t = 0, 1, \ldots,$ a single player k who has not traded by date t is recognized as the *proposer* with probability p_k.[21] Player k then chooses a bargaining partner h from the other side of the market and proposes a price (which determines how the surplus a_{kh} is divided between the two parties). If player h accepts the offer, then k and h trade at the agreed price and exit the market. Players have a common discount factor $\delta \in (0, 1)$.[22]

To gain some intuition for the competitive forces induced by this bargaining protocol, consider the monopoly scenario illustrated in Figure 26.8 in which a single seller, player 0, bargains with two buyers, players 1 and 2. Suppose that buyer 1 has a higher value than buyer 2 for the good supplied by the seller, i.e., $a_{10} > a_{20}$, and that each of the three players is recognized as the proposer with probability $1/3$. If $a_{10} > 2a_{20}$, then for δ above a certain threshold, there exists a unique MPE, in which the seller trades exclusively with buyer 1 and both players 0 and 1 obtain limit payoffs of $a_{10}/2$ as $\delta \to 1$. If instead $a_{10} \leq 2a_{20}$, then there exists a family of MPEs for high δ that yields limit payoffs of $a_{20}, a_{10} - a_{20}$, and 0 as $\delta \to 1$ for players 0, 1, and 2, respectively. We conclude that in both cases the limit MPE payoffs of players 0 and 1 are summarized by the formulae $\max(a_{10}/2, a_{20})$ and $\min(a_{10}/2, a_{10} - a_{20})$, respectively, whereas buyer 2 receives a limit MPE payoff of 0.

A different bargaining protocol proposed by Manea (2014c) in the context of an intermediation model generates the same predictions for the limit MPE payoffs. In

[21] Probabilities are not renormalized as players trade and exit the market, so there may be stages where no proposer is selected.

[22] This bargaining protocol is similar to the second protocol analyzed by Abreu and Manea (2012b), except that in the present setting a trader selected by nature automatically becomes the proposer in the bargaining round with his chosen partner, while in the model of Abreu and Manea either party serves as the proposer with probability $1/2$ after the selected trader activates a link.

the setting of the example above, the protocol stipulates that the seller gets a new opportunity to choose one of the buyers as a bargaining partner in every period, and both the seller and the chosen partner assume the role of the proposer with probability 1/2 after the choice is made. If $a_{10} < 2a_{20}$, then in either model the seller takes advantage of his outside option of trading with the low-value buyer with positive probability, but the probability of such an inefficient agreement converges to 0 as $\delta \to 1$.[23] As discussed in Manea (2014c), the common limit MPE payoffs can be explained as follows. When players are patient, the seller is able to pick his preferred outcome between two scenarios: (1) a bilateral monopoly situation corresponding to a two-player bargaining game between the seller and buyer 1 (where buyer 2 is absent); and (2) a second-price auction in which the seller extracts all the surplus created by the low-value buyer. Thus the outside option of trading with buyer 2 is "credible" in equilibrium only if $a_{10} \leq 2a_{20}$.[24]

Elliott and Nava explore the extent to which the conclusion of asymptotic efficiency of MPEs generalizes beyond the simple example above. To approach this issue, they restrict attention to assignment games that have a unique efficient assignment, denoted η. An agreement between buyer i and seller j is said to be *efficient* if $\eta(i) = j$. In this setting, Elliott and Nava call a family of MPEs for $\delta \in (0,1)$ *limiting efficient* if for every date t when a player k with $k \neq \eta(k)$ is selected as the proposer following a series of efficient agreements, k trades with $\eta(k)$ at date t with a conditional probability that converges to 1 as $\delta \to 1$.

Motivated by the example with two buyers and a single seller discussed above, define the *core outside option* o_k of player k as the maximum surplus that k can create by trading with a player on the opposite side of the market who is left unassigned by η,

$$o_k = \max \left\{ a_{kh} \mid \eta(h) = h \; \& \; h \text{ is on the opposite side of the market from } k \right\}.$$

In a limiting efficient MPE, all traders unassigned under η remain in the market indefinitely with limit probability 1 and receive limit payoffs of 0, so every player should obtain a limit payoff at least as high as his core outside option. Then an agreement between buyer i and seller $\eta(i)$ departs from their "Rubinstein shares" $(p_i a_{i\eta(i)}/(p_i + p_{\eta(i)}), p_{\eta(i)} a_{i\eta(i)}/(p_i + p_{\eta(i)}))$ in a two-player game in which they bargain in isolation only if the outside option of either player "binds." Based on this intuition, Elliott and Nava define the *shifted Rubinstein payoff* u_k of a player k assigned under η

[23] While the limit MPE payoffs coincide under the two bargaining protocols, the structure of equilibrium agreements for high δ is not exactly analogous when $a_{10} < 2a_{20}$. In the MPE of Manea's model for this case, the seller chooses to bargain with the low-value buyer 2 with positive probability and trades with buyer 2 with probability 1 conditional on this choice. In the model of Elliott and Nava, the seller makes an offer to buyer 2 with probability 0 conditional on being recognized as the proposer, whereas in the event that buyer 2 is the proposer, he forges an agreement with the seller with positive probability.

[24] See Shaked and Sutton (1984) for a classic study of outside options.

as follows:[25]

$$u_k = \begin{cases} o_k & \text{if } o_k \geq \frac{p_k}{p_k+p_{\eta(k)}} a_{k\eta(k)} \\ a_{k\eta(k)} - o_{\eta(k)} & \text{if } o_{\eta(k)} \geq \frac{p_{\eta(k)}}{p_k+p_{\eta(k)}} a_{k\eta(k)} \\ \frac{p_k}{p_k+p_{\eta(k)}} a_{k\eta(k)} & \text{otherwise.} \end{cases}$$

For any player left unassigned by η, the shifted Rubinstein payoff is simply defined to be 0. Elliott and Nava prove that limiting efficient MPEs must yield the shifted Rubinstein payoffs and provide a necessary condition for the existence of such equilibria.

Theorem 10 (Elliott and Nava 2014). *The payoffs for every family of limiting efficient MPEs converge to the shifted Rubinstein payoffs as players become patient. Limiting efficient MPEs exist only if the profile of shifted Rubinstein payoffs belongs to the core of the underlying assignment game.*

The first part of this result follows from an inductive argument on the number of buyers assigned under η. Consider a subgame for a family of limiting efficient MPEs where at least two buyers assigned under η are left. Then for each remaining buyer-seller pair $(i, \eta(i))$, there is positive limit probability that another efficient trade takes place. By the inductive hypothesis, $(i, \eta(i))$ obtain the shifted Rubinstein payoffs following that trade, and this property extends to the earlier stage. The induction base case, in which a single buyer is assigned under η, has the flavor of the example discussed above. The second part of the result follows from the first part, which establishes that any limiting efficient MPE must yield the shifted Rubinstein payoffs $(u_k)_{k\in B\cup S}$. If this payoff profile does not belong to the core of the assignment game, it must be that $u_i + u_j > a_{ij}$ for a buyer-seller pair (i,j). Then buyer i can obtain a limit payoff strictly greater than u_i by deviating from the underlying equilibria to make an offer slightly higher than u_j to j when he is selected as the proposer.

Elliott and Nava argue that Theorem 10 has negative welfare implications because the shifted Rubinstein payoffs and the core of the assignment game are not systematically related. However, limiting efficiency is a strong welfare criterion. A more natural efficiency notion in this setting would be a version of Abreu and Manea's (2012b) concept of asymptotic efficiency. Specifically, a family of MPEs for $\delta \in (0,1)$ with corresponding payoffs $(v_k^\delta)_{k\in B\cup S}$ is *asymptotically efficient* if $\lim_{\delta\to 1} \sum_{k\in B\cup S} v_k^\delta$ equals the maximum total surplus. Note that, under the assumption of a unique efficient assignment, a family of asymptotically efficient MPEs must eventually lead to the efficient assignment with a probability converging to 1 as $\delta \to 1$. Limiting efficiency requires that efficient agreements arise without any endogenous delay (in addition to the exogenous delay inherent to the bargaining protocol) as players become patient. Clearly, every family of limiting efficient MPEs is necessarily asymptotically efficient.

[25] The first two cases correspond to the outside options of k and $\eta(k)$ binding. The efficiency of the assignment η implies that it is not possible for both outside options to bind simultaneously.

At this time, it is not known whether a reverse relationship holds and whether Theorem 10 extends to also characterize asymptotically efficient MPEs.

26.6 Conclusion

This chapter surveys the growing body of research on bilateral trade in networks.[26] The theoretical models explored here advance our understanding of how local competitive forces shape trading outcomes and the balance of bargaining power in different parts of a network. The predictions of these models are qualitatively similar for some networks yet diverge sharply for others. We attempted to explain the modeling choices that account for discrepancies, but in many cases a complete comparison between models is intractable. We found that the nature of the matching process and the selection of the solution concept have substantially different implications for the dynamics of trade, the balance of bargaining power, welfare properties of market outcomes, and equilibrium multiplicity. In some of the models, the question of what types of networks and equilibrium concepts lead to inefficient trade or multiple equilibria requires further investigation. It would also be desirable to get a better grasp of which results are robust to the specification of the matching and bargaining process. Relatedly, it would be interesting to know what bargaining protocols offer more realistic predictions for markets in which participants meet and bargain "freely." Empirical analysis of real markets where network data is available and laboratory experiments can shed light on this issue.

We finally comment on a couple of restrictive assumptions that are prevalent in the existing literature. The work surveyed here deals exclusively with the case in which buyers have unit demand and sellers have unit supply. It would be useful to develop tractable models in which traders have multi-unit supply and demand. Another strong assumption maintained through most of the literature is that the underlying network is common knowledge among traders. Relaxing this assumption in plausible ways to allow for incomplete information about local network structure constitutes another important topic for future research.

Appendix

Proof of the First Part of Theorem 4

We start by showing that all matches result in first-period agreement in any MPE for every discount factor. Fix an MPE for the network G and the discount factor δ. Consider a link ij that is part of a maximum matching of G. Let u_i^{ij} and u_j^{ij} denote the expected payoffs of players

[26] We restricted attention to markets in which intermediation is not possible; a review of the literature on intermediation in networks can be found in Chapter 27.

i and j, respectively, in the MPE conditional on the event that they are matched with each other and do not reach an agreement in the first period. These payoffs are evaluated under the assumption that the other players conform to their equilibrium strategies, with discounting applied from the perspective of the first period. When i is selected to make an offer to j in the first period, his conditional expected payoff is at least $1 - u_j^{ij}$. Indeed, player j must accept any offer greater than u_j^{ij} from i since j's expected payoff in the MPE conditional on not reaching an agreement with i in the first period is u_j^{ij} (regardless of i's rejected offer). Furthermore, player j can secure a payoff of u_j^{ij} in the event that i is the proposer by rejecting any offer that i makes. Hence the sum of equilibrium continuation payoffs of i and j conditional on i being the proposer in the match (i, j)—an event which is commonly known by i and j—is at least 1. The same conclusion holds if instead j is selected to make an offer to i. If v_i^{ij} and v_j^{ij} denote the expected payoffs of players i and j, respectively, in the MPE conditional on i being matched to j in the first period (before a proposer is recognized), it follows that $v_i^{ij} + v_j^{ij} \geq 1$. Then for any maximum matching H of G, summing the previous inequality over the links of H, we obtain $\sum_{ij \in H}(v_i^{ij} + v_j^{ij}) \geq |H| = m$, where m is the maximum total surplus in G. Since the sum of expected payoffs of all players in the MPE is at least as large as the average of the expression $\sum_{ij \in H}(v_i^{ij} + v_j^{ij})$ over all maximum matchings H, it must be that the total sum of MPE payoffs is greater than or equal to m. Clearly, the total sum of MPE payoffs cannot exceed m, so it must be exactly equal to m. As $\delta < 1$, this is possible only if all matches result in agreement.

We can now prove that there exists exactly one MPE. When players i and j are matched in the original network, they correctly anticipate that all other matches result in agreement. If i and j fail to reach an agreement then they are left in a subnetwork in which the maximum matching has only one link. Hence the remaining subnetwork would consist of either the single link ij, a star with multiple spokes, or a three-player cycle (possibly with some isolated nodes). It is routine to verify that there exists a unique MPE in every such subnetwork. Players i and j share the same beliefs about the distribution of subnetworks that may arise in the second period. Since the sum of expected payoffs of i and j in any possible subnetwork cannot exceed 1, the sum of their discounted expected payoffs conditional on not reaching an agreement when matched with each other must be smaller than 1. Using the notation defined above, we have that $u_i^{ij} + u_j^{ij} < 1$. Then in any MPE, when i is selected to propose to j in the original network, i offers j a payoff of u_j^{ij} and j accepts the offer with probability 1. Therefore, there exists a unique MPE, in which every match materializes in an agreement in any subgame, as claimed.[27]

Acknowledgments

Section 26.5 benefited from discussions with Matt Elliott. I thank Ben Golub, Dilip Abreu, and the editors for comments as well as Aubrey Clark and Will Rafey for editing and proofreading.

[27] A reexamination of the argument shows that if the information structure is modified so that all players observe the entire matching activated by nature at each stage (but not the offers made in contemporaneous matches), then the resulting game also has a unique MPE, which generates the same expected payoffs as Polanski's benchmark model.

References

Abreu, D. and M. Manea (2012a). "Markov equilibria in a model of bargaining in networks." *Games and Economic Behavior* 75, 1–16.

Abreu, D. and M. Manea (2012b). "Bargaining and efficiency in networks." *Journal of Economic Theory* 147, 43–70.

Binmore, K. G., A. Shaked, and J. Sutton (1985). "Testing noncooperative bargaining theory: A preliminary study." *American Economic Review* 75, 1178–1180.

Charness, G., M. Corominas-Bosch, and G.R. Frechette (2007). "Bargaining and network structure: An experiment." *Journal of Economic Theory* 136, 28–65.

Corominas-Bosch, M. (1999). "On two-sided network markets." Ph.D. Thesis, Universitat Pompeu Fabra, Barcelona.

Corominas-Bosch, M. (2004). "Bargaining in a network of buyers and sellers." *Journal of Economic Theory* 115, 35–77.

Crawford, V. and E. Knoer (1981). "Job matching with heterogeneous firms and workers." *Econometrica* 49, 437–450.

Démange, G. (1982). "Strategy proofness in the assignment market game." Mimeo.

Démange, G., D. Gale, and M. Sotomayor (1986). "Multi-item auctions." *Journal of Political Economy* 94, 863–872.

Edmonds, J. (1965). "Paths, trees, and flowers." *Canadian Journal of Mathematics* 17, 449–467.

Elliott, M. (2015). "Inefficiencies in networked markets." *American Economic Journal: Microeconomics*, forthcoming.

Elliott, M. and F. Nava (2014). "Decentralized bargaining: Efficiency and the core." Mimeo.

Gale, D. (1986a). "Bargaining and competition, part I: Characterization." *Econometrica* 54, 785–806.

Gale, D. (1986b). "Bargaining and competition, part II: Existence." *Econometrica* 54, 807–818.

Gale, D. (1987). "Limit theorems for markets with sequential bargaining." *Journal of Economic Theory* 43, 20–54.

Gale, D. and L. Shapley (1962). "College admissions and the stability of marriage." *American Mathematical Monthly* 69, 9–15.

Gallai, T. (1964). "Maximale systeme unabhängiger kanten." *Magyar Tud. Akad. Mat. Kutató Int. Közl* 9, 401–413.

Gauer, F. (2014). "Strategic formation of homogeneous bargaining networks." Mimeo.

Gautier, P. and C. Holzner (2015). "Maximum matching in the labor market under incomplete information." Mimeo.

Hall, P. (1935). "On representatives of subsets." *Journal of the London Mathematical Society* 10, 26–30.

Jackson, M. O. and A. Wolinsky (1996). "A strategic model of social and economic networks." *Journal of Economic Theory* 71, 44–74.

Kelso, A. S. and V. P. Crawford (1982). "Job matching coalition formation and gross substitutes." *Econometrica*, 50, 1483–1504.

Koopmans, T. C., and M. Beckman (1957). "Assignment problems and the location of economic activities." *Econometrica* 25, 53–76.

Kranton, R. and D. Minehart (2000). "Competition for goods in buyer-seller networks." *Review of Economic Design* 5, 301–331.

Kranton, R. and D. Minehart (2001). "A theory of buyer-seller networks." *American Economic Review* 91, 485–508.

Lovász, L. and M. D. Plummer (2009). *Matching Theory.* AMS Chelsea Publishing, Providence, RI.

Leonard, H. B. (1983). "Elicitation of honest preferences for the assignment of individuals to positions." *Journal of Political Economy* 91, 461–479.

McLennan, A. and H. Sonnenschein (1991). "Sequential bargaining as a noncooperative foundation of general equilibrium." *Econometrica* 59, 1395–1424.

Manea, M. (2011a). "Bargaining in stationary networks." *American Economic Review* 101, 2042–2080.

Manea, M. (2011b). "Bargaining in stationary networks: Online appendix." https://www.aeaweb.org/aer/data/aug2011/20090320_app.pdf.

Manea, M. (2014a). "Bargaining in dynamic markets." Mimeo.

Manea, M. (2014b). "Steady states in matching and bargaining." Mimeo.

Manea, M. (2014c). "Intermediation in networks." Mimeo.

Nguyen, T. (2014). "Coalitional bargaining in networks." Mimeo.

Polanski, A. (2007). "Bilateral bargaining in networks." *Journal of Economic Theory* 134, 557–565.

Polanski, A. and F. Vega-Redondo (2014). "Markets, bargaining, and networks with heterogeneous agents." Mimeo.

Roth, A. E. and M.A.O. Sotomayor (1990). *Two-Sided Matching: A Study in Game-Theoretic Modeling and Analysis.* Econometric Society Monographs, Cambridge University Press, Cambridge.

Rubinstein, A. (1982). "Perfect equilibria of a bargaining model." *Econometrica* 50, 97–110.

Rubinstein, A. and A. Wolinsky (1985). "Equilibrium in a market with sequential bargaining." *Econometrica* 53, 295–328.

Rubinstein, A. and A. Wolinsky (1990). "Decentralized trading, strategic behaviour and the Walrasian outcome." *Review of Economic Studies* 57, 63–78.

Shaked, A. and J. Sutton (1984). "Involuntary unemployment as a perfect equilibrium in a bargaining model." *Econometrica* 52, 1351–1364.

Shapley, L.S. and M. Shubik (1971). "The assignment game I: The core." *International Journal of Game Theory* 1, 111–130.

Stahl, I. (1972). *Bargaining Theory.* Stockholm School of Economics, Stockholm.

CHAPTER 27

STRATEGIC MODELS OF INTERMEDIATION NETWORKS

DANIELE CONDORELLI AND ANDREA GALEOTTI

27.1 Introduction

A time-honored tradition in economics is to see markets as populated by a large number of agents interacting anonymously through the price system. The Walrasian paradigm, which culminated in the Arrow-Debreu-McKenzie model, has contributed to our understanding of such markets. However, a body of evidence accumulated over time shows that individual relationships and bonds of trust, arising from kinship ties, geographical proximity, joint investments, and so forth, affect economic outcomes in many relevant cases.[1]

A number of authors have taken up the challenge of studying the implications of decentralized trade.[2] Kirman (1997) and Tesfatsion (1997) stand as pioneering works in economics, for their modeling of markets as non-anonymous and the economy as a network. Corominas-Bosch (2003) and Kranton and Minehart (2000) developed the first strategic models of trading in networks. The key feature of these papers is that buyers and sellers have preferential relationships (i.e., links), that make trade among them feasible. The network structure becomes relevant because it determines the outside options of a buyer and a seller bargaining over the trade surplus. The literature

[1] For instance, the literature on social networks in sociology documents that personal relationships spill over into business relationships, and vice versa; also, the importance of ethnicity ties in shaping trade relations cannot be overemphasized, even in times of globalization, see Rauch (1999).

[2] There is a large literature that studies decentralized trade, in which the interaction across buyers, sellers, and intermediaries is modeled through search and random matching; see the seminal work of Rubinstein and Wolinsky (1987), and Duffie et al. (2005) for a more recent work. This survey focuses on models in which the interaction among traders is represented by a graph.

on buyer-seller networks has since grown to be relatively large, and it is surveyed in Chapter 26 of the handbook by Manea.

While buyer-seller networks are ubiquitous, trade in a wide range of markets involves a plethora of other subjects, such as intermediaries, dealers, brokers, market-makers, wholesalers, and retailers (see, e.g., Spulber 1999). This potentially long list of middlemen that connect producers to buyers all contribute to the value of the final product, or at least to its delivery. Long intermediation chains play a vital role in the market for agricultural goods in developing countries (e.g., see Fafchamps and Minten 1999), as well as in financial markets for the trade of assets sold over-the-counter (e.g., see Li and Schurhoff 2012). Complex processes of production and distribution lead to supply chains, a natural example of chains of intermediaries (e.g., see Hummels et al. 2001 and Antras and Chor 2013).

The presence of intermediate agents raises a whole new set of questions: how do different network structures affect the efficiency of trading outcomes? How does the position of a trader in the network affect his payoff? How does the interplay of horizontal competition among intermediaries for buyers and sellers and the vertical complementarities in completing a chain shape the final outcome of trade? The literature we survey in this chapter addresses these questions within a common framework, in which the network architecture of intermediation plays a central role in the determination of trading outcomes.[3]

The chapter is organized as follows. Section 27.2 focuses on pure *intermediation networks*: buyers and sellers are not directly connected and need the services of intermediaries to conclude an exchange. Intermediaries accomplish the task by buying from sellers or from other intermediaries and reselling to buyers or to other intermediaries. Intermediaries do not just match buyers and sellers or mediate the transaction; they are dealers. Intermediation networks are the norm in markets in which no centralized exchange exists, such as financial over-the-counter markets and the market for artworks and antiques. What distinguishes the papers in this section is the trading protocol through which interaction takes place: bilateral bargaining as in Condorelli, Galeotti, and Renou (2015) and Manea (2015), bid-and-ask prices posted by intermediaries as in Blume et al. (2007) and Gale and Kariv (2009), and auctions as in Kotowski and Leister (2014). On a more subtle level, these papers also differ in their information structure—complete *versus* incomplete information.

In the above models, it does not matter to buyers and sellers who intermediates the object, insofar as this does not alter the terms of trade. In Section 27.3, we relax this restriction and focus on network structure of supply chains more generally. In this case, the value of the object to the final buyers depends on who provides intermediate inputs, and the network describes how inputs from upstream firms can be combined with inputs of downstream firms.[4]

[3] We briefly discuss network formation in the concluding section.

[4] This also relates to recent work in macroeconomics and international trade on the role of production networks and firm-to-firm trade. The empirics of firm-to-firm trade and some modeling of production networks in macroeconomics are discussed in Chapter 28 of the handbook by Chaney.

Recent work in economics has focused on the optimal allocation of ownership rights along a supply chain, and on stable contracts along supply chains (e.g., Antras and Chor 2013; Ostrovsky 2008, and Hatfield et al. 2013). Some other work has focused instead on the role that production supply chains play in translating idiosyncratic shocks into volatility at the aggregate level (e.g., Acemoglu et al. 2012).[5] Here we survey a complementary body of work that studies oligopolistic pricing in competing supply chains. The discussion is based on the work of Choi et al. (2015) and Galeotti and Goyal (2015).

The models we discuss in Sections 27.2 and 27.3 differ in many aspects, but they generate recurrent equilibrium effects. First, the presence of intermediaries produces new forms of inefficiency when combined with information asymmetries or transaction costs. In particular, trade can occur via a long chain of intermediaries, even if shorter chains are feasible, and so intermediation costs can be too large. However, efficiency is re-established when intermediation stands as the only friction: reselling is sufficient to obtain allocative efficiency in equilibrium.

Second, market power is associated with network positions that give control and access to valuable parts of the networks. Betweenness centrality and variants of it are then adequate network proxies for market power in intermediated markets. These measures differ from classical network measures that have been used to describe power in networks, such as Bonacich and eigenvector centrality.[6] A corollary of this insight is that intermediaries can increase their market power by merging horizontally, and this comes at the expense of other traders, possibly also located further away from the merging intermediaries.

Finally, in models with complete information, the price at which the object is exchanged is increasing along the supply chain: this is necessary for intermediaries not to take a loss from buying and reselling. In contrast, when each trader's demand for the object is private information, trading conveys information, and the fact that the object remains in the market is bad news about its value. This implies that the price at which the object is exchanged decreases over time.

27.2 Intermediation

A finite set of traders is located in a directed network and the architecture of the network is common knowledge. A link from trader i to trader j represents the opportunity for i to sell the object to j. Traders can be buyers, sellers, or intermediaries. Each seller owns

[5] The effect of networks in translating idiosyncratic shock into volatility at the aggregate level is related to the topic of systemic risk, which is covered in Chapter 21 of the handbook by Acemoglu et al.

[6] The literature on network games shows a relation between equilibrium play and Bonacich and Eigenvector centrality. Network games are discussed in Chapter 5 of the handbook by Bramoullé and Kranton.

a single indivisible unit of a durable commodity. Buyers have consumption value for the good, whereas sellers and intermediaries wish to maximize their monetary payoff.

The various models that we consider in this section are developed in the context of this general environment, but they differ in terms of the information structure and the trading protocol used to exchange the objects. We first consider a model of bilateral bargaining with random selection of proposer developed by Manea (2015). Second, we discuss the bilateral bargaining model of Condorelli, Galeotti, and Renou (2015), in which each trader has a private value for the object and the seller makes all the offers. We then turn to Kotowski and Leister (2014), in which a seller uses auctions to bargain with multiple buyers or multiple intermediaries at once and to exploit competition among them. We conclude with a discussion of Blume et al. (2007) and Gale and Kariv (2009), in which intermediaries compete by posting bid-and-ask prices. The above papers occupy a central position in our survey, and we discuss related papers as we go along.

To discipline the discussion and to compare the results across model we focus, when possible, on a specific, yet rich, class of networks.

Definition 1. *A complete multipartite network is defined by a single seller s, single buyer b and a set $N = \{1,\dots,n\}$ of intermediaries, connected as follows:*

a. *the set of intermediaries is partitioned in L-subsets, $\{N_1,\dots,N_L\}$, $L \geq 1$;*
b. *the seller can only trade with all intermediaries in N_1; each intermediary in N_x can only trade with all intermediaries in N_{x+1}, for all $x = 1\dots L-1$; and each intermediary in N_L can only trade with the final customer b.*

Figure 27.1 illustrates complete multipartite networks. We refer to intermediaries in N_x as intermediaries in tier x. By convention, we say that the seller belongs to tier 0. The number of intermediaries in tier x is denoted by n_x. From the view point of an intermediary in tier x, intermediaries in tiers $\{x+1,\dots,L\}$ are downstream intermediaries, and those in tier $\{1,\dots,x-1\}$ are upstream intermediaries. We say that intermediary i in tier x is *critical* if he is the sole intermediary in that tier. For each tier x, we denote by k_x the number of critical downstream intermediaries. The total number of critical intermediaries in a network is denoted by k. The line network is a complete multipartite graph where there is only one intermediary in each tier (see Figure 27.1b). In a line network, each intermediary is critical. A competitive network is a complete multipartite network in which there are no critical intermediaries (see Figure 27.1a).

One exercise that this class of networks allows us to perform is to simulate the effect of horizontal mergers.[7] For a specific trading protocol, we compute the equilibrium outcome in a network in which there are L tiers with respective sizes $\{n_1,\dots,n_L\}$. We then allow a subset of intermediaries in a tier to merge: they choose their action in order to maximize their joint profits, while all other intermediaries still act to maximize

[7] Some of the papers we shall discuss also provide insights into vertical mergers. The formal discussion of these results requires introducing more notation and is beyond the scope of this survey.

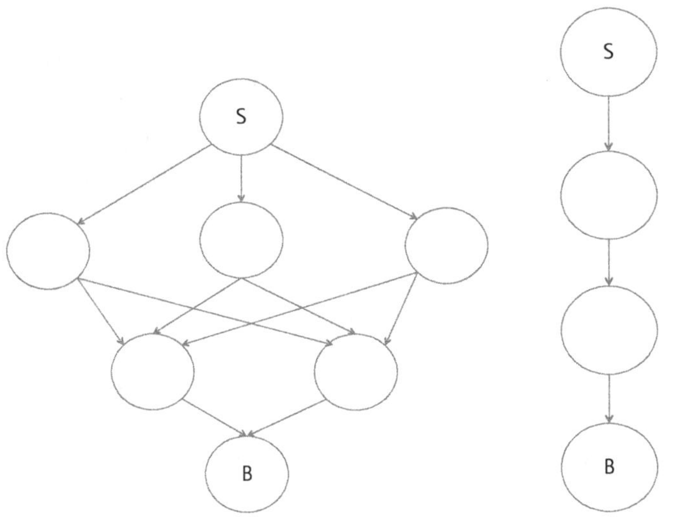

(a) Competitive network (b) Line network

FIGURE 27.1 Examples of complete multipartite networks.

their own individual payoff. We then compare the equilibrium outcomes in these two scenarios, focusing in particular on the changes in realized surplus and payoffs.

27.2.1 Bilateral Bargaining

Manea (2015) develops a bilateral bargaining model with complete information, in which a single object is exchanged until it reaches a buyer along one of the possible paths of intermediaries. We now introduce a simplified version of Manea (2015) and discuss equilibrium properties in the class of complete multipartite networks.[8] We assume that there is a single buyer with consumption value v for the good. The bargaining game in the network develops in an infinite number of rounds. Agents discount the future and have a common discount factor $\delta \in (0,1)$. At the beginning of round $t \geq 0$, let agent i in tier $x \in \{0,\ldots,L\}$ be the owner of the object (s is the owner at 0). Trade in period t develops as follows:

1. The owner i selects a trading partner, say j, among the intermediaries in tier $x+1$.
2. With probability p, the owner i makes an offer to j, who decides whether to reject or to accept; with the complementary probability, it is agent j who makes an offer to i.

[8] Manea (2015) allows for transaction costs, multiple buyers, and heterogenous valuations across buyers. As discussed below, Manea's (2015) analysis is for arbitrary acyclic networks.

3. If the offer is rejected, we enter period $t + 1$. If the offer is accepted, trade takes place and j becomes the new owner, in which case, if j is an intermediary, we enter $t + 1$; if j is a buyer, he consumes the good and the game ends.

All actions are observed and, therefore, the game has perfect information. The focus is on stationary Markov Perfect Equilibrium (MPE): a subgame perfect equilibrium in which, at any history, all actions taken in period t do not depend on actions taken in previous rounds. The main results of Manea (2015) characterize equilibrium payoffs in the limit case, in which agents become perfectly patient (i.e., $\delta \to 1$ in general acyclic networks). For an equilibrium and a subgame in which i owns the object, we define the resale value of i as his expected payoff in that subgame. Manea (2015) shows that for any family of stationary MPEs, resale values converge as $\delta \to 1$ and provide an elegant recursive characterization of the limiting vector of resale values.

This characterization is based on a decomposition of the network in a sequence of layers. This sequence is constructed recursively as follows: in layer 0, we first add buyers; then we add all intermediaries linked to at least two buyers; then we add all intermediaries linked to at least two agents already included in layer 0, and so on, until no more intermediaries are added to layer 0. To construct layer 1, we consider only agents that have not been assigned to layer 0. In layer 1, we first add intermediaries who have only one link to intermediaries in layer 0; then we add all intermediaries that have at least two links with intermediaries in layer 1; and we proceed untill we have no more intermediaries to add in layer 1. The algorithm continues until all agents have been allocated to one layer. For example, in Figure 27.2, the buyer is the only agent in layer 0; intermediary 3 is in layer 2; and all remaining agents are in layer 1. The obtained layer structure is, in turn, what determines limiting resale values. The limiting resale value of an intermediary in layer l is, in fact, $p^l v$.

In the case of complete multipartite networks, this characterization takes a very simple form. Recall that k_x is the number of critical intermediaries in tiers $\{x+1, \ldots, L\}$, while k is the total number of critical intermediaries.

Proposition 1. *Consider a complete multipartite network. As* $\delta \to 1$:

1. *the resale value of intermediary i in tier x converges to* $p^{k_x+1}v$;
2. *the equilibrium payoff of the initial seller converges to* $p^{k+1}v$, *and the equilibrium expected payoff of the buyer converges to* $(1-p)v$; *and*
3. *the equilibrium payoff of noncritical intermediaries converges to* 0, *and the equilibrium payoff of critical intermediary in tier x converges to* $(1-p)p^{k_x+1}v$.

Proposition 1 follows from the main equilibrium characterization in Manea (2015). We sketch a proof for Part 1 of Proposition 1 that uses a backward induction argument.[9] When the object reaches an intermediary in the last tier, we have a standard bargaining

[9] The proof of this result for general networks is more involved. To see why, note that a property of complete multipartite networks is that there are no paths connecting intermediaries in the same tier. Hence, whenever a seller in tier x sells to an intermediary in tier $x + 1$, the continuation payoff of all

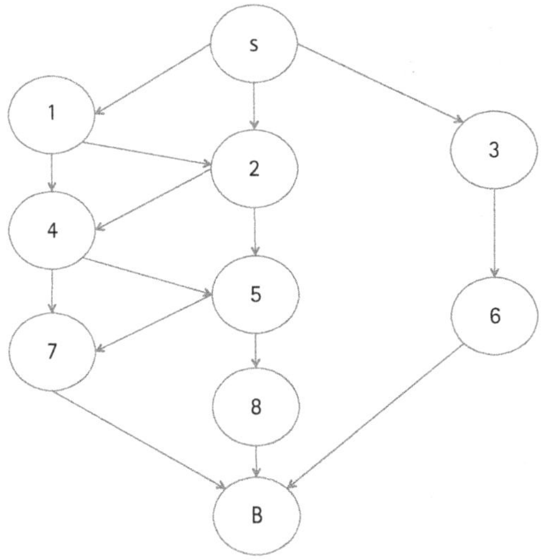

FIGURE 27.2 Trade does not minimize intermediation; Manea (2015).

game with random proposer and, in the limit equilibrium, the intermediary obtains payoff pv and the buyer obtains $(1-p)v$. The resale value of each intermediary in tier L is, then, pv. Suppose that the object has reached intermediary i in tier $L-1$. If tier L is monopolized by critical intermediary j, then we again have a standard bargaining game between intermediary i and intermediary j, where the total surplus of trade is j's resale value pv. In this game, intermediary i obtains an expected payoff of p^2v, which is his resale value. When tier L has more than one intermediary, we are in a bargaining model in which the owner, intermediary i, has multiple potential buyers—all the intermediaries in tier L, each with a resale of pv. Competition across intermediaries in tier L implies that intermediary i extracts all surplus, and so his resale value is pv. Part 1 of the proposition now follows, repeating this argument backwards. It is then straightforward to verify (parts 2 and 3).

To illustrate the economic contents of Proposition 1, first consider a network in which all traders are arranged in a line (see Figure 27.1b). In this case, the resale value of intermediaries is ranked according to their distance to the final customer: the closer the intermediary is to the final customer, the higher is his resale value. Furthermore, the equilibrium payoffs of intermediary i are simply $(1-p)$ times the resale value of i, and so the ranking of the equilibrium payoffs across intermediaries is the same as the ranking of their resale values. Finally, the equilibrium payoff of the initial seller is decreasing in the number of intermediaries, while the payoff of the final buyer is $v(1-p)$.

other intermediaries in tier $x+1$ is zero. In richer networks, the current seller may have neighbors i and j, and i and j may be connected via another path, so that intermediary j can still acquire the good after intermediary i has purchased it. We refer to Manea (2015) for the general proof that deals with these subtleties.

As the bargaining power shifts to upstream traders (i.e., p increases), the payoff of the initial seller increases and the payoff of the final buyer decreases; the seller extracts all the surplus in the limit case where he has full bargaining power (i.e., $p \to 1$). In contrast, the payoff of each intermediary changes nonmonotonically with p. It first increases with a shift of bargaining power to upstream traders, then eventually decreases. Interestingly, for moderate values of p, it can be the case that an increase in p increases the payoff of intermediaries who are close to the initial seller, but it decreases the rent for intermediaries that are close to the final customer.

Second, consider a competitive network (see Figure 27.1a). In this case, downstream intermediaries always compete for the object owned by the upstream intermediary. Competition at every level of the intermediation network implies that intermediaries have the same limiting resale value, regardless of their specific position. Furthermore, intermediaries obtain zero profit, and the seller and the buyer obtain the same payoff that they would obtain if they were to bargain directly (i.e., the seller obtains pv and the final buyer $v(1-p)$).

When the network is competitive, horizontal mergers can be very profitable, as they can create substantial market power. For example, if all intermediaries in one tier decide to merge, the sum of their expected payoff would jump from 0 to $pv(1-p)$, whereas the seller's profits would decrease from pv to $p^2 v$, *ceteris paribus*.[10]

In describing Manea's (2015) framework we have made two important simplifications. First, we have assumed that there are no transaction costs. In the context of bilateral intermediation, if the seller i has to pay a transaction cost c when trading with j, a classical hold-up problem can emerge, giving rise to inefficiencies.[11] Second, in complete multipartite networks, each intermediary in layer $x+1$ is identical from the viewpoint of a seller in tier x. Hence, all paths are equally efficient. However, consider the network in Figure 27.2, and assume that the final buyer has a valuation of 1 and that each trader has a transaction cost c, which is assumed to be small, but positive. In order to maximize aggregate surplus, the object should flow from the initial seller to the final buyer via the shortest path, from s to b via intermediaries 3 and 4. Instead, as Manea (2015) shows, for a high discount factor, in equilibrium, the seller trades with either intermediary 1 or intermediary 2 and the object reaches the final buyer via at least three intermediaries. The key aspect here is that the competition in the network that the seller accesses via intermediaries 1 and 2 is more intense, relative to the competition in the network that the seller accesses via intermediary 3. This implies that intermediary 1's and intermediary 2's resale values are higher than intermediary 3's resale value, and it is, then, more profitable for the seller to bargaining with intermediaries 1 and 2.

[10] These conclusions are specific to the competitive network considered here; see Manea (2015) for a more general analysis of horizontal mergers. Manea (2015) also studies vertical mergers.

[11] For example, suppose that the seller s is linked to intermediary i, who is linked to final buyer b; suppose, also, that the seller has a transaction cost of c and the valuation of the buyer is $v > c$. Since the resale value of intermediary i is pv, whenever $c > pv$, even if there are gains from trade, there is no trade in equilibrium. Wright and Wong (2014) study a similar model on a single chain of traders.

Siedlarek (2014) also studies a bargaining model in a directed network, but, building on Merlo and Wilson (1995), he focuses on a multilateral bargaining protocol. In each period, a path connecting the seller and the buyer is selected. Then, intermediaries in the path are ordered at random: the first intermediary makes a proposal, and, following the order, the other intermediaries sequentially accept or reject. The proposal is implemented if all intermediaries accept; otherwise, the game enters a new round of trade. Siedlarek (2014) characterizes stationary subgame perfect equilibria in a context in which different paths are allowed to generate different economic surplus. When the surplus of each path is homogeneous and agents become perfectly patient, the equilibrium characterization implies that only critical intermediaries obtain positive profit.

27.2.2 Bilateral Bargaining with Asymmetric Information

The framework in Section 27.2.1 makes a neat distinction between buyers and intermediaries, and the consumption value of each trader is common knowledge. Condorelli, Galeotti, and Renou (2015) propose a model of bilateral bargaining in networks with asymmetric information. All traders are potentially interested in consuming the object, and traders selling the good face incomplete information about the value of potential buyers. This framework is particularly appropriate for studying over-the-counter trading in financial markets where, depending on private information such as individual liquidity shocks, traders may buy for taking a certain market position or for reselling to other traders. Another interpretation is one of international trade, in which case the asymmetric information captures information frictions: the tastes of foreigns consumers may be very different from the tastes of domestic consumers.[12]

More formally, the network is populated by a single initial seller s, and a number of traders, each with either a high consumption value v_H, or a low value v_L, with $0 < v_L < v_H$. At the start of the game, the value of a trader is private information and the probability that trader i has high value is π_i. The game develops in an infinite number of trading rounds, and traders discount the future at a common rate $1 > \delta > 0$. In each round, the owner of the object either consumes the object or makes a take-it-or-leave-it offer to a neighbor of his choosing, who, in turn, decides whether to accept or reject.

Asymmetric information considerably complicates the equilibrium analysis in this model. Condorelli, Galeotti, and Renou (2015) provide a characterization of a specific class of stationary equilibria that they call "regular."[13] They show that all regular equilibria take the following form:

[12] The importance of informational frictions in international trade is discussed in Chapter 28 of the handbook by Chaney.

[13] Roughly speaking, these are Weak Perfect Markov equilibria in which the strategy of a high-value buyer is constrained to be monotone in the price asked, for a certain region of the price support.

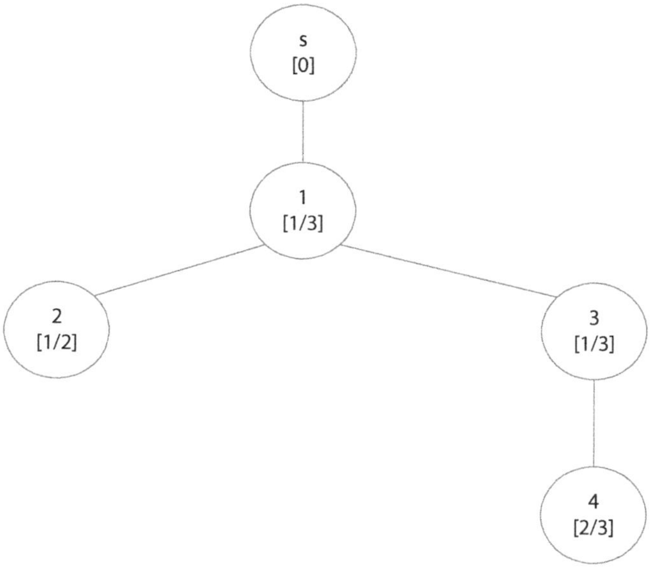

FIGURE 27.3 Examples of bargaining with asymmetric information.

a. If the owner of the object has low valuation, he makes a sequence of take-it-or-leave-it offers to a subset of his neighbors, each at a price that makes a trader with high value indifferent between accepting and rejecting.
b. If all these offers are refused, the owner makes an offer at a price that equals the resale value of the trader receiving the offer. This offer is accepted with probability 1 by both the high- and low-value trader.
c. When a high-value trader acquires the object, he consumes it; when a low-value trader acquires the object, he resells it.
d. The game continues in this fashion until a high-value trader acquires the object or a low-value trader decides to consume as the expected value of the object goes below v_L.

As an example, suppose that $v_H = 1$, $v_L = 0$, and consider the network in Figure 27.3; the number in the parentheses indicates the probability that the trader has high value. As agents become perfectly patient, one equilibrium path takes the following form: The initial seller offers the object to trader 1 at his resale value, which is $5/6$. Trader 2 accepts and consumes if he has high value; otherwise, he accepts and makes an offer of 1 to trader 2. Trader 2 accepts if he has a high value and otherwise rejects the offer. Upon rejection, trader 1 offers the object to trader 3 at his resale value, which is $2/3$. Trader 3 accepts the offer and consumes if he has a high value; otherwise, he makes an offer of 1 to trader 4. If trader 4 has high value, he accepts and consumes; otherwise, trader 3 consumes.

Two main insights can be derived from this characterization. First, equilibrium asked prices are nonmonotone in time: offers at the resale value decline over time, but the price asked between resale offers spikes upward. Offers at the resale value are declining over time because, as offers are rejected, traders learn that there are fewer and fewer potential customers in the network. The fact that prices spike from one resale offer to another reflects the attempt of intermediaries to exploit their local market power: they try to sell at a high price, and if they don't succeed, they lower the price and resell.[14] In the example in Figure 27.3, the sequence of prices asked is $\{5/6, 1, 2/3, 1\}$.

Second, intermediaries who are essential to connect other intermediaries to the initial seller will make, in expectation, higher expected profits. Since resale offers are declining over time, receiving an offer later in the game allows an intermediary with a high value to obtain a higher profit margin. However, the later an intermediary receives an offer, the higher is the ex-ante probability that the offer will not materialize, as earlier traders may consume the object. Condorelli, Galeotti, and Renou (2015) show that the latter effect dominates the former. In the example above, only traders 1 and 3 obtain a positive profit, and their ex-ante expected payoff, condition on having a high value, is $1/6$ for trader 1 and $1/9$ for trader 3.

27.2.3 Multilateral Bargaining through Auctions

We now move away from bilateral bargaining and consider trading protocols in which the seller more directly exploits competition across buyers. We build on Kotowski and Leister (2014), where in each round, the owner of the object sells it using a second-price auction. There is no asymmetric information about the valuation of final buyers, but each intermediary faces a random trading cost from purchasing an object. The trading cost can be either low, which we set equal to 0, or high, in which case it is higher than the consumption value to the buyer. Clearly, an intermediary can never recover a high cost of trade from buying and reselling; therefore, high-cost intermediaries will prefer not to trade. An intermediary's cost of trade is private information, and an intermediary has low cost with probability p.

We analyze the model within the class of complete multipartite networks.[15] Trading occurs via a sequence of second-price, sealed-bid auctions: the initial seller s runs an auction in which intermediaries in tier 1 bid; the winner then runs an auction in which intermediaries in tier 2 bid; and so on. We assume that the intermediary in the last tier L who eventually owns the object sells it to the final customer at v. Intermediaries can bid any positive price and can also abstain from the auction. Moreover, if only one intermediary makes a positive bid in one of the auctions, then he obtains the object at zero price. For simplicity, agents are assumed to be perfectly patient.

[14] This is reminiscent of hot-potato trading strategies in OTC financial markets.

[15] The analysis of Kotowski and Leister (2014) applies to nondirected networks. They show that, even without the assumption that the graph is directed, there is an equilibrium in which goods flow from upstream firms to downstream firms.

In this model of second-price auction with resale, there can be multiple equilibria. The focus is, then, on equilibria in which agents bid their expected resale value. Formally:

Proposition 2. *Consider a complete multipartite network. There is an equilibrium where, in each auction in which the owner is an agent in tier $x \in \{0,\dots,L-1\}$, high-cost intermediaries in tier $x+1$ do not participate in the auction, and low-cost intermediaries in tier $x+1$ bid their resale value, which is the asset's expected resale value conditional on all available information. Along the equilibrium path:*

1. *in the auction where the owner is an agent in tier $x \in \{0,\dots,L-2\}$, the resale value of each low-cost intermediary in tier $x+1$ is*
$$r_{x+1} = \Pi_{y=x+2}^{L}\delta(n_y)v,$$
where $\delta(n_y) = 1-(1-p)^{n_y} - n_y p(1-p)^{n_y-1}$ is the probability that at least two intermediaries in tier y have a low cost.
2. *The ex-ante expected equilibrium payoff of an intermediary in tier x is*
$$\pi_x = \Pi_{y=1}^{x-1}\left[1-(1-p)^{n_y}\right] \times p(1-p)^{n_x-1} \times r_x.$$

The proof of Proposition 2 follows from the main equilibrium characterization in Kotowski and Leister (2014). To understand the expression for the resale value, consider an intermediary that owns the object and that is located in tier $L-1$. All bidders in this auction have a resale value of v by assumption and, therefore, will bid v if they have a low cost. The expected profit of the intermediary owning the object is, then, the probability that at least two bidders have a low cost, $\delta(n_L)$, in which case he earns v. Proceeding backwards, we obtain the expression for the resale value.

The ex-ante payoff of an intermediary in tier x is affected by upstream concentration, horizontal concentration, and downstream concentration. First, the higher the number of intermediaries in each upstream tier, the higher the chance that the object will reach tier $x-1$, which is a necessary condition for an intermediary in tier x to make profit (i.e., first term in the expression). Second, the higher the number of intermediaries in each downstream tier, the higher downstream competition, and the higher the markup that the intermediary can obtain by buying and reselling (i.e., the last term in the expression). Finally, an intermediary in tier x is better off when there are fewer intermediaries in his tier, as this improves his terms of trade in the auction (i.e., the middle term in the expression).

Node criticality is also important in this context. In fact, when an intermediary is critical, he will purchase the object at zero price. The market power that critical traders have relative to other intermediaries is, however, confounded by the uncertainty that the object flows along the different tiers and the uncertainty of the level of competition within each tier. By letting the probability that an intermediary has a low cost going to 1, we eliminate these uncertainties and obtain that only critical traders obtain a positive profit.

We now consider the effect of an horizontal merger. We make the natural assumption that the merged set of intermediaries operates at a low cost of trade if at least one of the intermediaries has a low cost.

Corollary 1. *Suppose that in a complete multipartite network with tiers of size $\{n_1, \ldots, n_L\}$, a subset of intermediaries of size $\hat{n}_x \leq n_x$ in tier $x \in \{1, \ldots, L\}$ merge. By comparing the equilibrium characterized in Proposition 2 for the two networks, we obtain that:*

1. *The horizontal merger does not affect the aggregate expected surplus.*
2. *The horizontal merger increases the profit of each merged intermediary, it does not affect the profit of the other intermediaries in the same tier x or in downstream tiers $\{x+1, \ldots, L\}$, but it decreases the expected payoff of the intermediaries in upstream tiers $\{1, \ldots, x-1\}$.*

The aggregate surplus generated in equilibrium is v whenever there is at least a trading path of intermediaries, each with a low cost of trade; otherwise, the realized surplus is zero. The assumption that the merged set of intermediaries operates at a low cost of trade if at least one of the intermediaries has a low cost implies that horizontal mergers do not alter the expected surplus. However, horizontal mergers change the way the surplus is distributed. A merger in tier x reduces competition in that tier, and, therefore, each intermediary in tier $x+1$ anticipates that his resale value has decreased, and this extends to further upstream tiers.[16]

Polanski and Cardona (2012) consider exchange through first-price auctions in a multilevel intermediation network. They study a multilevel symmetric tree, with a single seller as the root of the tree, and buyers as terminal nodes. All buyers have independently distributed private values for the object held by the seller. Values are drawn from a uniform distribution. The intermediaries and the seller are interested only in their monetary payoff. In contrast to Kotowski and Leister (2014), their analysis is bottom-up: buyers with uncertain valuations submit bids to their intermediary. In turn, intermediaries submit their bids to the intermediary at the level above. Then, the process propagates until it reaches the seller, who awards the object to the highest bidder among the intermediaries connected to him. Polanski and Cardona (2012) rank networks in terms of payoff for the seller offered by different configurations. Intuitively, for the same number of potential buyers, the seller prefers networks with fewer tiers of intermediation.

[16] Kotowski and Leister (2014) also discuss the effect of vertical mergers. They assume that in a vertical merger, the probability that the partnership operates at a low cost is the probability that there is at least one member intermediary in each tier with a low cost. This defines an implicit cost to vertical mergers, as it can decrease the probability that the partnership operates at a low cost. In some circumstances, this cost will be sufficiently large to make vertical mergers unprofitable.

27.2.4 Multilateral Trading with Posted Bid-and-Ask Prices

We now turn to models in which intermediaries set a bid price to buy upstream and an ask price to sell downstream. Blume et al. (2007) introduce a bid and ask model in which each intermediary has no consumption value for the object and is connected to an arbitrary subset of buyers and sellers. Each seller has one unit of an homogeneous good, and each buyer demands one unit of the same good. The consumption value of buyers and sellers is arbitrary and commonly known.

The trading game in Blume et al. (2007) unfolds in two stages. First, each intermediary offers a bid price to each seller to whom he is connected, and an ask price to each buyer to whom he is connected. Second, sellers and buyers choose the best offer from the set of offers of traders connected to them. They may also choose not to sell or buy. A large penalty applies if intermediaries sell more units than they acquire, which guarantees that, in equilibrium, intermediaries will not default on their price commitment to buyers.

Blume et al. (2007) consider Nash equilibria of this game and show that all equilibria result in an efficient outcome (i.e., all possible beneficial trades are realized). They also show that a trader can make a positive profit if, and only if, the trader is essential. Here, a trader is essential if the social welfare decreases once the trader is removed from the network. When there are only one seller and one buyer, essentiality is equivalent to criticality.

By considering a setting in which there are only one buyer and one seller, it is possible to go beyond Blume et al. (2007) to consider complex networks, such as multipartite networks, in which all traders post bid-and-ask prices simultaneously. In this case, the object then flows from the initial seller to the highest bidder in tier 1, say i, and from tier 1 to the highest bidder in tier 2, provided that such a bid meets intermediary i's ask, and so on. The object stops flowing either when it is acquired by intermediary i and i's ask is strictly higher than the highest bid of intermediaries in the downstream tier, or when it reaches the final customer b. Ties are resolved with randomization. This framework builds on Gale and Kariv (2009), and most of the results we present in this section are proved in Choi et al. (2015).

Proposition 3. *Consider a complete multipartite network. There exists a Nash equilibrium that is efficient. In every efficient equilibrium:*

1. *Only critical intermediaries can obtain positive profit.*
2. *In the line network, all the surplus goes to intermediaries, whereas buyer and seller obtain zero profit. Indeed, any distribution of positive profits across intermediaries that sum up to v can be sustained as an equilibrium outcome.*
3. *In a competitive network, all the surplus goes to the seller and the buyer, whereas intermediaries obtain zero profit.*

The existence of efficient equilibria can be shown by construction. Consider the following bid/ask profile: each intermediary in tier $x = 2 \ldots L$ bids v and asks v; if there is only one intermediary in tier 1, then the intermediary bids 0 and asks v; if there is more than one intermediary in tier 1, then each intermediary bids v and asks v. Under this profile, if an intermediary acquires the object, he will resell it to v. So, each intermediary is willing to bid up to v. Since each intermediary asks v, for every intermediary in tier $x > 1$, it is a best reply to bid v. A monopoly intermediary in tier 1 will bid 0, while if there are multiple intermediaries, competition will push bids up to v.

In the line network, the intermediary connected to the seller has full bargaining power and will extract all the resale value that his connection generates; similarly, the intermediary connected to the final customer also has full bargaining power, and so must also extract all the surplus from the buyer. Equilibrium does not pin down, however, how the total surplus v is distributed across intermediaries. In fact, the problem faced by two consecutive intermediaries, x and $x+1$, is akin to a Nash demand game where, though, the surplus that can be shared is endogenous and depends on the resale value of intermediary $x+1$. In a competitive multipartite network, competition across intermediaries at every level destroys intermediation rents, and the seller is the sole agent extracting all the surplus. As in the models we have analyzed previously, intermediaries who merge horizontally may increase their market power.

Gale and Kariv (2009) provide an experimental analysis of this model that focuses on competitive networks. They find that, after a period of learning, the bid-and-ask prices converge to the competitive equilibrium prices, and the outcome of trade becomes efficient. So, despite trade requiring possibly long chains of intermediaries, and the impossibility of re-contracting, subjects are often able to coordinate on the efficient outcome. The experiment also points out that higher levels of competition across intermediaries, studied by increasing the number of intermediaries in different tiers, tend to speed up the learning and convergence towards efficient play.[17]

27.2.5 Static Trading Protocols

In all the papers we have surveyed so far, the market model contains an element of sequentiality, with sellers (or buyers) usually proposing the terms of trade and one or more buyers (or the seller) agreeing to trade according to their terms or refusing to trade. We now discuss three papers that, while maintaining that not everyone can trade with everyone else, exploit a more centralized/static equilibrium notion.[18]

Nava (2015) studies the exchange in a network of a single homogeneous commodity. Traders' preferences are heterogeneous, and their utility is concave in the quantity

[17] We refer to Chapter 17 of the handbook by Choi, Gallo, and Kariv for a detailed survey of experiments of networked markets. That chapter includes, among other related papers, an exhaustive discussion of Gale and Kariv (2009) and of Choi et al. (2015).

[18] Gofman (2014) builds a model of trading in a network without explicitly formulating a non-cooperative game. Each dealer receives an exogenous share of the gain from trade with the subsequent dealer and all gains from trade are endogenously determined. His main result is that trading protocols that do not assign all gains from trade to sellers result in inefficient outcomes.

of commodity consumed, while linear in money. The focus is on equilibrium in a simultaneous move game in which each trader decides how much quantity to sell to, or to buy from, other traders. As in a Cournot model, prices formed at every location reflect the local willingness to pay of a trader, given the demand and supply decisions in his local market. In equilibrium, there is intermediation, in that some traders buy some units of commodity in order to resell them to other traders. Prices strictly increase along any intermediation chain because a trader intermediates the commodity only if gains from trade are strictly positive. Efficiency is attained only in large economies and for specific network configurations. In particular, trade can be efficient only if intermediation is negligible in the economy. If intermediation is essential to clear markets, then a group of traders necessarily commands a rent that distorts trade.

Malamud and Rostek (2015) study a general model of asset trading where connected traders participate in the same market. Traders simultaneously submit demand schedules, and pricing in each market is attained via a uniform-price double auction. Linear Nash equilibria are analyzed (following Kyle 1989 and Vives 2011).

Babus and Kondor (2013) consider a setup closely related to that of Malamud and Rostek (2015). They focus on a less general connectivity structure, but they endow agents with idiosyncratic information on the asset's value and characterize the informational content of prices. They show that information diffusion is effective, but not informationally efficient. They also show that dealers with more trading partners are ex post better informed, so they tend to trade and intermediate more, and earn more profit per transaction.

27.3 Pricing in Supply Chains

Section 27.2 focuses on pure intermediation: the role of intermediaries is to buy and resell assets, and assets flow from sellers to final customers. However, one can envisage an environment in which a chain of intermediaries is a final product, and the value of consumers for different paths may depend on the particular intermediaries within the path. For example, in supply chains, intermediaries within a tier provide substitute inputs, and intermediaries across tiers provide complementary inputs. Chains are, therefore, final products, each produced through the combination of a set of complementary inputs. Similarly, in the context of transportation networks, one can interpret a path as a journey and the intermediaries along the path as transport service providers.

There is renewed interest in supply chains in economics. The important role of production supply chains in trade has been documented empirically in different studies, such as Hummels et al. (2001). Antras and Chor (2013) study the optimal allocation of ownership rights along a supply chain. Oberfield (2013) develops a theory in which the network structure of production—who buys inputs from whom—is the endogenous outcome of individual choices. Acemoglu et al. (2012) focus on the role that production supply chains play in translating idiosyncratic shocks into volatility at the aggregate level.

The framework developed in this section, due to Choi et al. (2015) and Galeotti and Goyal (2015), abstracts from the role of idiosyncratic shocks in the value chain and from the endogenous formation of these networks. It complements the literature on supply chains by providing a systematic study of strategic pricing in competing chains. This work is also related to Acemoglu and Ozdgar (2007a,b), who study the efficiency of oligopoly equilibria in congested markets, such as network flows in communication networks or traffic in transportation networks.[19]

Let T be the set of directed paths connecting s to b in a complete multipartite network. Following Galeotti and Goyal (2015), we use a Dixit-Stiglitz framework to model paths as differentiated products that buyers may demand, see, also, Singh and Vives (1984). Let p_i be the price for the service of intermediary i and let $p = \{p_1, \ldots, p_n\}$ denote a price profile. The cost of a path $q \in T$ is the sum of prices charged by the intermediaries along that path (i.e., $c(q;p) = \sum_{i \in q} p_i$). A representative consumer has quadratic utility over paths. That is, let x_q be the consumption of a representative consumer for path $q \in T$, and let

$$\hat{U} = \sum_{q \in T} x_q - \frac{\beta}{2} \sum_{q \in T} x_q^2 - \frac{\gamma}{2} \sum_{q \in T} \sum_{q' \in \mathcal{Q} \setminus \{q\}} x_q x_{q'}.$$

Then, the representative consumer maximizes $U = \hat{U} - \sum_{q \in T} c(q;p) x_q$, where $\beta > 0$ and $\gamma \in [0, \beta)$. Solving for optimum consumption leads to the following demand function for path q:

$$D(p,q) \equiv x_q = \frac{1}{\beta + \gamma(m-1)} \left[1 - \frac{\beta + \gamma(m-2)}{(\beta - \gamma)} c(q;p) + \frac{\gamma}{(\beta - \gamma)} \sum_{q' \neq q} c(q';p) \right],$$

where m is the total number of paths (i.e., $m \equiv |T|$). When $\gamma = 0$ paths are independent goods, when $\gamma > 0$ paths are substitute goods and, at the limit, when $\gamma \to \beta$, we have the case where paths are perfect substitutes. We are interested in Nash equilibria of the following simultaneous move game: intermediaries set prices simultaneously and wish to maximize their individual profits, where intermediary i's profits is $\Pi_i(p) = p_i \sum_{q \in T, i \in q} D(p,q)$.

The strategic relation between two intermediaries depends on their network location, together with the intensity of competition across paths. If intermediaries i and j share path q, then an increase in intermediary i's price decreases the demand of path q, and this creates incentives for intermediary j to decrease his price. So, sharing paths creates strategic substitutability between intermediaries' pricing strategies. In contrast, if intermediary i is located in path q and intermediary j is located in path q', an increase

[19] Acemoglu and Ozdgar (2007a,b) consider parallel paths networks. Another related paper is Hendricks, Piccione, and Tan (1999), who study the problem of competing airlines each designing their transportation networks.

in intermediary *i*'s price decreases the competitiveness of path q relative to path q', and intermediary j can then raise his profits by increasing his price. Hence, belonging to different paths creates strategic complementarities between intermediaries' pricing strategies. As intermediaries may share some paths and not others, whether the prices of two intermediaries are strategic complements or strategic substitutes will depend on their specific network location.

The following result provides a characterization of equilibrium pricing and profits across intermediaries in complete multipartite networks. The betweenness centrality of intermediary i is the ratio between the number of paths that i belongs to and the total number of paths. In a complete multipartite network, the total number of trading paths is $m = \Pi_{x=1}^{l} n_x$ and an intermediary in tier y belongs to $\Pi_{x=1,x\neq y}^{l} n_x$. Therefore, the betweenness centrality of an intermediary in tier y is simply $1/n_y$.

Proposition 4. *Consider a complete multipartite network. In equilibrium, the price and profit of an intermediary in tier x is higher than the price and profit of an intermediary in tier y if, and only if, the betweenness centrality of an intermediary in tier x is higher than the betweenness centrality of an intermediary in tier y.*

The result illustrates that betweenness centrality is an important determinant of market power for intermediaries. Betweenness centrality is formally related to the notion of node criticality. In fact, a node i is critical if, and only if, it has maximal betweenness centrality. When paths are imperfect substitute products, intermediaries enjoy rent even if they are not critical. As the next result illustrates, when paths become perfect substitutes, then only critical intermediaries charge a positive mark-up and obtain positive profits.

Proposition 5. *At the limit, where paths become perfect substitutes (i.e., $\gamma \rightarrow \beta$) the equilibrium price of each noncritical intermediary goes to* 0, *and the equilibrium price of each critical intermediary goes to*

$$p^* = \frac{\alpha}{k+1},$$

where k is the number of critical intermediaries. Hence, only critical intermediaries obtain positive profit.

Choi et al. (2015) provide a theoretical analysis of pricing in trading networks in which paths are perfect substitutes, and they provide experimental evidence of the model. In networks in which there are no critical intermediaries, their experimental results show that prices are generally very low, and the total cost of each chain is no larger than 20% of the consumers' willingness to pay. In contrast, when there are only critical intermediaries, intermediaries price similarly and, overall, the cost of the chain is above 90% of the consumers' willingness to pay. In the mixed cases, chains are priced similarly, and their cost is always above 60% of the consumer's valuation. Critical intermediaries price well above noncritical intermediaries and get most of intermediation's profits. Overall, the experimental results point to node criticality as an organizing principle for understanding pricing, efficiency, and the division of surplus in networked markets.

We have discussed oligopolistic analysis of supply chains. Ovstrosky (2008) studies a complementary model of supply chains in networks. A set of firms is arranged in an acyclic directed network of trading possibilities. Each firm's profit depends on the contracts it can sign with upstream and downstream firms. An outcome is a set of contracts signed by firms; an outcome is stable if no firm would like to break any of its relationships or establish a new one with a willing partner. Ovstrosky (2008) develops conditions on preferences that assure the existence of stable outcomes. These conditions generalize the gross-substitute condition in Kelso and Crawford (1982). Insights of two-sided matching theory (e.g., Adachi 2000 and Hatfield and Milgrom 2005) generalize besides existence: stable outcomes represent a lattice in the payoff space of firms at the top of the network (i.e., suppliers) and firms at the bottom (i.e., consumers). The generality of the model does not allow, though, to derive specific insights into the effect of the network structure on outcomes.[20]

27.4 Discussion and Open Questions

We conclude by discussing open questions in the literature. All the models discussed in Section 27.2 assume that there is only one seller with only one unit of the good. Extending the analysis to multiple units, heterogeneous supplies and heterogeneous demands can be a fruitful area for future research.

We have focused our attention on strategic intermediation in a given network and abstracted away from the equally important issue of trading network formation. Only a handful of papers are devoted to the study of network formation in an environment in which the benefits of linking are determined by an underlying game of trade. Goyal and Vega-Redondo (2007) characterize equilibrium networks under the assumption that only critical traders obtain positive payoff. Condorelli and Galeotti (2012) study network formation in a setting where, for every network, trade is efficient and there are no intermediation rents. In Section 27.3, we mentioned the work of Oberfield (2013), which develops a theory of endogenous input-output networks of production. Kotowski and Leister (2014), discussed in Section 27.2.3, characterize free-entry equilibrium networks. The formation of trading networks is a largely unexplored topic, and we believe it deserves further work.

The models we presented provide rich empirical predictions with respect to how the architecture of the network impacts efficiency, pricing, and distribution of economic surplus across traders. We discussed two recent experimental works on intermediation

[20] Hatfield et al. (2014) generalize Ovstrosky's (2008) to arbitrary networks (i.e., allowing for inter-dealer trading in addition to upstream and downstream trading).

networks (i.e., Gale and Kariv 2009 and Choi et al. 2015). Both experiments exploit the simultaneous move nature of these two models, and, therefore, they abstract away from the possibility of re-contracting among intermediaries. We believe that adding a dynamic aspect to these experiments can be useful to evaluate the insights that the theory provides on intermediation in networks.

Finally, it is well known that bargaining games are sensitive to the specification of the extensive form (i.e., the trading protocol). This is even more so when there are more than two bargaining parties as in intermediation networks. Understanding optimal trading protocols from the perspective of the various actors involved in the network is an open research agenda.

References

Acemoglu, D., V. M. Carvalho, A. Ozdaglar, and A. Tahbaz-Salehi (2012). "The network origins of aggregate fluctuations." *Econometrica* 80(5), 1977–2016.

Acemoglu, D. and A. Ozdaglar (2007a). "Competition in parallel-serial networks." *IEEE Journal of Special Areas of Communication*, Special Issue on Non-Cooperative Behavior in Networking 25, 1180–1192.

Acemoglu, D. and A. Ozdaglar (2007b). "Competition and efficiency in congested markets." *Mathematics of Operations Research* 32(1), 1–31.

Adachi, H. (2000). "On a characterization of stable matchings." *Economics Letters* 68(1), 43–49.

Antras, P. and D. Chor (2013). "Organizing the global value chain." *Econometrica* 81(6), 2127–2204.

Babus, A. and P. Kondor (2013). "Trading and information diffusion in OTC markets." Mimeo, CEU. Budapest.

Blume, L., D. Easley, J. Kleinberg, and E. Tardos (2007). "Trading networks with price-setting agents." In *Proceedings of the 8th ACM Conference on Electronic Commerce 2007*, New York, NY.

Choi, S., A. Galeotti, and S. Goyal (2015). "Trading in networks: Theory and experiment." Mimeo.

Condorelli, D. and A. Galeotti (2012). "Endogenous trading networks." Mimeo.

Condorelli, D., A. Galeotti, and L. Renou (2015). "Bilateral trading in networks." Mimeo.

Corominas-Bosh, M. (2004). "Bargaining in a network of buyers and sellers." *Journal of Economic Theory* 115, 35–77.

Duffie, D., N. Garleanu, and L. H. Pedersen (2005). "Over-the-counter markets." *Econometrica* 73, 1815–1847.

Fafchamps, M. and B. Minten (1999). "Relationships and traders in madagascar." *The Journal of Development Studies* 35(6), 1–35.

Fudenberg D, D. K. Levine, and J. Tirole (1985). "Infinite-horizon models of bargaining with one-sided incomplete information." In *Game Theoretic Models of Bargaining*, A. Roth, ed., 73–98. Cambridge, UK and New York: Cambridge University Press.

Gale, D. and S. Kariv (2009). "Trading in networks: A normal form game experiment." *American Economic Journal: Microeconomics* 1(2), 114–132.

Galeotti, A and S. Goyal (2015). "Competing chains." Mimeo.

Gofman, M. (2014). "A network-based analysis of over-the-counter markets." Mimeo Wisconsin-Madison.

Gul, F., H. Sonnenschein, and R. Wilson (1986). "Foundations of dynamic monopoly and the coase conjecture." *Journal of Economic Theory* 39(1), 155–190.

Goyal, S. and F. Vega-Redondo (2007). "Structural holes in social networks." *Journal of Economic Theory* 137, 460–492.

Hatfield, J. W. and P. R. Milgrom (2005). "Matching with contracts." *American Economic Review* 95(4), 913–935.

Hatfield, J. H., S. D. Kominers, A. Nichifor, M. Ostrovsky, and A. Westkamp (2013). "Stability and competitive equilibrium in trading networks." *Journal of Political Economy* 121, 966–1005.

Hummels, D., J. Ishii, and K.-M. Yi (2001). "The nature and growth of vertical specialization in world trade." *Journal of International Economics* 54(1), 75–96.

Kirman, A. (1997). "The economy as an evolving network." *Journal of Evolutionary Economics* 7(4), 339–353.

Kyle, A. S. (1989). "Informed speculation and imperfect competition." *Review of Economic Studies* 56, 517–556.

Kotowski, M. H. and C. M. Leister (2014). "Trading networks and equilibrium intermediation." Mimeo.

Kranton, R. and D. Minehart (2001). "A theory of buyer-seller networks." *American Economic Review* 91, 485–508.

Li, D. and N. Schurhoff (2012). "Dealer networks." Mimeo, University of Lausanne.

Malamud, S. and M. Rostek (2015). "Decentralized exchange." Mimeo, Wisconsin Madison.

Manea, M. (2015). "Intermediation in networks." Mimeo.

Merlo, A. and C. Wilson (1995). "A stochastic model of sequential bargaining with complete information." *Econometrica* 63(2), 371–399.

Nava, F. (2015). "Efficiency in decentralized oligopolistic markets." *Journal of Economic Theory* 157, 315–348.

Oberfield, E. (2013). "Business networks, production chains, and productivity: A theory of input-output architecture." Mimeo.

Ostrovsky, M. (2008). "Stability in supply chain networks." *American Economic Review* 98(3), 897–923.

Polanski, A. and D. Cardona (2012). "Multilevel mediation in symmetric trees." *Review of Network Economics* 11(3), 1–23.

Rauch, J. (1999). "Networks versus markets in international trade." *Journal of International Economics* 48, 7–35.

Rubinstein, A. (1982). "Perfect equilibrium in a bargaining model." *Econometrica* 50(1), 97–109.

Rubinstein, A. and A. Wolinsky (1987). "Middlemen." *The Quarterly Journal of Economics* 102(3), 581–593.

Siedlarek, J.-P. (2015). "Exchange with intermediation in network." Mimeo.

Spulber, D. (1999). *Market Microstructure: Intermediaries and the Theory of the Firm*. New York: Cambridge University Press.

Tesfatsion, L. (1997). "A trade network game with endogenous partner selection." In *Computational Approaches to Economic Problems*, H. Amman et al. eds., 249–269. Kluwer Academic.

Vives, X. (2011). "Strategic supply function competition with private information." *Econometrica* 79(6), 1919–1966.

Wright, R. and Y. Wong (2014). "Buyers, sellers, and middlemen: Variations on search-theoretic themes." *International Economic Review* 55(2), 375–397.

CHAPTER 28

NETWORKS IN INTERNATIONAL TRADE

THOMAS CHANEY

This chapter is organized as follows. In Section 28.1, I review theoretical work on the diffusion of information in networks, with a particular application on the geographic expansion of firm level trade. In Section 28.2, I review empirical evidence on the role of ethnic and social networks in international trade. In Section 28.3, I review theoretical and empirical work on production networks and firm to firm trade.

The theoretical models described in Sections 28.1 and 28.3 of this chapter are meant to address mostly macroeconomic questions. As such, those models do not go into great details about the interaction between agents connected through a network. The network itself is complex, but the interactions inside this network are rather simple. Several chapters in this handbook describe much more elaborate models of games and interactions within a network: Daniele Condorelli and Andrea Galeotti analyze models of intermediation within networks; Matt Eliott and Mihai Manea present models of bargaining within networks; and Ana Mauleon and Vincent Vannetelbosch study network formation games.

28.1 Information Networks in Trade

Introducing the notion of networks in international trade poses two main theoretical challenges. The first is conceptual, the second technical.

First, the empirical literature on networks in international trade identifies one key dimension along which networks matter for trade, the transmission of information. Information exchanges are inherently different from the exchange of physical goods: once someone knows something, that person can share this information with whomever they know and are willing to communicate with. So information has a tendency to diffuse along network connections.

Second, networks are inherently complex objects. The description of the properties of the network that connect together, say, importers and exporters from various countries, the characterization of the dynamics of this network, all require new analytical tools. So the second challenge for modeling networks in trade is technical: one has to develop new analytical tools to study those complex objects.

I now describe several theoretical contributions that address both of those challenges: modeling the concept of information, and modeling the complex dynamics of large-scale networks. Along both dimensions, new theoretical insights are gained from the study of networks. Note that while informational frictions do not necessarily imply that networks matter, conversely it is necessary to have a clear idea of what information frictions entail before introducing them into a model of information diffusion within a network.

28.1.1 Information Frictions

Two recent contributions introduce the notion of information frictions explicitly in international trade. Allen (forthcoming) uses a search and matching model to analyze the role of informational frictions in trade, and estimates that model using trade between islands in the Philippines. Dasgupta and Mondria (2014) model explicitly the cost of processing information for individual traders into an otherwise classical Ricardian trade model.[1]

Information frictions are prevalent in economics. They are arguably more severe in the context of international trade, where buyers and sellers are by construction far from each other. They are probably even more severe in developing countries where infrastructures that facilitate the flow of information are less developed. This is the context for Allen's (forthcoming) study of information friction in trade between islands in the Philippines. Allen models explicitly the cost for a potential buyer (exporter) to learn about demand conditions in remote markets (destinations). Formally, Allen models a costly sequential search by heterogenous producers. Heterogenous producers of a homogenous good are located in various islands. Because producers are heterogeneous in terms of productivity, there are potential gains from trade. Because these islands are far from each other, trade is costly. Moreover, because these islands are far from each other, information does not flow freely between them, and information about prices and quantities in other islands is not readily available to potential traders.

[1] The Ricardian trade model is a simple model where trade arises because countries differ in the technologies they have access to. In such a model, and even if one country is more efficient at producing all goods, and even in the presence of trade frictions between countries, trade will happen because of comparative advantages. Dornbusch, Fisher, and Samuelson (1977) extend David Ricardo's classic model to allow for a continuum of goods traded between two countries. Eaton and Kortum (2002) further extend this model to allow for any finite number of countries trading a continuum of goods, using a probabilistic approach to characterize the equilibrium.

Allen documents a series of stylized facts that would be incompatible with a model where only traditional trade costs exist, but that are consistent with a model that features information frictions. First, transportation costs cannot fully account for the negative impact of distance on trade. To the extent that information frictions increase with distance, they can explain the strong negative impact of distance on trade flows. Second, it is frequent two regions both import and export commodities. Again, in the case of the homogenous commodities considered here, a traditional model with only physical trade costs could not explain two-way trade. Information frictions can reconcile those facts with the theory. Moreover, the introduction of cell phones, which arguably lower information search costs, decrease the incidence of two-way trade. Third, the pass-through of price shocks between provinces declines when cell phones are introduced, again giving credence to the information friction channel. Fourth, larger farmers are more likely to "export" their output to other islands, but this size premium declines (small farmers start exporting) when cell phones are introduced. Fifth and last, the more heterogeneous the producers are in an origin island, the more elastic trade flows are to destination island prices. A conventional model with only physical trade frictions would predict the opposite (more heterogeneous producers, less elastic trade flows).

Note while networks do not play an explicit role in Allen's work, the joint presence of physical and informational trade frictions gives rise to two interesting networks: on the one hand, the conventional network of trade linkages between locations (in that case, islands in the Philippines); on the other hand, the information network of which farmer knows about prices in which set of locations. Those two networks, the observable network of trade linkages, and the notional network of information flows interact with each other in a subtle fashion: trade costs and prices affect which farmer learns about which market; this in turn determines the patterns of supply in each market, and hence equilibrium prices; equilibrium prices in turn affect trade flows and the incentives to acquire information.

Dasgupta and Mondria (2014) take the notion of information frictions in a somewhat different direction. Instead of assuming a direct cost for acquiring information, they model the cognitive cost of processing information. One interesting feature of such a model is that the cost of information becomes endogenous, as traders optimally choose how much information to process. Dasgupta and Mondria offer a simple and tractable model of trade with information frictions and rationally inattentive traders (importers in their case). As in Allen (forthcoming), the presence of information frictions allows for two-way trade even in a Ricardian trade model. In their model of rational inattention, Dasgupta and Mondrian can even explain why importers buy the very same good from different sources (and at different prices), as they optimally choose to randomly buy from different sources, not knowing exactly what price they will ultimately pay.

A nice feature of Dasgupta and Mondria is that they explicitly solve for a gravity equation type prediction for bilateral aggregate trade flows. Exports from country i to country j depend (mechanically) on the sizes of i and j, and on an index of the combined information and transportation frictions that are preventing the free flow of goods between countries. This index for the cost of trade has two components. The first one is exogenous, and depends essentially on the physical cost of trading goods between countries. The second component is more interesting: it is the endogenous outcome of the decision of rationally inattentive traders to acquire information or not. The presence of this second component generates subtle and nonstandard predictions.

First, in the presence of information frictions and rationally inattentive traders, the impact of changes in physical trade costs on trade flows is magnified: a small reduction in bilateral (physical) trade costs induces traders to optimally reallocate their attention. They will typically do so by paying more attention to markets where they are ex ante more likely to find profitable trading opportunities. If the cost of trading with one country decreases (keeping other costs constant), the likelihood of finding profitable trades with that country increase, and traders will optimally reallocate part of their (limited and hence costly) attention towards that country. Doing so, they de facto reduce the *informational* cost of trading with that country, and trade more with that country. The elasticity of trade with respect to physical trade costs is magnified.

Second, in the presence of information frictions, a reduction in the cost of processing information may have a non-monotone impact on trade flows with some countries. This is due to the endogenous decision of inattentive traders to allocate between different markets. When the cost of information increases, traders search for information less aggressively. There is an overall negative impact of information costs on the overall amount of information collected. But in addition to this global effect, there is a more subtle reallocation effect. Not only do traders reduce the overall amount of information they process, but they will also tend to allocate disproportionately more of their limited attention towards a priori more profitable markets. If this composition effect is strong enough, an increase of information processing costs may cause an *increase* in the amount of trade from some nearby countries, at the detriment of more remote countries. This result very much has a network flavor, where a shock in one part of the network has a very heterogeneous impact on other parts of the network.

Finally, Dasgupta and Mondria's model can easily explain the presence of zeros in bilateral trade matrices, the fact that many country pairs do not trade at all. When information processing costs become high, the expected gain from trade may not be able to compensate traders for the cost of processing information. As a consequence, for high information costs, traders endogenously decide not to trade (potentially not to trade with any foreign countries), even though they are fully aware that profitable trading opportunities exist.

A key takeaway from these models is that information frictions matter for trade. A direct corollary not explored in those papers is that due to the specific nature of information (once acquired, transmitting information is very cheap), information will tend to percolate along individual connections. In other words, the network of potential importers and exporters will evolve dynamically.

28.1.2 Information Diffusion in a Network of Traders

I now turn to the description of papers that explicitly take into account the network dimension of informational linkages, Chaney (2013) and Chaney (2014). If information is costly to acquire, then traders will optimally choose to have information only about a small number of trading partners. Subsequently, information will tend to diffuse among connected traders. The web of contacts that connect individual traders to each other can be described as a network. To understand the patterns of trade that emanate from the sharing of information among connected traders, one must describe how information diffuses along these connections between traders, as well as how this network endogenously evolves over time.

The diffusion of information about foreign trading partners shares many features of the propagation of financial crises, or financial contagion, described by Antonio Cabrales, Douglas Gale, and Petro Gottardi in this handbook.

In the case of international trade, one key information traders want to acquire is "*where* are potential trading partners located?". "Where" in international trade typically refers to which country those potential trading partners are located in, or even which city within a country. I present below a simple model of traders (exporting firms) gradually acquiring trading partners (customers) in various remote locations. But a priori, this "where" could represent their location in other spaces, such as a space of attributes, a space of cultural/linguistic/ethnic proximity, a space of prices. The technical tools I propose below can be used to model dimensions other than the geography of trade.

I now present a simple model of a dynamic network of importers and exporters. This model combines together elements of the models in Chaney (2013) and Chaney (2014). For simplicity, an individual exporter only sells to importers it knows. The question I seek to answer with the model is: "where are the firms an exporter sells to located?" I assume that the exporter gets to know importers in two ways: it can either search for importers at random (random search); alternatively, that exporter can meet the contacts of its existing contacts (remote search). Information about new trading partners percolates along existing connections. This model resembles the dynamic social network model in Jackson and Rogers (2008), where individuals meet new friends either at random, or via friends of friends. A key difference is that I characterize not only how many trading partners a firm has, but also where those trading partners are. Bramoulle, Currarini, Jackson, Pin, and Rogers (2012) propose an extension of Jackson and Rogers (2008) where they allow for agents of different types. Those types would correspond to what I call geographic location in my model. The model I propose shares many features of Bramoulle et al. (2012), but extends to as many as a continuum of types (geographic locations in the trade application), and offers further characterization of the aggregate properties of this dynamic network.

Consider a simplified setup with a continuum of locations along the real line, indexed by $x \in \mathbb{R}$. In each location, there is a continuum of firms. Time is continuous. When a firm is born, it randomly connects with a measure K of contacts, located in various locations of the real line. Consider a firm born at time $t = 0$ in the origin location $x = 0$. As the model is symmetric, this is without loss of generality. When the firm is born, the density of firms it is connected to located at coordinate x is given by $f(x)$, where f is symmetric, integrable, has a finite second moment, but is otherwise an arbitrary function that sums to K. After it is born, the firm meets new contacts through two channels. First, with a Poisson arrival rate ρ, it meets firms at random, according to the same density f: over a short time interval dt, it meets ρdtK firms overall, and $\rho dtf(x)\,dx$ firms in a neighborhood dx of location x. Second, with Poisson arrival rate β, any of the existing contacts of the firm reveals one of their own contacts.

Let us call f_t the function that describes the location of the firm's contacts at age t, so that at age t, the firm is in contact with $\int_a^b f_t(x)\,dx$ in an interval $[a,b]$; and $K_t = \int_{\mathbb{R}} f_t(x)\,dx$ the total measure of the firm's contacts at age t. The total measure/number of a firm's contacts evolves recursively according to the simple Ordinary Difference Equation,

$$\frac{\partial K_t}{\partial t} = \rho K + \beta K_t \tag{28.1}$$

with initial condition $K_0 = K$. This ODE admits a simple solution. The location of the firm's contacts (i.e., the function f_t) evolves recursively according to the more complex Partial Differential Equation,

$$\frac{\partial f_t(x)}{\partial t} = \rho f(x) + \beta \int_{\mathbb{R}} \frac{f_t(x-y)}{K_t} f_t(y)\,dy \tag{28.2}$$

with initial condition $f_t = f$. The first term on the right corresponds to random searches: with a Poisson arrival rate ρ, new contacts are formed in locations given by the exogenous density f. The second term corresponds to meeting the contacts of a firm's contacts: with Poisson arrival rate β, any contact in any potential location y (there are $f_t(y)$ of them in each $y \in \mathbb{R}$) reveals the name of one of her contacts, which happens to be in location x with a (density of) probability $f_t(x-y)/K_t$.

While the ODE (28.1) governing the dynamics of the number (measure) of contacts is simple, the PDE (28.2) governing the dynamics of the location of those contacts is more complex. How many contact a firm gains in any single location x depends on how many firms it knows in each and every locations in the world (i.e., on the entire function f_t). In mathematical terms, the complexity appears in equation (28.2) through the integral term, where the function $f_t(y)$ multiplied by $f_t(x-y)$ is integrated over all $y \in \mathbb{R}$: the entire function f_t becomes a state variable for the recursive evolution of $f_t(x)$ evaluated at any x.

This complexity is inherent to the network structure of the model. It would be present generically in any network model where information percolates along existing connections. In other word, any model which features connections between friends of friends will naturally display this kind of complexity: to know what information an agent acquires from one period to the next, one needs to know each contact this agent already has, as information can potentially percolate via any existing connection.

Fortunately, a simple mathematical tool allows to transform this seemingly complex problem, the partial differential equation (28.2), into a much simpler problem, and use instead a more tractable ordinary differential equation. The first trick is to recognize in the integral in equation (28.2) a convolution product.[2] Using $*$ for the convolution product, the recursive equation (28.2) can be written in a compact form as

$$\frac{\partial f_t}{\partial t} = \rho f + \beta \frac{f_t * f_t}{K_t} \tag{28.3}$$

The convolution product $*$ is per se just a notation, so equations (28.2) and (28.3) are one and the same. However, once a convolution product has been recognized, one can use its many useful properties. Among them, the convolution theorem is particularly useful. This theorem says that the Fourier transform of the convolution of two functions is the simple product of their respective Fourier transforms point-wise.[3] In other words, the Fourier transform replaces the rather intractable convolution product of functions into a much more conventional multiplication of numbers. So the second trick is to take a Fourier transform of the entire partial differential equation (28.3) to transform it into the following much simpler ordinary differential equation where I denote by $\hat{f}$ the Fourier transform of f,

$$\frac{\partial \hat{f}_t(s)}{\partial t} = \rho \hat{f}(s) + \beta \frac{\hat{f}_t^2(s)}{K_t}. \tag{28.4}$$

Plugging the solution to the ODE for the number of contacts K_t, this simple ODE for the Fourier transform of the entire distribution of contact $\hat{f}_t$ can easily be solved for any s independently. Instead of solving for the entire functions f_t's all at once, one instead solves for their Fourier transform point by point, a much simpler problem. Of course, one has to check ex post that for any t, the solution for $\hat{f}_t$ is continuous in s, at least in a neighborhood of zero. This is the case for the above recursive equation.

[2] The convolution product of two functions f and g is defined as,

$$(f * g)(x) = \int_{\mathbb{R}} f(y) g(x-y)\, dy$$

[3] The Fourier transform of a function f, denoted $\hat{f}$ by convention, is defined as,

$$\hat{f}(s) = \int_{\mathbb{R}} e^{-isx} f(x)\, dx$$

The convolution theorem simply says,

$$\widehat{f * g}(s) = \hat{f}(s) \times \hat{g}(s)$$

Once the Fourier transform $\hat{f}_t$ is known, it is possible to recover the underlying geographic distribution of contacts f_t.[4] But the beauty of the Fourier transform $\hat{f}_t$ is that it can directly be used to compute any moments of the function f_t, without having to ever recover the function f_t directly. This is because the Fourier transform is intimately related to the characteristic function.[5] For instance, if one is interested in knowing how far on average a firm of age t exports its output, one simply evaluates the second derivative of $\hat{f}_t$ at zero,

$$\hat{f}_t''(0) = \int_{\mathbb{R}} |x - 0|^2 f_t(x)\, dx,$$

which gives the average squared distance of exports. If one is interested in how much geographic dispersion there is in the exports of a firm of age t, one simply evaluates the fourth derivative of $\hat{f}_t$ at zero, and so on.

This mathematical method (the use of convolution products, of Fourier transforms to manipulate them, and derivatives of those transforms) has four main advantages.

First, it is fairly tractable. As long as a problem of information diffusion within a network can be modeled in such a way that some attributes (here geography) get added up, one can use those methods and derive very general results. Here, the additive part comes from the natural Euclidean metric of geographic space: if I am standing in location x and I want to meet someone in location z via one of my existing contacts in location y, I first need to "go" from x to y (acquire information about someone in y from x); this is represented by the vector $\overrightarrow{y-x}$; I then need my contact in y to have "gone" to z; this is represented by the vector $\overrightarrow{z-y}$; "going" from x to z via y is then represented by the vector $\overrightarrow{z-x} = \overrightarrow{y-x} + \overrightarrow{z-y}$; naturally in a Euclidian space, information diffusion is associated with adding up vectors, and Fourier transforms are then natural mathematical tools to analyze such a model.

Second, calculating moments in the data is straightforward, and the statistical properties of estimated moments are well understood. So the connection between the theory and the data can be made very tight.

[4] The inverse Fourier transform allows to recover the underlying function f from its Fourier transform $\hat{f}$,

$$f(x) = \frac{1}{2\pi} \int_{\mathbb{R}} e^{ixs} \hat{f}(s)\, ds.$$

[5] For a probability function, the Fourier transform and the characteristic functions are almost the same. If X is a random variable with p.d.f. f, the characteristic function φ_X is given by

$$\varphi_X(s) = \mathbb{E}\left[e^{isX}\right] = \int_{\mathbb{R}} e^{isx} f(x)\, dx = \hat{f}(-s).$$

The various moments of a random variable X with p.d.f. f can be calculated with the Fourier transform as easily as with the characteristic function, evaluating their successive derivatives at zero,

$$\mathbb{E}\left[X^k\right] = (-i)^k \varphi_X^{(k)}(0) = i^k \hat{f}^{(k)}(0).$$

Third, by allowing to solve for moments separately from each other, this mathematical method also allows to make sharp theoretical predictions without the need for very strong functional form assumptions. For instance, in the case of the model of international trade above, if one is interested primarily in the distance of exports of different types of firms, there is no need to characterize the exogenous function f beyond its first moment. In other words, one can simply make the reduced form assumption that firms, when they randomly look for foreign trading partners, meet partners at some average squared distance. There is no need to make any further assumption about the technology for this random search. Characterizing (and estimating) a single moment of this search technology is enough to generate rich prediction on firm-level trade.

Fourth, there exist various mathematical tools to study the limiting behavior of the Fourier transforms of functions, which I will not describe in detail in this chapter. There is a readily available mathematical toolkit for analyzing many other properties of any model of information diffusion within a network that resembles the one above. For example, Chaney (2013) uses such asymptotic tools to characterize the patterns of aggregate trade. Aggregate trade is calculated by adding up firm-level trade for a large number of firms. Using asymptotic results to solve for these large sums, Chaney (2013) shows the above model naturally predicts under weak conditions aggregate bilateral trade flows are inversely proportional to distance and proportional to county size, the so called gravity equation in international trade.

One should also note that as with any stylized model, mathematical elegance comes at the cost of strong simplifying assumptions. For instance, the tools described above are inappropriate when the model is not symmetric: in equation (28.2), if firms in location y have a different distribution of their own contacts than a firm located at the origin—in other words, if $f_{t,y}(x) \neq f_{t,0}(x-y)$ where $f_{t,y}$ is the distribution of contacts at time t for firms located in y and $f_{t,0}$ that for firms located at the origin—then the integral in equation (28.2) can no longer be treated as a convolution product, and most of the analytical tractability is lost. Numerical simulations can be useful in such cases to explore how differently the simplified (symmetric) model behaves compared to a more complex and analytically intractable asymmetric model. Chaney (2014) uses such numerical explorations to assess the robustness of a simplified model to relaxing some of the strong simplifying assumptions. Of course, numerical simulations offer only a partial and informal assessment of the robustness of this type of model, but they remain a useful complement to the formal study of simplified models.

It should be noted to conclude this section that similar mathematical tools, convolution products, and Fourier transforms to manipulate them have been used in other contexts to study similar problems of information diffusion. In particular, several works by Darrell Duffie and Gustavo Manso with various coauthors study problems of information percolation within financial markets. Since the questions asked are similar (how does information percolate within a network of traders?), it should not be surprising the same analytical tools are used.

Duffie and Manso (2007) study information percolation in a decentralized market with a large number of traders. Using the same tools, convolution products and Fourier transforms, they provide explicit analytical solutions at each point in time for the distribution of posteriors regarding some underlying asset value among market participants with diffused information that randomly meet each other.

Duffie, Malamud, and Manso (2009) study a similar setup and provide welfare analysis for various types of policy intervention: a subsidy to search costs (potentially welfare enhancing), or the provision of public signals (potentially welfare reducing).

Duffie, Giroux, and Manso (2010) characterize the speed of convergence towards a common posterior in a decentralized market where information percolates through random meetings, and provide an explicit formula for this speed of convergence.

Finally, Duffie, Malamud, and Manso (2015) study the dynamics of auctions taking place in over-the-counter markets. Over-the counter-markets differ from centralized markets in that trades occur between connected traders. This is therefore a natural network environment, where information about some underlying traded asset will gradually percolate as agents engage in bilateral trades and learn from each other. In this paper, unlike in the earlier ones, Duffie, Malamud, and Manso (2014) explicitly allow for a more interesting network structure: they consider cases where some agents are more connected than others; they even allow for traders to endogenously seek to trade with better connected agents.

All those papers use the same analytical tools as the one described in this section: convolution products (to describe the evolution over time of some attribute, in their case information about some underlying relevant payoff), and Fourier transforms to manipulate those convolution products and derive analytical solutions. In any setup where information diffuses gradually within a network of connected agents, such mathematical tools are natural candidates for deriving elegant and relatively simple solutions and characterizations of various equilibrium outcomes. I believe those tools are fairly easy to use, and can profitably be applied to a range of economic environments where networks play a central role.

28.2 Ethnic Networks and the Patterns of International Trade

In a seminal paper, Rauch and Trindade (2002) show that the presence of ethnic Chinese networks facilitates trade between countries, particularly so for differentiated goods. While this paper is not the first to look at the impact of migrant networks on international trade, the magnitude of the effects, as well as the relative precision with which the authors are able to quantify the ethnic proximity between two countries, explain why this paper has had a large and lasting impact on the study of networks in international trade. Even though the paper does not claim to address potential issues of endogeneity or reverse causality, it stands as a de facto benchmark in the study of the impact of social networks on international trade flows.

In this section, I review several empirical papers that have documented the quantitative impact of social networks on the patterns of international trade. Most of those use ethnic migrants to proxy for the presence of social ties. Few explicitly address concerns of endogeneity. After a brief description of the theoretical motivation behind those empirical studies, I describe in more details their empirical procedure and main findings.

There are two main reasons why social networks may affect the patterns of international trade: informal trade barriers and contract enforcement.

Impediments to trade take many forms, few of which are easy to measure. While transportation costs, explicit tariffs, or quotas are simple to quantify, and their impact on the cost of international transactions well understood, many other trade frictions hinder the flows of goods and services between countries. One indication informal barriers exist over and beyond traditional barriers to trade such as transportation costs, tariffs, and quotas is the fact that geographic distance has a strong negative impact on bilateral flows between countries, even after controlling for any measurable barrier to trade.

The first main category of informal trade barriers has to do with informational frictions: information about foreign products may be hard to acquire; the tastes of foreign consumers may differ significantly from domestic ones; in the case of trade in highly differentiated and customized intermediates, the precise instructions for customizing inputs according to an importer's specific needs may be hard to communicate. I collect all those informal trade frictions under the label of informational barriers.

The second main category of informal trade barriers has to do with contract enforcement. The inability of buyers and sellers to fully commit to pre-established contracts ex ante, and the inability of the justice system to perfectly enforce existing contracts ex post is not unique to international trade. However, it is arguably more salient for international trade than for domestic trade: differences in legal systems between countries, ambiguity of the extent of the jurisdiction of the national court system, and the mere distance between trading partners all contribute to making contract enforcement even harder for international than domestic trade. Greif (1989, 1993) shows that ethnic networks can mitigate the negative impact of incomplete contracts, by providing a punishment mechanism for traders that default on their promises.

For both categories the presence of migrant networks tends to mitigate informal trade frictions and facilitate trade. Migrant networks have the added advantage for the researcher of being relatively easy to identify, while other types of social networks are harder to quantify.

Kaivan Munshi in this handbook further describes community networks and migrations.

28.2.1 Ethnic Chinese Networks

Rauch and Trindade (2002) offer a reduced form test of the hypothesis that ethnic networks facilitate trade. They focus their attention on ethnic Chinese networks for three reasons. First, ethnically Chinese names are relatively easy to identify, and to distinguish from the names of other ethnic groups. Second, and more importantly, ethnic Chinese represent a large population, having penetrated countries all over the world. Third, Chinese emigration is both relatively recent, so that ties between ethnic Chinese may remain relatively strong, and not too recent, so that the determinants of ethnic Chinese migrations can plausibly be assumed to be independent of the contemporaneous barriers to international trade (even though Rauch and Trindade do not offer any direct test of that hypothesis).

Because of the size of ethnic Chinese networks, Rauch and Trindade do not only focus on trade to and from China, but also between third countries with varying ethnic Chinese populations. Their empirical strategy is straightforward: they estimate a conventional gravity equation[6] (the volume of bilateral trade between two countries is explained by their respective sizes, measured as GDP, and the bilateral distance between them), using as an additional control the product of the share of the population that is ethnic Chinese in the importing and exporting countries. In order to distinguish between the impact of ethnic Chinese networks on contract enforcement and reducing informational barriers, they estimate their model separately for trade in different industries. They use Rauch's classification (1999) of industries in three categories: commodities traded on organized markets, commodities with a reference price, and differentiated commodities. They argue that while contract enforcement via ethnic Chinese networks ought to be similar across all commodity types, informational barriers are likely to be mostly prevalent for differentiated commodities. Formally, they estimate via maximum likelihood the following equation separately for each commodity category,

$$V_{ijk} = controls_{ijk} \times \exp\left(\cdots + \psi_k CHINSHARE_{ij} + u_{ijk}\right)$$

where V_{ijk} is the volume of trade (imports plus exports) of type k commodities between countries i and j, and $CHINSHARE_{ij}$ is the product of the share of the population that is

[6] The gravity equation in international trade corresponds to the empirical regularity, first uncovered by Jan Tinbergen (1962), that relates the volume of bilateral trade $X_{A,B}$ between two countries A and B to their respective sizes, measured as GDP_A and GDP_B, and the bilateral distance between them $Distance_{A,B}$, according to a log-linear relationship:

$$X_{A,B} = constant \times \frac{GDP_A^{\alpha} GDP_B^{\beta}}{Distance_{A,B}^{\delta}}$$

Most estimates for the elasticities α, β and δ are close to 1. Since the early work of Tinbergen, empirical trade researchers have added a series of additional explanatory variables, such as tariffs, the existence of a trade agreement, the volatility of the exchange rate, colonial linkages etc. The measure of ethnic proximity used by Rauch and Trindade (2002) is one more example.

ethnic Chinese in i and j. This simple product of shares can be interpreted as a measure of the probability that if one selects at random an individual in each country, both will be ethnic Chinese.

Rauch and Trindade present three main results of interest. First, the presence of ethnic Chinese networks facilitates trade, raising trade by as much as 60% when one compares country pairs with Chinese networks as large as those prevailing in Southeast Asia to a counterfactual world without those networks. Second, ethnic Chinese networks facilitate trade significantly more for differentiated goods than for goods traded on organized exchanges. Rauch and Trindade interpret this difference as suggesting that ethnic Chinese networks not only improve contract enforcement (they do so for all categories of goods), but also serve as a conduit for information flows (which matter disproportionately more for differentiated goods). Third, Rauch and Trindade find that the trade enhancing effect of ethnic Chinese networks exhibits decreasing returns to scale: the marginal effect on trade of increasing the likelihood that two individuals are ethnic Chinese decreases with the size of the ethnic Chinese population.

While this paper does not attempt to deal with potential endogeneity or reverse causality issues (the same unobserved variables that facilitate trade between i and j may also encourage migrations of ethnic Chinese to those countries), it stands as a seminal contribution in the study of informal trade barriers, and in particular on how to use data on migrants to proxy for those informal barriers. I now turn to a series of papers that have used essentially the same idea to assess the impact of informal barriers on trade and investment flows both between and within countries.

28.2.2 Migrant Networks and Trade within Countries

While international migrants can act as facilitators of international trade, international migrations are far less frequent than migrations within countries. And while trade within countries is arguably subject to fewer frictions than international trade, within country trade frictions do exist as well: the negative impact of geographic distance, for instance, is about equally strong between as within countries (see the evidence starting with McCallum 1995). Interestingly, census data provides detailed information about the internal migrations within countries at least once a decade.

Combes, Lafourcade, and Mayer (2005) explore the hypothesis that networks may facilitate trade in their study of trade between regions within France. They offer two quantitative measures of the network of contacts that connects regions. The first one is similar to Rauch and Trindade's ethnic Chinese network: it is a measure of the number of people currently working in region j that were born in region i. The second one is a measure of the network of business linkages between regions, and it is formally close to the definition of ethnic Chinese networks in Rauch and Trindade. For each business group that owns plants in both regions i and j, Combes et al. take the product of the number of plants in i and in j, and then add up this product for all business groups with plants in i and j. Using an analogy to Rauch and Trindade's measure of ethnic Chinese networks, it is as if each business group represented a separate "ethnic" network, and Combes et al. simply add up all possible networks. This second measure is meant to capture the flow of information that transits within the boundaries of firms, but across geographic space, as well as the flows of workers between firms. Both measures of social and business linkages are expected to have a positive impact on trade flows.[7]

Combes et al. estimate again a simple gravity equation for bilateral trade flows between French regions, adding as control their measures of social and business networks. They use a somewhat more elaborate, or rather more theoretically founded, econometric procedure than Rauch and Trindade, either including importer and exporter fixed effects, or using ratios of trade flows to control for unobserved price effects.

Their main finding is in line with Rauch and Trindade. For two regions with a degree of either social or business connections equal to the average among French region pairs, migrants networks increase bilateral trade between 70% and 100% compared to a counterfactual world without migrant networks, and business networks increase trade by about 300% compared to a counterfactual world without business networks. Part of the reason why business networks have such a large impact is that, at least within France, the number of multi-plant firms and business groups is large. In other words, a counterfactual world with business networks is one where a very large fraction of business linkages would be severed.

Combes et al. perform an interesting decomposition of the impact of social and business networks on the traditional proxies for trade frictions. They find that accounting for this measure of informal trade barriers, proxied by social and business networks, reduces the estimated impact of all traditional measures of trade frictions: the estimated border effect (a measure of how much more trade happens within than between regions) is reduced by about 50%, the estimated negative effect of transportation costs (proxied by geographic distance) is reduced by about 60%, and the estimated positive effect of contiguity is reduced by about 20%. The estimated contiguity effect is also reduced, with all the reduction accounted for by social networks and not by business networks. This is mostly due to the fact that migrations typically occur over much shorter distances than business connections.

As is the case with the work of Rauch and Trindade, the interpretation of those numbers is subject to potential endogeneity issues. Combes et al. offer one attempt to deal with the problem of endogeneity or reverse causality: they instrument the stock of migrants in 1993 by the stock of migrants in 1978. This instrumentation strategy leaves the main results unchanged, or even slightly stronger.

[7] Note that contrary to Rauch and Trindade, Combes et al. do not directly control for the impact of size when building their proxies for the prevalence of social and business networks, as they use simple counts (of migrants or plants) instead of shares. They use a log specification which alleviates this concern partially. But instead of using $\ln(Migrants_{ij})$ to measure migrants networks, they use the nonhomogenous function $\ln(1 + Migrants_{ij})$. The use of $\ln(1 + \cdot)$ allows to deal with zeros, but it introduces a nonhomogeneity that makes the results unit dependent.

28.2.3 The Causal Impact of Migrant Networks on Trade

I now turn to several papers that deal with endogeneity concerns more forcefully. Migrations (as well as ownership of plants) is most often an endogenous decision. The same forces that determine the decision to migrate may also have a direct impact on trade flows. Several recent contributions offer solutions to this endogeneity problem.

Perhaps the most convincing evidence so far that social networks have a causal impact on economic outcome in general, and on trade in particular, is Burchardi and Hassan (2013). The authors show social ties that existed between East and West Germany pre-1989 can explain the unevenly distributed impacts of the German reunification on West German towns. West German towns where a large fraction of the population had lived in East Germany and kept social ties with the East benefited disproportionately from the German reunification, in the sense that income per capita rose relative to other towns following the fall of the Berlin wall. Part of income gains can be directly explained by increased entrepreneurial activity and increased investment of West German firms into East Germany.

Burchardi and Hassan address the two most salient issues of endogeneity related to the impact of social networks on economic outcome. First, social links can be formed among others for economic reasons: I become friends with Mr. X because I expect to do business with Mr. X in the future. Second, social links can be denser in some regions for the same reasons that trade towards those regions is large. To establish causal impact of social networks on economic outcomes, one must find social links that are not formed because of expected economic benefits, and plausibly exogenous variations over space of the density of the social network. Using the specificities of German post–World War II history, Burchardi and Hassan address both. First, they identify social ties between East and West Germany; until about the last minute, it was not expected that the Berlin wall would fall. People in the West who kept social links with the East did not expect any economic benefits from ties. Second, Burchardi and Hassan cleverly use the tumultuous history of Germany during World War II to identify plausibly exogenous variations in migrations from East to West. In particular, when a large number of ethnic German refugees and expellees arrived from Eastern Europe at the end of World War II, large swaths of Germany had been destroyed during battles, so that they had no choice but to temporarily relocate to the eastern parts of Germany. Those refugees and expellees formed social connections in the places where they settled. Many of those refugees eventually kept moving west, ultimately settling down in places that would become part of West Germany. To identify exogenous variations in the migrants that moved from East Germany to West Germany before the construction of the Berlin wall and the iron curtain (between 1945 and 1961), Burchardi and Hassan use as an instrument the bombing campaign of the Western Allies, and the uneven destruction of the housing stock in the western part of Germany that resulted from these bombing campaigns.

Having identified plausibly exogenous variations in the intensity of network connections between East and West Germany, Burchardi and Hassan assess the impact of those (exogenous) social connections on growth and investment post-reunification: following the German reunification, East Germans had access to relevant information about local demand conditions and the quality of the local productive assets, but no access to finance; West Germans on the other hand had the capacity to invest into the East, but a priori no good information about where valuable investment opportunities were. Those West Germans with social ties to East Germany were in a privileged situation to fully take advantage of the opportunities offered by investing in the East.

The quantitative effect of those social ties on macro-economic outcomes is sizable. A one standard deviation increase in the share of East German expellees settling in a West German region is associated with a 4.6% rise in income per capita in the years immediately after the fall of the Berlin wall (1989–1995), or a 0.7 percentage point higher growth. A one standard deviation increase in the share of East German expellees was also associated with a 3.4% increase in the likelihood that firms from that West German region would acquire a subsidiary in East Germany. This effect on investment persists at least until 2007. Finally, at the household level, the income of households with at least one relative in the East rises 4.9% in the years immediately after the fall of the Berlin wall (1989–1995). Overall, Burchardi and Hassan find a substantial *causal* impact of social networks on aggregate economic outcomes.

Several other papers have documented a *causal* impact of social networks on aggregate economic outcomes, and on the patterns of international trade in particular. Two papers in particular use innovative instrumental variable strategies to tease out this causal impact. Both Cohen, Gurun, and Malloy (2014) and Burchardi, Chaney, and Hassan (2014) evaluate the causal impact of migrants networks on U.S. trade and investment. Both papers study how the composition of ethnic migrants in different locations within the United States affect the pattern of trade of U.S. states (Burchardi et al. 2015), and the patterns of foreign investment of U.S. firms (Cohen et al. 2014 and Burchardi et al. 2015). To the extent that those papers look at trade and investment originating from the same country, the United States, some of the concerns of reverse causality are mitigated: the regulatory environment, most direct barriers to trade and investment, as well as the ease with which migrants from different countries can emigrate are all relatively uniform. Those two papers therefore isolate one single channel, the presence of ethnic migrant networks locally and the social network connection they bring with them, on the patterns of trade and investment.

Cohen et al. (2014) use the forced relocation of Japanese immigrants and Japanese Americans into internment camps during World War II to identify arguably exogenous variations in the local ethnic composition. They then assess the impact of those exogenous variations in the ethnic composition on the export performance of local U.S. firms. They find the local ethnic network affects the likelihood that a firm exports; firms that exploit this local network outperform other importers and exporters, they grow faster, and their (risk-adjusted) stock returns are 5–7% higher than other firms.

Burchardi, Chaney, and Hassan (2014) instead use the history of subsequent waves of migrants into the United States for over 150 years to instrument the current ethnic composition of U.S. states. They use decennial censuses to identify both the existing composition of migrants in different U.S. states at different times, and the arrival of new migrants from abroad. Under the assumption that new migrants from a given ethnic group are more likely to settle down in U.S. counties where there already exists a significant community from the same ethnic group, they predict for each decade the local ethnic composition using only past migrations. Doing so recursively from 1860 onward, they predict the local ethnic composition in 1990 taking as exogenous only the initial ethnic composition in 1860 and the size of inflows into the United States as a whole (not in any state in particular). Again, they find that the local composition of ethnic networks within the United States have a *causal* impact on the patterns of international trade of U.S. states, as well as the patterns of foreign investment of individual U.S. firms. This instrumentation strategy is data intensive, as it requires collecting data on migrations going back at least several decades, or even centuries. It is, however, very general and easy to replicate. As such, it differs from other instrumentation strategies that rely on specific historical accidents unlikely to be repeated.

Both papers isolate a *causal* impact of social networks (here proxied by ethnic networks) on the patterns of international trade and investment of the United States. These empirical contributions shed new light on the nature of some of the informal barriers to trade. They open up new questions on the role of various nonconventional policy instruments in reaping the gains from international trade and investment. By offering new empirical facts, they also open up new avenues for theoretical research on the dynamics of international trade and investment, international migrations, and the role of information frictions and contract enforcement. In the next section, I review recent theoretical advances in networks and international trade.

28.3 Production Networks and Firm-to-Firm Trade

I conclude this chapter with a description of recent work in macroeconomics and international trade on the role of production networks and firm-to-firm trade. Modern economies are characterized by complex production processes. Firms combine inputs with capital and labor to produce output; their output is often used as input by more downstream firms. Most firms have only a relatively small number of upstream suppliers and downstream customers, so that most of these firm-to-firm interactions are local in the sense they take place among a small number of firms. In other words, production is achieved by combining gradually more elaborate intermediate inputs along vertical production chains. Those chains form a complex network of input-output linkages. Understanding the aggregate properties of this web of input-output linkages requires a careful network analysis.

28.3.1 Theoretical Models of Production Networks

To gain intuition on the complexity of production processes with input-output linkages, consider the following simplified set-up, taken from Chaney (2013). Intermediates producers combine input from upstream suppliers with equipped labor, and sell their output to downstream customers. Formally, intermediate producer i combines equipped labor L_i with intermediates $q_{k\to i}$ ($q_{k\to i}$ stands for the sales from k to i) from a continuum of suppliers $k \in \mathcal{S}_i$ and sells its output to a continuum of customers $j \in \mathcal{C}_i$ using a constant returns to scale technology approximated by the following nested Cobb-Douglas-CES production function,

$$Q_i = \frac{1}{\alpha^{\alpha}(1-\alpha)^{1-\alpha}} \left(\int_{k\in\mathcal{S}_i} q_{k\to i}{}^{\frac{\sigma-1}{\sigma}} dk \right)^{\alpha\frac{\sigma}{\sigma-1}} L_i^{1-\alpha}. \tag{28.5}$$

Firms face the same iso-elastic demand from any customer $j \in \mathcal{C}_i$, $p_{i\to j} q_{i\to j} = \frac{p_{i\to j}{}^{1-\sigma}}{\int_{k\in\mathcal{S}_j} p_{k\to j}{}^{1-\sigma} dk} X_j$ with $p_{i\to j}$ the price charged by i to customer j, $q_{i\to j}$ the units sold by i to j, and X_j the total spending on intermediates by j. Given these iso-elastic demands, firm i charges all its customers the same constant mark-up, $\frac{\sigma}{\sigma-1}$, over its marginal cost,

$$p_{i\to j} = p_i = \frac{\sigma}{\sigma-1} w^{1-\alpha} \left(\int_{k\in\mathcal{S}_i} p_k{}^{1-\sigma} dk \right)^{\frac{\alpha}{1-\sigma}} \tag{28.6}$$

with w the competitive wage rate.

Equation (28.6) sheds light on the complex interactions between firms in a network of input-output linkages. The price firm i charges to any of its clients (p_i) depends on both the prices it pays to its suppliers (the p_k's) and the diversity of its suppliers (the set $\mathcal{S}_i$). With iso-elastic demand, prices are simply proportional to a firm's productivity. So from equation (28.6), a firm with either more efficient suppliers (lower p_k's) or with a more diverse set of suppliers (a larger measure for $\mathcal{S}_i$) will have a lower marginal cost of production. The productivity of each of firm i's suppliers in turn depends on the efficiency of their suppliers, so that the structure of the entire upstream production chain matters for firm i's efficiency.

Generically, given that the price set by any firm depends on the prices set by other firms, finding an equilibrium set of prices requires to jointly solve for all prices, taking as given the network of input-output linkages given by all the sets of suppliers and customers, the $\mathcal{C}_i$'s and $\mathcal{S}_i$'s. Generically, solving for prices in a complex network is hard. One should note that the expression $\int_{k\in\mathcal{S}_i} p_k{}^{1-\sigma} dk$ in equation (28.6) corresponds to the $(\sigma-1)^{th}$ moment of the prices of firm i's suppliers. Using the same analytical tools as those described in the previous section, Fourier transforms and the related moment generating function allow some insights into the solution to this problem. It is possible, however, to further simplify the problem to derive explicit solutions.

Chaney (2013) assumes the input-output network can partitioned in such a way that a given firm only competes with similar firms. Under this symmetry assumption, he solves explicitly for the equilibrium of this model, and further endogenizes the entire network of input-output linkages by explicitly modeling the choice of individual producers to seek suppliers and customers.

Oberfield (2013) goes a different route and assumes inputs are perfectly substitutable ($\sigma \to \infty$), so that in equilibrium, a firm only sources inputs from a unique supplier. He further allows firms to differ in their individual Total Factor Productivity, adding a firm-specific multiplicative term z_i to the production function in equation (28.5): $Q_i = \frac{1}{\alpha^\alpha (1-\alpha)^{1-\alpha}} z_i \left(\int_{k \in \mathcal{S}_i} q_{k \to i}^{\frac{\sigma-1}{\sigma}} dk \right)^{\alpha \frac{\sigma}{\sigma-1}} L_i^{1-\alpha}$. Under those two assumptions, Oberfield proposes an elegant fixed point solution concept for the endogenous structure of the network of input-output linkages. This network emerges semi-endogenously: Oberfield assumes that firms can *potentially* trade with an exogenously given set of firms (buy inputs from, sell their output to); among that exogenously given set, they decide endogenously to *actually* trade with only one of them. In other words, the actual flow of inputs along vertical production chains comes both from the exogenous distribution of potential network connections between suppliers and customers, and from the endogenous choice of each firm to pick only the best among its potential suppliers. Formally, the techniques used by Oberfield are related to the techniques used for the study of information percolation in the previous section. The one difference is instead of characterizing the behavior of the sums of random variables, Oberfield characterizes the maximum of random variables.

On this last note, I want to briefly comment on a series of recent work in macroeconomics that study the diffusion of knowledge, or of technologies, in a random network of firms. Luttmer (2007 and 2012); Alvarez, Buera, and Lucas (2008); Lucas (2009); Ghiglino (2011); Konig, Lorenz, and Zilibotti (2012); and Perla and Tonetti (2014) study the gradual diffusion of best techniques among firms that randomly meet each other. Lucas and Moll (214) and Luttmer (2014) further endogenize the choice of how intensively to seek information from other firms. Luttmer (2014) characterizes the long run growth rate in such an economy where information about technologies gradually diffuse among firms. All those models develop formal tools to study the asymptotic behavior of the maximum of many draws from a random variable (the best technique available after many meetings with other entrepreneurs).

In this handbook, Daron Acemoglu, Asuman Ozdaglar, and Alireza Tahbaz-Salehi describe the origins of systemic risk in a network model of input-output linkages. The model they describe shares some resemblance with the model above. They are interested in particular in the second moment properties of such a model, namely the variation of aggregate output emanating from local idiosyncratic shocks hitting each node in the input-output network.

28.3.2 The Empirics of Firm-to-Firm Trade

Few recent papers in international trade rigorously analyze the empirics of firm-to-firm trade.

In a series of papers with various co-authors, Andrew Bernard studies the importance of production networks, firm-to-firm trade, and the role of intermediaries in the context of international trade. Bernard, Grazzi, and Tomasi (2014) study the endogenous choice of potential exporters to serve foreign markets either directly or via an intermediary (wholesaler). They confront their model to detailed Italian firm level data on manufacturing and wholesale exporters, contrasting the modes of entry into foreign markets. Using detailed data on Norwegian firm-level exports, Bernard, Moxnes, and Ulltveit-Moe (2014) describe the network structure of firm-to-firm trade: for a given exporter, the details of their transaction-level data allows them to distinguish between exports towards different importers in the same destination country. Such detailed transaction data on firm-to-firm trade gives a unique perspective on the network of linkages between firms. Finally, Bernard, Moxnes, and Saito (2014) use an exhaustive data set of buyer-seller linkages among Japanese firms to study the importance of the network structure of production of firms' efficiency. Making a clever use of the arguably exogenous impact of the high-speed train (Shinkansen) on the ease with which firms can find new suppliers, they confirm firm performance is positively affected by the diversity and quality of its suppliers.

Antras, Fort, and Tintelnot (2014) incorporate a simplified version of the model described above into a conventional model of trade with heterogeneous firms. Firms that are able to gain access to more and/or better suppliers improve their efficiency and ultimately profits. Antras et al. assume a simple fixed cost of gaining access to input sources from new countries, and they solve for the equilibrium sourcing decision of heterogeneous producers. They propose a structural estimation of their model using data on U.S. importers.

This series of papers shows taking into account the network structure of production in models of international trade generates novel predictions. This is a promising avenue for future research.

References

Allen, Treb. Forthcoming. "Information Frictions in Trade." *Econometrica*.

Alvarez, Fernando E., Francisco J. Buera, and Robert E. Lucas, Jr. (2008). "Models of idea flows." NBER Working Paper 14135.

Antras, Pol, Teresa Fort, and Felix Tintelnot (2014). "The margins of global sourcing: Theory and evidence from U.S. firms." University of Chicago, unpublished manuscript.

Bernard, Andrew, Andreas Moxnes, and Yukiko Saito (2014). "Production networks, geography and firm performance." Dartmouth, unpublished manuscript.

Bernard, Andrew, Andreas Moxnes, and Karen Helene Ulltveit-Moe (2014). "Two-sided heterogeneity and trade." Dartmouth, unpublished manuscript.

Bernard, Andrew, Marco Grazzi, and Chiara Tomasi (2014). "Intermediaries in international trade: Margins of trade and export flows." Dartmouth, unpublished manuscript.

Bramoulle, Yann, Sergio Currarini, Matthew O. Jackson, Paolo Pin, and Brian W. Rogers (2012). "Homophily and long-run integration in social networks." *Journal of Economic Theory* 147(5), 1754–1786.

Burchardi, Konrad B. and Tarek A. Hassan (2013). "The economic impact of social ties: Evidence from German reunification." *Quarterly Journal of Economics* 128(3), 1219–1271.

Burchardi, Konrad B., Thomas Chaney, and Tarek A. Hassan (2014). "Migrants, ancestors, and investments." *Toulouse School of Economics*, unpublished manuscript.

Chaney, Thomas (2013). "The gravity equation in international trade: An explanation." *National Bureau of Economic Research*, Working Paper 19285.

Chaney, Thomas (2014). "The network structure of international trade." *American Economic Review* 104(11), 3600–3634.

Cohen, Lauren, Umit G. Gurun, and Christopher J. Malloy (2014). "Resident networks and firm trade." Harvard Business School, unpublished manuscript.

Combes, Pierre-Philippe, Lafourcade Miren, and Thierry Mayer (2005). "The trade-creating effects of business and social networks: Evidence from france." *Journal of International Economics* 66(1), 1–29.

Dasgupta, Kunal and Jordi Mondria (2014). "Inattentive importers." University of Toronto, unpublished manuscript.

Dornbusch, R., S. Fischer, and P. A. Samuelson (1977). "Comparative advantage, trade, and payments in a Ricardian model with a continuum of goods." *American Economic Review* 67, 823–839.

Duffie, Darrell, Gaston Giroux, and Gustavo Manso (2010). "Information percolation." *American Economics Journal: Microeconomic Theory* 2, 100–111.

Duffie, Darrell, Semyon Malamud, and Gustavo Manso (2009). "Information percolation with equilibrium search dynamics." *Econometrica* 77, 1513–1574.

Duffie, Darrell, Semyon Malamud, and Gustavo Manso (2015). "Information percolation in segmented markets." *Journal of Economic Theory* 157, 1130–1158.

Duffie, Darrell and Gustavo Manso (2007). "Information percolation in large markets." *American Economic Review, Papers and Proceedings* 97, 203–209.

Eaton, Jonathan and Samuel Kortum (2002). "Technology, geography, and trade," *Econometrica* 70(5), 1741–1779.

Ghiglino, Christian (2011). "Random walk to innovation: Why productivity follows a power law." *Journal of Economic Theory* 147(2), 713–737.

Greif, Avner (1989). "Reputation and coalitions in medieval trade: Evidence on the Maghribi traders." *Journal of Economic History* 49(4), 857–882.

Greif, Avner (1993). "Contract enforceability and economic institutions in early trade: The Maghribi traders' Coalition." *American Economic Review* 83(3), 525–548.

Jackson, Matthew O. and Brian W. Rogers (2007). "Meeting strangers and friends of friends: How random are social networks?" *American Economic Review* 97(3), 890–915.

Konig, M., J. Lorenz, and F. Zilibotti (2012). "Innovation vs. imitation and the evolution of productivity distributions." *CEPR* Discussion Papers 8843.

Lucas, Robert E., Jr. (2009). "Ideas and growth." *Economica* 76, 1–19.

Lucas, Robert E., Jr. and Benjamin Moll (2014). "Knowledge growth and the allocation of time." *Journal of Political Economy* 122(1), 1–51.

Luttmer, Erzo G. J. (2007). "Selection, growth, and the size distribution of firms." *Quarterly Journal of Economics* 122(3), 1103–1144.

Luttmer, Erzo G. J. (2014). "An assignment model of knowledge diffusion and income inequality." University of Minnesota, unpublished manuscript.

McCallum, J. (1995). "National borders matter: Canada–US regional trade patterns." *American Economic Review* 85, 615–623.

Oberfield, Ezra (2013). "Business networks, production chains, and productivity: A theory of input-output architecture." Princeton University, unpublished manuscript.

Perla, J. and C. Tonetti (2014). "Equilibrium imitation and growth." *Journal of Political Economy* 122(1), 52–76.

Rauch, J. and Trindade, V. 2002. "Ethnic chinese networks in international trade." *Review of Economics and Statistics* 84(1), 116–130.

Tinbergen, Jan (1962). "An analysis of world trade flows." In *Shaping the World Economy*, Jan Tinbergen, ed. New York: Twentieth Century Fund.

CHAPTER 29

TARGETING AND PRICING IN SOCIAL NETWORKS

FRANCIS BLOCH

29.1 Introduction

This chapter analyzes the optimal use of social networks by firms that wish to diffuse new products, rely on word-of-mouth communication for advertising, or exploit consumption externalities among consumers. Viral marketing—the exploitation of social networks to enhance firms' profits—has been discussed for the past 30 years, and the advent of the internet has tremendously increased the ease and scope of the use of social networks. The market capitalization of companies running digital social networks like Facebook, MySpace, or LinkedIn reflects the potential value of networking sites to firms targeting products to consumers.

The evolution of the academic literature on viral marketing has followed the expansion of social networking on the Internet. At the crossroads between different disciplines (marketing, computer science, and economics), the literature has grown so rapidly that it is impossible to survey all the relevant work. In addition, in some areas, like the use of recommendations and reviews, the literature is still growing fast, and the current state-of-the-art may very well be superseded in the next few months. In this chapter, I have chosen to be very selective, focusing attention on the targeting of individuals to diffuse information or opinions in a social network, and the pricing at different nodes of the social network when agents experience consumption externalities. In both cases, firms take the network of social interaction as given and consider how to optimally leverage social effects to introduce new products or maximize profits. In her chapter on marketing and social networks, Chapter 30 Dina Mayzlin considers the complementary problem, where firms take actions to stimulate social interactions and increase the effectiveness of social effects. Chapter 11 by Yves Zenou considers a class of targeting problems where the objective is not to seed the network to start a diffusion process but to remove a node in order to reduce or increase activity. Optimal targeting and pricing strategies are also examples of interventions which modify the outcome of actions chosen by consumers in a social network—relating this chapter to Chapter 9 by Yann Bramoullé and Rachel Kranton on general network games

Both topics covered in this chapter require an explicit modeling of the social network and emphasize how an existing social network shapes firms' decisions and performances.[1] The difficulty of obtaining analytical results on general networks has led to a multiplicity of methods and techniques to study targeting and pricing in social networks. Some papers rely on analytical mathematical results and others on numerical simulations. This chapter surveys both rigorous mathematical models and agent-based simulations.

For a firm acting in an environment where consumers are related by social links, the social network is either a vector of information or a vector of consumption externalities. In the first case, agents use their social connections to speak about new products or to convey their opinions on existing products. In the latter case, agents directly experience value from the fact that their neighbors consume a given product. Whether the social network is a vector of information or consumption externalities affect the way in which consumers interact in the social network, and the targeting and pricing decisions of the firm. We will discuss the two models separately.

Another important dividing line to organize the current literature on targeting and pricing in social networks is the degree of competition between firms. Some papers focus on the optimal choice of a monopolist while others consider a competitive environment, either in the form of perfect competition or in the form of oligopolistic competition. The outcome of the analysis is very different in the monopoly and competitive cases. Finally, the existing papers differ in the information structure of agents and firms on the social network. Some papers assume perfect knowledge of the network by consumers and firms. Other papers only consider limited information—consumers and firms only observe local neighborhoods and the degree distribution of the network. Differences in information structure typically result in large differences in the analysis of the optimal behavior of firms.

29.2 Information Externalities

We study how a firm uses social interactions to diffuse information about a new product. The firm's problem is to select a target in the social network in order to

[1] By contrast, a large part of the literature on word-of-mouth communication considers global interaction among consumers, without describing the exact network structure. The early literature on network externalities in consumption also considered only global interactions.

maximize diffusion of profits.[2] We first describe models of diffusion to explain how information travels in the social network. We then turn to the literature in computer science to analyze efficient targeting algorithms either in order to maximize diffusion of the new product (the influence maximization problem) or the revenue of the firm (the revenue maximization problem). In the next subsection, we review the literature on agent-based modeling whose objective is to uncover relations between the characteristics of the social network and/or of the nodes used as seeds and the speed and spread of diffusion of new products. Finally, we discuss the small body of literature in economics looking at analytical models of targeting and diffusion in social networks.

29.2.1 Information and Diffusion

We first analyze the diffusion of information in the social network. In the simplest model, information is a binary signal equal to 0 or 1 that can be interpreted as the awareness of a new product or the recommendation for an existing product. An agent is in state 0 if he is uninformed or has not received a positive recommendation, and in state 1 otherwise. As in models of contagion in epidemiology, agents' states evolve over time according to their interactions with other agents.

The social network g is fixed, and at any time t, a consumer may receive a message from one of her neighbors. Once an agent is informed, she remains informed forever. Diffusion of information is mechanical: a consumer does not control whether she sends information to her neighbors or not.[3] There are two main diffusion models:

- *The linear threshold model.* Agent i acquires information about the product if and only if the number of neighbors who are informed is greater than a number k or if the fraction of informed neighbors is greater than a fraction κ. If neighbors are heterogeneous, the numbers k and κ may be replaced by weighted averages of the status of the agent's neighbor reflecting the fact that some agents are more influential than others, independently of their location in the social network.
- *The independent cascade model.* Each of agent i's neighbors sends information with an independent probability p_j. Agent i is informed if and only if at least one of his neighbors has sent the information.

Targeting (or "seeding") the network amounts to choosing, given a network structure g, the agent or set of agents to whom the product should be given first in order to diffuse information to the entire network as quickly as possible. As the problem of targeting is mathematically intractable—there is no general analytical solution to the identification

[2] Optimal targeting problems arise in a number of areas in economics. For example, recent literature considers the optimal targeting of heterogeneous agents to induce coordination (Bernstein and Winter 2012, Sakovics and Steiner 2012).

[3] For models of information transmission where consumers control whether or not to pass on information in a network, see Chatterjee and Dutta (2010) or Bloch, Demange, and Kranton (2014).

of the targets in arbitrary networks—the literature has focused either on approximation methods and algorithms or numerical solutions to evaluate targeting policies. When firms have limited information on the network, the class of targeting strategies to choose from is reduced, and analytical results can be obtained.

29.2.2 Targeting Algorithms

29.2.2.1 Influence Maximization

Domingos and Richardson (2001) is the first paper raising the issue of targeting algorithms to maximize the probability of sales in a social network. They adopt a model where consumers can be of two types, 0 and 1, reflecting whether they buy the product or not. A consumer's probability of buying a product depends on two factors: marketing expenditures and the probability that her direct neighbors have bought the product.

The paper describes how a firm optimally targets consumers by directing marketing expenditures to specific agents in the network. It contrasts the performance of three algorithms: a single-pass algorithm which only looks at one iteration, a greedy algorithm which increases marketing expenditures wherever they increase payoffs, and a hill-climbing algorithm which increases expenditures where it matters most. Using data on an experimental program of movie recommendations, EachMovie, from 1996–1997, Domingos and Richardson (2001) compute the "multiplier affect" of marketing expenditures on a targeted agent and show that agent's market values may be as high as 20, meaning that one dollar spent on that consumer affects 20 other consumers. The distribution of network values appears to be very skewed so that targeted marketing strategies are very profitable. They compare three marketing strategies: mass marketing, where all agents receive uniform expenditures; directed marketing where agents are targeted according to their individual characteristics but ignoring their influence on others; and targeted marketing taking into account agents' market values. Domingos and Richardson (2001) quantify the profit increase due to the use of targeted strategies with respect to directed and mass strategies and demonstrate its importance.

In a follow-up paper, Richardson and Domingos (2002) consider the same model, assuming that the influence is measured linearly, through a matrix of weights $[w_{ij}]$ representing i's influence on j's choice and that marketing decisions are continuous. Using data from the knowledge-sharing site Epinions, the authors again quantify agents' network values and simulate the effect of different marketing policies. They also test the robustness of their results with respect to removal of nodes from the network.

Kempe, Kleinberg, and Tardos (2003) revisit the optimal targeting problem, casting it in the framework of standard diffusion processes. In their analysis, the objective of the firm is to select an initial set A of k nodes in the social network in order to maximize the total number of informed nodes when information is diffused according to a linear threshold or independent cascade model. Their first result shows that the optimal targeting problem is NP-hard. It is in general impossible to find a polynomial algorithm to compute the optimal target set A except in special cases—like Richardson and Domingos' (2002) linear model where the optimal target is a solution of a system of linear equations. The second result of the paper is an approximation bound on the efficiency of the hill-climbing algorithm, which selects agents to place in the set A by looking sequentially at those agents which result in the highest influence, where influence is measured by the number of additional "live" connections that would arise in the network if that agent was informed. The approximation bound derives from the analysis of submodular functions in integer programming.

A function f defined over subsets S of N is called submodular, if for any $S \subseteq T$ and any vertex $v \notin T$,

$$f(S \cup v) - f(S) \geq f(T \cup v) - f(T).$$

A result in integer programming due to Nemhauser, Wolsey, and Fisher (1978) shows the following:

The function f is monotone, positive, and submodular. Let S be the set of k elements obtained by adding one by one elements which maximize the increase in the function's value and S^* the set of k elements which maximizes the function's value. Then $f(S) \geq (1 - \frac{1}{e})f(S^*)$, where e is the base of natural logarithms.

Hence, if the objective function f is submodular, the hill-climbing algorithm produces an outcome that results in at least 63% of the efficient value. Kempe, Kleinberg, and Tardos (2003) prove that the influence function in the linear threshold and independent cascade models are submodular, thereby establishing the approximation bound for the hill-climbing algorithm.[4] The performance of the hill-climbing algorithm is tested using data on collaboration between high energy physicists. In the linear threshold model, the hill-climbing algorithm on average increases efficiency by 18% with respect to targeting based on the nodes' degree centrality, and by 40% with respect to a targeting based on closeness centrality. Interestingly, the first node in the set A_0 already accounts for 25% of the total number of nodes eventually informed.

29.2.2.2 Revenue Maximization

Hartline, Mirrokni, and Sundarajan (2008) introduce the revenue maximization problem. A monopolistic seller of a new product chooses the sequence in which buyers are approached and the prices charged to the buyers. In the revenue maximization problem, buyers choose whether to purchase the good or not, thereby indirectly controlling the spread of information. The value of a consumer i is drawn at random from a distribution $F_i(V)$ which depends on the set V of consumers influencing i. Hartline, Mirrokni, and Sundarajan (2008) assume that consumers are myopic and choose to purchase the good based on the current set of agents who have bought the good and not the final set of

[4] The proof of submodularity of the influence function is hard because the influence function is difficult to compute. Kempe, Kleinberg, and Tardos (2003) construct diffusion processes equivalent to the independent cascade and linear threshold models to compute the influence function, and show that their construction extends to more general diffusion models.

consumers purchasing the product. The revenue maximization problem adds a layer of complexity to the influence maximization problem: the seller has to compute optimal prices at each node taking into account the effect of prices on the diffusion of the good through the purchasing decisions of the buyers.

One case is particularly simple: if the network of interactions is the complete network and all agents are symmetric, the optimal pricing strategy only depends on the set of agents who have already bought the good and the set of agents who have not yet been approached, and can be computed in polynomial time as the solution to a linear programming problem. Beyond this simple case, the solution to the problem is NP hard. Hartline, Mirrokni, and Sundarajan (2008) assume that the revenue generated by consumer i is a monotone, submodular function of the set of agents V who influence agent i, and proceed to derive lower bounds on the efficiency of a two-step "exploit and influence" (EI) algorithm. This algorithm first searches for a set of initial agents A to whom the good is given for free using a hill-climbing strategy. The seller then visits agents in $N \setminus A$ in a random order and the algorithm selects for each buyer the optimal revenue maximizing price. In the linear threshold model the approximation bound is equal to $\frac{2}{3}$, and for general diffusion models this algorithm achieves at least $\frac{1}{3}$ of the total value. The marketing strategy proposed by the algorithm is rather simple, as it does not specify a fixed order of visiting buyers, and can be implemented with a small number of easily computed prices. It is adaptive, as the optimal prices chosen in the second part of the algorithm depend on the history of purchases by other buyers.

Arthur, Motwani, Sharma, and Xu (2009) propose a nonadaptive EI algorithm to solve the revenue maximization problem. The algorithm first computes a minimum cost-spanning tree of the graph with maximal number of leaves (this is an NP hard problem), and the set A is formed by the internal nodes, who receive a fixed cashback for each consumer they refer, while leaves are charged a fixed price. Arthur, Motwani, Sharma, and Xu (2009) show that this algorithm achieves a fraction of the total value, which depends on the complexity parameters of the problem, but the fraction is typically lower than the fraction computed by Hartline, Mirrokni, and Sundarajan (2008) for their adaptive algorithm.

29.2.2.3 Influence Maximization with Competition

The influence maximization problem becomes much more complex if one assumes that instead of one firm, there are two firms competing to seed the network. The firms are now players in a noncooperative game, and coordination problems may result in inefficiencies with respect to the policy adopted by a single firm maximizing influence. Hence, in addition to inefficiencies due to the approximation algorithm, new inefficiencies arise because firms play a game, and the "price of anarchy" measuring the worst ratio between the optimal value and the values obtained in the equilibrium of the noncooperative game becomes a relevant indicator. The influence model is now generalized by assuming that consumers can have three values: 0 (uninformed), A (buy product A) and B (buy product B). Faced with two competing diffusion models, the difficulty is to decide how a consumer exposed to both products A and B chooses the product she buys.

Bharathi, Kempe, and Salek (2007) and Carnes, Nagarajan, Wild, and Zuylen (2007) propose extensions of the independent cascade model to deal with a competitive environment. In Bharathi, Kempe, and Salek (2007), the consumer adopts the first product she becomes aware of. If she becomes aware of the two products at the same time, an exogenous tie-breaking rule decides which of the two products is chosen. Consider first the behavior of the follower—the reaction function of player A when player B has already chosen the set S_B of seeds. The influence function given the set S_B is a monotone submodular function of the set S_A. Hence, the hill-climbing algorithm provides an approximation of the efficient policy—with the same approximation bound of $1 - \frac{1}{e}$ as in the monopoly influence maximization problem. Carnes, Nagarajan, Wild, and Zuylen (2007) prove a similar result on the optimal policy of the follower under two different variants of the independent cascade model. In the first variant—the distance-based model—edges become activated according to a random process, and a consumer learns about the product with the minimal distance to a seed.[5] In the second variant—the wave propagation model—consumers become informed sequentially, and a consumer at period t picks at random one of his informed neighbors and adopts his product. Bharathi, Kempe, and Salek (2007) also compute a price of anarchy equal to 2—meaning that the number of nodes informed at any noncooperative equilibrium is at least equal to $\frac{1}{2}$ of the total number of nodes informed at the optimum.

The linear threshold model cannot as easily be extended to competitive diffusion in the network. Borodin, Filmus, and Oren (2010) analyze the follower's choice when firm B has already chosen the set of seeds S_B and note that the influence function of firms A is not necessarily monotone nor submodular. The hill climbing algorithm does not necessarily provide a good approximation of the optimal targeting policy. Goyal and Kearns (2012) study the optimal allocation of a fixed budget over nodes under a general threshold diffusion process characterized by two functions: the function f specifies, as a function of the proportion of informed neighbors, what is the probability that an agent is informed; and the function g specifies, as a function of the share of neighbors buying A and B the probability that the agent adopts one of the two products given that he is informed. Goyal and Kearns (2012) compute a price of anarchy of 4 when the function f is concave and the function g linear.[6]

Dubey, Garg, and de Meyer (2006) extend the linear model of Richardson and Domingos (2002) to multiple firms. Firms simultaneously choose marketing expenditures at every node. The proclivity of a consumer to buy from a given firm is a linear function of the number of neighbors buying from the firm and the relative marketing expenditures of the firm. Because the model is linear, the best response functions can be computed in polynomial time by solving systems of linear equations. The Nash

[5] If multiple seeds are at a minimal distance, the probability that the consumer consumes one of the two goods, say good A, is proportional to the number of seeds of good A at minimal distance.

[6] If the function f ceases to be concave, the price of anarchy jumps to infinity.

equilibrium is unique, can be computed analytically, and involves firms either spending zero resources on a consumer, or spending an amount which depends on the effective cost of marketing expenditures of the different firms.

29.2.3 Simulations on Targeting

Because optimal targeting has no analytical situation in general, numerical simulations can be used to assess the importance of different parameters on the diffusion of new products. Goldenberg, Libai, and Muller (2001) pioneered the use of agent-based modeling to study new product diffusion. They employ a cellular automata model, where each consumer forms a cell that is activated according to a fixed automaton, a set of rules relating the cell to its immediate environment. Since then, as reported by Rand and Rust (2011), agent-based modeling has proven particularly fruitful to study new product diffusion in complex social networks.

Watts (2002) studies conditions under which global cascades occur in random networks where diffusion follows a threshold rule. In addition to analytical derivations of the critical values of connectivity under which global cascades arise, Watts (2002) runs simulations to clarify the relation between the average connectivity in the network—measured by the fixed probability that a link is formed in a random Erdös Renyi network—and the percentage of nodes informed in the long run. He finds that the percentage of nodes informed is low when the graph has low connectivity and agents are likely not to receive information from neighbors, and high when agents have so many neighbors that the threshold condition becomes very difficult to satisfy. Hence global cascades—or full spread of new products across the social network—arise for intermediate values of the connectivity parameter.

Dodds and Watts (2007) enrich the model by distinguishing between two types of agents: influentials and imitators. Influentials are agents who belong to the top 10% of the degree distribution—namely, agents who are more connected than others. Dodds and Watts (2007) first study the likelihood that a global cascade arises, depending on the identity of the initial seed. As global cascades only arise in the "cascade window" when the connectivity is intermediate, the identity of the initial seed does not matter much in the likelihood of a global cascade and is clearly of lesser importance than the global network connectivity. However, interestingly, the size of the global cascade does depend on the identity of the seed: influentials trigger larger cascades even if the difference is not deemed to be very significant. The simulation results strongly depend on the diffusion process. If instead of a threshold diffusion process, the new product disseminates according to an independent cascade model, consumers with higher degree are more likely and not less likely to hear about the product. Higher connectivity in the network then favors diffusion rather than hampers it, and influentials are significantly more likely to trigger global cascades.

Recently, a flurry of papers have been using agent-based models to analyze the role of different characteristics of the seeds on diffusion in random and actual networks.

Goldenberg, Han, Lehmann, and Hong (2009) argue, contrary to Dodds and Watts (2009), that nodes with high connectivity (the "hubs") play an essential role in information diffusion. They use an independent cascade model, and their simulations are based on an actual network from a Korean social networking site. Stephen, Dover, and Goldenberg (2010) consider a model where transmission takes time, and emphasize the role played by two characteristics of the seed: her connectivity measured by the degree and activity measured by the number of times at which she is active. They show that both characteristics play a role in the diffusion process. Libai, Muller, and Peres (2013) consider competing firms, and study seeding strategies based on random targeting, targeting influential nodes and experts who are more likely to be believed by other consumers. They run simulations on 12 real-world networks. Haenlein and Libai (2013) add a demographic characteristic to the model, identifying consumers with high revenue value to the firm. Assuming homophily in the network—consumers with high value are more likely to be connected to other high-value consumers—they show that a revenue-targeting strategy performs better than a targeting strategy based on connectivity. They use a preferential attachment network generation model in order to generate networks with a given homophily structure.

In order to identify the best targeting policy, Stonedahl, Rand, and Wilensky (2010) run a genetic algortihm to select an optimal seeding strategy based on nodal characteristics like degree, two-step reach, clustering, and so on. They consider four stylized networks: random, lattice, small world, and preferential attachment, and one real network (a sample of Twitter users). Their analysis shows that a seeding strategy based on degree performs rather well with respect to the strategy generated by the genetic algorithm in all four stylized networks, but not in the actual network of Twitter users.

29.2.4 Analytical Models of Targeting

As the problem of targeting in a known fixed network is NP-hard, there is no hope to characterize analytically the optimal targeting policy when the firm has complete information on the network.[7] However, when information about the network is limited, analytical solutions of the influence maximization problem exist.

Galeotti and Goyal (2009) assume that the firm only knows the degree distribution of consumers in the social network, and the degree of each consumer. A targeted marketing strategy thus assigns a different advertising expenditure to each consumer on the basis of her degree. In their model, social links are directed, and consumers are characterized both by the number of agents who influence them (the out-degree) and the number of agents they influence (the in-degree). Assuming that local neighborhoods do not overlap, and that information is only transmitted at one step, the profit

[7] Recall, however, that analytical solutions can be found in the linear model of Richardson and Domingos (2002) and Dubey, Garg, and de Meyer (2006).

of reaching a consumer with (out)degree k and spending marketing expenditures x can be represented by a general function $\phi_k(x)$. The function $\phi_k(x)$ exhibits increasing (decreasing) marginal returns in degree if, for any $x > x'$, $\phi_{k+1}(x) - \phi_{k+1}(x') > (<)\phi_k(x) - \phi_k(x')$.

Two special models are investigated in the paper. The first model—a one-step diffusion version of the independent cascade model—supposes that consumers send information about the product to all their neighbors with a fixed probability depending on the expenditures x so that $\phi_k(x) = 1 - (1-x)^{k+1}$. The second model—a one-step version of the threshold models—supposes that an agent becomes informed if the fraction of informed neighbors is large enough and $\phi(x) = (1-x)xk^{1-\beta}$ where $\beta > 1$ is a parameter affecting the probability that the good is adopted as a function of the degree.

Galeotti and Goyal (2009) focus attention on the monotonicity of the optimal targeting policy: should consumers with a low degree (who are not influenced by many agents) receive higher advertising expenditures than consumers with a high degree (who are influenced by many other agents)? The answer to this question depends on the property of the profit function. If the profit function exhibits increasing marginal returns to degree (as in the threshold model), nodes with higher degree receive more advertising expenditures. If the profit function exhibits decreasing marginal returns to degree (as in the cascade model), nodes with lower degree receive more advertising. The intuition underlying the result is clear: in the first case, nodes with higher degree are less likely to receive the information from their neighbors, whereas in the second case nodes with low degrees are less likely to be informed through the social network. As word-of-mouth communication and advertising are substitutes, the firm compensates the poor information transmission through the network with more direct advertising.

Galeotti and Goyal (2009) also show that an increase in the dispersion of the degree distribution increases the value of targeting when the cost of expenditures is sufficiently low. They also obtain the intuitive result that consumers with higher in-degree (i.e., who influence more agents) receive more advertising expenditures.

Campbell (2013) uses the techniques developed for large random graphs to study monopoly pricing and targeting in the presence of word-of-mouth communication. A monopoly sets a price, and consumers decide whether to purchase or not, according to a random valuation. When a consumer buys the product, information flows to other consumers, but when a consumer does not buy, information stops.[8] The initial network thus becomes broken into smaller components as time goes, as in percolation diffusion processes.[9] The basic question asked by Campbell (2013) is to identify under which conditions, given an initial random graph, a giant component emerges. The emergence of a giant component depends on the price charged by the monopolist, and there exists a

[8] The fact that the diffusion process depends on prices controlled by the firm is reminiscent of the analysis of the revenue maximization problem by Hartline, Mirrokni, and Sundarajan (2008).

[9] See Callaway, Newman, Strogatz, and Watts (2000) on percolation processes on random graphs.

critical price P^{crit} such that a giant component emerges when price is below the critical price, $P < P^{crit}$, but not when the price is higher, $P > P^{crit}$.

In this framework, Campbell (2013) considers targeted advertising. The monopolist knows the degree distribution and chooses the consumer to whom the product is offered first, as a function of the consumer's degree. The revenue gained from the consumer depends on the size of the component the consumer belongs to. If she belongs to the giant component, advertising will be useless as word-of-mouth communication will eventually reach all other agents in the component, but if she does not belong to the giant component, targeted advertising is useful and its effectiveness is higher if the component is bigger. This observation translates into a simple targeting strategy given a fixed price P. If $P > P^{crit}$, the firm should target the consumer with the highest degree, and if $P < P^{crit}$, it should instead target the consumer with the smallest number of connections.

29.3 Consumption Externalities

Katz and Shapiro (1985) and Farrell and Saloner (1985) introduced the concept of network externalities—agents' consumption of a good is affected by the number of agents consuming the same good. Network externalities arise in telecommunications, when agents benefit from other agents using the same communication device, in the software industry where the development of application is driven by the number of users, etc. (See Shy 2001 for a thorough description of industries with network externalities.) Early models of consumption externalities modeled the externality as global rather than network based: the valuation of consumers was a function of the total number of users in the good. More recent models, starting with Jullien (2011) and Sundarajan (2007), analyze consumption externalities based on a given social network. In this section, we first study optimal pricing strategies of a monopolist, and then analyze equilibria of models of competitive pricing.

29.3.1 Monopoly Pricing with Consumption Externalities

Candogan, Bimpitis, and Ozdaglar (2012) and Bloch and Querou (2013) independently analyze the same model of linear consumption externalities.[10] Agent's utilities are given by the quadratic expression

$$U_i(q_1,\ldots,q_n) = a_i q_i - \frac{1}{2} b_i q_i^2 + q_i \sum_j g_{ij} q_j - p_i q_i,$$

[10] This model can be viewed as a foundation for the linear probability model of Richardson and Domingos (2002) and Dubey, Garg, and de Meyer (2006).

where $g_{ij} \in [0,1]$ measures the influence of agent j on i's consumption. The utility of agent i is quadratic in own consumption and includes an interaction term such that it increases with the consumption of direct neighbors. This positive externality in consumption implies that, given any vector of discriminatory prices $(p_1, \ldots, p_n)$, the consumption levels are computed as the equilibrium of noncooperative games played among consumers.

Suppose now that a monopolist sets the prices $(p_1, \ldots, p_n)$ in order to maximize profits $\Pi = \sum_i (p_i - c)q_i$. The optimal price vector is given by

$$\mathbf{p} = \mathbf{a} - (\Lambda - \mathbf{G})(\Lambda - \frac{\mathbf{G} + \mathbf{G}^T}{2})^{-1} \frac{\mathbf{a} - c\mathbf{1}}{2},$$

where Λ is a diagonal matrix with terms b_i on the diagonal, $\mathbf{G} = [g_{ij}]$ is the matrix of consumption externalities, and $\mathbf{a}$ the vector of a_is. Interestingly, when the matrix G is symmetric (influence is undirected), the monopoly sets a uniform price across the network. To understand this result, notice that when charging a price on a node, the monopolist balances two effects: if the node is more central, its overall value of the product is higher and the price should be higher to exploit this higher value, but consumption at that node also increases consumption at more neighboring nodes, so that prices should be lowered to increase consumption. In the linear model, the two effects exactly balance and the optimal price is uniform.[11] The equilibrium consumptions vary over nodes, and in fact are proportional to Katz-Bonacich centrality.[12] Bloch and Querou (2013) extend the model to allow different nodes to be served by different firms. In this oligopolistic setting, equilibrium prices are no longer uniform. Prices are increasing in degree and decreasing in the number of adjacent nodes served by the same firm.

Candogan, Bimpitis, and Ozdaglar (2012) compute the gain of a discriminatory pricing strategy with respect to uniform prices and provide an algorithm to compute the optimal uniform price. As in Hartline, Mirrokni, and Sundarajan (2008), they consider a two-price strategy where the good is given at a discount to some consumers and at a full price to others, and show that the problem of targeting is NP complete, but that an approximation algorithm returns at least 88% of the optimal profit. Corbo and Lin (2010) extend the analysis to k-pricing strategies for arbitrary values of k, and provide algorithms which achieve at least 90% of the optimal profit. Saaskhilati (2007) studies uniform monopoly pricing for specific network structures. He computes the optimal uniform price price for regular networks and stars.

Fainmesser and Galeotti (2013) analyze discriminatory pricing in the quadratic utility model when the monopoly only knows the degree distribution and observes the

[11] Bloch and Querou (2013) show that this is a knife-edge result. For example, if costs are quadratic, the result disappears and more central nodes are charged higher prices as production at these nodes becomes more expensive.

[12] Katz-Bonacich centrality is defined as the vector $(\mathbf{I} - \alpha\mathbf{G})^{-1}\mathbf{1}$ for some α such that $\alpha\rho(\mathbf{G}) < 1$ where $\rho(\mathbf{G})$ is the largest eigenvalue of G.

degree of each node. Consumers are assumed to have different in- and out-degrees. The in-degree of consumer i measures the number of agents influenced by i, and the out-degree the number of agents that influence i. Fainmesser and Galeotti (2013) compute the optimal price of a monopoly, which discriminates according to out-degrees or in-degrees. They show that prices are monotonic: a consumer with higher out-degree pays more, and a consumer with higher in-degree pays less.[13] Equilibrium consumption increases in out-degrees when the monopolist discriminates on out-degrees, and increases both in the in- and out-degrees when the monopolist discriminates on in-degrees.

The main result of the analysis in Fainmesser and Galeotti (2013) compares equilibrium profit and consumer surplus under uniform pricing, and discrimination based on in- and out-degrees. Discriminatory profits are increasing and convex in the variance of the out- (respectively in-) degree distributions. Consumers with low out-degrees benefit from discrimination based on out-degrees, whereas consumers with high out-degrees are harmed by discrimination. Consumers with high in-degrees are better off with a move from uniform pricing to discrimination based on in-degree, and total consumer surplus increases with the move. These welfare effects provide a full picture of the benefits and costs from discriminatory pricing in social networks.

29.3.2 Competitive Pricing in Social Networks

Suppose now that instead of monopoly pricing, we consider price competition with local network externalities. In an early paper, Jullien (2011) considers a model where consumers are divided into groups, with a fixed matrix of external effects across groups. Even though the interpretation offered by Jullien (2011) as competition among platforms on two-sided markets is somewhat different, the model can be viewed as a model of local network externalities where the matrix of external effects describes the social network. Two firms compete by setting prices for each group of consumers. The firms choose their prices sequentially, with firm A moving as the leader and firm B as the follower.

Focusing on the behavior of firm B, Jullien (2011) shows that the follower can use a "divide and conquer" strategy by targeting some consumer group to which it offers a low price, while raising the price of another consumer group which benefits highly from cross-group externalities with the first group. By choosing optimally its target group for cross-subsidy, the follower can in fact conquer the market even when it is less efficient than the leader. This advantage given to the follower translates into a lower bound on the follower's profit and an upper bound on the leader's profit in Stackelberg price competition with local network effects.

[13] As noted by Bloch and Querou (2013), the same result is obtained with complete information about the network, when externalities are sufficiently small.

Banerji and Dutta (2009) also assume that consumers are divided into groups, and explicitly specify a network of interaction measuring the externalities across groups. As opposed to Jullien (2011), prices are uniform across nodes and the two competing firms set prices simultaneously. Given Bertrand competition and network effects, a simple conjecture is to assume that in equilibrium, a single firm dominates the market. Banerji and Dutta's (2009) main objective is to characterize network structures under which a single firm operates, and network structures under which market segmentation may arise in equilibrium.

When the social network is complete and network externalities are global, price competition and network effects unambiguously lead to a single firm capturing the market. If two firms were active, they would have to charge the same price at any node and any firm would have an incentive to undercut his rival to capture the entire market. For other special networks, the circle and the star, Banerji and Dutta (2009) exhibit equilibria with market segmentation.

Duopoly pricing with social interactions is also discussed by Galeotti (2010) in a model where the social network plays a different role. Instead of capturing consumption externalities, the social network describes how consumers learn information about prices collected by other consumers. Galeotti's (2010) model incorporates social interactions into a standard model of consumer search. He shows that the presence of word-of-mouth communication affects the pricing strategies of the two firms. It implies that some consumers will always learn the prices of the two products, eliminating the possibility of an equilibrium where no consumer searches and all firms quote the monopoly price (the Diamond paradox). He also shows that prices and profits are not necessarily monotonic in the level of connectivity of the social network. If consumers sample more neighbors, they acquire more information but their incentive to search goes down. The two effects work in opposite direction. For high values of the search cost, the first effect dominates and the market becomes more competitive, reducing equilibrium prices and profits; for low values of the search cost, the second effect dominates and the market becomes less competitive, raising equilibrium prices and profits.

29.4 Open Questions

The literature in economics, computer science, and marketing on targeting and pricing has been very active in the past decade and is still growing very fast. As new data sets become available, theoretical models are increasingly being confronted to the data. Clearly, the empirical analysis of models of seeding and discriminatory pricing in social networks is a field of investigation that will grow in the near future. Other questions remain open and may be ripe for new theoretical investigations in the next few years. If privacy regulations prevent social networking sites from selling their data to third parties, advertisers will have to infer network data from consumers directly. Mechanisms to elicit information form consumers about their local neighborhoods—through referral discounts or rewards—still need to be studied. A closer look at consumers' incentives to propagate information in the network is also needed. If the new product is only available to a small select set of consumers, competition may give agents an incentive to keep silent about the product. Similarly, incentives to lie in recommendations about existing products need to be better understood. Finally, the analysis of competition among firms seeding the network remains sketchy. The revenue maximization problem with competing firms remains an open area of research, as are models of oligopolistic competition and product differentiation with local network externalities.

REFERENCES

Arthur, D., R. Motwani, A. Sharma, and Y. Xu (2009). "Pricing strategies for viral marketing on social networks." Mimeo, Stanford University.

Banerji, A. and B. Dutta (2009). "Local network externalities and market segmentation." *International Journal of Industrial Organization* 27, 605–614.

Bernstein, S. and E. Winter (2012). "Contracting with heterogeneous externalities." *American Economic Journal: Microeconomics* 4, 50–76.

Bharathi, S., D. Kempe, and M. Salek (2007). "Competitive influence maximization in social networks." *WINE 2007.*

Bloch, F., G. Demange, and R. Kranton (2014). "Rumors in social networks." Mimeo, Paris School of Economics and Duke University.

Bloch, F. and N. Querou (2013). "Pricing in social networks." *Games and Economic Behavior* 80, 263–281.

Borodin, A., Y. Filmus, and J. Oren (2010). "Threshold models for competitive influence in social networks." Mimeo, University of Toronto.

Callaway, D., M. Newman, S. Strogatz, and D. Watts (2000). "Network robustness and fragility: Percolation on random graphs." *Physical Review Letters* 85, 5468–5471.

Campbell, A. (2013). "Word of mouth and percolation in social networks." *American Economic Review* 103, 2466–2498.

Candogan, Bimpitis and Ozdaglar (2012). "Optimal pricing in networks with externalities." *Operations Research* 60, 883–905.

Carnes, T., C. Nagarjan, S. Wild, and A. van Zuylen (2007). "Maximizing influence in a competitive social network: A follower's perspective." *ICEC 07.*

Chatterjee, K. and B. Dutta (2010). "Word of mouth advertising, credibility and learning in networks." Mimeo, Penn State University and University of Warwick.

Corbo, J. and S. Lin (2012). "Optimal pricing with positive network effects: The big benefits of just a little discrimination." *ICIS 2012.*

Domingos, P. and M. Richardson (2001). "Mining the network value of customers." *Proceedings of the 7th Conference on Knowledge Discovery and Data Mining*, 57–66.

Dubey, P., R. Garg, and B. de Meyer (2006). "Competing for customers in a social network: The quasi-linear case." *WINE 2006*, long version in *Journal of Dynamics and Games* 1, 377–409.

Fainmesser, I. and A. Galeotti (2014). "The value of network information." Mimeo, Brown University and University of Essex.

Farrell, J. and G. Saloner (1985). "Standardization, compatibility and innovation." *RAND Journal of Economics* 16, 70–83.

Galeotti, A. (2010). "Talking, searching and pricing." *International Economic Review* 51, 1159–1174.

Galeotti, A. and S. Goyal (2009). "Influencing the influencers: A theory of strategic diffusion." *RAND Journal of Economics* 40, 509–532.

Goldenberg, J., B. Libai, and E. Muller (2001). "Talk of the network: A complex systems look at the underlying process of word-of-mouth." *Markting Letters* 12, 211–223.

Goldenberg, J., S. Han, D. Lehmann, and J. W. Hong (2009). "The role of hubs in the adoption process." *Journal of Marketing* 73, 1–13.

Goyal, S. and M. Kearns (2012). "Competitive contagion in networks." *STOC 2012*, long version forthcoming in *Games and Economic Behavior*.

Hartline, J., V. Mirrokni, and M. Sundarajan (2008). "Optimal marketing strategies over social networks." *Proceedings of WWW 2008: Beijing, China*, 189–198.

Jullien, B. (2011). "Competing in multi-sided markets: Divide and conquer." *American Economic Journal: Microeconomics* 3, 186–219.

Katz, M. and C. Shapiro (1985). "Network externalities, competition and compatibility." *American Economic Review* 75, 424–440.

Kempe, D., J. Kleinberg, and E. Tardos (2003). "Maximizing the spread of influence through a social network." *Proceedings of the 9th International Conference on Knowledge Discovery and Data Mining*, 137–146.

Libai, B., E. Muller, and R. Peres (2013). "Decomposing the value of word-of-mouth seeding programs: Acceleration versus expansion." *Journal of Marketing Research* 50, 161–176.

Nemhauser, G., L. Wolsey, and M. Fisher (1978). "An analysis of approximations for maximizing submodular set functions." *Mathematical Programming* 14, 265–294.

Rand, W. and R. Rust (2011). "Agent-based modeling in marketing: Guidelines for rigor." *International Journal of Research in Marketing* 28, 181–193.

Richardson, R. and P. Domingos (2002). "Mining knowledge-sharing sites for viral marketing." *KDDM 02*.

Saaskilahti, P. (2007). "Monopoly pricing of social goods." MPRA Paper 3526, University Library of Munich.

Sakovics, J. and J. Steiner (2012). "Who matters in coordination problems?" *American Economic Review* 102, 3439–3461.

Shy, O. (2001). *The Economics of Network Industries*. Cambridge University Press.

Stephen, A., Y. Dover, and J. Goldenberg (2010). "A comparison of the effects of transmitter activity and connectvity on the diffusion of information in social networks." Mimeo, INSEAD.

Stonedahl, F., W. Rand, and U. Wilensky (2010). "Evolving viral marketing strategies." *GECCO 10*.

Sundarajan, A. (2007). "Local network effects and complex network structure." *The BE Journal of Theoretical Economics* 7, art. 46.

Watts D. (2002). "A simple model of global cascades in random networks." *Proccedings of the National Academy of Sciences* 99, 5766–5771.

Watts, D. and P. S. Dodds (2007). "Influentials, networks and public opinion formation." *Journal of Consumer Research* 34, 441–458.

CHAPTER 30

MANAGING SOCIAL INTERACTIONS

DINA MAYZLIN

In the past decade and a half we have seen the rise of technologies that have enhanced peer-to-peer interactions: consumers now can check the status of their acquaintances on Facebook, browse through the photos of celebrities on Instagram, check Twitter for the latest updates on breaking news events, text their friends on their smart phones, and read hotel reviews written by strangers on TripAdvisor. One of the more exciting implications of this development from a marketing perspective is that these platforms enable the exchange of product-related information. For example, a Facebook friend may recommend a film that she recently saw, a recipe posted on Pinterest may mention a certain chocolate brand as an ingredient, and a blogger may comment on his experiences with a new digital camera. Moreover, not only is the information being shared to a greater extent on a greater variety of platforms, and among people who may not be as easily connected in the offline setting, but firms can now manage some of the interactions. For example, the firm can promote conversations among its customers and noncustomers by investing in online communities, or perhaps reaching out to certain bloggers.

This chapter summarizes recent research in marketing[1] that relates to the issue of management of social interactions, which we defined in Godes et al. (2005) as an action that is taken by an individual not actively engaged in selling the product or service, and that impacts others' expected utility for the product or service. In other words, social interaction is a broader concept than the traditional concept of word of mouth (WOM) since it encompasses new electronic types of communication such as email, consumer reviews, and Twitter posts. This chapter is complementary to Chapter 29 by Bloch, which focuses on the targeting of individuals to diffuse information or opinions in a social network, and the pricing at different nodes of the social network in a game with consumption externalities. While there is some overlap between the two chapters,

[1] The chapter also discusses a few related papers in economics, information technology, and finance.

the current work focuses more on social effects, while the chapter by Bloch takes the social effects as given and focuses on the diffusion of information in a network.

The first building block of social interactions is the motivation of the agents involved, which usually means the motivation of the sender. Berger (2014) lists the main motivations behind the generation of word of mouth as: (1) impression management (such as identity-signaling), (2) emotion regulation (for example, generating social support), (3) information acquisition, (4) social bonding, and (5) persuading others. A growing number of papers in the behavioral marketing literature are exploring these motivations and the implications that they have on the type of information that is shared between consumers. For example, if a sender is concerned about self-presentation, she is more likely to talk about products that are perceived as "cool." The second building block of social interactions is the shape of the social network of the agents, and the position in which they occupy in that network. Many of the papers that we will discuss in this chapter explore the impact of social networks on the diffusion of information.

We use a modified version of the framework developed in Godes et al. (2005) to classify the different roles that the firm can play in managing social interactions: observer, influencer, and participant. While the sender's motivation certainly has an important bearing on how the firm can optimally manage social interactions, we do not deal directly with the growing behavioral literature on word of mouth, for which Berger (2014) provides excellent coverage. Instead, we focus on the quantitative marketing literature as it applies to these topics.

30.1 Firm as Observer

The effectiveness of any management strategy relies in part on the ability to measure. However, there are two primary challenges to measuring social interactions: (1) How can one gather data on what are essentially private exchanges? (2) What aspects of conversations (highly unstructured data) should be measured and are managerially meaningful?

The lack of data availability in the past has implied that researchers traditionally have had two measurement techniques available to them: surveys (Reingen and Kernan 1986) and inference (Bass 1969). The emergence of online communication platforms has implied that conversations that were previously private are now not only public but can be collected by firms as well as by researchers. Below we summarize the various metrics that have been studied in the literature and the extent to which these metrics are relevant to the firm.

30.1.1 Volume

This is perhaps the most intuitive metric and represents the total amount of conversations about a product within a fixed period of time. There are a number of studies that have examined the relationship between volume and sales. For example, Godes and Mayzlin (2004) collected posts about new TV shows on public Usenet forums as measure of online conversations and tied them to TV ratings (firm sales in this setting). In that study *volume* measures the total number of posts across all user groups (n) about a show (i) during the course of the week (t),

$$POST_{it} = \sum\nolimits_{n=1}^{N} POST_{it}^{n}$$

In other words, here volume measures how many people mentioned the TV show within a week. The study estimates a model of TV ratings as a function of last week's word-of-mouth measures (including volume), controlling for previous TV ratings and the show fixed effect. Interestingly, Godes and Mayzlin (2004) find that the information provided by the volume metric (how many people mentioned the show) is already contained in the previous sales variable, and hence is not significant once the model includes previous ratings.

Chintagunta et al. (2010) examine the effect of online reviews on box office performance. In that study the volume measure is the number of reviews for a movie in the Yahoo! Movies website, and the dependent variables is the opening-day gross for a title in a local geographic market. They find that the volume of reviews does not have an effect on sales. In contrast, Gopinath et al. (2013) do find a positive relationship between blog volume (the number of blogs that mentioned the movie) and opening-day movie performance, but do not find a significant effect of volume on post-release movie performance. Some of the other studies that do find a significant effect of volume on sales include Dellarocas et al. (2007), Duan, Gu, and Whinston (2008), and Liu (2006).

In summary, the volume metric seems to be a valid measure of online word-of-mouth activity, especially soon after the product release and before sales data becomes available (Gopinath et al. 2013). What is less clear is the extent to which it captures new information about word of mouth above and beyond information contained in past sales (see Godes and Mayzlin 2004).

30.1.2 Valence

Valence measures the extent to which a conversation about a product is positive. In the context of online reviews, where a reviewer provides a numerical evaluation score of the experience, obtaining a valence measure is relatively easy, even though the text itself may still provide additional information above and beyond the numerical score. Measuring the valence of all other conversations involves text-processing which can be done either automatically using software or using human raters.

Table 30.1 illustrates the observed valence of conversations across various categories and platforms. Unless otherwise noted, all of the averages for reviews are on the 5-point scale (with "1" being the lowest possible review and "5" being the highest). One observation that we can make is that there is variation in mean valence across categories

Table 30.1 Valence of Conversations

Paper	Context	Valence distribution
Resnick and Zeckhauser 2001	Reviews of sellers and buyers on eBay	99% of buyers and 98% of sellers had positive feedback
Godes and Mayzlin 2004	Usenet posts on new TV shows	51% of posts were positive, 27% negative, and 22% mixed
Chevalier and Mayzlin 2006	Book reviews	Average rating on Amazon.com = 4.14 Average rating on BarnesandNoble.com = 4.45
Chintagunta et al. 2010	Movie user reviews on Yahoo! Movies website	Average rating of reviews until the movie is released in a new market = 9.9 out of 13
Moe and Trusov 2011	Product review on the website of national retailer of bath, fragrance, and beauty products	80% of all reviews were five-star
Godes and Silva 2012	Amazon book reviews	Average rating = 4.09
Ghose et al. 2012	Reviews of hotels	Average rating on Travelocity.com = 3.87 Average rating on TripAdvisor = 3.49
Mayzlin et al. 2014	Hotel reviews	Average TripAdvisor rating = 3.52 Average Expedia rating = 3.95
Tadelis and Nosko 2015	Reviews of sellers on eBay	More than 99% of sellers had positive feedback

and platforms. For example, the ratings on BarnesandNoble.com (average of 4.45) are higher than the rating on Amazon.com (average of 4.14), and the ratings for books are higher than the ratings for hotels (average between 3 and 4 stars) and movies (average of 3.8 when translated to 5-point scale). One robust observation has been the fact that reviews on eBay tend to be very high—a working paper by Tadelis and Nosko finds that 99% of sellers had positive feedback.

Book reviews on Amazon.com also tend to be quite positive—with an average of about 4.2 out of 5 stars. This could be due to self-selection (people may be able to effectively match themselves to a book based on available information) or could be possibly due to review manipulation. Hotel reviews appear to be more negative on average—with averages below 4 out of 5 stars. Mayzlin et al. (2014) point out that this effect could be due to differences in review manipulation: books on the same subjects may be complements whereas hotels in the same geographic area are substitutes. Hence, there is a more of an incentive for negative review manipulation by close competitors for hotels than for books.

In the context of reviews, several papers have found that valence (the book's average star rating) of reviews affects sales. For example, Chevalier and Mayzlin (2006) examine the effect of Amazon.com and BarnesandNoble.com reviews on book sales. They find that the valence of reviews drives a book's relative sales levels. Interestingly, they also find that the impact of very negative reviews is greater than the impact of very positive reviews. Chintagunta et al. (2010) find that movie reviews valence affects opening-day sales.

As we mentioned above, not all conversations contain a numerical rating. In that case, obtaining measures of valence must involve a more labor-intensive process. Godes and Mayzlin (2004) utilized two (human) coders to categorize Usenet postings as positive, negative, or mixed. A third coder was utilized when there was disagreement between the first two coders. In that paper the authors find that 51% of the relevant conversations are positive, 27% are negative, and 22% are mixed. A similar procedure was used by Gopinath et al. (2013).

Finally, there is a growing body of literature devoted to automated textual analysis. There is a variety of technical challenges associated with automatically mining text data. An excellent primer on the issues is Netzer et al. (2012). The various approaches include supervised machine learning and rule-based or dictionary-based text mining. The difficulty inherent in accurate text-mining can be illustrated by an example of a message board post from Netzer et al. (2012), "That's strange. I heard many people complaint[sic] about the Honda paint. I owned a 1995 Nissan Altima before and its paint was much better than my neighbor's Accord (1998+ model). I found the Altima interior was quiet [sic] good at that time (not as strange as today's)." Even a human coder might find it challenging to categorize this post's valence about the various brands mentioned.

Despite these challenges, a number of papers have used text mining techniques to extract additional information from online conversations. An early paper by Das and Chen (2007) extracts investor sentiment from online message boards and connects this information to stock performance. More recently, Archak et al. (2011) mine Amazon review data to obtain information on product features. Ghose et al. (2012) aggregate information collected from text analysis and user surveys (conducted on Amazon Turk) to create a hotel ranking system.

30.1.3 Variance

In the existing literature, the term variance has stood for two distinct types of measures. The first refers to the amount of disagreement in reviews on the evaluation of product quality. The second refers to the dispersion of conversations across the social network. We examine the two measures in turn.

The variance in the product evaluations has been measured by a few studies. Sun (2012) provides a theoretical foundation for why and how variance in ratings should affect sales. She presents a model where the consumer infers that the product with a high variance in ratings is in fact a niche product.This implies that a higher variance should only help product sales if the review average is low. The paper also provides some evidence for this effect using Amazon.com and BarnesandNoble.com book reviews. In contrast, Chintagunta et al. (2010) find no effect of review variance on movie sales. Moe and Trusov (2011) find that disagreement among review ratings is associated with a lower subsequent rate of posting of extreme reviews.

Godes and Mayzlin (2004) use entropy to measure the extent of information dispersion across a social network. This metric is motivated by Granovetter (1973) theory of the effect of social network structure on the flow of information. Granovetter characterizes relationships as being either "strong ties" or "weak ties." If we assume that communities or groups are characterized by relatively strong ties among their members, one implication of this model is that the only connections between communities are those made along weak ties. This has the important implication that information moves quickly within communities but slowly across them, and that the information that traverses a weak, as opposed to a strong, tie has the opportunity to reach more people.

Godes and Mayzlin (2004) use entropy to measure the degree to which conversations about TV shows are confined to few Usenet groups (low entropy) or are spread out across many Usenet groups (high entropy):

$$ENTROPY_{it} = -\sum_{n=1}^{N} \frac{POST_{it}^{n}}{POST_{it}} \mathrm{Ln}\left(\frac{POST_{it}^{n}}{POST_{it}}\right)$$

Based on the theories above, the authors hypothesize that, conditional on the same volume of word of mouth, conversations that are dispersed across many groups are more likely to result in higher awareness than conversations that are confined within fewer groups. The authors find that more dispersed word of mouth is associated with higher sales (viewership) in the next period, in a model that controls for other factors such as past sales and volume of word of mouth. Hence it is important to consider how information travels in a network when studying the effect of online conversations.

30.1.4 Review Dynamics

In the past few years, a number of papers have examined the effect of review dynamics on subsequent review behavior. That is, reviews are not simply independent draws of different opinions, but are in fact correlated across time in predictable ways. The existence of reviewer dynamics has several important implications. First, it calls into question the validity of simple metrics (such as valence, for example) to summarize the current state of word of mouth. Second, if there are consistent biases that arise in reviews due to the dynamics (if, for example, the early reviews are more positive predisposed to the product), and if consumers are not able to control for these biases, the usefulness of reviews may be compromised. Moreover, the firm may need to take actions that mitigate the biases that are introduced through review dynamics. For example, the firm may design more complicated summary metrics that attempt to de-bias review data or display information in a way that lessens potential biases.

The emerging literature in this area has pursued two objectives: (1) to document the existence of dynamic processes in reviews, and (2) to explain the mechanism behind these processes. We first turn to the question of existence of dynamics. Li and Hitt (2008) demonstrate the existence of an overall negative trend in Amazon book reviews over time, which is surprising given the fact that consumers who purchase the product later on should have more information available to them and are hence expected to make better purchase decisions. Godes and Silva (2011) argue that there are multiple and distinct dynamic processes that are present in consumer reviews. In particular, ratings change systematically over both order and time, and it is important to control for both of these effects. Moe and Trusov (2011) demonstrate the existence of negative autocorrelation in review ratings—an increase in average ratings tends to be associated with subsequent posting of negative ratings. They also show that an increase in rating volume is associated with an increased arrival of negative reviews, and that disagreement between reviewers is associated with fewer subsequent extreme ratings.

The drivers behind these dynamics have been debated. Li and Hitt (2008) argue that the downward trend in reviews arises due to reviewer self-selection, which exists if there is correlation between demand and quality perception. They argue that this correlation results in early product reviewers that are consistently biased compared to the general population. An example of positive correlation is the case where the early buyers of books are also fans of the author's previous books. They find that in a sample of Amazon book reviews the overall time trend is negative, which they interpret as evidence that for books the early reviews are positively self-selected. However, their theoretical model does not imply that the trend need always be negative. In fact, negative correlation between demand and quality perception, which they argue may occur in the case of software where the early buyers are particularly sensitive to defects, would result in a positive time trend.

Interestingly, Godes and Silva (2011) propose another explanation for the negative time trend found in Li and Hitt (2008): all reviews have become more negative over the past decade. In fact, when the authors control for this macro trend, they find that ratings at the book level in fact increase over time. The former (order) effect is driven by the hypothesized mechanism that purchase errors increase as more reviews arrive, and these errors lead to lower ratings. The authors provide support for this mechanism by showing that the sequential decline is bigger when reviewers are very dissimilar from each other. The review context here is also Amazon book reviews.

Moe and Trusov (2011) develop a modeling framework that allows them to separate product ratings into the following components: (1) consumers' independent (or socially unbiased) ratings, (2) the effect of social dynamics, and (3) noise. The review context here is a national retailer of bath, fragrance, and beauty products. Importantly, they show (in a simulation) that social dynamics in reviews may impact the evolution of sales. One interesting simulation result is that a high-quality product may actually benefit from early mixed ratings, which will result in social dynamics that quickly converge to the true high-level rating. Hence, it is not necessarily the case that marketers need to encourage only positive word of mouth early on.

The emerging literature on review dynamics demonstrates that instead of viewing reviews as independent reports of consumers' experience, it is more accurate to view reviews as pieces of an ongoing conversation between consumers. Note that these dynamics arise largely from the fact that the posting and the timing of each review is under the discretion of each consumer, which implies that reviews are prone to issues of self-selection as well as social dynamics. For example, consider a consumer who had a mildly negative experience at a restaurant. Whether or not she ever shares her opinion may depend on whether she sees a very positive review that contradicts her experience. From the firm's perspective these social dynamics in reviews introduce logistical challenges that relate to simple summary metrics, but, perhaps more importantly, introduce some additional risk since factors other than product quality drive online word of mouth.

In summary, the existing literature provides a rich set of guidelines to firms that want to invest in measuring online conversations. While the academic literature has been somewhat divided on the importance of volume of word of mouth on sales, volume-based measures of conversations have been embraced enthusiastically by industry. For instance, most measures of a company's success in social media are volume-based, such as the number of Facebook followers, number of Twitter followers, and so on. In fact, some company social media measurement communication efforts are exclusively focused on volume-based measures such as the number of mentions or followers on social media platforms. One possible reason behind the popularity of volume-based measures is that these measures are so intuitive—a company page with a million Facebook followers *sounds* like a very popular and successful page. However, studies such as Godes and Mayzlin (2004) call attention to the fact that volume-based measures may not be useful in all contexts. In particular, a volume-based measure may not provide any new information to the firm that is not already contained in sales. The research cited here suggests that a firm can benefit from collecting a wider range of metrics of word of mouth.

30.2 Firm as Influencer

In this role, the firm fosters and shapes social interactions among market participants. (Also see Chapter 12 by Fortin and Boucher on the empirics of network effects, and Chapter 15 by Aral on network experiments.) For example, in order to foster social interactions, the firm may choose to include consumer reviews on its site. The first fundamental question that needed to be addressed is determining whether the effect of social interactions is causal or not. The difficulty associated with identifying endogenous effects (the "reflection problem") was first pointed out by Manski (1993).[2] For example, suppose that we observe some students in a certain high school adopting a new app, and subsequently their friends adopt the same app. It is difficult to disentangle causality here—it could be that the adoption is due to word of mouth, but it could also be the case that all the students have similar preferences and demographics, and hence are making similar choices across time.

Note that the Manski's reflection problem is partially solved by obtaining direct word of mouth data. That is, in the example above, if we were to observe that the app adoption was preceded by a recommendation from one student to another, we could rule out that the students are independently adopting the app. However, we would still not be able to rule out that both the electronic communication between high school students and the adoption are influenced by an outside (and an unmeasured) factor. For example, it could be the case that the app targeted advertising at the high school students, which generated both adoption and conversations. Determining that WOM has a causal effect on sales and is not simply correlated with product success is especially important when one considers the firm's role in managing social interactions through a communication strategy.

30.2.1 The Causal Impact of Word of Mouth on Sales

There are three major approaches that studies have taken in order to demonstrate a causal link between word of mouth and sales. The first approach is a differences-in-differences approach that essentially compares word of mouth and sales across platforms (and across time). The second approach uses instrumental variables to identify the effect of word of mouth on sales. The third approach shows causality through natural experiments.

Let's first turn to the first approach which utilizes a cross-platform comparison. Chevalier and Mayzlin (2006) examine the effect of Amazon.com and BarnesandNoble.com reviews on book sales. This study is able to address the issue of causality of word of mouth directly. As an illustration, suppose that a new cookbook is heavily promoted by the publisher. This may generate a lot of word of mouth for the book as well as elevated sales. In order to conclude that it is WOM that is driving sales and not another factor such as the underlying quality of the book or the offline advertising campaign (that may be correlated with both WOM and sales), the paper utilizes a difference-in-differences approach: the authors examine the effect of user reviews across BarnesandNoble.com and Amazon.com and across time. That is, the authors examine whether a scathing review of a Julia Child cookbook on Amazon results in the book's lower popularity on Amazon relative to BN.com. The authors also rule out that

[2] In the marketing literature an influential paper that calls attention to the importance of correctly identifying social effects is Hartmann et al. (2007). Another paper that addresses these issues is Hartmann (2010).

the difference is driven by differences in preferences across sites by differencing across time. As mentioned before, the authors find that reviews have a causal effect on sales, and that the effect is asymmetric: very negative reviews have a bigger impact on sales than very positive reviews.

The second approach to disentangling causality issues when estimating social effects involves identifying the effects through instrumental variables. For example, Shriver et al. 2012 use wind speeds as exogenous variation on windsurfers' propensity to post content on a windsurfing social network platform. They find that social ties have a positive effect on content generation, and that content generation has a positive effect on obtaining social ties, even when the authors control for endogenous group formation and correlated unobservables.

Finally, a number of studies have used field experiments or natural experiments to demonstrate the causal effect of word of mouth. For example, Chen et al. (2011) use a natural experiment to compare the effect of observational learning (consumers observing what purchases are made by other consumers before them) and word of mouth on sales. The exogenous variation occurs due to the fact that from 2005 to 2007 Amazon removed and then reintroduced a platform feature that displayed what digital camera previous consumers bought after searching a particular model. One interesting result is that, unlike the effect of word of mouth, the positive effect of observational learning (a model sold well in the past) is greater than the negative effect of observational learning (a model did not sell well in the past) on future sales. Another study that utilizes the field experiment approach is Tucker and Zhang (2011). This paper examines the effect of providing popularity information (information on how many previous customers chose to purchase the product) on clicks in a platform that provided wedding service vendor listings. Some of the categories were randomly assigned to display popularity information and some were assigned not to show it. The authors find that the introduction of popularity information results in narrow-appeal vendors receiving more visits than equally popular broad-appeal vendors.

While these studies establish a causal link between social interactions and sales, all deal with "endogenous" or naturally-occurring word of mouth. This leaves open the question of whether it would be possible for the firm to encourage the creation of "exogenous" word of mouth. That is, it is still not clear that the firm would be able to create "buzz" that would result in additional product sales. Godes and Mayzlin (2009) implemented a field study within the context of a "buzz" campaign conducted by a promotional company on behalf of a national restaurant chain. In the field experiment the promotional company recruited a panel of "buzzers" who were encouraged to engage in conversation about the restaurant. The participants self-reported incidents of all interactions. The study also tracked weekly sales of the restaurant chain. This paper finds that indeed the firm is able to generate word of mouth that increases sales. In particular, the study finds that (consistent with Granovetter's theories), the impactful word of mouth is one between acquaintances (as opposed to conversations between friends). Hence, it is indeed possible for a firm to create word of mouth that meaningfully impacts sales.

30.2.2 Engaging the Right Sender

Consider a firm that plans to influence consumer interactions. The next managerially relevant issue for this firm is the optimal type of word of mouth that it wants to create. For example, does the firm want to engage only certain types of current users? Does it make sense for the firm to target a light or heavy user of the product, or is it perhaps better to target a potential user (and current non-user)? Similarly, does it make sense for the firm to invest in the effort to reach out to an influential user (or an opinion leader), and what is the best way to measure influence?

30.2.2.1 The Sender's Local Network

A number of papers have examined how the local network of the targeted node impacts the resulting diffusion of information. Interestingly, the results of some of the studies that focus on the network effects are quite counterintuitive in that it is not necessarily the most central and well-connected individuals that are most valuable from the firm's perspective. For example, consider two studies in this stream. A simulation study by Watts and Dodds (2007) shows that most information cascades are caused by "easily influenced individuals influencing other easily influenced individuals."

Yoganarasimhan (2012) finds that the size and structure of an author's local network is a significant driver of the popularity of YouTube videos seeded by her, even after the author controls for video characteristics, seed characteristics, and endogenous network formation. In contrast to Watts and Dodds (2007), this study finds that the marginal benefit of a second-degree friend (a friend of a friend) is higher than that of a first-degree friend (simply, a friend) in the spread of YouTube videos.

30.2.2.2 The Sender's Loyalty to the Firm

Another possible way to target senders is based on their loyalty to the firm. Godes and Mayzlin (2009) recruit different types of agents as part of the field study on word-of-mouth generation—while some "buzzers" were not initially aware of the restaurant chain (they are from the promotional firm's panel), others are part of the chain's loyalty program. In addition, within the loyal customers, there is heterogeneity on how loyal each customer is—while some have just recently signed up to be part of the loyalty program, others visit the restaurant on a regular basis. Interestingly, the paper finds that it is the less loyal customers (including the members of the panel who were not customers of the firm) whose incremental WOM leads to higher sales.

This may at first appear counterintuitive since the more loyal customers are more positively exposed towards the product. The explanation given by the authors for this result is network-based. That is, the less loyal customers are effective in generating incremental profit since their friends and acquaintances have not been previously exposed to information about the product. These results imply that the firm may optimally choose to concentrate on its new customers in spreading word of mouth as part of a viral campaign as opposed to the very loyal customers whose networks have already been saturated with word of mouth. Another implication of this study is the idea that new customers may be especially valuable to the firm from the perspective of incremental profit since they can spread information to previously unexposed potential customers.

In contrast, Iyengar et al. (2011) find a very different result from Godes and Mayzlin 2009. Iyengar et al. (2011) study the adoption by physicians of a new drug treating a chronic (and potentially lethal) condition. In addition to prescription data, they also collect network data, sales call data, and measures of opinion leadership. The network data is survey-based: each physician is asked to report the names of other physicians with whom she discusses the disease and to whom she refers patients. Each physician is assigned an "in-degree" measure based on nominations by others. Based on prescription volume, the authors are also able to trace product usage. The authors find that connections to heavy users are more influential in driving adoption than connections to light users.

Note that the two studies yield very different managerial implications: Godes and Mayzlin (2009) recommend targeting less loyal customers, who are more likely to be light users, for a viral campaign, while Iyengar et al. (2011) imply that heavy users are particularly important for contagion. There are several possible explanations for the discrepancy in the results. First, the two contexts are very different. The former study deals with restaurant recommendations with low awareness, where credibility of the source may be less important since trial is relatively cheap. The latter deals with a prescription drug with serious side effects where both credibility of the source may be important, and trial is relatively costly. Second, Godes and Mayzlin concerns itself with incremental word of mouth (exogenous word of mouth on top of existing endogenous word of mouth). In contrast, Iyengar et al. deals with endogenous word of mouth, since the authors do not deal with a buzz campaign but are measuring organic adoption. Hence, it could be the case that heavy users are crucial for adoption that is part of endogenous word of mouth and light users are an attractive target for generating incremental word of mouth.

30.2.2.3 Targeting Influential Senders

Another common targeting strategy is to seek out influential senders since they are, by definition, more likely to be persuasive. Of course, before one can answer this question, it is important to define what one means by "influential." The papers discussed below take different approaches in operationalizing this concept.

One common approach in marketing to measuring influence is to measure the extent to which the sender is an opinion leader using a sociometric scale. For example, the King and Summers' opinion leadership scale used in Godes and Mayzlin (2009) asks the respondent to fill out the following survey (on a 7-point scale):

1. In general, I like to talk to my friends and neighbors about category (7–very often to 1–never).

2. Compared with my circle of friends, I am ______to be asked about category (7–not very likely to 1–very likely).
3. When I talk to my friends about category, I (7–give a great deal of information to 1–give very little information).
4. During the past six months, I have told ____about category (7–no one to 1–a lot of people).
5. In discussions about category (7–my friends usually tell me about category to 1–I usually tell my friends about category).
6. Overall, in my discussions with friends and neighbors about category, I am (7–often used as a source of advice to 1–not used as a source of advice).

Despite the intuitive appeal of seeking out opinion leaders to spread information about the product, Godes and Mayzlin (2009) show that although opinion leadership is useful in identifying potentially effective spreaders of WOM among very loyal customers, it is less useful at identifying less loyal customers.This, along with their finding that it is the word of mouth from less loyal customers that is effective at driving sales, casts some doubt on the usefulness of opinion leadership as a targeting technique.

Iyengar et al. (2011) collect a survey-based measure that aggregates how many other physicians nominate the focal physician as someone with whom they discuss the disease or to whom they refer patients (this measure is referred to as "in-degree") in addition to a survey-based measure of own-opinion leadership. Hence, the former is based on others' reports of the focal doctor's opinion leadership, while the latter is self-reported. The authors also find that sociometric and self-reported measures of leadership are weakly correlated and associated with different effects: physicians with higher in-degree adopt earlier, while physicians who rank higher on the self-reported leadership measure of leadership adopt earlier but are less sensitive to contagion from peers.

Another interesting approach to identifying influential consumers is undertaken in Trusov et al. (2010). The authors propose to identify the influence of a social network member by examining the effect that the user has on her friends' behavior. In particular, if a member increases her usage and the people connected to her also increase their usage, they identify this person as influential. Conversely, if a member's usage does not impact her friends' behavior, the authors infer that this person is not influential. Interestingly, the authors find that an increase in the number of connections does not necessarily imply that the user is influential.

In summary, a firm that seeks to influence social interactions can be reassured that such an undertaking is possible—the literature has shown that organic word of mouth has a causal effect on sales and that it is possible for the firm to help create social interactions that will impact sales. In addition, the literature gives some guidance on the types of senders that should be targeted if the firm seeks to generate exogenous word of mouth. In particular, in cases where awareness is low, the firm needs to target its less loyal customers since their networks are less saturated with information about the product. Finally, while opinion leaders are more likely to adopt earlier, opinion leaders may be less useful if the firm seeks to generate word of mouth from less loyal customers.

30.3 Firm as Participant

30.3.1 Content Creation versus Linking

Finally, the most active role that the firm can play in managing social interactions is to actually participate in consumer conversations. For example, the firm can post its views on a corporate blog, inviting and responding to comments and feedback. One key question here is how the firm's posting and linking behavior affects the size of its audience. Mayzlin and Yoganarasimhan (2011) analyze the role of links to other blogs as signal of blog quality. This paper models bloggers as producers of information (or "breaking news"), and readers as consumers. By linking, a blog signals to the reader that it will be able to direct her to news in other blogs in the future. The downside of a link is that it is a positive signal about the rival's news-breaking ability. Hence, one action (a link) sends multiple signals to the reader: a positive signal about the quality of a focal blog as well as a positive signal about the quality of a potential rival. The paper shows that linking will be an equilibrium outcome when the heterogeneity on the ability to break news is low relative to the heterogeneity on the ability to find news in other blogs.

Several empirical papers have analyzed the value of links, and the connection between link formation and content generation. As mentioned before, Shriver et al. (2012) find that social ties have a positive effect on content generation, and that content generation has a positive effect on obtaining social ties. Stephen and Toubia (2010) find that in an online marketplace with links between various sellers, an increase in links creates economic value for the marketplace. In particular, the authors find that an increase in links between marketplace members is associated with an increase in commission revenue for the firm that runs the marketplace. In contrast, growth in the number of dead-end shops has a negative effect on marketplace performance.All of these papers suggest that both linking and content generation are important and productive activities.

Another option for firms is to manipulate conversations surreptitiously by exploiting the anonymity afforded by online communities. This approach raises some fundamental questions regarding the viability of online word of mouth since a large amount of this kind of promotion would undermine the credibility and, thus, usefulness of online conversations. Mayzlin (2006) examines this blurring of the lines between advertising and word of mouth. This paper develops a game-theoretic model where two products are differentiated in their value to the consumer. Unlike the firms, the consumers are uncertain about the products' quality. Firms have the option of posting anonymous, positive reviews about their product. One question that immediately arises is whether, given this anonymity and the firms' obvious self-interest, consumers would be influenced by online reviews. Broadly speaking, as more and more consumer purchases are being influenced by reviews posted by anonymous others and as the incentive grows for firms to surreptitiously manipulate these reviews, should consumers in equilibrium continue to place faith in them? In a unique equilibrium where online word of mouth is persuasive, the paper concludes that the answer is yes. In this equilibrium, firms spend more resources promoting inferior products: the firm with the better product optimally free-rides on unbiased word of mouth.

Dellarocas (2006) also develops an analytical model of strategic firm manipulation of online forums. This paper finds that manipulation may increase the quality of information provided by an online forum, which is the case if the amount of manipulation is increasing in the quality of the firm. Importantly, this paper points out that the development of "filtering" technologies makes it costlier for firms to manipulate online word of mouth.

Mayzlin et al. (2014) undertakes an empirical analysis of the extent to which manipulation occurs and the market conditions that encourage or discourage this activity. Specifically, this paper examines hotel reviews, exploiting the organizational differences between two travel websites: Expedia.com and TripAdvisor.com. That is, while anyone can post a review on TripAdvisor.com, a consumer could only post a review of a hotel on Expedia.com if she actually booked at least one night at the hotel through the website. Thus, the cost of posting a fake review on Expedia.com is quite high relative to the cost of posting a fake review on TripAdvisor.com. The authors show that the differences in the distribution of reviews for a given hotel between TripAdvisor.com and Expedia.com are affected by the firm's incentives to manipulate. Note that while several papers in marketing and computer science/IT journals have attempted to empirically document the existence of manipulated reviews, the methodology proposed in this paper avoids the challenge of classifying individual reviews as fake, and instead uses differences between sites to infer manipulation. The authors find evidence that supports the existence of manipulation, and the amount of manipulation is greater for small owner properties.

30.4 Conclusion and Future Directions

A firm that seeks to manage consumer social interactions can turn to a growing academic literature that directly addresses issues related to observing, influencing, and participating in consumer conversations. Despite the progress made so far, there are a number of outstanding issues that have yet to be addressed by researchers.

One potentially interesting area for exploration is the extent to which consumer motivation to talk has strategic implications on how the firm can manage consumer interactions. For example, Godes and Wojnicki (2014) demonstrate that consumers that are experts in a domain generate word of mouth that is positively biased since the valence of their experience signals their ability to make good choices. That is, a consumer that is a "foodie" may be more eager to share a positive experience ("a wonderful hole-in-the-wall Indian place I found") than a negative experience ("a terrible pizza joint"). One possible implication of this finding for a new product is that sampling by an expert is especially desirable—not only is the word of mouth generated potentially more persuasive due to the sender's expertise, but this consumer is positively biased.

Another area for exploration is the extent to which firm's traditional marketing actions impact organic word of mouth. Campbell et al. (2015) show that advertising for a new product may decrease the amount of organic word of mouth and information acquisition. In this model consumers engage in word of mouth in order to signal to others that they are high types who have lower costs of information acquisition. By advertising the firm reduces the asymmetry between high and low types, which reduces the signaling value of word of mouth, and hence decreases the amount of information acquisition.

Next, consider a firm that is considering response to negative word of mouth. To what extent does a public response lend legitimacy to the original consumer sentiment? Is the firm sometimes better off keeping silent on a negative rumor? How are these considerations affected by whether the rumor is true or false?

Finally, most of the papers reviewed here tend to examine firm optimal actions in isolation. Consider the targeting of particularly attractive senders by firms. A sender who is well-placed on a network is likely to appear to be a good target to multiple firms, which implies that the competition will be greater for certain types of senders. How does this increased competition affect optimal targeting? Does it make sense for firms to differentiate on whether they target the central versus the more fringe senders of information?

References

Archak, Nikolay, Anindya Ghose, and Panagiotis G. Ipeirotis (2011). "Deriving the pricing power of product features by mining consumer reviews." *Management Science* 57(8) 1485–1509.

Bass, Frank (1969). "New product growth model for consumer durables." *Management Science* 15(5), 215–227.

Berger, Jonah (2014). "Word of mouth and interpersonal communication: A review and directions for future research." *Journal of Consumer Psychology* 24(4), 586–607.

Campbell, Arthur, Dina Mayzlin, and Jiwoong Shin (2015). "Managing buzz." SOM working paper.

Chen, Yubo, Qi Wang, and Jinhong Xie (2011). "Online social interactions: A natural experiment on word of mouth versus observational learning." *Journal of Marketing Research* 48(2), 238–254.

Chevalier, Judith and Dina Mayzlin (2006). "The effect of word of mouth on sales: Online book review." *Journal of Marketing Research* 43(3), 345–354.

Chintagunta, Pradeep K., Gopinath Shyam, and Venkataraman Sriram (2010). "The effects of online user reviews on movie box office performance: Accounting for sequential rollout and aggregation across local markets." *Marketing Science* 29(5), 944–957.

Campbell, Arthur, Dina Mayzlin, and Jiwoong Shin (2014). "A model of buzz and advertising." SOM working paper.

Das, Sanjiv R. and Mike Y. Chen (2007). "Yahoo! for Amazon: Sentiment extraction from small talk on the web." *Management Science* 53(9), 1375–1388.

Dellarocas, Chrysanthos (2006). "Strategic manipulation of Internet opinion forums: Implications for consumers and firms." *Management Science* 52(10), 1577–1593.

Dellarocas, Chrysanthos, Xiaoquan Zhang, and Neveen F. Awad (2007). "Exploring the value of online product reviews in forecasting sales: The case of motion pictures." *Journal of Interactive Marketing* 21(4), 23–45.

Duan, Wenjing, Bin Gu, and Andrew B. Whinston (2008). "Do online reviews matter? An empirical investigation of panel data." *Decision Support Systems* 45(4), 1007–1016.

Ghose, Anindya, Panagiotis G. Ipeirotis, and Beibei Li (2012). "Designing ranking systems for hotels on travel search engines by mining user-generated and crowdsourced content." *Marketing Science* 31(3), 493–520.

Godes, David, Dina Mayzlin, Yubo Chen, Sanjiv Das, Chrysanthos Dellarocas, Bruce Pfeiffer, Barak Libai, Subrata Sen, Mengze Shi, and Peeter Verlegh (2005). "The firm's management of social interactions." *Marketing Letters* 16(3), 415–428.

Godes, David and Dina Mayzlin (2004). "Using online conversations to study word of mouth communication." *Marketing Science* 23(4), 545–560.

Godes, David, and José C. Silva (2012). "Sequential and temporal dynamics of online opinion." *Marketing Science* 31(3), 448–473.

Godes, David and Dina Mayzlin (2009). "Firm-created word-of-mouth communication: Evidence from a field study." *Marketing Science* 28(4), 721–739.

Gopinath, Shyam, K. Pradeep, and Sriram Venkataraman (2013). "Blogs, advertising, and local-market movie box office performance." *Management Science* 59(12), 2635–2654.

Granovetter, Mark S. (1973). "The strength of weak ties." *American Journal of Sociology* 78(6), 1360–1380.

Hartmann, Wes (2010). "Demand estimation with social interactions and the implications for targeted marketing." *Marketing Science* 29(4), 585–601.

Hartmann, Wesley R., Puneet Manchanda, Harikesh Nair, Matthew Bothner, Peter Dodds, David Godes, Kartik Hosanagar, and Catherine Tucker (2008). "Modeling social interactions: Identification, empirical methods and policy implications." *Marketing Letters* 19(3–4), 287–304.

Iyengar, Raghuram, Christophe Van den Bulte, and Thomas W. Valente (2011). "Opinion leadership and social contagion in new product diffusion." *Marketing Science* 30(2), 195–212.

King, Charles and John Summers (1970). "Overlap of opinion leadership across consumer product categories." *Journal of Marketing Research* 7(1), 43–50.

Li, Xinxin and Lorin M. Hitt (2008). "Self-selection and information role of online product reviews." *Information Systems Research* 19(4), 456–474.

Liu, Yong (2006). "Word of mouth for movies: Its dynamics and impact on box office revenue." *Journal of Marketing* 70(July), 74–89.

Manski, Charles F. (1993). "Identification of endogenous social effects: The reflection problem." *The Review of Economic Studies* 60(3), 531–542.

Mayzlin, Dina, Yaniv Dover, and Judy Chevalier (2014). "Promotional reviews: An empirical investigation of online review manipulation." *American Economic Review* 104(8), 2421–2455.

Mayzlin, Dina and Jiwoong Shin (2011). "Uninformative advertising as an invitation to search." *Marketing Science* 30(4), 666–685.

Mayzlin, Dina and Hema Yoganarasimhan (2011). "Link to success: How blogs build an audience by monitoring rivals." *Management Science*, forthcoming.

Mayzlin, Dina (2006). "Promotional chat on the Internet." *Marketing Science* 25(2), 155–163.

Moe, Wendy W. and Michael Trusov (2011). "The Value of Social Dynamics in Online Product Ratings Forums." *Journal of Marketing Research* 48(3), 444–456.

Netzer, Oded, Ronen Feldman, Jacob Goldenberg, and Moshe Fresko (2012). "Mine your own business: Market-structure surveillance through text mining." *Marketing Science* 31(3), 521–543.

Nosko, Chris and Steven Tadelis (2015). "The limits of reputation in platform markets: An empirical analysis and field experiment." University of Chicago, Working paper.

Reingen, Peter H. and Jerome B. Kernan (1986). "Analysis of referral networks in marketing—methods and illustration," *Journal of Marketing Research* 23(4), 370–378.

Resnick, Paul and Richard Zeckhauser (2002). "Trust among strangers in Internet transactions: Empirical analysis of eBay's reputation system." *The Economics of the Internet and E-Commerce*, Michael R. Baye, ed., volume 11 of *Advances in Applied Microeconomics*, 127–157. Amsterdam: Elsevier Science.

Shriver, Scott K., Harikesh S. Nair, and Reto Hofstetter (2013). "Social ties and user-generated content: Evidence from an online social network." *Management Science* 59(6), 1425–1443.

Stephen, Andrew T. and Olivier Toubia (2010). "Deriving value from social commerce networks." *Journal of Marketing Research* 47(2), 215–228.

Sun, Monic (2012). "How does the variance of product ratings matter?" *Management Science* 58(4), 696–707.

Trusov, Michael, Anand Bodapati, and Randolph E. Bucklin (2010). "Determining influential users in Internet social networks." *Journal of Marketing Research* 47(4), 643–658.

Tucker, Catherine and Juanjuan Zhang (2011) "How does popularity information affect choices? A field experiment." *Management Science* 57(5), 828–842.

Watts, Duncan J. and Peter Sheridan Dodds (2007). "Influentials, networks, and public opinion formation." *Journal of Consumer Research* 34(4), 441–458.

Wojnicki, Andrea and David B. Godes (2014). "Word-of-mouth as self enhancement." SSRN Working Paper 908999.

Yoganarasimhan, Hema (2012). "Impact of social network structure on content propagation: A study using YouTube data." *Quantitative Marketing and Economics* 10(1), 111–150.

CHAPTER 31

ECONOMIC FEATURES OF THE INTERNET AND NETWORK NEUTRALITY

NICHOLAS ECONOMIDES

31.1 Introduction and Organization

This chapter focuses on the issue of network neutrality on the Internet. Network neutrality means that content and applications of various types and from various providers are delivered to Internet users without prioritization for which the originators of the content and applications pay the Internet service providers (ISPs). Although the Internet developed under network neutrality, after 2005 practically all fixed-line ISPs to residential customers in the United States have demanded that they be paid to prioritize certain flows of Internet traffic. This issue is complex because the large collection of content and applications the Internet carries shows significant variation in the desirability of consumers for immediacy in their delivery. It is also complex because ISPs already get paid by users, and if they receive payment from the originators of the traffic, the resulting two-sided interaction has to be modeled.

We proceed as follows. Section 31.1 discusses the general structure of the Internet. Section 31.2 discusses the structure of the Internet. Section 31.3 discusses the issues that arise because of the potential abolition of network neutrality. Section 31.4 discusses the regulatory responses in the United States to the issue of network neutrality. Section 31.5 offer concluding remarks.

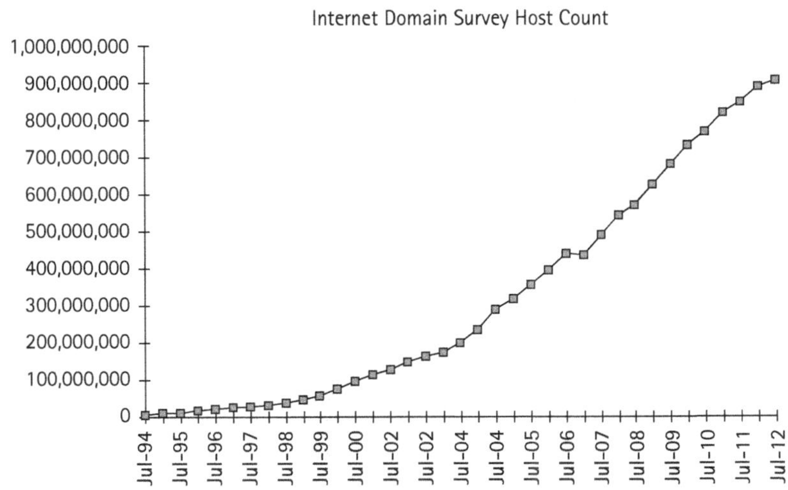

FIGURE 31.1 Internet growth. Source: Internet Systems Consortium (www.isc.org).

31.2 General Structure of the Internet

The Internet is a global network of interconnected networks that connect computing devices. The Internet allows data transfers as well as the provision of a variety of interactive real-time and time-delayed telecommunications services. Internet communication is based on common and public protocols.[1] Close to a billion computing devices are presently connected to the Internet. Figure 31.1 shows the expansion of the number of nodes connected to the Internet.

The vast majority of computing devices owned by individuals or businesses connect to the Internet through commercial ISPs. Educational institutions and government departments also connect to the Internet but typically do not offer commercial ISP services. Users typically connect to the Internet through cable modems, residential DSL, corporate networks, and, in rare cases, through satellite connection or dialup. Typically, routers and switches owned by ISPs send the caller's packets to a local Point of Presence (POP) of the Internet. Cable modems, DSL access POPs, and corporate networks' dedicated-access circuits connect to high speed hubs. High speed circuits leased from or owned by telephone companies, connect the high-speed hubs, forming an "Internet Backbone Network."

The Internet is based on three basic separate levels/layers of functions of the network:

(i) the hardware/electronics level of the physical network;
(ii) the (logical) network level where basic communication and interoperability are established; and
(iii) the applications/services level.

[1] See Bradner (1999).

Thus, the Internet separates the network interoperability level from the applications/services level. Unlike earlier centralized digital electronic communications networks, such as CompuServe, AT&T Mail, Prodigy, and early AOL, the Internet allows a large variety of applications and services to run "at the edge" of the network and not centrally.

Users pay ISPs for access to the whole Internet. Similarly, ISPs pay Internet backbones per month for a pipe of a certain bandwidth for access to the whole Internet. When digital content from provider A, for example, is downloaded by consumer B, both A and B pay their respective ISPs. Consumer B pays his ISP through his monthly subscription, and provider A pays similarly. In turn, ISPs pay to their respective backbones through their subscription.

31.3 Residential Broadband Access Networks and Network Neutrality

The present regime on the Internet does not distinguish in terms of price (or in any other way) between information packets depending on the services that these packets provide or on who is the sender. This regime, called "network neutrality" or "net neutrality," has prevailed on the Internet since its inception. Presently, information packets from a variety of services and providers are treated equally without discrimination or prioritization by the terminating ISP.

In 2005, taking advantage of a change in regulatory rules by the Federal Communications Commission that reclassified the Internet as an "information service" rather than a "telecommunications service" and therefore not subject to nondiscrimination provisions,[2] AT&T, Verizon, and cable TV networks advocate the introduction of price discrimination based on the originating provider of information packets.[3] Under the FCC classification, discrimination is generally prohibited in telecommunications services but allowed in information services. The local residential broadband access networks would like to abolish the regime of net neutrality and substitute for it a complex pricing schedule where the Internet local access network levies charges to the originating party (such as Google or Netflix) even when the originating party is

[2] See *Nat'l Cable & Telecomm. Assn. v. Brand X Internet Services*, 125 S. Ct. 2688 (2005).

[3] The issue arose first at an interview of Ed Whitacre, CEO of SBC with *BusinessWeek*, November 7, 2005.

Q. How concerned are you about Internet upstarts like Google, MSN, Vonage, and others?

A . How do you think they're going to get to customers? Through a broadband pipe. Cable companies have them. We have them. Now what they would like to do is use my pipes free, but I ain't going to let them do that because we have spent this capital and we have to have a return on it. So there's going to have to be some mechanism for these people who use these pipes to pay for the portion they're using. Why should they be allowed to use my pipes? The Internet can't be free in that sense, because we and the cable companies have made an investment and for a Google or Yahoo! or Vonage or anybody to expect to use these pipes [for] free is nuts!"

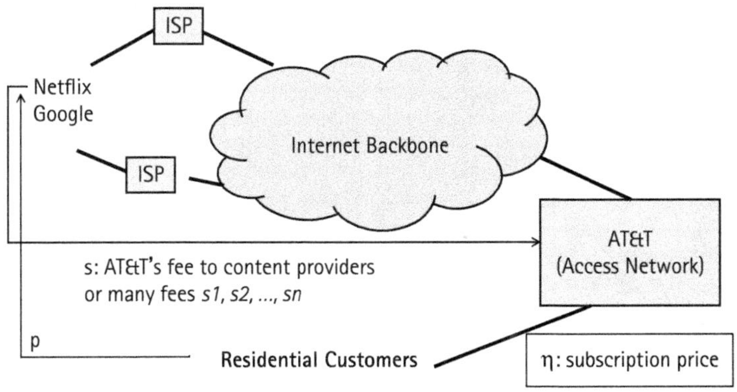

FIGURE 31.2 Schematic topology of the Internet showing possible violations of network neutrality.

not directly connected to and does not presently have a contractual relationship to the local access network. Notice local access networks propose to impose these charges to providers "on the other side" of the Internet and not on their ISPs. These providers would keep paying for transport of their information packets to their ISPs.

Figure 31.2 shows the basic elements of the problem. At the center of the figure, the Internet Backbone is considered effectively competitive.[4] In the lower-right corner is a residential ISP, such as AT&T. Residential ISPs collect a subscription price η from residential customers. At the top-left corner, there are content and applications providers, such as Google, Disney, and Netflix. They may receive a payment p from

[4] The Internet backbone market is considered effectively competitive. Although public information on this market is limited, during the preceding of the merger of AT&T with SBC and MCI with Verizon, these companies had to disclose their traffic to the Department of Justice, which is shown in the table below. We see that concentration is significant but not extreme. There are two additional reasons that increase competition on the Internet backbone. First, the (long distance) connection from point A to point B (points of presence) is a homogeneous good. Second, there is a tremendous amount of "dark fiber" that has been laid in the ground and only requires the addition of electronics to be functional. Thus, homogeneity of the good and overcapacity drive down prices on the Internet backbone.

Company	Traffic				Market Share Among All Providers
	1Q2004	2Q2004	3Q2004	4Q2004	4Q2004
A (AT&T)	37.19	38.66	44.54	52.33	12.58%
B	36.48	36.50	41.41	51.31	12.33%
C	34.11	35.60	36.75	45.89	11.03%
MCI	24.71	25.81	26.86	30.87	7.42%
E	18.04	18.89	21.08	25.46	6.12%
F	16.33	17.78	17.47	19.33	4.65%
G	16.67	15.04	14.93	15.19	3.65%
Total traffic—top 7 networks	183.53	188.28	203.04	240.38	57.78%
Total traffic—all networks	313	313	353	416	

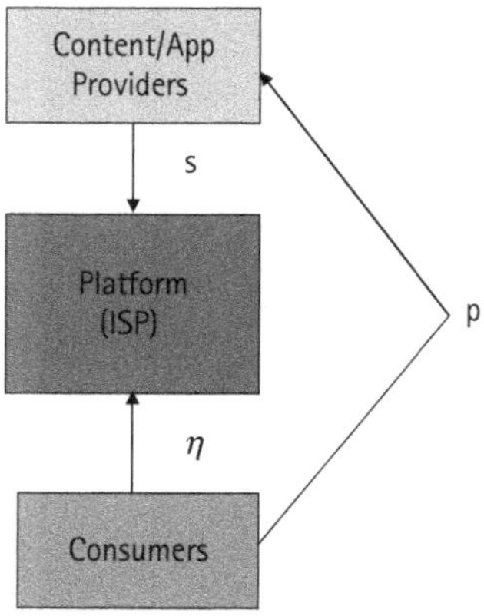

FIGURE 31.3 A residential ISP as a platform in a two-sided market.

residential customers, while some content and information are provided for free and supported by advertising. AT&T and other residential ISPs propose to collect fees *s1*, . . ., *sn* from content and applications providers. In return, residential ISPs propose to offer different degrees of prioritization to content and applications providers. At present, in the regime of network neutrality, all prices *s1*, . . . , *sn* are zero and there is no paid prioritization.

The imposition of price discrimination on the provider side of the market and not on the subscriber is a version of two-sided pricing. This is uniquely possible for firms operating in a network structure.[5] Figure 31.3 shows the setup of Figure 31.2 in an abstraction that may also be useful in analyzing other industries. The ISP is a platform that is paid η by consumers on one side of the market and *s* by content and applications providers on the other side of the market. Consumers may also pay a price *p* directly to content or applications providers. This analysis is based on the framework of Armstrong (2006).

It is expected that on the Internet content and applications providers would like more consumers, while each consumer would like more content and applications providers. Thus, there are two feedback loops or network effects in this setup.

The general two-sided setup of Figure 31.3 can result in positive or negative fee *s* assuming a positive price for η. For example, if the platform was a computer operating system, and the content/app side was third-party applications, we know that the platform typically subsidizes applications (so $s < 0$) by embedding in the OS subroutines that are only valuable to applications' developers. If the platform was a game platform, and the content/app side was third-party games, we know that the platform typically collects from software developers. If the platform is a search page on the Web (or Yellow Pages in older times), we know that users are typically offered a zero price *p*, while advertisers pay a positive price *s*. If the platform is an academic journal, typically readers/users pay a positive p while authors do not pay a positive *s*, although there

[5] For earlier work on two-sided pricing, see Armstrong (2006) and Rochet and Tirole (2003, 2006).

are journals that are freely distributed and charge the authors publication costs (after refereeing and acceptance). If the platform is a credit card network, typically issuers pay users (η <0) and get paid by the merchants (s >0). If the platform is a newspaper, typically users pay (η >0) and advertisers pay (s >0), but there are free newspapers (η = 0). In the particular case we have in mind, we observe AT&T, Verizon, Comcast, Time Warner, and other residential ISPs that demand a positive price s.

"Network neutrality" can have different definitions. Referring to pricing to the "other side" of the consumer market (that is, to content and applications providers) we may define network neutrality from the strictest to the weakest. The first, most extreme form of network neutrality is absolute nondiscrimination—that is, no quality of service variations offered for money or for free. A second, less strict form of network neutrality would allow ISPs to vary quality of service depending on the type of information packet or service, but the ISP could not charge any fees to upstream providers for these variations. This form of network neutrality would allow the ISP to implement variations in quality of service that consumers desire, but would still be prohibited from charging the upstream firms. All advocates of network neutrality accept that both of these definitions are consistent with network neutrality.

A third possibility for network neutrality is when tiered service is allowed but each tier is offered at the same price to all without exclusivity or identity-based discrimination. Academics and industry observers are divided on whether this is indeed network neutrality. On the one hand, within each tier, information packets are treated equally. On the other hand, across tiers, information packets are treated unequally, with prioritization given to packets in higher tiers. Thus, tiered service is generally considered a violation of net neutrality.

A fourth Internet regime could allow for identity-based discrimination, so that the same service is offered at different prices to different companies. An extreme version of this, a fifth Internet regime, would allow for full exclusivity to a particular content or application provider per industry segment. Essentially the ISP could go to an industry segment, such as to all search providers, and inform them that it would allow only one of them to be prioritized, and then auction the prioritized position. Clearly the last two regimes—identity-based discrimination and exclusivity—violate network neutrality.

Because of the very considerable market power of broadband Internet access networks, there is debate on the allocative efficiency properties of complex pricing strategies that violate network neutrality, as well as on their legality. Residential retail broadband Internet access customers may well have difficulty changing ISPs. Ninety-eight percent of US households are offered Internet access by at most two firms—a telephone company through digital subscriber line ("DSL"), and a cable TV company through a cable "modem"—and many households are facing a monopoly of either cable or DSL. Additionally, residential customers face switching costs, such as changing equipment and possibly the email service of the ISP. Finally, residential customers are affected by contracts that bundle broadband Internet access with other services such as telecommunications and cable television.

Network neutrality has allowed firms to innovate "at the edge of the network" without seeking approval from network operator(s). The decentralization of the Internet based on net neutrality facilitated innovation resulting in big successes such as Google and Skype. Net neutrality also increased competition among the applications and services "at the edge of the network" since they did not need to own a network to provide services.

Additionally, the existence of network effects on the Internet implies that efficient prices to users on both sides (consumers and applications) should be lower than in a market without network effects. Instead we see an attempt to increase prices that will reduce network effects and innovation.

At the most fundamental level, the problem of network neutrality can be analyzed as a two-sided market in either a static framework with fixed bandwidth or in a dynamic one, where the pricing regime changes the incentives to invest in bandwidth. Desirability of network neutrality can also be analyzed under the assumption of no congestion in residential Internet access or, alternatively, assuming congestion.[6] Finally, the effect of network neutrality on innovation may or may not be taken into consideration.

There are many advocacy papers written on the subject, but significantly less economic research. Among the academic analyses,[7] Economides and Tag (2012), assuming no congestion, showed that introducing a positive price to content providers is typically welfare-inferior to network neutrality. They assume a monopolist platform that charges a positive fee η to consumers. The crucial issue is the sign of the fee s from content and applications providers to the platform ISP. They show that the ISP would like to charge a positive s. They define total surplus $\text{TS}(\eta, s)$ as the sum of consumer surplus, profits of the ISP, and profits of content and applications providers. A regulator is given the job of choosing optimal pricing on the content side of the market. Economides and Tag show that a total surplus maximizing regulator in a two-sided market with network effects, when constrained to marginal cost pricing on the consumer market, chooses below-cost pricing in the content market—that is, that the maximizer of $TS(c, s)$ is $s^* < 0$, where c is the marginal cost of the ISP. This is essentially a consequence of network effects. In the presence of network effects, optimality implies pricing below cost for the end-to-end price $\eta + s$.[8] Since one side of the market was set to cost, intuitively, the other side of the market should be set at a price below cost.

Now consider a regulator setting fee s to content providers expecting the platform monopolist to set its profit-maximizing price to consumers $p(s)$, taking into account s. Then the regulator maximizes the constrained total surplus function $\text{TS}(p(s), s)$. The

[6] Evidence of presence of congestion has not been presented in the FCC proceedings so far.

[7] For a survey of academic papers, see Krämer, Wiewiorra, and Weinhardt (2013) and Lee and Wu (2009).

[8] In Economides and Tag (2012), $p = 0$.

regulator's optimal choice is a below-cost fee $s^{**} < 0$ to content providers.[9] The intuition is that same as before. Network effects imply below-cost end-to-end pricing $p(s) + s$. Since $p(s)$ is above cost, optimal s should be chosen below cost, here below zero. Since it is difficult to implement a negative price, the regulator can implement $s = 0$ as a second best. In summary, Economides and Tag (2012) show that there is a sharp divergence between private and public incentives, where the platform desires a positive fee from content providers while total surplus maximization implies a negative such fee, which is set to zero as a second best.

There are a number of other detriments to welfare that could result from departures from network neutrality. While the ISPs have promised an enhancement of the arrival time of information packets that originate from paying content and application firms, this is not necessary to generate value for themselves. For the latter, it is sufficient to degrade the arrival time of information packets that originate from nonpaying firms while keeping the arrival timing of the paying firms the same as before the violation of network neutrality. The present plan of access providers is to create a "special lane" for the information packets of the paying firms while restricting the lane of the nonpayers without expanding total capacity. By manipulating the size of the paying firms' lane, the access provider can guarantee a difference in the arrival rates of packets originating from paying and nonpaying firms, even if the actual improvement in arrival time for paying firms' packets is not improved over net neutrality.

If the access providers choose to engage in "identity-based" discrimination, they can determine which one of the firms in a content or applications industry, say in search, will get priority and therefore win. This can easily be done by announcing that prioritization will be offered to only one of the search firms; for example, the one that bids the highest. Thus the determination of the winner in search and other content or applications markets will be in hands of the access providers and not determined by innovative products or services on the other side. This can create very significant distortions since the surplus of the content and applications markets is a large multiple of the combined telecom and cable TV revenue from residential Internet access.

New firms with small capitalization (or those innovative firms that have not yet achieved significant penetration and revenues) are not very likely to be the winners of a prioritization auction. This is likely to lead to a calcification/freezing of industry structure and reduce innovation. Network externalities arise because a typical subscriber can reach more subscribers in a larger network. Under no network neutrality, access providers can limit the size and profitability of new firms in content and applications.

Typically, access networks also provide their own content and applications, or more generally, they provide substitutes to the content and applications independent firms. For example, Netflix's customers may use Comcast to download video and films from Netflix, while Comcast sells video services delivered through cable TV. Similarly, both telecom and cable TV ISPs provide their own phone services that are also provided by

[9] This holds provided that both consumers and content providers are sufficiently differentiated. Also, even paying the below-cost fee, the platform makes positive profits.

independent VOIP providers such as Vonage. The ISPs may favor their own services and degrade transmission of rivals that use their pipes. This is likely to distort competition and reduce total surplus.[10]

Since the Internet consists of a series of interconnected networks, any one of these, and not just the final consumer access ones can, in principle, ask content and application providers for a fee. This can result in multiple fees charged on a single transmission and lead to a significant reduction of trade on the Internet. Finally, there are political and news diversity concerns if content in newspapers and websites is delayed in comparison with sites and newspapers that pay for prioritization.

There are three main arguments supporting abolishing network neutrality. First, that if the ISP collects revenue from the content providers (for example, through paid prioritization), it will decrease prices to consumers. Second, that there is congestion on the local access network and paid prioritization can be way to alleviate it. Additionally, ISPs have claimed that the presence of congestion in the local access network automatically makes network neutrality suboptimal. Third, that if the ISP is allowed paid prioritization, its profits will be higher and it will invest more in network capacity, thereby decreasing distortions arising from network neutrality.

My assessment of these arguments is as follows. On the first issue, in many models, the ISP charging content providers leads to lower prices to users.[11] This is taken into account in all models, and the welfare results take it into consideration. On the issue of congestion, it should first be noted that the ISPs have not provided evidence in their numerous submissions to the FCC that there is congestion in the local access network. Second, paid prioritization is not the optimal way to deal with congestion because the incentives of the ISP differ from the incentives of society. The way the ISPs have proposed paid prioritization, it is provider-based and not volume-based. That is, if a provider pays, its content will be prioritized compared to providers that provide substitute services. The ISPs have never proposed a system that would prioritize some and delay other information packets based on the extent of congestion, which would naturally vary with the time of the day.

Most importantly, the ISP has an incentive to *create* artificial scarcity over and above the natural scarcity that may be there because of congestion, and paid prioritization gives the ISPs the opportunity to do so. Economides and Hermalin (2012) address the problem of network neutrality assuming congestion and examining for which class of additive utility functions network neutrality is optimal, and for which class prioritization is optimal. They assume homogeneous consumers, a monopolist ISP, and content and applications providers are indexed by θ, with consumers considering

[10] See, for example, a recent battle between Comcast and Level 3 Communications ("Comcast Fee Ignites Fight over Videos on Internet," *New York Times*, November 30, 2010) illustrates this point.

[11] Whether in fact ISPs such as AT&T and Verizon will reduce prices under paid prioritization is highly questionable. These companies are seen by Wall Street as "utilities" that are expected to pay a high dividend. If paid prioritization were to occur, key ISPs would be under pressure to distribute the added revenue as profits rather than to reduce prices to users.

content of higher θ to be more time-sensitive. Consumers have utility function

$$U = y + \int_{\underline{\theta}}^{\overline{\theta}} \left(\int_{\theta}^{x(\theta)} m\left(\frac{x}{\alpha\left(\tau\left(\theta\right),\theta\right)} \right) dx \right) dF\left(\theta\right)$$

where $x(\theta)$ is consumption of content of type θ, $\tau(\theta)$ is delay of content of type θ, $m(.)$ is the "adjusted" marginal utility of information packets (with $m > 0$, $m' < 0$), and $\alpha(\tau(\theta), \theta)$ reflects the extent that consumers care about time delay of packets of type θ. High α is equivalent to the consumer receiving fewer packets x, which means higher *marginal* utility.[12] As in Economides and Tag (2012), total surplus W is defined as consumer surplus plus profits of ISPs plus profits of applications. Because consumers are homogeneous, the monopolist ISP appropriates all of their consumer surplus.

Potentially, bandwidth can be divided into different lanes, with each lane given a different priority. Network neutrality then corresponds to no division of the bandwidth. A key result of Economides and Hermalin is that between two alternative divisions of the total bandwidth, one is welfare-superior to the other if and only if it results in more content being carried in equilibrium than the other. Therefore the *amount* of content carried is a "sufficient statistic" for welfare, and *how* bandwidth is allocated to content among applications does not matter. Content is analogous to income, and it has the maximum positive effect on welfare when it is allocated as the consumer sees fit without restrictions. Since welfare is increasing in bandwidth used, the problem of maximizing welfare under a bandwidth allocation constraint is reduced to maximizing throughput by choosing the appropriate bandwidth allocation.

The key question is, is it possible to divide bandwidth in classes (possibly charging different prices) and increase throughput (and therefore increase total surplus)? The answer of Economides and Hermalin is that it depends on an esoteric feature of demand: on the variation with respect to content type of the elasticity of content with respect to delay. In particular, maximizing total surplus subject to bandwidth constraint shows that higher θ content should be prioritized when the partial with respect to θ of the elasticity of content sales with respect to transmission delay τ is positive. Network neutrality is optimal when the elasticity of content with respect to transmission delay does not vary in content type θ. This is true for a wide class of utility functions.[13] Still, there are other utility functions for which prioritization is optimal.

Economides and Hermalin (2012) also find that, when network neutrality maximizes welfare, the profit-maximizing ISP will still discriminate and prioritize content. The ISP will choose prices that do not result in network neutrality. This is fundamentally because the ISP's profits do not coincide with total surplus, which also includes content providers' profits.

[12] Function α is decreasing in delay τ, $\partial a/\partial\tau < 0$ and therefore m is decreasing in τ, $\partial m/\partial\tau < 0$. Higher type θ content more time sensitive: $\partial^2\alpha/\partial\tau\partial\theta < 0$.

[13] In particular, network neutrality is optimal when $\alpha(\tau(\theta), \theta) = g(\tau)v(\theta)$, which means that the (lack of) preference for delay τ is "independent" of content type θ.

In a dynamic setup, when ISPs are allowed to deviate from network neutrality and invest the resulting profits in expanding bandwidth, Economides and Hermalin (2012) find that, in some examples, more bandwidth alleviates to some extent the static distortion, and sometimes it may reverse it. Whether the latter occurs depends on the bandwidth expansion technology and the relative contributions to welfare of consumer surplus and profits.

In some models, for example in Choi and Kim (2010), the ISP can have higher profits if it restricts bandwidth. In that model, lower bandwidth increases the difference between the price of a good in the fast lane in comparison to the one in the slow lane. Njoroge, Ozdaglar, Stier-Moses, and Weintraub (2013) find that network neutrality is optimal in a "priority lanes" model (as in Economides and Hermalin 2012) as well as in a "walled garden" model where ISPs effectively charge access fees to off-network CPs to deliver their content (as in Economides and Tag 2012).[14]

31.4 Regulatory Actions

The FCC passed rules in December 2010 that imposed, among others, the following requirements to ISPs:[15]

1. *Transparency*: Fixed and mobile broadband providers must disclose the network management practices, performance characteristics, and terms and conditions of their broadband services.
2. *No blocking*: Fixed broadband providers may not block lawful content, applications, services, or nonharmful devices; mobile broadband providers may not block lawful websites or block applications that compete with their voice or video telephony services.
3. *No unreasonable discrimination*: Fixed broadband providers may not unreasonably discriminate in transmitting lawful network traffic.

Even though these rules were weak (since they outlawed only unreasonable discrimination rather than all paid discrimination), Verizon appealed to stop them. In January 2014, Verizon won, and the FCC rules were nullified.[16] In October 2014, President Obama expressed strong support for net neutrality, "The president has

[14] The latter result holds only when the quality of content providers' heterogeneity is large.

[15] The NPRM (request for comments) of the FCC (2009) stated, "Subject to reasonable network management, a provider of broadband Internet access service must treat lawful content, applications, and services in a nondiscriminatory manner. We understand the term 'nondiscriminatory' to mean that a broadband Internet access service provider may not charge a content, application, or service provider for enhanced or prioritized access to the subscribers of the broadband Internet access service provider.... We propose that this rule would not prevent a broadband Internet access service provider from charging subscribers different prices for different services."

[16] The case was *Verizon v. F.C.C.*, D.C. Cir. No.11-1355 (Jan. 14, 2014). The essence of the decision is "[E]ven though the [FCC] has general authority to regulate in this arena, it may not impose requirements

made it abundantly clear that any outcome must protect net neutrality and ban paid prioritization—and has called for all necessary steps to safeguard an open Internet."

In the rules that were nullified, the FCC had classified Internet service under "Title I" as an "information service," over which it has only "ancillary jurisdiction." The FCC could have classified Internet service as a "telecommunications service" under "Title II," which would impose strict nondiscrimination. In fact, on March 12, 2015, the FCC reclassified the Internet as a telecommunications service falling under Title II.[17] This imposes strict nondiscrimination and bans paid prioritization. Essentially this rule adopts strict network neutrality. Industry group USTelecom (which includes AT&T and Verizon) sued the FCC on the validity of the rules on March 23, 2015. It is too early to judge the probability of success of this challenge.

31.5 Concluding Remarks

We identified salient features of the Internet and focused on the issue of network neutrality. Local broadband access residential ISPs possess market power and have proposed abolishing the present regime of no paid prioritization that has been called "network neutrality." We discuss how abolition of network neutrality and introduction of paid prioritization can result in significant reduction in the surplus created in the Internet ecosystem. In particular, we show how abolition of network neutrality can create an allocative distortion and reduce total surplus. However, profits accrued to the ISP from a violation of network neutrality, if used to expand bandwidth, can alleviate the allocative distortion caused by violating network neutrality. Additionally, we described how abolition of network neutrality could significantly reduce innovation at the edge of the network.

References

Armstrong, Mark (2006). "Competition in two-sided markets." *RAND Journal of Economics* 37, 668–691.

Bradner, Scott (1999). *The Internet Standards Process*, revision 3, Network Working Group, at ftp://ftp.isi.edu/in-notes/rfc2026.txt .

Cheng, Hsing Kenneth, Subhajyoti Bandyopadhyay, and Hong Guo (2011). "The debate on net neutrality: A policy perspective." *Information Systems Research* 22, 60–82.

that contravene express statutory mandates. Given that the Commission has chosen to classify broadband providers in a manner that exempts them from treatment as common carriers, the Communications Act expressly prohibits the Commission from nonetheless regulating them as such. Because the Commission has failed to establish that the anti-discrimination and anti-blocking rules do not impose per se common carrier obligations, we vacate those portions of the Open Internet Order."

[17] Federal Communications Commission, *Report and Order 14-28*, March 12, 2015.

Choi, Jay Pil and Byung-Cheol Kim (2010). "Net neutrality and investment incentives." *RAND Journal of Economics* 41, 446–471.

Economides, Nicholas (1996). "The economics of networks." *International Journal of Industrial Organization* 14(2), 675–699.

Economides, Nicholas and Benjamin Hermalin (2012). "The economics of network neutrality." *RAND Journal of Economics* 43(4), Winter 2012, 602–629, at http://www.stern.nyu.edu/networks/Economides-Hermalin_Economics_of_Network_Neutrality.pdf.

Economides, Nicholas and Joacim Tag (2012). "Network neutrality on the Internet: A two-sided market analysis." *Information Economics and Policy* 24, 91–104, at http://www.stern.nyu.edu/networks/Economides_Tag_Net_Neutrality.pdf.

Federal Communications Commission, *Report and Order 10-201*, Washington, DC, 2010.

Federal Communications Commission, *Report and Order 14-28*, Washington, DC, March 12, 2015.

Hermalin, Benjamin and Michael Katz (2007). "The economics of product-line restrictions with an application to the network neutrality debate." *Information Economics and Policy* 19, 215–248.

Kahn, Robert E. and Vinton G. Cerf (1999). "What is the Internet (and what makes it work)." At http://www.wcom.com/about_the_company/cerfs_up/internet_history/whatIs.phtml.

Katz, Michael and Carl Shapiro (1985). "Network externalities, competition and compatibility." *American Economic Review* 75(3), 424–440.

Krämer, Jan and Lukas Wiewiorra (2010). "Network neutrality and congestion-sensitive content providers: Implications for service innovation, broadband investment and regulation." NET Institute Working Paper 10-09.

Krämer, Jan, Lukas Wiewiorra, and Christof Weinhardt (2013). "Net neutrality: A progress report." *Telecommunications Policy* 37(9), 794–813.

Liebowitz, Stanley J. and Stephen E. Margolis (1994). "Network externality: An uncommon tragedy." *The Journal of Economic Perspectives* (Spring), 133–150.

Lee, Robin S. and Tim Wu (2009). "Subsidizing creativity through network design: Zero-pricing and net neutrality." *Journal of Economic Perspectives* 23(3), 61–76.

Njoroge, Paul, Asuman Ozdaglar, Nicolás E. Stier-Moses, and Gabriel Y. Weintraub (2014). "Investment in two-sided markets and the net neutrality debate." *Review of Network Economics* 12(4), 355–402.

Rochet, Jean-Charles and Jean Tirole (2003). "Platform competition in two-sided markets." *Journal of the European Economic Association* 1, 990–1029.

Rochet, Jean-Charles and Jean Tirole (2006). "Two-sided markets: A progress report." *RAND Journal of Economics* 37, 645–667.

Index

Figures and tables are denoted by an italicized *f* or *t*, respectively.

O

R

S